T0145101

Communications in Computer and Information Science 1534

More information about this series at https://link.springer.com/bookseries/7899

Isaac Woungang · Sanjay Kumar Dhurandher ·
Kiran Kumar Pattanaik · Anshul Verma ·
Pradeepika Verma (Eds.)

Advanced Network Technologies and Intelligent Computing

First International Conference, ANTIC 2021
Varanasi, India, December 17–18, 2021
Proceedings

Springer

Editors
Isaac Woungang
Ryerson University
Toronto, ON, Canada

Sanjay Kumar Dhurandher
Netaji Subhas University of Technology
New Delhi, India

Kiran Kumar Pattanaik
ABV-Indian Institute of Information Technology and Management
Gwalior, India

Anshul Verma
Banaras Hindu University
Varanasi, India

Pradeepika Verma
Indian Institute of Technology BHU
Varanasi, India

ISSN 1865-0929 ISSN 1865-0937 (electronic)
Communications in Computer and Information Science
ISBN 978-3-030-96039-1 ISBN 978-3-030-96040-7 (eBook)
https://doi.org/10.1007/978-3-030-96040-7

This Springer imprint is published by the registered company Springer Nature Switzerland AG
The registered company address is: Gewerbestrasse 11, 6330 Cham, Switzerland

Preface

The 1st International Conference on Advanced Network Technologies and Intelligent Computing (ANTIC 2021) was organized by the Department of Computer Science, Institute of Science, Banaras Hindu University, Varanasi, India, in online mode (due to the ongoing COVID-19 pandemic) and held during December 17–18, 2021. ANTIC 2021 aimed to bring together leading academicians, scientists, and researcher scholars, as well as undergraduate and postgraduate students, across the globe to exchange and share their research outcomes. The objective was to provide a state-of-the-art platform to discuss all aspects (current and future) of advanced network technologies and intelligent computing, enabling the participating researchers to exchange their ideas about applying existing methods in these areas to solve real-world problems.

ANTIC 2021 solicited two types of submissions: full research papers (equal to or more than 12 pages) and short research papers (between 8 and 11 pages), both identifying and justifying a principled advance to the theoretical and practical foundations for the construction and analysis of systems, where applicable supported by experimental validation. A total of 593 research papers were received through the EasyChair portal and 152 papers (25.63%) were accepted after the rigorous review process, comprising 119 (78.28%) were full research papers and 33 (21.71%) were short papers. Out of the 152 accepted papers, 61 papers (50 full papers and 11 short papers) were selected for publication in Springer's Communications in Computer and Information Science (CCIS). The papers are grouped into two thematic categories: Advanced Network Technologies and Intelligent Computing.

We would like to thank everyone who helped to make ANTIC 2021 successful. In particular, we would like to thank the authors for submitting their papers to ANTIC 2021. We are thankful to our excellent team of reviewers from all over the globe who deserve full credit for their hard work in reviewing the high-quality submissions with rich technical content. We would also like to thank the members of the Advisory Committee for their guidance and suggestions in making ANTIC 2021 a success. Finally, we would like to thank all the Track Chairs, Organizing Committee, and Technical Program Committee members for their support and co-operation.

December 2021

Isaac Woungang
Sanjay Kumar Dhurandher
Kiran Kumar Pattanaik
Anshul Verma
Pradeepika Verma

Organization

Chief Patron

Vijay Kumar Shukla	Banaras Hindu University, India

Patrons

A. K. Tripathi	Banaras Hindu University, India
Madhoolika Agrawal	Banaras Hindu University, India

Advisory Board

Anil Kumar Tripathi	Indian Institute of Technology (BHU), India
Anupam Shukla	Indian Institute of Information Technology, Pune, India
Bansidhar Majhi	Veer Surendra Sai University of Technology, India
Jagannathan Sarangpani	Missouri University of Science and Technology, USA
M. M. Gore	Motilal Nehru National Institute of Technology Allahabad, India
Pradeep Kumar Mishra	Jharkhand University of Technology, India
Pradip Kr. Das	Indian Institute of Technology, Guwahati, India
Rajeev Ranjan	Newcastle University, UK
Rajeev Srivastava	Indian Institute of Technology (BHU), India
Rajendra Sahu	ABV-Indian Institute of Information Technology and Management, India
Rajkumar Buyya	University of Melbourne, Australia
S. G. Deshmukh	Indian Institute of Technology, Delhi, India
Sanjay Kumar Madria	Missouri University of Science and Technology, USA
Sanjay Kumar Singh	Indian Institute of Technology (BHU), India
Shiv Mohan Singh	Science and Engineering Research Board (SERB), India
Sundaraja Sitharama Iyengar	Florida International University, USA

General Chairs

Isaac Woungang	Ryerson University, Canada
Sanjay Kumar Dhurandher	Netaji Subhas University of Technology, India

K. K. Pattanaik	ABV-Indian Institute of Information Technology and Management, India

Program Chairs

S. Karthikeyan	Banaras Hindu University, India
Pramod Kumar Mishra	Banaras Hindu University, India
Vivek Kumar Singh	Banaras Hindu University, India
Anshul Verma	Banaras Hindu University, India

Track Chairs

Himanshu Aggarwal	Punjabi University, India
Karm Veer Arya	ABV-Indian Institute of Information Technology and Management, India
Nanhay Singh	Netaji Subhas University of Technology, India
P. K. Singh	ABV-Indian Institute of Information Technology and Management, India
Sanjay Kumar	Pt. Ravishankar Shukla University, India
Udai Shanker	Madan Mohan Malaviya University of Technology, India
Akande Noah Oluwatobi	Landmark University, Nigeria
Alberto Rossi	University of Florence, Italy
Alireza Izaddoost	California State University, Dominguez Hills, USA
Andrea Visconti	University of Milan, Italy
Bhaskar Biswas	Indian Institute of Technology (BHU), India
Divakar Singh Yadav	National Institute of Technology, Hamirpur, India
Joshua D. Reichard	Omega Graduate School, USA
Lalit Garg	University of Malta, Malta
Pinar Kirci	Bursa Uludag University, Turkey
Pradeepika Verma	Indian Institute of Technology (BHU), India
Sukomal Pal	Indian Institute of Technology (BHU), India
Yousef Farhaoui	Moulay Ismail University, Morocco

Organizing Committee

Achintya Singhal	RGSC, Banaras Hindu University, India
Manoj Kumar Singh	Banaras Hindu University, India
Rakhi Garg	MMV, Banaras Hindu University, India
Manjari Gupta	Banaras Hindu University, India
Vandana Kushwaha	Banaras Hindu University, India
Gaurav Baranwal	Banaras Hindu University, India

Ankita Vaish	Banaras Hindu University, India
S. Suresh	Banaras Hindu University, India
Awadhesh Kumar	MMV, Banaras Hindu University, India
Manoj Mishra	RGSC, Banaras Hindu University, India
S. N. Chaurasia	Banaras Hindu University, India
Marisha	Banaras Hindu University, India
Sarvesh Pandey	MMV, Banaras Hindu University, India
Vibhor Kant	RGSC, Banaras Hindu University, India

Technical Program Committee

A. Senthil Thilak	National Institute of Technology, Surathkal, India
Abdus Samad	Aligarh Muslim University, India
Abhay Kumar Rai	Banasthali Vidyapith, India
Abhilasha Sharma	Delhi Technological University, India
Ade Romadhony	Telkom University, Indonesia
Afifa Ghenai	Constantine 2 University, Algeria
Ajay	Shree Guru Gobind Singh Tricentenary University, India
Ajay Kumar	Chitkara University, Punjab, India
Ajay Kumar	Central University of Himanchal Pradesh, India
Ajay Kumar Gupta	Madan Mohan Malaviya University of Technology, India
Ajay Kumar Yadav	Banasthali Vidyapith, India
Ajay Pratap	Indian Institute of Technology (BHU), India
Akande Noah Oluwatobi	Landmark University, Nigeria
Akash Kumar Bhoi	Sikkim Manipal Institute of Technology, India
Alberto Rossi	University of Florence, Italy
Aleena Swetapadma	Kalinga Institute of Industrial Technology, India
Ali El Alami	Moulay Ismail University, Morocco
Alok Kumar	CSJM University, Kanpur, India
Amit Kumar	BMS Institute of Technology and Management, India
Amit Kumar	Jaypee University of Engineering and Technology, India
Amit Rathee	Government College Barota, India
Angel D.	Sathyabama Institute of Science and Technology, India
Anil Kumar	London Metropolitan University, UK
Anirban Sengupta	Jadavpur University, India
Anita Chaware	SNDT Women's University, Mumbai, India
Anjali Shrikant Yeole	VES Institute of Technology, India

Ansuman Mahapatra	National Institute of Technology, Puducherry, India
Antriksh Goswami	Indian Institute of Information Technology, Vadodara, India
Anupam Biswas	National Institute of Technology, Silchar, India
Anuradha Yarlagadda	Gayatri Vidhya Parishad College of Engineering, India
Anurag Sewak	Rajkiya Engineering College, Sonbhadra, India
Arun Kumar	ABV-Indian Institute of Information Technology and Management, India
Arun Pandian J.	Vel Tech Rangarajan Sagunthala R&D Institute of Science and Technology, India
Ashish Kumar Mishra	Rajkiya Engineering College, Ambedkar Nagar, India
Ashutosh Kumar Singh	United University, Prayagraj, India
Aymen Jaber Salman	Al-Nahrain University, Baghdad
B. Arthi	SRM Institute of Science and Technology, India
B. S. Charulatha	Rajalakshmi Engineering College, India
B. Surendiran	National Institute of Technology, Puducherry, India
Balbir Singh Awana	Vanderbilt University, USA
Baranidharan B.	SRM Institute of Science and Technology, India
Benyamin Ahmadnia	Harvard University, USA
Bhabendu Kumar Mohanta	K L Deemed to be University, India
Bharat Garg	Thapar Institute of Engineering & Technology, India
Bharti	University of Delhi, India
Bhaskar Mondal	National Institute of Technology, Patna, India
Bhawana Rudra	National Institute of Technology, Karnataka, India
Binod Prasad	ABV-Indian Institute of Information Technology and Management, India
Boddepalli Santhi Bhushan	Indian Institute of Information Technology, Kottayam, India
Brijendra Singh	VIT Vellore, India
C. Selvi	Amrita Vishwa Vidyapeetham, India
Chanda Thapliyal Nautiyal	Dharmanand Uniyal Government Degree College, India
Chandrashekhar Azad	National Institute of Technology, Jamshedpur, India
Chetan Vyas	United University, Prayagraj, India
Chittaranjan Pradhan	Kalinga Institute of Industrial Technology, India
D. Senthilkumar	University College of Engineering, Anna University, Tiruchirappalli, India

Dahmouni Abdellatif	Chouaib Doukkali University, Morocco
Darpan Anand	Chandigarh University, India
Deepak Gupta	National Institute of Technology, Arunachal Pradesh, India
Deepak Kumar	Banasthali Vidyapith, India
Dharmendra Prasad Mahato	National Institute of Technology, Hamirpur, India
Dharmveer Kumar Yadav	Katihar Engineering College, India
Dhirendra Kumar	Delhi Technological University, India
Dinesh Kumar	Motilal Nehru National Institute of Technology Allahabad, India
Divya Saxena	Hong Kong Polytechnic University, Hong Kong
Ezil Sam Leni A.	JEPPIAAR SRR Engineering College, India
Gargi Srivastava	Rajiv Gandhi Institute of Petroleum Technology, India
Gaurav Gupta	Shoolini University, India
Gyanendra K. Verma	National Institute of Technology, Kurukshetra, India
Hardeo Kumar Thakur	Manav Rachna University, India
Harish Sharma	Rajasthan Technical University, India
Hasmat Malik	Netaji Subhas University of Technology, India
Inder Chaudhary	Delhi Technological University, India
Itu Snigdh	Birla Institute of Technology, Mesra, India
J. Jerald Inico	Loyola College, India
J. K. Rai	Defence Research and Development Organisation, India
Jagadeeswara Rao Annam	Gudlavalleru Engineering College, India
Jagannath Singh	Kalinga Institute of Industrial Technology, India
Jagdeep Singh	Sant Longowal Institute of Engineering and Technology, Longowal, India
Jainath Yadav	Central University of South Bihar, India
Jay Prakash	National Institute of Technology, Calicut, India
Jaya Gera	Shyama Prasad Mukherji College for Women, India
Jeevaraj S.	ABV- Indian Institute of Information Technology & Management, India
Jolly Parikh	Bharati Vidyapeeth's College of Engineering, New Delhi, India
Jyoti Singh	Banaras Hindu University, India
K. T. V. Reddy	Pravara Rural Education Society, India
Koushlendra Kumar Singh	National Institute of Technology, Jamshedpur, India
Lakshmi Priya G.	VIT Vellore, India
Lokesh Chauhan	National Institute of Technology, Hamirpur, India

M. Deva Priya	Sri Krishna College of Technology, India
M. Joseph	Michael Research Foundation, Thanjavur, India
M. Nazma B. J. Naskar	Kalinga Institute of Industrial Technology, India
Mahendra Shukla	LNM Institute of Information Technology, India
Mainejar Yadav	Rajkiya Engineering College, Sonbhadra, India
Manish Gupta	IPS College of Technology and Management, India
Manish K. Pandey	Banaras Hindu University, India
Manish Kumar	M S Ramaiah Institute of Technology, India
Manpreet Kaur	Manav Rachna University, India
Mariya Ouaissa	Moulay Ismail University, Meknes, Morocco
Mariyam Ouaissa	Moulay Ismail University, Meknes, Morocco
Meriem Houmer	Moulay Ismail University, Morocco
Minakhi Rout	Kalinga Institute of Industrial Technology, India
Mohit Kumar	National Institute of Technology, Jalandhar, India
Muhammad Abulaish	South Asian University, New Delhi, India
Mukesh Mishra	Indian Institute of Information Technology, Dharwad, India
Mukesh Rawat	Meerut Institute of Engineering and Technology, India
Nagarajan G.	Sathyabama Institute of Science and Technology, India
Nagendra Pratap Singh	National Institute of Technology, Hamirpur, India
Nandakishor Yadav	Fraunhofer Institute for Photonic Microsystems, Germany
Narendran Rajagopalan	National Institute of Technology, Puducherry, India
Neetesh Kumar	Indian Institute of Technology, Roorkee, India
Nisha Chaurasia	National Institute of Technology, Jalandhar, India
Nisheeth Joshi	Banasthali Vidyapith, India
Nitesh Bharadwaj	Indian Institute of Information Technology, Bhopal, India
Om Jee Pandey	SRM University, Andhra Pradesh, India
P. Manikandaprabhu	Sri Ramakrishna College of Arts and Science, India
Partha Pratim Sarangi	Kalinga Institute of Industrial Technology, India
Pavithra G.	Dayananda Sagar College of Engineering, India
Pinar Kirci	Bursa Uludag University, Turkey
Piyush Kumar Singh	Central University of South Bihar, India
Pooja	University of Allahabad, India
Prabhat Ranjan	Central University of South Bihar, India
Pradeeba Sridar	Sydney Medical School, Australia
Pradeep Kumar	University of KwaZulu-Natal, South Africa

Prakash Kumar Singh	Rajkiya Engineering College, Mainpuri, India
Prakash Srivastava	KIET Group of Institutions, India
Prasenjit Chanak	Indian Institute of Technology (BHU), India
Prashant Singh Rana	Thapar Institute of Engineering & Technology, India
Praveen Pawar	Indian Institute of Information Technology, Bhopal, India
Preeth R.	Indian Institute of Information Technology, Design and Manufacturing, Kurnool, India
Preeti Sharma	Chitkara University, Punjab, India
Priya Gupta	Jawaharlal Nehru University, India
Puneet Misra	University of Lucknow, India
Pushpalatha S. Nikkam	SDM College of Engineering and Technology, India
Raenu Kolandaisamy	UCSI University, Malaysia
Rahul Kumar Verma	Bennett University, India
Rahul Kumar Vijay	Banasthali Vidyapith, India
Ramesh Chand Pandey	Rajkiya Engineering College, Ambedkar Nagar, India
Rashmi Chaudhry	Netaji Subhas University of Technology, India
Rashmi Gupta	Atal Bihari Vajpayee University, India
Ravilla Dilli	Manipal Institute of Technology, India
Revathy G.	Sastra University, India
Richa Mishra	University of Allahabad, India
Rohit Kumar Tiwari	Madan Mohan Malaviya University of Technology, India
Rohit Singh	International Management Institute, Kolkata, India
S. Gandhiya Vendhan	Bharathiar University, India
Sachi Nandan Mohanty	College of Engineering Pune, India
Sadhana Mishra	ITM University, India
Sanjeev Patel	National Institute of Technology, Rourkela, India
Sanjeev Sharma	Indian Institute of Information Technology, Pune, India
Saumya Bhadauria	ABV-Indian Institute of Information Technology and Management, India
Saurabh Bilgaiyan	Kalinga Institute of Industrial Technology, India
Saurabh Kumar	LNM Institute of Information Technology, India
Seera Dileep Raju	Dr Reddy's Laboratories, India
Shailesh Kumar	Jaypee Institute of Information Technology, India
Shantanu Agnihotri	Bennett University, India
Shiv Prakash	University of Allahabad, India
Shivam Sakshi	Indian Institute of Management, Bangalore, India

Contents

Intelligent Computing

Advanced Network Technologies

Wirelessly Controlled Plant Health Monitoring and Medicate System Based on IoT Technology

Lijaddis Getnet Ayalew(✉), Channamallikarjuna Mattihalli, and Fanuel Melak Asmare

Bahir Dar University, Faculty of Electrical & Computer Engineering, Bahir Dar Institute of Technology, Bahir Dar, Ethiopia

Abstract. The availability of good quality fruits and vegetables is paramount in preventing starvation and minimizing outbreaks of diseases which leads to improving quality of life. One of the major obstacles of the mentioned availability is plant leaf disease. Although manpower plays a vital role in detecting such problems it is time-intensive, expensive, and very inefficient. Thus, developing a mechanism to vigorously monitor leaf's health and detect diseases of plant leaves at early stages is mandatory so that one can produce plenty. In this contribution, a system that detects leaf disease is developed using image processing algorithms, the k-nearest neighbor (KNN), support vector machine (SVM), and multilayer perception (MLP) machine learning algorithms are compared based on plant disease detection and classification systems performances. We also developed a prototype of simple-to-install technology that can recognize leaf diseases and allow medicine flow based on the results. This paper presents a smart plant health monitoring system that takes into account humidity, temperature, and soil contents.

Keywords: Image processing · IoT · KNN · MLP · SVM

1 Introduction

In many developing countries like Ethiopia, where more than 80% of the population lives in rural areas, their income is based on agriculture. These many farmers' living standard depends on the quality and yield of their products. Mostly this sector is challenged by plant diseases, pests, and weeds. Where traditional methods are still used to identify and threat. These traditional methods are time-intensive and need high manpower investment which leads to lesser quality outputs and wastage of resources. Thus, for effective farming auto health monitoring and disease detection systems are required.

Different parts of plants are infected by numerous kinds of diseases. The common plant diseases which result in huge production loss include viral, fungi, and bacterial disease and the causes of each are environmental changes, the presence of fungus, and the presence of germs respectively. Among these, leaf diseases highly disturb plants life cycle as it directly affects the course of photosynthesis. And bacterial blight, Cercospora Leaf Spot, Powdery Mildew, Anthracnose, and Rust are common leaf diseases in mango,

I. Woungang et al. (Eds.): ANTIC 2021, CCIS 1534, pp. 3–14, 2022.
https://doi.org/10.1007/978-3-030-96040-7_1

wheat, apple, grape, etc. For example, leaves attacked by bacterial blight show symptoms of dark brown surrounded by dark yellow.

We visited mango and apple farms here in Bahir Dar, Ethiopia and the mentioned problem is visible and threatening. According to experts on plant diseases, many of the observed leaves are affected by viruses and fungi. Because of leaf diseases, changes in the leaf's texture, color, shape, and size are noticeable. But still, experts are needed for proper identification and treatment suggestions. A Wrong diagnosis may lead to huge production loss.

To solve these problems the need for a detection system is unquestionable. Nowadays Internet of Things (IoT), machine learning, and Remote Sensing (RS) techniques are used in different areas of research for detecting, collecting, and analysis of data. Applying such techniques, many works were proposed to automate plant disease detection. Different machine learning algorithms along with image processing were used in [1–6] to detect and classify leaf diseases. G. Saradhambal, et.al [7] have also proposed a neural network classifier-based automatic plant diseases identification and classification system. Again Gurleen Kaur, et.al [8] have used SVM and Naïve Bayes classifiers for plant disease detection. In this work, SVM outperforms the Naïve Bayes classifier where achieved accuracies are 85% and 79% respectively. In [9], a web-based fruit diseases detection system is developed with an SVM classifier and 82% accuracy was achieved. In [10] a genetic algorithm and in [11] SVM and ANN are used. Channamallikarjuna M., et.al [6]

Table 1. Comparison of different previous works

Authors	Year	Algorithms used	System accuracy
M.Bhange et.al	2015	k-means clustering and SVM are used for segmentation and classification respectively	Using SVM they achieved 82% system accuracy
V. Singh et.al	2016	the search capability of the genetic algorithm has been used for segmentation where SVM and Genetic algorithms are used for discrimination	Using SVM the accuracy of 95.71% and Genetic Algorithm accuracy of 93.63%
R.Meena P et. al	2017	L*a*b color space, k-means clustering for segmentation, and SVM classifier	The accuracy achieved is 90% - 100% for the specified classifier
Pooja V et. al	2017	k-means clustering for segmentation and SVM for classifications	an average of 92.4% system accuracy is achieved
Gurleen K. and Rajbir K	2019	The major techniques employed were: BPNN, SVM, K-means clustering, Otsu's algorithm, CCM and SGDM	Using SVM and Naïve Bayes classifiers overall system accuracy of 85% and 79% respectively are achieved

proposed a statistical measure of distance, Bhattacharyya distance algorithm, based plant leaf disease detection system for real-time agricultural land automation. In this work Hue Saturation Value (HSV) color space representation is used for color perception.

K-means clustering is used for image segmentation in [7–11]. And GLCM texture feature extraction is also used in [4, 5, 7]. And in all cases developing accurate and reliable plants diseases detection systems is the ultimate challenge. Performances achieved in previous studies are compared in Table 1.

This work aimed to develop plant health monitoring and leaves disease detection systems at the early stages of the problem with possible maximum accuracy using leaves texture. Machine learning classifiers and image processing techniques are used for the identification and classification of healthy and unhealthy leaves. We compared the mango leaf diseases detection performance of K-Nearest Neighbor (KNN), Support Vector Machines (SVM), and MLP classifiers. Additionally, we vary image segmentation techniques for better accuracy of detection.

Then audio command-based wirelessly controlled plant health monitoring system based on IoT technology is presented. The system is developed applying image and audio signal processing techniques and the K-Nearest Neighbor (KNN) algorithm for the classification and identification of leaf diseases. Temperature, soil moisture, and contents, and humidity status of farming areas are remotely controlled using sensors. The presented method is easily configurable and cost-effective.

2 System Procedures and Architecture

2.1 System Procedures

In this work, the accuracy of different classifiers is compared by varying the number of datasets extracted and used for discrimination. The working principle of the proposed system involves two distinctive phases: the training phase and the testing phase. As can be seen from Fig. 1 below, leaf image acquisition, preprocessing, feature extraction, feature storage in the database, and train-test dataset split are the main activities during the training phase.

The testing phase starts from the train-test dataset split. Testing features' distance from features used to train classifiers is calculated where labeled class with minimum distance can be taken as a classification result.

Images of mango leaves are captured using picamera with the required resolution. We have used half of the training leaf images from the existing database which is available on the internet [12]. We have considered two classes of mango leaves where unhealthy category considers Bacterial Blight, Cercospora Leaf Spot, and Rust which are common in mango, wheat, apple, grape, etc. We have collected and included 95 healthy and 154 unhealthy mango leaf image samples which make a total of 249. As it is difficult to use captured images (RGB) directly, preprocessing is imposed as the next step to enhance image data.

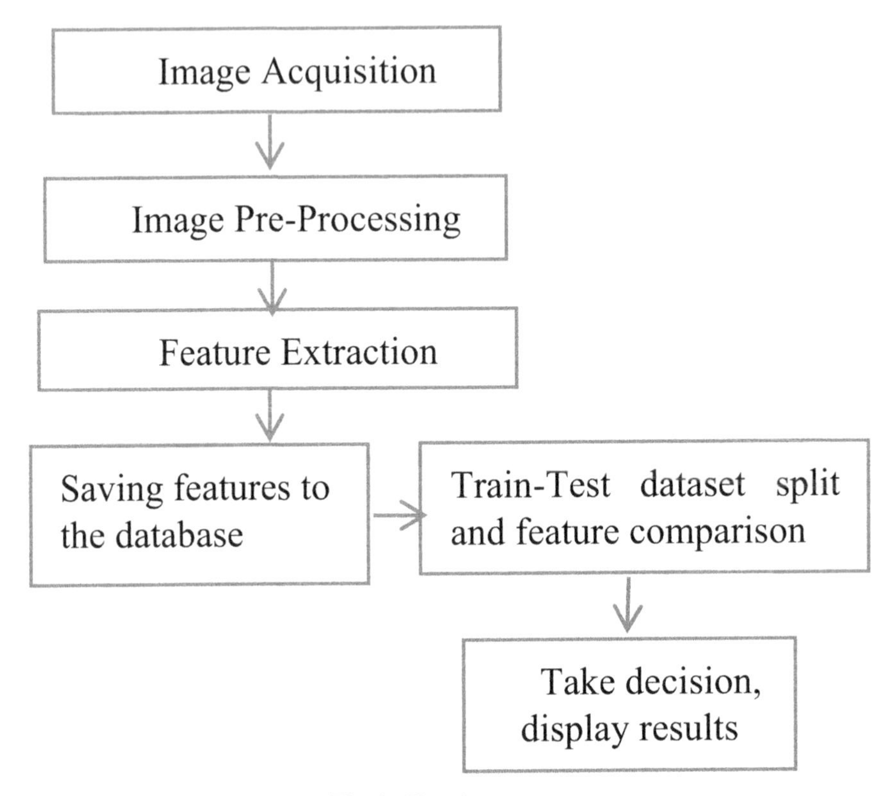

Fig. 1. Flowchart system

Image pre-processing is working on images to enhance image properties for further processing and analysis [13]. At the very beginning, input images are resized, grayscale clustered, and converted into the L*a*b color space. Color space representation is done for better color perception. Once images' noises, backgrounds, and undesired distortions are reduced, pre-processed images will be ready for segmentation.

Segmentation is applied to get an area of interest which is an infected region of leaves and various techniques like morphological image processing, k-means clustering, and Gaussian mixture model have been used. In our case, k-means clustering which accomplishes segmentation by minimizing squares of distances between image intensities and cluster centroids has been used. We also have applied morphological image processing, by taking the structure of the image into account, to remove imperfections that resulted during segmentation. Removal and addition of pixels to borders of objects in an image erosion and dilation respectively are the most common morphological operations.

Once an image is segmented, it is ready for feature extraction. Here Gray-Level Co-Occurrence Matrix (GLCM) features, which consider the spatial relationship of pixels, are generated and saved in the system database as comma-separated value (.csv) files. These stored features are used to train our system and as part of this proposed system, four features which include contrast, homogeneity, energy, and correlation are considered (Table 2).

Table 2. Features descriptions

Features	Representation	Possible values
Contrast	A measure of differences among neighbor pixels	Between 0 and 1
Homogeneity	Measures similarities among pixels	Between 0 and 1
Energy	Measures uniformity among squared elements summation in the GLCM	Between 0 and 1
Correlation	Measures association among neighbor pixels	Between -1 and 1

Contrast: Measures intensity contrast of a pixel and its neighbor pixel over the entire image. If the image is constant, contrast is equal to 0. Where Pij represents the (i, j)th entry in the normalized Gray-Level Co-Occurrence Matrix and N represents the number of distinct gray levels in the quantized image, the equation of the contrast is as follows [2].

$$\mathrm{Contrast} = \sum\nolimits_{\mathrm{i,j}=0}^{\mathrm{N}-1} (\mathrm{P_{ij}})(\mathrm{i}-\mathrm{j})^2 \tag{1}$$

Energy: Measures uniformity among squared elements summation in the GLCM. The range is between 0 and 1. Energy is 1 for a constant image. The equation of the energy is given by equation

$$\mathrm{Energy} = \sum\nolimits_{\mathrm{i,j}=0}^{\mathrm{N}-1} \left(P_{ij}\right)^2 \tag{2}$$

Homogeneity: Measures similarities among pixels. Its range is between 0 and 1. Homogeneity is 1 for a diagonal GLCM. The equation of the Homogeneity is as follows.

$$\mathrm{Homogeneity} = \sum\nolimits_{\mathrm{i,j}=0}^{\mathrm{N}-1} \frac{\left(P_{ij}\right)^2}{\left(1+(\mathrm{i}-\mathrm{j})^2\right)} \tag{3}$$

Correlation: Measures how correlated a pixel is to its neighborhood. Its range is between -1 and 1. Where μ, and σ represents the mean value of pixels contributed to Gray-Level Co-occurrence Matrix and variance respectively.

$$\mathrm{Correlation} = \sum\nolimits_{\mathrm{i,j}=0}^{\mathrm{N}-1} P_{ij}\left(\frac{(i-\mu)(j-\mu)}{\sigma^2}\right) \tag{4}$$

In this work, we have considered the KNN, SVM, and MLP algorithms for classification purposes. The KNN algorithm is a majority-vote-based supervised machine learning algorithm that assumes similar data points typically exist near each other. The value of k determines the boundary of clusters. We have tried different values of k and choose k that minimizes the error of classification. Usually, it is good to choose odd k to have a tiebreaker.

The support vector machine is also a supervised machine learning algorithm to get a hyperplane that distinctly classifies labeled data points. There can be many possible hyperplanes to separate two classes. And the ultimate goal is to find the hyperplane which minimizes misclassifications.

Multilayer Perception is the other classifier considered and it is a feedforward artificial neural network model that classifies a set of inputs to appropriate classes. Before applying any of the classifiers, feature vectors pre-stored in the database are divided into training and testing features. Classifiers use training features as experience to classify testing features. The performance of each classifier is determined by comparing predicted labels with corresponding actual values.

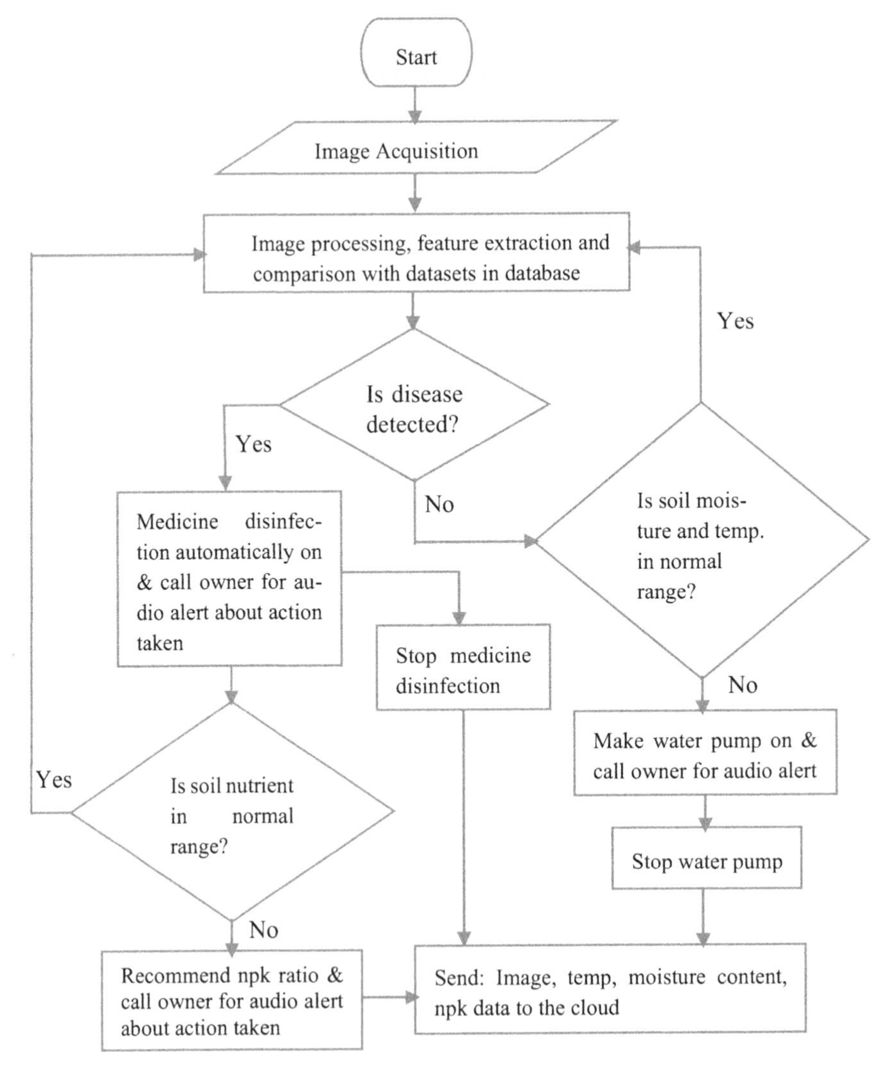

Fig. 2. Flowchart of the proposed system

If the detection result is diseased, the system is automated to take certain actions depending on the stage and type of disease. Here, if the fungal infection is only at the early stages, a disinfection system with less flow rate will be on for supplying the medicine. Providing information for farm owners and responsible officials about actions taken by the system through audio auto-play calls via GSM is the other system feature. When a disease is detected system is automated to call the owner and wait for an audio command. Owners can also check the status of their farm via the audio call of the system.

Plants need sufficient water for their existence. That is why they are highly affected at a dry time. And even with ample availability of water, pumping excess water is a wastage. Our system can address such situations as it includes a moisture sensor that makes water-pump on or off depending on soil moisture level. Responsible bodies can command the system for medicine, water flow depending on various scenarios on the farmland.

In parallel to leaf disease detection, our system sends real-time leaf images taken at the farm and related information to the cloud computing database. This feature of our system is valuable as stored information can be used by statisticians and other researchers for further analysis.

In general, the proposed real-time system shown as a flowchart in Fig. 2 is helpful as it can reduce labor effort and unnecessary time investments. And its cumulative effect can enhance agricultural yield and product quality.

2.2 System Architecture

The general block diagram of the proposed system is shown in Fig. 3. The raspberry-pi 3b+ is used as the main processor device and the database of leaf images (i.e. healthy and diseases) are pre-stored. A picamera with a required resolution is interfaced with the processor device and placed in an agricultural farm. The camera is set to capture leaf images continuously. Then the captured images have been processed and features are extracted and compared with the datasets of the leaves which are pre-stored in the device. In addition, soil moisture and temperature sensors are placed to avoid the spreading of diseases due to change in climatic conditions. Relays are used to turn on/off the valves for controlling water and medicine flow. Soil content sensors are also interfaced with the device for checking soil-content status and recommend actions to be taken. As part of the system, GSM is used to inform responsible bodies about the diseases stage and types and actions taken by the system. As compared with [6] and other similar works, our system includes additional features like soil content detector, cloud storage, and audio-based owner alert.

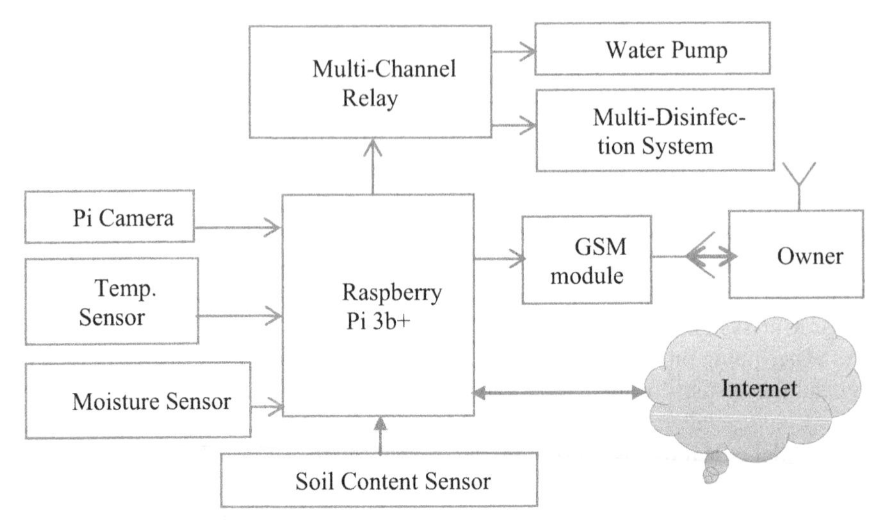

Fig. 3. System architecture

3 Experimental Setup and Results Discussions

In systems developed, we have considered two classes labeled as healthy and unhealthy. In unhealthy class leaves with Bacterial Blight, Cercospora Leaf Spot and Rust diseases which are common in mango leaves are included. To get accurate results for the variable number of input values, random numbers of datasets have been organized and tested. To be precise 60, 120, 180, and 249 data have been fed to the system which constitutes both healthy as well as unhealthy ones.

We applied image processing on RGB images and made them ready for feature extraction. During this process images are resized to 200 × 300 pixles, grayscale clustered, converted into L*a*b color space, and segmented applying k-means clustering and are shown in Fig. 4. Segmentation helps to identify a diseased part of leaves.

For all 249 images, GLCM texture features including contrast, homogeneity, energy, and correlation are pre-extracted and stored in the system database. SVM, KNN, and MLP algorithms are used for classification. Once features are ready, we applied the train-test split module with a test size of 0.2. We have tried different test size values and 0.2 is chosen as it results in good accuracy. Then testing datasets are compared with the templates included in the train category for the three classifiers. The closest match is displayed and the accuracy of algorithms used is also calculated and opted as performance metrics.

As in KNN, a certain test dataset is compared to k-neighbors, we have to choose k which can give optimum performance. And in this work, we have picked $k = 5$ and simulated the system with a KNN classifier. Similarly, SVM can be implemented with linear, polynomial, or other classifiers. In our case, we have used a linear kernel which results in good prediction accuracy. MLP classifier is also implemented with three hidden layers of size (20,180,200) and a maximum iteration of 5000 during performance

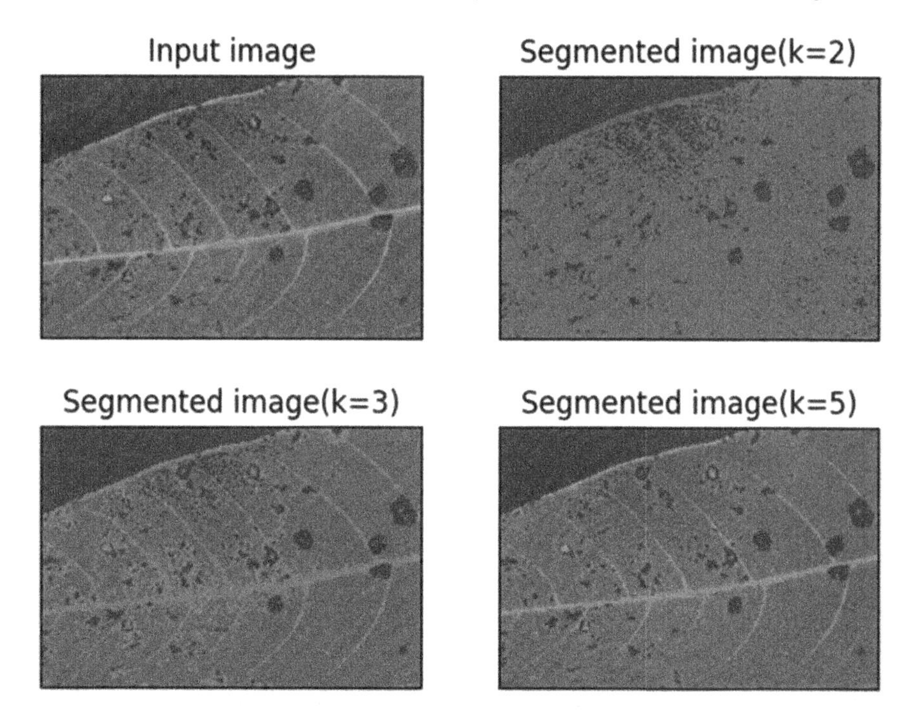

Fig. 4. Segmented mango leaf images with bacterial black spot

comparison. We have varied and tried different hidden layer combinations for good accuracy.

As shown in Table 3 below, we have achieved maximum accuracy of 98%, 90%, and 98% for KNN, SVM, and MLP algorithms respectively. We randomly choose different parameters for each classifier like the number of datasets in the training database to see its effect on classifiers' performance.

We can see that increment of datasets above some level doesn't significantly change that much the accuracy of the KNN classifier. In all cases, systems are developed using python. When we have compared these algorithms in terms of computational time MLP is time-intensive while SVM needs less computation time.

After selecting the components for the project, the fully assembled working prototype is designed for this purpose which is shown in Fig. 5. When the Raspberry pi 3b+ processor detects diseases with the aid of a camera, the medicine flow is automatically activated by turning on the valve. At the same time, auto-call will be activated where pre-stored audio about actions taken by the system is played and transferred to the farm owner. The system has also the possibility to receive audio commands back from the owner.

Table 3. Classifiers performance comparison

Classifier type	Number of datasets	Corresponding performance
KNN	62(k = 5)	85.71%
	124(k = 5)	92%
	187(k = 5)	94.74%
	249(k = 5)	98%
SVM	62(linear)	53.85%
	124(linear)	65%
	187(linear)	84.21%
	249(linear)	90%
MLP	62(20,180,200)	75%
	124(20,180,200)	84%
	187(20,180,200)	92.105%
	249(20,180,200)	98%

Fig. 5. Experimental system setup

4 Conclusions

The identification of plant leaf defects manually is inefficient and time-intensive. This can cause huge agro-production loss and ultimately brings nutrition shortage. Hence need for automatic plants leaves infection detection and classification system with maximum possible accuracy, in this technology-driven world, is demandable. For this purpose, we have implemented KNN, SVM, and MLP based leaf diseases detection and classification systems.

Systems considered are helpful to reduce labor effort and unnecessary time investments. And their cumulative effect can enhance agricultural yield and product quality. Systems developed are with optimum accuracy and can be used in plant leaves disease detection and classification. These systems can replace experts to identify plants' leaf diseases and simplify the tedious activity in the process.

Systems with different classifiers are implemented using python and achieved an accuracy of 98%, 90%, and 98% for KNN, SVM, and MLP respectively. And we recommend KNN classifier with 98% accuracy with optimum accuracy and computationally less intensive. As it can be seen from Table 1, the achieved performance of 98% is comparatively better. In all cases, the k-means clustering algorithm is used for segmentation and GLCM features are extracted and used for discrimination. Early detection of plant infections is supported by the proposed IoT platform. The proposed IoT platform would help with the early detection of plant infections. If any diseases are identified, the auto-medicine device is switched on, and medication is sprayed onto the rural ranch through a sprinkler. The installation of a platform for programmed medicine in the horticultural sector offers a possible solution to assist site-specific water system administration, allowing manufacturers to increase productivity by detecting diseases early on.

Similarly, soil humidity and temperature sensors are used to monitor disease transmission as climatic conditions change. The Raspberry pi 3b+ empowers the plants to take auto-medicine if the dampness/temperature levels reach the predetermined limit. It can be customized to any kind of leaf depending on the client's requirements. A high-speed processor is used to demonstrate the scheme. The client receives details about the discovery of the illness and the function of the valve via GSM. This structure is robust, easy to use, and needs little effort.

References

1. Pooja, V., Rahul, D., Kanchana, V.: Identification of plant leaf diseases using image processing techniques. In: 2017 IEEE International Conference on Technological Innovations in ICT for Agriculture and Rural Development (TIAR 2017). 978-1-5090-4437-5/17
2. Meena Prakash, R., Saraswathy, G.P., Ramalakshmi, G.: Detection of leaf diseases and classification using digital image processing. In: 2017 IEEE International Conference on Innovations in Information, Embedded, and Communication Systems (ICIIECS) (2017)
3. Tichkule, S.K., Dhanashri. H.: Plant diseases detection using image processing techniques. In: 2016 IEEE Online International Conference on Green Engineering and Technologies (IC-GET). 978-1-5090-4556-3/16
4. Suma, V., Amog Shetty, R., Rishab, F.T., Sunku, R. Triveni, S.P.: CNN-based leaf disease identification and remedy recommendation system. In: 2019 IEEE Proceedings of the Third

International Conference on Electronics Communication and Aerospace Technology [ICECA 2019], pp. 395–399 (2019). 978-1-7281-0167-5/19
5. Devaraj, A., Rathan, K., Jaahnavi, S., Indira, K.: Identification of plant disease using image processing technique. In: 2019 IEEE International Conference on Communication and Signal Processing, April 4–6, 2019, pp. 749–753 (2019). 978-1-5386-7595-3/19
6. Channamallikarjuna, M., Edemialem, G., Fasil, E., Adugn, N.: Real-time automation of agriculture land, by automatically detecting plant leaf diseases and auto medicine. In: 32nd 2018 IEEE International Conference on Advanced Information Networking and Applications Workshops, pp. 325–330 (2018)
7. Saradhambal, G., Dhivya, R., Latha, S., Rajesh, R.: Plant disease detection and its solution using image classification. Int. J. Pure Appl. Math. **119**(14), 879–884 (2018)
8. Gurleen, K., Rajbir, K.: Plant disease detection techniques: a review. In: 2019 IEEE International Conference on Automation, Computational and Technology Management (ICACTM), pp. 34–38 (2019). 978-1-5386-8010-0/19
9. Hingoliwala, M.B.H.A.: Smart farming: pomegranate disease detection using image processing. In: 2nd International Symposium on Computer Vision and the Internet, Vol. 58, pp. 280–288 (2015)
10. Vijai, S., Varsha, Misra, A.K.: Detection of an unhealthy region of plant leaves using image processing and genetic algorithm. In: 2015 IEEE International Conference on Advances in Computer Engineering and Applications, pp. 1028–1032 (2015). 978-1-4673-6911-4/15
11. Khirade, S.D., Patil, A.B.: Plant disease detection using image processing. In: 2015 IEEE International Conference on Computing Communication Control and Automation, pp. 768–771 (2018)
12. Chouhan, S.S., Kaul, A., Singh, U.P., Jain, S.: A database of leaf images: practice towards plant conservation with plant pathology. In: 2019 4th International Conference on Information Systems and Computer Networks (ISCON), pp. 700–707 (2019). https://doi.org/10.1109/ISCON47742.2019.9036158
13. Rastogi, A., Arora, R., Sharma, S.: Leaf disease detection and grading using computer vision technology &fuzzy logic. In: 2015 IEEE 2nd International Conference on Signal Processing and Integrated Networks (SPIN), pp. 500–505 (2015)

Multiple-Parameter Based Clustering for Efficient Energy in Wireless Sensor Networks

Ankita Srivastava(✉) and Pramod Kumar Mishra

Banaras Hindu University, Varanasi 221005, India
ankita.srivastava17@bhu.ac.in

Abstract. It is the smart technology era that has realizes the importance of research work in WSN's as it is the basic which can be applied to any technology for sensing and transmitting the information. While many research works has been done in this area but still many issues exists in WSN's such as efficient energy consumption, network lifetime, security, network connectivity, network coverage, optimal cluster heads selection and much more. As many solutions were provided by the researchers, clustering is one of the useful approach which enhances the network lifetime by making cluster and sending data via cluster heads. But selection of cluster heads in optimal way is need of today's era. Many works has been done regarding the selection of cluster heads b considering one or two attributes but for optimal solutions more attributes need to be considered. In this paper, MADM based TOPSIS approach has been used by taking four attributes for selection of optimal cluster heads and results of our simulation outperforms the other algorithms such as LEACH and EECS.

Keywords: WSN · Clustering · MADM · TOPSIS · Energy efficient

1 Introduction

Wireless sensor networks [1] has gained much recognition in recent years for its application in every fields. Wireless sensor networks have large numbers of small sensor nodes which can be deployed in healthcare [2, 3], agriculture, and military, smart homes, smart cities [4], etc. These sensors are small in size and have limited power battery. Their replacement is not so easy. Much issues in WSN's exists such as data aggregation, security, energy conservation, load balancing, and network lifetime much. Many researchers has given solutions for these issues, clustering is considered to be most important solutions which provide efficient energy consumption. In clustering, nodes were divided into clusters and after that from each clusters one cluster heads has been selected. These CHs (cluster head) collects the information or data from their sensor nodes and perform operation on them. After that cluster heads sends data to the base station. For clustering various algorithms has been proposed, LEACH [5] is the classical algorithm for the clustering. After this many improvements has been done over LEACH such as LEACH-C

I. Woungang et al. (Eds.): ANTIC 2021, CCIS 1534, pp. 15–24, 2022.
https://doi.org/10.1007/978-3-030-96040-7_2

[6], H-LEACH [7], EECS [8], EEHC [9], Random LEACH, etc. In clustering, many research work focusses on considering only one or two attribute for selection of cluster heads which leads to early die of the node in the networks. In certain applications [24, 25] where the sensor nodes were applied such as under water, military, healthcare where replacement of battery in sensor nodes is not possible and failure of one or more nodes leads to failure of networks thus for increasing the network lifetime such approaches needs to be applied where network lifetime increase with balanced load.

In our paper we have proposed multi-attributes decision matrix for selection of cluster heads by considering four conflicting factors and making co-ordination among them. Here we have used TOPSIS [10] method for cluster head selection. In today's era where everything is connected, technology driven in such sensor nodes needs optimal solution for increasing the network lifetime which will increase various applications performance too. In this paper, we have used TOPSIS methods for cluster heads selection for enhancing the network lifetime. MADM is an approach where multiple attributes for certain types of alternatives has been taken and co-ordination among has been made. This approach has been used in many fields for getting optimal solution. The rest of the paper has been organized as: Sect. 2 related work has been discussed. In Sect. 3 system model and proposed approach has been discussed. Section 4 the results and simulation part has been discussed giving the details of our work. In last section the conclusion of the paper together with future scope has been discussed giving the researchers an idea to work for issues in WSN (Fig. 1).

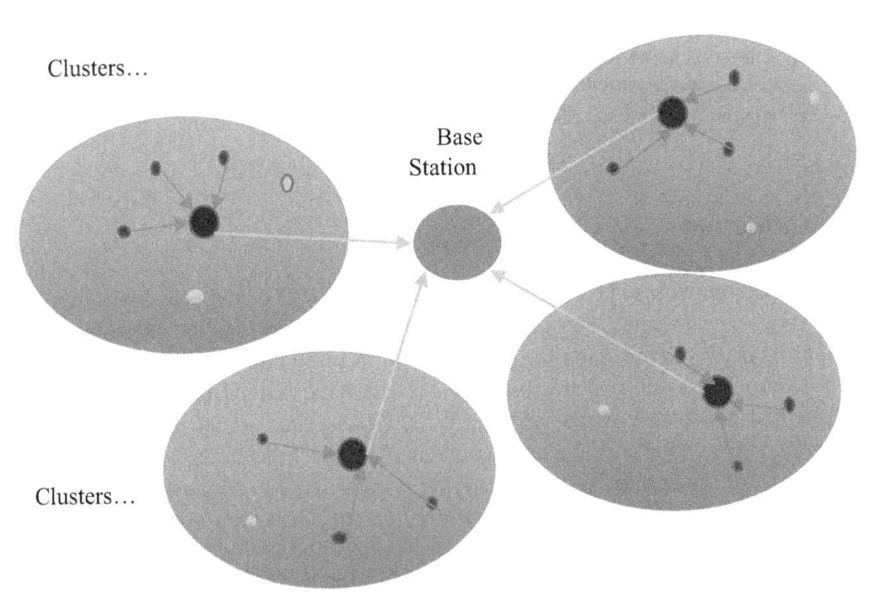

Fig. 1. Clustering in wireless sensor networks

2 Related Work

As various new emerging technologies were emerging among these WSN's plays an important role by collecting data through sensors, analyzing them, senses the environment, for wireless communication thus all of the works could not be performed without using WSN [11–13] and it directly related to IoT [13, 21–23, 26–28]. In this, section various works related to WSN has been discussed giving an understanding of the current scenario. Author [14] in this paper provides the detailed survey regarding various algorithms and protocols proposed for WSN's and current issues regarding this has been presented.

Papers also describes about the various applications of WSN's such as military, health, environment and commercial uses. Fault tolerance: is also known as the reliability which means the ability of sensor networks to survive even if any damaged caused by sensor node. Reliability depends upon the type of application of sensor networks. Power consumption: divided into three domain sensing, communication and data processing. [15] This paper the H-LEACH has been compared to LEACH and HEED algorithm. H-LEACH results shows that it is more efficient than these algorithms. H-LEACH is hybrid of LEACH and HEED algorithms which considers the residual energy and maximum energy of the nodes for cluster heads selection. Here the average and residual energy of nodes plays an important role in cluster heads selection and also it overcomes the problems of energy issues of nodes. HEED algorithm [16] has been introduced which is a hybrid approach which selects CH's based on their residual energy and on the basis of communication cost the nodes joins cluster. It also break tie by considering the proximity of nodes with its peer one's or degree of node. HEED is efficient for prolonging the network lifetime and also utilizes the various transmission power available at sensor nodes. As we know that clustering sensor nodes is an efficient technique for improving life time of sensor networks and its scalability. Cluster heads in clustering performs various tasks due to which it has extra load thus load balancing is crucial issue in WSN's in long run. Load balanced clustering is a NP-hard problem. Most of the time cluster heads were selected from normal sensor nodes which dies soon so some time special types of nodes can be used such as relay nodes or gateways having extra energy. But these are also battery operated thus improper formation of clusters may cause overloaded cluster heads. Two clustering algorithms were proposed [17], first algorithm considers equal load on the CHs while assigning the sensor nodes to it and second algorithm consider unequal load Balance on CHs while assigning sensor nodes to it. 2-Approximation and 1.5 Approximation have been proposed whose running time is (O nlogn) and compared with GLBCA, LBC, LDC and GA based clustering algorithms. But the proposed algorithms perform better than other. In this, considers two scenario one in which CH's have equal load and other in which CH's have unequal load balance. The results best shows its efficiency in terms of execution time, network life and balancing loads of CH'S.

As we know that in wireless sensor networks due to limited power resources the positioning of base station plays an important role thus dynamic positioning of base station could increase the lifetime of network. Positioning problem is a NP-Hard problem thus in this paper genetic algorithm has been used for solving the positioning problems. Each sensor nodes have limited energy capability due to which they can sense up to limited area thus WSN's requires more nodes to be deployed in a region. In this paper [18] the dynamic positioning of base station has been proposed by using genetic algorithm for optimal solution as positioning problems were NP-Hard problems. Here the author used same parameters as that of LEACH and HEED. It's simulate the results first with LEACH and HEED after that it combined DBSR algorithm with HEED, LEACH and compare the results. The results shows that DBSR outperforms significantly with the energy consumption and BS replacement dynamically. In this paper [19] lifetime of networks can be increased by node deployment has been shown. In this author consider random node deployment in square area, predetermined node deployment in rectangular area and hexagonal grid with mobile sink. The results shows that node deployment with mobile sink increases the network lifetime and also resolve energy hole problem of networks and hexagonal grid deployment of node has better coverage and connectivity than other two methods of node deployment. In this paper [19] energy-LEACH and Multi-hop-LEACH has been proposed and compared with LEACH algorithm. In this E-LEACH chooses cluster heads on the basis of residual energy of sensor nodes i.e. [20] nodes having higher residual energy as cluster heads in next round and Multi-hop-LEACH improves communication by using multi-hop instead of single hop. Simulation results shows that energy –LEACH and Multi-hop-LEACH outperforms than LEACH algorithm.

3 System Model and Proposed Approach

3.1 System Model

In wireless sensor networks, selected cluster heads (CHs) aggregate the data from the sensor node which sense the Data from the surrounding. Data collected by sensor nodes then they transmit it to respective ch, cluster heads sends data to the base station. For the transmission of data, energy or system model of energy is required here we have considered the basic energy model for our simulation [5] (Fig. 2). The model uses the following energy for ***k*** bit transmission and reception (Table 1).

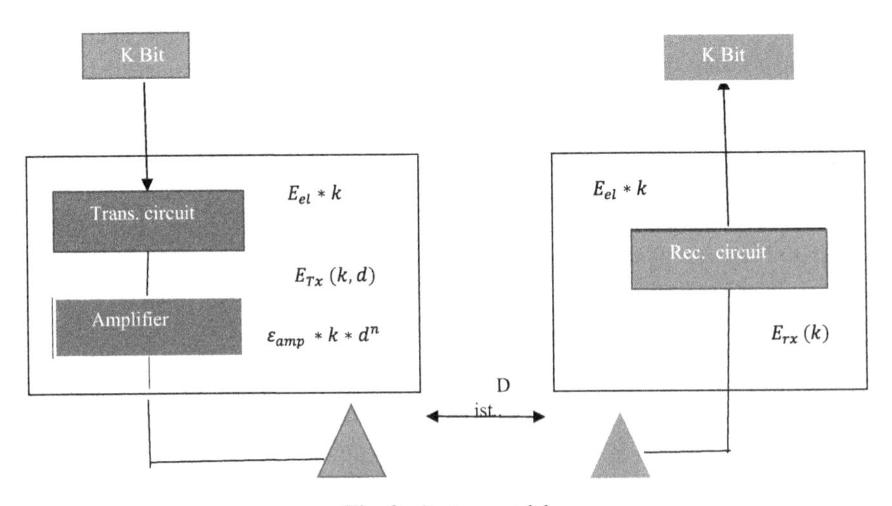

Fig. 2. System model

$$E_{tra} = E_{(tra-el)}(l) + E_{(tra-amp)}(l, d) \quad (1)$$

$$E_{tra} = \begin{cases} l * E_{el} + l * \varepsilon_{fs} d^2, ifd < d_0 \\ l * E_{el} + l * \varepsilon_{amp} d^4, ifd \geq d_0 \end{cases} \quad (2)$$

$$E_{rec} = E_{(rec-el)}(l) = lE_{el} \quad (3)$$

Table 1. Parameters values

Parameter name	Parameter value
Co-ordinates of origin	(0,0)
Deployment area	100*100
Sink position	Variable
Number of sensors	100
Initial energy of nodes	.2 J

3.2 Praposed Approach

We have proposed MADM based TOPSIS method for cluster heads (CHs) selection for the best deployment of the network by considering coverage, connectivity, distance from BS and residual energy. Here, the simulation has been done by using MATLAB. By using the proposed algorithm, efficient energy consumption can be achieved. The algorithm for the proposed work has been deliberated in Algorithm 1. Multiple parameters considered for the CH's selections were necessary for efficient energy consumption, here we want the maximum coverage of sensor nodes with better connectivity and each sensor nodes have efficient residual energy for becoming the CH's. These parameters were explained below briefly.

- Coverage: It means that the number of sensor nodes far away from their respective CH's, the less number of sensor nodes denotes more coverage.
- Connectivity: It means the number of CH's which are far from the base station, its lesser value indicate higher connectivity.
- Distance from BS: This factor tells us the maximum possible distance cluster heads can have, its lesser values is required for efficient energy consumption.
- Residual Energy: It indicates the remaining energy left for cluster heads or sensor nodes for the transmission of data, its higher value is needed for efficient energy consumption.

The proposed methodology procedure has been discussed step by step:

Step 1: First we have generate the 20 alternatives and calculated their values. After this, now we apply TOPSIS method for ranking the alternatives and finally we select the best alternative i.e. CH's for data transmission. The next step explains the TOPSIS process.

Step 2: Apply MADM approaches to rank the alternatives i.e. **TOPSIS** method:

$$M_{1ij} = \frac{M_{1ij}}{\sqrt{\sum_{i=1}^{m} M_{1ij}^2}} \text{ where } i = 1, 2, 3, 4, \ldots, n \text{ and } j = 1, 2, 3, 4, \ldots, m \quad (4)$$

Here M_{1ij} is the matrix of $n * m$.

Step 3: Now the weighted normalized matrix has been calculated M_{2ij}:

$$M_{2ij} = M_{1ij} * w_j (where\ i = 1, 2, 3, 4, \ldots, m \text{ and } j = 1, 2, 3, 4, \ldots n) \quad (5)$$

Step 4: Now compute v_b^+ and $v_{\overline{w}}$ which are the best and worst values for the attributes.

Step 5: s_k^+ and $s_{\overline{k}}$ for each alternatives will be calculated:

$$s_{ki}^+ = \sqrt{\sum_{j=1}^{m} (M_{2ij} - V_{bj}^+)^2} \text{ Here } i = 1, 2, 3, 4, \ldots, m \quad (6)$$

$$s_{\overline{k}i} = \sqrt{\sum_{j=1}^{m} (M_{2ij} - V_{\overline{w}j})^2} \text{ Here } i = 1, 2, 3, 4, \ldots, m \quad (7)$$

Step 6: P_{iw} means similarity to worst condition has been calculated:

$$P_{iw} = \frac{s_{\overline{k}}}{s_{\overline{k}} + s_k^+} \quad (8)$$

Step 7: In last step the rank to the alternatives has been given.

Algorithm 1: Proposed Algorithm

---

1. **Initialization:** $\boldsymbol{n} = number\ of\ sensor\ nodes$

   $\boldsymbol{CH's} = number\ of\ cluster\ heads$

2. Partition of the entire network into clusters number;
3. Generation of attributes and calculation of the factors;
4. Selection of CH's using the TOPSIS Method.

DATA: Sensor nodes and their positions
RESULTS: selection of CH's and nodes

---

4 Simulation and Result Analysis

For the simulation we have used the MATLAB software, in this simulation we have taken some alternatives for efficient energy consumption. Here the same model [5] has been for the simulation purpose for several rounds. Here first the co-ordination among all the some alternatives has been made by using the MADM methods and after that data were collected for the transmission. Here we have 2 cases for our simulation. The simulation results has been shown in Fig. 3, 4 and 5. Results shown in terms of FND means first node dead, LND means last node dead and CHD means cluster head dead. Figure 5 represents the comparative analysis of proposed algorithm in terms of residual energy.

CASE 1: n = 100, Base Station = (0, 0)
CASE 2: n = 100, Base Station = (50, 50)

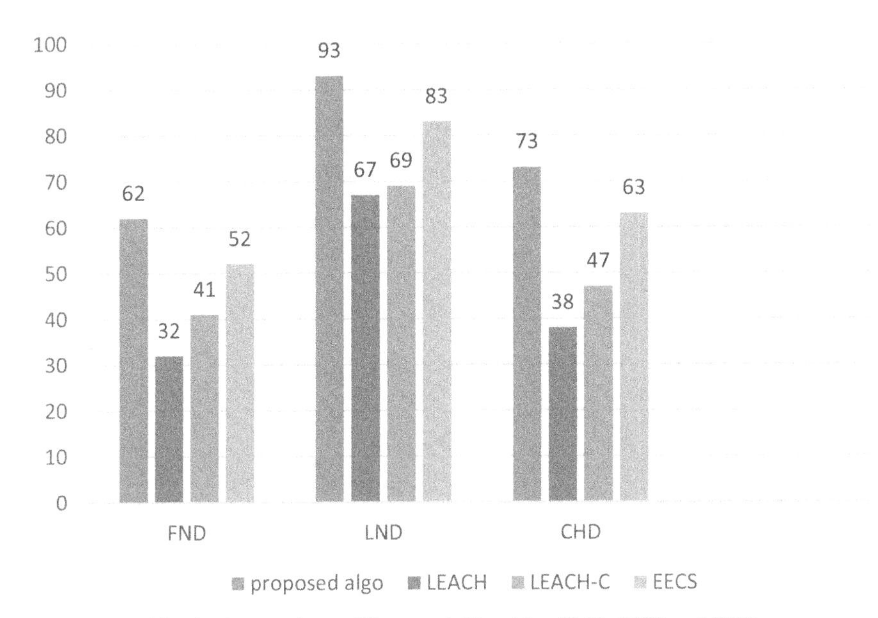

Fig. 3. Comparison of Proposed Algorithm FND, LND and CHD

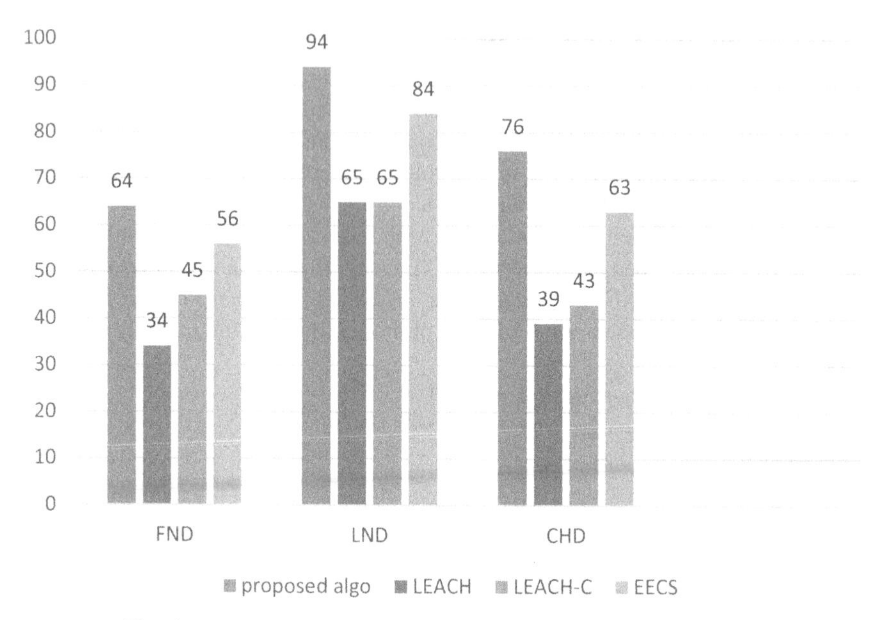

Fig. 4. Comparison of Proposed Algorithm in FND, LND and CHD

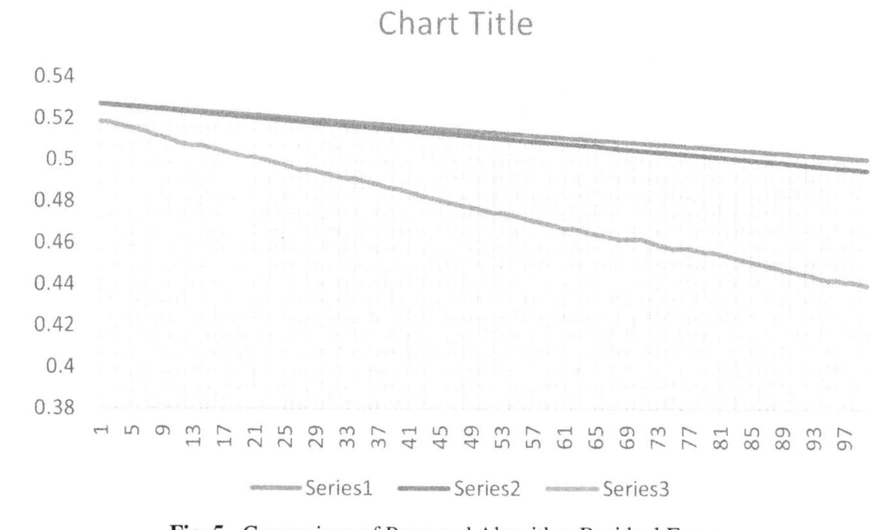

Fig. 5. Comparison of Proposed Algorithm Residual Energy

5 Conclusion and Future Direction

The proposed method gives the better performance in comparison to LEACH, LEAH-C and EECS. Here the results shows that FND, LND and CHD were better in proposed algorithm and its residual energy is higher than other algorithms. By using madm based approaches we get better performance such as increased lifetime of network and energy consumption in future we, will explore more attributes for the cluster heads selection which gives efficient energy consumption and also improved the network lifetime by the MADM based methods.

References

1. Srivastava, A., Mishra, P.K.: A survey on wsn issues with its heuristics and meta-heuristics solutions. Wireless Pers. Commun. **121**(1), 745–814 (2021). https://doi.org/10.1007/s11277-021-08659-x
2. Ketu, S., Mishra, P.K.: Enhanced Gaussian process regression-based forecasting model for COVID-19 outbreak and significance of IoT for its detection. Appl. Intell. **51**(3), 1492–1512 (2021). https://doi.org/10.1007/s10489-020-01889-9
3. Ketu, S., Mishra, P.K.: Scalable kernel-based SVM classification algorithm on imbalance air quality data for proficient healthcare. Compl. Intell. Syst. **7**(5), 2597–2615 (2021). https://doi.org/10.1007/s40747-021-00435-5
4. Dowlatshahi, M.B., Rafsanjani, M.K., Gupta, B.B.: An energy aware grouping memetic algorithm to schedule the sensing activity in WSNs-based IoT for smart cities. Appl. Soft Comput. **108**, 107473 (2021)
5. Heinzelman, W.B., Chandrakasan, A.P., Balakrishnan, H.: An application-specific protocol architecture for wireless microsensor networks. IEEE Trans. Wireless Commun. **1**(4), 660–670 (2002)
6. Tripathi, M., Gaur, M. S., Laxmi, V., Battula, R.B.: Energy efficient LEACH-C protocol for wireless sensor network (2013)
7. Vishal Gupta, M.N., Doja,: H-leach: Modified and efficient leach protocol for hybrid clustering scenario in wireless sensor networks. In: Lobiyal, D.K., Mansotra, V., Singh, U. (eds.) Next-generation networks, pp. 399–408. Springer Singapore, Singapore (2018). https://doi.org/10.1007/978-981-10-6005-2_42
8. Ye, M., Li, C., Chen, G., Wu, J.: EECS: an energy efficient clustering scheme in wireless sensor networks. In: PCCC 2005. 24th IEEE International Performance, Computing, and Communications Conference, 2005, pp. 535–540. IEEE (2005)
9. Kumar, D., Aseri, T.C., Patel, R.B.: EEHC: Energy efficient heterogeneous clustered scheme for wireless sensor networks. Comput. Commun. **32**(4), 662–667 (2009)
10. Pei, Z.: A note on the TOPSIS method in MADM problems with linguistic evaluations. Appl. Soft Comput. **36**, 24–35 (2015)
11. Srivastava, A., Mishra, P.K.: State-of-the-art Prototypes and Future Propensity Stem on Internet of Things (2019)
12. Ketu, S., Mishra, P.K.: Cloud, fog and mist computing in iot: an indication of emerging opportunities. IETE Techn. Rev. 1–12 (2021)
13. Ketu, S., Mishra, P.K.: Hybrid classification model for eye state detection using electroencephalogram signals. Cogn. Neurodyn. **1**, 18 (2021). https://doi.org/10.1007/s11571-021-09678-x
14. Sassi, H., Najeh, T., Liouane, N.: A selective 3-anchor DV-Hop algorithm based on the nearest anchor for wireless sensor network. Int. J. Electron. Commun. Eng. **8**(10), 1756–1760 (2014)
15. Razaque, A., Mudigulam, S., Gavini, K., Amsaad, F., Abdulgader, M., Krishna, G. S.: H-LEACH: hybrid-low energy adaptive clustering hierarchy for wireless sensor networks. In: 2016 IEEE Long Island Systems, Applications and Technology Conference (LISAT), pp. 1–4. IEEE (2016)
16. Younis, O., Fahmy, S.: HEED: a hybrid, energy-efficient, distributed clustering approach for ad hoc sensor networks. IEEE Trans. Mob. Comput. **3**(4), 366–379 (2004)

17. Kuila, P., Jana, P.K.: Approximation schemes for load balanced clustering in wireless sensor networks. J. Supercomput. **68**(1), 87–105 (2013). https://doi.org/10.1007/s11227-013-1024-6
18. Mollanejad, A., Khanli, L. M., Zeynali, M.: DBSR: Dynamic base station repositioning using genetic algorithm in wireless sensor network. In: 2010 Second International Conference on Computer Engineering and Applications, vol. 2, pp. 521–525 (2010)
19. Anand, V., Agrawal, D., Tirkey, P., Pandey, S.: An energy efficient approach to extend network life time of wireless sensor networks. Procedia Comput. Sci. **92**, 425–430 (2016)
20. Yektaparast, A., Nabavi, F. H., Sarmast, A.: An improvement on LEACH protocol (Cell-LEACH). In: 2012 14th International Conference on Advanced Communication Technology (ICACT), pp. 992–996. IEEE (2012)
21. Ketu, S., Mishra, P.K.: Performance analysis of machine learning algorithms for iot-based human activity recognition. In: Thangaprakash Sengodan, M., Murugappan, S.M. (eds.) Advances in Electrical and Computer Technologies. LNEE, vol. 672, pp. 579–591. Springer, Singapore (2020). https://doi.org/10.1007/978-981-15-5558-9_51
22. Sharma, A., Mishra, P.K.: Performance analysis of machine learning based optimized feature selection approaches for breast cancer diagnosis. Int. J. Inf. Technol. **1**, 12 (2021). https://doi.org/10.1007/s41870-021-00671-5
23. Sharma, A., Mishra, P.K.: State-of-the-Art in performance metrics and future directions for data science algorithms. J. of Scient. Res. **64**(02), 221–238 (2020). https://doi.org/10.37398/JSR.2020.640232
24. Mishra, S., Tripathi, A.R.: AI business model: an integrative business approach. J. Innov. Entrepreneur. **10**(1), 1–21 (2021)
25. Mishra, S., Tripathi, A.R., Singh, R.S., Mishra, P.: Design and Implementation of Internet of Everything's Business Platform Ecosystem (2021)
26. Mishra, S., Tripathi, A.R.: IoT platform business model for innovative management systems. Int. J. Financ. Eng. **7**(03), 2050030 (2020)
27. Mishra, S., Tripathi, A.R.: Literature review on business prototypes for digital platform. J. Innov. Entrepreneur. **9**(1), 1–19 (2020)
28. Mishra, S., Tripathi, A.R.: Platform business model on state-of-the-art business learning use case. Int. J. Financ. Eng. **7**(02), 2050015 (2020)

On the Performance Implications of Deploying IoT Apps as FaaS

Mohab Aly[1,2](✉), Foutse Khomh[1], and Soumaya Yacout[2]

[1] SWAT Lab., Département de Génie Informatique et Génie Logiciel (GIGL), Polytechnique Montréal, Montreal, QC H3T 1J4, Canada
{mohab.aly,foutse.khomh}@polymtl.ca

[2] Département de Mathématiques et de Génie Industriel (MAGI), Polytechnique Montréal, Montreal, QC H3T 1J4, Canada
soumaya.yacout@polymtl.ca

Abstract. Serverless computing provides a small run-time container to execute lines of codes; relieving developers from the management of the underlying infrastructure. In a serverless computing architecture, all administration operations are handled by cloud service providers. Thanks to its lightweight nature, ease of management, and ability to scale quickly, serverless computing is becoming a top choice for application development in the Internet of Things (IoT) community. However, like every new technology, there are limitations as well as advantages. For example, the highly-distributed, loosely coupled nature of Serverless applications can cause latency issues. Therefore, the advantages of a serverless architecture may come with a performance cost. To clear up this suspicion, this paper examines the performance implications of deploying IoT applications using a serverless infrastructure in comparison to deployments performed on an origin server. Our results show that different setups cannot always be assumed to have a straightforward overhead when deploying IoT applications even in the case of serverless architectures. Leveraging the results of our analysis, we provide some guidelines to help practitioners select optimal deployment strategies for their applications.

Keywords: Serverless · Eclipse · IoT · Hono · Container setups · OpenFaaS · Performance testing and evaluation · Empirical study

1 Introduction

Serverless computing, (or simply serverless), is an emerging Cloud service that enables event-driven computing for stateless functions executable on containers with small resource allocations. Nowadays, it is frequently used in the development of Cloud-based applications, thanks to the recent shift of enterprise applications architecture to microservices and containers [16]. With serverless computing, we can now deploy our applications in an environment where the Cloud provider manages all provisioning details. The focus on code has led to this concept often being referred to as Function-as-a-Service (FaaS).

Serverless computing makes it easy to patch, fix, or add new features to an application, since the application is broken up into separate, smaller functions. Using a serverless infrastructure, developers can run different functions

I. Woungang et al. (Eds.): ANTIC 2021, CCIS 1534, pp. 25–47, 2022.
https://doi.org/10.1007/978-3-030-96040-7_3

of the applications on different servers located closer to their users. They can easily scale resources up or down by controlling the number of functions that are deployed on different nodes. Due to its lightweight nature, ease of management, and ability to scale up and down quickly, serverless computing is becoming a development paradigm of choice for developers in the Internet of Things (IoT) community. However, the flexibility of the serverless architecture may come with a performance cost. In fact, because multiple parts of the code are not constantly running, they may need to 'boot up' every time they are invoked. This startup time may degrade the performance of the application. Also, the highly-distributed, loosely coupled nature of Serverless applications can induce latency issues. In this paper, we examine the performance implications of deploying an IoT application using a serverless infrastructure. We also examine the effect of that a serverless architecture can have on performance testing operations.

More specifically, we investigate the performance implications of deploying applications developed using Eclipse IoT [33] frameworks, such like the open-sourced IoT Eclipse Hono paradigm, that is getting a lot of traction in a number of businesses, i.e., industry, these days [4]. Eclipse IoT Hono can provision various types of protocol implementations, i.e., Hypertext Transfer Protocol REpresentational State Transfer `HTTP REST`, `MQTT`, Machine-to-Machine (M2M) administration, objects' authentication and control, over-the-air (OTA) update methodologies, security and enforcing it using authentication and authorization mechanisms. When developing an IoT application, developers try walk the fine line so that they are able to make a balance between choosing the proper deployments scenarios, resources consumption and the platform's quality of service (QoS). To find such a balance in a manual way is relatively a difficult challenge. Developers need guidelines, including instructions and–or recommendations, to help them choose effective design and deployment techniques.

To reduce performance issues and maintain software dependability and reliability, developers conduct performance tests on their IoT applications before deploying them in the field [38]. These performance tests are conducted by applying workloads (mimicking users' behavior in the field) on the software system [3,17,32] and monitoring some performance metrics, such as memory utilization, or latency. Because these performance tests are expensive to perform (they often need to run repeatedly over long periods of times, to build thorough statistical confidence on the obtained results [17]). Regression analysis are often performed on collected performance metrics to create estimators that can be used to make performance estimations in other contexts. To ensure that this process is accurate, it is important to understand performance variations across different contexts. Hence, in this work, we examine the effect of a serverless architecture on the relationship between different performance metrics. *To the best of our knowledge, our work is one of the first attempts that examines the discrepancy between performance testing results obtained in serverless and serverful contexts on top of open-sourced IoT frameworks using different container/orchestration setups.*

Exploring and identifying such discrepancy will help the practitioners, as well as the researchers, understand how to leverage performance testing results accurately. Our empirical studies are conducted using the open source Eclipse IoT platform, Eclipse Hono [13]. The performance tests are conducted using three container environments, i.e., Kubernetes, OpenShift and Docker Swarm. We compare testing results obtained in these environments along the following dimensions.

- Single performance metrics: the distributions and trends of each performance metrics.
- The relationship between the performance metrics: the correlation between every two performance metrics.

Results show that (1) performance metrics have different distributions shapes and trends in different container setups, and (2) relationships between different performance metrics are not stable across serverless and serverful environments. Practitioners should be careful and should not assume that performance results obtained in one environment will necessarily apply to another environment.

Organization of the Paper: The rest of this paper is organized as follows. Section 2 presents the background and discusses related works. Section 3 presents the set up of our case study. Section 4 presents the results of our case study, followed by a discussion. Section 5 discusses threats to the validity of our findings. Finally, Sect. 6 concludes the paper and outlines some avenues for future work.

2 Background and Related Works

This section provides background information about the technologies examined in this study and reviews related literature.

2.1 Serverless vs. Serverful Computing

A typical usage scenario on a serverless platform consists in writing a cloud function in a high-level programming language, specifying event(s) that should trigger the function's execution, such as pulling information about the platform being used (i.e., the configuration specs), or loading an image into cloud storage. Everything else is handled by the serverless system, including instance selection, scaling, and deployment, as well as fault tolerance, monitoring, logging, and security patches.

There are three crucial distinctions between serverless and serverful computing, that can be summarized as follows:

a. *Decoupled/not connected computation and storage.* The computation and storage scale separately. They are all provisioned and priced separately. The computation is usually stateless and the storage is handled by a distinct Cloud service.

b. *Executing code without managing resources allocation.* Instead of requesting resources, the user just gives a piece of code and the Cloud automatically creates the resources needed to run it.
c. *Paying in proportion to resources used instead of for resources allocated.* Billing is always linked to the execution, using indicators such as execution time, rather than metrics like the number and size of the virtual machines allocated (as it is the case in serverful computing).

2.2 Serverless Performance Prediction

Although Cloud functions have a noticeable lower startup latency rather than that of the traditional VM-based instances, some applications may experience substantial delays when establishing new instances. This cold start latency is influenced by three factors: (1) the time it takes to start a cloud function; (2) the time it takes to initialize the software environment of the generated function, such as loading node-information libraries, which contains the underlying platform, CPU counts and Uptime; and (3) application-specific initialization in the code. The latter two may appear small or unimportant compared to the former. While it can take less than one second to start a Cloud function, it also might take tens of seconds to load all application and–or function libraries.

Another obstacle to serverless predictable performance is the variability in the hardware resources that results from giving the container providers flexibility to pick up the underlying structure and–or architecture. Serverless computing relies on strong performance and security isolation to make multi-tenant hardware sharing possible. VM-like isolation is the current standard for multi-tenant hardware sharing for cloud functions [34]. In our prior experiment [2], we dealt with different underlying virtual-hardware technologies, including CPUs, network substates (i.e., latencies and actual network hardware influence the obtained results), memory configurations, etc. This uncertainty exposes a fundamental trade-off between the container providers desire to maximize the use of their resources and predictability.

Several serverless application developers conducted a variety of experiments to assess cold-start delay and–or latency, function instance lifetime, the maximum idle time before shutdown, and CPU consumption in Amazon Web Services (AWS) Lambda [1,5,8,9,29,35,36]. These experiments were ad-hoc and the results might be deceptive because they did not control the contention by other instances. On the other hand, a few research papers report about measured performance in Amazon Web Services (AWS). Authors in [15], assessed the desired latency and found it to have higher latency than Amazon Web Services (AWS) Elastic BeanStalk (a Platform as Service system). McGrath et al. [24] conducted early measurements on four serverless systems and discovered that Amazon Web Services (AWS) outperformed Google and Azure in terms of scalability, cold-start latency, and throughput. Recently, Lloyed et al. [20] investigated the elements that influence application performance in Amazon Web Services (AWS) and Microsoft Azure. They built a heuristic algorithm and function to identify the VM as a function that runs on Amazon Web Services (AWS) based on the VM uptime in /proc/stat. We assess the performance metrics for the underlying deployment setups, such as Kubernetes, OpenShift, as in our prior work [2], and Docker Swarm (i.e., to be considered and examined herein) and inspect their resource utilization when having Eclipse IoT-Hono and serverless application(s) on top of each.

2.3 Eclipse IoT Ecosystem

This sub-section explains the motivation behind the Eclipse IoT ecosystem and gives further background information on the platforms that make it up.

Eclipse Hono presents a Middleware layer between the back-end microservices and the different devices subscribed inside the framework to create a platform for scalable messaging in the IoT. The networking and communication with the back-end are handled via the Advanced Message Queuing Protocol (AMQP). Protocol adapters are also included within the Hono platform; they aid in the translation of the messages exist from the device's protocol to AMQP. Hono's core services are independent of the other protocols used by individual apps.

The Hono platform is comprised of various building components; the first is a set of protocol adapters that are required to link devices that do not support communications using AMQP. There are two protocol adapters to serve this purpose: Message Queuing Telemetry Transport (MQTT) and HTTP-based Representational State Transfer (REST) messages. In addition, there is also a dispatch router that is responsible for the routing AMQP messages appropriately between generating and consuming endpoints, within Hono. This router is based on the Apache Qpid project and is designed for scalability allowing it to handle connections from millions of connected objects. As a result, it does not assume ownership of the messages being exchanged instead passes the AMQP packets between accessible endpoints. This enables for horizontal scalability allowing for reliability and responsiveness [12].

Hono features a device registry that manages the registration, activation, and provisioning of credentials, as well as an Authentication Server that handles both the authentication and authorisation processes of such objects, ensuring and enforcing the security of routed communications within the ecosystem. By using a Grafana dashboard which is a Cloud front-end visualization tool, and an InfluxDB, Hono also comes with some monitoring infrastructure to visualize data using a number of different charts and diagrams that can be configured by its users, such as in form of series, time, histogram, stacks, bar/line charts, and other customization options. Other AMQP 1.0-compatible message brokers beyond Apache ActiveMQ Artemis can be used because of its modular design.

In the construction of big IoT systems, IoT performance assurance operations are critical. Such activities guarantee that the IoT, which consists of constructed platforms, meets the required performance standards [38]. However, their failures are usually due to performance issues rather than functional impairments [10,11]. Failures can lead to eventual quality rejection of targeted systems with reputational and financial consequences. In order to mitigate the resulting in performance hindrance and ensure systems' reliability at the same time, practitioners frequently conduct performance evaluations [38] to be applied to workloads, e.g., simulating users' behavior in the field, on software systems [32], and check the related performance metrics, i.e., CPU cores usage, that are generated according to those carried tests and evaluations. Traditionally, similar indicators have been used to evaluate system performance and identify potential performance issues, such as memory leaks [31], memory allocated usage as well as network throughput bottlenecks [21].

Furthermore, practitioners who use new systems seek advice or fresh directions on how to build such platforms and–or deploy their applications effectively on top of them. They must be able to select the appropriate configurations and frameworks since the participating objects are resources constrained; devices are not optimal in terms of resources utilization, such as CPU, memory, network, file systems, etc., and their misuse is likely to significantly degrade both the Quality of Service (QoS) and User Experiences (UEs). Prior research has proposed various ways of evaluating performance testing output, i.e., performance metrics. Conventionally, those techniques look at the following aspects metrics: (a) single performance metric and (b) performance metrics relationships.

2.4 Analyzing Performance Metrics from Performance Testing

Previous studies have offered a number of various approaches in order to analyze performance testing results. These methods look at distinct elements of performance metrics, such as: (i) single performance metrics, and (ii) relationship between them.

Single Performance Metrics. Nguyen et al. [25] pioneered the idea of using control graph/charts [28] to detect performance regressions. These control charts employ a preset threshold to detect performance abnormalities. Such graphs/charts, on the other hand, require that the output follows a uni-model distribution, which may be inappropriate assumption for performance. Authors in [25] presented a method for normalizing performance metrics and indicators across varied settings, environments, and workloads to produce robust control charts. Malik et al. [22,23] provided several approaches for clustering performance metrics using Principal Component Analysis (PCA), in which each component created is linked to performance metrics by a weight value. Weight values measure how a metric could contribute to a specific component. To detect performance regressions, a comparison of the weight values of various components is conducted for each performance indicator.

Heger et al. [14] proposed a method for diagnosing the fundamental causes of performance regression using software development history and unit tests. They used Analysis of Variance (ANOVA) to compare the system's reaction time in order to discover performance regressions as the first stage in their technique. Similarly, Jiang et al. [19] were able to derive response times from systems logs.

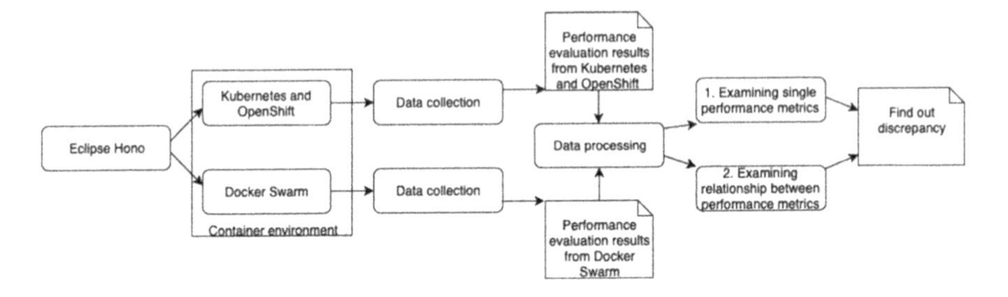

Fig. 1. An overview of our case study setup

Instead of conducting statistical tests, they visualized the trend of the response time during the performance tests to be able to identify performance issues.

Relationship Between Performance Metrics. Malik et al. in [21] leveraged Spearman's rank correlation to capture the likely linkages, i.e., probable connections, between the performance metrics and–or indicators. The deviance of the correlation is examined to determine which subsystem is responsible of the performance fluctuation. In addition, Foo et al. [11] proposed an approach that uses association rules to address the difficulties of manually detecting performance deterioration in large-scaled software systems. Association rules capture the historical correlations between performance metrics and generate rules depending on the outcomes of the previous performance tests. Deviations in the association rules are considered to be signs of performance regressions. Moreover, Jiang et al. [18] used the normalized mutual information as a similarity measure to clusters' correlated performance metrics. Since metrics could be highly correlated in one cluster, hence, the uncertainty among metrics can tend to be low. In addition, the authors leveraged the information theory entropy, a logarithmic measure of the rate of transfer of information of a particular message, to monitor the uncertainty of each cluster. A noticeable, i.e., significant, change in the entropy is considered to be a sign of performance degradation.

3 Design of the Study

The aim of our study is to understand the performance implications of different serverless application deployments scenarios under various container setups. In particular, we are interested in understanding the impact of a serverless architecture on the performance of applications deployed on Eclipse Hono using different containers technologies. A load driver is used to exercise our subjected platform; after collecting and processing the performance metrics, we analyze and draw the conclusion based on (1) single performance metrics, and (2) relationship between performance metrics. An overview of our case study setup is depicted in Fig. 1. As previously stated, we selected three container technologies (i.e.,

Table 1. Setup design on Amazon AWS

Criteria deployments	Experimental designs Kubernetes, OpenShift–(KO), and Docker Swarm- (DS)	Version
Basic Version	Bare metal deployment of Hono on Kubernetes, OpenShift, and Docker Swarm	KO-0 & DS-0
Hono-Serverless app	Deploying serverless apps on top of Kubernetes, OpenShift, and Docker Swarm	KO-1 & DS-1

Kubernetes, OpenShift, and Docker Swarm) which are recommended for applications deployment in the Eclipse Community. We answer the following research questions:

- **RQ1. Does Eclipse Hono display similar performance(s) when deployed on top of Kubernetes, OpenShift, or Docker Swarm setups?**
- **RQ2. Do serverless applications deployed using Eclipse Hono display similar performance(s) in Kubernetes, OpenShift and Docker Swarm setups?**
- **RQ3. To what extent does the relationship between the performance metrics change across Kubernetes, OpenShift, and Docker Swarm setups?**

To address these three research questions and attain our objectives, we performed a series of experiments with different deployments configurations to collect and analyze differences between performance metrics. Table 1 provides details about our configuration(s); the serverless application has a workload generator that can be used to generate a synthetic container workload(s), it attempts to generate performance records at a relatively steady rate across the specified container configuration(s). All deployments were deployed from scratch every time a new analysis is performed, the findings were gathered by running a series of stress tests on the platform (i.e., fixing the number of issued requests and applications developed on top of it) and tracing back their executions. The same set of configurations and–or tests were utilized for all experiments to ensure comparable results. The rest of this section goes over our conducted experiments in further depth.

3.1 Environmental Setup

The overall performance assessment is carried out on a Linux machine, i.e., Ubuntu 18.04, in a lab environment; this machine has an Intel i7-870 Quad-Core 2.93 GHz CPU with 16 GB of RAM, 630 GB SATA storage, 8MB Cache and is linked to a local gigabyte Ethernet connection. The hosting machine is configured to have available Kubernetes, OpenShift, and a cluster of Docker

Engine nodes running in Swarm mode. The same configurations were used in our prior study [2]. The main deployment target for the Swarm mode, in this study, is a multi-node cluster running on Amazon Web Services (AWS). We created the clusters as follows.

Amazon Web Services (AWS): platform that helps running Docker Swarm Clusters on top of the Cloud. Eclipse Hono components are distributed by means of `Docker images` which can be deployed to arbitrary environments where Docker is available (i.e., Docker Swarm on AWS).

Since our goal is to compare the performance metrics in such created clusters, we set up the Swarm stack using five Swarm manager nodes and ten (10) worker nodes; this was the limit for the free tier we were using. On the other hand, previously in our study [2], for both Kubernetes and OpenShift, we setup the VM instances with sufficient resource capacity so that the deployment is successful, namely four (4) CPUS, ten (10) GB of Memory and thirty (30) GB of disk-space. The default instance settings only provide a small part of the host machine's resources, which may not be enough for them to start successfully. That's why, based on the capabilities of the real physical machine, it is recommended to scale up the resources wherever feasible; "the more resources used, the better".

Container Environments. We opted for the following configurations:

- Minikube: tool that is used to run a single-node Kubernetes cluster inside a VM locally. This makes it easier to try Kubernetes or develop with(in) it.
- Minishift: tool that is used to run OpenShift locally by running a single-node OpenShift cluster inside a VM.
- Docker Community Edition (CE) for Amazon Web Services (AWS) – stable – that uses existing Virtual Private Cloud (VPC) on our Cloud infrastructure for the Docker Swarm setup.

3.2 Design and Procedure

To analyze the benefits and the trade-offs of the different installations explored, the tests we setup were organized using two separate forms of issued requests, as previously indicated `(REST HTTP and MQTT)`. For each type, we simulated the registered devices sending both requests at the same time in a telemetry fashion[1]. Each experiment was carried out five times (with the number of registered devices being gradually incremented after each performed sub-experiment: starting with 10 devices up to 100 registered devices – 100 is the limit for this platform by default, at the time of the experiments). The goal of such setup is to acquire the minimum, maximum, and average values of the resources been utilized. We opted to repeat each experiment five times to reduce the impact of variabilities (which are frequent in virtual settings/environments) on our obtained results. Tables 1 shows the three deployment versions of the platform, the basic versions

[1] Expecting no response in return.

KO-0 & DS-0 do not use any additional overhead, just the deployment of Hono on top of Kubernetes, Openshift, and a cluster of Docker Engine nodes running in Swarm mode with the aforementioned set up configurations.

3.3 Performance Tests and Evaluation

Minikube, Minishift as well as Docker Swarm are dispatched with `DOCKER_HOST` environment variable to point to the Docker daemon running within the virtual instances; this daemon is used to make the final Docker images visible and accessible within the setups and prepare them for Hono's deployment. Deployment provides access to the platform via a number of services, the main and most important of which are:

1. dispatch router: router network allowing business applications to consume data.
2. rest-adapter: protocol adapter that uses the HTTP protocol to broadcast telemetry data and events.
3. mqtt-adapter: protocol adapter that uses the MQTT protocol to broadcast telemetry data and events.
4. service-device-registry: component for registering and managing devices.

After each round of experimentation, we delete all the environments, specifically the VMs created, and restart them to guarantee consistency between the performance tests. This ensures that the clusters that have been built are healthy enough to deliver meaningful result(s).

3.4 Data Collection and Pre-processing

Performance Metrics. We utilized heapster[2] and Prometheus[3] to record the acquired values of the performance metrics (i.e., CPU, memory usage, and network consumption). Heapster enables container cluster monitoring and performance analysis for Kubernetes. It collects and interprets various signals, such as compute resources consumption, whereas, Prometheus (i.e., for OpenShift and Docker Swarm), is a service monitoring system. It gathers metrics from configured targets at certain time intervals, assesses, shows the findings, and sends out warning/alarms if certain conditions are fulfilled or observed to be true. We ran Prometheus on each cluster we built and then performed statistical analysis on the collected data. Furthermore, we captured the metrics throughout a time span corresponding to "starting the cluster within the VM until the destruct phase".

[2] https://github.com/kubernetes/heapster.
[3] https://prometheus.io.

System Throughput. We utilized the measurements obtained to determine the minimum, maximum and average values of the systems' resources by counting the amount of telemetry messages, `HTTP and MQTT`, sent from each of registered device while using the serverless applications on top of the messaging architecture, Amazon Web Services (AWS). The goal is to integrate the performance indicators and system throughput while reducing the amount of noise collected. The combination is based on the time stamp – on a minute-by-minute basis; a similar method was used to solve mining performance metrics issues in [11].

Hypotheses. To answer our research questions, we formulated the following null hypotheses, KO-x & DS-x ($x \in \{0, 1\}$), KO-0 & DS-0 are the basic versions of the platform described in Table 1:

- HR_x^{1-0}: there is no difference between the values of the performance metrics, *CPU Cores*, obtained for Hono's design when deployed on Kubernetes, OpenShift, and Docker Swarm setups.
- HR_x^{2-0}: there is no difference between the values of the performance metrics, *Memory Usage*, obtained for Hono's design when deployed on Kubernetes, OpenShift, and Docker Swarm setups.
- HR_x^{3-0}: there is no difference between the values of the performance metrics, *Network Consumption*, obtained for Hono's design when deployed on Kubernetes, OpenShift, and Docker Swarm setups.
- HR_x^{1-1}: there are no differences in the utilized resources, *CPU Cores*, consumed by the serverless workload generator app(s) deployed on top of Hono's Kubernetes, OpenShift, and Docker Swarm setups.
- HR_x^{1-2}: there are no differences in the utilized resources, *Memory Usage*, consumed by the serverless workload generator app(s) deployed on top of Hono's Kubernetes, OpenShift, and Docker Swarm setups.
- HR_x^{1-3}: there are no differences in the utilized resources, *Network Consumption*, consumed by the serverless workload generator app(s) deployed on top of Hono's Kubernetes, OpenShift, and Docker Swarm setups.

Analysis Method. We performed the Wilcoxon test [30] to accept or reject HR_x^{y-0} and HR_x^{y-1}, where y $\in \{1,2,3\}$. We also estimated the Cliff's δ effect size [26] to measure the significance of the changed found between metrics values. All of the tests were carried out with a 95% confidence level (i.e., p-value ≤ 0.05).

A p-value that is less than or equal to 0.05 indicates that the results are statistically significant. In such case, we reject the null hypothesis (i.e., two populations come from the same distribution) and accept the alternative hypothesis which helps to determine if the performance metrics in each of Kubernetes, OpenShift and–or Docker Swarm environments have the same distribution. We picked the Wilcoxon test since it make no assumption(s) about the distribution of the metrics. *When it happens to have a statistically significant value(s) $\leq$ (or merely very near) to the value calculated, we depict both its median and effect size (EF) values to demonstrate which environment performs best and which one is the worst for such sort of deployment(s).*

Table 2. p-value of Wilcoxon Test (p-VAL) – *(Median Kubernetes, Median Docker Swarm)* – *(Cliff δ Effect Size (ES))*

HONO AWS-Swarm Deployment DS-0									
# Devices	CPU cores usage (p-value)			Memory usage (p-value)			Network consumption (p-value)		
	Min	Max	Avg	Min	Max	Avg	Min	Max	Avg
10	0.1508	0.1508	0.1508	0.8413	0.834	0.6905	0.4206	0.5476	0.6905
20	**0.007937** ***(65, 1.77)*** ***(ES = 0.68)***	0.1193	**0.007937** ***(58.5, 7.32)*** ***(ES = 0.68)***	0.3095	0.2222	0.3095	0.1508	0.6905	0.6905
30	**0.01167** ***(67, 8.13)*** ***(ES = 0.68)***	**0.007937** ***(88, 32.18)*** ***(ES = 0.68)***	**0.007937** ***(78, 21.22)*** ***(ES = 0.68)***	0.8413	0.8413	0.8413	0.2222	0.1508	**0.03175** ***(82, 39.06)*** ***(ES = 0.68)***
40	**0.007937** ***(58, 18.3)*** ***(ES = 0.68)***	**0.007937** ***(82, 37.28)*** ***(ES = 0.68)***	**0.007937** ***(70, 28.98)*** ***(ES = 0.68)***	0.4206	0.2222	0.3095	0.09524	0.1508	0.1508
50	**0.01193** ***(80, 26.02)*** ***(ES = 0.68)***	**0.007937** ***(93, 45.86)*** ***(ES = 0.68)***	**0.007937** ***(86.5, 34.98)*** ***(ES = 0.68)***	**0.03615** ***(68, 46.98)*** ***(ES = 0.68)***	**0.03175** ***(80, 64.26)*** ***(ES = 0.68)***	**0.03175** ***(74, 54.84)*** ***(ES = 0.68)***	0.05556 *(76, 40.98)* *(ES = 0.68)*	**0.03175** ***(95, 63.27)*** ***(ES = 0.68)***	**0.03175** ***(85, 53.72)*** ***(ES = 0.68)***
60	**0.007937** ***(78, 35.11)*** ***(ES = 0.68)***	**0.03175** ***(84.18, 53.63)*** ***(ES = 0.68)***	**0.007937** ***(84.18, 44.01)*** ***(ES = 0.68)***	0.05556 *(75, 3.61)* *(ES = 0.68)*	0.1508	0.1425	0.09369	0.05556 *(103, 74.07)* *(ES = 0.68)*	0.09524
70	0.6905	0.5476	0.6905	0.1508	0.05556 *(95, 72.65)* *(ES = 0.68)*	0.09524	**0.007937** ***(85.5, 46.26)*** ***(ES = 0.68)***	**0.007937** ***(98, 62.03)*** ***(ES = 0.68)***	**0.007937** ***(90.17, 52.75)*** ***(ES = 0.68)***
80	0.1587	0.1587	**0.03175** ***(86, 71.44)*** ***(ES = 0.68)***	0.1508	0.2222	0.3095	0.09524	**0.03175** ***(97, 53.55)*** ***(ES = 0.68)***	**0.03175** ***(90.82, 39.57)*** ***(ES = 0.68)***
90	0.6905	0.2087	0.3095	0.09524	0.05556 *(100, 39.59)* *(ES = 0.68)*	0.05556 *(91.17, 25.48)* *(ES = 0.68)*	0.4206	0.2222	0.4206
100	0.4206	0.3095	0.4206	0.1508	0.09524	0.09524	0.05556 *(87, 73.37)* *(ES = 0.68)*	0.2222	0.2222

The Wilcoxon test is a non-parametric statistical test that examines whether two independent distributions and–or trends are the same. Cliff's δ is a non-parametric effect size measure that shows how much two sample trends overlap [26]. It ranges from -1 (when all selected values in the first group are larger than those in the second) to $+1$ (if all selected values in the first group are smaller than that of the second group). It is zero when the two sample trends are equal [6]. *A Cliff's δ effect size is deemed insignificant, i.e., negligible, if it is < 0.147, small if < 0.33, medium if < 0.474, and large if ≥ 0.474.*

4 Case Study Results

This section discusses the outcomes of our aforementioned research questions. Tables 2, 3, 4 and 5 summarize the findings of the Wilcoxon test, median values and Cliff's δ effect size values for each performance metric. Significant results are marked in **bold**.

RQ1. Does Eclipse Hono display similar performance(s) when deployed on top of Kubernetes, OpenShift, or Docker Swarm setups?

Table 3. p-value of Wilcoxon Test (p-VAL) – *(Median OpenShift, Median Docker Swarm) – (Cliff δ Effect Size (ES))*

HONO AWS-Swarm Deployment DS-0									
# Devices	CPU cores usage (p-value)			Memory usage (p-value)			Network consumption (p-value)		
	Min	Max	Avg	Min	Max	Avg	Min	Max	Avg
10	**0.007937** ***(34.682, − 5.04)*** ***(ES = 0.68)***	**0.007937** ***(43.212, 13.38)*** ***(ES = 0.68)***	**0.007937** ***(38.914, 4.46)*** ***(Es = 0.68)***	1	0.4206	0.4206	0.402	0.6752	0.6752
20	0.1508	0.1508	0.09524	0.6905	0.6905	0.6905	0.4206	1	1
30	0.1508	0.5556 *(66.327, 32.18)* *(ES = 0.68)*	0.05556 *(56.849, 31.22)* *(ES = 0.68)*	0.4206	0.3095	0.3095	1	1	0.4206
40	**0.007937** ***(42.991, 18.3)*** ***(ES = 0.68)***	0.4206	0.3095	**0.03175** ***(2.09, 47.44)*** ***(ES = 0.68)***	**0.01587** ***(16.62, 63)*** ***(ES = 0.68)***	**0.01587** ***(9.58, 53.21)*** ***(ES = 0.68)***	0.6905	0.6905	0.8413
50	0.1587	**0.01587** ***(64.305, 45.86)*** ***(ES = 0.68)***	**0.007937** ***(59.507, 34.98)*** ***(ES = 0.68)***	0.6905	0.6905	0.6905	0.8413	1	1
60	0.1508	0.3095	0.1508	0.1508	0.2222	0.2222	1	1	0.8413
70	0.1587	0.09524	0.05556 *(72.536, 55.57)* *(ES = 0.68)*	0.6905	0.5476	0.6905	0.1508	0.2222	0.2222
80	1	0.2222	0.8413	0.4206	0.3095	0.3095	0.8413	**0.03175** ***(75.42, 53.55)*** ***(ES = 0.84)***	0.8413
90	0.5476	0.8413	0.6905	0.4206	0.5476	0.4034	0.5476	0.6905	0.6905
100	0.8413	0.5476	0.6905	1	0.8413	1	0.3095	0.5309	0.5476

Motivation. In this research question, we want to understand the performance implications of scaling the Eclipse Hono framework using different container setups. If there is a significant performance implication, in the next research question, we will examine how the deployment of serverless applications is impacted by it.

Approach. Since performance metrics are likely to have different value ranges because of differences in the architecture of the platforms (Kubernetes/OpenShift environment(s) may have a higher network I/O consumption than Docker Swarm), instead of comparing the values of each performance metrics across the different environments, we instead study whether the trends of the metrics are the same.

Results. Tables 2 and 3 demonstrate that there are statistically significant differences between all the **CPU cores usages**, **Memory usages** and **Network throughput** when deploying Hono on top of both Kubernetes and OpenShift rather than on top of Docker Swarm. We therefore reject HR_x^{y-0} for KO-x & DS-x ($x \in \{0..1\}$), where y $\in \{1, 2, 3\}$ for all the performance metrics identified. Both container setups are greedy while consuming their resources to (1) keep up with the processes being issued within the system, such as setting-up/spinning the VMs, building existing `Docker` images, and deploying Hono platform. (2) delivering telemetry messages from the participating devices.

Observation: in both Kubernetes and OpenShift, containers are not allowed to be used above their CPU and–or memory limits. If a container, in such setups, allocates more resources than their limits, it becomes a candidate for termination. Thus, if the container continues to seize resources beyond its limits, it is then terminated to ensure that the environment is properly configured and formulated. And in our case here, both of the above mentioned setups hit the limits several times leading to the excessive resources utilization until the complete deployment has occurred successfully by adding up more resource specs. *This was not the case while dealing with Docker Swarms on Amazon Web Services (AWS), where the Cloud infrastructure is scaled in–out based on the workloads generated by the platform and deployed application(s).*

Findings: Performance metrics typically do not follow the same distribution in OpenShift, Kubernetes, and Docker Swarms environments.
Actionable implications: Practitioners should consider deploying Hono on top of Docker Swarm for better performance.

RQ2. Do serverless applications deployed using Eclipse Hono display similar performance(s) in Kubernetes, OpenShift and Docker Swarm setups?

Motivation. Kubernetes, OpenShift and Docker Swarm are three leading containers technologies in the Industry nowadays. This research question aims to inspect and understand potential performance differences when serverless applications are deployed in the three different container setups.

Approach. After executing and collecting the performance metrics, we compare each performance metric between the container environments. Since the performance tests are carried out in diverse containers' behavior, the scales of performance metrics are not obviously the same. The Kubernetes environment, for example, may consume more Memory than OpenShift. Hence, rather than comparing the values of each performance metric in each environment, we investigate if the distributions of the performance metric have the same shape and trend across the multiple container settings and–or environments. We perform Wilcoxon tests and compute Effect Size (ES) metrics, to capture differences between the distributions of different corresponding performance metrics.

Results. Tables 4 and 5 show differences between CPU/Memory usages and network consumption when our serverless application is deployed on top of the scaled out Hono in the container setups, respectively. There is a statistically significant difference between the amount of consumed **Memory** for OpenShift. Therefore, we reject HR_x^{1-2} for KO-x & DS-x ($x \in \{0..1\}$). We explain this difference in memory consumption as follows. When OpenShift nodes are memory overcommitted, they run out of resources that successful deployment is hindered, such situation is called resource pressure. We experienced such behavior several times while experimenting our scenario. A number of Hono pods[4] where stuck

[4] *hono-adapter-http-vertx-1-deploy.*
hono-adapter-kura-1-deploy.
hono-service-device-registry-1-deploy.
hono-service-messaging-1-deploy.

Table 4. p-value of Wilcoxon Test (p-VAL) for Serverless Apps – *(Median Kubernetes, Median OpenShift) – (Cliff δ Effect Size (ES))*

HONO-Serverless-App KO-DS									
# Devices	CPU cores usage (p-value)			Memory usage (p-value)			Network consumption (p-value)		
	Min	Max	Avg	Min	Max	Avg	Min	Max	Avg
10	0.8345	0.6761	1	0.6761	0.8345	1	0.2963	0.2963	0.2101
20	0.2101	0.1437	0.2101	1	0.8345	1	1	1	1
30	0.2963	0.6761	0.5309	0.5309	0.4034	0.5309	0.1437	0.09469	0.09469
40	0.1437	0.09469	0.1437	0.4034	0.5309	0.4034	0.5309	0.8345	0.8345
50	0.1437	0.2101	0.1437	1	1	1	0.2101	0.2101	0.2101
60	0.1437	0.09469	0.09469	0.5309	0.6761	0.5309	1	0.6761	0.8345
70	0.09469	0.2101	0.09469	0.6761	0.6761	0.6761	0.5309	0.345	0.6761
80	0.09469	0.5309	0.4034	**0.03671** *(19.06, 59.14)*-*(**ES = −0.84**)*	**0.03671** *(38.9, 73.29)*-*(**ES = −0.84**)*	**0.03671** *(28.26, 66.43)*-*(**ES = −0.84**)*	0.2963	0.2101	0.2963
90	0.1437	0.09469	0.2101	0.1437	0.09469	0.09469	0.8345	0.6761	0.8345
100	0.8345	0.8345	1	0.8345	0.6761	1	0.6761	0.8345	0.6761

being created leading to the "freeze" state of the whole experiment, hence, we had to destroy such deployment and start a new creation to heal such phenomenon. When OpenShift node service realizes that it is under resource pressure, it then stops accepting new pods formulations requests to try complete what is already there, in terms of processes.

Findings: Performance metrics typically do not follow the same distribution in container environments when deploying serverless applications on top of the framework and after scaling it out using the Eclipse Hono messaging infrastructure.
Actionable implications: Developers should be aware of the performance implications of choosing Kubernetes/OpenShift containers to deploy their IoT serverless applications. They should ensure that enough memory is available for proper deployments. Moreover, they should consider deploying their IoT serverless applications on top of Docker Swarm, for better performance.

In the case of CPU cores usage and Network consumption, we did not notice any significant difference between Kubernetes and OpenShift. Hence, in these cases, we cannot reject HR_x^{1-1} and HR_x^{1-3} for the CPU cores usage and network consumption for such environments. A similar behaviour was also observed in our previous study [2] for OpenShift; it also consumed more **Memory** than Kubernetes, when regular applications were deployed on top of the scaled Hono.

In addition to the previous observation, the trend of the Memory usage, in both container setups, are slightly larger than that of the Docker Swarm environment. The trends tend to fall down, yet maintaining a statistically significant difference with the distribution of memory consumed by the Docker Swarm environment. Both Kubernetes and OpenShift needs larger amounts of memory to allow the proper deployments of the serverless application.

Table 5. p-value of Wilcoxon Test (p-VAL) – *(Median Kubernetes/OpenShift, Median Docker Swarm) – (Cliff δ Effect Size (ES))*

HONO-DS-AWS-Serverless simulating workload app DS-1									
# Devices	CPU cores usage (p-value)			Memory usage (p-value)			Network consumption (p-value)		
	Min	Max	Avg	Min	Max	Avg	Min	Max	Avg
10	1	1	1	0.8413	0.8413	0.8413	0.2222	0.1508	0.2222
20	0.6905	0.5476	0.4206	0.8413	0.6905	0.8413	**0.01587** ***(16.9, 47.48) (ES = 0.36)***	**0.03175** ***(35.83, 62.02) (ES = 0.36)***	**0.03175** ***(20.75, 53.19) (ES = 0.36)***
30	0.8413	0.6905	0.6905	0.05556 *(76.22, 22.01) (ES = 0.36)*	0.1508	0.05556 *(88.83, 26.76) (ES = 0.36)*	0.8413	0.8413	0.8413
40	**0.007937** ***(64.244, 44.92) (ES = 0.68)***	0.09524	**0.01587** ***(73.242, 57.67) (ES = 0.68)***	1	0.6905	0.6905	0.1508	0.6905	0.6905
50	0.6905	**0.007937** ***(92.953, 68.76) (ES = 0.68)***	**0.007937** ***(83.413, 63.33) (ES = 0.68)***	1	1	0.8413	0.05556 *(73.71, 24.14) (ES = 0.36)*	0.05556 *(87.64, 36.31) (ES = 0.36)*	0.09524
60	0.6905	0.6905	0.6905	0.6905	0.8413	0.6905	0.3095	0.4206	0.4206
70	0.1508	0.1508	0.1508	0.05556 *(72.62, 41.93) (ES = 0.36)*	**0.03175** ***(96.27, 56.95) (ES = 0.36)***	**0.03175** ***(84.59, 47.45) (ES = 0.36)***	0.8413	0.4206	0.8413
80	0.3095	0.3095	0.3095	0.3095	0.3095	0.3095	0.2222	0.3095	0.2222
90	0.6905	0.1508	0.6905	0.6905	1	1	0.8413	1	1
100	**0.007937** ***(64.775, 90.19) (ES = 0.68)***	**0.007937** ***(108.31, 82.725) (ES = 0.68)***	**0.007937** ***(74.732, 100.92) (ES = 0.68)***	0.8413	0.8413	0.8413	0.2222	0.5476	0.3095

RQ3. To what extent does the relationship between the performance metrics change across Kubernetes, OpenShift, and Docker Swarm setups?

Motivation. The relationship between performance metrics can fluctuate dramatically and varies considerably between different environments, which may be an indicator of system regression or performance problems. According to [7], combinations of performance metrics are more predictive of performance difficulties than a single metric. A change in correlation values between a collection of performance metrics frequently shows performance difficulties in an application and–or discrepancies between its performance across different platforms. Practitioners frequently depend on correlation analysis of performance indicators to determine the behavioral changes of a system between different environments. For example, the CPU may be highly correlated with network usage in one system (e.g., when network's operations are high due to the workload being generated, eventually CPU must eventually increase to accommodate this increase); on the other hand, in the same system, the correlation between CPU and memory may become low. Such a shift might indicate performance difficulties (for example, high CPU without memory or network I/O operations might be due to a performance failure). Also, if there is a significant difference in the correlations just due to the platform being utilized, such as Kubernetes vs. OpenShift vs. Docker Swarm, then practitioners may need to be informed of the potential consquences of picking one platform over another. In this research question, we examine whether the relationship between performance metrics differ among the three studied environments we look at: Kubernetes, OpenShift, and Docker Swarm.

Table 6. Spearman's rank correlation summary of performance metrics in Docker Swarm on top of Amazon AWS

# Devices	CPU–Memory			CPU–Network			Memory–Network		
	Min	Max	Avg	Min	Max	Avg	Min	Max	Avg
HONO AWS-Swarm Deployment DS-0									
10 -> 100	0.1	0.5	0.2	0.1	**0.6 (strong)**	−0.3	0.2	0.3	0.3
HONO-DS-serverless app DS-1									
10 -> 100	**0.9 (v. strong)**	−0.4	0.4	−0.6	**0.6 (strong)**	−0.9	−0.7	−0.2	−0.7

Table 7. Spearman's rank correlation summary of serverless performance metrics in Kubernetes deployments

# Devices	CPU–Memory			CPU–Network			Memory–Network		
	Min	Max	Avg	Min	Max	Avg	Min	Max	Avg
HONO Bare metal Deployment									
10 -> 100	−0.1025978	−0.1	−0.3077935	**0.6 (strong)**	**0.7 (strong)**	**0.6 (strong)**	0.2051957	0.1	0.2051957
HONO-Serverless-App KO									
10 -> 100	−0.9	−0.7	−0.9	−0.6	−0.4	−0.6	**0.8 (v. strong)**	**0.9 (v. strong)**	**0.8 (v. strong)**

Approach. We assessed the Spearman's rank correlation coefficient between all performance metrics, i.e., in the container configurations and Docker Swarm stack, and studied if they differed. For instance, in one environment, the CPU cores usage may be significantly associated, i.e., highly correlated, with the Network I/O; whereas in another environment, such correlation may be minimal. In this situation, we consider that there is a discrepancy in the correlation coefficient between the CPU and the Network I/O.

Spearman's correlation coefficient is a statistical strength measure of a monotonic relationship between paired data. It is used to figure out far apart the performance metrics are [27]. In a given sample, Spearman's correlation is indicated by r_s and limited to $-1 \leq r_s \leq 1$ and is interpreted as the closer r_s to ± 1 the stronger is the monotonic relationship.

Correlation is an effect size (ES) measure and the strength of the correlation for the r_s *is described as follows: 0.00–0.19 "very weak", 0.20–0.39 "weak", 0.40–0.59 "moderate", 0.60–0.79 "strong", and 0.80–1.0 "very strong".*

We picked the Spearman's rank correlation coefficient because it does not need any assumptions on the distribution of the variables. This is required because load test data contains traces that do not follow a normal distribution.

Results. There exists differences in correlation between the performance metrics in each of Kubernetes, OpenShift and Docker Swarm; i.e., are not all correlated in the same way. Tables 6, 7, and 8 demonstrate the variations in the correlation coefficient on the resources used in each environment. By examining them closely, we discover that (CPU–Network) in Kubernetes and Docker Swarm have a stronger correlation than that of OpenShift (i.e., noticeable network's operations due to Hono's deployment workload); whereas (Memory–Network) coefficients in OpenShift are stronger than those in the other two environments[5].

When the serverless application was added on top of Kubernetes, the (Memory–Network) coefficient has showed a very strong bond rather than the other two setups. Whereas, while deploying the serverless application on top of the Docker Swarm platform on AWS, both the (CPU–Memory) and (CPU-Network) have shown very strong and strong relationships, respectively. This is becuase the workload simulator serverless app is used to determine system performance and response time to evaluate network design, and to simulate the actions of a number of different events. OpenShift was neutral and has not shown any bond between it's metrics. For the sake of brevity, we do not show the detailed analysis here, but it is available in our replication package indicated in the next section, i.e., Sect. 5. Box-plots, as well as Spearman's correlation trend-lines, are depicted therein to show the correlation changes among related metrics.

Findings: The correlations between performance metrics may change considerably between container environments when serverless apps are deployed on top of them using Eclipse Hono.
Actionable implications: Practitioners should pay attention to these performance deviations across the three studied platforms when planning the migration of their IoT apps. They should also avoid reusing blindly any performance benchmarking result obtained in a different environment.

5 Threats to Validity

This section investigates the threats to the validity of our carried out study based on the recommendations suggested by Wohlin et al. [37].

Construct validity threats pertaining the relationship between theory and observations, such as the measurements errors. We instrumented the various versions of deployments mentioned in the previous Sect. 3 to obtain execution readings from which we derived the performance metrics values; the minimum, maximum and average. To limit the possibility of biases that may be caused by changes in the network, hardware, and our tracing, we repeated each experiment five times and averaged the results, i.e., obtained their median values. We are certain that such repetitive measures enhanced the accuracy of our calibration and results. We tracked the performance of the containers from the moment they were built until they were destroyed, then averaged the performance metrics for every minute together as a median value.

[5] A thorough statistical study and analysis for Hono's bare deployment and other serverful applications on top of both Kubernetes and OpenShift has been carefully investigated in our previous work [2].

Table 8. Spearman's rank correlation summary of serverless performance metrics in OpenShift deployments

# Devices	CPU–Memory			CPU–Network			Memory–Network		
	Min	Max	Avg	Min	Max	Avg	Min	Max	Avg
HONO Bare metal Deployment									
10 -> 100	**0.6 (strong)**	0.1	0.3	−0.1	0.1	0.1	−0.3	**0.8 (v. strong)**	**0.6 (strong)**
HONO-Serverless-App KO									
10 -> 100	0.3	0.3	0.3	−0.9	−0.7	−0.6	−0.4	−0.3	−0.1

Internal validity threats concern our analysis method. Our empirical analysis is based on the performance testing results acquired from the examined subject systems. The manner in which the performance tests were conducted as well as their quality might jeopardise the validity of our findings. Our technique is particularly reliant on the recorded performance metrics, and the quality of these metrics might affect the internal validity of our study. Our performance evaluations lasted for a duration of, roughly, six → eight months, while the length of the assessments may influence the outcomes of the performed case study, we believe that we have observed and investigated performance readings throughout a realistic period of time. The statistical analysis that is carried out is considered to be another internal validity threat. To mitigate it, we paid attention not to violate the assumptions of the statistical tests. Specifically, we utilized non-parametric tests, which do not require making any assumptions about the distribution of our data.

External validity threats concern the prospect of generalizing our findings. More testing with various settings and–or configurations is required to gain better understanding of the influence of Eclipse Hono deployment techniques on the resources consumption and to give practitioners guidance on how to use such platform while developing, building and deploying IoT serverless applications.

Reliability validity threats concern the prospect of replicating this study. We make every effort to offer of the information and details needed to duplicate our conducted study. All the data used in the study are accessible online in our replication package[6].

Finally, the **conclusion validity** threats which pertain to the relationship between the treatment and the outcome had no bearing on our study since we took care not to violate the assumptions of the statistical tests employed in our analysis. However, more complicated clusters and various applications should be used in future replications of our work to make our findings more general.

[6] Complete results, box plots, data and scripts are shared online on *osf.io* scientific data repository: http://bit.ly/benchmarking-hono-FaaS-deployments.

6 Conclusion

Performance assurance activities are crucial in ensuring platform–software reliability. Virtual environments nowadays are used to conduct performance tests. However, the discrepancy between testing results between different virtual environments, including container setups, are still under evaluated. We aimed at highlighting whether a discrepancy present between different container environments will impact the studies and tests carried out in the IoT domain. In this paper, we tested and evaluated such discrepancy by conducting performance tests on the open-sourced IoT platform, Eclipse IoT, in different container/stack environments. By examining the performance testing results, we find that there exists discrepancies between performance testing results in each of the environment involved when examining single performance metric, the relationship, and building statistical performance models among different performance metrics.

The major contribution of this paper includes:

- Our paper is one of the first research works that attempts to evaluate the discrepancy in the context of analyzing the performance testing results in container environments.
- We find that relationships among related performance metrics have large differences between different setups. Practitioners, including developers, cannot always assume a straightforward overhead from deploying an open-sourced IoT platform, specifically IoT–Hono, with serverless applications on top of different container/Cloud technologies.
- We evaluated serverless computing environment invoking functions in parallel to demonstrate the performance and throughput of serverless computing for open-sourced framework. We compared their performance regarding CPU, memory and network I/O between a sequential and a concurrence invocations that helps understanding performance and function behaviors on serverless computing environment.
- We distributed the deployment of the Hono platform on top of AWS Cloud infrastructure and assessed such performance implications while adding an additional layer of a serverless app on top of it.

Our results highlight the need to be aware of and to reduce the discrepancy between performance testing results in container environments, for both practitioners and researchers.

7 Future Work

This work is the first step to lay a concrete ground for a deeper understanding of the discrepancy/differences between performance test results in different container setups while adopting open-sourced IoT platforms, deploying serverless apps on top of them, and the impact of detecting performance issues with such kind of discrepancies. Having such knowledge of the discrepancies, we can better comprehend the existence and magnitude of impact on detecting real-world performance degradation in the future. Furthermore, future research effort can focus on generating comparable performance testing results from different setups with different workloads.

References

1. Akin, M.: How does proportional CPU allocation work with AWS lambda? (2018). https://engineering.opsgenie.com/how-does-proportional-cpu-allocation-work-with-aws-lambda-41cd44da3cac. Accessed 26 Mar 2019
2. Aly, M., Khomh, F., Yacout, S.: Kubernetes or OpenShift? Which technology best suits eclipse Hono IoT deployments. In: 2018 IEEE 11th Conference on Service-Oriented Computing and Applications (SOCA), pp. 113–120. IEEE (2018)
3. Bukh, P.N.D.: The art of computer systems performance analysis, techniques for experimental design, measurement, simulation and modeling (1992)
4. Cabé, B.: Key trends from the IoT developer survey 2018 (2018). https://blogs.eclipse.org/post/benjamin-cab%C3%A9/key-trends-iot-developer-survey-2018. Accessed 12 Feb 2018
5. Chapin, J.: The occasional chaos of AWS lambda runtime performance (2017). https://blog.symphonia.io/the-occasional-chaos-of-aws-lambda-runtime-performance-880773620a7e. Accessed 26 Mar 2019
6. Cliff, N.: Dominance statistics: ordinal analyses to answer ordinal questions. Psychol. Bull. **114**(3), 494 (1993)
7. Cohen, I., Chase, J.S., Goldszmidt, M., Kelly, T., Symons, J.: Correlating instrumentation data to system states: a building block for automated diagnosis and control. In: OSDI, vol. 4, pp. 16–16 (2004)
8. Cui, Y.: How does language, memory and package size affect cold starts of AWS lambda? (2017). https://read.acloud.guru/does-coding-language-memory-or-package-size-affect-cold-starts-of-aws-lambda-a15e26d12c76. Accessed 26 Mar 2019
9. Cui, Y.: How long does AWS lambda keep your idle functions around before a cold start? (2017). https://read.acloud.guru/how-long-does-aws-lambda-keep-your-idle-functions-around-before-a-cold-start-bf715d3b810. Accessed 26 Mar 2019
10. Dean, J., Barroso, L.A.: The tail at scale. Commun. ACM **56**(2), 74–80 (2013)
11. Foo, K.C., Jiang, Z.M., Adams, B., Hassan, A.E., Zou, Y., Flora, P.: Mining performance regression testing repositories for automated performance analysis. In: 2010 10th International Conference on Quality Software (QSIC), pp. 32–41. IEEE (2010)
12. Foundation, E.: Eclipse Hono (2017). https://www.eclipse.org/hono/. Accessed 01 Feb 2018
13. Foundation, E.: Open source software for industry 4.0, an eclipse IoT working group collaboration (2017). https://iot.eclipse.org/resources/white-papers/Eclipse%20IoT%20White%20Paper%20-%20Open%20Source%20Software%20fo%20Industry%204.0.pdf. Accessed 20 Oct 2017
14. Heger, C., Happe, J., Farahbod, R.: Automated root cause isolation of performance regressions during software development. In: Proceedings of the 4th ACM/SPEC International Conference on Performance Engineering, pp. 27–38. ACM (2013)
15. Hendrickson, S., Sturdevant, S., Harter, T., Venkataramani, V., Arpaci-Dusseau, A.C., Arpaci-Dusseau, R.H.: Serverless computation with openlambda. In: 8th {USENIX} Workshop on Hot Topics in Cloud Computing (HotCloud 2016) (2016)

16. NGINX, Inc.: Nginx announces results of 2016 future of application development and delivery survey (2016). https://www.nginx.com/press/nginx-announces-results-of-2016-future-of-application-development-and-delivery-survey/. Accessed 31 May 2019
17. Jain, R.: The Art of Computer Systems Performance Analysis: Techniques for Experimental Design, Measurement, Simulation, and Modeling. Wiley, Hoboken (1990)
18. Jiang, M., Munawar, M.A., Reidemeister, T., Ward, P.A.: Automatic fault detection and diagnosis in complex software systems by information-theoretic monitoring. In: IEEE/IFIP International Conference on Dependable Systems and Networks, DSN 2009, pp. 285–294. IEEE (2009)
19. Jiang, M., Munawar, M.A., Reidemeister, T., Ward, P.A.: System monitoring with metric-correlation models: problems and solutions. In: Proceedings of the 6th International Conference on Autonomic Computing, pp. 13–22. ACM (2009)
20. Lloyd, W., Ramesh, S., Chinthalapati, S., Ly, L., Pallickara, S.: Serverless computing: an investigation of factors influencing microservice performance. In: 2018 IEEE International Conference on Cloud Engineering (IC2E), pp. 159–169. IEEE (2018)
21. Malik, H., Adams, B., Hassan, A.E.: Pinpointing the subsystems responsible for the performance deviations in a load test. In: 2010 IEEE 21st International Symposium on Software Reliability Engineering (ISSRE), pp. 201–210. IEEE (2010)
22. Malik, H., Hemmati, H., Hassan, A.E.: Automatic detection of performance deviations in the load testing of large scale systems. In: Proceedings of the 2013 International Conference on Software Engineering, pp. 1012–1021. IEEE Press (2013)
23. Malik, H., Jiang, Z.M., Adams, B., Hassan, A.E., Flora, P., Hamann, G.: Automatic comparison of load tests to support the performance analysis of large enterprise systems. In: 2010 14th European Conference on Software Maintenance and Reengineering, pp. 222–231. IEEE (2010)
24. McGrath, G., Brenner, P.R.: Serverless computing: design, implementation, and performance. In: 2017 IEEE 37th International Conference on Distributed Computing Systems Workshops (ICDCSW), pp. 405–410. IEEE (2017)
25. Nguyen, T.H., Adams, B., Jiang, Z.M., Hassan, A.E., Nasser, M., Flora, P.: Automated detection of performance regressions using statistical process control techniques. In: Proceedings of the 3rd ACM/SPEC International Conference on Performance Engineering, pp. 299–310. ACM (2012)
26. Romano, J., Kromrey, J.D., Coraggio, J., Skowronek, J.: Appropriate statistics for ordinal level data: should we really be using t-test and cohen'sd for evaluating group differences on the NSSE and other surveys. In: Annual Meeting of the Florida Association of Institutional Research, pp. 1–33 (2006)
27. Rosner, B.: Fundamentals of biostatistics. Nelson Education (2015)
28. Shewhart, W.A.: Economic Control of Quality of Manufactured Product. ASQ Quality Press, Milwaukee (1931)
29. Smith, C.: Understanding AWS lambda performance-how much do cold starts really matter? (2017). https://blog.newrelic.com/technology/aws-lambda-cold-start-optimization/. Accessed 26 Mar 2019
30. Stapleton, J.H.: Models for Probability and Statistical Inference: Theory and Applications, vol. 652. Wiley, Hoboken (2007)
31. Syer, M.D., Jiang, Z.M., Nagappan, M., Hassan, A.E., Nasser, M., Flora, P.: Leveraging performance counters and execution logs to diagnose memory-related performance issues. In: 2013 29th IEEE International Conference on Software Maintenance (ICSM), pp. 110–119. IEEE (2013)

32. Syer, M.D., Shang, W., Jiang, Z.M., Hassan, A.E.: Continuous validation of performance test workloads. Autom. Softw. Eng. **24**(1), 189–231 (2016). https://doi.org/10.1007/s10515-016-0196-8
33. The Eclipse Foundation: Open source software for industry 4.0, an eclipse IoT working group collaboration. Made available under the Eclipse Public License 2.0 (EPL-20) (2017). https://iot.eclipse.org/resources/white-papers/Eclipse%20IoT%20White%20Paper%20-%20Open%20Source%20Software%20for%20Industry%204.0.pdf. Accessed 20 Oct 2017
34. Wang, L., Li, M., Zhang, Y., Ristenpart, T., Swift, M.: Peeking behind the curtains of serverless platforms. In: 2018 {USENIX} Annual Technical Conference ({USENIX} {ATC} 2018), pp. 133–146 (2018)
35. Willaert, F.: AWS lambda container lifetime and config refresh (2016). https://www.linkedin.com/pulse/aws-lambda-container-lifetime-config-refresh-frederik-willaert/. Accessed 26 Mar 2019
36. Windisch, E.: Understanding AWS lambda coldstarts (2017). https://read.iopipe.com/understanding-aws-lambda-coldstarts-49350662ab9e. Accessed 26 Mar 2019
37. Wohlin, C., Runeson, P., Höst, M., Ohlsson, M.C., Regnell, B., Wesslén, A.: Experimentation in Software Engineering. Springer, Heidelberg (2012). https://doi.org/10.1007/978-3-642-29044-2
38. Woodside, M., Franks, G., Petriu, D.C.: The future of software performance engineering. In: 2007 Future of Software Engineering, pp. 171–187. IEEE Computer Society (2007)

Method of Constructing a Graphic Model of the Regulatory and Legal Framework in the Sphere of Information Security

A. V. Manzhosov(✉) and I. P. Bolodurina(✉)

Orenburg State University, Victory Avenue, 13, 460018 Orenburg, Russia

Abstract. Data visualization in the field of information security (IS) is becoming an urgent task. The more informative the visual display for the analyst, the faster and better he will be able to get the result. The visualization method described in the article allows to reduce the time needed for determination the actual IS threats when processing the IS threat model. In the classic way of displaying, threat models are a list or a table of several tens of printed pages. To process such a volume of data, information security analysts need to spend a lot of time. It remains likely that the analyst will lose sight of important data. The article proposes a method for visualizing the spatio-temporal model of IS threats in three-dimensional and two-dimensional form. When visualizing a threat model, a surface with axes is used: time, risk, information asset. The proposed visualization of the IS threat model allows the analyst to identify current threats in less time. Scope - management of information security. The result of this work is a way to visualize the IS threat model, taking into account the spatio-temporal factor.

Keywords: Information security · Spatio-temporal model of information security threats; visualization · Graphic model

1 Introduction

An analysis of the documents of the IS regulatory framework made it possible to identify patterns in the use of links to documents, in other words, some tacit citation system. This system assumes the use of certain words and phrases indicating the possible mention of another regulatory legal act. The citation system is used to visualize the relationship between regulations.

The graph model is defined as the main visualization element. The pinnacle of the graph model is represented by normative legal acts, the edge - the links between normative legal acts.

The implementation of the method involves the solution of the following tasks: a review of research on methods of visualizing the regulatory framework, as well as the use of a graph model in the field of information security, classification of existing information retrieval models, development of a formalized model of a method for constructing a graph model of a regulatory framework, development of a functional model of a construction

I. Woungang et al. (Eds.): ANTIC 2021, CCIS 1534, pp. 48–62, 2022.
https://doi.org/10.1007/978-3-030-96040-7_4

method a graph model of the regulatory framework, development of an algorithm for the operation of a software tool that implements a method for constructing a graph model of a regulatory framework.

2 Review of Research on Ways to Visualize the Regulatory Framework, as Well as the Use of the Graph Model in Information Security

The article by Jiansong Zhang [1] presents a method for automated processing of regulatory documents in the construction sector based on semantic analysis, the result of which is a tree-like model of the main requirements presented in them. This development became possible due to the peculiarities of the English language, namely the strict order of words in sentences. Application of the method allows to increase the speed of reading, and, consequently, the application of the stated requirements to building structures.

The article by Innar L. and Anton V. [2] describes a method for visualizing normative documents in the form of graphs, where the vertex is a paragraph in the document. This is possible due to the peculiarities of the structure of the normative legal acts of Estonia and the United States. They are based on a huge number of paragraphs, numbering in hundreds, which, among other things, refer to each other, which is the basis for constructing a graph that visualizes the internal relations of an act.

The authors of the article "Citation Analysis: An Approach for Facilitating the Understanding and the Analysis of Regulatory Compliance Documents" [4] offer a more global approach to the system of defining relationships between documents. They propose to build a directed graph, based on the analysis of citations found in documents, where the vertices are a quotation or a cited document. The line between the vertices exists if at least one of these two documents "quotes" the other.

The article by the authors Josef K., Luka S. [3] does not describe the method used for determining and visualizing the links between regulatory documents, but demonstrates the structure of a program that could implement a method for constructing visualization of a regulatory legal framework.

The authors of the article "Applying Graph Centrality Metrics in Visual Analytics of Scientific Standard Datasets" [5] suggest using a graph model to reflect citation networks presented in text form, this allows the network user to focus on higher priority aspects of the research.

In the article "Applying Graph Centrality Metrics in Visual Analytics of Scientific Standard Datasets" [5], the authors propose to use a graph model to reflect citation networks presented in text form, this allows the network user to focus on higher priority aspects of the research.

The authors of the article [7] Weihua C., Lijun C., Xiaotang C., Kaiqi H. propose to use the graph model to visualize the possible path of the object being tracked in the video image. Several types of vertices are used: real and assumed positions of the object, and as edges - possible displacements between them.

The main tool for visualizing relational data types is the graph model, which is also used in other areas of visualization. A graph model represents not only data, but also relationships between data. The presented publications do not form a holistic picture for solving the problem of automated construction of a graph model of the regulatory framework.

The main contribution of our work is the presented additional method for analyzing the regulatory framework. In particular, this method can help certain groups of analysts gain insight into the complex relationships between regulations.

3 Classification of Existing Models of Information Search in a Regulatory Legal Act

Search models have been actively used for a long time, however, user needs are constantly growing and changing, and some algorithms can no longer provide the same speed and efficiency when performing any narrowly focused tasks, which include the subject of this work.

Consider the classification of the existing, most common search models [8], presented (see Fig. 1) and highlight their main features.

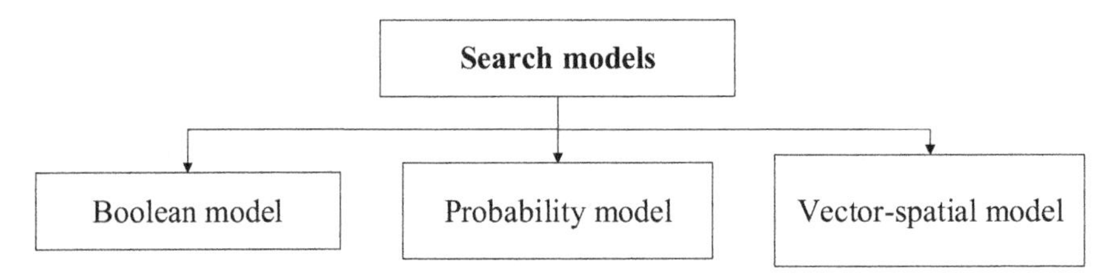

Fig. 1. Classification of search models.

The Boolean model is the simplest one of the above list. Its meaning is to handle boolean expressions (eg the query "search engine" is represented as "search AND system"). Pre-indexing occurs - it is determined in which documents various terms are found, based on the indexed pages (documents), the search is carried out as follows: the query entered by the user is divided into terms, then each of these terms is found in the dictionary and, based on Boolean laws, it is determined which of the indexed pages satisfy the stated condition. The resulting list is presented to the user. The obvious disadvantage of this method is the lack of ranking in accordance with relevance (semantic match to the request).

The disadvantage inherent in the Boolean model is absent in the probabilistic search model, which is based on Bayes' theorem on conditional probabilities of events. For the model to function, each document is associated with a certain vector

$$l = (t_1, t_2, \ldots, t_n) \tag{1}$$

where $t_i = 1$, if this term is included in the document, and $t_i = 0$, if it is not included. Designating for W_1 the event that signals that the document in which the search is carried out is relevant, and for W_2 – the opposite event W_1, the event of document irrelevance, then calculating $P(W_i|l)$ we will determine the probability of occurrence of the event W_i

for the document being checked [9]. Applying Bayes' theorem can lead to probabilities whose values are easier to estimate, namely:

$$P(W_i|l) = P(l|W_i)P(W_i)/P(l) \tag{2}$$

the canonical model uses a simplification that assumes the independence of any pair of terms in the document. The main disadvantage of this model is its low scalability (as the number of documents grows, the processing time grows exponentially).

Within the framework of a vector-spatial model, a document is described by a vector in Euclidean space, in which each term used in the document is assigned its weight value, which is determined on the basis of statistical information about its appearance both in a separate document and in the entire documentary array. The user's query is also a vector in the same Euclidean term space. The dot product of the corresponding query and document vectors is used to estimate the proximity of the query and the document. This model operates with large amounts of data and high-dimensional arrays, which makes it difficult to process large requests [10].

The developed method offers an alternative model for bypassing the use of terms. The Boolean model is used in the basis, but the search is carried out not by a full request, but only by a limited set of predefined sets. Then the found sections are matched with the already complete set of requests (document names) and, based on the probabilistic approach, the relevance of this fragment to the corresponding document is determined, which is initially useless due to the presence of a large array of words in each of the documents.

Each of the described methods is based on a mathematical apparatus that describes a set of functions used to implement it, which is also called a formalized model of the method.

4 Formalized Model for Visualization of the Regulatory Framework

It is necessary to define a set of data defining the structure of the graph before proceeding to its construction.

The Boolean search model using the associative-majority model of pattern recognition was chosen as a mathematical model for determining the coherence of regulatory legal acts.

The following conventions are introduced:

D = {d1, d2, d3, ..., dm} – a plurality of rules that make up the legal framework;
T = {t1, t2, t3, ..., tm} – a plurality of titles of regulatory legal acts **D**;
ti,k – ordinal k heading word of the regulatory legal act **di**, **k = 1, kmax**;
t^{x}i,q – ordinal q heading word of the regulatory legal act, found in the text, **q = 1, qmax**;
F = {f1, f2, f3, ..., fm} – a set of textual information contained within each of the regulatory legal acts of a closed set **D;**
fj,r – ordinal j word belonging to the text of a regulatory act **dj**, **r = 1, rmax**;
Vx,ij – a measure of proximity between a title **ti** and text **fj**;

W = {w1, w2, w3, …, wn} – a plurality of sets of keywords by which the search is carried out;
wi, p – ordinal p keyword from ordinal i set, by which the search is carried out, **p = 1, pmax**;
Vz,ij – a measure of proximity between a set **wi** and text **fj**;
yij - mention of regulatory legal act **dj** in ordinal **i** document of the plurality **D**, **i = 1, m**.

The definition of a reference to any act from the current set that is present in a regulatory legal act is made by searching for all sentences containing heading keywords. The array of received document links is matched with titles from the current set, and matches between the received link and the title are identified using keywords. In this case, it is possible to obtain many different situations of mentioning the heading in the text. All of them can be represented by means of Euler circles [11]. The options for the entry of the heading into the text of the normative legal act are shown (see Fig. 2).

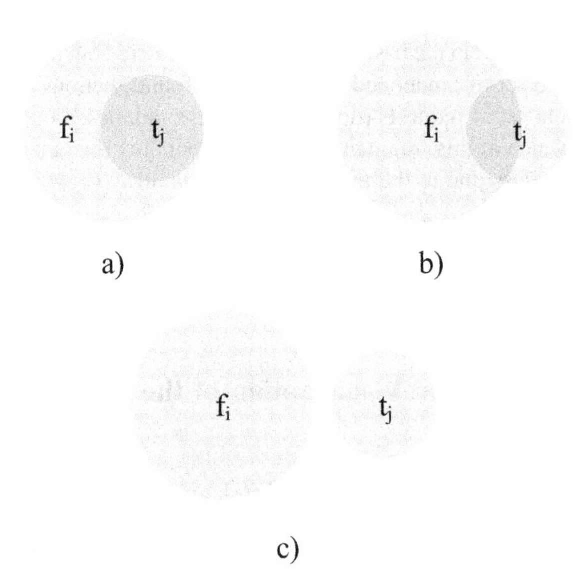

Fig. 2. Options for the title to be included in the text of a regulatory legal act.

There are 3 different options for the presence of the heading in the text of the document, but option "b" may have different provisions, since the percentage of heading words in the text may vary.

In other words, there are 3 different situations:

1 heading is fully presented in the text of the document (a);
2 the title is partially presented in the text of the document (b);
3 heading is missing in the current document (c).

From the point of view of the boolean search model, it is necessary that most of the query was contained in the text of a document d_x.

$$t_{i,1} \wedge t_{i,2} \wedge t_{i,3} \wedge \cdots \wedge t_{i,k} \tag{3}$$

The task of searching for a large array of text can be simplified: the search is performed only in sentences in which the probability of meeting a reference to a normative legal act tends to one. The presence of predefined conditional expressions in a sentence is taken as a factor that increases the probability.

Based on this statement, the following measure of proximity between words in an expression and words in a text is determined.

$$V_{z,ij} = \begin{cases} 1, \mathit{if}\ w_{i,p} = f_{j,r} \\ 0, \mathit{if}\ w_{i,p} \neq f_{j,r} \end{cases} \tag{4}$$

Word order is important for correct search. This means that the sum of the measure (array) of proximity must be maximum, which can be expressed by the following expression

$$w_i \in f_j \leftrightarrow \begin{cases} 1, \mathit{if} & \sum_{z=1}^{p_{max}} V_{z,ij} = p_{max} \\ 0, \mathit{if} & \sum_{z=1}^{z_{max}} V_{z,ij} \neq p_{max} \end{cases} \tag{5}$$

In this way, all sentences that may contain document titles are found.

When the names of the documents are determined, they are compared with those documents that are present in the current set D. This can be done using the following condition.

$$V_{x,ij} = \begin{cases} 1, \mathit{if}\ t_{i,q}^{x} = t_{j,k} \\ 0, \mathit{if}\ t_{i,q}^{x} \neq t_{j,k} \end{cases} \tag{6}$$

The condition for a document reference is represented by the following expression

$$t_{i,q}^{x} \in f_j \leftrightarrow y_{i,j} = \begin{cases} 1, \mathit{if}\ \sum_{x=1}^{q_{max}} V_{x,ij} \rightarrow q_{max} \\ 1, \mathit{if}\ V_{\{0\},ij} + V_{\{1\},ij} = 2 \\ 0, \mathit{if}\ \sum_{x=1}^{q_{max}} V_{x,ij} \rightarrow 0 \end{cases} \tag{7}$$

Where, $V_{\{0\},ij}$ и $V_{\{1\},ij}$ are the conditions for the coincidence of the date and document number.

Based on the values obtained from expression (7), an adjacency matrix is constructed and presented in Table 1. An adjacency matrix is a way to represent a graph as a boolean matrix {0, 1}, where 0 – is the absence of an edge between two vertices, and 1 – the presence of this rib [12]. The direction is defined as follows: the vertex defined in the row is the beginning, and the vertex defined by the column is the end, in other words, the adjacency matrix is not symmetric for a directed graph.

Table 1. Adjacency matrix.

d_i	d_j				
	d1	d2	d3	…	dm
d1	y1,1	$y_{1,2}$	$y_{1,3}$	…	$y_{1,m}$
d_2	y2,1	$y_{2,2}$	$y_{2,3}$	…	$y_{2,m}$
…	…	…	…	…	…
d_m	ym,1	$y_{m,2}$	$y_{m,3}$	…	$y_{m,m}$

A graph is defined, which formally has the following form

$$G = (D, E). \tag{8}$$

where D – the set of vertices of the graph (this set coincides with the above);

E – set of ordered vertices d_i, $d_j \in$ D.

According to Table 1, the graph model of the regulatory framework has a formal representation, but it is also necessary to take into account those regulatory legal acts that are not present in the current closed set D, but references to which are present in it.

All these links are found in the set of titles of normative legal acts found earlier and remaining after, among them, links to documents of the set D have been identified.

After taking into account such links, new vertices may appear, the adjacency matrix will change (updated), and now has the form presented in Table 2.

Table 2. Updated adjacency matrix.

d_i	d_j							
	d1	d2	d3	…	dm	dm+1	…	dm+a
d1	y1,1	$y_{1,2}$	$y_{1,3}$	…	$y_{1,m}$	$y_{1,m+1}$	…	$y_{1,m+a}$
d_2	y2,1	$y_{2,2}$	$y_{2,3}$	…	$y_{2,m}$	0	…	$y_{2,m+a}$
…	…	…	…	…	…	…	…	…
d_m	ym,1	$y_{m,2}$	$y_{m,3}$	…	$y_{m,m}$	$y_{m,m+1}$	…	$y_{m,m+a}$

Cells whose value turned out to be empty are filled with zeros.

After the adjacency matrix has taken its final form, the graph can be built. As an example, consider a schematic representation of the graph model of a part of the regulatory framework (see Fig. 3).

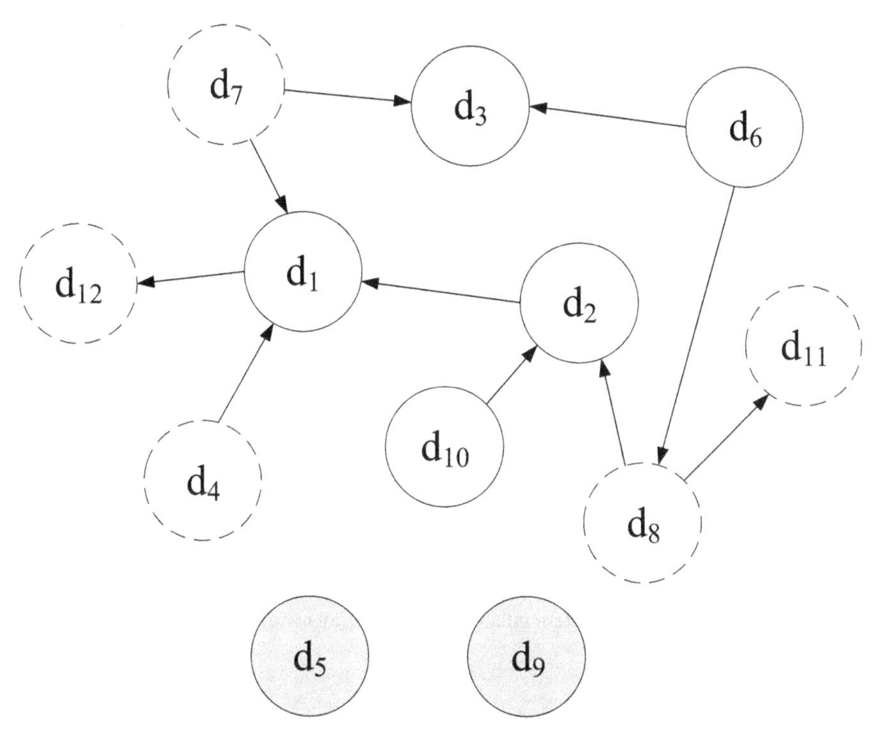

Fig. 3. Schematic representation of the graph model of a part of the regulatory framework.

The figure shows a graph consisting of 12 vertices. It is possible to single out some conditional groups of normative legal acts into which it is divided.

1 belonging to the closed set D and participating in the construction of the main graph ($d_1 - d_4$, $d_6 - d_8$, d_{10}).
2 not belonging to the closed set D and participating in the construction of the main graph (d_{11}, d_{12}).
3 belonging to the closed set D and not participating in the construction of the main graph (d_5, d_9). Vertices containing such documents are also called isolated.

A real example is the graph model of a part of the regulatory and legal framework for the safety of fuel and energy complex facilities (see Fig. 4).

To build an example, 10 documents were used, among which 7 belong to group 1 and 3 more documents - to group 3.

The next step after describing the mathematical basis of the future model is its functional design.

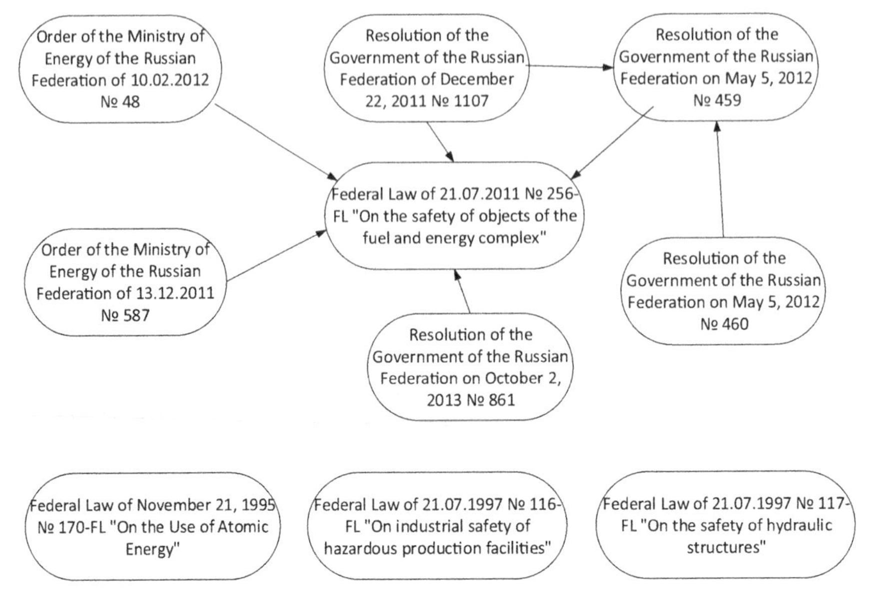

Fig. 4. Graph model of a part of the regulatory and legal framework for the safety of objects of the fuel and energy complex.

5 Functional Modeling of the Developed Method

The functionality of the system is described by the IDEF0 model. They most accurately and fully reveal the essence of the process due to their clarity, understandable language of description and simplicity. The system gradually builds up as the design process proceeds using this methodology [13].

Figure 5 presents a contextual diagram of the process of constructing a graph model of the regulatory framework.

To divide the process into its constituent functions, we will carry out its decomposition. Figure 6 shows the decomposition of the context diagram presented in Fig. 5

When decomposing the process of constructing a graph model of the regulatory framework, 6 child functions are obtained.

1. Preliminary preparation of documents.

   The list of normative legal acts necessary for an analyst to work is fed to the input of this function, and headers are determined on their basis. This function can be implemented both programmatically and manually, by first renaming the files in accordance with the title of the regulatory legal act. The output is a lot of headers.
2. Search for matches.

   The function determines which of the headings are found in documents, whose heading differs from the current one, based on the previously received set of headings and the entered set of normative legal acts using the closed set search algorithm. Thus,

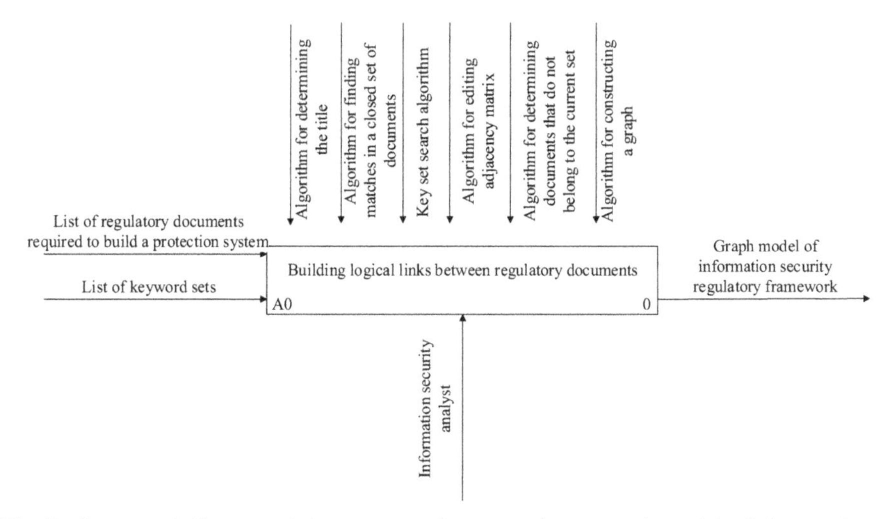

Fig. 5. Contextual diagram of the process of constructing a graph model of the regulatory framework.

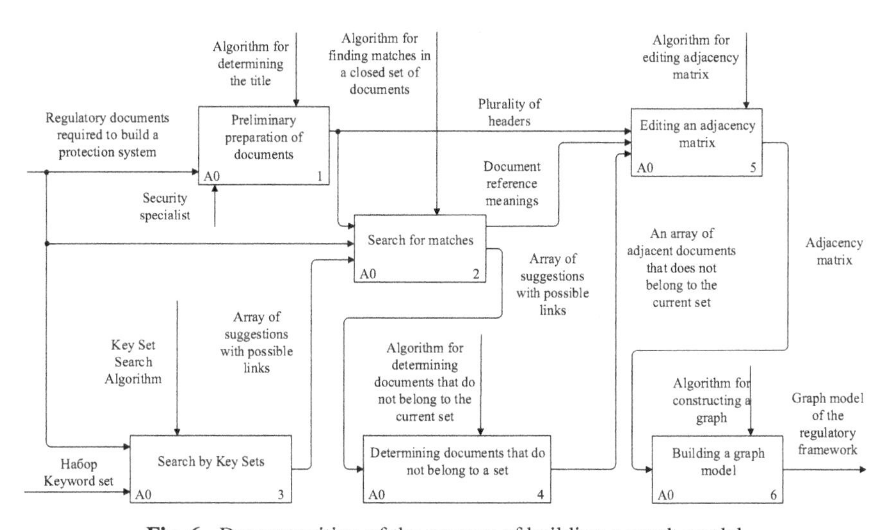

Fig. 6. Decomposition of the process of building a graph model.

at the output of the function, the values of references to documents are obtained, as well as a selection from sentences in which no matches were found for the current set of documents.

3. Search by Key Sets.

   The input values of the function are plurality regulations and sets of keywords. The search algorithm for key sets in all regulatory legal acts searches for possible references to documents. The sentences with found matches are copied into an array and passed to the search function.

4. Definition of documents that do not belong to a set.
   The function by the skeleton of known headers determines the name of a possible mention of a regulatory legal act after receiving an array of sentences and, in case of successful detection, transfers the corresponding name to the adjacency matrix editing function.
5. Editing adjacency matrix.
   The function is the heart of the whole method of constructing a graph, since it forms an adjacency matrix, based on the values of references to regulatory legal acts and a set of headers obtained from functions 2 and 4 using the editing algorithm, which is the output value of this function.
6. Building a graph model.
   The graphical construction of a graph model of the regulatory framework occurs in the last function of the method using a predetermined algorithm, based on the adjacency matrix obtained from function 5, which is the final result of the process.

The above functions are the basis for the development of the algorithm for the operation of the software tool.

6 Algorithm for Constructing a Graph Model of the Regulatory Framework

One of the main tasks that need to be solved in the development of this method is the development of an algorithm for the functioning of the software tool. The result of this stage is shown in Fig. 7.

This algorithm includes several stages.

1. Adding regulations.
   At this stage, the user adds normative legal acts, on the basis of which the current model is built. This is done by choosing a directory with the required set of files.
2. Definition of titles of regulations.
   Headings are determined from regulatory legal acts and placed in a separate array. This assumes that the files found in the directory may or may not have names that match the pattern (%document type% % publisher% from% date% % number % % title%).
3. Cycle processing regulation.
   In this cycle, the main work is carried out to determine the links between normative legal acts as follows: a search is carried out for all sentences containing headings keywords. The array of received document links is matched with titles from the current set, and matches between the received link and the title are identified using keywords. If no matches are found for a closed set, then according to predefined templates it is determined which normative legal act the link was found to. The detected referenced normative legal act is added to the adjacency matrix, which is subsequently edited.
4. Building a graph model of the regulatory framework.
   The adjacency matrix obtained in the cycle is the basis for the development of the graph, with which the user subsequently works.

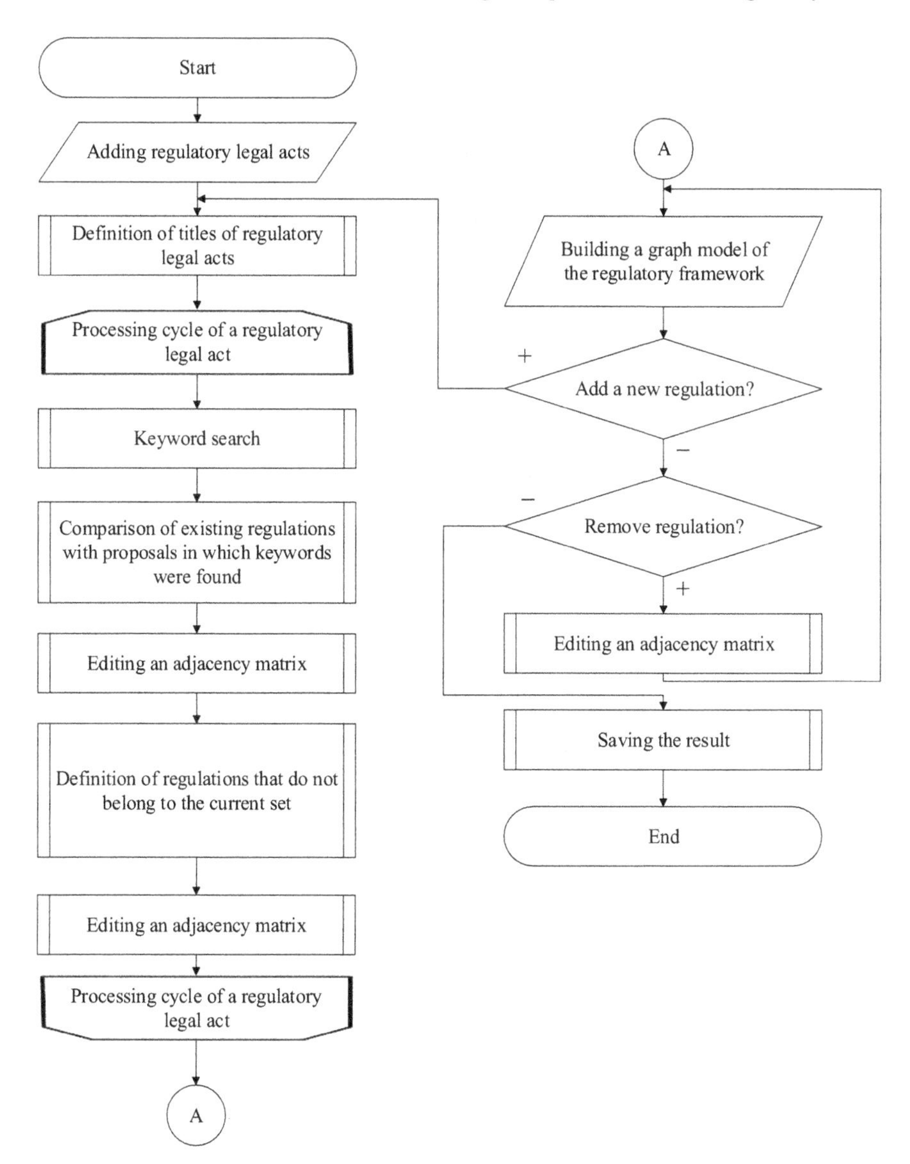

Fig. 7. Algorithm of the software tool.

5. Adding or removing regulation.

   After the graph has already been built, the user can either add a new normative legal act for processing by the program, or delete it. If added, the program will return to the loop and check for references to this regulatory legal act in others, as well as detect links directly in this regulatory legal act to identify new dependencies. If deleted, the matrix will be edited and the graph visualization will be updated.

6. Saving the result.

   After the graph is built, the user has the opportunity to save it in order to continue working in the future.

Prospects for using the means of automated construction of a graph model of the regulatory and legal framework

The purpose of using the means of automated construction of the regulatory framework is to reduce the complexity of the task of analyzing the requirements of the regulatory framework. The analysis of the requirements of the regulatory framework is carried out by information security analysts of the organization periodically. Issues faced by an information security analyst when analyzing the requirements of the regulatory framework:

- are all information security requirements met?
- have all the regulations taken into account when analyzing the legal framework?
- if not all regulatory legal acts are taken into account, then what are the names of the unrecorded documents?
- have all the necessary documents been developed for the organizational and methodological support of the organization in the field of information security?
- is there consistency of documents, organizational and methodological support of the organization in the field of information security?
- is the coherence of documents in the organizational and methodological support of the organization in the field of information security correct? etc.

Analysts in the field of information security were interviewed in order to find out their satisfaction with the proposed means of automated construction of a graph model of the regulatory framework. The survey was conducted twice: after the first use of the product and after testing for 7 days. The results of testing the means of automated construction of a graph model of the regulatory and legal framework are presented in Fig. 8.

The results of testing the means of automated construction of a graph model of the regulatory framework provided suggestions and recommendations for further improvement and development of the tool. Information security analysts have shown interest in the automation of information security management processes. Analysts changed their minds about the tool after a week of testing. This demonstrates interest in the proposed solution and reusability.

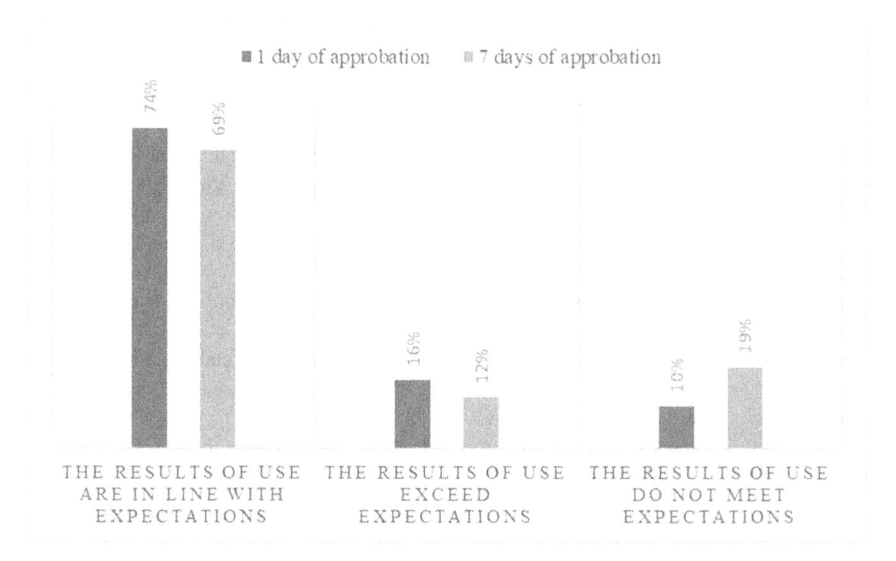

Fig. 8. The results of approbation the means of automated construction of a graph model of the regulatory and legal framework.

7 Conclusion

The presented developments offer a model and an algorithm for automated visualization of relations between regulatory legal acts by building a graph model, which allows the information security analyst to accelerate the perception of information. The main advantage of the proposed method is the speed of data analysis, since its software implementation is capable of reducing the time spent on manual search, analysis and systematization of information by tens of times. The study was carried out with the financial support of the Russian Foundation for Basic Research within the framework of scientific project No. 20-07-01065, as well as a grant from the President of the Russian Federation for state support of leading scientific schools of the Russian Federation (NSh-2502.2020.9).

References

1. Zhang, J.: A logic-based representation and tree-based visualization method for building regulatory requirements. Visual. Eng. **5**(1), 1–14 (2017)
2. Innar, L., Anton, V., Ermo, T.: Visualization and structure analysis of legislative acts: a case study on the law of obligations. In: ICAIL 2007: Proceedings of the 11th International Conference on Artificial Intelligence and Law, pp. 189–190 (2007)
3. Josef, K., Luka, S.: Visualization of Inter-Communication in Normative-Legal Documents and Automated Design of Analysis in Reference-Legal Systems. Moscow State University, 8–10 (2016)
4. Mohammad, H., Abdelwahab, H.: Citation analysis: an approach for facilitating the understanding and the analysis of regulatory compliance documents. In: Sixth International Conference on Information Technology: New Generations (2009)

5. Applying Graph Centrality Metrics in Visual Analytics of Scientific Standard Datasets. https://www.mdpi.com/2073-8994/11/1/30/htm. Accessed 20 Aug 2021
6. Timo S.: Anomaly detection in log data using graph databases and machine learning to defend advanced persistent threats. In: Lecture Notes in Informatics (2017)
7. Weihua, C., Lijun, C., Xiaotang, C., Kaiqi, H.: An equalized global graph model-based approach for multicamera object tracking. IEEE Trans. Circuits Syst. Video Technol. **27**(11), 2367–2381 (2017)
8. Baeza-Yates, R., Ribeiro-Neto, B.: Modern Information Retrieval: The Concepts and Technology behind Search. ACM Press Books, New-York (1999)
9. Manning, K.D., Raghavan, P., Schütze, H.: An Introduction to Information Retrieval. Williams Publishing House, Moscow (2011)
10. Greiff W.R.: A theory of term weighting based on exploratory data analysis. In: Proceedings of the 21st annual international ACM SIGIR conference on Research and development in information retrieval. ACM, pp. 11–19 (1998)
11. Kuzichev A.S: Venn Diagrams. History and Application, Publishing house "Science", Moscow (1968)
12. Bondy, J.A., Murty, S.R.: Graph Theory with Applications. The Macmillan Press Ltd., Ontario, Canada (1976)
13. Information technology to support the life cycle of products. Functional Modeling Methodology. https://docs.cntd.ru/document/1200028629. Accessed 230 Aug 2021

Decision Factor Based Modified AODV for Improvement of Routing Performance in MANET

Priyanka Pandey(✉) and Raghuraj Singh

Harcourt Butler Technical University, Kanpur, India

Abstract. Routing in Mobile Ad Hoc Network (MANET) suffers from route failure issue. Major causes of failure are unstable link, low battery power of participating nodes and frequent topological changes which further leads to performance degradation of routing protocol in terms of packet delivery ratio, delay and routing overheads. Therefore, to overcome from these consequences, a new mechanism is proposed namely Modified AODV (Mo-AODV) which selects the path on the basis of Decision Factor (DF) metric. The DF is estimated by considering received signal power and node's residual energy. Further, a node updates its neighbourhood information on the basis of neighbors' energy level through hello message. Simulation results prove that the proposed approach gives better output in terms of packet delivery ratio, delay, normalized routing load, throughput and control message overhead.

Keywords: AODV · Energy · Signal power · Routing · MANET

1 Introduction

In MANET [1] environment mobile nodes move independently. Communication in this kind of network is performed without any access point and centralized control. Routing data packets in such environment is one of the crucial tasks due to node mobility which causes frequent link failure. Several routing protocols are proposed by many researchers and they can be broadly classified as reactive and proactive.

Reactive routing protocols such as AODV [2], DSR [3] and TORA [4] construct a route only when data is required to be transmitted from sender to receiver. Thus, these protocols are also known as on demand routing protocol. Among these, AODV is one of the most widely used routing techniques in MANET as it performs better routing operation in large and dynamic network. However, the protocol still faces challenges due to limited MANET characteristics.

Another category of routing is proactive protocols. OLSR [5] and DSDV [6] are some of its examples. These protocols are also known as table driven as every node has maintained list of route to every other node in its routing table.

I. Woungang et al. (Eds.): ANTIC 2021, CCIS 1534, pp. 63–72, 2022.
https://doi.org/10.1007/978-3-030-96040-7_5

In this work, we proposed improvement over standard AODV routing technique. The scheme Mo-AODV makes decision about forwarding of RREQ packet on the basis of Decision Factor metric which is estimated using signal strength and residual energy parameters. Simulation results of the proposed approach gives better output in terms of all essential performance metrics such as delay, routing overhead and packet delivery ratio.

Rest of the sections is organized as follows: Sect. 2 describes related work done in order to improve AODV performance along with their limitations. In Sect. 3 working of proposed approach is discussed. Section 4 analyzes the results obtained after simulation and in Sect. 5 conclusion is drawn.

2 Related Work

AODV constructs a route on the basis of number of hops and sequence number. However, other important factors such as signal strength; node energy, delay, distance etc. are also required to be considered during route discovery operation. Avoidance of aforementioned factors may lead to route failure and thus degrade overall routing performance. This section presents some of the works done related to improvement over AODV scheme.

In [7] Ant Colony Optimization technique is utilized for selection of an optimal path. The scheme considers factors such as number of nodes, velocity, residual energy and travel time during route discovery operation. Performance of the proposed scheme is better in terms of packet delivery ratio, delay and energy consumption as compared to AODV. However, routing overhead has not discussed in result analysis section.

Enhanced Energy AODV (EE-AODV) [8] estimates energy consumption rate and residual energy factors during route request (RREQ) forwarding operation. The im-proved version of AODV (EE-AODV) has shown better performance in terms of packet delivery, overhead, delay and energy consumption by varying number of connections. However, the scheme is not effective when number of connection is less than 10 in simulation.

Sofian Hamad et al. [9] proposed Average Link Stability with Energy Aware protocol which takes into account parameters such as LLT (Link Lifetime) and Residual energy during RREQ packet forwarding operation for construction of stable route. The protocol shows significant improvement in terms of network lifetime, delay and total packet received. However throughput performance has not shown much improvement.

Route Constancy and Energy Aware routing protocol [10] estimates residual energy of a node while receiving RREQ packet. If the energy is low, delay forwarding tech-nique is applied which holds the received packet for certain time and later forwards it. Network lifetime, routing load, throughput and packet delivery ratio showed significant improvement. Though, there is a possibility of increase in delay due to delay forwarding technique.

Pratik Gite [11] applied a Runge Kutta method to estimate stability time of a link. In the scheme signal strength is measured as a function of distance and time.

Once the source node receives RREP packets from all neighbours, it predicts route stability. The proposed AODV is compared with standard AODV and results are represented in the form of graphs which show significant improvement in terms of throughput, packet delivery ratio and delay. Though, analysis in terms of routing overhead is out of scope.

In [12] Ant Colony Optimization with Bacterial Foraging Algorithm (ACO-BFA) for AODV is proposed. The technique is applied in order to enhance the route selection process based on signal strength, energy and hop value. The performance has been improved in terms of delay, overhead, network lifetime and energy consumption. However, packet delivery ratio performance has not been analyzed in the analysis section.

Fuzzy-Hybrid-AODV [13] focuses on improvement of throughput by considering hybrid bio inspired approach based on Genetic Algorithm and Hill climbing tech-nique. The inputs parameter to the fuzzy inference engine are start to finish delay, number of time hubs leave the machine, parcel dropped and number of RERR packet generated. The scheme achieves higher throughput. However, other metrics such as delay, overhead and packet delivery performance have not been discussed in result analysis section.

Bonu Satish et al. [14] proposed a scheme to decide Active Route Timeout value by considering different parameters such as number of control packet sent, hop count and number of nodes. These parameters are fed into fuzzy inference engine which helps in deciding ART value for routing operation with AODV protocol. Simulation results show good improvement in delay and throughput. However, routing overhead and packet delivery ratio have not been taken into account during analysis.

In [15] the proposed technique considers number of nodes and node speed to suggest value of hello interval. Results show significant improvement. However, as number of nodes increases end to end delay performance is not shown much improvement.

Kumari Praveen et al. [16] proposed cluster based energy efficient AODV technique which works on residual energy and QoS for reliable communication. Results are improved in terms of energy consumption, throughput, overhead and packet delivery. However, end to end delay is not discussed in analysis.

QoS-SM [17] measures the probability of link availability using factors such as loca-tion information, speed and direction of nodes. Further in order to measure QoS support parameters such as traffic, neighbours of each node and frequency of transmission are combined as well.

Summary of related work is also represented in the Table 1.

From literature review it can be concluded that researchers have mainly utilized signal strength and energy parameters as criteria factors for selection of final path. Different approaches such as delay forwarding [10], Runge Kutta method [11], bio inspired techniques [7,12] and Fuzzy technique [13,14], selection of hello interval [15], cluster based [16] and prediction of link support [17] have been applied for performance enhancement of AODV routing algorithm. However, the proposed works have not shown significant improvement over all

Table 1. Summary of related work

Reference	Protocol	Approach of work
[7]	ORA_AODV	ACO technique is applied by considering travel time, velocity, energy and number of nodes in the path
[8]	EE-AODV	Energy consumption rate is considered as a selection criteria for route establishment
[9]	A-LSEA	Residual energy and Link Lifetime factors are considered for route selection
[10]	RCEARP	Link stability and residual energy factors have been considered for route establishment
[11]	PAODV	Runge-Kutta method is applied to approximate upcoming strength
[12]	ACO-BFA	ACO and Bacterial foraging techniques are applied during route discovery with main focus to minimize energy consumption
[13]	Fuzzy- Hybrid GA	Genetic Algorithm and Hill climbing approaches are applied combinely
[14]	FBARAODV	Fuzzy approach is applied using sent control packets, number of nodes and hop count factors
[15]	ANFISHIAODV	Hello interval time is suggested on the basis of number of nodes and node speed
[16]	EE-AODV	Residual energy and QoS are considered for reliable communication
[17]	QoS-SM	Measures possibility of link support at specific route

considered performance metrics such as normalized routing load, delay, control overhead and packet delivery ratio or some of them are not evaluated and analyzed. Hence, considering these flaws, the proposed work aims to improve routing performance in terms of all aforementioned metrics and results are represented and analyzed through graphs.

3 Proposed Algorithm

The proposed Modified AODV (Mo-AODV) considers signal quality and energy factors to decide next hop neighbors. The scheme utilizes hello mechanism for monitoring neighborhood nodes. The node receiving hello message updates its neighbor table on the basis of residual energy of the sender i.e. if residual energy is below threshold, the node does not update its neighbor table. Consideration of received signal power and residual energy during route discovery process restricts unnecessary broadcasting of RREQ packets and thus makes protocol more efficient in performing routing operation.

3.1 Parameters

Following parameters have been utilized in route discovery operation.

Residual Energy: In MANET environment, nodes operate with the help of battery power. If active node has less energy level then it may cause break in routing operation. This may further lead to unwanted delay, packet loss and message overheads. Residual energy of a node is calculated as the difference between initial energy and total consumed energy after certain time interval as shown below in Eq. 1

$$ENE_{re} = ENE_i - ENE_c \tag{1}$$

Where ENE_i is initial energy ($60J$), ENE_{re} is residual energy and ENE_c is total energy consumed in transmitting, receiving and in idle mode after certain time.

Signal Strength: It is utilized for examining link stability between two nodes. If the signal strength between two nodes is strong, probability of link lifetime is more as compared to the link with weak signal strength. For estimation of received signal power, Two Ray Ground [18] propagation model is used in this approach.

3.2 Route Discovery Operation

Whenever an intermediate receives RREQ packet, it evaluates Reward Factor RF on the basis of parameters such as residual energy (ENE_{re}), initial energy (ENE_i), current received signal power (Rec_p) and received signal power threshold (Rec_{th}). Equation 2 is used to estimate Reward Factor RF.

$$RF = ({}^{ENE_{re}}/_{ENE_i} + (1 - {}^{Rec_p}/_{Rec_{th}})) \tag{2}$$

Further the node calculates hop score H_{sc} using Eq. 3 as shown below.

$$H_{sc} = {}^{RF}/_{2} \tag{3}$$

Afterwards, the receiving node compares the evaluated hop score H_{sc} against hop score threshold (H_{scth}) in order to take decision about routing as mentioned in the Eq. 4.

$$DecisionFactor = \begin{cases} ForwardPacket, & \text{if } H_{sc} > H_{scth} \\ Discard, & \text{Otherwise} \end{cases} \tag{4}$$

Based on estimated Decision Factor (DF), an intermediate node either forwards or discards the RREQ packet. The overall operation is also represented through the flow chart as mentioned in Fig. 1.

As shown in the flowchart (Fig. 1), whenever an intermediated node receives route request packet, it first determine received signal power and node's energy

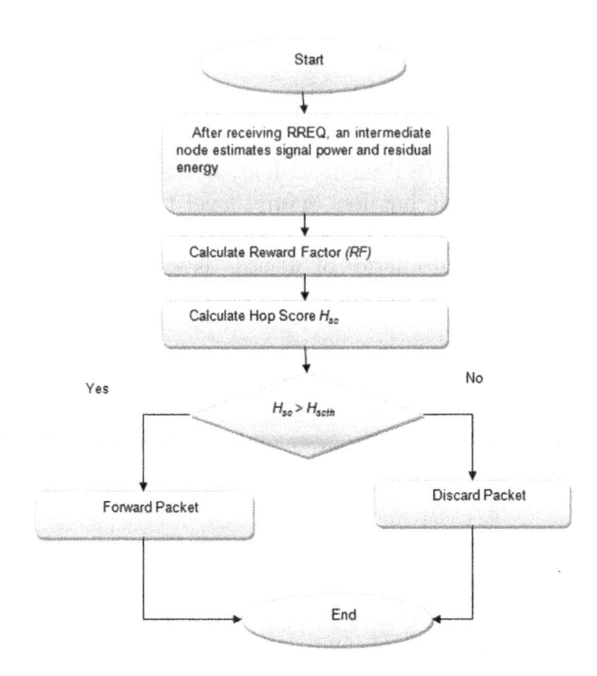

Fig. 1. Route discovery operation

and estimates node's reward factor using Eq. 2. Further, the hop score H_{sc} is estimated and compared with the threshold H_{scth} to take decision for forwarding or discarding of control packet.

Moreover, the scheme continuously monitors energy level of its neighbors using hello message and if the level is found below the threshold the node does not update its neighbor table. Along with this the node can receives multiple duplicate RREQ packets which helps in giving fresher route information.

4 Simulation Environment and Result Analysis

The proposed scheme is implemented using network simulator NS 2.35 [19]. It is a discrete event simulator. The algorithm has deployed in 900×1000 simulation region and Random Way point [20] is chosen as mobility model. In this scheme, Constant Bit Rate (CBR) traffic type is considered. Main simulation parameters of experimental set up is represented below in Table 2.

Performance of the proposed scheme is evaluated in terms of packet delivery ratio, Normalized Routing Load, Control overhead, Throughput and delay by varying simulation time. Results obtained after simulation are represented in the graph as shown below from Figs. 2, 3, 4, 5 and 6 along with discussion about their performance analysis in terms of aforementioned performance metrics.

Table 2. Main simulation parameters

Simulator	NS2.35
Simulation Area	900×1000
Propagation Model	Two Ray Ground
MAC Type	IEEE802.11
Antenna	Omni Antenna
Mobility Model	RWP
Simulation Time	100, 200, 300, 400, 500 s
Number of Nodes	62
Max. Connection	12
Pause Time	5.0 s
Protocols	Mo-AODV, AODV

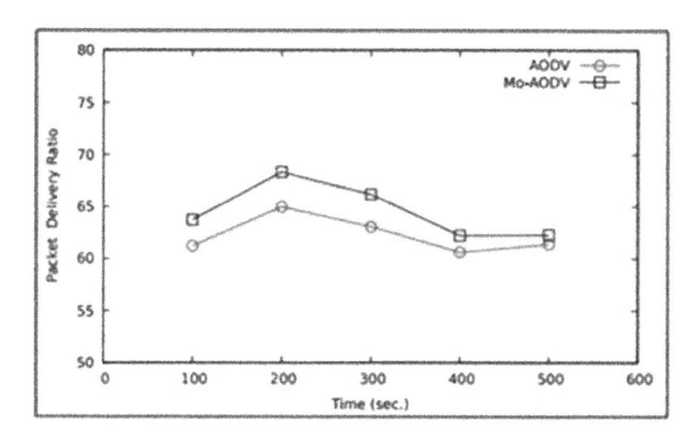

Fig. 2. Packet delivery ratio vs Simulation time

Figure 2 shows packet delivery ratio performance of the proposed approach. When simulation time is 200 s the packet delivery ratio is highest. As the simulation time increases the performance decreases gradually. But overall the proposed approach has 3.6% more packet delivery ratio as compared to AODV.

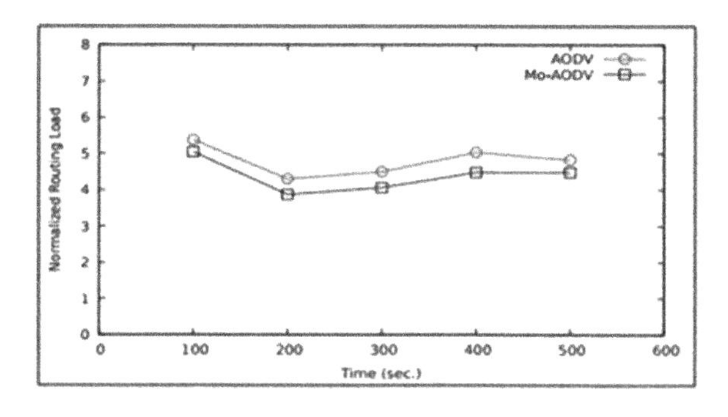

Fig. 3. Normalized routing load vs Simulation time

Normalized routing load is the total number of control packets required to transmit each data packet. From the graph as shown in Fig. 3, the proposed scheme requires 9.14% of less control packets than AODV. This is because, the scheme filters weak nodes and links during route discovery process which prevents unnecessary broad-casting of route request packets.

Control overhead is the total number of packets required during routing operation. From the Fig. 4, it is clear that as the simulation time increases control overhead also increases. But the overall control packets required by proposed approach is 5.3% less than the standard AODV.

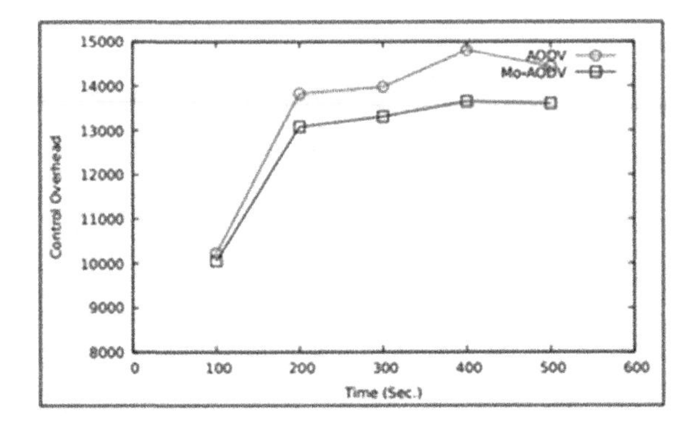

Fig. 4. Control overhead vs Simulation time

Throughput performance of Mo-AODV is represented in the Fig. 5. The proposed scheme has achieved 3.30% higher throughput as compared to standard AODV. This is because consideration of node's residual energy and signal strength parameters makes the path more stable and clear which allows a destination node to receive more number of data packets per seconds.

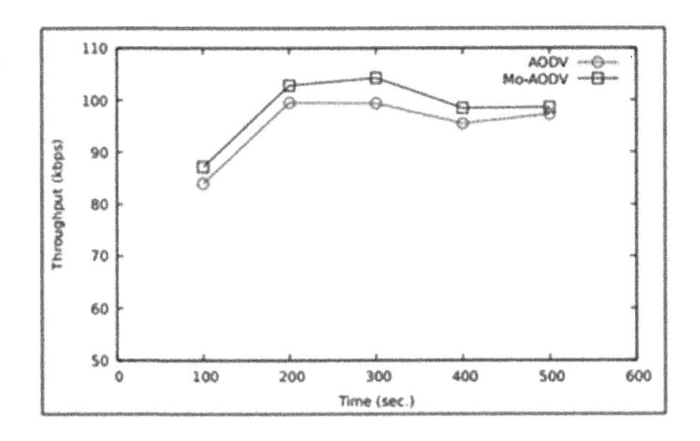

Fig. 5. Throughput vs Simulation time

Figure 6 represents end to end delay performance. From the graph it can be concluded that the proposed approach has 31.25% less end to end delay as compared to AODV. This is because; the scheme searches a path that has less probability of failure during data transmission.

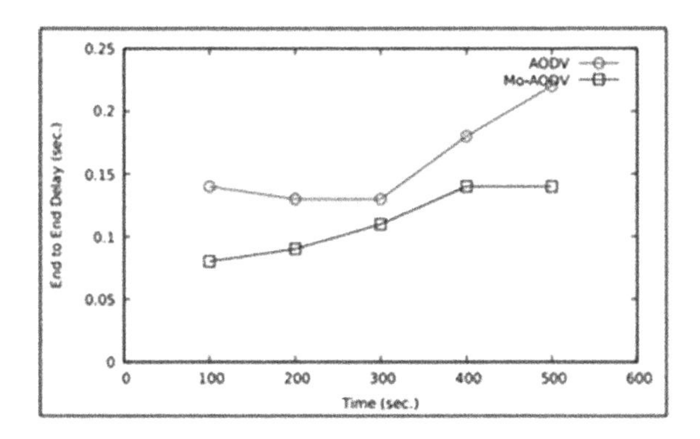

Fig. 6. End to end delay vs Simulation time

5 Conclusion

The proposed Mo-AODV scheme is an enhancement over standard AODV routing protocol that forwards an RREQ packet based on Decision Factor. The factor is estimated using parameters such as signal strength and energy. Final route created by this technique is more efficient as it has higher throughput, better packet delivery ratio, less delay and less routing load as compared to standard AODV. Simulation results are analyzed and represented in the form of graphs. In future, we try to expand the solution by applying Reinforcement Learning technique.

References

1. Rappaport, T.S.: Wireless Communications: Principles and Practice, vol. 2. Prentice Hall PTR, New Jersey (1996)
2. Chakeres, I.D., Belding-Royer, E.M.: AODV routing protocol implementation design. In: Proceedings of the 24th International Conference on Distributed Computing Systems Workshops, pp. 698–703. IEEE (2004)
3. Johnson, D.B., Maltz, D.A.: Dynamic source routing in ad hoc wireless net-works. In: Imielinski, T., Korth, H.F. (eds.) Mobile Computing, pp. 153–181. Springer, Boston (1996). https://doi.org/10.1007/978-0-585-29603-6_5
4. Gupta, A.K., Sadawarti, H., Verma, A.K.: Performance analysis of AODV, DSR & TORA routing protocols. Int. J. Eng. Technol. **2**(2), 226 (2010)
5. Clausen, T., et al.: Optimized link state routing protocol (OLSR) (2003)
6. He, G.: Destination-sequenced distance vector (DSDV) protocol. Networking Laboratory, Helsinki University of Technology, pp. 1–9 (2002)

7. Zant, M.A., Jabareen, S., Rattrout, A., Hamarsheh, M.: Enhancing AODV routing protocol to predict optimal path using ant colony algorithm in MANET. Int. J. Appl. Eng. Res. **13**(18), 13637–13646 (2018)
8. Er-Rouidi, M., Moudni, H., Faouzi, H., Mouncif, H., Merbouha, A.: Enhancing energy efficiency of reactive routing protocol in mobile ad-hoc network with prediction on energy consumption (2017)
9. Hamad, S., Belhaj, S., Muslam, M.M.: Average link stability with energy-aware routing protocol for MANETs. Int. J. Adv. Comput. Sci. Appl. **9**(1), 554–562 (2018)
10. Khan, S.M., Khan, M.M., Khan, N.A.: Route constancy and energy aware routing protocol for MANET. Int. J. Comput. Sci. Netw. Secur. **17**(5), 56–64 (2017)
11. Gite, P.: Link stability prediction for mobile ad-hoc network route stability. In: 2017 International Conference on Inventive Systems and Control (ICISC), pp. 1–5. IEEE, January 2017
12. Divya, K., Srinivasan, B.: Energy-aware-AODV: optimized route selection process based on ant colony optimization-bacterial foraging algorithm (ACO-BFA). Eur. J. Mol. Clin. Med. **8**(3), 3938–3950 (2021)
13. Kuppuchamy, N.K.: Qos improvement using hybrid genetic algorithm based AODV in manet. Turk. J. Comput. Math. Educ. (TURCOMAT) **12**(8), 2128–2135 (2021)
14. Kumar, B.S., et al.: Soft computing approach to enhance the performance of AODV (ad-hoc on-demand distance vector) routing protocol using active route TimeOut (ART) parameter in MANETs. Turk. J. Comput. Math. Educ. (TURCOMAT) **12**(2), 3060–3068 (2021)
15. Raju, K.N., et al.: An Adaptive Hello Interval for AODV through ANFIS to Improve the Performance of MANETs, pp. 235–242 (2021)
16. Parveen, K., Gaurav, A., Sugandha, S.: Reliable communication using energy efficient protocol in cluster based manet. JCR **7**(12), 4384–4392 (2020). https://doi.org/10.31838/jcr.07.12.627
17. Sankar Ganesh, S., Ravi, G.: Real time link quality based route selection and transmission in industrial Manet for improved QoS. J. Ambient. Intell. Humaniz. Comput. **12**(7), 6873–6883 (2020). https://doi.org/10.1007/s12652-020-02331-1
18. Kanthe, A.M., Simunic, D., Prasad, R.: Effects of propagation models on AODV in mobile ad-hoc networks. Wireless Pers. Commun. **79**(1), 389–403 (2014)
19. Issariyakul, T., Hossain, E.: Introduction to network simulator 2 (NS2). In: Issariyakul, T., Hossain, E. (eds.) Introduction to Network Simulator NS2, pp. 1–18. Springer, Boston (2009). https://doi.org/10.1007/978-0-387-71760-9_2
20. Bettstetter, C., Hartenstein, H., Pérez-Costa, X.: Stochastic properties of the random waypoint mobility model. Wireless Netw. **10**(5), 555–567 (2004)

Detection and Separation of Transparent Objects from Recyclable Materials with Sensors

Saruhan Furkan Senturk, Huseyin Kaan Gulmez, Mehmet Faruk Gul, and Pinar Kirci(✉)

Computer Engineering Department, Bursa Uludag University, Bursa, Turkey
pinarkirci@uludag.edu.tr

Abstract. Environmental pollution poses a threat to the ecological balance, economy and budgets of states. For this reason, many studies have been carried out recently to reduce environmental pollution, increase recycling, protect the environmental order and prevent economic deterioration. Many municipalities all over the world put recycling bins at certain points in cities, districts and even neighborhoods. However, these boxes are only designed to dispose of materials that can be recycled according to certain chambers with human hands and eyes. People separate their garbage and come and throw their garbage into that bin. Despite this, it can be seen that different wastes are in different bin and a manpower is needed to separate them again. In the project, recyclable materials (metals, transparent materials and others) will be separated into different bins by using sensors and smart systems without using human power.

Keywords: Sensor · Transparent objects · Recyclable material · Seperation

1 Introduction

Elements of various types of sensors and executions are used in optical automatic extraction systems. Considering the sensor elements: Camera, Optical Sensor (Optical Obstacles, Color Sensors), Spectrum Sensors, X-Ray Technologies (Used to detect various types of content depending on their targets).

Actors used in separation systems can include various types of motion: Pneumatic, Electric (Engine, Surveillance Motors, (Separation Pallets, Mobile)), Mechatronics (Robot Arm).

Optical separation systems use different criteria for classification according to the visual properties of objects: shape, size, color, texture, etc.

Sorting by Object Length: in the study, two optical barriers (X1, X2) were placed to detect objects at different heights. Photoelectronic sensor is used here.

Sorting by Object Size: in the study, electronic items are classified according to their size, and this process is applied to a conveyor belt. To realize the application, FX3G type PLC is used.

Sorting by Object Color: In the study, a booth was created to sort items according to experimental level color properties. TCS230 color sensor is used in classification system

I. Woungang et al. (Eds.): ANTIC 2021, CCIS 1534, pp. 73–81, 2022.
https://doi.org/10.1007/978-3-030-96040-7_6

application. In this study, separation processes performed in 3 different methods were performed with 3 different sensors [1].

In this study, capacitive sensors are used to measure the paper height. However, since the sensor is sensitive to humidity, measuring the volume does not give reliable results.

The detection system is based on infrared with four lines of sight. Each emitted beam is detected by photodiodes located on the other side of the container. A photodiode of at least 33 KHz is used for detection. This study tested how sensors work in such a system in case of contamination and whether they can be trusted. It has been found that sensors in this area are not as sensitive as expected and are suitable for most conditions [2].

Waste is processed in recycling facilities and useful materials are separated for recycling. Examples include glass, metal, paper and plastic. These are separated and processed in compressed form. Infrared rays are used to separate paper or cardboard. A difference is created here if wavelengths in the range of 0.78–2.5 μm are reflected. Ballistic machines can detect delicate, two-dimensional objects. Large boxes can be separated by disk analysis. Metals can be separated using induction sensors. Also, ferrous materials are absorbed by a magnetic flux and non-ferrous materials are separated by placing them under a welding flux [3].

This work has a three-layered architecture. The layers use some sensors to measure waste levels, GSM and Wi-Fi technologies to deliver the received data to its destination. In the third layer, it is desired to store these data obtained from the sensors in a cloud-based system. In this application, the ultrasonic sensor srf05 and the distance sensor VL53L0X were also tested to detect different types of waste. And it detected the tested materials (paper, aluminum, glass). Distance sensor and ToF sensor gave precise and fast results [4].

The general purpose of this project is to ensure the separation of waste. The return on this recyclable waste is huge if it can be recycled properly. Existing waste is pushed into the said system through a cover. It detects this situation with the IR (Infrared) distance sensor and the system starts. It is then sequentially dropped into the metal detection system. Then it is dropped into the compartment where the capacitive sensor is located. Here, wet and dry wastes are separated from each other. The separation process is completed from the compartment where the foldable covers are located [5].

Capacitive sensors are used to non-contactly detect the distance, proximity and position of a particular object. The capacitive sensor in this research was used to detect recyclable materials. Glass and plastic were detected with a capacitive sensor. The main purpose of the study is to make a distinction based on material permeability. Different types of paper and plastic were used for the experiment [6, 7].

In [8, 9], the study worked on to isolate hard-to-distinguish materials such as paper and plastic. According to researches, capacitance is an electrical variable and the detection of each substance is tried to be made according to the level of change in the sensor. Capacitance value differs according to distance and object properties. In this study, it is aimed to solve the materials passing over a conveyor platform with a capacitive proximity sensor and an application called LABVIEW.

2 Photoelectric Sensors

Substances that can detect objects through optical properties are called photoelectric sensors. Photoelectric sensors have an emitter to emit light and a receiver to receive light again. When the light emitted from the transmitter is interrupted or reflected again by the object, the change in the amount of light received by the receiver can provide the detection of the object. This beam, which interacts with the object, is detected by the receiver and converted into an electrical output. The classification of photoelectric sensors is shown below.

Features:

Long sensing distance: The reciprocal sensor can detect objects more than 10 m away. It may be impossible to do this with other detection methods.

Almost no detection object restrictions: Sensors with this feature work according to whether the object interrupts the light. Therefore, many objects can be detected.

Fast response time: Since the sensor uses light, the detection time is very fast and the sensor does not perform any mechanical action.

Non-contact detection: Since there is no physical contact with the objects, it is very unlikely that the light will damage the objects.

Working Principles-Features of Light:

Reflection: Surfaces such as glass and mirrors can reflect light as it comes. This is called regular reflection. Matte surfaces such as white paper can scatter light in different directions. This scattering of light is called diffuse reflection.

Polarization of Light: photoelectric sensors usually use LEDs as a light source. LED lights are not polarized. They can spread in vertical and horizontal directions.

-Light Sources:

Pulse Modulated Light: These sensors generally use pulse modulated light. This pattern emits light repeatedly at fixed intervals.

Unmodulated Light: These types are known as continuous light beams. Although it has advantages such as fast detection time, it has disadvantages such as short distance detection.

Classification-Classification by Detection Method

1. Reciprocal Sensors

   There is a transmitter and a receiver. In order for the beam coming out of the transmitter to be detected from the receiver, the receiver and transmitter must be positioned opposite each other. An object passing between these two positions interrupts the light and with this interruption the object can be perceived.

Features

- They have long sensing distances.
- They have a detection position that is not affected by changes.
- It is a study that is not generally affected by the physical properties of the object.

2. Diffuse Sensors
   The transmitter and the receiver are in the same mechanism. As this is the case, the light does not return to the receiver in a normal state. If the beam from the transmitter hits an object, the reflected beam from the object returns to the receiver and detection is made.

Features:

- Long sensing distance
- Easy assembly process
- The intensity of the beam reflected from the object may vary according to the characteristics of the object.

3. Retroreflective Sensors
   The transmitter and receiver are placed in the same mechanism as diffuse reflective sensors. In order for the receiver to receive the beam from the transmitter, a reflector is attached to a certain distance and the reflected beam returns to the receiver. The sensor responds when the object that enters between these two positions interrupts the light sent by the receiver.

Features:

- Detection distance may vary between cm-m.
- Simple to install.
- Detection process is not affected much by the physical properties of the object.
- Transparent objects can be detected with this sensor.
- Objects that can reflect rays from their surfaces may not be detected with this sensor.
- This type of sensors has dead zones at close ranges [10].

3 Structure of the Project

In the project, the detection of transparent objects with the photoelectric sensor, the detection of substances entering and leaving the platform with the ultrasonic sensor, and the presence of substances with the capacitive sensor can be detected.

The reason why sensors are preferred over image processing in the system is to produce more realistic results and to operate them at less cost. At the same time, very large data sets are needed in image processing. Even as a result of training, sensors are used instead of image processing, as the validation of the model may still return low levels or overfitting may be encountered in case of overtraining.

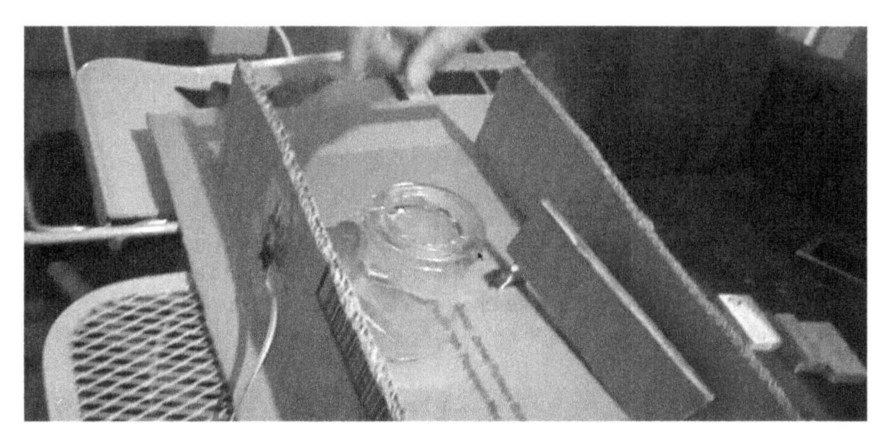

Fig. 1. Glass substance entry to the platform

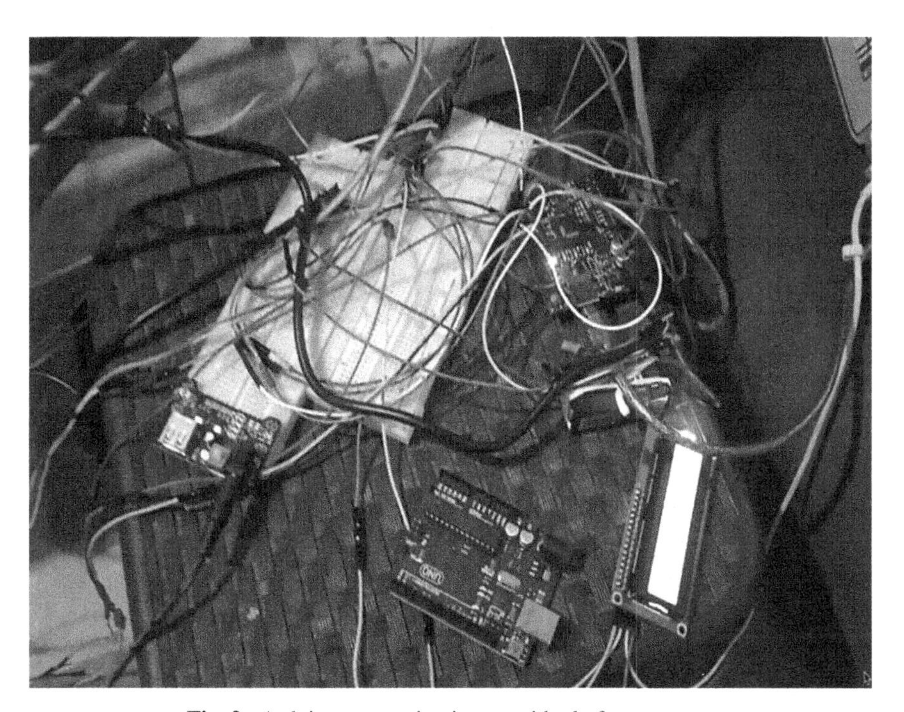

Fig. 2. Arduino connection image with platform sensors

As shown in Figs. 1 and 2, ultrasonic sensor, capacitive sensor and optical sensor are placed on an inclined (30°) platform, respectively. There are doors fixed with servo motors under each sensor. After the sensors detect, these doors open and the substance reaches the next sensor. At the last door below the optical sensor, the waiting time is slightly increased. According to the results, the separation box under the platform is rotated according to the information received and accumulates the substance in a separate section according to the type of substance.

4 Results

4.1 Non-transparent Substance Test

Photoelectric sensors can detect objects with optical properties. It has a transmitter to diffuse the light and a receiver to detect when the light returns. There are more than one type of photoelectric sensors. The photoelectric sensor to be used in this project is a reflector sensor.

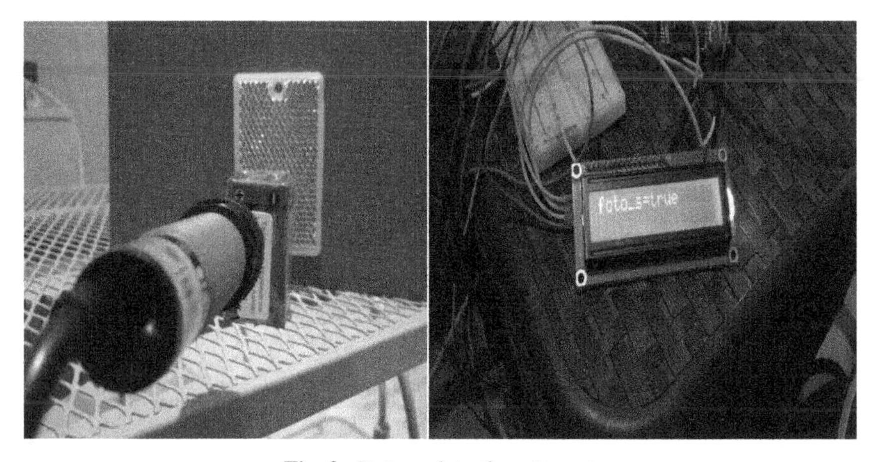

Fig. 3. Battery detection attempt

Fig. 4. Paper detection trial

A reflector is placed some distance from the sensor, and the light passing through the material is reflected from the reflector and returns to the sensor's receiver, so whether the object is transparent or not can be detected as shown in Figs. 3, 4, 5 and 6. If the object is not transparent, it absorbs the light and does not return it, the sensor receiver cannot receive the light again and outputs as shown in Figs. 3 and 4.

4.2 Transparent Substance Test

Fig. 5. Clear container detection experiment

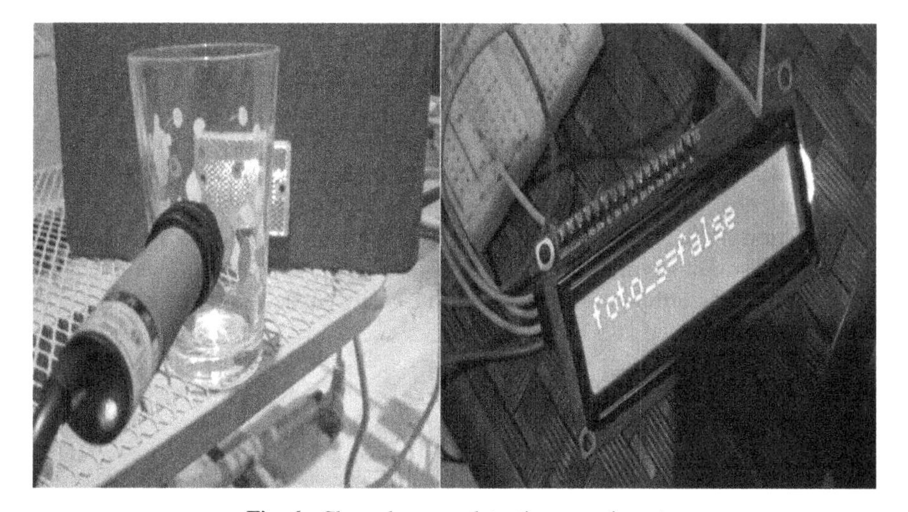

Fig. 6. Clear glass cup detection experiment

If the object is transparent, it does not absorb the light and returns it back, the sensor receiver receives the light again and outputs as shown in Figs. 5 and 6 accordingly.

Table 1 shows the substance detections made with the photoelectric sensor. Experiments were made with batteries, paper and transparent containers. Accuracy detection

Table 1. Transparency detection accuracy table with photoelectric sensor

Item Type	Accuracy detection	DetectionTime (seconds)
Battery (Fig. 3)	No Transparent Substance Detected	0–1
Paper (Fig. 4)	No Transparent Substance Detected	0–1
Transparent Container (Fig. 5)	Transparent Substance Detected	0–1
Transparent Glass Cup (Fig. 6)	Transparent Substance Detected	0–1
	%100	**0–1**

and detection times are given in seconds in cases where transparent material is detected in the trials.

5 Conclusion

As a result of the research, sensors were determined for the separation of substances. It was determined that it is appropriate to use optical sensors for transparent objects. It has been concluded that the fastest and most precise solution for separation can be done with sensors. Optical sensors have a mechanism that works by returning the beam emitted from the transmitter to the receiver, so that transparent objects can be easily detected. The use of ultrasonic sensors was deemed appropriate in order to check whether there was any substance entering the mechanism. Capacitive sensors also detect whether there is an approaching object. Ultrasonic sensors have been determined as sensors that can detect all kinds of substances. In this project, ultrasonic sensors detect if there is any substance entering the platform and trigger other components.

In the future study, it has been determined that Arduino is necessary to use different types of sensors in this type of mechanism. The sensors can be easily integrated with the arduino. With an arduino set, sensors can be easily connected to the computer and data flow can be provided.

References

1. Romul, C., Holonec, R., Drăgan, F.: The PLC Implementation of an automated sorting system using optical sensors. Acta Electrotehnica, **58**(4) (2018)
2. Leitão, P.O., Marques, M.B.: Reliability of a line-of-view sensor for recycling point waste containers. In: World congress, Lisbon (2009)
3. Bonello, D., Saliba, M.A., Camilleri, K.P.: An exploratory study on the automated sorting of commingled recyclable domestic waste. Procedia Manuf. **11**, 686–694 (2017)
4. Likotiko, E., Petrov, D., Mwangoka, J., Hilleringmann, U.: Real time solid waste monitoring using cloud and sensors technologies. Online J. Sci. Technol. **8**(1) (2018)

5. Chandramohan, A., Mendonca, J., Shankar, N.R., Baheti, N.U., Krishnan, N.K.: Automated waste segregator. In: 2014 Texas Instruments India Educators' Conference (TIIEC), Bangalore, India, 4–5 April 2014 (2014)
6. Ahmad, I.K., Harun, S.N.H., Azmi, M.R.: An application of capacitance proximity sensor for identification of recyclable materials. J. Kejuruteraan SI **1**(5), 37–41 (2018)
7. Ye, Y., Zhang, C., He, C., Wang, X., Huang, J., Deng, J.: A review on applications of capacitive displacement sensing for capacitive proximity sensor. IEEE Access **8**, 45325–45342 (2020)
8. Ahmad, I.K., Mukhlisin, M., Basri, H.: An application of capacitance proximity sensor for identification of paper and plastic from recyclabling materials. Res. J. Appl. Sci. Eng. Technol. **12**(12) (2016)
9. Agovino, M., Ferrara, M., Faezah, A.N., Marchesano, K., Garofalo, A.: The separate collection of recyclable waste materials as a fywheel for the circular economy: the role of institutional quality and socio-economic factors. Econ. Polit. **37**, 659–681 (2019)
10. Technical Explanation for Photoelectric Sensors. https://www.ia.omron.com/data_pdf/guide/43/photoelectric_tg_e_8_3.pdf. Accessed 27 Feb 2021

Integration of Fog Computing to Internet of Things for Smart Sensitive Devices

Zaheer Khan Hussainkhel[1] and Md Motaharul Islam[2](✉)

[1] Khana-e-Noor University, G5GW+HCW, Kabul, Afghanistan
[2] United International University, Dhaka 1212, Bangladesh
motaharul@cse.uiu.ac.bd
http://cse.uiu.ac.bd/profiles/motaharul/

Abstract. Fog computing is gaining tremendous importance as a near to network edge storage and connectivity service to user. Fog computing has a big scope with Internet of Things. Integration of IoT and Fog might be a reliable pair for improved performance in term of latency. Mostly IoT devices are used to provide real-time and latency-sensitive service requirements. Due to the distance between client devices and data processing center the integration of IoT and cloud affect the performance of IoT devices. This paper introduces Fog-IoT model for integrating IoT devices with Fog computing. This integration model improves the performance of the IoT devices in term of latency. The focus of this paper is to introduce model for IoT integration with Fog, Harmonization of IoT in Fog to enhance the performance of IoT latency sensitive devices and future research opportunities for latency-sensitive environment in fog computing.

Keywords: Fog computing · Cloud computing · Internet of Things · Fog-IoT integration · Fog-IoT harmonization

1 Introduction

A revolutionary smart environment for the future connectivity and reachability between the Internet of Things (IoT) devices has come to an action due to the rapid usage of IoT devices or objects for example (Radio Frequency Identification (RFID) tags, Sensors, Laptops, Mobile Phones and actuators). In order to establish a smooth connectivity and actuating smart environment IoT devices are harmonized [1,2] with cloud storage. Internet Protocol (IP) addresses are assigned to IoT devices in order to harmonize IoT devices with cloud [3,4]. In this harmonization process cloud server is responsible to store the sensed data. Almost all IoT objects have adopted cloud computing [5]. Sad to say that, no technology is immaculate and cloud technology has its own shortcomings especially when used with IoT. Due to the distance between data processing centers and client devices, cloud cannot guarantee low latency for IoT. The physical location of cloud server is not known to the end devices/users, thus the integration of IoT with cloud affects the performance of IoT objects [6]. Cloud is a

I. Woungang et al. (Eds.): ANTIC 2021, CCIS 1534, pp. 82–93, 2022.
https://doi.org/10.1007/978-3-030-96040-7_7

centralized system which leads to single-point-failure. Furthermore, cloud cannot establish an earthly-level connectivity promised by IoT. In earthly-level connectivity smart internet devices are connected to every workplace, home and market. Moreover, the industrial control systems which require low delay response time are not satisfied with cloud technology [7]. The integration of IoT and fog is a perfect solution. Moreover, fog computing deals with IoT based services and applications [15]. An infrastructure paradigm named as fog computing is announced in 2012 by Cisco [8]. It is a computing concept in order to overcome the limitations of cloud computing. Fog computing is the extended version of cloud computing with a new feature, servers being near to edge networks. Computing, networking and storage facilities of the cloud computing is extended by Fog paradigm, while the cloud data centers are offloaded and service latency to the end users are reduced [14]. Fog is a distributed decentralized infrastructure which overcomes the single-point-failure, by putting the servers distributed over the network unlike cloud where servers are located in a centralized approach. Fog server runs at edge network. It reduces the delay for real time and latency-sensitive services in cloud-fog migrating model. Fog computing storage is near-to-network edge, thus fetching and storing the sensed data is faster comparing to far-to-network edge storages like cloud. In this paper, we have designed Fog-IoT model to enhance the security of forests in term of fire and natural disaster. In our proposed model fire/smoke and heat detection sensors are used to fetch the input generated by input sensors (fire/smoke and heat detectors). The input sensors are programmed in Arduino to monitor the forest area and send the sensed input value to the fog layer. A raspberry Pi is installed in fog layer to store the input value passed by input sensors each two seconds and pass the average value each 5 min to cloud layer. A threshold value for the highest heat bearable by forest is assigned. A status sensor is used to continuously monitor the value generated by input sensors. Smart fire extinguisher and water pump are installed in the smart forest environment and are triggered by the status sensor. Second Arduino is used to handle the status sensor in triggering smart output devices. The input sensors keep monitoring the smart environment even though the output devices are triggered. If the input devices detect that, the smart environment heat is less or equal to half of the defined threshold value, then sensor status stops the output devices by sending the signal.

The core contribution for this paper are summarized as follow:

- We have proposed Fog-IoT model for Integration of Fog computing with latency-sensitive, real time services and IoT devices.
- In our proposed model, the performance of IoT devices are increased in term of latency
- We have analyzed and compared the integration of IoT-Cloud computing and IoT-Fog computing using CoAP and HTTP as communication protocols.
- We have proposed research opportunities for all latency-sensitive environments and harmonizing IoT in fog.

The rest of the paper structure: Sect. 2 discuss related works. Proposed model and Algorithmic analysis for fog-IoT environment are described in Sect. 3 and 4 respectively. Analytical model is in Sect. 5, Future research trend is discussed in Sect. 6, finally Sect. 7 covers future work and concludes the paper.

2 Related Works

Researchers have proposed an integrated fog cloud IoT (IFCIoT) architectural paradigm [9] which promise scalability, better localized accuracy, quicker response time, increased performance and energy efficiency for future IoT applications. Edge servers, Smart routers, and base stations are fog nodes which receive the offloading computational requests and sensed data from different IoT devices. In order to promise enhanced energy efficiency, performance and real time responsiveness of applications a reconfigurable and layered fog node architecture is proposed. The proposed architecture reconfigure the architecture resource and analyzes application characteristics to better meet the workload demands.

For computing, storage, control and networking fog is an emergent architecture where these services are distribute closer to end users along the cloud-to-things continuum[10]. Furthermore, it covers both mobile and wireline scenarios. It traverses across the software adn hardware and resides in the network edge. Moreover, it over access networks among end users and includes both control and data plane. A growing variety of applications, including those in the fifth generation (5G) wireless systems, embedded artificial intelligence (AI) and Internet of Things (IoT) is supported by proposed architecture. Researchers have summarized the challenges and opportunities of fog and main focus is the networking context of IoT.

In order to share resources of the physical network, Network Function Virtualization (NFV) provides a good paradigm. The focus of this paper [11] is the deployment of service function chains (SFCs) which is composed of a specific order of virtual network functions (VNFs). Furthermore, to solve the challenges faced by centralized cloud computing a distributed fog computing mechanism is proposed. As the mobile users move among different fog-based radio access networks SFC must be migrated respectively. Therefore, the problem caused by the user movement in cloud-fog computing environment for SFCs migration and remapping is addressed. Firstly the migration problem of SFCs is modeled as an integer linear program. Afterward, two SFC migration strategies are proposed in order to reduce the reconfiguration cost, downtime and migration time of SFCs.

The requirements of the IoT applications which are not met by today's solutions are beard by fog computing. Different initiative has been presented in order to drive the development of fog and much work has been done for the improvement purpose of the certain aspects. Even though a detail and in-depth analysis of the different solutions and pointing out how they can be integrated and applied in order to meet specific requirements, is still needed. Researchers have presented a new taxonomy and unified architectural model [12], comparing vast number of solutions. Eventually, they have drawn some guideline and conclusions for the development purpose of fog-based IoT applications.

A rapid growth in the number of connected devices with the Internet of Things (IoT) is predicted, which becomes part of our environment and daily life. It is expected to connect billions of devices and humen through IoT to initiate promising advantages. Fog along with edge computing paradigms (e.g. multi-access edge computing (MEC), cloudlet) are meant to be the promising solution in order to handle the large number of time sensitive and security critical data produced by IoT devices. Firstly, a tutorial on fog and its related computing paradigm considering their differences and similarities is provided [13]. Afterwards, a classification of research topics in fog computing is proposed. Finally, the researchers have summarized and categorized the endeavor on fog and its related computing paradigms using a comprehensive survey.

A new era is brought in wireless networking platform through wireless sensor network. Due to its intense usage the researchers has been led to design new network architecture paradigm. Furthermore, wireless network and traditional ad hoc do not properly suit in this scenario. The architecture of wireless sensor network are different from application to application. The researchers have proposed a health care application which cannot achieve the eventual goal of the sensor network through energy awareness [17]. The readability of links plays an important role. In this context many protocols like CTP, RPL are proposed by the researchers. The proposed routing protocol by the researchers is meant to enhance the throughput of data packet transmission in healthcare application.

Wireless sensor network (WSN) plays a vital role while stepping forward towards the age of Internet of Things (IoT) in the overall networking architecture. Applications like smart-city and healthcare extensively use WSN framework to collect and analyse real-time data considering actions required actions based on the application requirements. Considering various scenarios like generalizing the networking framework while ignoring applications-specific requirements would lead towards designing an efficient routing protocol. Considering the healthcare applications the researchers [18] have proposed an application specific routing protocol for better increase in the network life time and maintaining the maximum throughput in data transmission.

3 Proposed Model

A typical architecture for integrating IoT devices with fog computing is depicted in Fig. 1. The proposed model is categorized into three different layers known as IoT layer, Monitoring layer (Fog), and Cloud layer. IoT layer is located at the edge. It is responsible to detect the smart environment through IoT input devices and trigger the IoT output devices to actuate in the smart environment. In our proposed model we have divided the IoT layer into two parts.

1. Input Part: It detects the smart environment through installed sensors (Heat and Fire/Smoke detectors) as shown in Fig. 1. An Arduino board is

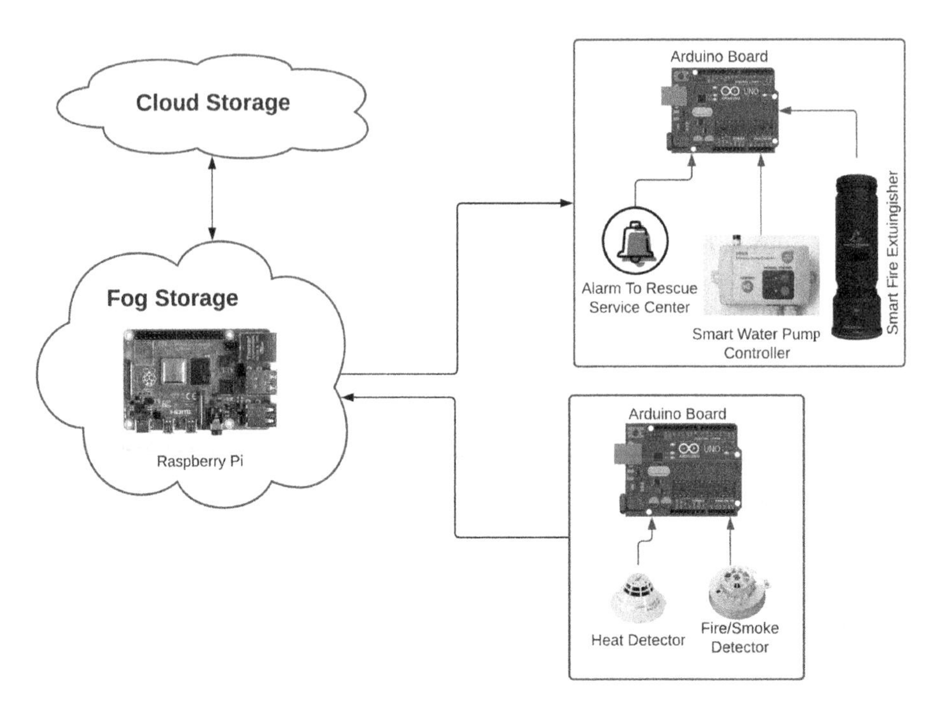

Fig. 1. Proposed model for fog-IoT integration.

programmed to control the fetched input and send the data to the next layer known as Fog.

2. Output Part: This part continuously monitors the input generated by the input sensors using status sensor to check the defined threshold value for the heat. An alarm buzzer is installed to alert the Rescue Service Center (RSC) team regarding the risk of burning the forest due to high heat detected from the smart environment. Smart water pump is also installed and it is triggered automatically to pour the water to reduce the heat and fire from the environment. A smart fire extinguisher is also triggered if the heat threshold value is met.

The input sensors are continuously checking the forest smart environment. If a normal heat threshold value is detected from the environment it is passed to the fog layer. In the output part the status sensors keep checking the heat threshold value and if the value is found to be normal then immediately engaged IoT devices (e.g. Smart Water Pump Controller, Smart Fire Extinguisher) as shown in Fig. 1 are kept idle. An Arduino is programmed to control the overall scenario of output part. Second layer is the monitoring node also known as fog. A raspberry Pi is installed to process the data passed by input devices and store the average heat threshold value each five minutes to the cloud permanent storage which is the third layer of the proposed model. The installed raspberry Pi is also responsible to communicate with the Arduino installed in output part of the proposed model. Through this communication the output devices are triggered based on the defined heat threshold value. Cloud storage is located in the third layer of the proposed model which stores the detected data permanently. This layer is rarely used and thus the latency generated between the IoT devices and Cloud is eliminated by introducing an intermediate monitoring layer. We have analyzed the security issues of cloud and fog respectively. Cloud security level is not defined because the physical location of the data server is not defined. End users do not have a clue regarding the transmission of data. IoT input devices continuously send the detected data from the smart environment and the output devices monitors the status throughout the internet. There is no doubt that the data is vulnerable to cyberattack and a risk of data loss exist as well. Fog improves security by keeping data close to the edge. The geographical distribution of cloud use centralized approach and fog is based on distributed approach.

4 Algorithmic Analysis

Two algorithms are proposed for the implementation purpose, first for controlling the input part and second for the output part. Algorithm 1 describes the working mechanism of the input part. Here four local variables are used namely, heatSensor, fire/smokeSensor, heatThresholdValue, fireStatus to detect the smart forest environment. Heat Sensor detects the heat; fire/smoke sensor detects the existence of fire from the smart forest environment. heatThresholdValue defines the max heat bearable by the forest. fireStatus defines the existence of fire which is

Algorithm 1. Input Part

1: **procedure** InputDevices
2:     $heatThresoldValue \leftarrow the\ standard\ heat\ value\ bearable\ by\ forest$ ▷ if the heat generated by the input sensors exceed the predefined heat threshold automatically the output devices are triggered by setting **fireStatus** variable value to **TRUE**.
3:     $heatSensor \leftarrow detects\ the\ heat\ from\ forest\ smart\ environment$
4:     $fireSmokeSensor \leftarrow detects\ the\ existence\ of\ fire$
5:     $fireStatus \leftarrow False$
6:     **if** heatSensor **is greater than** heatThresholdValue **OR** fireSmokeSensor == **True then**
7:         $fireStatus \leftarrow True$
        break
8:     **end if**
9:     **if** heatSensor **is less than** heatThresholdValue/2 **AND** fireSmokeSensor == **False then**
10:         $fireStatus \leftarrow False$
        break
11:     **end if**
12: **end procedure**

by default false. The fireStatus value is changed to true if the generated heat by the heat sensors are greater comparing to defined threshold value. Algorithm 1 is triggered each 2 s and it is responsible to set fireStatus value to false once the environment is under control as shown in Algorithm 1.

Algorithm 2. Output Part

```
1: procedure OUTPUTDEVICES
2:     statusValue ← continuously checks the fireStatus value in fog layer   ▷
   if this value is true it means existence of fire is detected from the smart forest
   environment.
3:     if statusValue == True then
4:         Smart water pump controller is triggered
5:         Smart Fire Extinguisher is triggered
6:         Notify RSC through an Alarm Buzzer
       break
7:     end if
8:     if statusValue == False then
9:         Stop smart water pump controller
10:        Trigger smart fire extinguisher to its idle state
       break
11:    end if
12: end procedure
```

The output part is controlled by Algorithm 2. Here we have a variable statusValue which continuously checks the fireStatus in the fog layer. If the existence of fire is detected, Notification alarm is given to Rescue Service Center (RSC), smart water pump and smart fire extinguisher are triggered. Algorithm is executed each 2 s and checks the fireStatus. If it is detected that, the fire is controlled; smart water pump and fire extinguisher are returned to their idle states.

5 Analytical Model

Analytical model of the proposed system is depicted in Fig. 2. Here we are integrating IoT sensing devices with fog. In this integration model, the IoT input sensors fetch the real data from the forest smart environment and store it in the fog platform. The stored data is continuously monitored by the IoT actuating devices/sensors each 2 s in order to trigger the output devices (Fire Extinguisher, Water Pump Controller, Alarm). Two connection protocols (e.g. Hyper Text Transfer Protocol, Constrained Application Protocol) are used for connectivity purpose. Nowadays IoT devices can easily communicate with IP networks. Since a computer can communicate with computer in a Network Address Translation (NAT) environment, IoT devices can communicate as well [3]. There will be no congestion in the proposed network model, because IoT devices require minimal

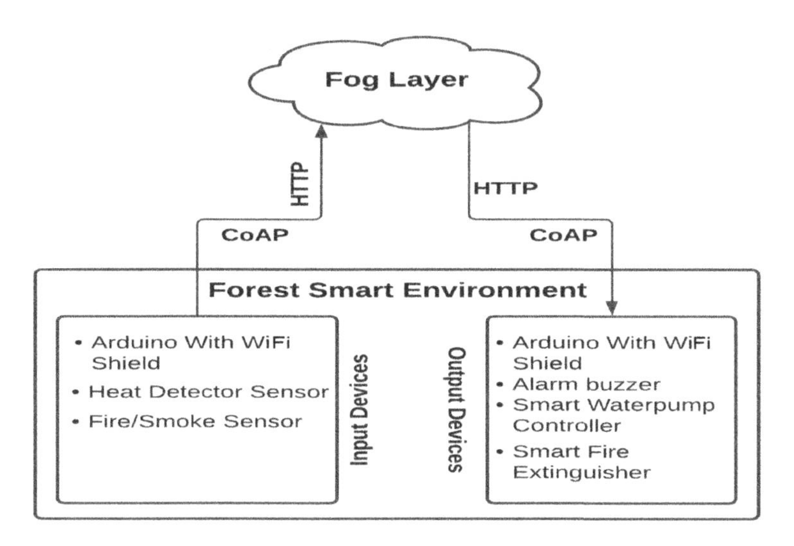

Fig. 2. Proposed system analytical model.

bandwidth to operate. Advanced congestion control mechanisms will be used to control congestion if any occurs. Performance evaluation can be performed based on latency, throughput, reliability, etc. measurement techniques. In this section we have measured the performance of the proposed system based on generated latency. Latency is the total time consumed by a message to completely reach at the destination from the first bit sent out from the source. If latency is considered to be Lat then,

$$Lat = Prop_t + Trans_t + Que_t + Proc_d \tag{1}$$

Equation 1 can be further expanded to,

$$Lat = dist/Sp + Msg/Bnd_{w_{Fog}} + Que_t + Proc_d \tag{2}$$

Considering Eq. 2, the total time required for a bit to travel from the source to destination is the propagation time ($Prop_t$), distance is measured by dist, Sp shows the speed. Transmission time ($Trans_t$) is measured on the basis of ratio of message size (Msg) and bandwidth (Bnd_w) of the channel. Since queueing delay and processing delay are very negligent we may ignore them, finally the Eq. 2 becomes,

$$Lat = dist/Sp + Msg/Bnd_w \tag{3}$$

Distance plays an important role in calculating the latency of a network specially in case of IoT devices. Latency generated by Fog-IoT integration model is lower comparing to latency produced by Cloud-IoT integration model. We have done an analysis on latency calculation of Fog-IoT integration model and Cloud-IoT integration model. In this analysis we have used Hypertext Transfer Protocol (HTTP) and Constrained Application Protocol (CoAP) for the communication purpose of Fog-IoT integration model. For the latency generation of the Cloud-IoT we used Google Cloud as our analyzing model. We analyzed the latency of Fog-IoT and Cloud-IoT model in the following ways.

Latency for google cloud is calculated as shown in Eq. 4.

$$Lat_{googlecloud} = dist_{googleCloud}/Sp + Msg/Bnd_{w_{googleCloud}} \tag{4}$$

Latency for Fog computing can be calculated as given in Eq. 5.

$$Lat_{fog} = dist_{fog}/Sp + Msg/Bnd_{w_{fog}} \tag{5}$$

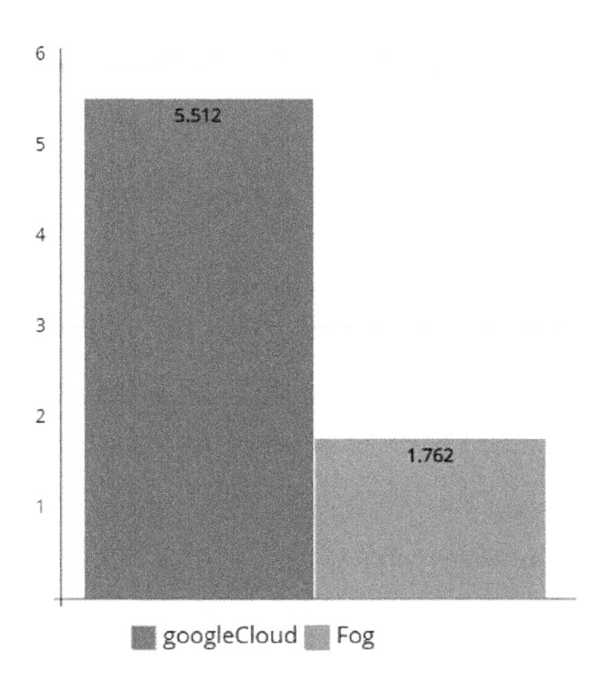

Fig. 3. Latency consumed by fog-IoT and googleCloud-IoT using HTTP.

A calculation is performed for latency consumption of Fog-IoT and Cloud-IoT integration based on HTTP and CoAP. The analysis shows that, integrating IoT with Fog decreases the latency of the network compared to integrating Cloud with IoT devices.

The total latency generated by fog based platform and cloud based platform using HTTP is depicted in Fig. 3. The analysis shows that integrating IoT with fog generates low latency. Total latency generated by fog-IoT integration model using HTTP is 1.762 ms. Furthermore, 5.512 ms is the total latency generated by google cloud using HTTP. We have analyzed and calculated total latency consumed by Fog-IoT and Cloud-IoT integration model as depicted in Fig. 4 using CoAP. The calculations show that, GoogleCloud consumes 5.1152 ms latency and total latency generated by fog is 1.3652 ms.

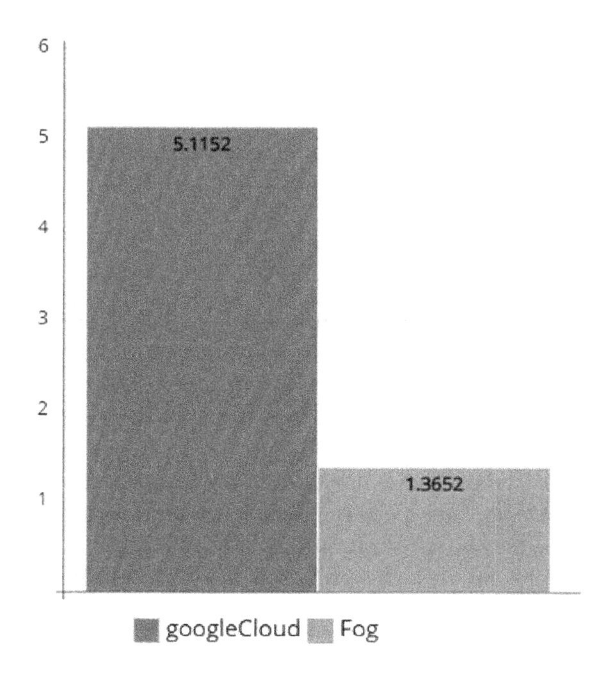

Fig. 4. Latency consumed by fog-IoT and googleCloud-IoT using CoAP.

Wrapping up the analysis we can conclude that IoT devices work faster with Fog. Moreover, Fog-IoT integration model generates less latency compared to cloud-IoT platform.

6 Future Research Trends

An improvement in quality of daily life through billions of IoT devices, numerous IoT applications (e.g. Virtual Reality, Haptic Technology, Augmented Reality and autonomous driving) and connected IoT sensors is expected through Internet of Things. Emerging technologies such as Internet of Things (IoT) require latency-aware computation for real-time application processing. For processing real-time applications latency-aware computation is required by emerging technologies such as IoT. In order to analyze collected data and meet latency requirements an improvement in IoT applications in term of energy efficiency and scalability is enabled with the use of fog/edge computing. Moreover, to fully utilize fog computing for IoT a series of challenging problems need to be addressed [16].

In fog computing most of the computational nodes are battery-operated, which makes fog computing unreliable. Moreover, further study is required to efficiently provide computing resource to latency sensitive IoT applications. Fog computing is prevailing component of cloud computing and it does not replace cloud computing. In fog computing the processing of information is performed at the edge [19], though still providing the option to communicate with the data center of cloud. Moreover, As a promising solution for the IoT and next generation networks fog computing has emerged. Fog computing enables service provisioning along te continuum from the cloud to things to reduce bandwidth demands and latency. Moreover, it empower end users in their neighboring as an extension to cloud computing [20].

7 Conclusions and Future Works

In this paper, we propose a model for integrating IoT devices with Fog computing in a smart forest environment. IoT devices are responsible to fetch the input from the smart environment through sensing or input devices and further store it in the Fog layer. The actuating IoT devices (output devices) act based on the input fetched by input devices. We use Fog as main storage for IoT sensed/input devices which makes the proposed model fog based implementation. In analysis phase we have shown that the performance of the proposed model is increased by producing much lower latency compared to the existing system. In our future work, we are going to enhance the proposed model considering different measurement techniques and communication protocols.

References

1. Islam, M.M., Khan, Z., Alsaawy, Y.: An implementation of harmonizing internet of things (IoT) in cloud. In: Pathan, A.-S.K., Fadlullah, Z.M., Guerroumi, M. (eds.) SGIoT 2018. LNICSSITE, vol. 256, pp. 3–12. Springer, Cham (2019). https://doi.org/10.1007/978-3-030-05928-6_1
2. Islam, M., Khan, Z., Alsaawy, Y.: A framework for harmonizing internet of things (IoT) in cloud: analyses and implementation. Wireless Netw. **27**(6), 4331–4342 (2021)
3. Islam, M., Huh, E.N.: Sensor proxy mobile IPv6 (SPMIPv6)–a novel scheme for mobility supported IP-WSNs. Sensors **11**(2), 1865–1887 (2011)
4. Islam, M.M., Hassan, M.M., Huh, E.N.: Sensor proxy mobile IPv6-a framework of mobility supported IP-WSN. In: 2010 13th International Conference on Computer and Information Technology (ICCIT), pp. 295–299 (2010)
5. Bilal, K., Malik, S.U.R., Khan, S.U., Zomaya, A.Y.: Trends and challenges in cloud datacenters. IEEE Cloud Comput. **1**(1), 10–20 (2014)
6. Mutlag, A.A., Abd Ghani, M.K., Arunkumar, N.A., Mohammed, M.A., Mohd, O.: Enabling technologies for fog computing in healthcare IoT systems. Future Gener. Comput. Syst. **90**, 62–78 (2019)
7. Mubeen, S., Nikolaidis, P., Didic, A., Pei-Breivold, H., Sandstrom, K., Behnam, M.: Delay mitigation in offloaded cloud controllers in industrial IoT. IEEE Access **5**, 4418–4430 (2017)

8. Elmisery, A.M., Rho, S., Aborizka, M.: A new computing environment for collective privacy protection from constrained healthcare devices to IoT cloud services. Cluster Comput. **22**(1), 1611–1638 (2019)
9. Munir, A., Kansakar, P., Khan, S.U.: IFCIoT: integrated fog cloud IoT: a novel architectural paradigm for the future internet of things. IEEE Consum. Electron. Mag. **6**(3), 74–82 (2017)
10. Chiang, M., Zhang, T.: Fog and IoT: an overview of research opportunities. IEEE Internet Things J. **3**(6), 854–864 (2016)
11. Zhao, D., Sun, G., Liao, D., Xu, S., Chang, V.: Mobile-aware service function chain migration in cloud-fog computing. Future Gener. Comput. Syst. **96**, 591–604 (2019)
12. Bellavista, P., Berrocal, J., Corradi, A., Das, S.K., Foschini, L., Zanni, A.: A survey on fog computing for the internet of things. Pervasive Mob. Comput. **52**, 71–99 (2019)
13. Yousefpour, A., et al.: All one needs to know about fog computing and related edge computing paradigms: a complete survey. J. Syst. Archit. **98**, 289–330 (2019)
14. Mukherjee, M., et al.: Security and privacy in fog computing: challenges. IEEE Access **5**, 19293–19304 (2017)
15. Sen, A.A.A., Yamin, M.: Advantages of using fog in IoT applications. Int. J. Inf. Technol. **13**(3), 829–837 (2021)
16. Naha, R.K., et al.: Fog computing: survey of trends, architectures, requirements, and research directions. IEEE Access **6**, 47980–48009 (2018)
17. Rahman, M.R., Islam, M.M., Shahaz, E.A., Alsaawy, Y.: Application specific energy aware and reliable routing protocol for wireless sensor network. In: 2019 7th International Conference on Smart Computing and Communications (ICSCC), pp. 1–5 (2019)
18. Rahman, M.R., Islam, M.M., Pritom, A.I., Alsaawy, Y.: ASRPH: application specific routing protocol for health care. Comput. Netw. **197**, 108273 (2021)
19. Sabireen, H., Neelanarayanan, V.: A review on fog computing: architecture, fog with IoT, algorithms and research challenges. ICT Express **7**(2), 162–176 (2021)
20. Chen, N., Yang, Y., Zhang, T., Zhou, M.T., Luo, X., Zao, J.K.: Fog as a service technology. IEEE Commun. Mag. **56**(11), 95–101 (2018)

Exploiting Vulnerabilities in the SCADA Modbus Protocol: An ICT-Reliant Perspective

Ayush Sinha(✉), Saurabh Singh Patel, Abhishek Kumar, and O. P. Vyas

Department of IT, Indian Institute of Information Technology, Allahbad, Prayagraj, India
{pro.ayush,mit2019027,mit2019040,opvyas}@iiita.ac.in

Abstract. Industrial plants like power, gas, water, and transport are controlled remotely and managed by a protocol like Modbus-TCP which has a significant contribution towards Industrial Control systems (ICS) and Supervisory control and data acquisition (SCADA) systems. In the previous years, occurrences of cyber-attacks influenced the SCADA structures and their associated protocols though few in numbers but lethal. The attack may affect the confidentiality and integrity of the Modbus/TCP module and unauthorized control of coils and registers has the potential for appalling conditions in some unacceptable situations. The proposed work investigates the security of an industrial framework utilizing the Modbus transmission convention in an ICS to build up a particular security test framework for the discovery attack, Man in the middle (MIMT) attack, Denial of Service (DoS), and Metasploit attack. This work focuses to execute the attack results and show the interaction in the virtual climate of Conpot and Rapid SCADA and presents an analysis using CVSS 3.1 score to compare the Metasploit, DOS, MITM in terms of vulnerabilities and threat levels. Finally, the severity of ease of happening for different attacks is mentioned as a conclusion of this study.

Keywords: SCADA · Modbus-TCP · Attack detection · CVSS score · SCADA vulnerability

1 Introduction

SCADA system software uses real time data for making decision in various part of industrial process like monitoring and controlling communication in transportation, industrial plant like water and oil system, manufacturing, energy management and other different real time automation systems. The present work focuses on the widely used industrial protocol MODBUS [24] which is primarily used for communication in Scada systems.

Modbus is the industrial world's de facto standard communication protocol [25]. It was created while considering serial communications protocols like

I. Woungang et al. (Eds.): ANTIC 2021, CCIS 1534, pp. 94–108, 2022.
https://doi.org/10.1007/978-3-030-96040-7_8

Modbus RTU/ASCII. However, due to the growing popularity of the TCP/IP stack, it was recently changed to TCP. To be more precise, Modbus itself does not have any security features because it was built for controlled serial lines in the first place. Thus, the objective of this work is to investigate vulnerabilities associated with SCADA Modbus protocol, as an outcome of this work, various benchmarks as in Table 3 is being mentioned to evaluate the performance of networking devices that use Modbus TCP. MODBUS is the most extensively used communication protocol in modern ICS systems for bi-directional sensor data transfer between data acquisition servers and Intelligent Electronic Devices (IED) such as Programmable Logic Controllers (PLC) or Remote Telemetry Units (RTU). The safety and security of ICS systems is a crucial concern in these plants' safe and secure operations. Because security precautions were not considered when the Modbus protocol was designed, it is more vulnerable to cyber security attacks. One of the most common attacks for MODBUS is a denial-of-service (DoS) or flooding attack, which impacts the control system's availability. In this study, a new approach for detecting user application-level flooding or DoS attacks is described, which activates an alarm annunciator and displays appropriate alarms in a Supervisory Control and Data Acquisition system (SCADA) to alert administrators or engineers to take corrective action. When compared to other methods, our method recognised the largest percentage of attacks in the shortest amount of time. This solution took into account all of the conditions that cause flooding attacks in the MODBUS protocol.

To study the overall perspective of the SCADA system and it's related protocol, work in [17] emphasis that the SCADA systems are the monitoring and control components that underpin key infrastructures such as power, telecommunications, transportation, pipelines, chemicals, and manufacturing plants. Because legacy SCADA systems ran on separate networks, they were less vulnerable to Internet attacks. However, as SCADA systems become more interconnected with the Internet and corporate networks, serious security concerns arise. As the number of security incidents against critical infrastructures rises, as a consequence to this, security considerations for SCADA systems are receiving more attention. This work gives an overview of the overall SCADA architecture, as well as a full description of the SCADA communication protocols. The work also talks about some high-profile security incidents, goals, and dangers. Furthermore, we provide a thorough examination of the security approaches and techniques aimed at securing SCADA systems. While observing the flooding attack scenario, paper [18] presents that Industrial Control Systems (ICS) are commonly used to monitor and control a variety of industrial plants around the world, including oil and gas refineries, nuclear reactors, power production and transmission, and other chemical plants. To continue the experiments with the machine learning (ML) models, dataset availability is of very importance. So, to overcome the dataset availability issues, authors in [16] states that aggressors attempt to overwhelm the endpoint by either sending transmissions that are too fast for the endpoint to handle or by sending bundles designed to cause programming errors that result in exceptions with an accident the organisation stack, the running project, or the working arrangement of the targeted device.

While considering another type of attack called Man-in-The-Middle(MiMT), the paper [19] says that MiTM attacks pose several risks to a smart grid. In a MiTM attack, an intruder inserts itself into a dialogue between two devices to either eavesdrop or impersonate one of the devices, causing the discussion to appear to be normal. As a result, the intruder can carry out fake data injection (FDI) and false command injection (FCI) attacks on power system activities such as state estimation, economic dispatch, and automatic generation control (AGC). Few studies have concentrated on MiTM approaches that are difficult to detect in a smart grid. To demonstrate how such attacks might induce physical contingencies such as mistaken operation and incorrect measurements, authors have implemented multi-stage MiTM incursions in an emulation-based cyber-physical power system testbed against a large-scale synthetic grid model. The work described detection algorithms that were built using numerous alarms from intrusion detection systems and network monitoring tools that enables stakeholders to protect against the stealthy attacks. To consider the signature based intrucsion detection for Modbus basec ICS, the work in [20] shows attacks like response injection, command injection, Reconnaissance, and denial of service, are possible on ICS communication networks. These type of attacks make difficulties in monitoring and can lead to system failure and that can lead to financial loss and can create concern for economy and safety. The MODBUS application layer network protocol is used by industrial control systems, and the work discusses a series of 28 cyber attacks against it. On the top of attack simulation and detection, authors in [21] presents the performance evaluation in case of Distributed Denial of Service(DDoS) Attack. While these all are examples of outside attack on smart grid (SG), some authors [28] also focuses on an insider attack arises due to presense of malicious devises and presented a solution based on blockchain technology. As an overall approach to handle cyber security, authors in [26] attempted to give a complete framework based on NIST guidelines.

For the rest of the paper, the organization of the manuscript is as: Sect. 2 refers the Modbus TCP introduction and it's related vulnerabilities along with possible Cyber attack scenarios. Section 3 states the major contribution of the previous work in the same related domain. Section 4 is about the problem statement that tell the objective of this work. Section 5 presents the lab setup needed to carry out the experiment related to present work. Finally Sect. 6 mentions the experiment carried out with detailed discussion about the outcome achieved and Sect. 7 is about the concluding remarks along with future directions of the present work.

2 Modbus-TCP and Related Vulnerabilities

MODBUS-TCP is an application-layer protocol, positioned at level seven of the Open Systems Interconnection model(OSI). It gives client correspondence between devices associated with different sorts of transports or organizations.

Now days, assistance for the straightforward rich construction of MODBUS-TCP, Fig. 1, keeps on developing. The more detailed discussion for the Modbus-TCP Header and it's Data Model is presented in the subsequent subsections along with the explanation of Vulnerabilities and Cyber Attack for Modbus-TCP protocol.

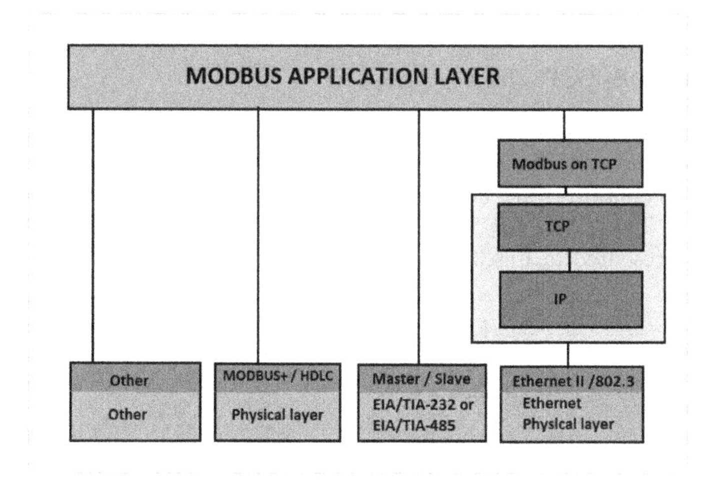

Fig. 1. MODBUS-TCP communication stack

2.1 MODBUS-TCP Header

The MODBUS-TCP header consists of four parts that is Transaction Identifier, Protocol Identifier, Length and Unit Identifier. The Fig. 2 explains the structure of the header format.

Transaction Identifier: 2 bytes are set by the Master to uniquely identify any request. These bytes are repeated by the Slave device in the response, as the responses of the Slave device may not always be received in the same order as the requests.

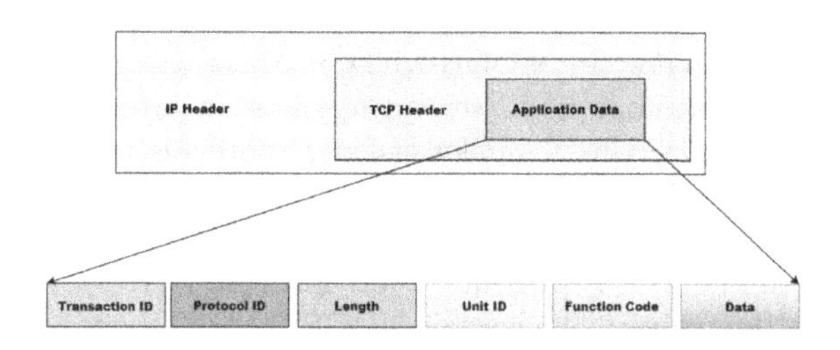

Fig. 2. MODBUS-TCP header

Protocol Identifier: 2 bytes are set by the Master, will always be 00 00, which corresponds to the Modbus protocol.

Length: 2 bytes are set by the Master, identifying the number of bytes in the message that follow. It is counted from Unit Identifier to the end of the message.

Unit Identifier: 1 byte is set for Master. It is repeated by the Slave device to uniquely identify the Slave device.

2.2 MODBUS-TCP Data Model

The MODBUS-TCP describe a protocol data unit called as PDU which is not dependent on the underlying structure of TCP-IP layers where as PDU application data unit contains some additional fields when mapping with different buses and network. Thus, MODBUS-TCP defines info-model with relative to known qualities as presented in Fig. 3.

Primary tables	Object type	Type of	Comments
Discrete Input	Single bit	Read-Only	This type of data can be provided by an I/O system.
Coils	Single bit	Read-Write	This type of data can be alterable by an application program.
Input Registers	16-bit word	Read-Only	This type of data can be provided by an I/O system
Holding Registers	16-bit word	Read-Write	This type of data can be alterable by an application program.

Fig. 3. MODBUS-TCP data model table

2.3 Vulnerabilities of MODBUS-TCP

The MODBUS-TCP protocol has some weaknesses that may give an ability for a malicious user to perform different security attacks. The major exposed weaknesses as of now are as:

1. **Absence of Secrecy:** The data that are sent across the transmission media may be possibly seen by other entity in the network.
2. **Absence of Integrity:** In MODBUS-TCP, there are less trustworthiness registers worked and complied with an application layer. As an output, it thus relies upon underlying TCP lower-layer protocols for saving the trustworthiness.
3. **Absence of Validation:** Any device can freely access at any level because no validation mechanism is present on any layer of TCP.
4. **Shortsighted Framing:** The core of MODBUS-TCP protocol is connection oriented mechanism. While such connections are generally good, it may also have critical downside.

5. **Absence of Session Structure:** Each transaction counts as one unit in MODBUS-TCP not like other protocols as (SMTP,HTTP. etc.). When joined with the absence of validation and worst TCP (ISN) in many inserted devices, it gets conceivable for assailants to infuse orders with no information on the current meeting.

2.4 Attacks Against Modern Control Frameworks

In past few years, several instance of cyber attacks affected the SCADA frameworks and their protocols. These attack led to harm the financial situation of people and organization. The below subsection gives explanation of three cyber attack named as Metasploit, Man-In-The-middle and Denial of Service (DoS).

1. **Metasploit Attack:** The Metasploit framework are consist of many several modules which can be used to exploit the Industrial Control System. There are 3 main modules which made for monitoring and penetrating the MODBUS-TCP. Using the Kali Linux system for deploying this Metasploit tool to perform attack using the three modules of Metasploit tools that are: **modbus_findunitid, modbusclient, modbusdetect**.
2. **Man in The Middle Attack (MITM):** There is no security on transport layer for SCADA protocols, so it has to be protected from intruders. An intruder can easily intercept the master, slave in the middle of communication with method like ARP Poisoning. The detail steps that is adopted for the present work for the intrusion are as:
   - Using ARP poising, poison the ARP table of Master and Slave. This will force them to send all the traffic using Attackers system.
   - Now using the Ettercap filter attacker can modifies the communication going towards the client computer.
   - While leaving the communication, Attacker will mostly clear the ARP table. This will make reverse engineering hard to trace back.
3. **Denial of Service Attack (DoS):** In ICS, DoS attack is used to disturb the proper functioning of some modules and can cause the whole system to the complete shutdown state. A DoS attack may target either the digital or physical framework. In digital framework, DoS attack targets the programs running on framework endpoint by crashing them, that controls the log data, data exchange. In Physical framework, DoS attack focuses on changing the arguments of actual cycle that prevents the activity. In Modbus/TCP setup, master and client can be identified by using IP address and an Attacker can use different IP for organization devices for launching attack against victim by using a malware. Using Flooding (a type of DoS attack) a large amounts of traffic are sent from an organization's endpoint. Aggressors try to exhaust the end point by sending transmissions that are too fast to handle for the endpoint or by sending bundles designed to cause programming errors that result in exceptions with an accident for the organisation stack, the running project, or the working arrangement of the targeted device.

3 Related Work

Industrial plants like water, gas and power are control remotely and monitored by SCADA Modbus-TCP which have a important work in Industrial Control System [1]. Due to the unavailability of security in communication protocol of Scada System leads to various security threats. This paper show the various types of threats and defects in Scada system by taking real time incident which arises in databases and also focused on the improvement of the Scada system. SCADA systems is using Modbus-TCP since last few years [2]. Auhtors have shown that in the past, Scada system have been vulnerable to various cyber threats. Scada mostly used Modbus TCP for its communication protocol which is not secure. Authors in [3] developed a Secure Modbus-TCP based on the original Modbus-TCP to provide various security features like transmission of data with authentication, protect from exploiting various function code because it uses cryptic key like symmetric algorithms to give secure confidentiality to save data. In the work [4], authors highlighted that the SCADA system relies on susceptible communication protocols and lacks a cyber-security strategy, putting the CI's general operation at risk. Here the important subject of matter is Modbus-TCP that is widely used in electrical grid. Authors in the survey article [27], presented the overall possible scenario for the energy infrastructure related attack protection, detection and applicable countermeasures. However in the work like [5] a successful operation of (SCADA) deployed around the different part of world which is crucial to many of today's most critical infrastructures. It was further explained that as technology has progressed, networks have been able to expand in terms of both space and complexity, allowing system operators to operate these systems more efficiently. Authors in [6] states that at the inception of SCADA system in 1979, there was not much need of security in Modbus. The ascent of organization protocols, for example, Modbus is frequently joined by the extraordinary risks related to its application in genuine frameworks, particularly the ones which are basic in their environment. In the work [7], authors have highlighted the importance of IoT devices that is an integral part of Sg infrastructure. They also demonstrated that urgency of regular updation of IoT devices. The framework in the work can be utilized to look through the honeypot framework that may exist in the provincial organization, and plan a keen honeypot framework to befuddle the attack of the programmer and ensure the inward arrangement of the venture.

Authors in paper [8] showed that how the SCADA system is affected by open-source gadgets in the different manners. In continuation to this, the work in [9] composed a few benchmarks to assess the exhibition of systems administration devices that run Modbus TCP. Boundaries announced by this benchmarks incorporate (1) response time for Modbus requests, (2) most extreme number of requests effectively took care of by Modbus devices in a particular measure of time, and (3) observing of Modbus devices when enduring a Distributed Denial of Service attack. To demonstrate the modern computerization control frameworks, it is shown in [10] that the Modbus TCP/IP protocol is widely used. This study discusses and investigates the use of interruption detection and prevention systems (IDPS) with advanced bundle assessment capabilities and advanced parcel inspection current firewalls that can detect and halt extremely precise attacks hidden somewhere in the communication stream.

The cyber attack arises when alternate Distributed Energy Resources (DERs) enter dispersion networks [11]. While in the work [12], authors says that because it monitors and regulates the computerization cycles of modern hardware, SCADA frameworks play an important role in Critical Infrastructures (CIs). In many event, SCADA is based on insecure communication protocols with no network security component, thus making it possible to jeopardise the CI's whole operation. In this paper, the authors focused on the Modbus TCP protocol, which is widely used in CIs and, in particular, in the electrical systems. The work analyse and improve the cyber attacks that the SMOD pen-testing gadget generates. In paper [13], authors presents an understanding for the attack on SCADA frameworks explicitly zeroing in on frameworks that utilization the Modbus TCP protocol. An entrance testing approach is embraced utilizing a novel infiltration testing device to (1) test the viability and effectiveness of the apparatus, (2) inspect the insider danger just as the outside danger through the interior and outer entrance testing individually, and (3) rate the weaknesses distinguished through the entrance tests as indicated by the Common Vulnerability Scoring System. The investigation likewise analyzes and tests the current security countermeasures that are novel to SCADA frameworks and blueprints a few papers that may improve security in SCADA frameworks. Their test results showed that a portion of the attack may seriously affect respectability and accessibility. The work in [14] shows that SCADA is the obvious depiction of Cyber Physical Systems (CPS). As SCADA is updated with cutting-edge registration and communication innovations, the risks associated with receiving/updating to new technology should be thoroughly reviewed and approved. All CPS frameworks that are currently in use and attached to actual cycles cannot be scheduled for penetration testing and inspection. Using Colored Petri Nets (CPN) apparatus, this work describes the design and implementation of a current consistent SCADA testbed, as well as the protocol evaluation of semantics and security of the Modbus TCP protocol. In the work titled as Cyber Threat Investigation of SCADA Modbus Activities, [15] the utilization availability of SCADA and ICS networks in shrewd advances have presented them to an enormous assortment of safety dangers. Moreover, not many examinations are done in this field from the Internet (digital) point of view. Along these lines, the paper researches unapproved, malevolent, and dubious SCADA exercises by utilizing the dark net address space. Specifically, this work explores Modbus administration, which is a true standard protocol for correspondence and it is the most accessible and used interface electronic devices in basic and modern foundations.

To this end, we can say that a lot of researchers have explored the SCADA system vulnerabilities and impact of different cyber attack on it. Some more common vulnerabilites highlighted in the work [22] that states that Process Control Systems (PCS) or Industrial Control Systems provide the plant automation systems. They eliminate the need for human involvement and boost the plant's efficiency by decreasing errors. Previously, these systems were housed in a control room with no access to the internet or the outside world. However, as technology advances, these systems are now connected to the internet for remote monitoring and data sharing with other systems such as ERP. As a result, these systems are vulnerable to cyber-attacks too. To be more specific, some researchers like [13] handled the real time attack scenario and developed a simulators which are used to show the way to consider security in client/server model, and parameters which are required to reduce security risks while enhancing efficiency, decreasing transmission effects, ensuring protocol independence, and attaining significant performance using cryptography techniques. As the present work focuses on IEC 60870 protocol, that is also explored in the work [14]. This study looks into cyber attacks on industrial control systems (ICS) that use IEC -60870-5-104 for communications. The study focuses on MITM attacks, including command alteration, as well as capture and replay attacks. On a local software simulated laboratory, an initial series of attacks is carried out. In collaboration with a power distribution operator, final trials and verification of a MITM attack are carried out in a thorough testbed environment.

4 Problem Statement

SCADA/MODBUS is a legacy communication protocol and lags in protection mechanism, authentication mechanism and exploit of function codes. These days, security is viewed as a major test for the MODBUS protocol as a component of SCADA system. Genuine consideration is expected to determine these potential security issues of SCADA/MODBUS system, and an efficient system is needed for security execution to essentially improve the overall performance. The threat impact of cyber attacks are different in different situation and thus a proper threat level monitoring is required under different cyber attack scenario.

5 Experimental Setup

In our setup we are using Modbus master and Modbus slave setup simulated using open source tool QmodMaster [22] and ModbusPal [23]. QmodMaster can perform various function code given in Table 1.

Table 1. QmodMaster function codes

READ COILS	0x01
Operation	Function Code
READ HOLDING REGISTERS	0x03
READ DISCRETE INPUTS	0x02
WRITE SINGLE COIL	0x05
WRITE INPUT REGISTERS	0x04
WRITE MULTIPLE COILS	0x0f
WRITE SINGLE REGISTER	0x06
WRITE MULTIPLE REGISTERS	0x10

ModbusPal receives Function code from master and within a defined time interval, it sends status of sensors to the master. Attacker workstation is using kali linux as its primary operating system and attacking communication in different scenarios. For attack simulation, we are using open source tools like Metasploit, SMOD and Ettercap as well as for analysis we are using Wireshark. Metasploit contains pre-build vulnerabilities available in Modbus/TCP communication and these vulnerabilities can cause severe security threat to CIA triad that is confidentiality, availability and integrity.

Once successfully manipulated the coil state at address, it is needed to cross-verify this with the RTU client.

Using Ettercap, we have performed a MITM attack on Master and client PLC systems. The Ettercap tool focused on two flaws of Modbus TCP: (1) sending queries with explicit content and (2) the lack of authentication inside the Modbus TCP protocol. The Ettercap tool was originally used on the Kali Linux system by putting the communication interface into unified sniffing mode.

DoS attack chocked the communication and increased request response time between PLC devices and Master server. This is properly visible in the Wireshark snapshot as depicted in Figs. 4 and 5.

```
SMOD >use modbus/dos/writeSingleCoils
SMOD modbus(writeSingleCoils) >show options
 Name     Current Setting  Required  Description
 ----     ---------------  --------  -----------
 Output   False            False     The stdout save in output directory
 RHOST                     True      The target IP address
 RPORT    502              False     The port number for modbus protocol
 Threads  24               False     The number of concurrent threads
 UID                       True      Modbus Slave UID.
SMOD modbus(writeSingleCoils) >set RHOST 192.168.43.84
SMOD modbus(writeSingleCoils) >set uid 3
SMOD >exploit
SMOD >use modbus/dos/writeSingleCoils
SMOD modbus(writeSingleCoils) >show options
 Name     Current Setting  Required  Description
 ----     ---------------  --------  -----------
 Output   False            False     The stdout save in output directory
 RHOST    192.168.43.84    True      The target IP address
 RPORT    502              False     The port number for modbus protocol
 Threads  24               False     The number of concurrent threads
 UID                       True      Modbus Slave UID.
SMOD modbus(writeSingleCoils) >set UID 3
SMOD modbus(writeSingleCoils) >set RHOST 192.168.43.84
SMOD modbus(writeSingleCoils) >exploit
[+] Module DOS Write Single Coil Start
```

Fig. 4. DOS attack using SMOD

No.	Time	Source	Destination	Protocol	Length	Info
333	31.737059	192.168.43.84	192.168.43.222	Modbus...	63	Response: Trans: 124; U
334	31.749299	192.168.43.222	192.168.43.84	Modbus...	66	Query: Trans: 125; U
335	31.751645	192.168.43.84	192.168.43.222	Modbus...	63	Response: Trans: 125; U
336	31.764476	192.168.43.222	192.168.43.84	Modbus...	66	Query: Trans: 126; U
337	31.765539	192.168.43.84	192.168.43.222	Modbus...	63	Response: Trans: 126; U
338	31.777666	192.168.43.222	192.168.43.84	Modbus...	66	Query: Trans: 127; U
339	31.778612	192.168.43.84	192.168.43.222	Modbus...	63	Response: Trans: 127; U
340	31.790812	192.168.43.222	192.168.43.84	Modbus...	66	Query: Trans: 128; U
341	31.791752	192.168.43.84	192.168.43.222	Modbus...	63	Response: Trans: 128; U
342	31.804599	192.168.43.222	192.168.43.84	Modbus...	66	Query: Trans: 129; U
343	31.806005	192.168.43.84	192.168.43.222	Modbus...	63	Response: Trans: 129; U
344	31.816243	192.168.43.222	192.168.43.84	Modbus...	66	Query: Trans: 130; U
345	31.817924	192.168.43.84	192.168.43.222	Modbus...	63	Response: Trans: 130; U
346	31.830187	192.168.43.222	192.168.43.84	Modbus...	66	Query: Trans: 131; U
361	36.838049	192.168.43.222	192.168.43.84	Modbus...	66	Query: Trans: 132; U
372	41.843368	192.168.43.222	192.168.43.84	Modbus...	66	Query: Trans: 133; U
589	131.333743	192.168.43.222	192.168.43.84	Modbus...	66	Query: Trans: 2; U
590	131.335913	192.168.43.84	192.168.43.222	Modbus...	64	Response: Trans: 2; U
1156	301.598639	192.168.43.101	192.168.43.84	Modbus...	66	Query: Trans: 144; U
1157	301.600279	192.168.43.84	192.168.43.101	Modbus...	66	Response: Trans: 144; U

Fig. 5. Communication under DOS attack

6 Results and Discussions

In the present simulation, we are successfully update the coil value with metasploit module. This attack affects the confidentiality and integrity of the Modbus/TCP module and this control of coils and registers has the potential for appalling conditions in some unacceptable situation. From CVSS, the threat level is high and metasploit based attack also affects the authenticity as they can bypass the UID scan based attacks.

Attackers system is using ettercap tool for performing MITM attack on Modbus/TCP communication using ARP poisoning. The coils write request was sent from Ettercap using the filter to change the coil order from on to off between PLC and Master. In regular communication, Wiresharks operations were normal between Master and PLC and requests were not dropping, but with the MITM attack, Wireshark uncovered anomalies in Modbus TCP like re transmission error, unknown addresses and duplicate acks. In DOS attack, It is properly visible in wireshark simulation that a network is sending a continuous write request to a PLC device on an illegal data address and this can be seen in wireshark frame capture Figs. 5 and 6. Due to continuous requests, PLC devices are spending most time in exception handling, and it is delaying the Master's request for Data read. This will eventually affect availability. After comparing the threat level of all these attacks performed, we reached to the result as given in Table 2.

```
> Frame 35814: 63 bytes on wire (504 bits), 63 bytes captured (504 bits) on interface \Device\NPF_{161E075F-3AD1-4457-B6D7-76D6E70F0859}, id 0
> Ethernet II, Src: HonHaiPr_24:aa:6b (5c:ea:1d:24:aa:6b), Dst: IntelCor_90:b7:bc (e0:94:67:90:b7:bc)
v Internet Protocol Version 4, Src: 192.168.43.84, Dst: 192.168.43.222
    0100 .... = Version: 4
    .... 0101 = Header Length: 20 bytes (5)
  > Differentiated Services Field: 0x00 (DSCP: CS0, ECN: Not-ECT)
    Total Length: 49
    Identification: 0x73db (29659)
  > Flags: 0x40, Don't fragment
    Fragment Offset: 0
    Time to Live: 128
    Protocol: TCP (6)
    Header Checksum: 0xae68 [validation disabled]
    [Header checksum status: Unverified]
    Source Address: 192.168.43.84
    Destination Address: 192.168.43.222
> Transmission Control Protocol, Src Port: 502, Dst Port: 54906, Seq: 1, Ack: 13, Len: 9
v Modbus/TCP
    Transaction Identifier: 1789
    Protocol Identifier: 0
    Length: 3
    Unit Identifier: 3
v Function 5: Write Single Coil. Exception: Illegal data address
    .000 0101 = Function Code: Write Single Coil (5)
    Exception Code: Illegal data address (2)

0000  e0 94 67 90 b7 bc 5c ea  1d 24 aa 6b 08 00 45 00
0010  00 31 73 db 40 00 80 06  ae 68 c0 a8 2b 54 c0 a8
0020  2b de 01 f6 d6 7a af 83  12 ad 06 02 0d 70 50 18
0030  01 00 1b a8 00 00 06 fd  00 00 00 03 03 85 02
```

Fig. 6. Illegal data write under DOS attack

From CVSS 3.1 score, Table 3, we are getting some clarity about MITM attack that this is complex than DOS attack and require some hardware privileges. In our case, Network sniffing is easy but changing the coil value is hard and for that we need to have the proper idea of memory location of the coil addresses. DOS attack has highest threat level due to easy of implementation and less hardware knowledge and DOS is mostly targeting system availability.

Table 2. Vulnerability analysis for various attacks

Attack type	Confidentiality (Read Coil/Register)	Integrity (Write Coil/register)	Authenticity (Bypass UID)	Availability
Metasploit	Yes	Yes	Yes	No
MIMT	Yes	Yes	Yes	No
DOS	No	No	Yes	Yes

Table 3. CVSS 3.1 scoring for DOS, MITM and Metasploit

	Metric value for DoS attack	Metric value for MIMT attack	Metric value for MetaSploit
CVSS Base Score	**8.2(High)**	**6.8(Medium)**	**7.4(High)**
Attack Complexity	Low	High	High
Attack Vector	Network	Network	Network
User Interaction	None	None	None
Privileges Required	None	Low	None
Confidentiality	Low	High	High
Scope	Unchanged	Unchanged	Unchaged
Availability	High	None	None
Integrity	None	High	High
CVSS Temporal Score	**7.8(High)**	**6.4(Medium)**	**7.2(High)**
Remediation Level	Work Around	Work Around	Work Around
Exploit Code Maturity	Functional	Functional	Functional
Report Confidence	Confirmed	Confirmed	Confirmed
CVSS Enviornmental Score	**7.6(High)**	**6.4(Medium)**	**7.2(High)**
Confidentiality Requirement	Low	Low	Low
Modified Attack vector	Local	Adjacent Network	Adjacent Network
Modified User Interaction	None	None	None
Availability Requirement	High	Low	Low
Modified Privileges Required	None	Low	Low
Integrity Requirement	High	High	High
Modified Scope	Unchanged	Unchanged	Unchanged
Modified Attack Complexity	Low	High	Low
Modified Integrity	None	High	High
Modified Confidentiality	None	Low	Low
Modified Availability	High	Low	None

7 Conclusion

The present work analyse the vulnerabilities present in SCADA modbus TCP protocol while performing the different attack categories like MiTM, DoS, Meta-exploit etc. The experimental setup used in the work would eventually helps the organizations to test their safeguards and improve the defence of an expected attack on one's ICS organizations. The work presents an analysis using CVSS 3.1 score to compare the Metasploit, DOS, MITM in terms of vulnerabilities and threat levels. After comparing these attacks based on CVSS score, it is concluded that DOS attack is more easy to performed on Modbus TCP and it is more harmful than Metasploit attack as well as MIMT that has shown low threat level than Metasploit and DOS. For future scope IDS may be used for lowering the CVSS 3.1 score. We are planning zeek as network layer IDS and Samhain host-based intrusion detection system (HIDS) as host IDS simultaneously. We will use wireshrak network to sniff the data changes in pcap file generated by wireshark and then compare their value using Scapy modules to generate a graph value before and after the attack performed. We will also analyse the scenario where network is sending a continuous write request to a PLC device on an illegal data address. Due to this, continue request PLC devices is spending of its most time in exception handling, and it is delaying the Master's request for Data read. This use case for performing attack can also be made using multiple devices, and in that case, the communication will be blocked between PLC and Master devices. These types of attacks can be detected using standard IDS because they make an attack pattern.

Acknowledgement. The work is funded by Department of Science and Technology(DST), India for the Cyber Physical Security in Energy Infrastructure for Smart Cities (CPSEC) project under Smart Environments theme of Indo-Norwegian Call.

References

1. Upadhyay, D., Sampalli, S.: SCADA (supervisory control and data acquisition) systems: vulnerability assessment and security recommendations. Comput. Secur. **89**, 101666 (2020)
2. Stranahan, J., Soni, T., Heydari, V.: Supervisory control and data acquisition testbed vulnerabilities and attacks. In: SoutheastCon 2019, pp. 1–5 (2019). https://doi.org/10.1109/SoutheastCon42311.2019.9020436
3. Xuan, L., Yongzhong, L.: Research and implementation of Modbus TCP security enhancement protocol. J. Phys: Conf. Ser. **1213**, 052058 (2019). https://doi.org/10.1088/1742-6596/1213/5/052058
4. Radoglou-Grammatikis, P., Siniosoglou, I., Liatifis, T., Kourouniadis, A., Rompolos, K., Sarigiannidis, P.: Implementation and detection of modbus cyberattacks. In: 2020 9th International Conference on Modern Circuits and Systems Technologies (MOCAST), pp. 1–4 (2020). https://doi.org/10.1109/MOCAST49295.2020.9200287
5. Wilson, P.L.: ModSec: A secure Modbus protocol. SMARTech Home (2018). https://smartech.gatech.edu/handle/1853/62615
6. Parian, C., Guldimann, T., Bhatia, S.: Fooling the master: exploiting weaknesses in the Modbus protocol. Procedia Comput. Sci. **171**, 2453–2458 (2020)
7. Chou, C.-H., et al.: Modbus packet analysis and attack mode for SCADA system. J. ICT Des. Eng. Technol. Sci. **2**, 30–35 (2018). https://doi.org/10.33150/JITDETS-2.2.1
8. Parcharidis, M.: Simulation of cyber attacks against SCADA systems - Thesis presentation (2018)
9. Gamess, E., Smith, B., Iii, G.: Performance evaluation of Modbus TCP in normal operation and under a distributed denial of service attack. Int. J. Comput. Netw. Commun. **12**, 1–21 (2020). https://doi.org/10.5121/ijcnc.2020.12201
10. Nyasore, O.N., Zavarsky, P., Swar, B., Naiyeju, R., Dabra, S.: Deep packet inspection in industrial automation control system to mitigate attacks exploiting modbus/TCP vulnerabilities, pp. 241–245 (2020). https://doi.org/10.1109/BigDataSecurity-HPSC-IDS49724.2020.00051

11. Siddavatam, I.A., Parekh, S., Shah, T., Kazi, F.: Testing and validation of Modbus/TCP protocol for secure SCADA communication in CPS using formal methods. Scalable Comput. Pract. Exp. **18**(4), 313–330 (2017). https://doi.org/10.12694/scpe.v18i4.1331
12. Fachkha, C.: Cyber threat investigation of SCADA Modbus activities. In: 2019 10th IFIP International Conference on New Technologies, Mobility and Security (NTMS), pp. 1–7 (2019). https://doi.org/10.1109/NTMS.2019.8763817
13. Peng, T., Leckie, C., Ramamohanarao, K.: Protection from distributed denial of service attacks using history-based IP filtering. In: IEEE International Conference on Communications, ICC 2003, vol. 1. IEEE (2003)
14. Zaballos, A., Vallejo, A., Selga, J.M.: Heterogeneous communication architecture for the smart grid. IEEE Network **25**(5), 30–37 (2011)
15. Gawande, A.R.: DDoS detection and mitigation using machine learning. Dissertations, Rutgers University-Camden Graduate School (2018)
16. Ullah, I., Mahmoud, Q.H.: An intrusion detection framework for the smart grid. In: 2017 IEEE 30th Canadian Conference on Electrical and Computer Engineering (CCECE). IEEE (2017)
17. Wei, L., et al.: Review of cyber-physical attacks and counter defense mechanisms for advanced metering infrastructure in smart grid. In: 2018 IEEE/PES Transmission and Distribution Conference and Exposition (TD). IEEE (2018)
18. Zafar, R., et al.: Applications of ZigBee in smart grid environment: a review. In: Proceedings of the 2nd International Conference on Engineering and Emerging Technologies (ICEET). Superior University, Lahore (2015)
19. Chen, T.M., Sanchez-Aarnoutse, J.C., Buford, J.: Petri net modeling of cyber-physical attacks on smart grid. IEEE Trans. Smart Grid **2**(4), 741–749 (2011)
20. Kundur, D., et al.: Towards modelling the impact of cyber attacks on a smart grid. Int. J. Secur. Netw. **6**(1), 2–13 (2011)
21. Emmanuel, M., Seah, W.K., Rayudu, R.: Communication architecture for smart grid applications. In: 2018 IEEE Symposium on Computers and Communications (ISCC). IEEE (2018)
22. GitHub. https://github.com/zhanglongqi/qModMaster. GitHub - zhanglongqi/qModMaster: The maintainer's repo. https://github.com/ed-chemnitz/qmodbus/. Accessed 27 Aug 2021
23. ModbusPal. ModbusPal - Java MODBUS simulator. http://modbuspal.sourceforge.net/. Accessed 27 Aug 2021
24. Huitsing, P., Chandia, R., Papa, M., Shenoi, S.: Attack taxonomies for the Modbus protocols. Int. J. Crit. Infrastruct. Prot. **1**, 37–44 (2008)
25. Gamess, E., Smith, B., Francia, G.: Performance evaluation of Modbus TCP in normal operation and under a distributed denial of service attack. Int. J. Comput. Netw. Commun. (IJCNC) **12**(2), 1–21 (2020)
26. Sinha, A., et al.: Cyber physical defense framework for distributed smart grid applications. Front. Energy Res. **8**, 407 (2021)
27. Sinha, A., et al.: Critical infrastructure security: cyber-physical attack prevention, detection, and countermeasures. In: Quantum Cryptography and the Future of Cyber Security, pp. 134–162. IGI Global (2020)
28. Singh, J., et al.: Insider attack mitigation in a smart metering infrastructure using reputation score and blockchain technology. Int. J. Inf. Secur. 1–20 (2021)

Privacy Preserving of Two Collaborating Parties Using Fuzzy C-Means Clustering

Latha Gadepaka[1(✉)] and Bapi Raju Surampudi[1,2(✉)]

[1] School of Computer and Information Sciences, University of Hyderabad, Hyderabad, India

[2] Cognitive Science Lab, International Institute of Information Technology, Hyderabad, India

Abstract. In this smart and rapidly growing computing world, most of the organizations and companies needs to share necessary information (data) to third parties for their analysis which plays major role in decision making. Any data generally consists of sensitive (personal) information about people and organizations and when the sensitive information to be revealed to third parties, there is always chance of privacy loss. In this regard any organization or an individual requires the application of privacy preserving techniques to ensure data privacy, that means the data to be exchanged between two or more sites, without data loss or privacy violation. When data to be shared between multiple organizations or sites (parties) the privacy issues will be arises more. In this line we choose collaborative fuzzy clustering that gives better solutions to preserve privacy of data while sharing information between multiple parties to achieve combined results. Collaborative clustering is a concept or mechanism to find a common data points along with relationships within data residing at various individual data sites. We propose two ways of collaborative clustering using Fuzzy C-Means Clustering approach, one is Horizontal-PPFCM and other is Vertical-PPFCM. In both proposed methods the information is securely shared between two parties to compute the combined results and privacy also preserved. In fuzzy collaborative clustering two parties generate clusters with mutual collaboration of information they own at each individual sites and a collaborative objective function does all the necessary computations with granular information collected from each party. The overall process flow requires only the secured information and a common objective function for final outcomes. The proposed methods are working well for both horizontal and vertical collaborations and showing better results with assured data privacy.

Keywords: Privacy · Preserving · Clustering · Partitioning · Fuzzy C-Means · Collaboration

1 Introduction

In this rapidly growing distributed computing world usually to share sensitive data to others requires strong techniques to ensure privacy of data. The information about people and organizations needs to be shared when a general outcomes

I. Woungang et al. (Eds.): ANTIC 2021, CCIS 1534, pp. 109–122, 2022.
https://doi.org/10.1007/978-3-030-96040-7_9

are to be collected, hence the privacy concerns of their data must be addressed. There are many methods and models are being developed in order to preserve the privacy of distributed data. Collaborative scheme of exchanging information between multiple parties has been giving secured outcomes from distributed databases. The main objective of fuzzy clustering to be applied at input level is to share or communicate inputs to other parties in a secured way and also make sure that there will not be any privacy loss at input level. The following section gives a complete idea of fuzzy c-means clustering applied through a collaborative clustering.

2 Fuzzy C-Means Clustering

Fuzzy c-means (FCM) is a method of clustering that allows one single attribute value of data to belong to two or more clusters. The Fuzzy clustering method developed by Dunn in 1973 and improved by Bezdek in 1981 [1]. FCM clustering builds fuzzy partitions with c clusters where all the elements become members of more than one clusters and aimed for solution of minimizing the membership value of elements using the objective function given below.

$$J_{FCM}(U,V) = \{\sum_{i=1}^{c}\sum_{k=1}^{n}(u_{ik})^m ||x_k - v_i||^2\} \tag{1}$$

In the above equation, partition matrix is $u_{ik} \in [0,1]$ and v_i denotes cluster center of the $i-th$ cluster and m is a fuzzy coefficient. The general FCM method has two main steps, first step estimates the optimal membership functions of elements to clusters and second step estimates the centroids for each cluster. Fuzzy C-Means Clustering is presented in the following algorithm:

Algorithm 1. FUZZY C-MEANS CLUSTERING ALGORITHM

Input: Data set X, No of clusters C and fuzzy coefficient m.
Step 1: Initialize the membership matrix $U = [u_{ij}]$, $U(0)$.
Step 2: Compute cluster centers $C(k) = [c_j]$, using membership matrix $U(k)$ and data objects using

$$C_j = \frac{\sum_{i=1}^{n} u_{ij}^m x_i}{\sum_{i=1}^{n} u_{ij}^m} \tag{2}$$

Step 3: Update the membership matrix $U(k)$ to $U(k+1)$ using

$$u_{ij} = \frac{1}{\sum_{j=1}^{C}(\frac{||x_i - c_j||}{||x_i - c_k||})^{\frac{2}{m-1}}} \tag{3}$$

Repeat step 2 and step 3 until termination condition usually the threshold of changing membership matrix.

2.1 Process of Fuzzy C-Means Clustering

The process of Fuzzy C-Means Clustering algorithm explained in following steps [6].

1. Initializing c centers for clusters and Computing membership matrix u_{ik}.
2. Updating each cluster center as weighted averages to every cluster center until termination.

3 Collaborative Clustering

When data to be shared or distributed, the collaboration scheme is introduced to establish any interaction between one site with another. When data is in distributed manner, and the results to be combined then based on the requirements of the parties and their expected solutions, the collaboration between data sites can give better combined solutions without violating or compromising privacy of each data site in collaborative clustering method [3]. In collaborating approach there will be two or more number of data sites and at each data site a structure can be revealed by forming c number of clusters; assuming same number of clusters for each data site [2]. For a specific data site first a local level structure is discovered by receiving and considering the results of clustering, reported by other data sites, then the local site can update the local findings at their specific data sites [11].

3.1 Modes of Collaboration

In collaborative scheme because of (a) privacy reasons and (b) technical constraints and their feasibility, each data site behaves like a separate computing entity and restricted to share the data outside this site. The data sites can exchange only local findings or outcomes and then use them in further development of the collaborative scheme. There are two fundamental modes of interaction between data sites in collaboration scheme [9]:

- (a) centralized mode: this mode of collaboration reconcile the local outcomes with results shown at other remaining data sites.
- (b) distributed mode: in this mode all data sites are allowed to interact between each other and the resulting local outcomes or local models can be shared.

4 Collaborative Fuzzy C-Means Clustering

The collaborative fuzzy clustering was originally introduced by pedrycz, that helps in finding the ways of improving the quality of unsupervised learning, and clustering in particular. In fuzzy based clustering, there are two fundamental facts of granularity of data: one is prototypes and other is partition matrices [5]. These two outcomes of granular information could be used as a communication vehicle in collaboration process. Here, data sets resides in same feature space, so that communication can be done by exchanging prototypes computed at each data site [10]. The structure of a granular interface and their communication flow in collaboration presented in diagrams below (Fig. 1).

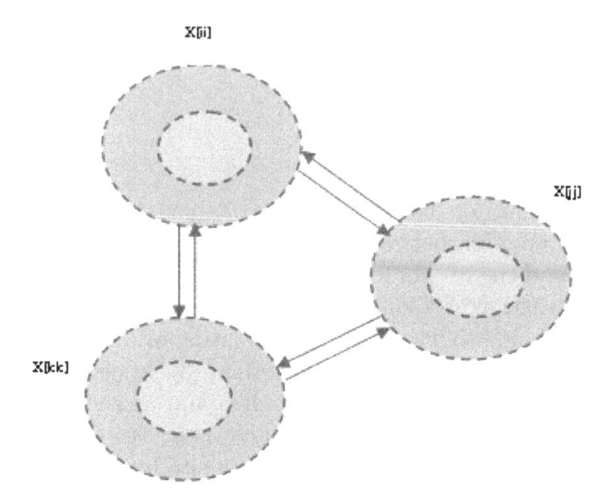

Fig. 1. The collaboration between granular interfaces of the numeric data

5 Privacy Preserving Collaborative Fuzzy C-Means Clustering-Horizontal

In this section we discuss our proposed method, privacy preserving collaborative FCM (PP-HFCM). In this method the dataset is horizontally partitioned and distributed between n parties to form C number of clusters. All the parties follows the horizontal way of clustering at the time collaborating. All the parties share only local outcomes (prototypes) and prepare the required parameters for next phase of the FCM clustering in collaboration process. The algorithm of privacy preserving collaborative fuzzy c-means clustering for horizontal data distribution is given below.

5.1 Process of Horizontal-PPFCM

In first phase, FCM algorithm executed at each individual data site (no collaboration phase) [11] and after FCM has terminated at each site, process stops and the data sites communicate their local findings to collaboration phase. The effectiveness of the collaboration depends on the way in which one data site communicates to others in terms of what has been discovered so far in the form of information granules (Prototypes). The process of privacy preserving collaborative FCM clustering for horizontal data distribution is presented in a process flow diagram (Fig. 2).

Algorithm 2. Privacy Preserving Collaborative FCM Clustering - Horizontal

Partition: Dataset is horizontally partitioned and distributed between n parties.
Initialize: Parties decide on number of clusters c, initialize the membership matrix matrix u_{ik}, objective function prototype v_{ij}, fuzzy coefficient $m = 2$ and collaboration matrix $\alpha[ii, jj]$ for collaborative FCM clustering process.
Non Collaboration:
1. FCM Clustering: For All Parties
The first party p_1) computes prototype v_{ij} vectors for its data and partition matrix u_{ik} for all C clusters, using FCM algorithm.
2. Collaboration Phase:
2.a: Now party p_1 sends it's prototype v_{ij} vectors and partition matrix u_{ik} to next party p_2 through collaboration matrix $\alpha[ii, jj]$
2.b: Now party p_2 computes partition matrix u_{ik} and prototype v_{ij} vectors, then sends to the next party p_n through $\alpha[ii, jj]$.
2.c: Now party p_n computes prototype $V[ij]$ and partition matrix $U[ij]$, then minimizes final collaborative function $Q[ii]$ using

$$Q[ii] = \sum_{i=1}^{c}\sum_{k=1}^{n} u_{ik}^2[ii]d_{ik}^2[ii] + \sum_{jj=1, jj\neq ii}^{P} \alpha[ii, jj]\sum_{i=1}^{c}\sum_{k=1}^{n}(u_{ik}[ii] - u_{ik}[jj])^2 \quad (4)$$

Repeat Until Termination (End of Collaboration)

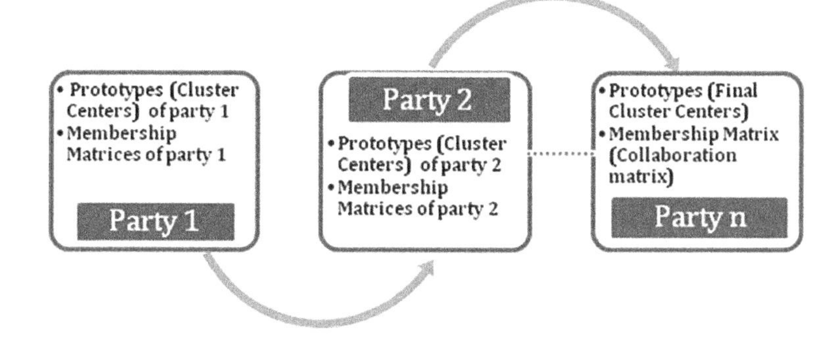

Fig. 2. Process flow of privacy preserving collaborative FCM clustering - horizontal

6 Privacy Preserving Collaborative Fuzzy C-Means Clustering-Vertical

When dealing with vertical clustering, the collaboration process concerned with different patterns (usually disjoint subsets of patterns) but in the same feature space. Here the collaborative clustering carried out based on the prototypes and the same distance function and same fuzzy coefficient is being used. Individual prototypes are being computed in the entire clustering process [4,8]. The collaborative FCM clustering algorithm proposed for vertical data distribution is given below. In this algorithm, the ith prototypes $v_i[1], v_i[2], ...v_i[j]$ and the corresponding membership degrees $u_i[1], u_i[2], ..., u_i[P]$ are computed and joined through a collaborative matrix $\beta[l,m]$. The collaborative objective function is used to get final structure of the $l-th$ data set of any party involved in process, and collaboration matrix $\beta[l,m]$ combines the prototypes received from previous data sites, then the differences between the prototypes $v_i[l]$ and $v_i[m]$ is computed.

Algorithm 3. Privacy Preserving Collaborative Fuzzy C-means clustering algorithm - vertical

---

Partitioning: Data set partitioned into P number of vertical partitions.
For all Parties: each one holding subset of inputs $X_1, X_2, X_3, ..., X_p$.
Initialize : Objective function, number of clusters C, Termination condition, and collaboration matrix $\beta[l,m]$.
randomly initialize all partition matrices $U[1], U[2], ..., U[P]$.
For each data item:
Compute individual prototypes: $Vi[l], i = 1, 2, ..., C$ and individual partition matrices $U[l], l = 1, 2, ..., p$ for all subsets of input patterns.
Until termination condition.
Collaboration Phase:
Collaboration Matrix: $\beta[l,m]$ computes combined prototypes $Vi[l]$ and combined partition matrices $U[l]$, and minimize the collaboration index $Q[ii]$ using following equation.

$$Q[ii] = \sum_{k=1}^{X[l]} \sum_{i=1}^{c} u_{ik}^2[l] d_{ik}^2[l] + \sum_{m=1, m \neq l}^{P} \beta[l,m] \sum_{i=1}^{c} \sum_{k=1}^{X[l]} u_{ik}^2[l] ||v_i[l] - v_i[m]||^2 \quad (5)$$

Until termination.

6.1 Process of Vertical-PPFCM Algorithm

The Privacy preserving model and process of vertically collaborative FCM clustering is given in diagram below (Fig. 3).

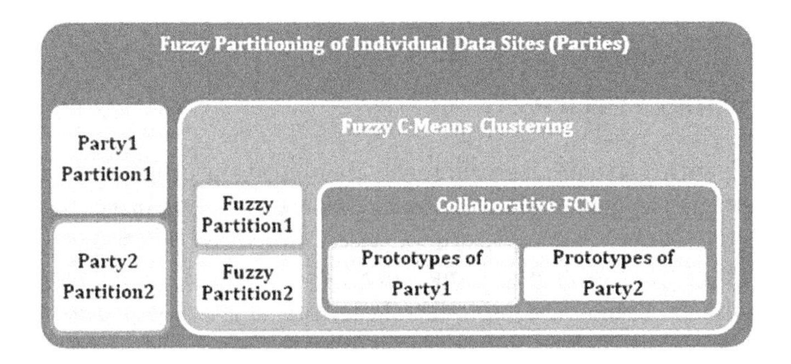

Fig. 3. Privacy preserving collaborative FCM clustering model - vertical

1. For all the parties of their given data sites $D[1], D[2], ..., D[P]$, define the number of clusters C, set the termination criterion, collaborative coefficient β, that optimizes the level of collaboration.
2. **Initial phase:** for each individual data site compute fuzzy c-means clustering results individually in the form of prototypes $v_i[ii], i = 1, 2, ..., c$.
3. **Collaboration phase:** allow the individual results (prototypes)to interact to form collaboration between the sets.
4. For each data site (ii), minimize the collaborative objective function $\beta[l, m]$ at each data site by iterative proceeding with the iterative calculations of the partition matrix and the prototypes.
5. Continue until termination condition of the collaboration.

7 Experiments and Results

All the experiments are undertaken for horizontal and vertical collaboration for fuzzy c means clustering using Matlab2013a with help of fuzzy clustering tool box. The results are presented for collaboration (preserving privacy) and non collaboration (non privacy) based fuzzy clustering for Iris, Glass Identification data sets. For all the experiments, fuzzy coefficient $m = 2$ and the number of clusters $C = 3$.

7.1 Results of Horizontal-PPFCM

The results of collaboration in Horizontal-PPFCM are presented for two data sets (iris and Glass identification). The performance of objective function is plotted and compared between non collaborative (non privacy) and horizontal collaborative(privacy preserving) fuzzy C-Means Clustering.

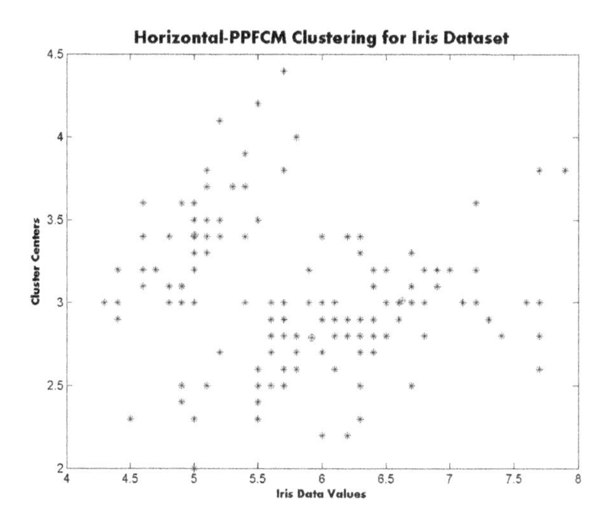

Fig. 4. Horizontal-PPFCM clustering for iris dataset

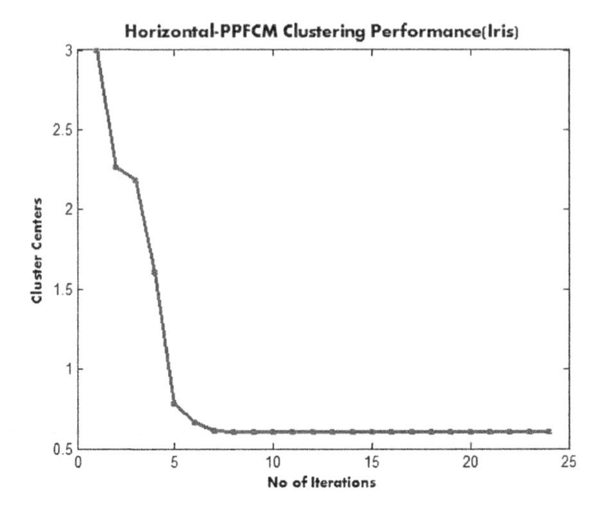

Fig. 5. Horizontal-PPFCM collaborative clustering performance for iris dataset

Iris Data Set: Figures shows the results of non collaborative FCM and the performance of the objective function for iris data set. The next two figures shows the results for Horizontal-PPFCM when the iris data set is horizontally partitioned, where each party holds 75 samples (Figs. 4 and 5).

Glass Identification Data Set: Figures shows the non privacy preserving FCM clustering and its performance of glass identification data set of 214 instances. Next two figures gives details of privacy preserving horizontal fuzzy c means clustering performed (Figs. 6 and 7).

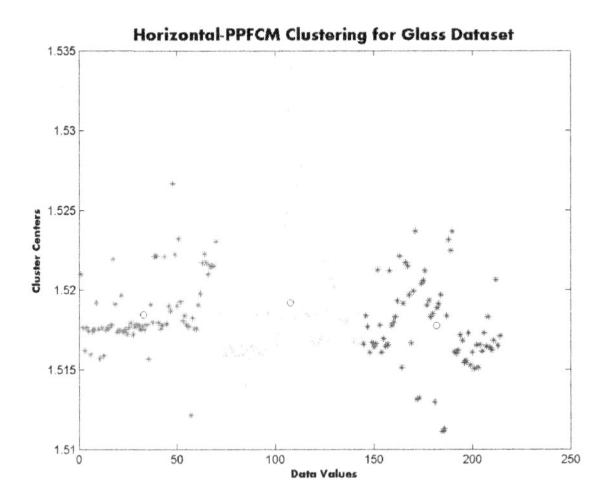

Fig. 6. FCM clustering performance of glass identification dataset

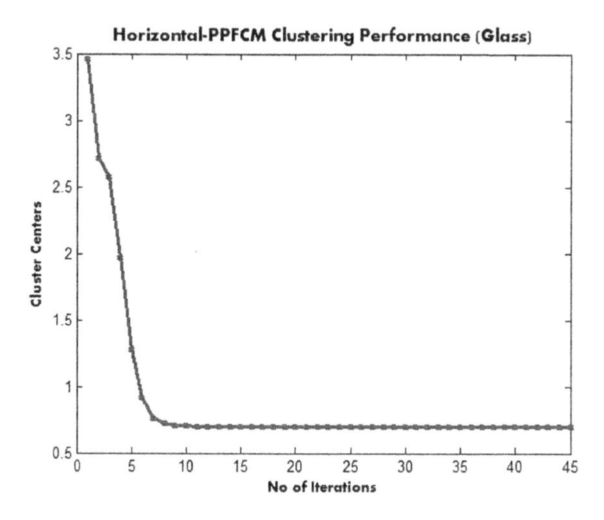

Fig. 7. Horizontal-PPFCM clustering performance for glass dataset

7.2 Privacy Analysis of Horizontal-PPFCM

The above Table 1 shows the privacy levels of Horizontal-PPFCM with respect to three different privacy measures for three levels of PPFCM process. The privacy levels are explained below.

Table 1. Privacy metrics results and analysis of horizontal-PPFCM

Privacy metric	Range & level	Metric result	Privacy level
priv_CS	[0, 1] Low	0.25	**High**
priv_TC	[0, 8] Low	1.11	**High**
priv_PS	[0, 8] Low	1.00	**High**

- ***Privacy of Granular Information:*** The granular information means the input level outcomes to share with other party in collaboration process. In PPFCM method the inputs are shared in the form of granular information of the input data. The cluster centers and the partition matrices are shared between parties instead if direct input attribute values. The fuzzy partitions and their cluster centers does not carry any original or sensitive information to other party, hence the privacy is well preserved in this phase. The privacy measured at this stage using cluster similarity $priv\_CS$ where the similarity between the clusters formed by the parties are compared and if the difference is Low then the privacy is said to be high. The privacy level of Horizontal-PPFCM has shown high level privacy in this input phase.

- ***Privacy of Parties in Collaboration:*** The main phase of collaboration process, where privacy of parties to be highly preserved is the collaboration phase. Collaborating parties expect high privacy level in collaboration phase as they share the information by exchanging prototypes to each other. The privacy metric t-Closeness $priv\_TC$ is used to evaluate the privacy level of parties, where the distribution of original input values must be close to the distribution of the shared information in collaboration. The difference (distance) between two parties must be small to gain high privacy. The result shows low value and the high privacy level for Horizontal-PPFCM in this phase.
- ***Privacy of Outputs in Collaboration:*** The output level is the final phase where the combined results of the collaboration are published. The collective output from any collaborative process must ensure privacy of parties involved. Privacy score $priv\_PS$ measures the privacy level assured in collaboration. Privacy score indicates the privacy risk increases with the sensitivity of information granules and their visibility in collaboration. Low visibility decreases the privacy risk and gives high privacy level. Horizontal-PPFCM shows the high privacy level by showing low visibility of information in collaboration

7.3 Results of Vertical PPFCM

Experiments are carried out for Iris and seeds data sets. Vertical-PPFCM algorithm has shown following results with collaboration coefficient $\beta = 2.0$ for two parties that are derived before and after collaboration. The results are shown for iris dataset in graphical and tabular formats below (Figs. 8 and 9).

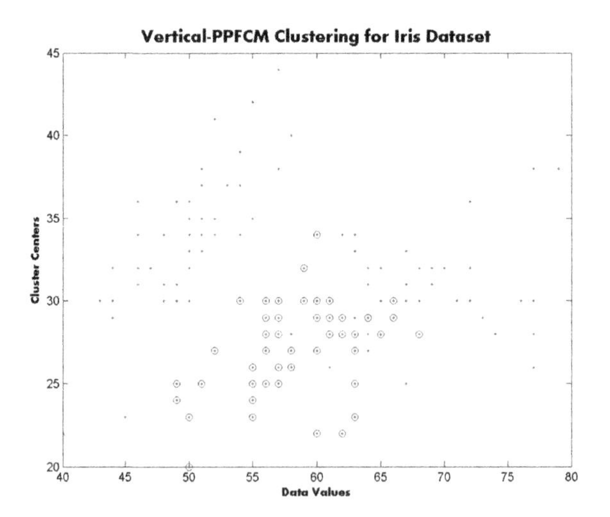

Fig. 8. Vertical-PPFCM clustering for iris dataset

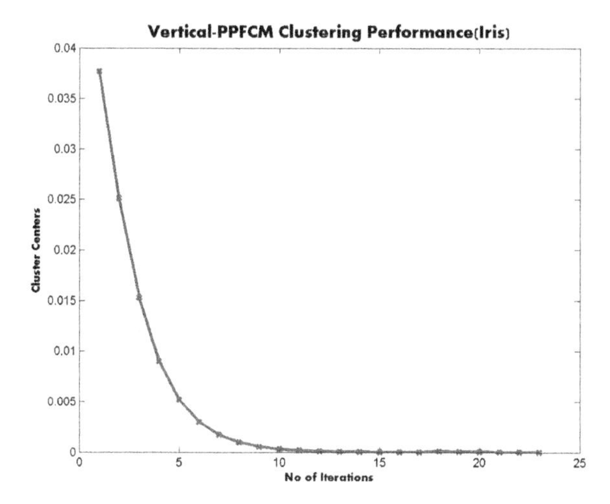

Fig. 9. Vertical-PPFCM clustering performance for iris dataset

Glass Identification Data Set: Figures shows FCM clustering and its performance of glass identification data set, that holds 214 instances. Next two figures present the cluster centers in collaborative process when data is vertically partitioned between two parties (Figs. 10 and 11).

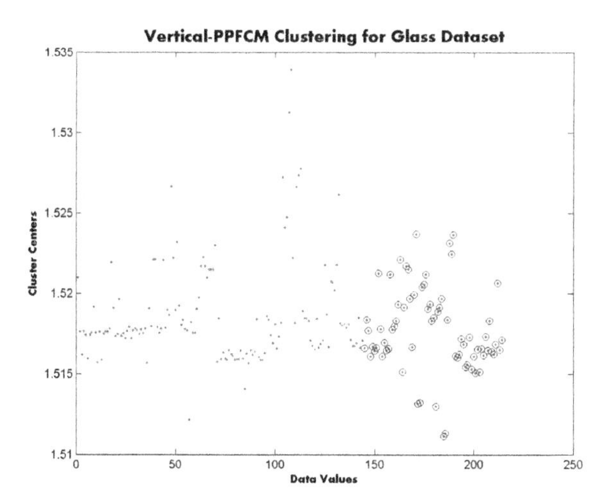

Fig. 10. Vertical collaborative FCM clustering for glass dataset

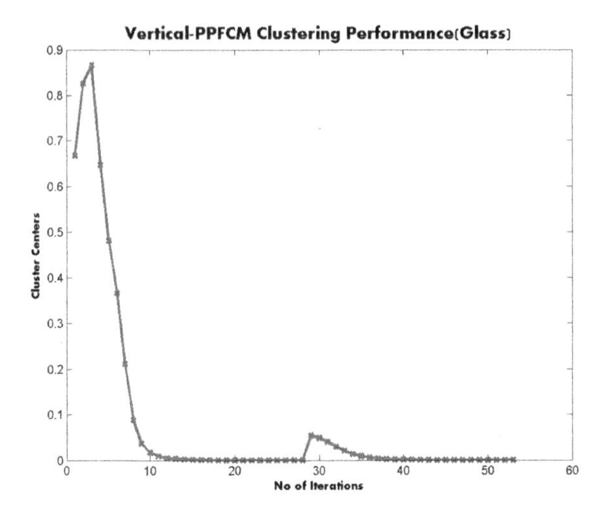

Fig. 11. Vertical-PPFCM clustering for seeds dataset

Table 2. Privacy metrics results and analysis of vertical-PPFCM

Privacy metric	Range & level	Metric result	Privacy level
priv_CS	[0, 1] Low	0.18	**High**
priv_TC	[0, 8] Low	2.45	**High**
priv_PS	[0, 8] Low	0.01	**High**

7.4 Privacy Analysis of Vertical-PPFCM

The above Table 2 shows the privacy levels of Vertical-PPFCM with respect to three different privacy measures for three levels of PPFCM process. The privacy levels are explained below.

- ***Privacy of Granular Information:*** The input level outcomes in the form of granular information are shared with other parties in collaboration process. The cluster centers and the partition matrices are shared between parties instead of direct input values. The privacy measured at this stage using cluster similarity $priv\_CS$ where the similarity between the clusters formed by the parties are compared and if the difference is Low then the privacy is said to be high. The privacy level of Vertical-PPFCM has shown high level privacy in this input phase.
- ***Privacy of Parties in Collaboration:*** Vertical collaborating parties expect high privacy level in collaboration phase as they share their information (partial) through prototypes to each other. The privacy metric t-Closeness $priv\_TC$ is used to evaluate the privacy level of parties, where the distribution of original input values must be close to the distribution of the shared information in collaboration. The difference (distance) between two

parties must be small to gain high privacy. The result shows low value and the high privacy level for Vertical-PPFCM in this phase.
- ***Privacy of Outputs in Collaboration:*** The collective output from any collaborative process must ensure privacy of parties involved. Privacy score *priv_PS* measures the privacy level assured in collaboration. Privacy score indicates the privacy risk increases with the sensitivity of information granules and their visibility in collaboration. Low visibility decreases the privacy risk and gives high privacy level. Vertical-PPFCM shows the high privacy level by showing low visibility of information in collaboration

8 Summary

In this paper one can understand the way of collaboration for preserving privacy of individual parties while exchanging the intermediate results with each other. The paper presented the proposed methods and algorithms to en-corporate the collaboration between data sites. The collaborative clustering process has been undertaken and implemented for horizontal and vertical data distribution mechanisms for two parties, that can also be used for multiple parties in future. Experiments and Results of the proposed algorithms along with the privacy analysis of both Horizontal-PPFCM and Vertical-PPFCM are presented.

9 Conclusions

We adopted the core work done by [7,8], and implemented a sequential collaboration method when exchanging internal outputs between parties to perform collaborative clustering. We proposed algorithms for horizontal and vertical data distribution for collaborative fuzzy c-means clustering between two parties. The results were produced for two data sets (iris and seeds). The aim of achieving privacy preserving was successful in collaborative clustering using Fuzzy C-Means Clustering between two parties with acceptable privacy level.

10 Future Scope

We aim to undertake further investigations on multiple number of parties (more than two) in large scale distributed systems environment. After adopted for horizontal and vertical partition methods in this work, we are working on arbitrarily partitioned data between two parties and then the same to be implemented for more number of parties. We are working towards the challenge of building fully privacy preserved model with privacy assured at each level of data transfer. Then we have plan for extending the same model to give solutions for other data mining techniques like, classification, regression and deep learning.

References

1. Bezdek, J.C., Ehrlich, R., Full, W.: FCM: the fuzzy c-means clustering algorithm. Comput. Geosci. **10**(2), 191–203 (1984). https://doi.org/10.1016/0098-3004(84)90020-7. http://www.sciencedirect.com/science/article/pii/0098300484900207
2. Bilge, A., Polat, H.: A comparison of clustering-based privacy-preserving collaborative filtering schemes. Appl. Soft Comput. **13**(5), 2478–2489 (2013)
3. Cano, I., Ladra, S., Torra, V.: Evaluation of information loss for privacy preserving data mining through comparison of fuzzy partitions. In: 2010 IEEE International Conference on Fuzzy Systems (FUZZ), pp. 1–8. IEEE (2010)
4. Estivill-Castro, V.: Private representative-based clustering for vertically partitioned data. In: Proceedings of the Fifth Mexican International Conference on Computer Science, ENC 2004, pp. 160–167. IEEE (2004)
5. Inan, A., Kaya, S.V., Saygın, Y., Savaş, E., Hintoğlu, A.A., Levi, A.: Privacy preserving clustering on horizontally partitioned data. Data Knowl. Eng. **63**(3), 646–666 (2007)
6. Kannan, S., Ramathilagam, S., Chung, P.: Effective fuzzy c-means clustering algorithms for data clustering problems. Expert Syst. Appl. **39**(7), 6292–6300 (2012)
7. Pedrycz, W.: Collaborative fuzzy clustering. Pattern Recogn. Lett. **23**(14), 1675–1686 (2002)
8. Pedrycz, W.: Distributed and collaborative fuzzy modeling. Iran. J. Fuzzy Syst. **4**(1), 1–19 (2007)
9. Pedrycz, W., Rai, P.: Collaborative clustering with the use of fuzzy c-means and its quantification. Fuzzy Sets Syst. **159**(18), 2399–2427 (2008)
10. Strehl, A., Ghosh, J.: Cluster ensembles–a knowledge reuse framework for combining multiple partitions. J. Mach. Learn. Res. **3**, 583–617 (2003)
11. Yu, F., Tang, J., Cai, R.: Partially horizontal collaborative fuzzy c-means. Int. J. Fuzzy Syst. **9**(4), 198 (2007)

A Multi Agent Framework to Detect in Progress False Data Injection Attacks for Smart Grid

Ayush Sinha(✉), Ashrith Reddy Thukkaraju, and O. P. Vyas

Indian Institute of Information Technology, Allahabad, Prayagraj, India
pro.ayush@iita.ac.in, {icm2016005,opvyas}@iiita.ac.in

Abstract. With the rapid growth in the use of the internet and the rapid digitization of the energy services, telecommunication and other critical infrastructures, there is a proportionate growth in the magnitude and the complexity of the cyber attacks on these services. Attacks on such services can cause a disruption of a magnitude that can affect the lives of millions of people and cause financial loss in billions of dollars. This necessitates ensuring the security of these services and systems to be of the utmost priority. The Cyber Physical defence is a concept that deals with ensuring and enforcing security mechanisms against different possible attacks on these critical infrastructures. This paper discusses some of the possible attacks on the smart grids, describes various detection and mitigation techniques against these attacks and also proposes a defence mechanism that follows the NIST framework for critical infrastructures. In particular, this paper focuses on one particular type of attack called the False Data Injection (FDI) Attack. FDI attack is one of the most common and lethal attacks on the smart grids with the ability to inflict both financial damage and cause essential service disruptions. We discuss the possible ways to identify and defend against FDI attacks and propose a reinforcement learning based approach using Deep neural networks to detect and defend against these FDI attacks. We propose the learning algorithm and the reward system that ensures the convergence of the learning algorithm.

Keywords: FDI attack · Smart grid · Reinforcement learning · Deep Q learning · Attack detection

1 Introduction

Utilization of the network became a trend present with the development of technology, especially the Internet. It is a fact that the majority of the individuals around the world are using the Internet for their personal needs, or for their business needs. This trend is growing at an exponential pace due to the digitization of every service in every sector and the growth in the pace of innovation in and around the web. However, this trend of web use is directly proportional

I. Woungang et al. (Eds.): ANTIC 2021, CCIS 1534, pp. 123–141, 2022.
https://doi.org/10.1007/978-3-030-96040-7_10

to the crime rate on the internet, or that is better known as cyber crime. This sort of crime has various levels of motivation behind it ranging from mischief to terrorism.

With the digitization of the power production and distribution sector, regular electric grids were improved upon to smart grids. Smart grid utilizes the advancements of information technology to smartly produce, monitor, manage, and deliver energy to consumers by using two-way communication. Although the smart grid was originally designed as an improvement to the traditional grid, the security aspects are failed to be taken into consideration while it was developed. This causes many attackers to perform lethal attacks [5] that can not only disrupt the power supply, but also cause damage to lives in the worst case scenario.

The term "Industrial Control System" refers to a wide range of control systems, including Supervisory Control and Data Acquisition systems, Programmable Logic Controllers (PLC) and other control systems like Distributed Control Systems (DCS), which are common in the industrial sector and smart grid critical infrastructure. Reports from the Industrial Control Systems Cyber Emergency Response Shows that Cyber attacks on critical infrastructure in the US have rocketed over the past few years. Several previous attacks like Blackenergy, Industroyer, and Triton were specially targeted on large power supply stations as a result of which millions of people were left in freezing temperatures without power supply. Eventually, security became a major threat and a very vital part of the ICS networks. While both physical and cyber attacks are possible on smart grids, cyber attacks are more dangerous because of their surreptitious nature and the magnitude of damage that can be inflicted. This is the reason we focus on the Cyber attacks on smart grids in this paper.

1.1 Goals

The main motto of the paper can be observed from the following points. Discuss the architecture of Smart Grid Discuss the need for Cyber Security in Smart Grid Describe the possible attacks on smart grids Define a mechanism to identify and defend against these attacks using the NIST framework Propose a novel approach of using Reinforcement learning to detect false data injection attacks.

1.2 NIST Framework

NIST framework is a Cyber Security standards framework that defines and describes a set of recommended security practices, standards and mechanisms to ensure the security of an organization or critical infrastructures. It defines a 5 step process to ensure that cyber physical security of any organizational network (Fig. 1).

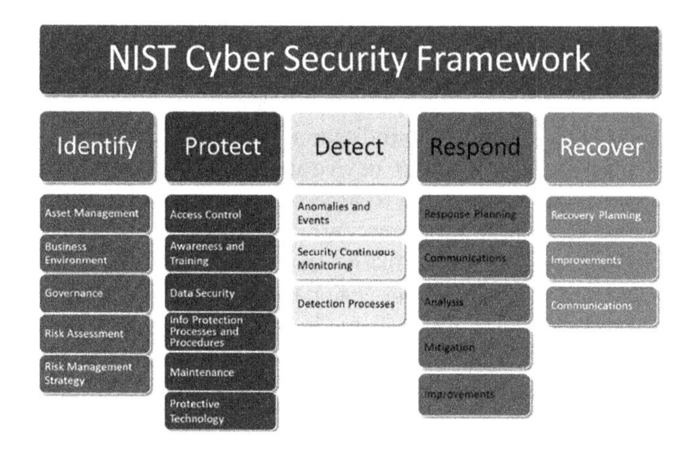

Fig. 1. NIST CyberSecurity framework [5]

1.3 Identify

The first step is identification, which involves an organizational perspective of cybersecurity. It takes into account the various security risks possible to the overall system and the assets that these risks will affect. The cybersecurity attacks on a smart grid can be broadly classified as attacks on the following [5].

1. Confidentiality of Data: The confidentiality of data being sent through the network is essential because otherwise, attackers would be able to determine usage patterns, which could reveal the user's personal information. The pricing data on the other hand does not need to be kept confidential as it is public information.
2. Integrity of Data and Commands: The integrity of data such as pricing information and demand details must be maintained to ensure the smooth functioning of the smart grid. Incorrect pricing information can lead to situations such as unexpected excessive consumption and can result in blackouts in some areas.
3. Availability of information and electricity: DDoS attacks on the network can lead to the unavailability of pricing information which can have huge financial repercussions. It can also lead to an incorrect amount of power being generated, thus leading to either wastage or scarcity of Power.

Attacks of these three types can hinder communication services. In addition to these, several physical attacks can be performed such as stealing the smart meter, tampering with sensors etc.

1.4 Protect

Protection involves ensuring that adequate safeguards are in place so that critical communication is not hindered.

1. Access Control: Customers will only have access to their own smart meter information and public information such as the pricing information. Access to another customer's smart meter information is prohibited. NAN Control centres will be able to access the individual smart meter data, however, they will not have access to any smart device information. The Control and Distribution Centre will also be able to access the individual NAN data but will not have access to individual smart meter information as aggregation would have taken place at the NAN Control Centre.
2. Awareness and Training: Customers using the smart grid must be made aware of possible attacks that can be performed on their smart devices. They should also be advised regularly to not download content from unverified sources and to install anti-virus software and malware detectors on their devices.
3. Data Security: [6]
   (a) Encryption: Encryption of data from the smart devices till the power generation station is essential so that attackers cannot infer any information about usage patterns. Encryption can be performed using either symmetric or asymmetric key cryptography. Some of the commonly used symmetric key encryption algorithms are AES, DES and RC4, and some commonly used asymmetric key encryption algorithms are El Gamal and RSA. Encryption will ensure that the attacker cannot directly access the usage data of a customer.
   (b) Integrity and Authentication: It is necessary to ensure that data such as the smart meter data and the pricing data do not get modified when communicated through the network. It should also be ensured that this information has come from an authentic source and not an attacker and non-repudiation of the data should also be ensured to hold accountability. Hence each communicating device in the smart grid can have its public-private key pair that is used for signing the data that they send. Some signature algorithms that can be used are the RSA digital signature scheme and ECDSA.
   (c) Key Exchange: For symmetric key exchange methods such as Diffie Hellman Key, exchange protocol can be used or a Key Distribution centre can be used. For public key exchange, a Public Key Infrastructure (PKI) consisting of various Certification Authorities can be used.
4. Protective Technology:
   (a) Load Balancing of Servers: Multiple servers can be provided at the NAN Control Centre and WAN Control Centre to prevent congestion.
   (b) Channel hopping on wireless networks: This technique can be used in case a particular communication channel frequency is jammed.
   (c) Limiting packet size: There is a possibility of attackers jamming the network communications by transmitting excessively large quantities of data which can lead to disruptions to the essential communications within the smart grid network. This can be prevented or avoided by setting the limits based on the volume of data requests and responses within the Smart grid network.

1.5 Detect

This step involves identifying the occurrence of an attack on the smart grid. Detection methods could involve continuously monitoring the grid for anomalies or probing at frequent intervals to ensure that everything is functioning smoothly. Anomaly-based and Signature-based Intrusion Detection are two approaches for detecting intrusions in a smart grid [7]. An anomaly-based IDS can identify any deviation from the normal functioning of a system while a signature-based IDS will check whether the functioning of the system matches any predetermined attack signatures. Initially, anomaly-based methods are a better method to check whether any anomalous activity is occurring in the smart grid. When an attack is detected and identified in the smart grid the signature of the attack can be added to the list of detection mechanisms to ensure more thorough detection of repeated attacks.

1.6 Respond

This involves implementing measures to take action against the attacker once an attack has been discovered. If the source of the attacks can be identified the smart grid can block the attacker in some way to stop the attack. However, this is not an easy task and hence it is important to provide resilience to the smart grid.

1.7 Recover

In the recovery phase services that were hindered by the attack are restored. It is also the phase used to ensure that the functioning of the smart grid continues irrespective of an attack. The recovery phase will usually involve modifying the architecture of the network and making use of DERs to ensure the smart grid functionality. We can make use of the Tri-level optimal hardening plan [8] to strengthen the system's resilience and avoid potential blackouts. This optimal hardening plan makes use of distributed generators to keep power flowing throughout the network in the event of an attack.

2 Background Work

The Smart Grid Architecture Model [1], is a frequently used reference framework was created in response to the European Commission's Standardisation Mandate M/490. It is depicted in Fig. 2. It is a three-dimensional reference architecture comprising of Zones, Domains and Interoperability Layers that may be utilised to create a contemporary smart grid system. The smart grid is covered by five interoperability layers, which are made up of five domains that corresponds various components of the electricity and energy conversion and transmission, and the six zones that reflect hierarchical levels organization of power system. This architecture provides a solid and well-structured foundation for the design, analysis, and validation of innovative solutions and technologies built for smart grids. It helps in understanding the structure and the complexities involved in a smart grid.

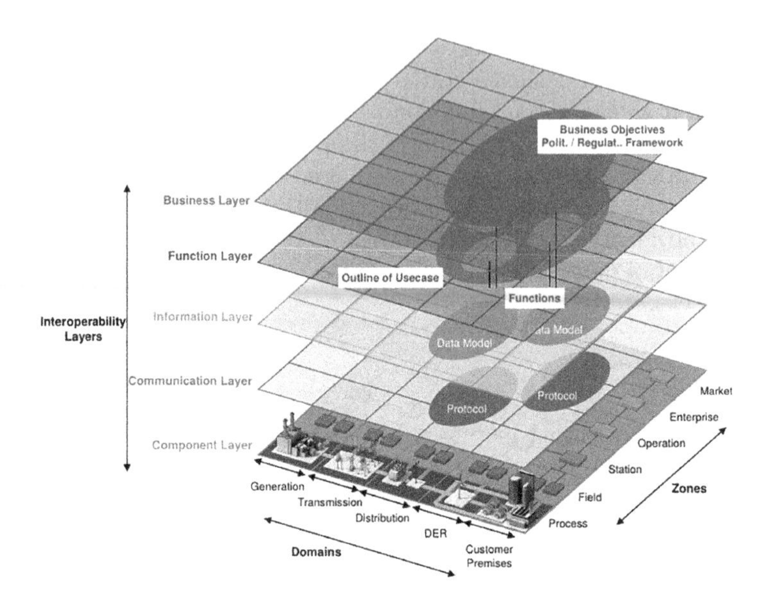

Fig. 2. SGAM structure [1]

Literature Review on Smart Grid Cyber Security [2] discusses the various attacks that can occur on a smart grid network and classifies the smart grid security into 4 classes: Smart Meter Security, Process Control System Security, Smart Grid Communication Protocol Security and, Power System State Estimation Security. We will focus on the Power System State Estimation security primarily as we explore a technique to protect smart metres against False Data Injection Attacks.

Snort [3] is an example of an Network Intrusion Detection System (NIDS). It monitors several devices in a local area network and analyses network data to detect intrusions. To tap and collect network traffic, it is connected to either a hub or a network switch through port mirroring. It is one of the approaches that can be used to monitor and detect intruisons into a smart grid or any organizations networks.

Prelude [4] is an example of a Hybrid Intrusion Detection System. Data from host-based IDS agents (monitors system calls, application log files, and modifications to files such as binary executables to identify intrusions) is combined with network information from NIDS to provide a full view of the network and system.

National Institute of Standards and Technology [5] is a cyber security standards framework defined to ensure the security of organizational networks, cyber physical systems and critical infrastructures. IT defines a 5 step process that we follow in our work while designing the cyber security mechanism to defend against the FDI attacks on smart grids.

Cryptography and Network Security [6] defines and discusses the concepts and methods of Cryptography and network security both in breadth and depth. It discusses each layer and aspect of security in detail. We can use these cryptography techniques to ensure the access control, confidentiality of the data being transmitted in our network to make it secure.

An Intrusion Detection Framework for the Smart Grid [7] proposes a method using multiple Intrusion Detection Systems (IDS) at HAN, NAN and WAN layers and uses a centralized system to analyze and detect the anomalous activity. The central system correlates the information from all the sensor nodes to detect and react to the attacks on the network. It shows how IDS systems can be used to detect and respond to cyber attacks.

Tri-level optimal hardening plan for a resilient distribution system considering reconfiguration and DG islanding [8] is a Defender-Attacker-Defender reconfiguration based strategy to increasing smart grid resilience. It first hardens the lines to reduce DG islanding, then performing the worst possible attacks on it, and lastly reconfiguring the network to reduce DG islanding once again.

Cyber security in the Smart Grid: Survey and challenges [9] discusses the security requirements of a smart grid network as Confidentiality, Availability and Integrity. This paper also classifies the possible attacks on the Smart grid as DDoS attacks, Data manipulation attacks and attacks on confidentiality to gain unauthorized access to the services and devices in a smart grid. It discusses these threats in detail and also suggests countermeasures to deal with the same.

Cyber Security Requirement for Smart City - Model Framework [10] The Indian government established a framework that highlights the different security parameters and requirements that should be taken into account when designing and developing smart city architectures. This framework provides data exchange information and it's security requirements for a smart grid network.

In **Cyber-Physical Security of a Smart Grid Infrastructure** [11], in this paper, the fundamental information security criteria, which are integrity, confidentiality, and availability, were considered to establish particular security needs for a smart grid network. It also detects potential attacks and access points, as well as the repercussions of these on the grid and it's security. It also proposes and discusses defence mechanisms against these potential threats.

The **NIST Guidelines for Smart Grid** [12] in this framework, smart grid's cybersecurity needs are defined as confidentiality, integrity, and availability. It also delineates what each of these requirements implies. It consists of a 5 step approach to deal with the security threats on a smart grid. These steps are discussed in detail in the coming sections.

Machine learning in cybersecurity: A review [13] this paper outlines a number of machine learning based approaches that may be used to identify intrusions. When it comes to SCADA networks, one of the recommended ways of intrusion detection is to discover aberrations in the system using a amalgamation of OC-SVM and k-means clustering. This technique is touted for having excellent accuracy, little overhead, and decent performance.

Detection of False Data Injection Attacks in Smart Grid Utilizing ELM-Based ECON Framework [14] makes use of a one class one network, machine learning approach to distinguish between false data and normal data. This paper also provides a prediction recovery technique for correcting erroneous data by analyzing the spatial correlation of data.

Communication Architecture for Smart Grid Applications [15] outlines three tiers of networks in a generic smart grid infrastructure-WAN, HAN, and NAN- and provides both the wired and wireless communication infrastructures in this network.

Towards modelling the impact of cyber attacks on a smart grid [16] gives a way for analysing the impact using graphs. This entails using directed graph for modelling the smart grid's infrastructure and then executing attacks on it to determine the predicted impact on the system.

A Cyber-Physical Modeling and Assessment Framework for Power Grid Infrastructures [17] discusses and proposes an architecture for a CPMA framework that synthesises a Markov Decision Process (MDP) based analysis model infiltration of the attackers into the network using a threat model (attack tree), power topology, and cyber topology. It also employs a cyber security state estimation of the entire system and power system analysis to pinpoint the locations of the attacks.

The **SOCCA** [18] infrastructure uses all of these factors to do a contingency analysis and rating for the power infrastructure. This framework is only a quantitative measure since it takes into account only the number of loads that can be controlled directly from each node as the parameter to estimate the risk. This fails to account for some of the packet flooding attacks like DOS, DDoS and SYN Flooding attacks and also fails to account other coordinated attacks.

A Machine Learning-Based Technique for False Data Injection Attacks Detection in Industrial IoT [19] discusses the machine learning algorithm based methods like Support Vector Machines (SVM) and Auto Encoders for identifying the FDI attacks on the Industrial IOT devices which encapsulates the smart grid devices and services. They propose an Auto Encoder based real-time detection mechanism that significantly out performs the SVM based approach.

Power System State Estimation: Theory and Implementation [20] is a book that proposes the standard implementation of the weighted least squares (WLS) approach which is the standard System state estimator for smart grid networks. This book also uses a simple 3 bus example grid for the statistical and performance analysis of this WLS algorithm. In this paper, we use this 3 bus example grid as one of the grids to test the RL based approach against.

So, in a nut shell, there is no standardised security framework for a smart grid that takes into account different types of attacks. Most research works focus on particular attacks and defence and hence it is necessary to develop a standard defence framework that can be incorporated across all smart grids. The standard defence framework should provide the basic structure of how to tackle a potential cyber attack- from detection of the attack to recovery and should also include protective measures to thwart an attack before it occurs.

Detection of False Data Injection Attacks in Smart Grids Based on Forecasts [24] proposes using the data co-linearity that exists in the smart meter readings to make forecasts for the networks smart meter observations which can be then used to predict and detect the attacks on smart grids. This is a predictive approach and we are looking to propose a real time framework for attack detection instead.

Hybrid Multilayer Network Traceback to the Real Sources of Attack Devices [25] proposes a mechanism to traceback both the network layer source (IP address) and the Data link layer source (MAC address) with just a single traceback operation. It also analyzes the performance efficiency and contrasts other such traceback approaches. This can be used in the respond and recovery phase of our approach to detect the attack source and deal with in the appropriate way.

A Novel Data Analytical Approach for False Data Injection Cyber-Physical Attack Mitigation in Smart Grids [26] proposes using a data analysis based approach which employs the margin setting algorithm (MSA) to detect two different scenarios of the false data injection attacks. It shows that it is better than the SVM and ANN bases approaches when it comes to accuracy of attack detection. However, in our approach, we propose a RL bases agent to detect the attacks in real-time as they are happening to reduce the damage caused by the attacks. This helps us minimize the delay of detection of the attacks.

HPSIPT: A high-precision single-packet IP traceback scheme [27] proposes another highly accurate and precise single packet IP traceback method that has a very less storage and computational overhead which can be used for real-time attacker tracing when the IP addresses have been spoofed. This approach has a very less computational overhead and can be used to find the attacker sources of a DDOS attack in real-time.

3 Common Attacks on Smart Grids

3.1 DoS/DDoS Attacks

DoS and DDoS are some of the most common and lethal attacks on the Smart Grid networks. This is when a single device or multiple devices floods the smart meters or any other device in the ICS network with packets at a rate the devices can not process them. This can disrupt the entire smart grid and cut off the necessary power transportation. A DDoS attack is one in which the attacker uses numerous devices or Source IP addresses to make the host network inaccessible, therefore interrupting service for normal, legitimate users. More than 65% of DDoS attacks are volumetric or flooding based attacks, which are carried out by overloading and inundating the target device or service with excessive requests [21] (Fig. 3).

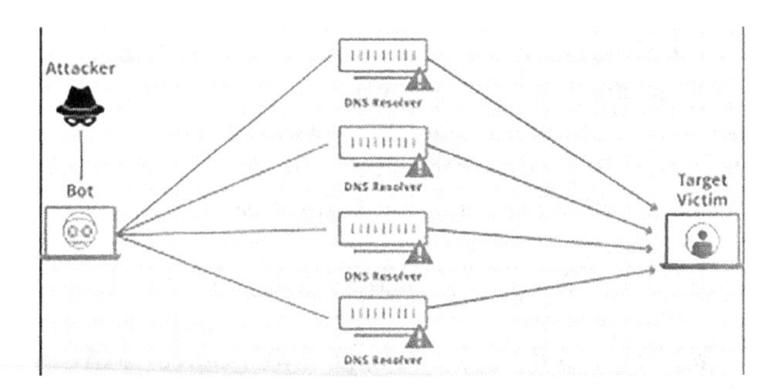

Fig. 3. DDOS attack [23]

3.2 SYN Flooding Attacks

This is another form of packet flooding attack that is very common. A SYN flooding attack occurs when a specific device receives TCP packets with the SYN flag set at an extremely high rate from a single source (Fig. 4).

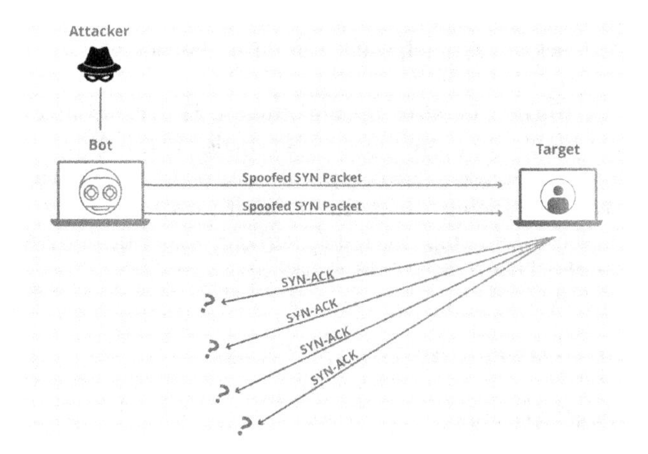

Fig. 4. SYN Flooding attack [23]

3.3 False Data Injection (FDI) Attack

FDI attacks include the injection of misleading data into the smart grid via a hacked sensor in order to influence its operation. Such erroneous data may create surges or decreases in power production, both of which would be problematic for the entire smart grid network. It can cause power supply disruptions at a very large scale that could prove to be very costly (Fig. 5).

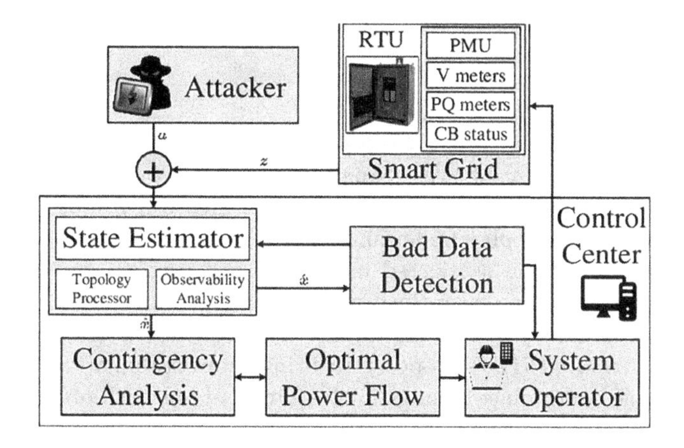

Fig. 5. False Data Injection (FDI) attack [22]

4 Proposed Model

In the past, many methods have been proposed to detect False data injection attacks including clustering techniques like k-Means, machine learning based approaches like Support Vector Machines (SVM) and other neural network-based approaches. Through this work, we explore the possibility of using Reinforcement Learning (RL) to detect an FDI attack in progress. We have proposed RL based model for detecting the FDI attack in progress. The model uses the deep Q learning instead of traditional Q learning to overcome the issues of space complexity.

Furthermore, we employ techniques like Experience Replay and Target networks to increase and improve the training stability and also propose the metrics that can be used to evaluate the performance of the agent. These metrics tell us about the performance of the agent in real-time along with the accuracy of the detection model. We also propose a reinforcement learning reward system that seems to ensure convergence of the RL agent. It has been designed and fine tuned in a way the model learns to do the necessary job quickly and with stability.

4.1 Deep Q-Learning

To overcome the space complexity problem of storing the Q-table in the Q-learning method, we can use Deep Q-Learning. Rather than remembering all of the action value pairs of the table, we generalise the approximation of the Q-table function in this approach. The inputs and target values in deep reinforcement learning are continually changing, making the training process unstable making it an extremely difficult task to learn a mapping f for a continually changing input and output. However, because both input and output can converge, we may be able to describe f while allowing it to evolve if we slow down the changes in both.

4.2 Experience Replay

To overcome the problem of ever-changing inputs and outputs, we can use a method called a replay. To achieve this, we can create a list of the past few observations called the replay buffer and then generate the training set for each step by randomly picking a selected number of observations from this buffer. Since we sample at random from the replay buffer, the data is more independent of one another and is dispersed uniformly. This follows the Independent and Identically Distributed (IID) property to bring more stability to the training. We store a tuple containing 4 things in this buffer about each observation. The current state s_t, the current action a_t, the reward received r_t and the next state s_{t+1}. The tuple at a time step t e_t is given by is $e_t = (s_t, a_t, r_t, s_{t+1})$.

4.3 Target Network

This is another method designed to overcome the problem of training instability due to regularly changing input and outputs in the deep Q-learning training algorithm. In this, we have a reinforcement learning agent called the target network which is updated only at regular intervals and not at every training step. We have the normal model updated at every step but when we take the decision, we only use the target models output to update the weights of the normal model. After every few steps, the target model weights are made same as that of the normal model. This way the training is much more stable.

4.4 State

$S_t \in S$ represents the system state, and S is the set of potential states in which the agent can be found. We define the state space as $[S_n, S_a]$, where S_n implies that the system is in good working order and S_a implies that the system is under attack. The state is not known to the agent, which means that the agent cannot be certain whether the system is functioning normally or under attack; the agent can only guess the state of the system based on it's observations of the present state.

4.5 Action

$a_t \in A$ denotes the agent's action at any time step t, whereas A denotes the collection of actions the agent may do. The action space is defined as $[A_c, A_s]$, where A_c signifies that based on the agents observation of the present state of the system, the agent infers that it is in normal operation and that the system should be let to function normally, and A_s denotes that the that the system is under attack and it should be halted to avoid further damage from happening.

Deterministic Actions: $T : S \times A \longrightarrow A$ We define a new state for each state and action.

Stochastic Actions: $T : S \times A \longrightarrow prob(A)$ We define a probability distribution $P(s'|s, a)$ for the next states given any state and action.

4.6 Probability

$p_t : S_t \times a_t \longrightarrow P(s_{t+1})$ is the probability that the agent enters s_{t+1} when it performs the action at the state s_t. Because we propose a model-free deep reinforcement learning technique, the agent is unaware of this probability.

4.7 Reward

r_t denotes the reward obtained by the agent while doing the action a_t at the state s_t. This function has to be defined in such a way that we punish the agent for taking unintended actions and reward it for taking the intended ones. This is the key for the reinforcement learning agent to learn.

4.8 Goal of the Reinforcement Learning Agent

We have 2 possible states that our system can be in:

1. Normal functioning (S_n)
2. Under FDI Attack (S_a)

We have 2 possible actions our RL agent can take:

1. Continue the normal functioning of the grid (Do not stop the simulation) (A_c)
2. Stop the simulation (A_s)

Our objective is to create an agent to identify the attacks as soon as they begin (not sooner, not later) in order to avert severe grid damage. We have four different possibilities as a result of our agents' actions. They are as follows:

1. The simulation is terminated by the agent before the attack occurs.
2. After the attack starts, the agent terminates the functioning of the simulation.
3. After the attack starts, the agent does not halt the functioning of the simulation.
4. The simulation is not terminated by the agent before the attack occurs.

In the above four outcomes, only 2 and 4 are desired, whereas actions 1 and 3 are unintended.

5 Case Study on IEEE-3, IEEE-9, IEEE-14 and IEEE-30 Bus System

The evaluation will be will be carried out on the three bus example grids defined at [14] and some of the standard IEEE grids like the IEEE 9 bus, IEEE 14 bus and the IEEE 30 bus systems. The current system state vector (voltage magnitudes and phase angles) is calculated using the State Estimation functions available in the PANDAPOWER liberty of python. We run the simulation initializing the networks to the default values of the simple 3 bus example grid, IEEE 9 bus, IEEE 14 bus and the IEEE 30 bus systems and run it for a certain number of steps in each episode. We start the FDI attack randomly at any step in an episode and continue till a random amount of time. The goal is to teach our model to stop the episode as soon as the attack begins (Not later, Not earlier) (Figs. 6 and 7).

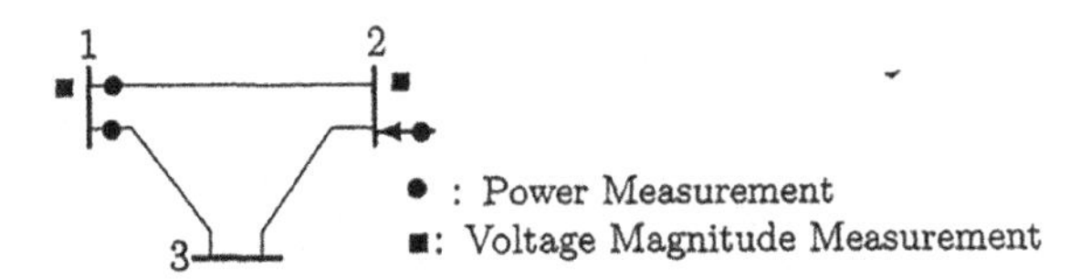

Fig. 6. 3 bus grid

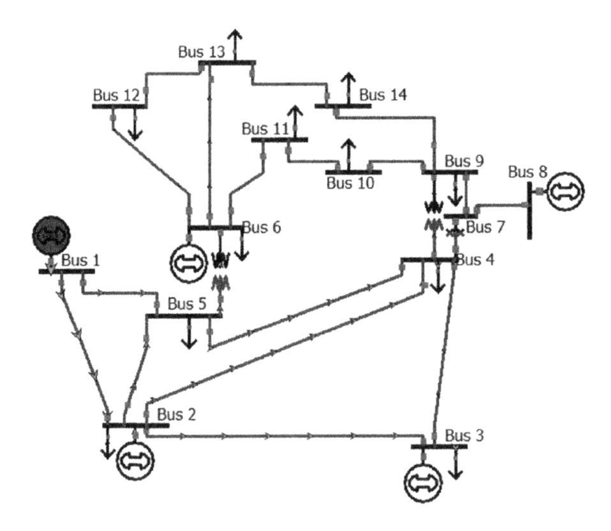

Fig. 7. IEEE 14 bus grid network

5.1 Simulating Attacks

We simulate the attack by dividing the simulation into episodes. In each episode that spans 500 timesteps, the attack starts at a random time step(t). We train our Reinforcement Learning model on several episodes and use a predetermined reward system to compute the reward and take the appropriate action.

5.2 Predict Action

In each episode at each step, the neural network model receives the state value and predicts two values, the estimated reward for each of the actions (stop or continue) for that given state. Using this prediction, we take the action with the maximum reward and proceed to the next state. The reward system ensures that the model is punished for taking the wrong actions at the right time.

5.3 Reward System

For each of the above 4 consequences of our agent's action, we reward it in such a way that we punish the unintended consequences and reward the intended ones. Suppose the current state is defined as St and the Current action as At, we can define a possible reward policy as:

$$Reward_t = C_1, \text{ if } S_t = S_a \text{ and } A_t = A_s \quad (1)$$

$$Reward_t = -k_1 * (t - start), \text{ if } S_t = S_a \text{ and } A_t = A_c \quad (2)$$

$$Reward_t = C_2, \text{ if } S_t = S_n \text{ and } A_t = A_c \quad (3)$$

$$Reward_t = -k_2 * (k_3 - ||noise||), \text{ if } S_t = S_n \text{ and } A_t = A_s \quad (4)$$

Where C_1 and C_2 can be small positive values to ensure positive reward, k_1, k_2 and k_3 are constants that we can fine-tune states to improve performance. The start is the timestep when the attack begins. Equation (2) is the reward when the agent fails to stop the grid while the attack is happening. In this case, the reward is based on the time elapsed since the attack began (t-start). Equation (4) is the reward when the agent stops the grid when there is no attack (False positive). We want to punish this consequence and thus the reward can be a huge negative value when this happens. We use the mean of the noise vector as the reward term at that state to incorporate it into the agent's learning process and impact its decision when a similar observation occurs at a future point in time. Hence, this reward policy should theoretically ensure that our agent learns to avoid unintended actions.

6 Results and Discussions

The sections explains the result that has been achieved after applying the proposed model on IEEE-3 bus and IEEE-14 bus system. To understand that the model is learning to do the intended task, we can use the plots of the value of the loss to check if it is converging to ensure this. We show the plot of the loss function as the training progresses as in Fig. 8:

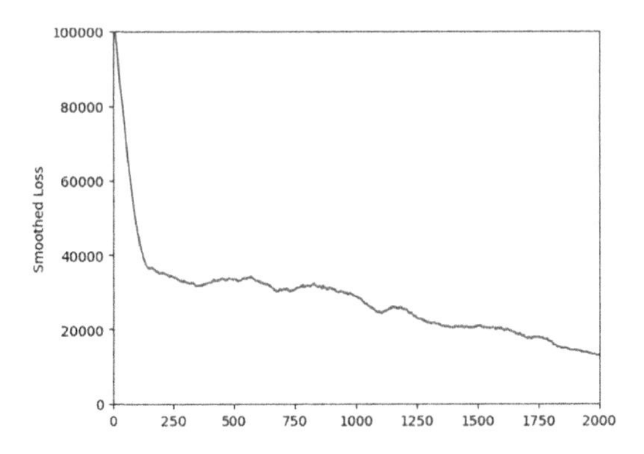

Fig. 8. Loss curve

6.1 Evaluation Metrics

We present several evaluation criteria to assess the performance of our proposed RL agent in real-time. The following are the evaluation metrics:

1. Perfect Calls Percentage
2. Good Calls Percentage
3. Late Calls rate
4. False Alarm rate
5. Detection Failure Percentage

1. **Perfect Calls Percentage:** The number of attacks that our RL agent can detect as soon as they start. Let the number of episodes where the attack start time step is the same as the attack detection time step be N_p and the total number of episodes be T. This can be computed by the equation:

$$Perfect\ Calls\ Percentage = \frac{N_p}{T} \times 100$$

2. **Good Calls Percentage:** We can define good calls as detecting attacks before a certain number of time steps after it starts. We can call this threshold as. Let the number of episodes where the attack detection is within time steps after it starts to be N_g and the total number of episodes be T. This can be computed by the equation:

$$Good\ Calls\ Percentage = \frac{N_g}{T} \times 100$$

3. **Delayed Calls rate:** We can define Delayed calls as detecting the attack anytime after the attack begins. Let start be the time step when the attack has begun and t be the time step when the agent detects and the total number of episodes be T. For all the episodes where start t this can be computed by the equation:

$$Delayed\ Calls\ rate = \frac{\sum |t - start|}{T}$$

4. **False Alarm rate:** False Alarms are the calls that occur before the attack begins. We would want to avoid these as much as possible to avoid disruptions in power supplies. Let start be the time step when the attack has begun and t be the time step when the agent detects and the total number of episodes be T. For all the episodes where start t this can be computed by the equation:

$$False\ Alarms\ rate = \frac{\sum |start - t|}{T}$$

5. **Detection Failure Percentage:** When our agent is unable to detect that the attack occurred by the end of the episode, we call this a detection failure. Let the number of episodes this happens to be called as Nd fand the total episodes be T. This can be computed by the equation:

$$Detection\ Failure\ Percentage = \frac{N_{df}}{T} \times 100$$

6.2 Evaluation Results

We use the proposed evaluation metrics for the different models trained for the different power grid systems to estimate the performance of our Reinforcement Learning Agents. Here are the findings for the test conducted for 100 episodes for the 2 grids. The threshold for good calls was set to 10 time steps after the attack began (Table 1).

Table 1. Evaluation results

Bus type	Perfect calls percentage	Good calls percentage	Delayed calls rate	False alarms rate	Detection failure percentage
3 Bus Grid	82	97	0.74	1.87	0
IEEE 9 Bus	95	97	0.01	0.50	0
IEEE 14 Bus	89	100	0.53	0.00	0
IEEE 30 Bus	93	93	0.00	4.74	0

7 Conclusion

Through this word, we explore that there are many problems associated with the security of critical infrastructure, especially the smart grids. We have seen that it is of utmost importance to ensure the smart grid networks has comprehensive security measures in place to avoid the possible magnitude of damage that can be caused by any cyber or physical attacks on them. Since is it very hard to employ a monolithic solution that takes care of all the possible threats, we focus on the individual attacks and propose comprehensive detection and recovery mechanism for them. Hence, in this word, we focus primarily on the False Data Injection attack. We explore the possibility of using a Deep Reinforcement Learning agent to detect and act against this attack. In this work, we have proposed a Deep reinforcement learning based approach for responding to the FDI attacks. We explored methods like target network and experience replay on top of regular Deep reinforcement learning algorithm to improve the stability of training and speed up the convergence. The evaluation has been done on 4 simulated smart grids (simple 3 bus grid, IEEE 9 bus, IEEE 14 bus and the IEEE 30 bus).

We have a separate model (RL agent) to detect attacks for each of these grids. However, it would be much more intuitive and convenient to have a single model (RL agent) that detects the attack on any given smart grid network configuration. This could be an interesting problem that can be explored in the future. Furthermore, an as a future scope of this work, an attack recovery mechanism combined with the proposed work would make it a comprehensive strategy that follows NIST guidelines framework for FDI attack detection.

8 Compliance with Ethical Standards

8.1 Funding

The work is funded by Department of Science and Technology(DST), India for the Cyber Physical Security in Energy Infrastructure for Smart Cities(CPSEC) project under Smart Environments theme of Indo-Norwegian Call.

8.2 Conflict of Interest

All authors declares that they has no conflict of interest for the presented work.

8.3 Ethical Approval

This article does not contain any studies performed by any of the authors with human participants or animals.

References

1. Smart Grid Coordination Group: Smart Grid Reference Architecture. Technical report, CEN-CENELEC-ETSI (2012)
2. Baumeister, T.: Literature Review on Smart Grid Cyber Security (2016)
3. Snort. https://www.snort.org/
4. Prelude OSS. https://www.prelude-siem.org/
5. National Institute of Standards and Technology: Framework for Improving Critical Infrastructure Cybersecurity (2018)
6. Forouzan, B.: Cryptography and Network Security (2011)
7. Ullah, I., et al.: An intrusion detection framework for the smart grid. In: IEEE 30th Canadian Conference on Electrical and Computer Engineering (CCECE) (2017)
8. Lin, Y., et al.: Tri-level optimal hardening plan for a resilient distribution system considering reconfiguration and DG islanding. Appl. Energy **210**, 1266–1279 (2018)

9. Wang, W., Lu, Z.: Cyber security in the smart grid: survey and challenges. Comput. Netw. **57**, 1344–1371 (2013). https://doi.org/10.1016/j.comnet.2012.12.017
10. National Security Council Secretariat, New Delhi: Cyber Security Requirement for Smart City - Model Framework (2016)
11. Mo, Y., et al.: Cyber-physical security of a smart grid infrastructure. Proc. IEEE **100**(1), 195–209 (2012)
12. NIST: NISTIR 7628 Revision 1 Guidelines for Smart Grid, Cybersecurity Volume 1 - Smart Grid Cybersecurity Strategy, Architecture, and High-Level Requirements (2014)
13. Shukla, S., et al.: Machine learning in cybersecurity: a review. Wiley (2019)
14. Xue, D., et al.: Detection of False Data Injection Attacks in Smart Grid Utilizing ELM-Based ECON Framework (2019)
15. Emmanuel, M., et al.: Communication architecture for smart grid applications. In: IEEE Symposium on Computers and Communications (ISCC) (2018)
16. Kundur, D., et al.: Towards modelling the impact of cyber attacks on a smart grid. Int. J. Secur. Netw. **6**(1), 2–13 (2011)
17. Davis, K., et al.: A cyber-physical modeling and assessment framework for power grid infrastructures. IEEE Trans. Smart Grid **6**(5), 2464–2475 (2015)
18. Zonouz, S., et al.: SOCCA: a security-oriented cyber-physical contingency analysis in power infrastructures. IEEE Trans. Smart Grid **5**(1), 3–13 (2014)
19. Aboelwafa, M.M., et al.: A machine learning-based technique for false data injection attacks detection in industrial IoT. IEEE Internet Things J. **7**(9), 8462–8471 (2020). https://doi.org/10.1109/JIOT.2020.2991693
20. Abur, A., Expósito, A.G.: Power System State Estimation: Theory and Implementation. Marcel Dekker, New York (2004)
21. Cao, Y., Gao, Y., Tan, R., Han, Q., Liu, Z.: Understanding internet DDoS mitigation from academic and industrial perspectives. IEEE Access **6**, 66641–66648 (2018)
22. Konstantinou, C., Maniatakos, M.: A case study on implementing false data injection attacks against nonlinear state estimation, pp. 81–92 (2016). https://doi.org/10.1145/2994487.2994491
23. Cloudflare: What is a DDoS attack? https://www.cloudflare.com/learning/ddos/what-is-a-ddos-attack/
24. Kallitsis, M.G., Bhattacharya, S., Michailidis, G.: Detection of false data injection attacks in smart grids based on forecasts. In: 2018 IEEE International Conference on Communications, Control, and Computing Technologies for Smart Grids (SmartGridComm), pp. 1–7 (2018). https://doi.org/10.1109/SmartGridComm.2018.8587473
25. Yang, M.-H., Luo, J.-N., Vijayalakshmi, M., Shalinie, S.M.: Hybrid multilayer network traceback to the real sources of attack devices. IEEE Access **8**, 201087–201097 (2020). https://doi.org/10.1109/ACCESS.2020.3034226
26. Wang, Y., Amin, M.M., Fu, J., Moussa, H.B.: A novel data analytical approach for false data injection cyber-physical attack mitigation in smart grids. IEEE Access **5**, 26022–26033 (2017). https://doi.org/10.1109/ACCESS.2017.2769099
27. Murugesan, V., Selvaraj, M.S., Yang, M.-H.: HPSIPT: a high-precision single-packet IP traceback scheme. Comput. Netw. **143**, 275–288 (2018). https://doi.org/10.1016/j.comnet.2018.07.013. ISSN 1389-1286

Fake News Detection Using Ethereum Blockchain

Akanksha Upadhyay(✉) and Gaurav Baranwal

Department of Computer Science, Banaras Hindu University, Varanasi, Uttar Pradesh, India
akankshaupadhyay2016@gmail.com, gaurav.baranwal@bhu.ac.in

Abstract. The progression of technology, Internet availability, and rapid adoption of social media unintentionally paved the way for the viral spread of fake news and hoaxes. Around the world, people are using fake news for their profit either by presenting information in the wrong manner for gaining more viewers or by manipulating the publics' opinion for influencing their behavior in major events of society, for example, elections or view towards someone's part in an unfortunate event. In this paper, we present a blockchain-based platform where one can get the news contents authenticated by anonymous individuals around the globe independently as genuine or fake. It gives people a secure and anonymous channel for verifying content's reliability and further limiting the reckless sharing and spreading of news content by people. Inherited properties of blockchain by the platform, i.e., immutability and traceability, combat the spread of fake news.

Keywords: Blockchain · Fake news detection · Decentralized · Immutability

1 Introduction

The Internet's inception and availability unintentionally aided in the viral spread of fake news and hoaxes. The dissemination of misinformation being witnessed right now is a consequence of the uncontrolled usage of these platforms. It enlightens the limitations of the currently working systems and the loopholes of the administration controlling them. Also, the progressions and revolutions being made in technology have unintentionally opened many doors to increase this spread of misinformation. With this rapid spread, it is impossible to manually assess all the information being shared on digital platforms, as multiple aspects must be explored before judging the reliability and integrity of an article and its source.

Various solutions involving artificial intelligence and machine learning methods that involve natural language processing, ensemble methods, classification, etc., have been proposed to tackle this problem. Despite all these efforts, the system fails to stop creating and spreading fake news due to corrupt systems, authorities, and centralized databases. These systems are limited to risking over-blocking, prone to false positives or false negatives, and fail to catch emotional concepts. Also, the variations in language, culture, and political behavior add up to this challenge.

I. Woungang et al. (Eds.): ANTIC 2021, CCIS 1534, pp. 142–152, 2022.
https://doi.org/10.1007/978-3-030-96040-7_11

One more approach that is still under study and is fairly a new area of research to be explored for this specific issue is the blockchain. The characteristics of blockchain can help in proving the authenticity and reliability of media content. Though blockchain may not necessarily prevent people from posting misinformation, it can foster trust in the viewers by making it a lot easier to track and verify the information. In this paper, blockchain technology-based solutions to combat fake news content issues over the internet are explored. We provide a platform where one can get the news contents authenticated by anonymous individuals independently as genuine or fake, giving people a secure and anonymous channel for verifying content's reliability and further limiting the reckless sharing and spreading of news content by people. The platform also allows the users to help in verifying the authenticity of news content giving their bid to improve society.

2 Literature Review

A significant number of initiatives aiming to counter the broadcasting of fake information have been developed worldwide. Most of the work focusing on detecting fake news stories has concluded social media as the primary cause of spread. Shu et al. 2017 [1], in a comprehensive study of fake news detection from the perspective of data mining, explored the problem in two phases' viz. characterization and detection. Karimi et al. 2018 [2] combined multiple sources of information to distinguish the degrees of fakeness, and further proposed a framework named, Multi-source Multi-class Fake news Detection framework MMFD. With the introduction of blockchain, a more effective approach can be developed. Chen et al. 2020 [3] developed a method for preventing fake news using Proof of Authority consensus focusing on leveraging blockchain to enroll and monitor news organizations. Fraga-Lamas et al. 2020 [4] recommended crowdfunding utilizing tokens to stimulate the investigation of fraudulent media making users participate in peer-to-peer transactions without a third party. Shang et al. 2018 [5] used a ledger of timestamps and connections between distinct blocks to track the provenance of information. But very few works used blockchain to detect fake news. This paper presents an Ethereum blockchain-based platform where users can share news content and get their content verified by an anonymous and unrelated group of people in simple steps. This paper is different from other papers using blockchain for solving fake news problems in many different ways. This paper is based on the Ethereum blockchain which is the most popular blockchain for decentralized applications. Ether is the second-highest cryptocurrency, enhancing Ethereum's market value which will provide us a greater reach and more users base once the application is implemented on Ethereum Mainnet. Ethereum has a huge team working on it, making it more trustworthy, giving long life and greater assurance of quality improvement. This paper also presents a working prototype. Whereas many previous papers like Chen et al. 2020 talk about designing a dedicated blockchain with a new consensus algorithm which has its disadvantages of a smaller user base and the time it will take to ultimately make a working model for it.

2.1 Fake News

Fake news is false or misleading information represented as news [6]. The objective of such content is to damage the reputation of an entity or a person or to gain money with the help of advertising revenue. The idea of fake news is not new and has been present there since long ago, but it has achieved a rapid increase with the rise of online platforms and the availability of the internet [7]. Fake news has produced conflicts and confusion and has reduced the trust and reliability of the real news. Social media algorithms are also one of the causes of the spreading of fake news.

2.2 Blockchain

A blockchain is a continuously increasing stack of blocks, each block containing a cryptographic hash of the preceding one, the transaction data, and a time-stamp [8]. This technology assures data integrity once the transaction data is stored in a blockchain. In the blockchain, decentralized digital ledger stores transactions over multiple computers in blocks. No single block can be altered separately without any modification in all other blocks, allowing the participants in independent and comparatively cheap verification and audit process of transactions. Typically managed by a peer-to-peer network, the nodes of blockchains mutually adhere to a protocol for communication and validation of new blocks. Centralized systems have limitations like the single point of failure or corruption leading to total shutdowns. The decentralization feature of blockchain eliminates this limitation as the data is stored in each node of the blockchain and also the public-key cryptography makes the data incorruptible. There are various time-stamping schemes, like proof-of-work, or proof-of-stake, used by blockchains to serialize changes.

2.3 Proposed Work

This paper presents a blockchain-based platform where users can share news content that a panel of auditors can review for its authenticity verification. Auditors in this panel are users who have earned a good credit score by giving authentic feedback over time. This paper presents a distributed platform for sharing and reviewing news content based on blockchain technology using smart contracts eliminating third-party interaction instead of centralized client-server technology.

Participating Entities

The system involves three different types of participating entities. Their roles are discussed as follows:

- *Customers*: Customers can be anyone and have the ability to post content.
- *Reviewers*: Any Customer that gives feedback to content is marked as a Reviewer.
- *Validators*: Reviewers can be further designated as Validators by smart contract without human interference if their credibility score based on their past activities exceeds a defined threshold.

Workflow of System

As shown in Fig. 1, the workflow of the system can be described as follow,

1. Once a customer posts news content (in the form of image and description), that content goes to InterPlanetary File System(IPFS) which stores that file providing a unique hash for that file.
2. As saving content on Ethereum blockchain is costly, IPFS generated hash is saved on Ethereum blockchain instead of the large content file.
3. This decentralized application(dApp) takes hash from Ethereum blockchain and access the content from IPFS using this hash.
4. Users can see all the uploaded content initially with two feedback buttons: Real and Fake.
5. If users want, they can give their feedback about the post, which will be saved along with their addresses for future reference, and these users will be marked as Reviewers.
6. Addresses marked as Validators are the credible; the result for a certain post is decided based upon their vote.
7. Initially, some trusted memberswith a score more than threshold value will be added to function as Validator. As the users connect to the dApp, there will be more Validators making system more and more robust.
8. Once required numbers of Validators have voted for a post, no more voting for that post will be accepted. Buttons for voting will be replaced by a result button.
9. The result is decided on the majority of Validators votes and is displayed alongside the content.
10. A threshold is defined to designate a customer as a Validator or Reviewer.
11. When the result is declared for a post, the rating of the customers who voted on that post is updated according to the vote they gave for that certain post.
12. If anyCstomer's score C_{Score}crosses a certain threshold,in future their vote will be counted as Validator's vote V_{val} as they have given authentic feedback for a long time to cross that threshold. Similarly, if a Customer's score whose vote was earlier counted as Validator's vote goes below the threshold, then their vote will be saved as Reviewer's vote V_{Rew} and will not be counted in result calculation for future posts until they regain the authority of being a Validator.

Apart from this, no address can vote twice and once the result is announced, feedback buttons will be replaced with the Result button.

Algorithm for Voting

Let Customer C votes vote V on a post P.

a) If C already Voted for P
   a. Reject V
b) Else
   a. If No. of Validator Votes V_{Val}< Required Validator Votes (Voting Allowed)
      i) If Customer Score C_{Score}> Threshold, t (C is Validator)
         Add V to Validators Votes V_{val}
         (a) If V == Real
            (i) Increment P_{Real}i.e., Total Votes claiming P as Real.
         (b) Else
            (i) Increment P_{Fake}i.e., Total Votes claiming P as Fake.
      ii) Else (C is Reviewer)
         (a) Add V to Reviewer Votes $V_{Rev.}$
   b. If No. of Validator Votes V_{Val}= Required Validator Votes
      i) Voting Not Allowed

Algorithm for Declaring Result

Let Post P have No. of Validator Votes V_{val} = Required Validator Votes

a) If P_{Real}>P_{Fake}
   a. Result for Post P = Real
b) Else
   a. Result = Fake

Algorithm for Updating Score of Costumer

After Result is calculated for Post P,
   For all Costumers C, who voted V on Post P,
   a) If V == Result
      a. Increment C_{Score}
   b) Else
      a. Decrement C_{Score}

All these calculations are done simultaneously by Smart Contract.

Design Architecture

The proposed system is a decentralized platform for news/information sharing and verification. The system is designed for the Ethereumblockchain. The back-end is built using truffle and node.js. A smart contract in solidity was written in which all the functioning was specified, which will send data to the blockchain and emits events in response to transactions. This smart contract was then deployed on a local blockchain using Ganache. Keeping the functionality in mind, the front-end was designed. A truffle template was used for designing the front-end. React and JavaScript is used majorly in designing of front-end, and the web3 JavaScript library is used for communicating between the

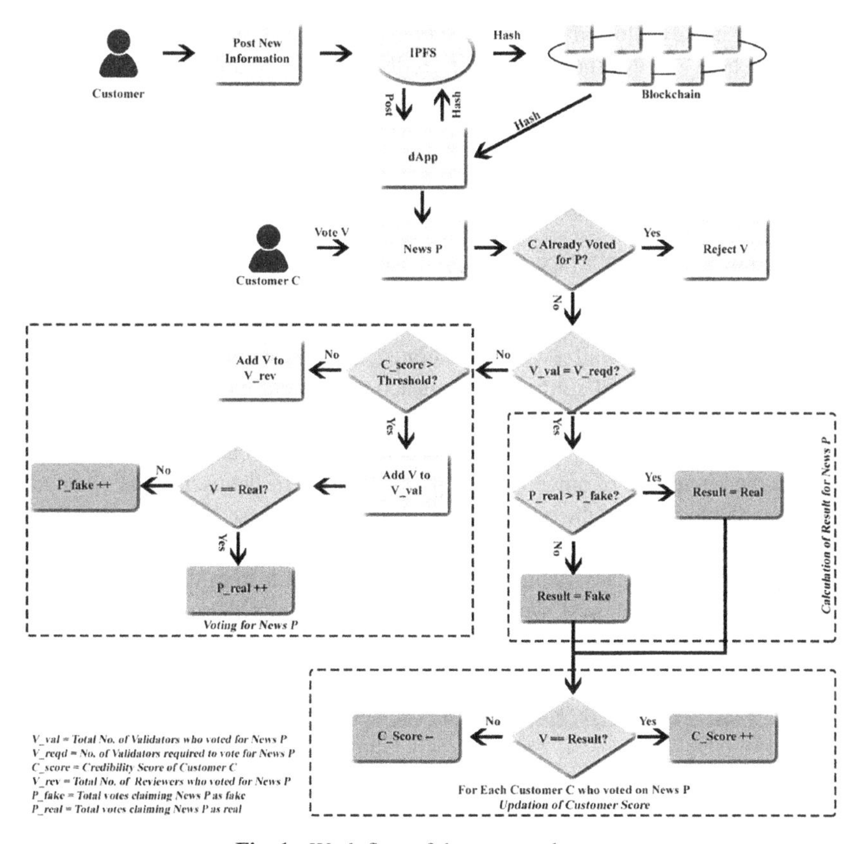

Fig. 1. Work flow of the proposed system

front-end and back-end. The Web3.js library instantiated the smart contract inside the front-end JS file. IPFS is also used for storing content that was accessed by Infura's gateway. Using the web3.js library, the front-end was made accessible via Metamask Chrome Extension, which users need to use to interact with dApp using a Chrome web browser.

3 Result

The proposed system is a functional dApp running over a local Ethereumblockchain.This system includes a simple user interface containing the following features and functionalities:

1. A Form at the top of the page for users to submit their content.
2. Display of content sorted according to time of submission.
3. A navigation bar at the top displaying the user's public key with an icon associated with it.
4. Each post is displayed with buttons for voting them real or fake.

5. Every user is only allowed once to vote on a particular post.
6. After reaching the threshold value for feedbacks from validators, a button for the display of results is visible in place of voting buttons.
7. The result is displayed along with posts once declared.

3.1 User Interface of Proposed System

Figure 2 shows the user interface of dApp it contains a form for submission of news content with an image and written description related to it on top. Content submission is very simple and easy. Users only have to choose a file and write a description and upload it. A pop-up of Metamask wallet will be generated, asking for confirmation for spending ether, and once the user will confirm the transaction, content will be uploaded. Next to it is the latest uploaded news content. Any user connected to the distributed network, which can be done simply by logging into Metamask wallet, can scroll and read all of the uploaded content free of any cost. The public address of the user who uploaded the news is also visible at top of the post. The two voting buttons viz. "FAKE" and "REAL" for recording the user's vote on that post are present at the bottom of the post. A user who wants to give feedback has only had to click any one button from the two given below with that a pop-up of Metamask will be generated, and after confirming the transaction the feedback is recorded.

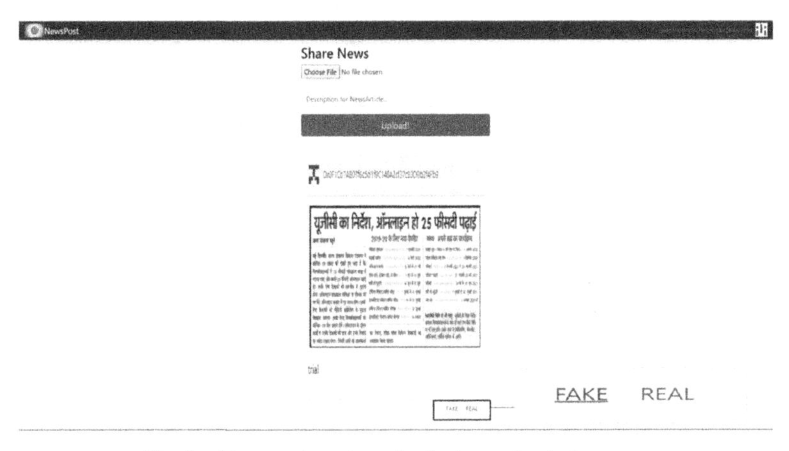

Fig. 2. The user interface displaying uploaded content

The User Interface displays the User's Public Address (Fig. 3(a)) on the top of the screen, and on top of each news, the public address of the user who uploaded it is displayed (Fig. 3(b)). Figure 3(c) shows the result button in the bottom right with the displayed result for that post once result is declared. Buttons for feedback will automatically be removed from the post once the result is declared.

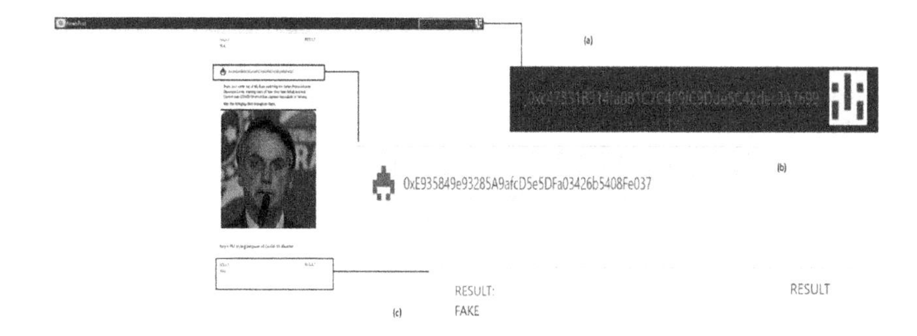

Fig. 3. (a) Address of user (b) Address of the Source of Post (c) Result Declared

Once a user is connected to decentralize web using their Metamask wallet the user interface of Dapp is simple and handy for any user.

3.2 Result Analysis

The result of the proposed paper is a prototype for a decentralized application working on the Ethereum blockchain. The resultant system has a simple interface. The Proposed model uses an Ethereum based wallet which makes it easily accessible. This system can run on any normal desktop or laptop. The user does not need any special high power machine, also making it easy to use as once the user is connected to the application with a fewer number of steps to be performed.

Privacy Analysis
Ethereum provides pseudonymity to its users, making it difficult to find the person's real identity behind the screen. This feature of Ethereum is helping our model to keep users' real identity hidden. Users can only be identified by their public address, which states nothing about their real-world identity. Since Ethereum is a public blockchain, all transactions made by any public address are recorded on the chain, which anyone can see. With effort, a pattern can be found out about users' posting and reacting habits, but that does not lead to the user's real-life identity. This feature allows users to post and react without any fear of judgment and hatred, and online bullying, as online bullying is a big issue nowadays in the centralized social media system. Also, since it's pseudonymous in case of serious threat or illegal activity with the help of traceability, authorities can try to find out IP address locating the user's physical address.

Security Analysis
Anything in the cyber world is prone to attacks. Smart contracts can also be attacked. Keeping this in mind this system has been designed to avoid all these known attacks.

Reentrancy: When a smart contract calls an external contract, the external contract can take over and change the data unexpected by the calling contract.

As in this model, we are not calling any other smart contract, we have prevented our system from reentrancy attack. There is no time stamp dependence in our contract, neutralizing it against any manipulation of timestamp by miners. Front running is also an issue in Ethereum, but our system does not benefit anyone just by having any information in advance, again neutralizing the effect of the front running attack. While implementing the system on Ethereum Mainnet problem of integer overflow might occur after prolonged use of application for which slight modification will be beneficial and probably easy in the future as Ethereum is also working to provide better solutions for these issues.

Local v/s Real-Time Implementation
The proposed system is a prototype running on local Ethereum blockchain provided by Ganache which brings certain limitations such as in Ganache, ten different accounts were provided on one blockchain. Therefore, the number of participants, including validators and all other users, was limited, leading to a limited number of validators validating the news content. As greater the number of validators will be, the more secure and bias-free network will be, therefore while implementing the system on Ethereum Mainnet, slight modification and verification for scalability will be needed. As discussed in security analysis problem of integer overflow may occur in the present system after prolonged use of the application. In the real-time implementation, the system will be free from downtime situations due to a large number of nodes working on Ethereum.

4 Advantages of Proposed System

The proposed system acts as a self-regulating platform with equal rules for everyone and is free from any third-party intervention. The developed model is for tackling fake news issues without limiting people's right to free expression and without making boundaries thrashing the global village concept. Primarily, the system along with the information provides whether the information is genuine or fake, which in the current scenario of uncontrolled flow of information is very necessary. The platform presented in this paper will stop the spread of fake news and save the time of users in today's busy world. Apart from that, anyone can also use this platform as an application to verify the authenticity of some news received, just by uploading it to the platform. The Validators available on the platform will verify it for you, keeping you safe from false content. Ethical journalists can also use this platform to share information they can't share anywhere else due to the fear of getting blackmailed or threatened as it provides anonymity. The general public can also use this platform for putting forward complaints about institutions or people in power. It can gain others' support as well, all staying anonymous making it safe for everyone. Since the validators are random, widespread people who don't know each other, the chances of bias and manipulation will also decrease tremendously. And since there is no central authority to control the system, no one can promote or censor content available on this platform to manipulate the contents' visibility.

Being based on Ethereum instead of any other blockchain gives the system an advantage. Ethereum is the largest platform for Dapp nowadays. In the cryptocurrency market, Ethereum is in the second position, giving it a huge user base worldwide, which guarantees us quality, future success, and greater security with long life for application.

5 Limitations of Proposed Work

The problem of fake news spread is a multileveled issue.The most obvious limitation of this model is its reach to the general public as the blockchain based dApps are not that user-friendly in comparison to the traditional social media apps.

The primary limitation of this prototype is, it's based on the goodwill of people who want to give time as validators and help in the verification of content. Secondly, since this prototype is running on Ethereum local blockchain Ganache which only provides ten accounts for a blockchain, i.e. there is a limitation on number of validators required for validating the news. Small modifications can work for accommodating a large number of participants while implementing this model in the real world on main net Ethereum.

6 Conclusion and Future Work

Fake new circulation is a major problem of the current time. It has become really easy to create and circulate false information without getting caught. Blockchain is an emerging technology with features like traceability, transparency, and immutability, making it a very good candidate for the detecting and preventing fake news circulation. This paper presented a distributed application for users to help find out if the posted news is real or fake. All information is stored on the blockchain, making it traceable, immutable, and transparent and eliminating the idea of central authority that might influence the results.

Blockchain is a new emerging technology with lots of potentials. In this paper, we used this potential to solve major issues of fake news detection providing a simple approach. Although apart from existing features there is a lot of scope of further enhancement of this system, some of them are included below:

- Monetary incentives can be added to motivate users for being more responsible.
- This system can also be adopted in other fields as per for registering complaints and supporting issues anonymously without any interference of a central authority.
- A dedicated blockchain can be designed for this system making it more resilient.
- This system using blockchain can also be integrated with existing social media platforms running over client-server technology, making this widely applicable.
- There is always a scope for more user-friendly features for better user satisfaction.

References

1. Shu, K., Sliva, A., Wang, S., Tang, J., Liu, H.: Fake news detection on social media: a data mining perspective. ACM SIGKDD Explor. Newsl. **19**(1), 22–36 (2017)
2. Karimi, H., Roy, P., Saba-Sadiya, S., Tang, J.: Multi-source multi-class fake news detection. In Proceedings of the 27th international conference on computational linguistics, pp. 1546–1557 (2018)
3. Chen, Q., Srivastava, G., Parizi, R.M., Aloqaily, M., Al Ridhawi, I.: An incentive-aware blockchain-based solution for internet of fake media things. Infor. Proc. Manag. **57**(6), 102370 (2020)

4. Fraga-Lamas, P., Fernández-Caramés, T.M.: Fake news, disinformation, and deepfakes: leveraging distributed ledger technologies and blockchain to combat digital deception and counterfeit reality. IT Prof. **22**(2), 53–59 (2020)
5. Shang, W., Liu, M., Lin, W., Jia, M.: Tracing the source of news based on blockchain. In 2018 IEEE/ACIS 17th International Conference on Computer and Information Science (ICIS), pp. 377–381. IEEE (2018)
6. Tufekci, Z.: It's the (Democracy-Poisoning) Golden Age of Free Speech. Wired. https://www.wired.com/story/free-speech-issue-tech-turmoil-new-censorship/ (2018)
7. Guardian News and Media.: What is Fake News? How to Spot it and What You Can Do to Stop It. The Guardian. https://www.theguardian.com/media/2016/dec/18/what-is-fake-news-pizzagate (2016)
8. Narayanan, A., Bonneau, J., Felten, E., Miller, A., Goldfeder, S.: Bitcoin and Cryptocurrency Technologies: A Comprehensive Introduction. Princeton University Press (2016)

Multi-criteria Based Cloud Service Selection Model Using Fuzzy Logic for QoS

Mohammad Faiz(✉) and A. K. Daniel

MMM University of Technology, Gorakhpur, U.P., India

Abstract. Due to the expansion of web users and increase demand for Cloud in the public domain, a cloud service provider (CSP) tries to fulfill the exclusive functionalities and specifications required by different users which is a challenging task. There are numerous challenges such as migration of all the applications and data from one Cloud to another can drop-off the computing capabilities offered by the Cloud platform because different Cloud service providers (CSP) provides cloud services at different pricing categories with a different set of choices where one cloud charge low cost for storage but high cost for computation or vice-versa. In this paper, a Fuzzy based Cloud selection model is proposed using cost, capacity, performance, security and maintenance as key parameters to improve QoS experience of the cloud users.

Keywords: Fuzzy cloud · QoS · Ranking · Multi-criteria

1 Introduction

Cloud Computing technology is one of the most important inventions of the current decade because it has increased computing capabilities to a great extent. Clouds allow companies to target their base businesses, instead of worrying about computing technology and maintenance of resources. Cloud computing helps companies to upgrade their software more efficiently, with better storage, less maintenance, and resources to meet changeable and unpredictable demands [1]. Cloud offers robust, versatile, reliable, elastic cloud services to the users. Cloud users can choose from a variety of services such as Software-as-a-Service (SaaS), Platform-as-a-Service (PaaS), and Infrastructure-as-a-Service (IaaS) to upgrade their system's performance such as storage, speed, availability, durability, etc. In traditional computation systems, the installation and deployment of computing devices are much expensive and not easily available to individual users. In Cloud computing, the users have to pay user charges based on the duration they used the cloud service. The rapid increase in the number of web users more CSPs is available in the market that leads a tuff competition among CSP. It is a challenging task for a CSP to provide Quality oriented services to the users in the present dynamic compelled environment and for the users to pick the best suited cloud that fits to satisfy their demand for Cloud computing [1, 2]. Various types of cloud service models with their popular cloud services are shown in Fig. 1.

I. Woungang et al. (Eds.): ANTIC 2021, CCIS 1534, pp. 153–167, 2022.
https://doi.org/10.1007/978-3-030-96040-7_12

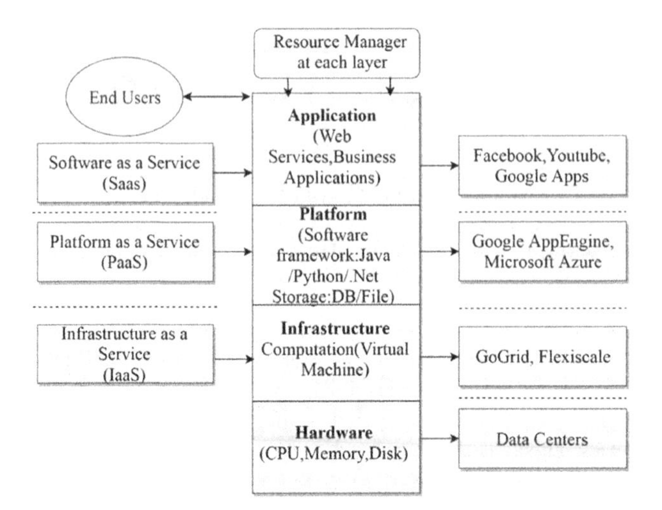

Fig. 1. Cloud computing services

Cloud computing offers services to users to use different applications such as Facebook, Google.etc. The users can build their robust and versatile softwares using PaaS cloud services such as Microsoft Azure. Cloud even offers the infrastructure to create users own virtual machine with desired hardware (C.P.U, memory, disk) and software configurations [3].

The paper is divided into the following sections: Sect. 2 parameters in cloud resource ranking, Sect. 3 related work, Sect. 4 proposed fuzzy cloud selection model, Sect. 5 simulation, and Sect. 6 conclusion.

2 Parameters in Cloud Resource Ranking

There are various parameters for ranking the cloud resource are as follows:

I. **Availability**
It is the amount of time that a service is in the operating state. It is also represented by the extent to which the cloud service is in the specified functional and committable state [3].

II. **Reliability**
Reliability is the metric for evaluating a cloud service performance in compliance with its Service Level Agreement (SLA) requirements. It represents the operating behavior of service without any failure in a specified time and condition [4].

III. **Stability**
Stability is defined by variations in the efficiency of the provided software instances or variations in the performance of the provided platform.

IV. **Response time**
The time span between receiving requests for cloud service and the beginning of transmitting response messages. The response is represented by the time taken to

create a virtual machine instance + time taken to initialize virtual machine + time of reply of a request [5].

V. **Flexibility**
It refers to the adoption of the design when there is any variation occurs internally or externally. Dynamic discovery and Dynamic adoption are the two things that are applied to cloud services. Dynamic discovery refers to the effectiveness and accuracy of cloud service given at the run time for needed service specification. Dynamic adoption (DA) refers to the ability of the CSP to update the modifications of services on user requests.

VI. **Scalability**
Scalability is referred as the ability of the cloud to handle a large number of simultaneous requests. It is also referred as the expanding ability of cloud-based service during large request times. It denotes the scalable nature of cloud software instances, the scalable nature of the cloud platform instances and the virtual machine/host/storage instances scalability.

VII. **Transparency**
It is indicated by the degree to which Cloud computing usability is exaggerated with any kind of changes in the service. It can also be determined for the rate at which these effects occur. It is denoted by the transparency in cloud software requests, the transparency of instances on the cloud platform and transparency in terms of instances in virtual machine/host/storage [6].

3 Related Work

The selection of cloud depends on the parameters that need to be considered to prioritize the resources of different CSPs. In order to select the suitable cloud for a cloud user the selection parameters are designed in the model to get the best cloud that matches with the requirements of the customer and suitable SLA can be made between the CSP and cloud user [7]. This section gathers the related work of different authors with respect to the algorithms and techniques used in their model, different QoS parameters used in their model, which cloud service model (SaaS, PaaS, IaaS) is addressed by their model and what approach they had used in their cloud selection model. The related work of different authors is shown in Table 1.

Table 1. Existing cloud service ranking models

Reference	Technique/Algorithm	Framework	QoS Parameters	Approach
Goscinski and Brock (2010) [8]	Dynamic Programming	Resource Via Web Services Framework	Time and effort	Optimization Technique
Menzal et al. (2011) [9]	AHP	Framework (MC2)2	Cost, Benefits, opportunities and Risk	Multi-criteria Decision Model(MCDM)

(continued)

Table 1. *(continued)*

Reference	Technique/Algorithm	Framework	QoS Parameters	Approach
He et al. (2012) [10]	Greedy algorithm	MSS Optimizer (Multi-tenant SaaS Optimizer)	Cost, response Time, Availability, Throughput	Optimization Technique
Klein et al. (2012) [11]	Genetic Algorithm	Network alert approach	Scalability	MCDM
Lo et al. (2013) [12]	TOPSIS	Calculating weights of diverse criteria and provides rating of alternate service	Scalability, Capability, Performance, Reliability and Availability	MCDM
Ghosh et al. (2014) [13]	SELCSP Framework	Risk between cloud provider and cloud consumer	Risk, Trustworthiness	Trust Approach
Supriya, M. et al. (2015) [14]	Fuzzy-AHP	Sugeno Fuzzy inference system (FIS)	Agility, Assurance, Cost and Performance	MCDM
Pan et al. (2016) [15]	Trust Approach	Trust Modelling Method	Capacity, Cost, Security	Trust Approach
Supriya, M. et al. (2020) [16]	Genetic Algorithm	Trust Estimation Method	Security, Dependability, Availability, and Ability	Multi-criteria
Eisa, M. et al. (2020) [17]	Hybrid Approach	QoS aware selection model	QoS attributes: Security, Usability, Performance	MCDM
Faiz, M. et al. (2020) [18]	Fuzzy Technique	Trust based Model	Capacity, Cost, Performance	MCDM
Tiwari, R et al. (2021) [19]	RE-TOPSIS	Modified-TOPSIS	Cost, Performance, Scalability,	MCDM

4 Fuzzy Cloud Selection Model

Fuzzy logic is a multi-valued logic based on fuzzy set theory. [8]. Let an input point ‘k’ lies in the universe of discourse ‘X’ which belongs to a fuzzy set ‘F’ with the membership degree as:

$$0 \leq \mu_F(k) \leq 1$$

Using flow control systems, a fuzzy logic technique develops a series of fuzzy rules to alleviate the ambiguity that arises in trust based management systems such as decision making and pattern recognition. Fuzzy logic is used to address a flow control problem by translating crisp inputs into fuzzy values and the fuzzy values are used to deduce Fuzzy rules (IF-THEN-ELSE). Fuzzy based Cloud selection model is proposed to select the best-suited cloud to the users as per the requirements of the users based on different selection parameters. The proposed Fuzzy model for Cloud ranking using five QoS parameters is shown in Fig. 2 where 'Mamdani' FIS applied for the flow control of the proposed model.

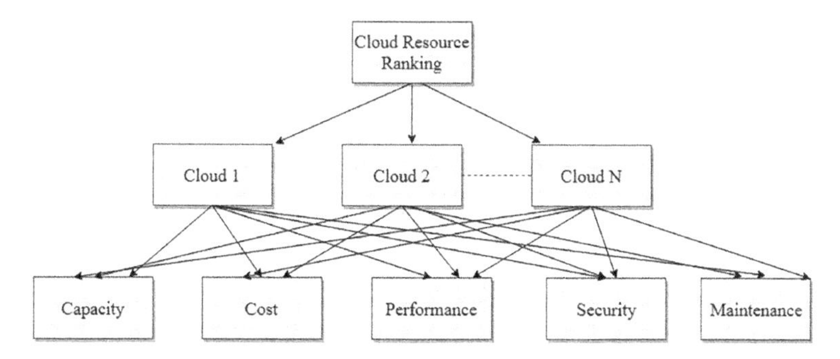

Fig. 2. Cloud ranking attributes

The dataset for reference model is shown in Table 2 where C_1, C_2,......C_5 are different Clouds.

Table 2. Cloud ranking attributes [7]

Criteria	Parameters	C1	C2	C3	C4	C5
Capacity	RAM (GB)	15	14	15	13	14
	C.P.U (GHz)	9.6	12.8	8.8	9.3	10.2
Cost	Vm Cost ($/month)	162.7	97.1	141	110.2	117.1
	Storage (TB/month $/GB)	0.02	0.03	0.03	0.04	0.03
Performance	Database performance (%)	55.0	68.0	65.0	75.0	72.0
	Max. C.P.U performance Score	2500	3100	3600	3800	4000
	Max. Network performance (Mbps)	3200	2100	2200	2700	3000
Security	Avg. Strength	0.9	0.5	0.8	0.7	0.8
Maintenance	Up-time (%)	99.90	99.59	99.92	99.45	99.97
	Free Support (Boolean)	Y	Y	N	N	Y

The next step is modeling the linguistic terms and selecting the membership functions (MF). The membership function plays a vital role in the performance representation of the fuzzy model [8]. There are different types of membership functions such as Gaussian, Triangular, Trapezoidal, etc. The membership function range varies between 0 to 1. In our model, the selected membership functions are used for the fuzziness representation are triangular and trapezoidal due to the simplicity in representation, easily represent fuzzy intervals. The Triangular MF has higher performance compared to Gaussian MF. The proposed model used Trapezoidal MF as the main shape for output because it better captures the vagueness of linguistic inputs [9].

Trapezoidal MF

$$Trapezoid(x : a, b, c, d) = 0 \quad if \ x \leq a,$$

$$\frac{(x-a)}{(b-a)} \quad if \ a \leq x \leq b,$$

$$1 \quad if \ b \leq x \leq c,$$

$$0 \quad if \ d \leq x$$

$$\mu_{trapezoid} = \max\left(\left(min\frac{x-a}{b-a}, 1, \frac{d-x}{d-c}\right), 0\right)$$

Where x is the input, a and b are lower and upper threshold values.

Triangular MF

$$\mu_{triangular}(x) = \begin{cases} 0 if & x <= a \\ \frac{x-a}{b-a} if & a \leq x \leq b \\ \frac{c-x}{c-b} if & b \leq x \leq c \\ 0 if & x \geq c \end{cases}$$

A. **Fuzzification of Input**
In this process, crisp values are transformed into fuzzy values. In the proposed model the fuzzification process used triangular and trapezoidal membership function because these membership functions are simple and precisely represents the crisp input values. The range of crisp inputs is represented by defining the range of fuzzy variables for the successful execution of a fuzzier program [10].

i. **Capacity**
The parameter used for the capacity (Gigabyte) is the amount of RAM required by the user. The linguistic terms and corresponding crisp input range values using trapezoidal MF are shown in Table 2. The dataset of Table 2 is used to consider the range value. The linguistic terms used to represent capacity are 'Reasonable', 'Good' and 'Extensive' where reasonable represents low amount of RAM, good represent a generic amount of RAM and extensive represents large amount of RAM capacity required by the cloud user[11].

ii. **Cost**
In our research, we found that cost is one of the most important factors for the selection of cloud service providers. The users desire to get suitable cloud services within the least amount of payment. The parameter used for the cost is the amount charged for Virtual Machine (Vm) required by the user. The linguistic terms and corresponding crisp input range values using trapezoidal MF are shown in Table 3 where cost of a cloud service 'x' offered by the CSP ranges between low to very high cost.

iii. **Performance**
The performance of the cloud is computed using performance score of the C.P.U and the database performance. The parameter used for the performance is the database performance of the cloud service provider. The linguistic terms and corresponding crisp input range values using the same trapezoidal MF are shown in Table 3 where the linguistic term used for performance ranges from poor to excellent. The universe of discourse is selected as 100 to represent the database performance on the scale (0–100%).

iv. **Security**
Security is a major aspect of each cloud user because the precious data is stored in a location far away from the end-user. If a cloud system suffers from security flaws it may lead to heavy loss for both service provider and end-user. There are multiple layers of security such as tier-1, tier-2 provided by the different CSP A proper SLA is needed to be signed between CSP and the end-user [12]. The parameter used to represent security is the average strength of the cloud system offered by the provider. The linguistic terms and corresponding crisp input range values using the triangular MF to more precise representation of values are shown in Table 3 where the security level ranges from low security to very high security.

v. **Maintenance**
The different services offered by the Cloud require proper maintenance of the system over time. We have considered the maintenance service parameter using up-time and free support provided by different service providers. Up-time denotes the portion of the cloud that remains active during maintenance. Free support is denoted using binary representation (0 or 1) which means whether the cloud offers free support in a purchased plan or not. The linguistic terms and corresponding crisp input range values using the trapezoidal MF are shown in Table 3. The universe of discourse is selected 100 to represent the maximum up-time (%) of the system under maintenance. Table 3 represents the fuzzufication of different selected parameters.

Table 3. Linguistic terms and corresponding range values for different parameters for QoS

No	Parameter	Linguistic term	Range(input values)
1	Capacity	Reasonable	[0, 0,6,8]
2		Good	[7, 15, 20,25]

(continued)

Table 3. (*continued*)

No	Parameter	Linguistic term	Range(input values)
3		Extensive	[24,28, 36, 40]
4	Cost	Low	[0 0 90 100]
5		Reasonable	[90 100 110 120]
6		High	[110 120 140 150]
7		Very high	[150 170 180 200]
8	Performance	Poor	[10, 20, 30,40]
9		Satisfactory	[35, 40, 50, 60]
10		Good	[55, 65,75,85]
11		Excellent	[80, 85, 90,100]
12	Security	Low	[−0.1 0 0.2]
13		Medium	[0.1 0.3 0.4]
14		High	[0.3 0.5 0.7]
15		Very high	[0.6 0.8 1]
16	Maintenance	Poor	[10, 20, 30,40]
17		Satisfactory	[35, 40, 50, 60]
18		Good	[55, 65,75,85]
19		Excellent	[80, 85, 90,100]

B. **Output Membership Function**To represent each of the five input parameters that are capacity of memory (R.A.M), cost of services ($/month), performance of cloud server, security offered(low to high) and maintenance(support and updates), the membership functions are $\mu A(Cp)$, $\mu B(Co)$ $\mu C(P_f)$ $\mu D(Se)$ and $\mu E(M_t)$ respectively. Each variable is associated with Upper and Lower threshold values. The associated thresholds are TH_1, TH_2, $TH_{3...}TH_{10}$ respectively. The activation point of the model is represented by a lower threshold value and the model works amid the upper and lower regions of the threshold [13].

$$\mu_x(Capacity) = \begin{Bmatrix} 1 & \text{if } Cp \le TH_1 \\ TH_1 - Cp/TH_1 - TH_2 & \text{if } TH_1 < Cp < TH_2 \\ 0 & \text{if } Cp \ge TH_2 \end{Bmatrix}$$

$$\mu_y(Cost) = \begin{Bmatrix} 0 & \text{if } Co \le TH_3 \\ Co - TH_3/TH_3 - TH_4 & \text{if } TH_3 < Co < TH_4 \\ 1 & \text{if } Co \ge TH_4 \end{Bmatrix}$$

$$\mu_z(Performance) = \begin{Bmatrix} 0 & \text{if } Pf \le TH_5 \\ Pf - TH_5/TH_5 - TH_6 & \text{if } TH_5 < Pf < TH_6 \\ 1 & \text{if } Pf \ge TH_6 \end{Bmatrix}$$

$$\mu_z(Maintenance) = \begin{Bmatrix} 0 & \text{if } Mt \le \mathrm{TH}_7 \\ \mathrm{Mt} - \mathrm{TH}_7 / \mathrm{TH}_7 - \mathrm{TH}_8 & \text{if } \mathrm{TH}_7 < \mathrm{Pf} < \mathrm{TH}_8 \\ 1 & \text{if } Pf \ge \mathrm{TH}_8 \end{Bmatrix}$$

$$\mu_z(Security) = \begin{Bmatrix} 0 & \text{if } Se \le \mathrm{TH}_9 \\ \mathrm{Se} - \mathrm{TH}_9 / \mathrm{TH}_9 - \mathrm{TH}_{10} & \text{if } \mathrm{TH}_9 < \mathrm{Pf} < \mathrm{TH}_{10} \\ 1 & \text{if } Pf \ge \mathrm{TH}_{10} \end{Bmatrix}.$$

Let Co, Cp, P_f, S_e, and M_t be represented for the five inputs as cost, capacity, performance, security and maintenance respectively. The Cloud ranking values are expressed linguistically in the output variables. The defined fuzzy sets used for the output membership function for the Cloud ranking are represented as 'VLOW', 'LOW', 'MEDIUM', 'HIGH' and 'VHIGH' to define the range between very low to very high where higher value represents a higher rank shown in Table 4.

Table 4. Fuzzy output MF variables

No	Linguistic Term	Values
1	[6.8 7.5 9 10]	VHigh
2	[5.2 6 6.5 7]	High
3	[3.2 4.5 5 5.5]	Medium
4	[1.5 2.5 3 3.5]	Low
5	[0 0 1 2]	VLow

The membership functions are chosen using a pre-defined Fuzzy rule set. The following is the order in which the input parameters for selecting a CSP is considered:

$$\mathrm{Cp} > \mathrm{P_f} > \mathrm{Co} > \mathrm{S_e} > \mathrm{M_t}.$$

Fuzzy Inference Rules:

1. IF Cp Extensive, and P_f High and Co Low and S_e High and M_t High THEN the Rank is 'VHigh'.
2. IF Cp Reasonable, and P_f Low and and Co High and S_e Medium and M_t Satisfactory THEN the Rank is 'Medium'.
3. IF Cp Reasonable, and P_f High and Co Reasonable and S_e Medium and M_t High and THEN the Rank is 'High'.
4. IF Cp Extensive, and P_f Medium and Co Reasonable and S_e Medium and M_t High THEN the Rank is 'Medium'.
5. IF Cp Good and, Co High and, P_f Excellent, S_e High and M_t Good THEN the Rank is 'High'.

6. IF Cp Extensive, and Co Reasonable and P_f High and S_e Medium and M_t Good THEN the Rank is 'High'.

Similarly, 27 rules created in the FIS to get the rank value of CSP using the requirements provided by different customer.

5 Simulation

The simulation is carried out with the help of the Matlab fuzzy toolbox. The Cloud ranking parameters used for the simulation are taken from the defined cloud instances of Table 1 as reference data. Random data are tested to deduce the rank value of different Clouds (C_1, C2,......C_5) for different users(U_1, U_2, U_3 and U_4). The flow control mechanism used in the simulation is 'Mamdani' Fuzzy Inference System (FIS) due to its improved simulation representation compared to 'Sugeno' FIS. The input and output membership used in the simulation process is shown in Fig. 3.

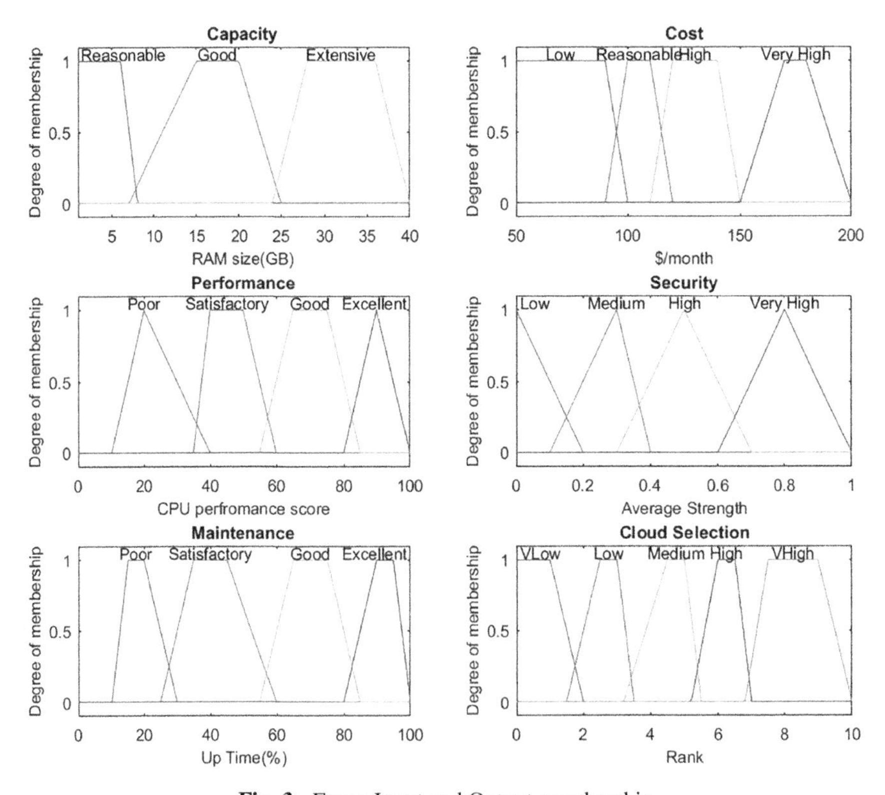

Fig. 3. Fuzzy Input and Output membership

Giving the input values for User1, the priority index for different parameters used for User1 is:

$$Cp > P_f > S_e > M_t > Co.$$

The proposed fuzzy cloud ranking model computes the rank value for User1 is 0.82 that represents a very high chance of selecting the cloud by the cloud User1. The rank value for User1 is shown in Fig. 4.

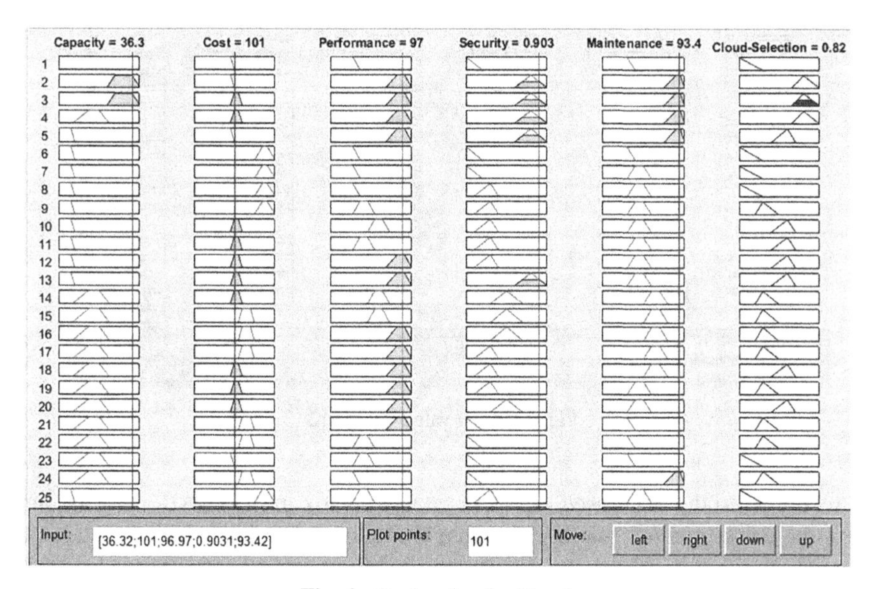

Fig. 4. Rank value for User1

Similarly, giving the input values of Table 5, the priority index for User2 is:

$$Cp > Co > S_e > M_t > P_f.$$

The proposed fuzzy cloud ranking model computes the rank value for User2 as 0.5 that represents a moderate chance of selecting the cloud by the cloud User2.

Similarly, the priority index for User3 is:

$$M_t > S_e > Co > Cp > P_f.$$

The model computes the rank value for User3 as 0.34 that represents a low chance of selecting the cloud by the cloud User3. The priority index for User4 is:

$$M_t > S_e > Cp > P_f > Co.$$

The model computes the rank value for User4 as 0.72 that represents is a very high chance of selecting the cloud by the cloud User4.

In order to compute the fuzzy relation and rank value, AND (Λ) operator is applied using the Fuzzy Inference rules on membership value of different parameters as:

$$A(x, y) \cup B(x, y) = \max(A(x, y), B(x, y));$$

Where A(x,y) and B(x,y) are the two relations that are define over two crisp sets $x \in X$ and $y \in Y$.

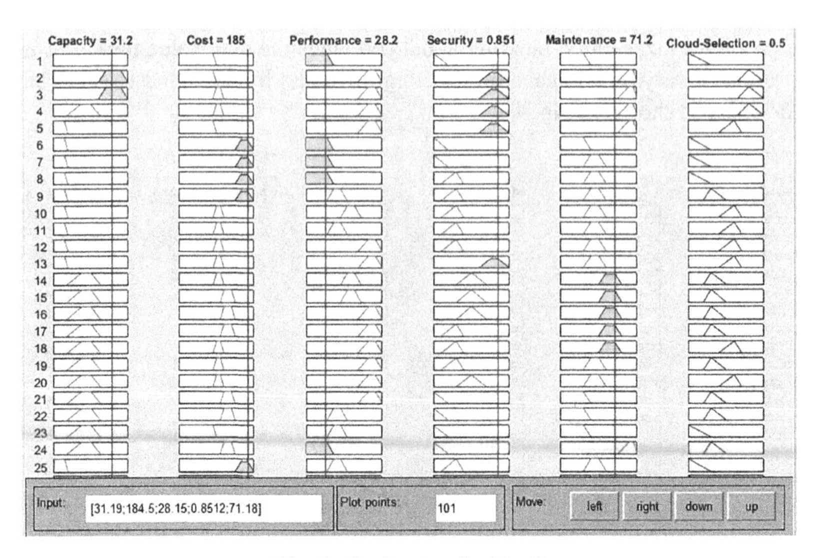

Fig. 5. Rank value for User2

According to the rule based of FIS in Fuzzy toolbox in MATLAB using the input values the corresponding rule is fired from the defined rule base and respective rank value output is computed in the defined range of MF variables in Table 3 and shown in Fig. 4 and Fig. 5. The rank value of different users are shown in Table 5.

Example: IF Cp Extensive (Λ) P_f High (Λ) Co Low (Λ) S_e High (Λ) M_t High THEN the Rank is 'VHigh'.

Table 5. Ranking of cloud users

Sr. No	User	Input [Capacity, Cost, Performance, Security, Maintenance]	Rank
1	U1	[36.3,101,97,0.90,93.4]	0.82
2	U2	[31.2,185,28.2,0.85,71.2]	0.5
3	U3	[16,135,64.8,0.63,75.9]	0.34
4	U4	[30.2,101,83.2,0.79,80.6]	0.72

The following is a generalized comparison chart of attributes for different types of clouds and users (Fig. 6).

The proposed model for cloud ranking observes that the parameters that are considered for different clouds based on the collected information in Table 2, it is observed that offering quality service at low cost cloud is a major factor for selecting the cloud with very high chances by the user1.

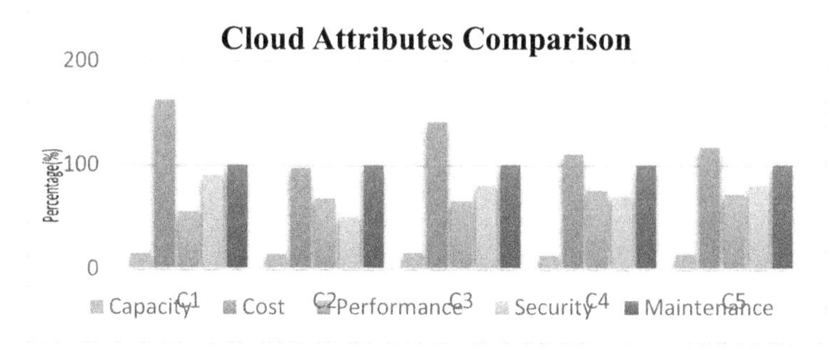

Fig. 6. Cloud attribute comparison

According to the obtained rank value the cloud priority recommendation for User 1 is shown in Fig. 7 is as follows:

$$C_4 > C_5 > C_2 > C_3 > C_1.$$

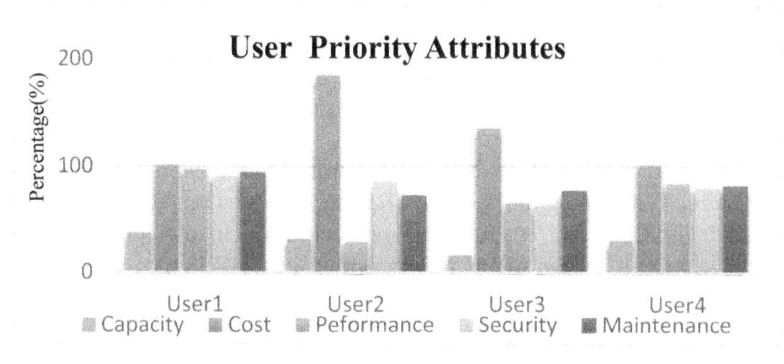

Fig. 7. User attributes comparison

The User 2 pays a higher charges for cloud service looking for high memory capacity, and performance. The CSP even offers high security and good maintenance but suffers with low performance that leads to downfall the rank in moderate rank.

According rank value the cloud recommendation priority for User 2 is as follows:

$$C_1 > C_3 > C_2 > C_4 > C_5.$$

The User 3 pays high charges for Cloud service where the CSP offered moderate level of service in terms of capacity, performance, security and maintenance leads to decrease the rank with great extent.

According to the obtained rank value the cloud recommendation priority for User 3 is as follows:

$$C_3 > C_1 > C_2 > C_4 > C_5.$$

Similarly, the User4 gets valuable cloud service offered by CSP that provides high quality service in low cost that lands the ranking in high category.

According to the obtained rank value the cloud recommendation priority for User 4 is as follows:

$$C_4 > C_5 > C_3 > C_2 > C_1.$$

6 Conclusion

The proposed model computes the rank of different Clouds using the selected parameters. The model tested for four users. The model enables the Cloud users to select the Cloud optimally according to their desired features of cloud. The proposed ranking methodology effectively ranks Cloud services based on a set of criteria such as capacity, pricing, performance, security, and maintenance. The simulation result shows that among the five reference cloud the recommended cloud is C_4 that fulfills service requirements of the user and the ranking priority of the clouds are $C_4 > C_3 > C_1 > C_5 > C_2$. We believe that the proposed methodology is a significant step in providing accurate cloud service measurement and selection for Cloud users. The Cloud service selection problem can be solved by this model.

References

1. Neelakanteswara, P., Suryanarayana Babu, P.: Prioritized rank based technique for resource allocation in cloud computing. Int. J. Innov. Technol. Explor. Eng. **8**(6), 520–523 (2019)
2. Mohammadkhanli, L., Jahani, A.: Ranking approaches for cloud computing services based on quality of service: a review. ARPN J. Syst. Softw. **4**(2), 50–58 (2014)
3. Elmubarak, S.A., Yousif, A., Bashir, M.B.: Performance based ranking model for cloud SaaS services. Int. J. Inf. Technol. Comput. Sci. **9**(1), 65–71 (2017)
4. Faiz, M., Shanker, U.: Data synchronization in distributed client-server applications. In: 2016 IEEE International Conference on Engineering and Technology (ICETECH), pp. 611–616. IEEE (2016)
5. Supriya, M., Sangeeta, K., Patra, G.K.: Comparison of AHP based and fuzzy based mechanisms for ranking cloud computing services. In: Proceeding - 2015 Int. Conf. Comput. Control. Informatics its Appl. Emerg. Trends Era Internet Things, IC3INA 2015, pp. 175–180 (2016)
6. Luo, C., et al.: CloudRank-D: benchmarking and ranking cloud computing systems for data processing applications. Front. Comput. Sci. China **6**(4), 347–362 (2012)
7. He, Q., et al.: QoS-driven service selection for multi-tenant SaaS. In: Cloud Computing (cloud), 2012 International Conference on. IEEE (2012)
8. Daniel, A.K.: A cloud trusting mechanism based on resource ranking. In: Shanker, U., Pandey, S. (eds.) Handling Priority Inversion in Time-Constrained Distributed Databases, pp. 130–155. IGI Global (2020). https://doi.org/10.4018/978-1-7998-2491-6.ch008
9. Ardagna, D., Casale, G., Ciavotta, M., Pérez, J.F., Wang, W.: Quality-of-service in cloud computing: modeling techniques and their applications. J. Int. Serv. Appl. **5**(1), 1–17 (2014)
10. Garg, S.K., Versteeg, S., Buyya, R.: A framework for ranking of cloud computing services. Futur. Gener. Comput. Syst. **29**(4), 1012–1023 (2013)

11. Sirohi, P., Agarwal, A., Maheshwari, P.: A comparative study of cloud computing service selection. Int. J. Eng. Adv. Technol. **8**(5), 259–266 (2019)
12. Alhamad, M., Dillon, T., Chang, E.: A Trust-Evaluation Metric for Cloud applications. Int. J. Mach. Learn. Comput. **1**(4), 416–421 (2011)
13. Paunovic, M., Ralevic, N.M., Gajovic, V., Vojinovic, B.M., Milutinovic, O.: Two-stage fuzzy logic model for cloud service supplier selection and evaluation. Math. Prob. Eng. **2018**, 1–11 (2018). https://doi.org/10.1155/2018/7283127
14. Srivastava, R., Daniel, A.K.: Efficient model of cloud trustworthiness for selecting services using fuzzy logic. In: Abraham, A., Dutta, P., Mandal, J.K., Bhattacharya, A., Dutta, S. (eds.) Emerging Technologies in Data Mining and Information Security. AISC, vol. 755, pp. 249–260. Springer, Singapore (2019). https://doi.org/10.1007/978-981-13-1951-8_23
15. Sirohi, P., Agarwal, A., Maheshwari, P.: Framework for cloud service selection and ranking. SSRN Electron. J.https://doi.org/10.2139/ssrn.3394093
16. Supriya, M.: Cloud service provider selection using non-dominated sorting genetic algorithm. In: 2020 4th (ICOEI) (48184), pp. 800–807. IEEE (2020)
17. Eisa, M., Younas, M., Basu, K., Awan, I.: Modelling and simulation of qos-aware service selection in cloud computing. Simulat. Model. Pract. Theory **103**, 102108 (2020)
18. Faiz, M., Daniel, A.K.: Fuzzy cloud ranking model based on QoS and Trust. In: 2020 Fourth International Conference on I-SMAC (IoT in Social, Mobile, Analytics and Cloud) (I-SMAC), pp. 1051–1057. IEEE (2020)
19. Tiwari, R.K., Kumar, R.: A robust and efficient MCDM-based framework for cloud service selection using modified TOPSIS. Int. J. Cloud Appl. Comput. (IJCAC) **11**(1), 21–51 (2021)

Improving the Aggregate Utility of IEEE 802.11 WLAN Using NOMA

Badarla Sri Pavan(✉) and V. P. Harigovindan

Department of E.C.E., National Institute of Technology Puducherry, 609 609 Karaikal, India
hari@nitpy.ac.in

Abstract. The IEEE 802.11 WLAN standards are very popular since three decades due to its flexibility and operation in unlicensed frequency bands. It provides physical and MAC layer specifications. We present an analytical model with a channel access scheme to improve the aggregate utility of WLAN using non-orthogonal multiple access (NOMA). The NOMA is used at physical layer and in MAC layer distributed coordination function (DCF) is considered. With the inclusion of NOMA in WLAN, multiple transmissions are possible in a resource block. Moreover, it mitigates the contention process to enhance the performance of WLAN. With the analytical model, we show the performance of NOMA based WLAN improved compared to conventional WLAN. The analytical results are validated with simulations.

Keywords: Internet of things · IEEE 802.11 WLAN · NOMA

1 Introduction

In recent days, the Internet of things (IoT) getting more attention due to the applications of E-Learning, smart environment, E-Health, video streaming, etc. This increases the number of Internet users connected to Internet. Besides, the network parameters such as high throughput and connectivity are very much essential [1]. IEEE 802.11 wireless local area networks (WLANs) is widely used because of simplicity and flexible deployment for high throughput. For next-generation WLANs, the high throughput and connectivity are key challenges. The IEEE 802.11 provides specifications for physical and medium access control (MAC) layers [2]. In physical layer, depending on channel conditions, the nodes use various data rates based on rate adaptation [3]. The widely used IEEE 802.11b standard defines data rates (11, 5.5, 2, 1 Mbps). In the MAC layer, DCF is used for channel access. Although the nodes in IEEE 802.11 WLAN share the unlicensed wireless spectrum, the available radio resources are limited and crowded with users, which lead to more contention, resulting in low throughput [4]. Thus, resource allocation is key for the effective utilization of the wireless spectrum.

I. Woungang et al. (Eds.): ANTIC 2021, CCIS 1534, pp. 168–176, 2022.
https://doi.org/10.1007/978-3-030-96040-7_13

The sharing of resources among the users is possible with multiple access techniques. NOMA technique has proved its potentiality for future networks. In NOMA, several nodes can complete transmissions in one block. It uses spectrum efficiently and provide high connectivity [5]. Among existing NOMA variants, power domain NOMA (PD-NOMA) is efficient and compatible [6]. In this, the nodes can transmit the data with different powers in same block. It offers superposition coding (SC) to serve multiple users along with successive interference cancellation (SIC). Different sections are as follows. Section 2 presents realated work. Section 3 gives system model to find the throughput for conventional and NOMA based WLAN with channel access scheme. Results are provided in Sect. 4. The paper ends with Sect. 5.

2 Related Work

Due to the limited radio spectrum, the existing multiple access techniques in IEEE 802.11 WLAN are unable to support dense networks [4]. Thus, there is a necessity of supporting dense IoT networks for data-intensive applications. These requirements will be fulfilled with the inclusion of NOMA because it allows multiple transmissions. Very few works are proposed for NOMA based WLANs [4,7–9]. The author [4] improved the throughput by proposing NOMA in WLAN. In [7], the authors implemented a scheduling approach using NOMA for scheduling the resources in Wi-Fi network. The authors of [8] proposed a client pairing on NOMA in CSMA/CA to provide more connectivity for clients in WLANs. In [9], the authors built a prototype using NOMA to improve the weighted sum rate of WLAN.

3 System Model

Here, the channel access scheme is discussed using NOMA. We consider the IEEE 802.11 network with an access point (AP) and N nodes. The coverage area of the AP is divided into E regions as shown in Fig. 1. These are deployed uniformly over E regions. The transmission rate of a node $\mathcal{R}_e \,|e \in [1, E]$ in region-e depends on distance from the AP and nodes use rate adaptation for transmission [3]. Here, the nodes exhibit multi-rate nature. The near region nodes obtain high data rate, while far region nodes have low rate which depends on SNR. The channel access scheme is explained as follows. (1) We form the user-clusters with one node from each region as shown in Fig. 2 [10,11]. (2) Here, the region-1 nodes of $\lceil \frac{N}{E} \rceil$ contend for channel access. (3) After DIFS, the region-1 nodes gets channel access. (4) Region-1 node shares request to send (RTS) frame with AP and it returns with clear to send (CTS) frame having a destination address, which is received by every node. (5) Now, the respective user-cluster nodes send the data concurrently along with near region node with PD-NOMA. (6) Received data of user-cluster nodes is decoded using SIC technique at the AP.

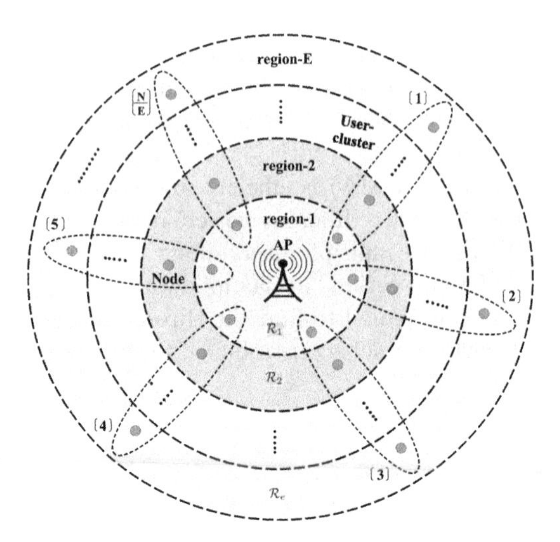

Fig. 1. Network architecture

We assume the network in saturated conditions and no hidden nodes. At first, the nodes sense the channel using DCF [2]. Once the channel in ideal state, the node is ready for transmission after DIFS duration. For busy channel condition due to collision/another node's transmission, the node initiates a back-off process

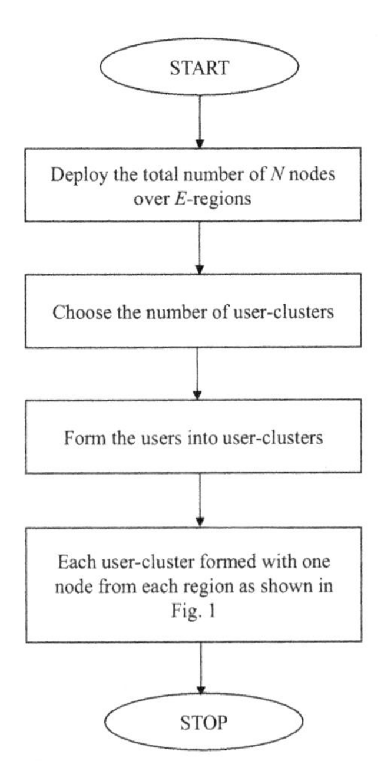

Fig. 2. Flow chart for formation of user-clusters based on data rates

and choose initial back-off over $[0, W_0 - 1]$, where W_0 gives minimum contention window (CW). The counter decrements in ideal channel state and frozen in the busy state. The back-off stage doubles till the maximum CW of $W_m = 2^m W_0$ for unsuccessful transmission/collision. When m approaches maximum CW, then it stays till retry limit (L) [2]. The node propagate data if counter reaches zero. The node's transmission probability [2,10] in region-e of data rate $\mathcal{R}_e \mid e \in [1, E]$ is expressed as,

$$\tau_e^{(\mathcal{R}_e)} = \frac{2(1 - 2\mathbb{P}_{c,e}^{(\mathcal{R}_e)})(1 - (\mathbb{P}_{c,e}^{(\mathcal{R}_e)})^{L+1})}{\left(\begin{array}{c} W_0(1 - (2\mathbb{P}_{c,e}^{(\mathcal{R}_e)})^{m+1})(1 - \mathbb{P}_{c,e}^{(\mathcal{R}_e)}) + (1 - 2\mathbb{P}_{c,e}^{(\mathcal{R}_e)})(1 - (\mathbb{P}_{c,e}^{(\mathcal{R}_e)})^{L+1}) + \\ 2^m W_0 (1 - 2\mathbb{P}_{c,e}^{(\mathcal{R}_e)})(\mathbb{P}_{c,e}^{(\mathcal{R}_e)})^{m+1} \times (1 - (\mathbb{P}_{c,e}^{(\mathcal{R}_e)})^{L-m} \end{array} \right)}, \tag{1}$$

where $\mathbb{P}_{c,e}^{(\mathcal{R}_e)}$ is the collision probability.

3.1 Conventional IEEE 802.11 WLAN

This subsection presents the computation of throughput for conventional WLAN. The node's collision probability in region-e with rate $\mathcal{R}_e \mid e \in [1, E]$ is expressed as [10],

$$\mathbb{P}_{c,e}^{(\mathcal{R}_e)} = \left(1 - (1 - \tau_e^{(\mathcal{R}_e)})^{n^{(\mathcal{R}_e)} - 1}\right) \times \prod_{y=1, y \neq e}^{E} (1 - \tau_y^{(\mathcal{R}_y)})^{n^{(\mathcal{R}_y)}}. \tag{2}$$

The success probability with rate $\mathcal{R}_e \mid e \in [1, E]$ in region-e is expressed as,

$$\mathbb{P}_{s,e}^{(\mathcal{R}_e)} = \frac{\left(n^{(\mathcal{R}_e)} \tau_e^{(\mathcal{R}_e)} (1 - \tau_e^{(\mathcal{R}_e)})^{n^{(\mathcal{R}_e)} - 1}\right)}{\mathbb{P}_{tr,e}^{(\mathcal{R}_e)}} \times \left(\prod_{y=1, y \neq e}^{E} (1 - \tau_y^{(\mathcal{R}_y)})^{n^{(\mathcal{R}_y)}} \right), \tag{3}$$

where $\mathbb{P}_{tr,e}^{(\mathcal{R}_e)} = 1 - (1 - \tau_e^{(\mathcal{R}_e)})^{n^{(\mathcal{R}_e)}}$. Further, the throughput with rate $\mathcal{R}_e \mid e \in [1, E]$ of region-e is expressed as,

$$A_t^{(\mathcal{R}_e)} = \frac{E[P] \; \mathbb{P}_{tr,e}^{(\mathcal{R}_e)} \; \mathbb{P}_{s,e}^{(\mathcal{R}_e)}}{\mathbb{E}[\mathbb{T}_{slot}]}, \tag{4}$$

where $\mathbb{E}[\mathbb{T}_{slot}]$ is average slot time and payload is $E[P]$. The below equation shows the $\mathbb{E}[\mathbb{T}_{slot}]$ for two regions ($e = 2$) with rate $\mathcal{R}_1$ and $\mathcal{R}_2$ [12].

$$\begin{aligned} \mathbb{E}[\mathbb{T}_{slot}] = (1 - \mathbb{P}_{tr,e})\, \delta + \sum_{i=1}^{2} \mathbb{P}_{tr,e}^{(\mathcal{R}_e)} \mathbb{P}_{s,e}^{(\mathcal{R}_e)} \mathbb{T}_{sw}^{(\mathcal{R}_e)} + \\ \mathbb{P}_{tr,1}^{(\mathcal{R}_1)} (1 - \mathbb{P}_{s,1}^{(\mathcal{R}_1)})(1 - \mathbb{P}_{tr,2}^{(\mathcal{R}_2)}) \mathbb{T}_{cw}^{(\mathcal{R}_1)} + \mathbb{P}_{tr,2}^{(\mathcal{R}_2)} (1 - \mathbb{P}_{s,2}^{(\mathcal{R}_2)}) \times \\ (1 - \mathbb{P}_{tr,1}^{(\mathcal{R}_1)}) \mathbb{T}_{cw}^{(\mathcal{R}_2)} + \mathbb{P}_{tr,1}^{(\mathcal{R}_1)} \mathbb{P}_{tr,2}^{(\mathcal{R}_2)} \max\,(\mathbb{T}_{cw}^{(\mathcal{R}_1)}, \mathbb{T}_{cw}^{(\mathcal{R}_2)}). \end{aligned} \tag{5}$$

where δ is empty slot time. In Eq. (5), the 1st term signifies the average idle slot length. The 2nd term represents average successful slot due to success of any one of the nodes of rate $\mathcal{R}_1$ and $\mathcal{R}_2$. The 3rd term signifies the mean collision slot duration due to the collisions among rate $\mathcal{R}_1$ nodes (region-1), the 4th term gives the collisions among rate $\mathcal{R}_2$ nodes (region-2). The 5th term accounts for the average length of the collision slot when the rate $\mathcal{R}_1$ and $\mathcal{R}_2$ nodes collide with each other. Similarly, $\mathbb{E}[\mathbb{T}_{slot}]$ can be computed for E-regions, but skipped here due to cumbersome expressions.

The success and collision slots duration are expressed as [2],

$$\mathbb{T}_{sw}^{(\mathcal{R}_e)} = t_d + t_{RTS} + t_{CTS} + t_H^{(\mathcal{R}_e)} + t_{E[P]}^{(\mathcal{R}_e)} + 3t_s + t_a + 4\sigma,$$

$$\mathbb{T}_{cw}^{(\mathcal{R}_e)} = t_d + t_{RTS} + \sigma,$$

where propagation delay is σ; t_d is DIFS duration, t_{RTS} is RTS frame duration, t_{CTS} is CTS duration, $t_H^{(\mathcal{R}_e)}$ gives header length, $t_{E[P]}^{(\mathcal{R}_e)}$ is payload length, t_s is SIFS length, and t_a is acknowledgment duration. The aggregate throughput is expressed as [10],

$$A_t = \sum_{e=1}^{E} A_t^{(\mathcal{R}_e)}. \tag{6}$$

3.2 NOMA Based IEEE 802.11 WLAN

In NOMA based WLAN, region-1 nodes alone compete for channel access. So the contending nodes is $\lceil \frac{N}{E} \rceil$ instead of N nodes. Therefore, the collision probability with data rate $\mathcal{R}_1$ is expressed as [10],

$$\mathbb{P}_{c,1}^{(\mathcal{R}_1)} = 1 - (1 - \tau_1^{(\mathcal{R}_1)})^{n^{(\mathcal{R}_1)}-1}. \tag{7}$$

The probability of success that the node transmits with data rate $\mathcal{R}_1$ is expressed as,

$$\mathbb{P}_{s,1}^{(\mathcal{R}_1)} = \frac{n^{(\mathcal{R}_1)}(1 - \tau_1^{(\mathcal{R}_1)})^{n^{(\mathcal{R}_1)}-1}}{\mathbb{P}_{tr,1}^{(\mathcal{R}_1)}(\tau_1^{(\mathcal{R}_1)})^{-1}}, \tag{8}$$

where, $\mathbb{P}_{tr,1}^{(\mathcal{R}_1)} = 1 - (1 - \tau_1^{(\mathcal{R}_1)})^{n^{(\mathcal{R}_1)}}$. The region-1 node's throughput with rate $\mathcal{R}_1$ is expressed as [10],

$$S_{nw}^{(\mathcal{R}_1)} = \frac{E[P]\ \mathbb{P}_{tr,1}^{(\mathcal{R}_1)}\ \mathbb{P}_{s,1}^{(\mathcal{R}_1)}}{E[\mathbb{T}_{nslot}]}, \tag{9}$$

The average slot time length $E[\mathbb{T}_{nslot}]$ is expressed as [10,12],

$$\mathbb{E}[\mathbb{T}_{nslot}] = (1-\mathbb{P}_{tr,1})\delta + \mathbb{P}_{tr,1}^{(\mathcal{R}_1)}\mathbb{P}_{s,1}^{(\mathcal{R}_1)} \times \max(\mathbb{T}_{sw}^{(\mathcal{R}_1)}, \mathbb{T}_{sw}^{(\mathcal{R}_2)}, \cdots, \mathbb{T}_{sw}^{(\mathcal{R}_E)}) + \mathbb{P}_{tr,1}^{(\mathcal{R}_1)}(1 - \mathbb{P}_{s,1}^{(\mathcal{R}_1)})\mathbb{T}_{cw}^{(\mathcal{R}_1)}. \tag{10}$$

Here, the 1st term gives mean ideal slot duration. The 2nd term provides the mean successful slot duration of user-cluster nodes. The 3rd term gives average collision slot duration, where the collisions are due to the region-1 nodes. Finally, the aggregate throughput [10] is expressed as,

$$A_{nt} = \sum_{e=1}^{E} \frac{E[P]\, \mathbb{P}_{tr,e}^{(\mathcal{R}_e)}\, \mathbb{P}_{s,e}^{(\mathcal{R}_e)}}{\mathbb{E}[\mathbb{T}_{nslot}]}. \tag{11}$$

4 Results and Discussion

We provide the analytical (A) and simulation (S) results using MATLAB 2020a and ns-3 respectively [13]. The WLAN consists of 480 nodes and an AP. We consider equal nodes in every region. The user-clusters are formed as shown in Fig. 1. The system parameters are shown in Table 1. For result analysis, we divide the coverage area into four regions. The length of each region and its data rates are shown in Table 2 using IEEE 802.11b [12]. The region-1 nodes attain 11 Mbps, while region-4 nodes exhibit 1 Mbps rate.

The aggregate throughput for NOMA based IEEE 802.11 WLAN is shown in Fig. 3 using RTS/CTS. The results are compared with the conventional WLAN. From Fig. 3, the throughput in both the methods is reduced with rise in network

Table 1. System parameters

Parameter	Value
$E[P]$	512 bytes
δ	20 μs
Header	52 bytes
RTS frame	44 bytes
CTS frame	38 bytes
DIFS	50 μs
SIFS	10 μs
ACK	38 bytes
Propagation delay, σ	1 μs
Back-off stage, m	5
L	6
W_0	32

Table 2. Region lengths and data rates

Region	1	2	3	4
Length d_e (m)	60	40	30	25
Data rate $\mathcal{R}_e$ (Mbps)	11	5.5	2	1

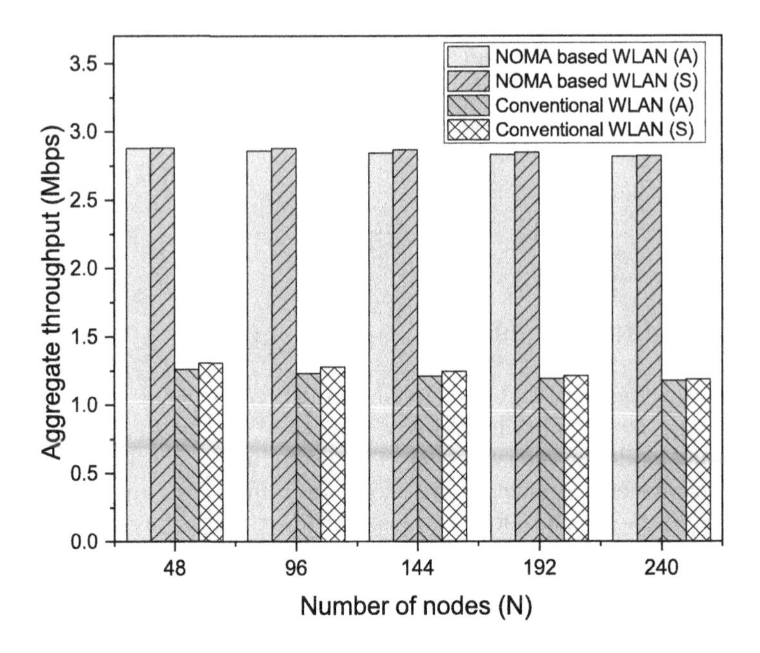

Fig. 3. Aggregate throughput performance

size. Because, collisions increases as the contention among the nodes increases. It makes the contention process tedious and reduces the throughput. Besides, we can observe the increase in aggregate throughput performance using NOMA. Here, the multiple transmissions of user-cluster nodes are possible using NOMA in a resource block. Hence, the spectrum resources are efficiently utilized. Also, due to the reduced number of nodes in the contention process, which further mitigates the collisions among nodes, because the near regions nodes only contend for channel access.

The connectivity analysis is illustrated in Table 3. We consider the different network sizes $N \in \{50, 150, 250, 350\}$. For NOMA based WLAN, the aggregate throughput for $N = 350$ is significantly improved in comparison with conventional

Table 3. Connectivity

Network size	Aggregate throughput (Mbps)			
	NOMA based WLAN		Conventional WLAN	
	A	S	A	S
50	2.8768	2.8788	1.2597	1.3146
150	2.8422	2.8747	1.2089	1.2690
250	2.8150	2.8426	1.1741	1.2209
350	2.7903	2.7975	1.1425	1.1785

WLAN for $N = 50$. This is due to the simultaneous transmissions possible with NOMA. Thus, from all the results, we conclude that the significant improvement of throughput and connectivity due to the inclusion of NOMA in WLAN.

5 Conclusion

We have presented an analytical model for IEEE 802.11 WLAN using NOMA for future WLANs. The NOMA based WLAN provides multiple transmissions in a resource block by user-cluster nodes. It also reduces the contending nodes during the process of contention. Using the model, we have evaluated the throughput and connectivity of IEEE 802.11 WLAN. From results, the NOMA based WLAN provides significant advantage in throughput and connectivity compared to conventional WLAN. The analytical results validated with simulation works.

References

1. Foerster, J.R., Costa-Perez, X., Prasad, R.V.: Communications for IoT: connectivity and networking. IEEE Internet Things Mag. **3**(1), 6–7 (2020)
2. Chatzimisios, P., Boucouvalas, A.C., Vitsas, V.: IEEE 802.11 packet delay-a finite retry limit analysis. In: GLOBECOM 2003, vol. 2, pp. 950–954, December 2003
3. Babu, A.V., Jacob, L.: Fairness analysis of IEEE 802.11 multirate wireless LANs. IEEE Trans. Veh. Technol. **56**(5), 3073–3088 (2007)
4. Uddin, M.F.: Throughput performance of NOMA in WLANs with a CSMA MAC protocol. Wireless Netw. **25**(6), 3365–3384 (2018). https://doi.org/10.1007/s11276-018-1730-3
5. Dai, L., Wang, B., Ding, Z., Wang, Z., Chen, S., Hanzo, L.: A survey of non-orthogonal multiple access for 5G. IEEE Commun. Surveys Tuts. **20**(3), 2294–2323 (2018)
6. Islam, S.M.R., Avazov, N., Dobre, O.A., Kwak, K.s.: Power-domain non-orthogonal multiple access (NOMA) in 5G systems: potentials and challenges. IEEE Commun. Surveys Tuts. **19**(2), 721–742 (2017)
7. Khorov, E., Kureev, A., Levitsky, I., Tutelian, S.: Scheduling for downlink non-orthogonal multiple access in Wi-Fi networks. In: 2018 International Scientific and Technical Conference Modern Computer Network Technologies (MoNeTeC), pp. 1–6 (2018)
8. Tian, Z., Wang, J., Wang, J., Zhang, C., Tang, Z., Wu, F.: More clients connected by NOMA in the downlink transmission of WLANs. In: 2017 13th International Wireless Communications and Mobile Computing Conference (IWCMC), pp. 1968–1973 (2017)
9. Kheirkhah Sangdeh, P., Pirayesh, H., Yan, Q., Zeng, K., Lou, W., Zeng, H.: A practical downlink NOMA scheme for wireless LANs. IEEE Trans. Commun. **68**(4), 2236–2250 (2020)
10. Pavan, B.S., Harigovindan, V.P.: A novel channel access scheme for NOMA based IEEE 802.11 WLAN. Sādhanā **46**(3), 1–6 (2021). https://doi.org/10.1007/s12046-021-01669-2
11. Ali, M.S., Tabassum, H., Hossain, E.: Dynamic user clustering and power allocation for uplink and downlink non-orthogonal multiple access (NOMA) systems. IEEE Access **4**, 6325–6343 (2016)

12. Harigovindan, V.P., Babu, A.V., Jacob, L.: Proportional fair resource allocation in vehicle-to-infrastructure networks for drive-thru Internet applications. Comput. Commun. **40**, 33–50 (2014)
13. nsnam: ns-3 Design Documentation Wi-Fi Module (2018). https://www.nsnam.org/docs/models/html/wifi.html

Extension of Search Facilities Provided by 'CoWIN' Using Google's Geocoding API and APIs of 'CoWIN' and 'openweathermap.org'

Sudipta Saha[1], Saikat Basu[1(✉)], Koushik Majumder[1], and Debashish Chakravarty[2]

[1] Maulana Abul Kalam Azad University of Technology, Haringhata, West Bengal, India
saikat.basu@makautwb.ac.in
[2] Indian Institute of Technology, Kharagpur, West Bengal, India
dc@mining.iitkgp.ac.in

Abstract. Pandemic COVID-19 creates devastating impacts on healthcare, education, economy, and society. Herd immunity against COVID-19 would permit society to return to normal. To achieve herd immunity against COVID-19 harmlessly, a significant portion of a population would need to be vaccinated. India is the second most populated country globally, already administered the second-highest number of doses to its citizen successfully. The Government of India's popular web portal/app 'CoWIN' is used to register and monitor vaccinations. It helps Indians to find the nearest available vaccination slots. In the present work, an Android application is developed to enhance the search facility provided by 'CoWIN' to make it more convenient. It helps the users to find all available vaccination slots within a 15 km radius in one search. This application also provides a weather forecast for the searched location and search date. Users are also able to know how much fees are charged by a center during search time. This application has tried to make the 'CoWIN's search by map facility more informative. It is expected that this application will improve the user's vaccination slots search experiences and make the searchings more convenient.

Keywords: CoWIN · Vaccination · India · Slot · Search facility · API

1 Introduction

Severe acute respiratory syndrome coronavirus-2 (SARS-CoV-2), previously called 2019 novel coronavirus, was first identified in Wuhan city of Hubei Province in China in December 2019. This virus creates flu-like symptoms with dry cough, fever, sore throat, and severe breathing problems among humans. Nearly 221 countries and territories are affected by this virus [3]. The World health organization (WHO) declared this epidemic a 'Public Health Emergency

I. Woungang et al. (Eds.): ANTIC 2021, CCIS 1534, pp. 177–189, 2022.
https://doi.org/10.1007/978-3-030-96040-7_14

of International concern' on 30th January 2020 and Pandemic on 11th March 2020. To date, 225,119,797 people worldwide diagnosed with SARS-CoV-2 positive, among them, 4,638,720 had already died [3]. The country that suffers the highest impact of this virus is the United States [3]. It losses 677,737 citizens among 41,816,668 positive cases. Unfortunately, India is in the second-highest position; 442,688 Indians passed away among 33,236,921 positive cases [3]. This pandemic has had a profound impact on the education system. According to United Nations Educational, Scientific and Cultural Organization (UNESCO), more than 800 million students worldwide have been suffered; one-fifth of students unable to attend school, one-fourth of students unable to attend higher education classes. Most countries imposed stringent restrictions on the mobility of people to restrict the spread of the virus; as a result, most of the economic activity remained nearly shut down for an extended period [6]. India's Gross Domestic Product (GDP) contracted by 7.3% in 2020–21, according to the National Statistical Office. This performance is the worst for the Indian economy since independence. Lakhs of Health Care Workers (HCW) were infected, and thousands were dead worldwide due to COVID-19. The Pan American Health Organization (PAHO) reported 570,000 HCW were infected, and 2500 were dead before September 2, 2020, due to COVID-19 [12]. Indian Medical Association (IMA) also reported near about 800 doctor's deaths in the second wave of COVID-19 [11]. COVID-19 hospitalization costs in India in the case of private hospitals are too high for an average Indian.

Herd immunity from covid-19 is an essential requirement to restore society into a normal situation. It can be achieved by immunizing the whole community by either vaccination or previous infection. Some of the vaccine development efforts have already started and have created some vaccines successfully. The popular vaccines have shown positive results in preventing disease among the community, minimization of hospital admission, minimization in disease complication, and of course, minimization of deaths among patients [14]. To date, 5,692,346,079 vaccine doses have been applied across 184 countries; the rate is 33,488,886 per day [18]. India's drug regulator has approved three vaccines Covishield (Indian name for the Oxford-AstraZeneca vaccine), Covaxin (India's Indigenous vaccine produced by Bharat Biotech), and Sputnik V (Gamaleya National Research Institute of Epidemiology and Microbiology, Moscow, Russia), for Indians [14]. 73,89,05,342 vaccine doses are already applied in India, 17,64,42,772 Indians are fully vaccinated [7]. Eight vaccines had already undergone or are currently undergoing clinical trials in India [4]. These are Covishield by the Serum Institute of India (phase III human clinical trial completed), Covaxin by Bharat Biotech Ltd (phase III human clinical trial completed), ZyCoV-D by Zydus Cadila (phase III human clinical trial ongoing), novel COVID-19 vaccine by Biological E. Limited (phase I and II human clinical trials ongoing), Sputnik V by Dr. Reddy's Laboratories(phase II human clinical trial ongoing), BBV154 - Intranasal vaccine (phase I human clinical trial ongoing), Covovax by join venture of Indian council of Medical Research (ICMR) and Serum Institute of India (phase II/III human clinical trial ongoing), and mRNA based vaccine HGCO19 by Gennova Biopharmaceuticals Ltd (phase II/III human clinical trial ongoing). India has an estimated population of 1380 million (as of 2020), and the Government of India has planned to administer the COVID-19 vaccine to all willing Indians [14].

'CoWIN' [1] is a web portal/app used by Indians to vaccinate successfully. Indians use this portal to find suitable slots for themselves. They can register themselves in this portal, book slots at their convenience, get vaccinated by visiting centers physically, and finally download a vaccination certificate. It is owned and operated by India's Ministry of Health and Family Welfare (MoHFW). People have a natural tendency to prefer the nearest vaccination center around them. In the cases of senior citizens, physically weak people, and differently-abled people, this requirement is more crucial. Vaccination of children, which yet not stated in India, but will be started shortly. It is inconvenient to carry small children to a far-vaccination center to vaccinate them. In 'CoWIN', search by pincode is not equally beneficial for each pincode. at all. Sometimes, slots are all filled in a particular user's pincode, but slots are empty very near to him/her. In that case, users have to search more than once using different pincodes of his/her vicinity, which is very cumbersome. On the other hand, the district-wise search covers a large geographical area. Users may not agree to visits the vaccination centers, which are geographically located in his/her district but far away from his/her home. In the present work, an Android application is developed to enhance the 'CoWIN' portal's [1] search facility. Android platform is chosen by considering its immense popularity among people. This application implements a search facility where all the vaccination slots under all the pincodes with 15 km of searching pincode are listed. This application also tried to improve and clarify the search by Map facility provided by 'CoWIN'.

Weather forecasts have a profound effect on people's planning. Weather and weather forecasts affect people's convenience highly, lifestyle mediumly, and travel extremely [15]. Therefore, this application provides weather highlights when a user searches for a vaccination center for a particular pincode and a particular date. It is expected that this additional information will make the user's vaccination plan more realistic.

2 Related Works

India has linked the vaccine campaign into 'CoWIN' portal/app [1], allowing the government to monitor immunization coverage, analyse the efficacy of the program, and plan future tactics to address the difficulties presented in a dynamic COVID environment [21]. Apart from providing vaccination slot finding and booking facilities, the 'CoWIN' [1] portal/app also provides dashboard facilities [7]. It shows the total number of first and second doses vaccinations administered and the number of sites conducting vaccination. It gives information about the state-wise and district-wise number of doses one and doses two applied. The total number of sites all over the country conducting vaccination and also state-wise and district-wise vaccination site details are provided by 'CoWIN'. It displays live data of the number of persons vaccinated partially and fully in every Indian state and union territory. It provides the same data district-wise also. 'CoWIN' displays vaccination provides by each center of a district of a state or union territory. It also updates information about the newly administered doses for the current day. It gives a graphical representation (using pie charts) to represent the percentage of individual age group vaccinated and individual gender vaccinated. 'CoWIN' also displays the percentage of individual vaccine doses administered using pie charts. It also gives a graphical representation of the rural vs. urban trend of vaccination. The number of government and private sites conducting vaccination in each district can also be known from 'CoWIN'. 'CoWIN' portal/app displays the total number of registered people in 'CoWIN', the number of new registrations today. It also displays the percentage of the individual age group registered on this site. Dr. R.S. Sharma, CEO of National Health Authority (in 2nd Public Health Summit 2021organized by the Confederation of Indian Industry (CII)) said that the 'CoWIN' portal is unique in the world in terms of portability, scalability, and inclusivity. It is a citizen-centric platform and can provide granular details of each individual. More than 50 countries from Latin America, Africa, and Asia have shown interest in the 'CoWIN' system, and the Government of India will share the technology with them free of cost.

The 'CoWIN' portal/app, on the other hand, has received a few criticisms [1]. The technology behind this portal/app was initially not scalable enough to accommodate the growing user demand. As a result, it had suffered from certain server problems, such as inconsistencies among the information displayed in the portal and actual slot allocations [16]. The 'CoWIN' app is also prone to freezing and crashing at the beginning [16]. Initially, those aged between 18–45 were barred from registering for vaccination on-site. A person in this age range would need to use 'CoWIN' to register and reserve slots [17]. There were numerous problems with this policy. This approach ignored the approximately 18 million Indian homes without access to mobile phones or the Internet [19]. People in remote regions who were already afflicted by inadequate healthcare and a scarcity of medical supplies have to withstand even more institutional ignorance because of the e-registration way to be vaccinated [19]. Differently-abled people were also impacted. According to some surveys, 86% of differently-abled individuals are unaware of how to use the 'CoWIN' portal/app, and 61% of them believe that a smartphone is too expensive for them to purchase [19]. However, on June 15, 2021, the Indian government modified its policy. This makes it optional to pre-register and book slots in advance using 'CoWIN.' Any adult, regardless of age, may visit the convenient vaccination location, register on-site, and receive the vaccine all in one visit [20].

Some research also shows that the 'CoWIN' site might be improved. According to [13], the 'CoWIN' platform may be utilized to disseminate vaccine-related educational and communication resources. Additionally, the 'Cowin' platform may be utilized to train additional workers for the inoculation effort [21]. Statistics on adverse responses to various CoVID-19 vaccinations should be included in this portal/app [16]. [16] proposes a required QR code scan-based batch number registration (identifying) in the 'CoWIN' before opening a vial to safeguard Indians against fraudulent vaccinations.

3 Application Details

3.1 Functional Description

The name of the application is 'Protishedhak Bondhu.' The term 'Protishedhak' is a Bengali word that means 'prohibitive' or anything that can prevent disease (i.e., vaccine), and the term 'Bondhu' is a Bengali translation of the word 'friend.' This is an Android app and can be installed on any smartphone running the Android Operating System. This app provides vaccination slot search facilities of India like 'CoWIN'[1]. In the app 'Protishedhak Bondhu,' a user is needed to provide a particular pincode and pick a date from a popup calendar. In response to these inputs, this app displays the details of all vaccination slots (available and booked) under this particular pincode in a listed view like 'CoWIN'. In 'CoWIN', when searching for slots, a user can not see the corresponding weekdays of the dates. She/he needs to check an external calendar if necessary. Since this app uses a popup calendar as a date picker, a user is always aware of the weekdays for which he/she searches. Figure 1 shows how a user needs to pick a date from the popup calendar at the searching time. 'CoWIN' does not show the charges of vaccination in the case of paid vaccination centers. This app, 'Protishedhak Bondhu,' indicates the fee type (free or paid) of vaccination along with vaccination charges in INR. Figure 2 shows that this application listed vaccination charges along with other slot information. In addition to slot information, this application indicates the place name of the pincode in boldface top of the listed view. The most interesting factor of this app is that it is capable of showing weather details like maximum temperature, minimum temperature, humidity, overall weather description (i.e., sunny, cloudy, moderate rain, little rain), along with a matching weather icon for the particular date and particular pincode user searches. This facility is available for the current date and the next seven days. Figure 3 represents the screen where 'Protishedhak Bondhu' displays the exact place name and weather details of searching pincode. In 'CoWIN', there are three types of search facilities- i.e., Search by Pin, Search by District and Search by Map. Sometimes no vaccination centers are present under a particular pincode, but vaccination centers are present under the other pincodes, which are very close to the searched pincodes. Districtwise search facilities indicate the vaccination centers under a very long geographic area. Users may not find it is convenient to visit far distanced vaccination centers under his/her district. For this purpose, this app has a facility to search vaccination centers under a particular pincode along with all vaccination centers under all pincodes, which are within a 15 km radius of the searching pincode. This app is also capable of showing the vaccination center on a map view on a date basis. This application never asks for user location; a user does not need to use the location service of his/her device. This app asks the user for a pincode and date; the application sets markers on all vaccination centers under this pincode for that date in map view. Figure 4 shows vaccination centers under a pincode with cyan-colored markers. This app provides detailed information for all vaccination slots of the center on that date when a user clicks on a marker. Figure 5 shows that this application shows all the slot information of a particular date as Toast message on clicking a marker. In 'CoWIN', when the user clicks on a particular marker, it will show very brief details of empty slots only for a week at once for that particular center. Using this application, a user can search for all vaccination centers under all pincodes within 15 km of a particular pincode in map view. This app can set the markers on all vaccination centers with a 15 km radius of a particular pincode and provide details of all slots of a particular date on clicking the marker. Figure 6 shows that this app sets the magenta coloured markers on all the vaccination centers under all the pincodes, which are inside a 15 km radius of a given pincode. In Table 1, the search facilities provided by 'CoWIN' [1] and 'Protishedhak Bondhu' are compared.

Table 1. Comparison of search facilities of 'CoWIN' and 'Protishedhak Bondhu'

Search facilities	CoWIN [1]	Protishedhak Bondhu
Pincode wise Search present?	Yes	Yes
District Wise Search present?	Yes	No
Search inside all the pincodes within 15 km of a given pincode present?	No	Yes
Weekday of the date shown in search time?	No	Yes
Exact Place Name of pincode is shown in searching time?	No	Yes
Weather forecast available for the pincode at searching time?	No	Yes
Search by Map using pincode available?	Yes	Yes
Search by Map using current location available?	Yes	No
Search by Map using pincode contains marker only on centers under this pincode?	No, it contains markers on centers that are outside of the pincode, misleading users	Yes
Search by Map using pincode to find vaccination centers under all pincodes within a 15 km radius of a given pincode?	No	Yes
Vaccination slot details information are complete in map view?	No, contains only age group and available doses if only slots available	Yes, complete information about slot details irrespective of slot availability
Filters like "search by age", "search by vaccine type," and "search by cost" are present while searching?	Yes	No, very soon it would be included

3.2 Needed API Details

This application uses some Application Programming Interfaces (APIs) for its purpose. 'Apisetu//gov.in' [8] is a web portal that provides many APIs for multiple purposes. Ministry of Electronics & IT, Government of India, is the owner of this portal. 'Apisetu//gov.in' has a section for listing CoWIN's API [2]. This section listed all public, protected APIs provided by web portal 'CoWIN' [1]. The appointment availability API of 'CoWIN' (a public API) is used in this application. It collects vaccination slot information. It is a REpresentational State Transfer (REST) API version 2 and responds to a request Uniform Resource

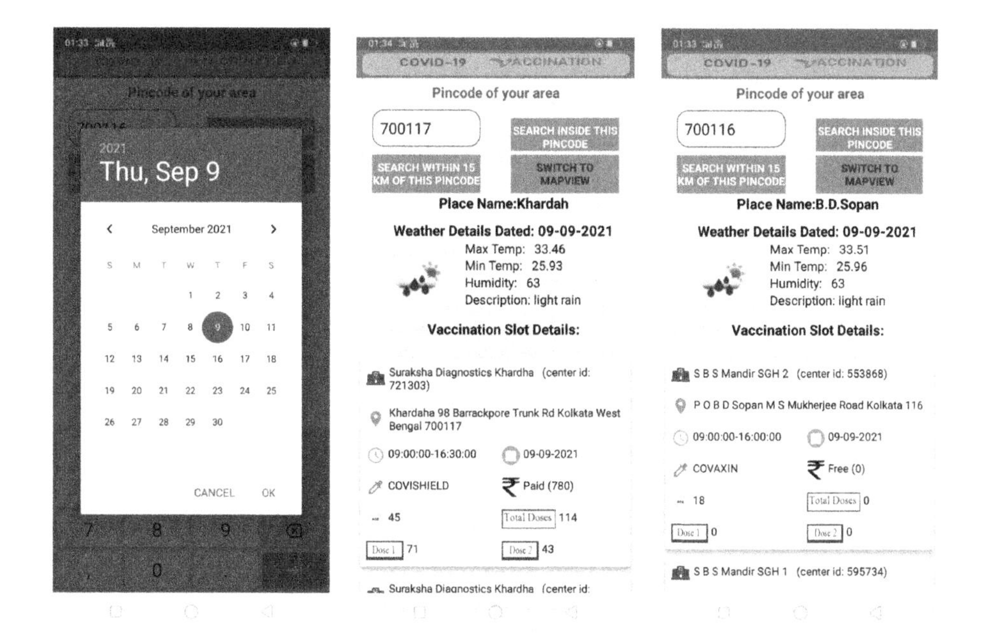

Fig. 1. Date picker is used to pick date at searching time.

Fig. 2. Vaccination fee is shown in slot information.

Fig. 3. List view of vaccination slot and weather details of pincode.

Locator (URL) in JavaScript Object Notation (JSON) format. The base URL used in this app for accessing vaccination slot details is 'https://cdn-api.co-vin.in/api/v2/appointment/sessions/public/findByPin? pincode=⟨*pincode*⟩ & date =⟨*Date*⟩'. The application fetches all the vaccination center's details under a particular pincode using this API. Fetched fields are vaccination center's id, center's name, center's address, vaccination timing, vaccine name, fee type (free or paid), amount of fee, total available doses, dose 1's availability, dose 2's availability, minimum age, latitude and longitude of the center. The application presents the selected fields among them in front of users.

This application uses free APIs of the web portal 'openweathermap.org' [9] to collect weather-related data. 'openweathermap.org' provides REST API version 2.5 and responds to a request URL in JSON and Extensible Markup Language (XML) format. To use APIs provided by 'openweathermap.org', a key that they generate is necessary. Key generation is free on this website. This application uses JSON format responses of request URL. Since this app provides vaccination slot availability data pincode-wise and date-wise, weather data needed to be fetched according to particular pincode and date. However, 'openweathermap.org' has no free API, which provides weather data date-wise and pincode-wise together. This application uses 'One Call' APIs of 'openweathermap.org,' which provides weather data of current date and following seven days for a particular latitude and longitude. 'One Call API's have no facility of pincode-wise search. For that reason,

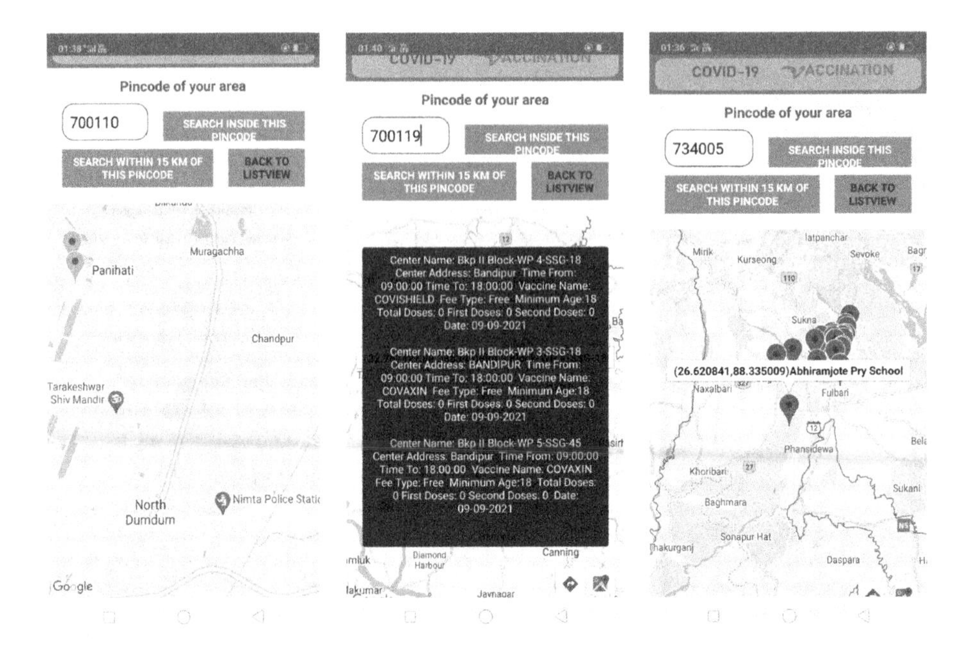

Fig. 4. Cyan markers indicate vaccination center in Map view.

Fig. 5. Clicking on a marker give all slot details of this particular center in Toast message.

Fig. 6. Searching for all vaccination center inside 15 km of given pincode in Map view.

this application uses two free APIs of 'openweathermap.org.' It uses 'current weather data' API -'https://api.openweathermap.org/data/2.5/weather?zip =⟨*pincode*⟩,⟨*countrycode*⟩&appid=⟨*APIkey*⟩' to fetch exact place name and accurate latitude and longitude of the searched pincode. Now using accurate latitude and longitude of searched pincode, this application calls the 'One call' API of 'openweathermap.org'. This 'One call' API 'https://api.openweathermap.org/ data/2.5/onecall?lat=⟨*latitude*⟩&lon=⟨*longitude*⟩ &exclude=current,minutely , hourly,alerts&appid=⟨*APIkey*⟩'. Since weather details APIs for the next 16 days or 30 days are not free from 'openweathermap.org', this app only provides weather details for the current and next seven days. Latitude and longitude values get from the 'CoWIN' API is very crude, contain only integer parts. So, using these values, accurate weather condition fetching is not possible.

Finally, to display a map, this application uses Google Maps Platform's APIs. This application generates a key for this purpose by logging in 'https://console.cloud.google.com.' 'Maps SDK for Android' API and 'Geocoding' API were needed to be enabled for this key. 'Maps SDK for Android' was needed for showing google maps. Latitude and longitude values from 'CoWIN'

API are very crude; these cannot be used to set the markers on Google Map. If it is used, then markers for multiple vaccination centers will overlap. To solve this problem, 'Geocoding' API is used. 'Geocoding' API is used to fetch the exact latitude and longitude of the center using the vaccination center name. Then the exact latitude and longitude values of centers set markers on those positions of google map.

3.3 Design Details

This application is developed using Android Studio version 4.2.2. Android Studio is an official Integrated Development Environment (IDE) for developing apps for Android OS. This application is compiled against API level 30 (Android 11) and it is capable to run any android device uses API level 19 (Android 4.4) or above. The Android PacKage (APK) size of this application - 'Protishedhak Bondhu' is 8.6 MB, and it requires 28.8 MB storage upon installation.

This application provides search results in two views- list view and map view. Users can also switch between two views. In both views, a user can search for vaccination slot details under a particular pincode and slot details under all the pincodes within a 15 km radius of a particular pincode. The second search facility is now available for the state of West Bengal only. Soon, we will cover this facility all over India. An adjacency matrix is prepared to implement searching all slot details under all the pincodes within a 15 km radius of a particular pincode. In this matrix, each row corresponds to every pincode of West Bengal, and columns hold the pincode numbers within a 15 km radius of the particular pincode. The developers of this app have taken the help of the website 'www.mapsofindia.com' [10] to collect all the pincodes of West Bengal. To find the pincodes within 15 km of a particular pincode, developers take the help of the website 'https://www.freemaptools.com/' [5].

Some of the vital algorithms of this application are given below. Algorithm 1 is used to provide vaccination slot details in list view and weather details under a particular pincode area. Algorithm 2 is used to deliver vaccination slot details under a particular pincode area and vaccination slot details under all pincodes areas within 15 km of the searching pincode in list view, along with weather details. Algorithm 3 is used to provide vaccination slot details in map view under a particular pincode area. Algorithm 4 is used to deliver vaccination slot details in map view under a particular pincode area and vaccination slot details under all pincodes areas inside 15 km of the searching pincode. Algorithm 5 is used to display all vaccination slot details of a center in toast message when the marker on this center is clicked.

Algorithm 1 – Provide vaccination slot details in list view and weather details under a particular pincode area

Input:	Date, pincode
Output:	Brief weather details and vaccination slot details under this pincode in list view
Step 1:	Take a valid 6-digit pincode from the user.
Step 2:	Take a date input from the user using a date picker.
Step 3:	Use 'current weather data' API of 'openweathermap.org' to fetch the place name, latitude, longitude of this pincode
Step 4:	Use 'One Call' API of 'openweathermap.org' to fetch current and following seven days weather data using latitude and longitude of this pincode
Step 5:	Save weather data of 8 days in a 2-dimension Weather matrix, where each row represents weather data for a particular date
Step 6:	Use a for loop to match input date with Weather matrix 1st column and display the rest of the columns data of this matched row in device screen
Step 7:	Create a Vaccination_Model class with necessary fields like center's id, center's name, center's address, vaccination timing, vaccine name, fee type (free or paid), amount of fee, total available doses, dose 1's availability, dose 2's availability, minimum age.
Step 8:	Create an empty Array List named Vaccine_Center of object type Vaccination_Model
Step 9:	Use 'CoWIN' API with the date and pincode to fetch vaccination slot details
Step 10:	For each fetched vaccination slot, create a Vaccination_Model object with the fetched data and add this object to the Array List Vaccine_Center
Step 11.	Display the Array List Vaccine_Center in the screen using List View

Algorithm 2 – Provide vaccination slot details under a particular pincode area and vaccination slot details under all pincodes areas within 15 km of the searching pincode in list view, along with weather details

Input:	Date, pincode, an Adjacency matrix in which each row represents a pincode and columns represent pincodes with a 15 km radius of that particular pincode
Output:	Brief weather details and vaccination slot details under this pincode and all pincodes with 15 km radius of this pincode in listview
Step 1:	Take a valid 6-digit pincode from the user.
Step 2:	Take a date input from the user using a date picker.
Step 3:	Use 'current weather data' API of 'openweathermap.org' to fetch the place name, latitude, longitude of this pincode
Step 4:	Use 'One Call' API of 'openweathermap.org' to fetch current and following seven days weather data using latitude and longitude of this pincode
Step 5:	Save weather data of 8 days in a 2-dimensional Weather matrix, where each row represents weather data for a particular date
Step 6:	Use a for loop to match input date with matrix 1st column and display the rest of the columns data of this matched row in device screen
Step 7:	Create a Vaccination_Model class with necessary fields like center's id, center's name, center's address, vaccination timing, vaccine name, fee type (free or paid), amount of fee, total available doses, dose 1's availability, dose 2's availability, minimum age.
Step 8:	Create an empty Array List named Vaccine_Center of object type Vaccination_Model
Step 9:	Use a for loop to match input pincode with a row of the adjacency matrix
Step 10:	For each pincodes which are presents in the columns of the matched row of adjacency matrix, use 'CoWIN' API with the date to fetch slot details
Step 10.1:	For each fetched vaccination slot details under a pincode, create a Vaccination_Model object with the fetched data and add this object to the Array List Vaccine_Center
Step 11:	Display the ArrayList Vaccine_Center in the screen using List View

Algorithm 3 – Provide vaccination slot details in map view under a particular pincode area

Input:	Date, pincode
Output:	Vaccination slot details under this pincode in map view
Step 1:	Take a valid 6-digit pincode from the user.
Step 2:	Take a date input from the user using a date picker.
Step 3:	Create a 2D array of string type named Vaccine_Center with 2 columns
Step 4:	Initialize a variable rowno with 0
Step 5:	Use 'CoWIN' API with the date and pincode to fetch slot details what include the center's id, center's name, center's address, vaccination timing, vaccine name, fee type (free or paid), amount of fee, total available doses, dose 1's availability, dose 2's availability, and minimum age
Step 6:	For each fetched vaccination slot
Step 6.1:	Keep slot_details data in all columns of Vaccine_Center[rowno][1], left Vaccine_Center[rowno][0] blank
Step 6.2:	Use the vaccination center's name and 'Geocoding' API to fetch the latitude and longitude of the center.
Step 6.3:	Use this latitude and longitude to set a marker on google map, and set the title of the marker as $(\langle latitude \rangle, \langle longitude \rangle)\langle centername \rangle$
Step 6.4:	Keep $(\langle latitude \rangle, \langle longitude \rangle)\langle centername \rangle$ in the Vaccine_Center[rowno][0]
Step 6.5:	Increment rowno by 1

Algorithm 4 – Deliver vaccination slot details in map view under a particular pincode area and vaccination slot details under all pincodes areas inside 15 km of the searching pincode

Input:	Date, pincode, an Adjacency matrix in which each row represents a pincode and columns represent pincodes with a 15 km radius of that particular pincode
Output:	Vaccination slot details under this pincode and all pincodes with 15 km radius of this pincode in map view
Step 1:	Take a valid 6-digit pincode from the user.
Step 2:	Take a date input from the user using a date picker.
Step 3:	Create a 2D array of string type named Vaccine_Center with 2 columns
Step 4:	Initialize a variable rowno with 0
Step 5:	Use a for loop to match input pincode with a row of the adjacency matrix
Step 6:	For each pincodes which are presents in the columns of the matched row of adjacency matrix, use 'CoWIN' API with the date to fetch slot details
Step 6.1:	For each fetched vaccination slot details under a pincode
Step 6.1.1:	Keep slot_details data in all columns of Vaccine_Center[rowno][1], left Vaccine_Center[rowno][0] blank
Step 6.1.2:	Use the vaccination center's name and 'Geocoding' API to fetch the latitude and longitude of the center.
Step 6.1.3:	Use this latitude and longitude to set a marker on google map, and set the title of the marker as (⟨*latitude*⟩ ,⟨*longitude*⟩) ⟨*centername*⟩.
Step 6.1.4:	Keep (⟨*latitude*⟩, ⟨*longitude*⟩) ⟨*centername*⟩ in the Vaccine_Center[rowno][0]
Step 6.1.5:	Increment rowno by 1

Algorithm 5 – Algorithm of vaccination center's Marker's Onclick event

Input:	int rowno, 2-D String array Vaccine_Center
Output:	Vaccination slot details of a center
Step 1:	Get the title on the marker in a string variable temp1
Step 2:	Set a string variable temp2=Null
Step 3:	For int i=0 to rowno-1
Step 3.1:	Compare temp1 with Vaccine_Center[i][0]
Step 3.2:	If both matches
Step 3.2.1:	concatenate temp2 with Vaccine_Center[i][1]
Step 4:	Display temp2 in Toast Message

4 Benefits Provided by 'Protishedhak Bondhu'

'Protishedhak Bondhu' is a COVID-19 vaccination slot searching application. Indians can use this application to search for the nearest vaccination slots in any place inside India. This application provided an enhanced search facility compared to 'CoWIN'.

1. Most significant addition of 'Protishedhak Bondhu' is to provide a vaccination slot searching facility within a 15 km radius of a given pincode, not just under the pincode.
2. Second significant addition is, this app can deliver weather forecasts for searched pincode and particular search date.
3. It provides the "Search by Map" facility to display the vaccination centers under a given pincode or within a 15 km radius of a given pincode in mapview.
4. This app makes the "Search on the Map" facility of 'CoWIN' more informative. 'CoWIN' provides little information regarding vaccination slots when "Search on the Map" is used. Depending on the user's preference, this app can place a marker on every vaccination center under a given pincode or within a 15-kilometer radius of the given pincode. This app provides sufficient details of the vaccination center in toast view when a user clicks on a marker.
5. This app makes the "Search on the Map" facility of 'CoWIN' more accurate. While searching on the map using a pincode, 'CoWIN' set markers on the vaccination center outside the given pincode. This can mislead the user. This app can set markers exactly the vaccination centers under a given pincode.

6. This app can show the exact fees of vaccination by fetching real-time data from the 'CoWIN' database. 'CoWIN' does not provide this information to the users.
7. This app displays a popup calendar while a user searches for a vaccination slot using a date. Sometimes users decide on slot booking depending on their preferred weekdays. However, 'CoWIN' does not provide this facility yet.

5 Conclusions and Future Scopes

This application is designed to enhance the search facility provided by 'CoWIN' (India government web portal for COVID-19 vaccination registration) by introducing a new kind of search where a user can search for all available vaccination slots within a 15 km radius of his/her pincode at once. Along with this facility, this app also provides weather forecasts for a particular pincode and a particular date. Not only that, this app does 'Search by Map' facilities of 'CoWIN' more informative. Child (age group 2–18) COVID-19 vaccination will start shortly. This app may be equally beneficial for them also. This app uses some public and free APIs for providing these facilities to the users. However, the search facility for the vaccination slots within 15 km of his/her pincode at once is only available for the state of West Bengal only. Very soon, this facility for all over India will be available. Presently, all the free-of-cost vaccination slots are booked just after immediate release. So most ordinary people feel trouble finding free-of-cost vaccination centers in their vicinity. This app will shortly add a service to check free slot availability every 5 min in the background and immediately notify the users.

References

1. Check Your Nearest Vaccination Center And Slots Availability. https://www.cowin.gov.in/. Accessed 12 Sept 2021
2. Co-WIN Public APIs. https://apisetu.gov.in/public/marketplace/api/cowin. Accessed 12 Sept 2021
3. Countries where COVID-19 has spread. https://www.worldometers.info/coronavirus/countries-where-coronavirus-has-spread/. Accessed 12 Sept 2021
4. COVID-19 Vaccines under trials in India. https://vaccine.icmr.org.in/covid-19-vaccine. Accessed 10 Sept 2021
5. Find Indian pincodes inside a radius. https://www.freemaptools.com/find-indian-pincodes-inside-radius.htm. Accessed 28 Apr 2019
6. The Global Economic Outlook During the COVID-19 Pandemic: A Changed World. https://www.worldbank.org/en/news/feature/2020/06/08/the-global-economic-outlook-during-the-covid-19-pandemic-a-changed-world. Accessed 8 June 2020
7. Total Vaccination Doses. https://dashboard.cowin.gov.in/. Accessed 12 Sept 2021
8. Use APIs as Building Blocks for Innovative Applications. https://apisetu.gov.in/. Accessed 12 Sept 2021
9. Weather API. https://openweathermap.org/api. Accessed 12 Sept 2021

10. West Bengal Pin Code. https://www.mapsofindia.com/pincode/india/west-bengal/. Accessed 26 Apr 2021
11. ANI: Nearly 800 doctors died amid second wave of COVID-19, says Indian Medical Association. https://zeenews.india.com/india/nearly-800-doctors-died-amid-second-wave-of-covid-19-says-indian-medical-association-2372775.html. Accessed 30 June 2021
12. Erdem, H., Lucey, D.R.: Healthcare worker infections and deaths due to Covid-19: a survey from 37 nations and a call for who to post national data on their website. Int. J. Infect. Diseases. **102**, 239 (2021)
13. Gupta, M., Goel, A.D., Bhardwaj, P.: The COWIN portal-current update, personal experience and future possibilities. Indian J. Community Health **33**(2), 414–414 (2021)
14. Kumar, V.M., Pandi-Perumal, S.R., Trakht, I., Thyagarajan, S.P.: Strategy for Covid-19 vaccination in India: the country with the second highest population and number of cases. NPJ Vaccines **6**(1), 1–7 (2021)
15. McCarthy, P.: Defining the impact of weather. In: Proceedings 22nd Conference on Weather Analysis and Forecasting/18th Conference on Numerical Weather Prediction, American Meteorological Society, Utah (2007)
16. Mukherjee, D., Maskey, U., Ishak, A., Sarfraz, Z., Sarfraz, A., Jaiswal, V.: Fake Covid-19 vaccination in India: an emerging dilemma? Postgrad. Med. J. (2021)
17. PTI: CoWin portal: registration must for those between 18–45 years to get COVID-19 vaccine shot. https://www.indiatvnews.com/news/india/cowin-portal-online-registration-mandatory-covid-19-vaccination-for-18-to-45-years-cowin-gov-in-700359. Accessed 25 Apr 2021
18. Randall, T., Sam, C., Tartar, A., Murray, P., Cannon, C.: More than 5.66 billion shots given: Covid-19 tracker. https://www.bloomberg.com/graphics/covid-vaccine-tracker-global-distribution/. Accessed 12 Sept 2021
19. Sharma, R.: India: digital divide and the promise of vaccination for all. South Asia@ LSE (2021)
20. Singhal, A.: Online registration on CoWIN not mandatory for Covid vaccination: health ministry. https://www.indiatoday.in/coronavirus-outbreak/story/online-registration-booking-appointment-cowin-not-mandatory-covid-vaccine-1815219-2021-06-15. Accessed 15 June 2021
21. Verma, N.: Reassess, reorganize & realign vision for effective inoculation: a holistic review of SARS covid-19 vaccination drive in India

Detection of DDoS Attack in IoT Using Machine Learning

Naveen Kumar[1], Abdul Aleem[1(✉)], and Sachin Kumar[2]

[1] Siksha 'O' Anusandhan Deemed to be University, Bhubaneswar, Odisha, India
{naveenkumar,abdulaleem}@soa.ac.in

[2] Motilal Nehru National Institute of Technology Allahabad, Prayagraj, Uttar Pradesh, India

Abstract. Distributed Denial-of-Service (DDoS) attacks have recently increased exponentially in numbers. With the advent of the Internet-of-things (IoT), it becomes easy for attackers to perform a DDoS attack. IoT devices are resource constraints; therefore, their security is compromised. Adversary takes the benefit of compromised security and captures the IoT devices; these captured devices are used to perform a DDoS attack. This article proposes machine learning-based techniques for the detection of DDoS. The research involves performing a DDoS attack in an IoT network and then capturing the attack traffic and regular traffic using Wireshark. The captured traffic is processed to fetch its various features, and machine learning is applied for classification that can distinguish the attack traffic from the regular traffic. Four different machine learning-based approaches have been applied to the collected data to detect malicious traffic. Naïve Bayes turns out to be the better performing algorithm for this purpose.

Keywords: DDoS · DoS · IoT · Network security · Security attack

1 Introduction

In the late 20th century, the Internet was developed to make sharing of information easy among all connected machines in a network. However, at that time, while designing the Internet, designers did not keep security in mind; this may be the reason that something equivalent to a virus began along with the birth of the Internet. In ARPANET (the first wide area network consists of around 60000 machines), a self-replicating worm caused DoS on 10% of the connected devices [1]. With the advancement of technology, for over 1 billion users, the Internet has become a basic need for today's life. Business, shopping, communication, everything needs a secure Internet connection. The downside of this convenience is a vulnerability that caused the problem if explored with the wrong intention. Malicious users often interrupt normal operations and steal information.

An attack aiming to disrupt the services of a connected website server is called an availability-based attack, which is one of the most severe security attacks. When performed using different protocols, variable packet sizes, and different time intervals, such attacks are more threatful. Such attacks are commonly referred to as DoS attacks [2]. When the attack is carried out via more than one attacking machine, it is termed

I. Woungang et al. (Eds.): ANTIC 2021, CCIS 1534, pp. 190–199, 2022.
https://doi.org/10.1007/978-3-030-96040-7_15

a DDoS attack. There have been recent reports of new attacks using IoT devices [3]. IoT devices are resource constraint devices that have low power, low memory, and low processing. The security of IoT devices is compromised due to these constraints. An adversary can easily capture IoT devices and use them to perform DDoS. These attacks are called botnet-based attacks.

Most of the previous works focus on detecting and mitigating DoS and DDoS attacks in a normal network scenario. Very few works have shown the practical implementation of DDoS attacks using IoT devices. In this article, we have described the successful execution of a DDoS attack in an IoT scenario. After, DDoS the traffic corresponding to regular and attack scenarios has been captured using the Wireshark tool. The pre-processing of the data is performed to fetch the attributes that can be used to detect the DDoS attack. This pre-processed data has been used for attack detection. Four machine learning approaches, i.e., Support Vector Machine (SVM) [4], Multi-Layer Perceptron (MLP) [5], Naïve Bayes (NB) [6], and Random Forest (RF) [7], are used for the DDoS detection. The performance evaluation has been done through metrics like accuracy, precision, sensitivity, and F-measure.

The rest of the paper is summarized as follows. Section 2 describes the work done by others for DDoS detection. Section 3 discusses the proposed work. Section 4 analyses the results obtained during DDoS detection; the paper is concluded in Sect. 5.

2 Related Work

In 2010, over three billion malware attacks were reported, and the number of DoS attacks skyrocketed by 2013 [8]. Many different ways to detecting and mitigating DDoS assaults have been presented. In this section, some of the most important DDoS detection research works have been briefed.

Apthorpe et al. [9] demonstrated that IoT data could even be used to infer consumer usage behaviors; behavior includes food habits, likes, dislikes, etc. It can be performed by reflecting finite states of IoT devices in the repeated temporal structure of incoming/outgoing traffic rates. SCADA anomaly detection focuses on unique data traffic patterns in sensors and controllers in the system [10, 11].

Miettinen et al. [12] demonstrated the use of machine learning to identify distinctive patterns in IoT networks. A variety of anomaly detection algorithms based on supervised, unsupervised, and clustering methods had been provided [13, 14], but with no emphasis on network anomaly detection. No and Ra [15] proposed a rapid entropy-based technique for DDoS detection. The suggested technique combines the FOEC method's lossless compression entropy with entropy calibration based on the number of packet kinds and packets. The fast Entropy technique lowered the computing time by 90% while maintaining the same accuracy as the standard entropy method.

The Rank Correlation-based Detection (RCD) algorithm, developed by Wei et al. [16], detects Distributed Reflection Denial of Service (DRDoS) attacks. This strategy assumes that the resulting flows will have the same properties if an attack is carried out using the same attacking pattern. RCD calculates the rank correlation between two flows on identifying the suspicious flows, which is then used for DRDoS detection. They have not, however, tested against complex scenarios or real-world DRDoS on the Internet.

Zhou et al. [17] have suggested an approach to differentiate two types of low-rate DDoS attacks, the constant attack, and the pulsating attack. This approach uses packet size distribution difference as a parameter to differentiate attack traffic from regular traffic. Real datasets are used in the simulations to show that the false-negative rate is minimal. The approach is unaffected by network topology, attack packet arrival patterns, or pulse patterns. The delay induced by network congestion is not considered. Network middlebox has limitations as limited power, limited storage, and processing.

Sivanathan et al. [18] suggested a solution to meet these constraints using a Software-Defined Network (SDN) to find an anomaly in the traffic patterns at flow-level granularity. Dynamic Attack detection is proposed by Lim et al. [19] using a scheduling-based architecture that detects attack traffic during the DDoS attack. This frame is based on a Software-Defined Network (SDN) controller.

Mousavi and St-Hilaire [20] proposed an entropy-based DDoS detection technique to secure the SDN controller. This approach used the entropy of the destination IP address to detect the attack. The approach detected the DDoS within the arrival of 250 packets of attack traffic in the SDN-based network. Kokila et al. [21] used SVM to detect DDoS attacks in the SDN environment. The authors used the DARPA intrusion detection dataset of the evaluation. SDN have been also experimented with advanced machine learning models like LRCN [22].

The application of machine learning-based techniques to detect DDoS is an emerging area. The area of DDoS detection and mitigation has been widely explored. However, very few works have been done regarding DDoS attack execution on a real IoT network using resource-constrained devices. This article focuses on executing DDoS, finding features that can be used to differentiate actual traffic from the attack traffic, and then using those features to detect DDoS attacks.

3 Proposed Work

This research aims to detect the attack based on the features extracted from the traffic of real-network. In order to achieve this, first, a network is set up that has various IoT devices and other network devices attached. Then the traffic is captured using a suitable tool, and the dataset is created that has extracted features of the captured traffic. Finally, machine learning techniques are applied to the dataset of captured traffic so that attack and non-attack scenarios can be distinguished based on the values of the various network traffic attributes. Hence, the proposed detection methodology can be summarized into three major phases, namely, network setup, dataset preparation, and attack detection. These phases are discussed in the subsequent subsections.

3.1 Network Setup

Figure 1 shows the network used for the proposed work. The network consists of five devices in which one laptop with low configuration acts as a gateway router to which all the other devices are connected. Another laptop act as an attacker, which is used to attack the gateway router and other devices in the network. There are three legitimate hosts, out of which two are mobile devices, and one is another laptop machine. One of the mobile devices is connected to a Bluetooth smartwatch (IoT device). Wireshark is installed on the gateway machine, which can collect all the communicating packets that will be analyzed further.

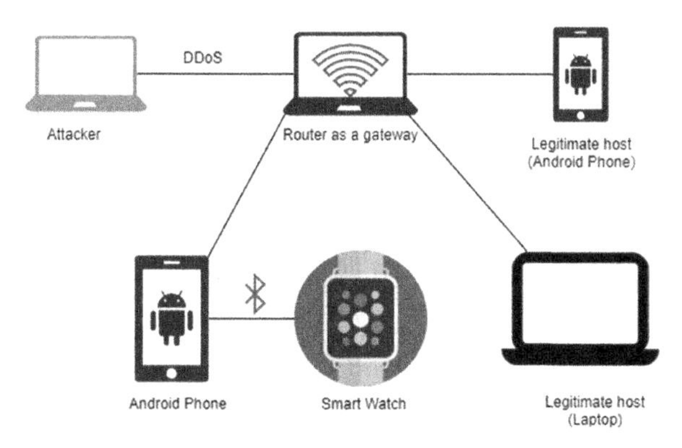

Fig. 1. Network setup

The attack traffic is detected at the router placed in a LAN of IoT devices and communicates with the public Internet. There are few assumptions about the consumer IoT network that a gateway router or any middlebox can observe all traffic between IoT edge device and network. Data traffic from the WiFi device should traverse through a gateway where the detection module would be installed. The attack traffic generated from devices within the network is inspected by the detection module. The victim may be a machine within LAN or outside the network. Any device can send regular traffic as well as attack traffic at any time.

3.2 Dataset Preparation

After the network setup, the next step is to run the scenario and capture the traffic so that it can be used for extracting the features. These features are chosen so that the regular traffic can be distinguished from the attack traffic. The packets have been captured using the Wireshark tool, which is installed on the gateway machine. Wireshark generates PCAP files for each interface present in the device. These PCAP files contain the headers of the captured packets. Thus, PCAP files cannot be used directly by the machine learning approaches. Features are extracted from the PCAP files in order to use them for further classification. The subsequent subsections discuss the generation of regular and attack traffic and the description of features.

3.2.1 Generating Regular Traffic

The dataset contains a mixture of data corresponding to attack traffic and regular traffic. Regular traffic is collected when the normal usage of the network is going on. Data is collected through the Wireshark tool. To capture regular traffic, we interacted with smartphones, smartwatches, and laptops for 3 min and collected PCAP data for all packets during this time interval. Device interaction for regular traffic includes application update, storing data to the cloud, pinging the router, sending an email, etc., performed many times.

3.2.2 Generating Attack Traffic

The remaining data of the dataset contains attack traffic that has been captured during the attack. The DDoS attack was launched with the destination IP address of the victim machine. In this experiment, UDP flood attack, TCP SYN flood attack, and ICMP flood attack were performed to capture the data traffic in the attack scenario. These attacks have been performed using an Ubuntu 16.04 virtual machine running on a laptop using the hping3 utility. A total of 560088 packets has been produced, which includes both malicious packets and benign packets.

3.2.3 Dataset Description

The final dataset contains legitimate traffic of 37903 data packets and the attack traffic of 522185 data packets. Captured data is in string form, so first, all string data is converted to numeric data. Dataset is then processed to make an interval of 5 s and 10 s. The dataset consists of 5 features, namely, packet count, unique protocol, average packet size, unique source IP, and unique destination IP. The target variable has two classes corresponding to attack and not attack. The features are described next.

1. **Packet Count**: Number of captured packets during a duration. This feature is generally high during the attack.
2. **Unique Protocol**: The number of unique protocols in all the captured packets during a duration. This feature is generally low during the attack.
3. **Average Packet size**: The average packet size of the packets captured during a duration. An attacker tries to maximize the number of connections in an interval by sending small-sized packets during the attack.
4. **Unique Source IP**: The number of source IPs is almost the same in regular data traffic because the source IP address does not frequently change with time, and there would almost fixed a number of the device with fixed IP addresses in a time interval. When an attack is performed, the number of source IPs increases rapidly due to random IP generators at the attack source.
5. **Unique Destination IP**: IoT devices have limited devices connected to them, so their destination IP address changes within a limit. However, when an attack is launched, many packets are received at the gateway (in this model laptop is acting as a router), and most of the legitimate traffic gets blocked. Hence there is only one destination IP of attack and very little legitimate traffic.

3.3 Attack Detection

After preparing the dataset, the next task is to use this dataset for the classification of traffic scenarios as regular or attack. Although in terms of detection time, the classical statistical threshold-based approaches are more suitable than machine learning approaches.

However, these approaches can only be used when there are one or two features in the dataset. Handling more than two features is difficult for statistical approaches. Machine learning does not require manual threshold finding and manual configurations. Therefore, we switch to machine learning approaches for the detection of the attack. The description of machine learning approaches utilized in this research is described next:

3.3.1 Support Vector Machine

Both classification and regression problems can be solved with the SVM. SVM divides data into classes by drawing a line or hyperplane through it. It detects the Support vectors, which are the points closest to the line from both classes. The gap between the line and the support vectors is known as the margin. The hyperplane with the biggest margin is the ideal hyperplane.

3.3.2 MLP

Three layers of nodes make up a multilayer perceptron: input, hidden, and output. The count of nodes in the input layer and output layer is the same as the number of features and the number of classes used in the dataset. Depending on the situation, the number of hidden layers and nodes in the hidden layer may vary. Backpropagation is used to train the weights between the input-hidden and hidden-output layers to minimize the error between anticipated output and targets.

3.3.3 Random Forest

Random Forest is a decision tree-based classifier. Internal nodes and leaf nodes make up the decision tree. Each internal node assigns a condition to an attribute, and one of the branches is chosen depending on that condition. The leaf nodes represent the class. The random forest algorithm produces numerous decision trees and links them together to boost accuracy.

3.3.4 Naïve Bayes

The Bayes theorem is used to develop the naïve Bayes classifier, a probabilistic classifier. The Bayes theorem is used to compute the posterior probability for each class. The tuple in question belongs to the posterior probability class with the highest likelihood.

4 Experimental Details and Results Analysis

Scikit-learn Python framework has been used for applying machine learning. The machine learning algorithms implemented are SVM, MLP, RF and NB. All hyper-parameters were the default values unless otherwise stated. Classifier training has been performed on a training set of 85% of the dataset having an attack as well as non-attack data items. After training, the testing has been done using the remaining 15% of the dataset. The comparative results of four machine learning approaches for 5 s interval, and 10 s interval is shown in Table 1.

Table 1. Performance measure of machine learning approaches on both datasets

Algorithm name	Accuracy	Precision	Recall	F1-Score
SVM (5 s)	0.9991	0.998	0.989	0.993
SVM (10 s)	0.9934	0.981	0.921	0.950
MLP (5 s)	0.9915	0.997	0.995	0.995
MLP (10 s)	0.9895	0.989	0.966	0.977
Random Forest (5 s)	0.9998	0.997	0.995	0.995
Random Forest (10 s)	0.9892	0.994	0.991	0.992
Naive Bayes (5 s)	0.9992	0.999	0.998	0.998
Naive Bayes (10 s)	0.9907	0.987	0.944	0.965

The accuracy of the four classifiers ranged from roughly 0.98 to 0.99. The efficiency of the decision tree classifier is good, indicating that data can be segmented in a higher function space. MLP classified data with an accuracy of 0.9915 in the 5 s interval dataset and 0.9895 in the 10 s interval dataset. Naïve Bayes, which performs classification by calculating probability, shown an accuracy of 0.9987 in the 5 s interval dataset and 0.9907 in the 10 s interval dataset. Support vector machine-classified dataset with the efficiency of 0.9991 and 0.9934 for 5 s interval and 10 s interval, respectively. The precision of naïve Bayes is best in the case of 5 s interval, whereas in the case of 10 s interval, the precision of random forest is best. Recall and F1-measure values show a similar pattern. The graphical comparative analysis of four machine learning approaches for 5 s interval, and 10 s interval is shown in Fig. 2 and Fig. 3, respectively.

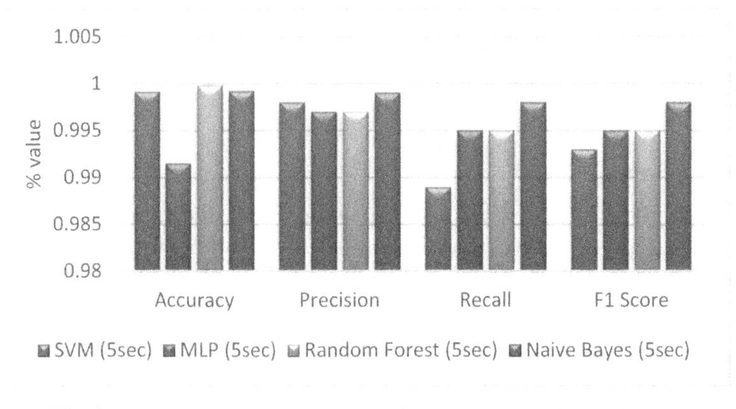

Fig. 2. Comparison of classifiers' performance on 5 s interval data

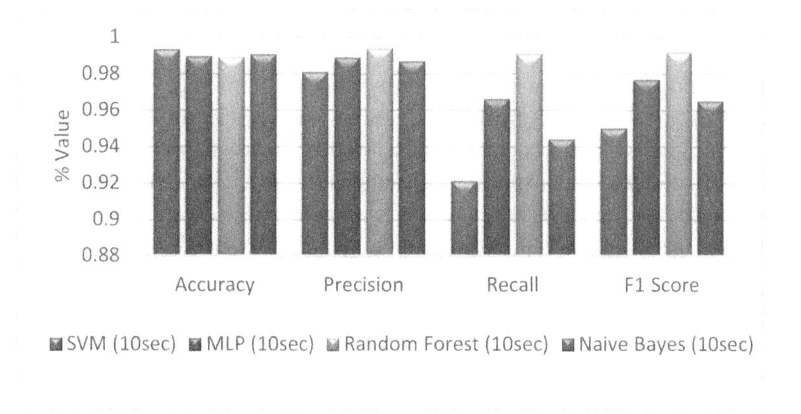

Fig. 3. Comparison of classifiers' performance on 10 s interval data

Another metric which is average detection time is significant for fast detection of the attack. The average detection time for 5 s interval and 10 s interval is shown in Table 2 and Table 3, respectively. SVM takes the least time as compared to other machine learning approaches. However, the accuracy of SVM is low. Thus, the naïve Bayes approach is most suitable for DDoS attacks as it has high accuracy, precision, recall, F1-measure, and it also takes less time to detect the attack.

Table 2. Average detection time for different machine learning approaches on 5 s interval data

Algorithm name	Average time taken (μs)
SVM	620.67
MLP	852.365
Random Forest	13902.32
Naïve Bayes	628.36

Table 3. Average detection time for different machine learning approaches on 10 s interval data

Algorithm name	Average time taken (μs)
SVM	785.78
MLP	925.26
Random Forest	14854.36
Naïve Bayes	798.25

5 Conclusion and Future Work

DDoS attack aims to take down the server by flooding unnecessary packets in a huge amount. All resource of the victim machine gets busy in handling this flood of the attacked packets. Hence the actual users cannot get service from the victim machine. In order to analyze the DDoS attack, an experimental setup is formed. The traffic corresponding to attack and nonattack scenarios are collected and processed to create a dataset. A limited set of features has been used to minimize computation overhead, which is an essential requirement for real-time installation and working. Machine learning algorithms are used to differentiate attack traffic from legitimate traffic. The naïve Bayes approach is most suitable as it has low detection time and higher accuracy, precision, recall, and F1-measure. This experiment was done in a synthetic lab environment. More complicated scenarios can be created to mimic the real DDoS situations. The deployment of the detector in IoT devices to check the detector's performance can be explored in the future.

References

1. Kelty, C.: The morris worm. Limn 1.1 (2011)
2. Us-cert.cisa.gov. 2009. Understanding Denial-of-Service Attacks | CISA. https://us-cert.cisa.gov/ncas/tips/ST04-015. Accessed 5 Sept 2021
3. cpomagazine.com. 2020. IoT-Based DDoS Attacks Are Growing and Making Use of Common Vulnerabilities. https://www.cpomagazine.com/cyber-security/iot-based-ddos-attacks-are-growing-and-making-use-of-common-vulnerabilities/. Accessed 5 Sept 2021
4. Noble, W.S.: What is a support vector machine? Nat. Biotechnol. **24**(12), 1565–1567 (2006)
5. Gardner, M.W., Dorling, S.R.: Artificial neural networks (the multilayer perceptron)—a review of applications in the atmospheric sciences. Atmosph. Environ. **32**(14–15), 2627–2636 (1998)
6. Rish, I.: An empirical study of the naïve Bayes classifier. In: IJCAI 2001 Workshop on Empirical Methods in Artificial Intelligence, vol. 3. no. 22 (2001)
7. Biau, G., Scornet, E.: A random forest guided tour. Test **25**(2), 197–227 (2016). https://doi.org/10.1007/s11749-016-0481-7
8. Symantec internet security threat report. http://www.symantec.com
9. Apthorpe, N., Reisman, D., Feamster, N.: A smart home is no castle: Privacy vulnerabilities of encrypted iot traffic. arXiv preprint arXiv:1705.06805 (2017)
10. Bigham, J., Gamez, D., Lu, N.: Safeguarding SCADA systems with anomaly detection. In: International Workshop on Mathematical Methods, Models, and Architectures for Computer Network Security. Springer, Heidelberg (2003)
11. Shirazi, S.N., et al. Evaluation of anomaly detection techniques for scada communication resilience. In: 2016 Resilience Week (RWS). IEEE (2016)
12. Miettinen, M., et al.: IoT sentinel: automated device-type identification for security enforcement in IoT. In: 2017 IEEE 37th International Conference on Distributed Computing Systems (ICDCS). IEEE (2017)
13. Patcha, A., Park, J.-M.: An overview of anomaly detection techniques: existing solutions and latest technological trends. Comput. Netw. **51**(12), 3448–3470 (2007)
14. Hodge, V., Austin, J.: A survey of outlier detection methodologies. Artif. Intell. Rev. **22**(2), 85–126 (2004)

15. No, G., Ra, I.: An efficient and reliable DDoS attack detection using a fast entropy computation method. In: 2009 9th International Symposium on Communications and Information Technology. IEEE (2009)
16. Wei, W., Chen, F., Xia, Y., Jin, G.: A rank correlation based detection against distributed reflection DoS attacks. IEEE Commun. Lett. **17**(1), 173–175 (2013)
17. Zhou, L., et al.: Low-rate DDoS attack detection using expectation of packet size. Secur. Commun. Netw. **2017**, 1–14 (2017)
18. Sivanathan, A., et al.: Low-cost flow-based security solutions for smart-home IoT devices. In: 2016 IEEE International Conference on Advanced Networks and Telecommunications Systems (ANTS). IEEE (2016)
19. Lim, S., et al.: A SDN-oriented DDoS blocking scheme for botnet-based attacks. In: 2014 Sixth International Conference on Ubiquitous and Future Networks (ICUFN). IEEE (2014)
20. Mousavi, S.M., St-Hilaire, M.: Early detection of DDoS attacks against SDN controllers. In: 2015 International Conference on Computing, Networking and Communications (ICNC). IEEE (2015)
21. Kokila, R.T., Thamarai Selvi, S., Govindarajan, K.: DDoS detection and analysis in SDN-based environment using support vector machine classifier. In: 2014 Sixth International Conference on Advanced Computing (ICoAC). IEEE (2014)
22. Rai, A., Aleem, A., Gore, M.M.: Employing LRCN model for application classification in SDN. In: Tiwari, A., Ahuja, K., Yadav, A., Bansal, J.C., Deep, K., Nagar, A.K. (eds.) Soft Computing for Problem Solving. AISC, vol. 1393, pp. 347–359. Springer, Singapore (2021). https://doi.org/10.1007/978-981-16-2712-5_29

Efficient IaC-Based Resource Allocation for Virtualized Cloud Platforms

Nirmalya Mukhopadhyay and Babul P. Tewari(✉)

Department of Computer Science & Engineering,
Indian Institiute of Information Technology, Bhagalpur 813210, India
{nirmalya.cse.2103004,bptewari.cse}@iiitbh.ac.in

Abstract. Cloud computing has become an inevitable part of information technology (IT) and other non-IT businesses. Every computational facility is now provided as computing services by the cloud service providers (CSPs). While providing these services, CSPs try to maintain efficiency by keeping the performance index as high as possible. Although virtualization technology has made this possible by applying resource provisioning techniques, but this approach is still hectic and expertise-dependent. In this paper, we propose an efficient infrastructure as code (IaC) based novel framework for optimizing resource utilization percentage through an automatic provisioning approach. This framework maximizes the resource utilization and performance metrics of virtualized cloud platforms. In this context, we have presented some mathematical formulations and considering those, we addressed our designed programming model for the proposed IaC-based framework. Extensive simulations have been performed to establish the novelty of the proposed approach. We have also presented a comparative study by considering two data centers, one with IaC based proposed model and the other is a conventional contemporary model. Result analysis confirms the performance of our proposed IaC-based framework.

Keywords: Infrastructure as code (IaC) · Virtualization · Performance analysis in cloud · Performance optimization technique · Cloud computing · Resource allocation in cloud

1 Introduction

Provisioning the traditional computing resources is costly and time consuming process. Moreover, this process is normally carried out by the expert personnel. With the advancement in virtualization technology, it has been possible to remove the hectic installation and configuration process through cloud native development. This has enabled the developers to provision the resources as per their convenience [18]. However, this has diverted the main role of the developers in cloud as resource provisioning is needed with every new deployments. In addition to this, as the process was manual, the changes in operating environments, execution models and inconsistent performances were difficult to monitor

I. Woungang et al. (Eds.): ANTIC 2021, CCIS 1534, pp. 200–214, 2022.
https://doi.org/10.1007/978-3-030-96040-7_16

in timely manner. There is a need to automate the provisioning method for the emerging application programming interface (API) driven cloud platform and the time-to-live (TTL) enabled information technology (IT) infrastructure shrinking [19]. In this context, infrastructure as code (IaC) has played a significant role in creating, managing and provisioning hardware resources in cloud data centres through machine-readable scripts. This approach effectively reduces the time of installation, configuration and management of resources. This helps the developers to concentrate on writing applications without worrying about resource provisioning [13].

With the use of IaC, the speed of provisioning the infrastructure has been dramatically increased for development, deployment, testing, production and scaling of cloud services. IaC prevents ad-hoc configuration drifts by resolving mismatched development environment, faulty deployment and security vulnerabilities to meet standard compliance and governance of cloud platforms [18]. IaC ensures provisioning intelligence stays linear in the enterprises even if the core responsible team is no more available for a project as everything is ordered-up, fully documented and versioned during the development stage [19].

In this paper, we investigate the issue of virtual resource utilization percentage with the objective of maximize the performance index in virtualized cloud platforms through efficient use of IaC. We have designed a novel IaC-based framework to show that the performance metrics of a virtualized cloud platform can be analyzed and significantly improved if the resource selection and utilization can be efficient. Based on this IaC enabled framework, we have formulated a mathematical model. Then, we have addressed a programming model that enables us to implement the proposed framework in virtualized cloud platform. Our proposed approach is used for increasing the percentage of active CPU and memory cycle utilization. We have established the novelty and efficiency of our framework through extensive simulations. We have applied our framework in a simulated cloud platform with two different data centers, where two customers are sending in their huge tasks to process. The data center that is customized with our framework shows better performance index than the other. This event was staged to prove the efficiency of our proposed model.

We have classified the rest of the paper in the following sections. We elaborated related works in Sect. 2 followed by the proposed system model in Sect. 3. Section 4 signifies our motivation for this research. The proposed framework along with it's programming model is described in Sect. 5. We have analysed the performance of the proposed framework in Sect. 6. At last, Sect. 7 concludes the paper.

2 Related Works

Cloud services are becoming essential for education, research, business and day-to-day life. This is mainly because of the good performance metric of the cloud platforms. However, performance related issues were not sufficiently considered for large-scale systems at the early stage. Therefore, a continuous growth of research is noticed in this direction. In this section we will highlight some of the prominent studies that are relevant to the problem addressed in this paper.

VM consolidation (VMC) has been proved as an effective strategy for minimizing performance-energy trade-off. Beloglazov A. et al. [1] and Ren R. et al. [8] used semi-online multi-dimensional bin packing to resolve VM placement problem. They further investigated overloaded host detection by Markov chain model. On contrary, Esfandiarpoor S. et al. [3] concentrated on arrangement of racks, cooling structures, server distribution and network topologies for VMC. In a similar context, Li Z. et al. [4] conducted their research to verify dynamic workload, processor utilization, VM migration timeline and scopes for consolidating VMs. Considering the cons of centralized VMC, Khani H. et al. [6] came up with a new distributed model for VMC using a non-cooperative game for tumbling power consumption in large data centers. The authors in [2], was looking to achieve burstiness-aware resource reservation policy to reduce the under-utilization of cloud resources. To make this policy even more robust, a queuing theory based model has been proposed in [14]. On contrary, the authors in [7,9] have addressed high cloud resource utilization index through developing elastic algorithms for migrating VMs to appropriate location.

Sayadnavrad M. H. et al. [5] focuses on reliability constraints of VMC. So, they prepared a novel framework to predict and analyze PM reliability. The authors in [10,12] also designed novel algorithms to obtain VMC in an efficient way. This technique reduces the server sprawl count in a data center. A six phase based VMs consolidation of cloud platform has been addressed in [11], which provides easy management of the entire process. Rahmani S. et al. in [13] have proposed a burst-aware algorithm to reduce resource wastage and SLA violation. In an almost similar research, Rajabzadeh M. et al. [15], came up with a new energy-aware model to create virtual cluster for classification. They experimented VM migration time based on critical conditions through Markov chain and population based simulated annealing algorithm for placement of VMs. Further research on energy-efficient VMC was conducted in [5,16]. While authors in [5] have came up with a reliable algorithm-based approach for achieving energy-aware VMC method, [16] established an adaptive fuzzy threshold-based approach to obtain even better consolidation state.

Eventually, it has been observed that due to the colossal size of cloud computing platforms, the size of the state space of the fundamental Markov process stands too large for numerical difficulties. In this context, Ever E. [20] proposed a new approximation solution with two state variables that are capable of incorporating numerous traffic loads for large numbers of facility servers. Horri A. et al. [21] claimed to achieve a QoS-aware VMC based on resource utilization history of VMs. This SLA-aware algorithm is capable of finding the underutilized hosts in the data center and then participates in the decision of VM placement. The authors in [17] also proposed a resource-aware VM placement algorithm to achieve optimal resource utilization in Infrastructure as a Service (IaaS) cloud.

On contrary, our research focuses on a novel IaC-based framework that will reduce the under-utilization of the virtual resources and increase the performance of the cloud platform without violating the SLA.

3 The Proposed System Model

The ability to allocate several virtual machines in a single physical server (generally underutilized) is referred to as server consolidation. It aims to save up-front cost on hardware, server management and infrastructure management. We consider a virtualized cloud platform with two data centers having heterogeneous resources placed arbitrarily to serve user requests. Let us consider, a set of n number physical servers are installed inside the data center which is represented by $P = \{P_1, P_2 \ldots P_n\}$. Also consider, each physical server belonging to the set P is accessed with index i, $1 \leq i \leq n$. Let, there is m number of virtual servers is installed in each physical server P_i, $i \in \{1, 2, \ldots n\}$. Each virtual server deployed inside a physical server is represented with the notation V_{i_j}, $i \in \{1, 2, \ldots n\}$, $j \in \{1, 2, \ldots m\}$. Hence, the total number of virtual servers running inside a physical server is $n \times m$.

The performance of the physical machines depends on the throughput and utilization percentage of central processing unite (CPU) cycles, random access memory (RAM), network bandwidth and disk space. However, the dynamic variation percentage of storage is safely disregarded due to the huge requirement of the disk space. At any point of time, a cloud data center handles many user requests for provisioning heterogeneous resources to complete their tasks. So, the available computer resources has to be picked and placed in an efficient manner. Figure 1, shows the basic hierarchical model of resource allocation in a virtualized cloud platform.

The overall CPU utilization index (CUI) of the data center is the ratio of CPU utilization of virtual servers to the CPU utilization of physical servers. This can be formulated as follows:

$$CUI = \frac{\sum_{i=1}^{n} \sum_{j=1}^{m} UCC_{ij}}{\sum_{i=1}^{n} ACC_i}. \tag{1}$$

Note that, UCC_{ij} is utilized CPU cycles of jth virtual server running inside i^{th} physical server and ACC_i is available CPU cycles of ith physical server. The higher the CUI is, the higher the performance of the platform. Furthermore, if we consider that T_s is the time for which all the virtual servers under a physical server are utilizing 100% of their capability without violating the SLA and T_a is the time for which all the virtual servers under the same physical server are active, then we can formulate the fraction of time T_P for which the physical server will reach 100% CPU utilization as below:

$$T_P = m \times \frac{T_s}{T_a}. \tag{2}$$

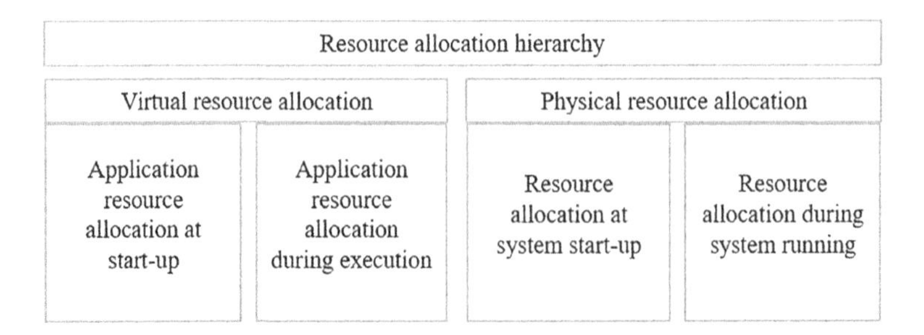

Fig. 1. Basic hierarchical resource allocation model

Now, to optimize Eq. 1 using Eq. 2, we replace ACC_i with T_P for i^{th} physical server and rewrite the Eq. 1 as follows:

$$CUI = \frac{\sum_{i=1}^{n}\sum_{j=1}^{m} UCC_{ij}}{mn \times (\frac{T_s}{T_a})}. \tag{3}$$

Note that, we have considered the existence of n physical servers each having m number of virtual servers to model ACC of all the physical servers. Again, UCC can be considered as the ratio of actual running time of all the virtual servers T_r and the time T_s, for which all the virtual servers were utilizing 100% of their capability without violating the SLA. So, we can rewrite Eq. 3 using the running time notations as follows:

$$CUI = \frac{T_r.T_a}{(T_s)^2}. \tag{4}$$

From Eq. 4, it is clearly observed that the entire performance of the data center can be expressed in terms of actual running time and the time for which all the virtual servers under all the physical server utilizes 100% of their capability without violating the SLA. In addition to that, we can measure memory utilization index (MUI) and the active network bandwidth utilization index (BUI) in the virtual servers using the same formulae. Thus, Eq. 4 can be considered as a generic global performance optimization formulae for virtualized cloud platforms.

4 Motivation

All the computing resources of a data center is fully virtualized. Therefore, analyzing the performance of the VMs is necessary in order to identify the efficiency, throughput and energy consumption of the virtual servers in a data center. A high performance data center ensures the best cloud services to the customers without effective delay. Optimizing the performance metric of a data center depends on many parameters. One of the essential parameter is the infrastructure of computation. It has been observed that the performance of the data center of a CSP deeply depends on how efficiently they are handling their hardware resources.

Infrastructure can be created, configured, manipulated, managed, altered and deleted through several processes. There are some management console GUIs available in public cloud domains for the handing infrastructures to obtain infrastructure as a service. Beside, these CSPs also provide some command line interfaces (CLIs), which are capable of providing same facilities like the GUIs. There are some third party vendors, that also provide some state-full APIs specially designed to access the cloud infrastructure from remote devices.

All these approaches to access cloud infrastructure becomes limited in the context of performance improvement. Writing hard-coded scripts provides the opportunity to get feasible solutions that allow the researchers to analyze and verify the credibility of performance metrics for the cloud platforms.

Motivated by these facts, we develop an efficient IaC-based framework to improve the performance of the cloud platforms. The fundamental objective is to enhance the resource utilization in cloud data centers.

5 The Proposed IaC-Based Framework for Optimizing the Performance of the Virtualized Cloud Platform

In this section, we described the working principles of our proposed framework and the process of analysing the performance of the system. Then we propose a programming model that works behind our framework. More formally, we have formulated a mathematical model to justify the proposal.

5.1 IaC-Based Framework for Performance Analysis and Optimization

Infrastructure as Code is a convenient way to analyze the overall performance of a virtualized system because it helps converting all the physical infrastructure into programmable modules. These modules can be executed within the native or non-native environments to eliminate the dependencies of the physical hardware. These software modules are easy to implement, monitor, modify and maintain. Moreover, there are tools and methods, which can evaluate the performance of these codes just like other logical entities. This section highlights how the performance of the entire virtualized system can be analyzed.

Software Analysis. The old analysis approaches for software components were not consistent because of two main reasons; first, the program interacts with hardware in unpredictable and less-controllable ways and second the run-time was additive in nature. Hence, the dependency of accessibility of memory causes instability of performance. IaC provides a stable software performance analysis because we have been able to make valid interaction cycles between software and modern hardware components.

To achieve this, a piece of performance data is collected and queued or processed for later display in a simple model. For example, in a large scientific computer cluster of 1024 computers, the entry and exit times of each programme feature could be collected. Instead, the entry and exit times could be subtracted and added up locally in each function to provide a total of the processing time spent in each software function. Individual programmes would submit the execution time histogram to a central server when they finished so that output reports could be prepared offline.

VM Performance Analysis. Several virtual machines were tested operating on a single physical host in the second dimension to assess the efficiency in an environment where resource contention may be a factor. One of the most difficult aspects of modelling virtual machine performance is that, the program no longer has platform resources devoted exclusively to its consumption. The transition from a dedicated execution to a shared execution results in cache contention between the VMs, which has an impact on efficiency. The resource contention of two VMs, running through resource sharing model, can have a significant impact on VM performance analysis. It is therefore critical to accurately define all shared resource contention effects in the virtualized platform in order to model appropriate VM results. The objective is to completely calibrate a model in order to forecast virtual machine performance on a potential server architecture or configuration. The aim is to create a model that can predict the performance of a virtual machine on a different platform.

5.2 The Proposed Framework

Our proposed framework consists of seven major modules. They together ensure the performance-optimized solution when a cloud consumer sends in a task-request to a specific cloud service of a desired CSP. For any category of task, the proposed framework selects configuration parameters rationally. The logic of data center selection and VM selection is closely associated with load management. The Load balancer takes the initial and in-process load distribution responsibility. During the selection of VMs for a process, the CUI and MUI are taken into consideration. On the other hand, network and storage parameters are mainly taken care by the load balancer. Figure 2, shows different components and working principles of our proposed framework. The performance monitoring module keeps track of the utilization percentages of the CPU, RAM and network bandwidth. If performance drops below a predefined percentage, this module re-initiate the resource allocation policy efficiently.

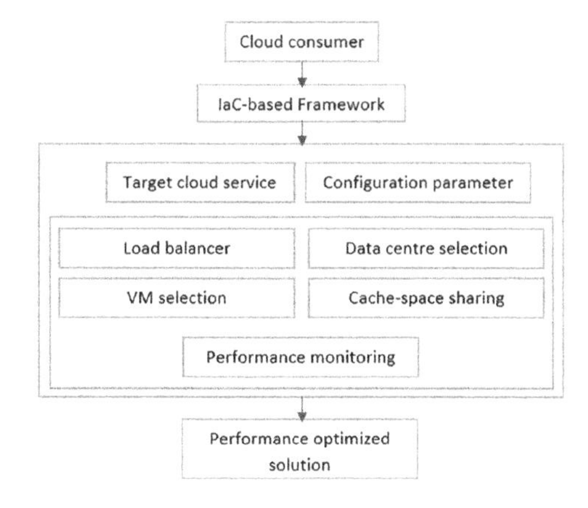

Fig. 2. Proposed IaC-based framework

5.3 Programming Model

We have designed a programming model to support the above mentioned IaC-based framework. This has two integral parts such as the front-end and the back-end. The front-end handles the application level access to the cloud user and the back-end actually processes the efficient resource allocation methodology.

The VM is an integral part of the interpreter architecture. It handles the code that we write to deploy our applications. The code and data structure for our VM resides inside a pre-specified module. Initially the VM is empty, then gradually it includes the pile of stack according to the state change. So, we have created a structure with initial chunk of parameters that the VM needs to boot and configure itself. We have incorporated $\#ifndefclox\_vm\_h$ and $\#defineclox\_vm\_h$ to create the basic functions of the VM and then initialized them using $initVM()$.

Interestingly, the defined module is going to move the functions obliquely and it will be a chore to allocate a pointer to the VM for further referencing. So, we have decided to declare a single global VM object to do the job. Before we end up the session, we need to use $freeVM()$ function to disassemble the initial "chunk" and free the VM from the main memory to keep the performance up. Now, when we run the "clox", it starts the VM before it executes the initial chunk and hence the probability of memory-overlapping is less. As a result, the CPU cycles can be utilized for actual execution and we will not face any performance degradation. The compiler detects the static errors (if any) and the VM itself detects the run-time errors. The interpreter uses these information to set the exit code of the process.

We are using a global VM object which is behaving like a byte pointer pointing towards a byte-code array. It is faster than integer indexing pointers and also holds up during the initialization of the VM functions. When we run the entire code for allocating resources through the IaC-based framework, the system spends more than 90% of its time to execute $run()$ function, which is a clear indication that the resource utilization percentage is satisfactory.

As the VM core is visible and consolidated, the VMM can monitor the virtual platform architecture (VPA) core usage. This can be obtained through minimizing the ratio of available virtual servers and total VM utilization. This can be formulated as follows:

$$CUI_{min} = \min_{\forall i,j \in n,m}(VM_{AU_{ij}}, \frac{VM_{AU_{ij}} \times m \times n}{VM_{AU_i}}) \tag{5}$$

Where, $VM_{AU_{ij}}$ is the single core utilization of the jth VM running under i^{th} physical server, VM_{AU_i} is the core utilization of all VMs running under i^{th} physical server, n is the number of all the physical servers and m is the number of virtual servers. So, from Eqs. 4 and 5, we can write

$$\frac{T_r.T_a}{(T_s)^2} = \min_{\forall i,j \in n,m}(VM_{AU_{ij}}, \frac{VM_{AU_{ij}} \times m \times n}{VM_{AU_i}}) \tag{6}$$

On the other hand, as the shared cache is an invisible resource, so its performance can not be monitored directly through the hypervisor. So, we model the following formulae to estimate the cache-space utilization index (CSUI). This can be computed easily by keeping in mind the VM execution profiles that how many cores are sharing the same cache space.

$$CSUI_{VM_y(x)} = \frac{VM_yCUI}{VM_{all}CUI - VM_xCUI} \tag{7}$$

where, $CSUI_{VM_y(x)}$ is the cache-space utilization index of y VM sharing the cache-space with x VM, VM_yCUI is the CPU utilization index of y VM, VM_xCUI is the CPU utilization index of x VM, $VM_{all}CUI$ is the CPU utilization index of all the VMs running inside a physical server.

Cache contention is the only tiny issue that a cache-space shared model faces. In our proposed model, we therefore, have formulated the performance loss index to get a proper estimation of the loss in overall efficiency of the model. It can be calculated as follows:

$$CCPLI_{VM_y(x)} = \frac{VM_y(x)ACPI}{VM_y(x)CCPI} \tag{8}$$

Where, $CCPLI\_VM_y(x)$ is the cache contension performance loss index, $VM\_y(x)ACPI$ is the actual cycles per instruction for y VM sharing cache-space with x VM and $VM\_y(x)CCPI$ is the consolidated cycles per instruction for y VM sharing cache-space with x VM.

These mathematical formulations are synthesized along with the programming model and the IaC-based framework we developed.

Table 1. Parameter list for data centers

Parameter/Heads	*Value/Description*
VM *Allocation policy*	*Single threshold*
Architecture	*x86*
Operating system	*Linux*
Hypervisor	*Xen*
Number of hosts	3
Number of vCPUs	12
RAM	120 GB
Bandwidth	1000 MBPS
VM *migration*	*Enabled*
Monitoring interval	180 ms
Scheduling interval	30 ms
VM *scheduling type*	*Time shared*
Provisioning style	IaC – *based for Datacenter*1 *simple for Datacenter*2

Table 2. Parameter list for customers

Parameter/Heads	*Value/Description*
Broker policy	*Round Robin*
Processing elements	1
Input image size	1000 MB
Input file size	500 GB
Output file size	500 GB
Resource utilization model	*Full*

6 Performance Evaluation

In this section we present the simulation results related to the performance evaluation of the proposed framework. Through extensive simulations, we rationally prove that our proposed framework is efficient enough to improve the overall performance of the virtualized cloud platform.

6.1 Simulation Environment

We have used CloudSim simulator for deploying and testing our experiments. We have developed a customized Java code that can mimic the cloud data center environment. We also have used CloudReports GUI toolkit to generate the simulation reports. For implementing IaC, we have written scripts using hashicorp configuration language (HCL) in Terraform platform and deployed them using

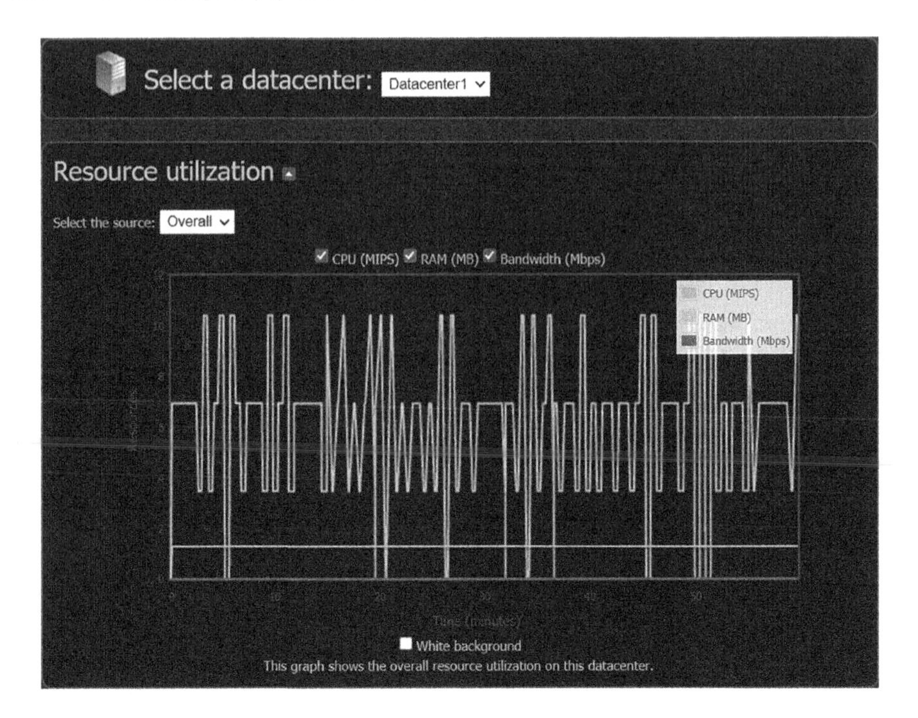

Fig. 3. Resource utilization in Datacenter1

Iac-supported API. We have considered two different cloud data centers with two different cloud customer. One data center (Datacenter1) is enabled with our proposed IaC-based framework and the other one (Datacenter2) is the conventional non-IaC data center. We created two cloud customers with same input data set to check the comparison of performance indexes of these two data centers. The performance graphs are generated and captured using CloudReports toolkit. Different cloud parameters that were used in this simulations are listed in Table 1 and Table 2. We have run the simulations for 100 times and taken the average results for comparison.

6.2 Simulation Results

IaC plays an important role in automating the resource provisioning technique. Use of IaC reduces the number of performance drops during computation in a cloud platform. This eventually increases the overall performance of the system. This performance optimization also impact throughput and reliability (in a positive manner) of the cloud services that a cloud customer will adapt.

Figure 3 and Fig. 4 demonstrate a comparison between cloud resource utilization indices of Datacenter1 and Datacenter2. As mentioned earlier, Datacenter1 is enabled with our proposed IaC-based framework whereas Datacenter2 follows the conventional one. From the comparative study of Fig. 3 and Fig. 4 respec-

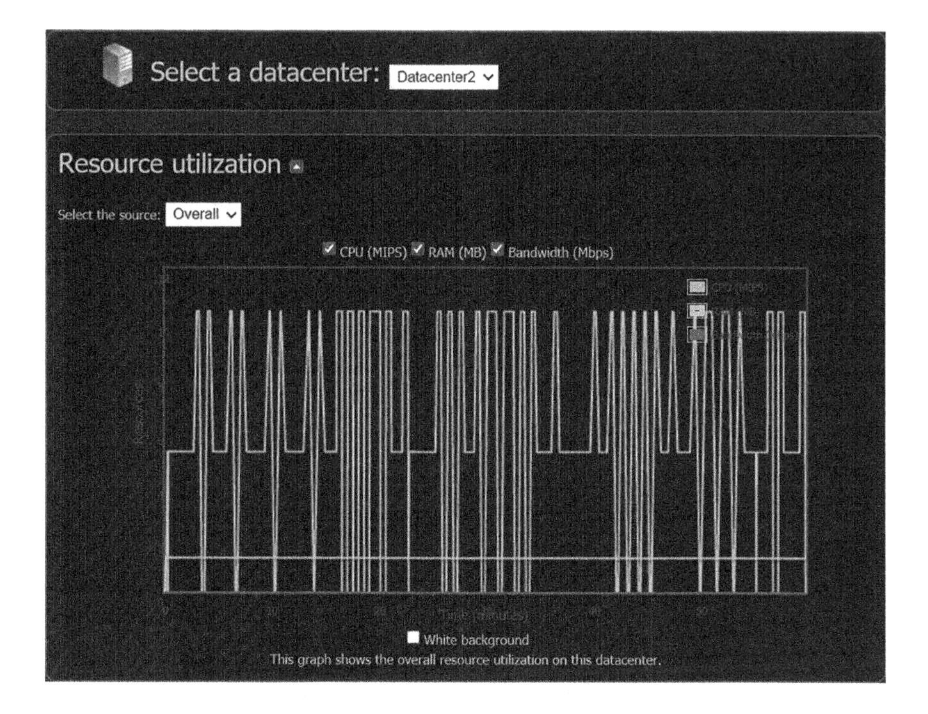

Fig. 4. Resource utilization in Datacenter2

tively it is observed that there is a significant number of CPU performance drops in Datacenter2 in comparison to Datacenter1.

The number of performance drops is directly proportional to the loss of performance. So, this comparison proves that, Datacenter1 has performed better than that of Datacenter2 in terms of CUI. This is due to the fact that in Datacenter1, provisioning of resources was taken care by the Iac-based framework for the new deployments and changed environments. The total number of task provided by the customers were same, but for every new task, the deployment environment was changed and re-provisioning of resources were needed. Now, for Datacenter2, there was no automated process re-provisioning. Therefore, the load balancer (LB) and distributed resource allocator (DRA) took initiative to reassign new resources to each and every new deployment. The use of CPU cycles during this resource reallocation phase was little and the system experiences performance loss. On contrary, for Datacenter1, the proposed IaC-based framework automatically re-provisioned the cloud resources and the overall performance stands better.

We have also compared the resource utilization percentages of Customer1 and Customer2 in Fig. 5. It is evident from the figures that, Customer1 needed lesser amount of CPU cycles to complete the given task. This is due to the fact that our proposed framework results in increased percentage of resource utilization for each resource. On contrary, Customer2 faces more performance drops and

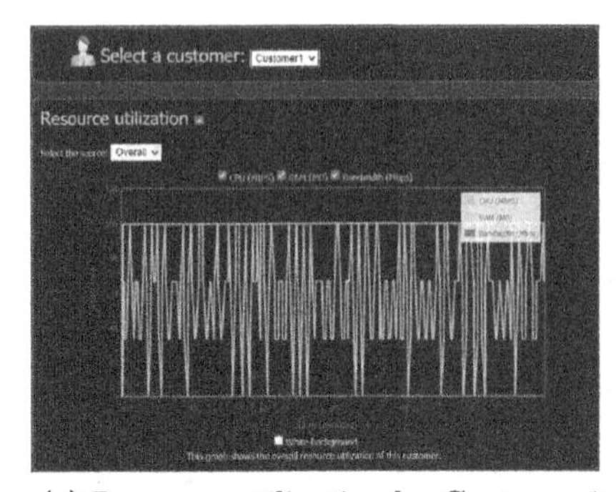

(a) Resource utilization by Customer1

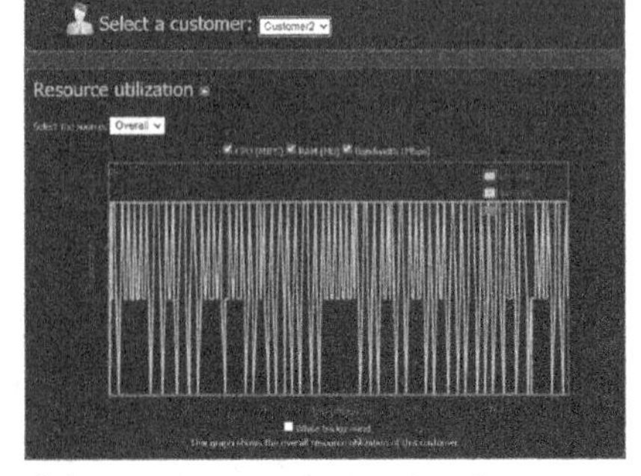

(b) Resource utilization by Customer2

Fig. 5. Resource utilization comparisons of 2 Customers

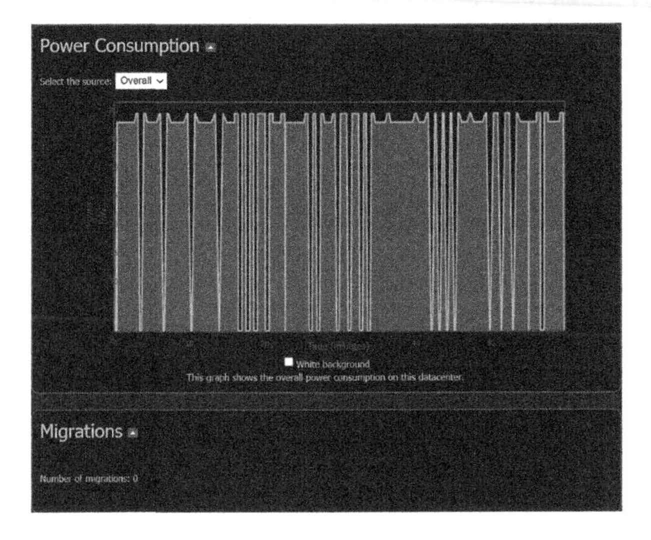

Fig. 6. Power consumption level during simulation in Datacenter1

it leads to configuration drift and lesser throughput of the system hence causes reduced performance. This once again establishes our argue that our proposed framework optimizes the system performance in a virtualized cloud platforms.

In Fig. 6, we represent the simulation results of power consumption in Datacenter1. As evident, the proposed framework results in an uniform power consumption with time. Power consumption is an important parameter for the design of energy efficient cloud system. The uniformity in the power consumption from the proposed framework establishes its energy efficiency.

7 Conclusion

Performance optimization in cloud platform is extremely significant research topic for recent times because almost every business has been migrated to cloud. Moreover, computation has become cloud service-oriented which indicates that the cloud service providers have to maintain the performance, efficiency, quality of service (QoS), reliability and availability indexes for all resource-intensive computational services. In this regard, we have proposed an efficient IaC-based framework for virtualized cloud platforms which will optimize the performance index of the cloud data centers. We have designed a programming model to implement the framework in simulated data center environment and proved the efficiency of the proposed framework through appropriate simulation. We have also formulated a mathematical model in this context. Finally, all the simulations we have conducted in this study were framed to emphasize the performance analysis of our proposed framework with contemporary systems and we found that the simulation results satisfy the formulated mathematical model.

References

1. Beloglazov, A., Buyya, R.: Managing overloaded hosts for dynamic consolidation of virtual machines in cloud data centers under quality of service constraints. IEEE Trans. Parallel Distrib. Syst. **24**(7), 1366–1379 (2013)
2. Zhang, S., Qian, Z., Luo, Z., Wu, J., Lu, S.: Burstiness-aware resource reservation for server consolidation in computing clouds. IEEE Trans. Parallel Distrib. Syst. **27**(4), 964–977 (2016)
3. Esfandiarpoor, S., Pahlavan, A., Goudarzi, M.: Structure-aware online virtual machine consolidation for datacenter energy improvement in cloud computing. Comput. Electr. Eng. **42**, 74–89 (2015)
4. Li, Z., Yan, C., Yu, X., Yu, N.: Bayesian network-based Virtual Machines consolidation method. Futur. Gener. Comput. Syst. **69**, 75–87 (2017)
5. Sayadnavrad, M.H., Toroghi, A.H., Rahmani, A.M.: A reliable energy-aware approach for dynamic virtual machine consolidation in cloud data centers. J. Supercomput. **75**, 2126–2147 (2018)
6. Khani, H., Latifi, A., Yazdani, N., Mohammadi, S.: Distributed consolidation of virtual machines for power efficiency in heterogeneous cloud data centers. Comput. Electr. Eng. **47**, 173–185 (2015)
7. Nashaat, H., Ashry, N., Rizk, R.: Smart elastic scheduling algorithm for virtual machine migration in cloud computing. J. Supercomput. **75**(7), 3842–3865 (2019). https://doi.org/10.1007/s11227-019-02748-2
8. Ren, R., Tang, X., Li, Y., Cai, W.: Competitiveness of dynamic bin packing for online cloud server allocation. IEEE/ACM Trans. Netw. **25**(3), 1324–1331 (2017)
9. Ashry, N., Nashaat, H., Rizk, R.: AMS: adaptive migration scheme in cloud computing. In: Hassanien, A.E., Tolba, M.F., Shaalan, K., Azar, A.T. (eds.) AISI 2018. AISC, vol. 845, pp. 357–369. Springer, Cham (2019). https://doi.org/10.1007/978-3-319-99010-1_33
10. Chang, Y., Gu, C., Luo, F., Fan G., Fu, W.: Energy efficient resource selection and allocation strategy for virtual machine consolidation in cloud datacenters. IEICE Trans. Inf. Syst. **101**(7), 1816–1827 (2018)

11. Arianyan, E., Taheri, H., Sharifian, S., Tarighi, M.: New six-phase on-line resource management process for energy and SLA efficient consolidation in cloud data centers. Int. Arab. J. Inf. Technol. **15**(1), 10–20 (2018)
12. Bermejo, B., Juiz, C.: Virtual machine consolidation: a systematic review of its overhead influencing factors. J. Supercomput. **76**(1), 324–361 (2019). https://doi.org/10.1007/s11227-019-03025-y
13. Rahmani, S., Khajehvand, V., Torabian, M.: Burstiness-aware virtual machine placement in cloud computing systems. J. Supercomput. **76**(1), 362–387 (2019). https://doi.org/10.1007/s11227-019-03037-8
14. Luo, Z., Qian, Z.: Burstiness-aware server consolidation via queuing theory approach in a computing cloud. In: 2013 IEEE 27th International Symposium on Parallel & Distributed Processing (IPDPSW2013), pp. 332–341. ACM Digital Library (2013)
15. Rajabzadeh, M., Toroghi Haghighat, A., Rahmani, A.M.: New comprehensive model based on virtual clusters and absorbing Markov chains for energy-efficient virtual machine management in cloud computing. J. Supercomput. **76**(9), 7438–7457 (2020). https://doi.org/10.1007/s11227-020-03169-2
16. Salimian, L., Safi Esfahani, F., Nadimi-Shahraki, M.-H.: An adaptive fuzzy threshold-based approach for energy and performance efficient consolidation of virtual machines. Computing **98**(6), 641–660 (2015). https://doi.org/10.1007/s00607-015-0474-5
17. Gupta, M.K., Amgoth, T.: Resource-aware virtual machine placement algorithm for IaaS cloud. J. Supercomput. **74**(1), 122–140 (2017). https://doi.org/10.1007/s11227-017-2112-9
18. Riti, P., Flynn, D.: Infrastructure as Code. In: Beginning HCL Programming, 1st edn., pp. 65–78. Apress, Berkly (2021)
19. Hüttermann, M.: Infrastructure as code. In: DevOps for Developers, 1st edn., pp. 135–156 Apress, Berkly (2012)
20. Ever, E.: Performability analysis of cloud computing centers with large numbers of servers. J. Supercomput. **73**(5), 2130–2156 (2016). https://doi.org/10.1007/s11227-016-1906-5
21. Horri, A., Mozafari, M.S., Dastghaibyfard, G.: Novel resource allocation algorithms to performance and energy efficiency in cloud computing. J. Supercomput. **69**(3), 1445–1461 (2014). https://doi.org/10.1007/s11227-014-1224-8

Patient Feedback Based Physician Selection in Blockchain Healthcare Using Deep Learning

Narendra Kumar Dewangan(✉) and Preeti Chandrakar

National Institute of Technology, Raipur, India
{nkdewangan.phd2019.cse,pchandrakar.cs}@nitrr.ac.in

Abstract. In the healthcare system, selecting the correct physician is very important. Physicians have specialization and repute also for the treatment of the patient. Blockchain is reliable, anonymous, immutable and distributed ledger technology, best for the healthcare system. In blockchain privacy of the patient is maintained. So for selecting the correct physician, the feedback of previously treated patients by the physician also matters. We proposed a blockchain-based healthcare system that takes feedback from the patients. This proposed system uses a deep learning model based on RNN (Recurrent Neural Network), which optimizes the patient's feedback and suggests selecting the recommended physician for the next cycle of patient appointment. This proposed system is implemented using PHP-based blockchain and uses a customized hash solving consensus algorithm. This system is implemented in a private blockchain. The deep learning part is implemented using LSTM(Long Short Term Memory) RNN using TensorFlow. Feedback of patients is collected in the blockchain and mark as transactions. The reputed physicians are a miner in the blockchain network. The accuracy of the proposed system is about 70% as the blockchain da does not have many features as the plain dataset.

Keywords: Blockchain · Deep learning · Healthcare · Patient feedback · RNN · LSTM

1 Introduction

1.1 Healthcare, Blockchain and Data Privacy

In the healthcare sector, the privacy of patient data is essential. Blockchain is a way to provide secure, immutable, structured, verified and trusted data storage. Blockchain used distributed ledger technology to store data and send a copy of data to all over the nodes in the blockchain network. According to [1] authentication between the data stored in the blockchain and nodes of the blockchain is essential.

I. Woungang et al. (Eds.): ANTIC 2021, CCIS 1534, pp. 215–228, 2022.
https://doi.org/10.1007/978-3-030-96040-7_17

Security of blockchain from Sybil attack and man in the middle attack in case of cloud-based blockchain, DDoS (Distributed Denial of Services) attack prevention in distributed node, and many other attacks should be resistible in the blockchain to protect patient data. In the healthcare feedback system, private information about the patient should not be revealed in any condition. However, the patient can give feedback about the treatment and hospital after completion of therapy.

1.2 Elliptic Curve Cryptography and Elliptic Curve Digital Signature

Elliptic curve cryptography(ECC) is public-key cryptography. For ease of calculation, operations are performed on finite fields. The finite fields used in software and hardware applications may differ. The equation,

$$y^2 = x^3 + ax + b \ (modp)$$

Which is known as the Weierstrass equation, where a and b are constants integers less than$p$ that is a prime number and satisfies the following condition.

$4a^3 + 27b^2 \neq 0(modp)$

Let define an elliptic curve$E(a, b)$ over prime field E_p. ECDSA was proposed in 1992 by Scott Vanstone, which is based on the implementation of the digital signature algorithm (DSA) on the elliptic curves. Its security is based on the elliptic curve discrete logarithm problem (ECDLP). ECDSA consists of three phases: key pair generation, signature generation and signature verification [2].

1.3 Deep Learning

Deep learning is an unsupervised feature learning method that is used to extract high-level features from data. Since feature identification is time-consuming and expensive, Deep learning (DL) is used. Unsupervised learning does not need the labeled data. Most of the patient's feedback cases from the doctors and hospital do not contain any labeled data. DL can also be used with supervised learning, which is used to feature labeled data. There are various DL techniques like RNN, ANN (Artificial Neural Network), DBN (Deep Belief Network), CNN (Convolutional Neural Network), etc. In this paper, we take feedback data from the patient and extract the features from that data to give rating doctors and hospitals the rating by the patients in terms of feedback transaction in blockchains. For the feature extraction on DL, we use an RNN because RNN contains multiple hidden layers and has a success rate in different applications. Traditionally in machine learning, the process of extraction of knowledge from a large dataset loaded into the machine are done. Machine learning professionals rectify errors made by machines. This is removed in deep learning. In the machine learning approach, human contribution needs to do the data extraction and applying the machine. Deep learning models can create new features by themselves. In deep learning, problems are solved on an end-to-end basis. In deep understanding, approach cost can be eliminated, and labelling is not required. The deep learning approach provides high-quality results [3].

In previously developed systems, the blockchain-based data set is not taken for the recommender system. From [4], we can see that the proposed a good recommender system in deep learning methods for the cardiac patients, but their system not used blockchain in any data sets. From [5], the author selected the random forest method as the best approach for a machine learning-based recommender system. However, their system has not analysed any blockchain-based generated datasets. [6] propose a deep learning-based recommender system for movie lens dataset but not have any encrypted dataset in their proposed system. [7] presented a deep learning-based hybrid model for recommendation generation and ranking with a hybrid Bayesian stacked auto-denoising encoder model. In their system multilayer neural network to manipulate the non-linearities between the user and item communication from data. [8] proposed multi-model trust-based system for movie recommendations. Their proposed system worked in machine learning. They obtain an accuracy of 0.83% with 0.74 MSE. In [9] compares two models for a recommendation for datasets with MovieLens, but not used any blockchain-based dataset for recommendations. Our proposed system uses blockchain to collect the feedback data and as a recommender for the next cycle of patient registration and appointment. Our proposed system can help patients to find the best doctor in their area of expertise. Our proposed system goals are explained in the next subsection.

1.4 Goal and Paper Organization

According to the above situation, we are designing a patient feedback collection in a blockchain system that measures the patient's data and uses deep learning to extract the features of unlabelled feedback data. The main goals of this articles are as follows:

1. To develop a patient feedback collection blockchain with consent from the patient to share data.
2. To extract the features from the feedback and process for making a ranking of hospitals and doctors using deep learning.
3. To propose the doctor and hospital selection system based on the result of deep learning and ranking.

The literature review of this paper is described in Sect. 2. Section 3 explained the proposed system and proposed algorithm for patient feedback in the blockchain. Deep learning used for the feature extraction and ranking of doctors and hospitals are explained in Sect. 4. Section 5 described the implementation and experimental setup of proposed system. Section 6 described the results and outcomes from the deep learning and blockchain transactions. Section 7 concluded the paper with future scope.

2 Literature Review

Rahoof et al. [10] described the secure, scalable, low storage for the blockchain in the healthcare management system. It explained that it could use public social networks to share data through secure blockchain nodes. They developed an attribute-based scheme to combine with blockchain and use IPFS as scalable data storage. Tu et al. [11] explained rapid retrieval of large-scale data in the consortium blockchain, which is based on a B++ tree. It is used an indexing mechanism with user files and block files. Ko et al. [12] aim to predict the next block in the blockchain using machine learning, which covers the 91% accuracy in the terms predict the hash and content of the next block. For this, they extract the features of 100000 blocks and calculate the transaction fees. Abbas et al. [13] explained blockchain and machine learning-based pharmacy and drug recommendation system for the patients. They used Hyperledger fabric for the blockchain implementation, a private blockchain and for machine learning, they used N-gram and LightBGM methods. Albanese et al. [14] described dynamic consent management in the blockchain in clinical trials using Hyperledger fabric. In this system, consent is automatic generated as the trial properties. Cao et al. [15] described storage of EMR (Electronic Medical Records) in Hyperledger calliper blockchain-based in IPFS and CNN. CNN used to recognize relations in clinical medical records and predict consequences in big data environments. Bagchi et al. [16] used an artificial neural network using an adaptive learning algorithm to handle the diverse type of cardiovascular clinical data. They used them in blockchain for secure sharing between nodes. It uses ethereum for blockchain implementation. This paper did not mention the dataset repository from which they are taking the data and what kind of machine learning modelling they are using. Shankar et al. [17] presented a system to store face recognition data in the blockchain and use that data using VGGface deep neural network for recognition. It provides logistics and feature extraction for face data. It converts the face data in blockchain hash. It does not implement any blockchain and no security details are provided by the authors in this paper. Zhang et al. [18] explained device to device cache transfer and store them into the blockchain. They provide a miner selection algorithm, mining time, and feature selection based on deep learning with the FEL model. This paper does not implement any real-time blockchain also does not provide any concept of dealing blockchain with security. Lobo et al. [19] described the different types of deep learning used for feature extraction and optimization. The blockchain model of this paper is based on the Exonum blockchain platform with a byzantine fault tolerance algorithm. It uses artificial intelligence for feature extraction from patient data by analysis and optimization. In this paper, DNN (Deep Neural Network) is used for the deep learning model. This paper uses cryptocurrency as a reward system to generate corruption in the system while sharing data in the blockchain network. The signature algorithm used in this blockchain is not precise and the validation mechanism is not explained. Singh et al. [20] explained the drone registration, verification and storage of surveillance data in blockchain with V2V communication. It uses a zero-knowledge protocol for security in the blockchain model. The deep learning algorithm for miner selections and dataset results are not shown in the paper. Saif et al. [21] proposed big health data storage in blockchain technology, which deals with collecting health data through IoT devices and storing them in the blockchain. This system uses CNN for the deep learning implementation. No blockchain platform mention in this paper and no real blockchain is implemented.

3 Proposed System: Patient Feedback

The proposed system designed for the patient feedback system in the blockchain contains nodes, patients and doctors/hospital, where doctors are approver for the transaction and blocks. Diagnosis of the patient is made online through blockchain-based patient-centric clinic methods. From Fig. 1 we can understand the proposed system, which starts from the doctor's treatment certificate generation. When the treatment of a patient is completed, a treatment certificate is generated by the doctor or by the hospital. Once the patient received the treatment certificate, he/she can give feedback on the feedback portal by using their public identity and a 256-bit hash address generated using the ECC key pair generation during patient registration. While sending the feedback and data sharing consent, he/she signed the transaction using the ECDSA(Elliptic Curve Digital Signature Algorithm) and select any miner node available at that time in the mining pool. The miner doctor verifies the signature and treatment history from the blockchain transaction pool and confirms that the feedback is about the same doctor/hospital who generated the treatment certificate. If this condition is valid, then the miner approves the transaction. After few transactions, the miner node validates the block and the block is added into the blockchain. The transaction format for feedback and consent can be seen from Fig. 2. The algorithm used for searching and matching the treatment certificate generator and receiver, signature validation and block addition can be understood from Algorithm 1. Since this blockchain is implemented in the private blockchain, so we are using a proof-of-authentication consensus algorithm. More specifically, Fig. 1 can be explained as the first step is to obtain a certificate for treatment. After getting a certificate of treatment second step is to give feedback to the doctor and consent for data sharing. The third step is to approver verified the feedback and add this transaction to the block. After mining of block, the block is added to the blockchain. In the next cycle of patient registration, when a patient search for the doctor, a deep learning model processed the feedback data and found the most valued doctor according to feedback. This system recommends the doctor in patient search.

3.1 Blockchain Formation

When a patient want to register for the treatment to the doctor, he/she sends the registration request with the problem statement and public address. This transaction is signed by the ECDSA private key of the patient. At approver node, this transaction is verified by the proof-of-authentication consensus algorithm [22]. After approval the transaction is added into the block and the blockchain is mined by the approver. In this process, patient diagnosis and treatment is done in online mode only. After treatment the doctor generates the treatment certificate and consent request. This certificate and request is sent to the patient as transaction along with ECDSA signature on this transaction by the doctor. The transaction verification and block approval is done by the approver. Consensus algorithm and consent transaction can given in Algorithm 2.

Algorithm 1: Patient feedback and miner selection algorithm

Result: Transaction and Block to be added in Blockchain
Input : Patient feedback
Output: Transaction & Block

1 $B \leftarrow$ Blockchain;
2 $T \leftarrow$ Transaction;
3 $Pt \leftarrow$ Patient;
4 $Dr \leftarrow$ Doctor;
5 $Fb \leftarrow$ Feedback;
6 $n \leftarrow$ Total blocks in blockchain;
7 $Cr \leftarrow$ Treatment certificate;
8 $Pt[Fb] \leftarrow$ Patient feedback;
9 **while** $B[i] \leftarrow n-1$ **do**
10   Search transaction between Dr and Pt **if** $Pt \leftarrow Dr[Cr]$ **then**
11         $Dr \leftarrow Pt[Fb]$
12   **else**
13   Wait for Treatment Certificate;
14   **end**
15 **end**
16 Apply RNN LSTM on $Pt[Fb]$;
17 Prepare New list of Dr;
18 **if** $T[Dr \leftarrow Pt]$ **then**
19   Valid$[T]$;
20     $B \leftarrow B[T]+1$;
21 **else**
22   Transaction Rejected
23 **end**
24 **return** B

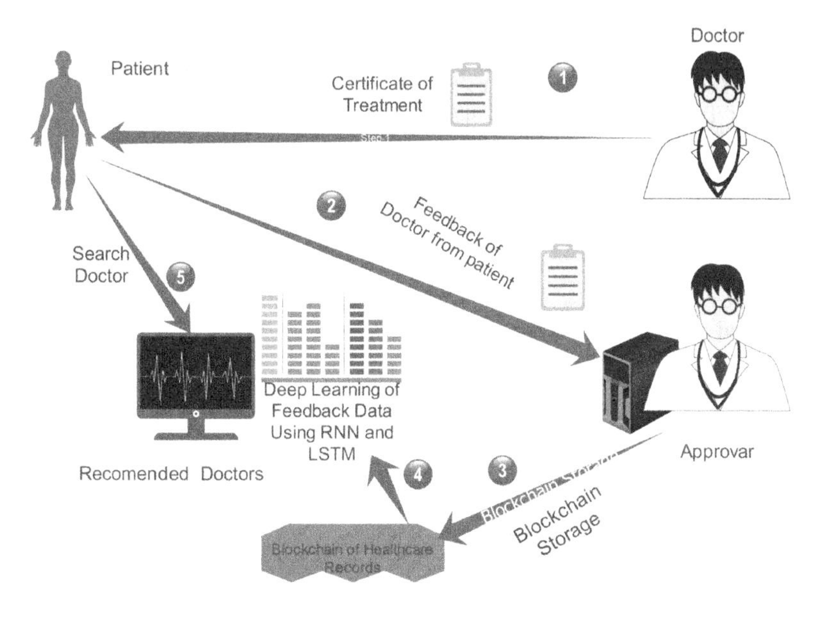

Fig. 1. Overall proposed system

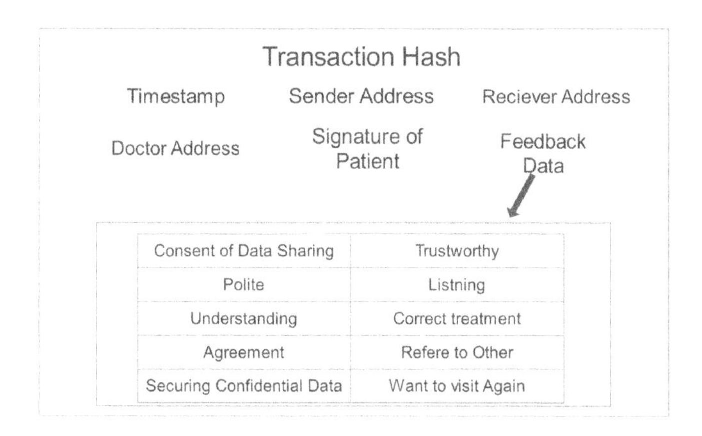

Fig. 2. Transaction format for patient feedback

Algorithm 2: Transaction approval and consent

Result: Transaction and Block to be added in Blockchain
Input : Transaction from patient/doctor
Output: Transaction and Block

```
1  B ← Blockchain;
2  T ← Transaction;
3  Pt ← Patient;
4  Dr ← Doctor;
5  Fb ← Feedback;
6  n ← Total blocks in blockchain;
7  Cr ← Treatment certificate;
8  Pt[Fb] ← Patient feedback;
9  Data_Consent ← Consent from patient;
10 X_privk ← ECDSA private key of X;
11 X_pubk ← ECDSA public key of X;
12 Sig(X)_Y ← ECDSA signature on X by Y;
13 if Dr ← Pt(T_registration) then
14 |   verify(Sig(Pt(T_registration))_{Pt_privk}) if verify==1 then
15 |   |   "Transaction is verified"
16 |   else
17 |   |   v
18 |   end
19 |   erify(B_new) B ← B[T] + 1;
20 else
21 |   Transaction Rejected
22 end
23 return B
24 Consent Transaction:
25 if Pt ← Dr(Cr&ConsentRequest) then
26 |   verify(Sig(Dr(Cr&ConsentRequest)))_{Dr_privk}) if verify==1 then
27 |   |   "Transaction is verified"
28 |   else
29 |   |   v
30 |   end
31 |   erify(B_new) B ← B[T] + 1;
32 else
33 |   "Transaction Rejected"
34 end
35 Dr ← Pt(Consent_Y) return B
```

4 Proposed System: Deep Learning for Feature Extraction

As explained in [23], in RNN non-neural hypothesis can be expressed, but CPU consumption is very high since the blockchain is already designed to be run in the high computation. Feature extraction using data-driven methods is more accurate, and RNN is a useful technique for healthcare because it supports streaming data and can be analyzed further. Fixed-size input vectors are used here also. Data such as text or DNA sequences can be provided as input where output depends on previous input. In the architecture of RNN, perceptions are interconnected with themselves, which act as a memory for consecutive inputs. In RNN for each timestamp t, the activation a_t and the output y_t are expressed as $a_t = g_1(W_{aa}a_t + W_{ax}x_t + b_a)$ and $y_t = g_2(W_{ya}a_t + b_y)$, where W_{ax}, W_{aa}, W_{ya}, b_a, b_y are coefficients that are shared temporally and g_1, g_2 activation functions.

5 Implementation of Proposed System

The proposed system is broadly divided in two parts for implementation: 1) Blockchain implementation and 2) RNN implementation. The system requirements can be seen in the Table 1. Detailed of this implementation are as follows:

Table 1. System remarks for implementation of proposed system

Property	Details
Hardware	Intel i3 processor, 4 GB RAM
Operating System	Ubuntu 20.04 64-bit
Blockchain	PHP
Deep Learning	Google Colab, RNN, LSTM

Blockchain Implementation. We are using a PHP-based blockchain with a custom consensus algorithm to implement the proposed system in the blockchain. In this system, doctors play the role of miner and ECDSA is used as the signature algorithm. For implementing the model, we take a total number of patients as 80, doctors as 20, and the complete transaction in feedback is 10000.

Deep Learning in RNN. RNN based deep learning model is implemented in Google Colab with LSTM (long short-term memory). The dataset generated from feedback in the blockchain is taken as a CSV file for input to the training and test process model. It is splitting the CSV file into train and test data. 70% as train and 30% as test data. Train the machine with weight. It implemented using Keras [23] and TensorFlow [24]. For deep learning implementation [25–27] are helpful. The proposed system uses LSTM RNN and epoch of 100 with resulting produced loss of 0.0950 in adam optimizer. Feedback collected from the patients are in form of the integer values. The error generated in the proposed system is mean squared error.

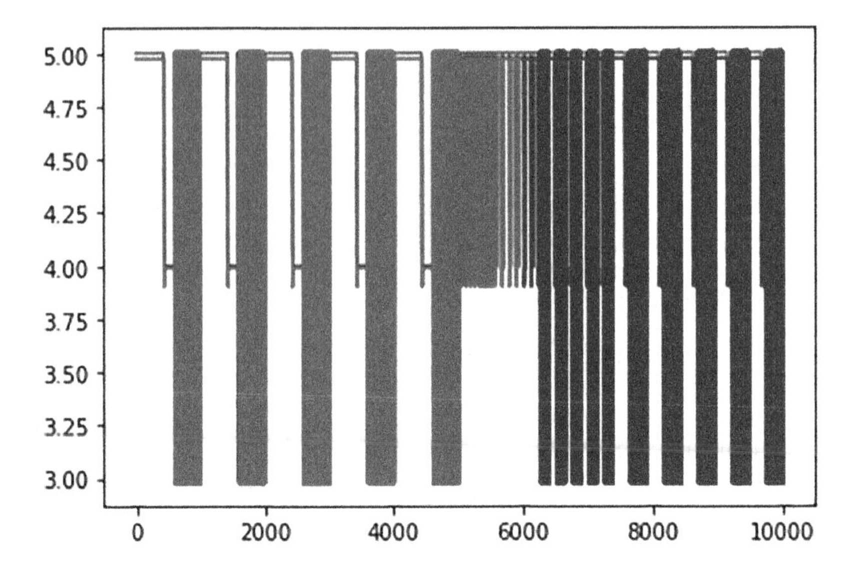

Fig. 3. RNN test, train and accuracy graph with 69% accuracy

6 Results

In the result section, there are two parts. In the first part, we measure the accuracy of our feedback system. In the second part, the blockchain implementation results show the transaction and CPU utilization during the transactions.

Deep Learning Results: From the Table 2 the test data percentage and train data percentage with the accuracy obtained in each run are shown. With 70% and 30% of the train and test, accuracy is 69%, with 80% and 20% of the train and test accuracy is 70% and with 90% and 10% of the train and test, accuracy is 70%. Since data is generated by blockchain and many fields are hard to scale. So we are talking only a few features of feedback form like politeness of doctor with a patient, providing the right treatment or not, reliable to protect private data or not, taking consent about data sharing, revisiting the same doctor, and referring this doctor to others. The graph generated by the deep learning model can be seen in Fig. 3 and Fig. 4. In these figure red lines indicates the training dataset and blue lines indicates the test run dataset.

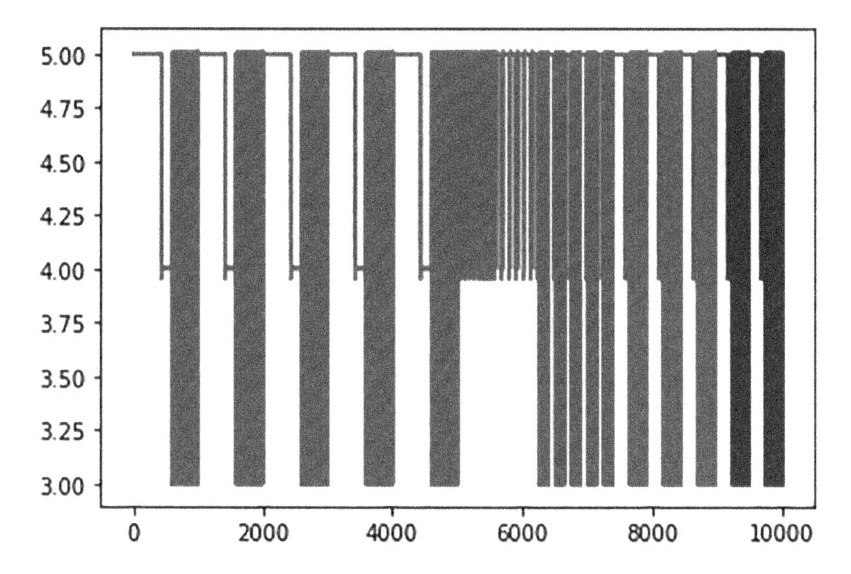

Fig. 4. RNN test, train and accuracy graph with 70% accuracy

Table 2. Deep learning training, test and accuracy

Training data	Test data	Accuracy
70%	30%	69%
80%	20%	70%
90%	10%	70%

Blockchain Results: The blockchain is implemented in the PHP. As we transact the data from the patient to doctor or doctor to patient the CPU consumption is increasing as the deep learning is working for suggesting doctors based on the feedback provided by the patients. From Fig. 5 we can see that the as number of patient is increased by the time the transaction is also increasing linearly, so the deep learning model for prediction consumes more CPU power. The figure sows that the in one second there is 17 transactions are possible in our proposed system.

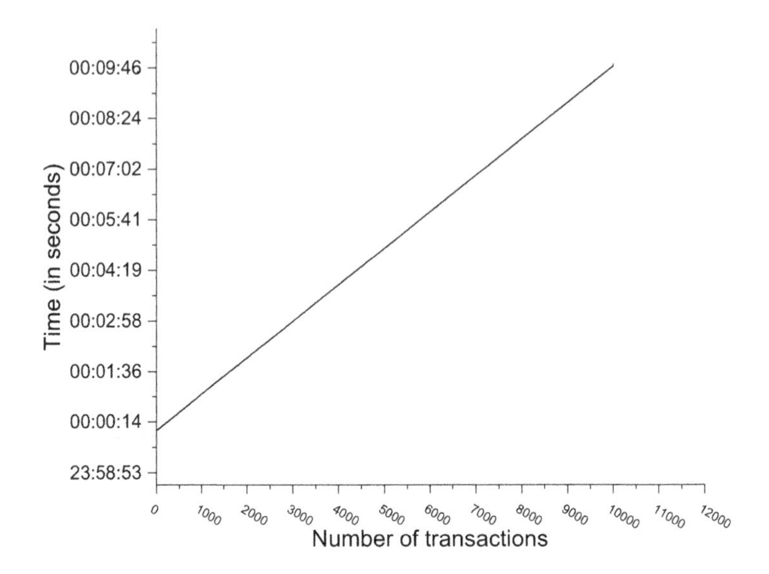

Fig. 5. Graph between transactions and mining time

7 Conclusion and Future Scope

In the healthcare field, the privacy and security of data are essential. Blockchain is truly reliable for storing data and sharing it with others in the healthcare network while complying with the GDPR (Global Data Protection Regulation). The proposed system gives the patient an opportunity to select the right doctor for the treatment with the other patients' help. The proposed system also categorizes the doctors by specialization and fees to treat patients by the doctor of their own choice. This system deals with only the hash code addresses of patients, doctors, and miners, so the identity of any node is not revealing compared to the traditional system.

In the future, this system can be made more accurate by providing more data to the deep learning system, training the system for more time, and scaling the data in a correct format. This system can be used for drug and test suggestions in the future. The asset management and financial system can also use the deep learning-based blockchain system for the blockchain network's asset and financial schemes recommendation. The limitations of this system are that there is a possibility of recommending the same doctor at all time if he/she got good feedback from most of the patients. Also, the proposed system is based on a limited number of features, so if the feature selection is expanded in future, we can get more accurate results.

References

1. Tang, F., Ma, S., Xiang, Y., Lin, C.: An efficient authentication scheme for blockchain-based electronic health records. IEEE access **7**, 41678–41689 (2019)
2. Goodfellow, I., Bengio, Y., Courville, A.: Deep Learning. The MIT Press (2016)
3. Genç, Y., Afacan, E.: Design and implementation of an efficient elliptic curve digital signature algorithm (ECDSA). In: IEEE International IOT. Electronics and Mechatronics Conference (IEMTRONICS), vol. 2021, pp. 1–6 (2021). https://doi.org/10.1109/IEMTRONICS52119.2021.9422589
4. Mustaqeem, A., Anwar, S.M., Majid, M.: A modular cluster based collaborative recommender system for cardiac patients. Artif. Intell. Med. **102**, 101761 (2020)
5. Sharma, R., Rani, S.: A novel approach for smart-healthcare recommender system. In: Hassanien, A.E., Bhatnagar, R., Darwish, A. (eds.) AMLTA 2020. AISC, vol. 1141, pp. 503–512. Springer, Singapore (2021). https://doi.org/10.1007/978-981-15-3383-9_46
6. Chavare, S.R., Awati, C.J., Shirgave, S.K.: Smart recommender system using deep learning. In: 2021 6th International Conference on Inventive Computation Technologies (ICICT), pp. 590–594 (2021). https://doi.org/10.1109/ICICT50816.2021.9358580
7. Sivaramakrishnan, N., Subramaniyaswamy, V., Viloria, A., et al.: A deep learning-based hybrid model for recommendation generation and ranking. Neural Comput. Applic. **33**, 10719–10736 (2021). https://doi.org/10.1007/s00521-020-04844-4
8. Choudhury, S.S., Mohanty, S.N., Jagadev, A.K.: Multimodal trust based recommender system with machine learning approaches for movie recommendation. Int. J. Inf. Tecnol. **13**, 475–482 (2021). https://doi.org/10.1007/s41870-020-00553-2
9. Gupta, A., Sharma, A.: Implementation of recommender system using neural networks and deep learning. In: Sharma, M.K., Dhaka, V.S., Perumal, T., Dey, N., Tavares, J.M.R.S. (eds.) Innovations in Computational Intelligence and Computer Vision. AISC, vol. 1189, pp. 256–263. Springer, Singapore (2021). https://doi.org/10.1007/978-981-15-6067-5_28
10. Abdul Rahoof, T.P., Deepthi, V.R.: HealthChain: a secure scalable health care data management system using blockchain. In: Hung, D.V., D'Souza, M. (eds.) ICDCIT 2020. LNCS, vol. 11969, pp. 380–391. Springer, Cham (2020). https://doi.org/10.1007/978-3-030-36987-3_25
11. Tu, J., Zhang, J., Chen, S., Weise, T., Zou, L.: An improved retrieval method for multi-transaction mode consortium blockchain. Electronics **9**(2), 296 (2020)
12. Ko, K., Jeong, T., Maharjan, S., Lee, C., Hong, J.W.-K.: Prediction of bitcoin transactions included in the next block. In: Zheng, Z., Dai, H.-N., Tang, M., Chen, X. (eds.) BlockSys 2019. CCIS, vol. 1156, pp. 591–597. Springer, Singapore (2020). https://doi.org/10.1007/978-981-15-2777-7_48
13. Abbas, K., Afaq, M., Ahmed Khan, T., Song, W.C.: A blockchain and machine learning-based drug supply chain management and recommendation system for the smart pharmaceutical industry. Electronics **9**(5), 852 (2020)
14. Albanese, G., Calbimonte, J.-P., Schumacher, M., Calvaresi, D.: Dynamic consent management for clinical trials via private blockchain technology. J. Ambient. Intell. Humaniz. Comput. **11**(11), 4909–4926 (2020). https://doi.org/10.1007/s12652-020-01761-1
15. Cao, Y., Sun, Y., Min, J.: Hybrid blockchain-based privacy-preserving electronic medical records are sharing scheme across medical information control system. Measur. Control **53**(7–8), 1286–1299 (2020)

16. Bagchi, S., Chakraborty, M., Chattopadhyay, A.K.: APDRChain: ANN based predictive analysis of diseases and report sharing through blockchain. In: Chakraborty, M., Chakrabarti, S., Balas, V.E. (eds.) eHaCON 2019. AISC, vol. 1065, pp. 105–115. Springer, Singapore (2020). https://doi.org/10.1007/978-981-15-0361-0_8
17. Shankar, S., Madarkar, J., Sharma, P.: Securing face recognition system using blockchain technology. In: Bhattacharjee, A., Borgohain, S.K., Soni, B., Verma, G., Gao, X.-Z. (eds.) MIND 2020. CCIS, vol. 1241, pp. 449–460. Springer, Singapore (2020). https://doi.org/10.1007/978-981-15-6318-8_37
18. Zhang, R., Yu, F.R., Liu, J., Huang, T., Liu, Y.: Deep reinforcement learning (DRL)-based device-to-device (D2D) caching with blockchain and mobile edge computing. IEEE Trans. Wireless Commun. **19**(10), 6469–6485 (2020)
19. Lobo, V.B., Analin, J., Laban, R.M., More, S.S.: Convergence of blockchain and artificial intelligence to decentralize healthcare systems. In: 2020 Fourth International Conference on Computing Methodologies and Communication (ICCMC), pp. 925–931. IEEE, March 2020
20. Singh, M., Aujla, G.S., Bali, R.S.: A deep learning-based blockchain mechanism for secure internet of drones environment. IEEE Trans. Intell. Transp. Syst. **22**, 4404–4413 (2020)
21. Saif, S., Biswas, S., Chattopadhyay, S.: Intelligent, secure big health data management using deep learning and blockchain technology: an overview. In: Dash, S., Acharya, B.R., Mittal, M., Abraham, A., Kelemen, A. (eds.) Deep Learning Techniques for Biomedical and Health Informatics. SBD, vol. 68, pp. 187–209. Springer, Cham (2020). https://doi.org/10.1007/978-3-030-33966-1_10
22. Puthal, D., Mohanty, S.P., Nanda, P., Kougianos, E., Das, G.: Proof-of-authentication for scalable blockchain in resource-constrained distributed systems. In: IEEE International Conference on Consumer Electronics (ICCE), vol. 2019, pp. 1–5 (2019). https://doi.org/10.1109/ICCE.2019.8662009
23. Chollet, F., et al.: Keras. GitHub (2015). https://github.com/fchollet/keras
24. Zhifeng Chen, et al.: TensorFlow: large-scale machine learning on heterogeneous systems (2015). Software available from tensorflow.org
25. Arora, M., Chopra, A.B., Dixit, V.S.: An approach to secure collaborative recommender system using artificial intelligence, deep learning, and blockchain. In: Choudhury, S., Mishra, R., Mishra, R.G., Kumar, A. (eds.) Intelligent Communication, Control and Devices. AISC, vol. 989, pp. 483–495. Springer, Singapore (2020). https://doi.org/10.1007/978-981-13-8618-3_51
26. Sak, H., Allauzen, C., Nakajima, K., Beaufays, F.: Mixture of mixture N-gram language models. In: 2013 IEEE Workshop on Automatic Speech Recognition and Understanding, pp. 31–36. IEEE, December 2013
27. Waqar, M., Majeed, N., Dawood, H., Daud, A., Aljohani, N.R.: An adaptive doctor-recommender system. Behav. Inf. Technol. **38**(9), 959–973 (2019)

Analyzing the Behavior of Real-Time Tasks in Fog-Cloud Architecture

Pratibha Yadav(✉) and Deo Prakash Vidyarthi

Computer and Systems Sciences, Jawaharlal Nehru University, New Delhi, India

Abstract. The growing number of IoT devices generates a huge amount of data that are generally processed by the Cloud datacenter. However, it results in inordinate delay for time-critical applications due to network intricacies. Fog computing has evolved in recent which provides similar facilities as of Cloud though in a reduced manner. In order to provide the desired quality of service to the IoT users, it is essential to classify and allocate Fog-Cloud resources optimally to the time-critical requests. In this work, we have developed an analytical model focusing on the design mechanism approach and optimal policies for the allocation and offloading of real-time tasks that results in overall time minimization.

Keywords: Cloud computing · Fog computing · Fog-aggregation · Offloading

1 Introduction

The increasing complexity and heterogeneity of interconnected devices over the communication network results in massive challenges for the network architect and researchers to meet efficient resource and bandwidth utilization. In this regard, cloud computing has emerged as a viable solution to handle this considerable data demand and efficiently handle connected network resources. Fog computing [1] can be thought of as a scaled-down version of cloud computing that provides the same functionality as a cloud which ultimately constitutes storage, computing, and other resources. However, it does not possess as huge infrastructure as a Cloud. The proximity of the Fog nodes with the end-users provides the advantage in processing the data for the delay-sensitive and time-critical applications. These end-users are usually the IoT devices that communicate with Fog infrastructure for information processing and decision-making. Fog-integrated Cloud architecture [2] can provide a better approach to solve these problems by optimally utilizing and scheduling Fog and Cloud resources. Also, it can provide a highly flexible and reliable platform for the computation of IoT data. Further, it optimizes the objectives such as latency, delay, cost, and power consumption for IoT devices.

I. Woungang et al. (Eds.): ANTIC 2021, CCIS 1534, pp. 229–239, 2022.
https://doi.org/10.1007/978-3-030-96040-7_18

There are colossal mission-critical IoT applications with different data processing needs based on their size or tolerable delay. For example, for driverless vehicles, information regarding the shortest available route to the destination, route change instructions, immediate and future maintenance needs, and many more may be required. Smart home equipment might be managed remotely using data from deployed sensors. Various IoT applications, such as real-time systems, healthcare, virtual and augmented reality (VR/AR), real-time control loops, streaming analytics, and so on [3], are highly delay-sensitive and are referred to as mission-critical [4], not only because of the traditional concept of "life risk" but also because of the dangers of public services being disrupted. Some of the real-time tasks (soft real-time and hard real-time) are displayed in Table 1 with their tolerable delay in milliseconds.

Table 1. Various application and related delays.

Task	IoT application	Tolerable delay (ms)
Hard real-time	Industrial control in real-time	0.5
	Augumented reality	2
	Virtual reality	1
	Road safety	10
	Robotics/Tele-presence	3
	Remote healthcare	20
Soft real-time	Gaming	60
	Education and culture	70
	Smart grid	100

High delay sensitivity is a common feature of many applications. The end-to-end delay is determined by several factors, including network latency, computer processing time, and placement of nodes. Moreover, these are also of two types: hard delay-sensitive tasks that must provide resources in given time and soft delay-sensitive tasks that can proceed with some penalty.

The server's placement creates a bottleneck in the network. Various studies has been carried out in the literature to place the servers deep into the network such as [5]. In other words, regardless of the size of the task, a time delay is unavoidable due to the distance between the data generating device and server, as data must have to travel back and forth. Rather than transferring a massive amount of data produced by the IoT to the cloud, it is more effective to analyze the data at the multiple layers (Fig. 1), right from where it has been generated.

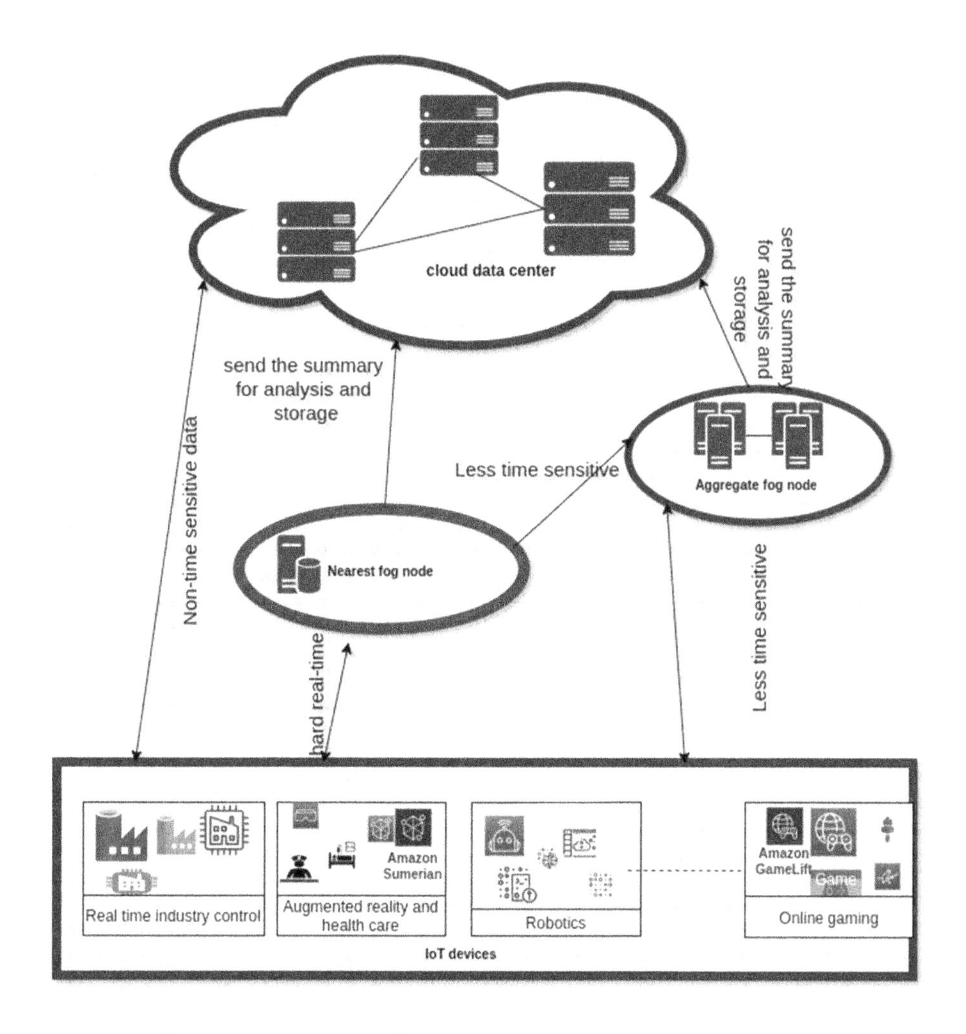

Fig. 1. Generalized IoT-fog-fog aggregation-cloud architecture.

2 Related Work

A number of algorithms and strategies for job offloading in fog-cloud environment have been published. For IoT applications, Fan and Ansari [6] suggested an allocation policy of workload. For mobile cloud, they designed an $M/M/1$ queuing model that considers offloading time and computational delay with the minimize resource cost and response time. Nan et al. [2] developed a comprehensive analytic framework for minimization of energy cost and time to support the IoT operation as well as to deal with trade-offs in terms of average monetary, average response time, and energy charges using the Lyapunov optimization approach. Guo et al. [7] designed ECTCO (Energy-efficient Collaborative Task Computation Offloading) algorithm based on semidefinite relaxation and stochastic mapping approach to obtain task computation offloading strategies for IoT sensors. Deng et al. [8] developed a framework for workload allocation with optimum task distribution in Fog and Cloud with confined service delay and reduced power consumption. To reduce task latency, Yousefpour et al. [9] established an analytical model for the three-tier architectural Fog-Cloud system. They used a threshold technique for work offloading decisions.

To minimize power consumption and offload the task, Rahbari and Nickray [10] created a module placement strategy based on classification and regression tree methods. Liu et al. [11] presented a hybrid computational offloading technique to reduce latency while maximizing resource uses in a diverse Fog-Cloud environment. The performance of fog offloading has been modelled by Majeed et al. [12]. They proposed methods for calculating offloading time using containers in Fog computing. There are four approaches to container-based offloading that have been investigated. Three are stateless techniques (Save and Load, Push and Pull, Export and Import), and one stateful technique (CRIU-based Live Migration). The offloading time is estimated using four estimation methods based on Ridge regression, Multivariate Linear Regression, Random Forest Regression, and Polynomial Multivariate Regression. Zhang et al. [13] proposed the FEMTO (fair and energy-minimized task offloading) method, which is based on a fairness scheduling metric that considers three main characteristics: task offloading energy intake, fog node priority and historical average energy. Furthermore, the proposed FEMTO algorithm achieves high and consistent level of fairness in the energy utilization of FNs.

L. Gu et al. [14] investigate virtual machine placement in fog-computing by integrating fog computing and medical -cyber physical systems because of growing healthcare trend in society. Moreover, task distribution reduces total costs and meets QoS requirements. In an SDN-enabled IoT-Fog, A. Akbar et al. [15] introduced a novel technique to provide reliable and adaptive communication. To determine the reliability level of links, they employed the k-nearest neighbor approach of machine learning. Five months of real-life network traces were used to train the model. They also employed a well-known multi-objective optimization (MOO) Technique (NSGA-II) to generate Pareto-optimal routes that optimize the trade-off value between two goals. Furthermore, their technique is built on an adaptive decision mechanism in which the SDN controller chooses the appropriate route for various types of applications depending on packet characteristics (TCP and UDP).

In this work, we have developed an analytical model for workload allocation in a three-tier fog-cloud system (i.e., fog-fog aggregation-cloud) and developed an offloading strategy to maximize network utility and time minimization.

3 System Model

The task or workload allocation problem in fog computing is similar to task offloading, which involves allocating tasks or workloads to fog nodes or cloud servers while reducing energy, cost, or delay. The input to the problem is a collection of tasks, and the problem is modelled as a static optimization problem that specifies how the jobs or workloads are assigned.

The entire network is modelled as an undirected graph $G = \{V; E; W\}$, where the node V is the collection of nodes consist of IoT, Fog, Fog aggregation, and Cloud nodes, i.e., $V = \{I \cup F \cup FA \cup C\}$ Moreover, communication links between nodes are represented by edges E; W is the set of weight which consist of upload and download time of the tasks. Thus $W = \{W^{up} \cup W^{dn}\}$.

In hierarchical fog-cloud architecture, the collection of I set of IoT devices are represented as $I = \{1, 2, 3, ..., i\}$ and set of F locally distributed fog nodes as $F = \{1, 2, 3, ..., j\}$. Furthermore, we consider FA fog aggregation nodes defined as $FA = \{1, 2, 3, ..., k\}$. Additionally, a set of C cloud data servers is denoted as $C = \{1, 2, 3, ..., l\}$. An intelligent controller will offload the task to either of the nodes based on the requirement of the tasks or the availability of the node. Also, we consider CPU frequency at different layers (Fog, Aggregated Fog, and Cloud) as $C^{cpu} > FA^{cpu} > F^{cpu}$ and memory of these are represented as $C^M > FA^M > F^M$. Various types of computing, depending upon their place of execution, are defined as follows:

3.1 Native Computing

Tasks computed locally at IoT devices due to low computing power and storage requirement. In this case, the total time for the task mainly depends on local execution time (IoT devices' frequency and data size), and no network delay (i.e., Uptime and Downtime) is involved. So, the total time taken for task completion is expressed as t_n, represented as below:

$$t_n = \sum_{I=1}^{i} \left(\frac{D_i}{f_i} \right) \quad \forall I \in \{1, 2, 3, ..., i\} \tag{1}$$

Where D_i is the data size of the request and f_i is the frequency of the IoT device.

Table 2. Parameter definitions.

Symbols	Definition	Symbols	Definition
I	Total number of IoT devices	P_r^f	Probability that fog nodes accept incoming request
F	Total number of fog nodes	P_r^{fa}	Probability that fog-aggregation nodes accept incoming request
FA	Total number of fog aggregation nodes	P_r^c	Probability that cloud nodes accept incoming request
C	Total number of cloud nodes	UP_{ij}^{IF}	Time taken to upload i^{th} IoT request to the jth fog node
D_i	Data size of the ith task	EX_{ij}^F	Time taken by j^{th} fog node to execute i^{th} IoT
F_i	CPU frequency of ith computing device	DW_{ji}^{FI}	Time taken to send from the jth fog node to i^{th} IoT node, after processing
t_n	Total time taken in native computing	UP_{ik}^{IFA}	Time taken to upload i^{th} IoT request to the kth fog node
t_r	Execution time of task at remote computing	EX_{ik}^{FA}	Execution time to execute i^{th} IoT request on k^{th} fog-aggregation node
tt_r	Total time taken in remote computing	DW_{ki}^{FAI}	Time taken to send from the kth fog-aggregation node to i^{th} IoT node, after processing the assigned task
C_M, FA_M, F_M	Main memory of Cloud, Fog Aggregation, and Fog nodes, respectively	UP_{il}^{IC}	Time taken to upload i^{th} IoT request to j^{th} fog-aggregation node
t^{up}	Upload time	EX_{il}^C	Execution time to execute i^{th} IoT request on l_{th} cloud node
t^{dn}	Download time	DW_{li}^{CI}	Time taken to send from l_{th} cloud node to i^{th} IoT node, After processing the assigned task
BW^{up}	Upload bandwidth	BW^{dn}	Download bandwidth
$C^{cpu}, FA^{cpu}, F^{cpu}$	CPU frequency of Cloud, Fog Aggregation, and Fog nodes, respectively		

3.2 Remote Computing

Because IoT devices have limited processing and storage capacity, processing of most of the jobs are not possible here, and these are offloaded to locally distributed fog nodes, fog aggregation nodes, or centralized cloud servers depending on the task's criticality, the status of the nodes. The communication delay of the devices to other computing devices (fog, fog-aggregation, and cloud) serves an essential role in making the offloading decision. Here, we have assumed communication delay as upload and download time of the tasks. Upload time is defined as uploading the task from IoT to the most feasible computing nodes like fog, fog aggregation, and cloud. Download time sends the task to the IoT device requested after processing the assigned tasks to various computing nodes.

Thus, the total time in remote computing relies on the execution time taken by a particular node and the network delay.

t_{up} is the upload time of the tasks defined as below:

$$t_{mn}^{up} = \frac{\sum_{m=1}^{i} D_m}{BW_{mn}^{up}} \tag{2}$$

Where $m = I$ and $n \in \{F \cup FA \cup C\}$. D_i is the data size, BW_{up} is the upload bandwidth.

Similarly, t^{dn} is the downtime of tasks represented as below:

$$t_{nm}^{dn} = \frac{\sum_{m=1}^{i} D_m}{BW_{mn}^{dn}} \tag{3}$$

Where $m \in I$ and $n \in \{F \cup FA \cup C\}$. D_i is the data size, BW^{dn} is the download bandwidth.

Execution time of tasks at remote nodes t_r is computed as given in Eq. (4).

$$t_r = \sum_{m=1}^{i} \left(\frac{D_m}{f_n}\right) \tag{4}$$

Where $m \in I$ and $n \in \{F \cup FA \cup C\}$. D_m is the data size of the tasks and f_n is the frequency of nodes.

The notation used in this work is listed in Table 2.

4 Model Analysis for Total Time in Fog-Cloud Architecture

In this work, we have categorized the tasks based on their response time (tolerable delay) into three classes: hard real-time tasks, soft real-time tasks with the penalty, and soft real-time tasks without penalty. The proposed controller will offload the task to either fog, fog aggregation, or cloud.

tt_r is the total time taken for remote computing. It is described as below:

$$\begin{aligned} tt_r = P_r^f(UP_{ij}^{IF} + EX_{if}^{F} + DW_{ji}^{FI}) + \\ P_r^{fa}(UP_{ik}^{IFA} + EX_{ik}^{FA} + DW_{ki}^{FAI}) + \\ P_r^c(UP_{ij}^{IC} + EX_{C}^{il} + DW_{li}^{CI}) \end{aligned} \tag{5}$$

Where, $\forall I \in \{1, 2, 3, ..., i\}$, $\forall F \in \{1, 2, 3, ..., j\}$, $\forall FA \in \{1, 2, 3, ..., k\}$ and $\forall C \in \{1, 2, 3, ..., I\}$.

The sum of probability of task execution at the fog, fog-aggregation, and cloud is 1, i.e., $P_r^f + P_r^{fa} + P_r^c = 1$ which means that the task will be executed at either of the nodes.

5 Node Offloading Policy

An offloading strategy at different nodes is introduced in this section. The task will be offloaded to other nodes based upon the minimum upload time and the status of that node, where status is the flag variable having the value of either one or zero. Value "zero" means the node is free, i.e., no execution is going on, and "one" means it is busy with other tasks execution. New tasks will be offloaded only when the node has the status value "zero" and minimum upload time. The nodes frequently update the status to the controller. As we can see in Table 3, the first node having lower upload time and status field represents "one", which means it is busy with other task execution so, the controller will offload the task to the second node. By applying this offloading policy, one can minimize the offloading and backtracking time from various nodes to the controller.

Table 3. Example of policy offloading.

S. No.	Upload time (ms)	Status	Node ID
1	20	1	121.110.75.16
2	25	0	121.110.75.18
–	–	–	–
–	–	–	–
N	–	–	–

6 Performance Evaluation and Discussion

Experiments to assess the performance of the proposed analytical model are presented in this section. Furthermore, it is shown how the proposed model works well for time-critical tasks. First, the simulation setup and parameters for the evaluation of the model are described. Next, the simulation results are presented, and the framework's behavior is analyzed under varying parameters.

For the analysis, a four-level hierarchical network model, as shown in Fig. 1, is used, where the cloud node is situated at the highest level and IoT nodes at the last level. Informed by earlier works, we have assigned simulation parameters to the nodes defined in Table 4. Each IoT node will generate 100 random requests; thus, together, all IoT nodes generate ten thousand requests. Also, each request is combined with a random delay. As discussed, the tasks are categorized into three types: hard real-time, soft real-time with the penalty, and soft real-time without any penalty.

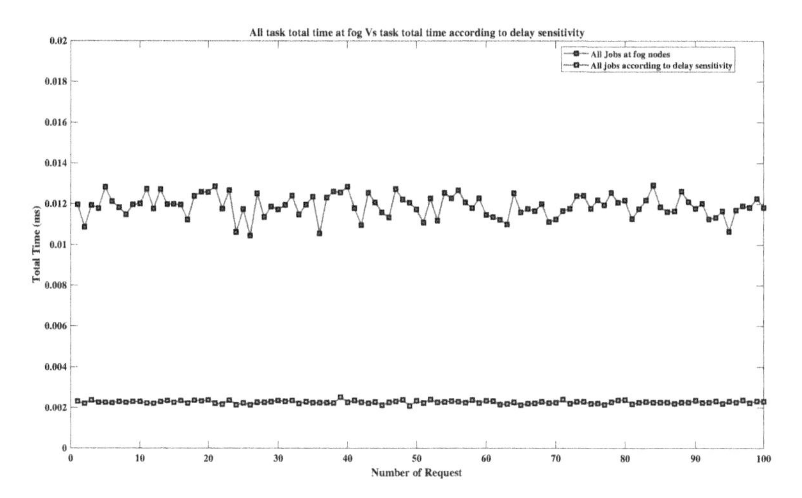

Fig. 2. Comparison on total time taken when all tasks are offloaded to fog vs. proposed model.

Table 4. Simulation parameters

Parameters	Values
Number of IoT devices	100
Number of fog nodes	30
Number of fog aggregation nodes	10
Number of cloud nodes	1
Upload and download bandwidth at fog nodes	54 Mbps
Upload and download bandwidth at fog aggregation nodes	100 Mbps
Upload and download bandwidth at cloud nodes	10 Gbps
CPU frequency of fog nodes	600 MHz
CPU frequency of fog aggregation nodes	1 GHz
CPU frequency of cloud nodes	10 GHz

Further, we have also considered the criticality of jobs for the task classification. We offload the tasks at three different layers, such as tasks whose tolerable delay is less than or equal to 50 ms will be sent to fog, tasks whose tolerable delay lies between 51 to 200 ms will be offloaded to fog-aggregation node, and the tasks whose tolerable delay is greater than 200 ms will be offloaded to cloud node. For offloading at a particular layer, they will follow the offloading policy defined in Sect. 3. The real-time tasks tolerable delay generated randomly between [1–10000] in millisecond. The request length is exponentially distributed between the range [100 kb–70 kb]. Cloud servers' CPU frequency and memory are more than fog-aggregation, and memory and CPU frequency of fog-aggregation are more than fog nodes.

In Figs. 2 and 3, the performance of the proposed model is shown by comparing with the situation where all requests have been served by the Fog only and Cloud only, respectively. From Fig. 2, one can see that the proposed model, based on the tolerable delay method, performs better than the model where all requests are served by the fog nodes. Although all the fog nodes are in near proximity, the total time taken by the fog nodes is still more than the total time taken by the nodes where the request has been processed by the nodes according to the delay tolerance. Further, one can also say that the size of the task as well as the processing capability of the nodes plays a vital role in providing better response time and quality of service for the user. Figure 3 is self-explanatory with one significant factor that the total time is far greater than our model. It is because of the network delay. Thus, one can conclude that the cloud nodes are inefficient for time-critical tasks.

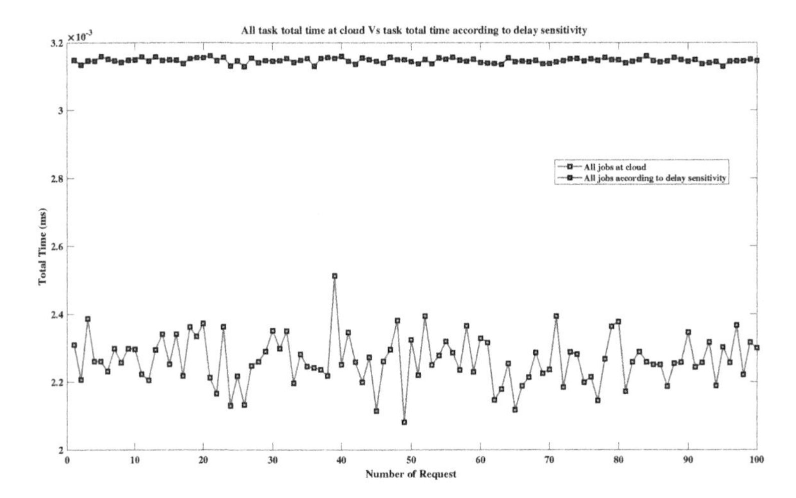

Fig. 3. Comparison on total time taken when all tasks offloaded to cloud vs. proposed model.

7 Conclusion

In this paper, we have proposed a scheme to minimize the total time taken by the IoT tasks in a hierarchical fog-fog aggregation-cloud system. Further, we have developed an offloading strategy that can minimize offloading time and backtracking time. We have also shown that our design approach can minimize the overall time taken to process the requests. The result shows that offloading real-time tasks according to tolerable delay and using the proposed analytical model can be a better approach to provide quality of service.

As a part of future work, we intend to add extra dimensions to IoT requests which will be helpful for proper task classification. It is also intended to include parameters such as security, costs, and energy minimization.

References

1. He, J., Wei, J., Chen, K., Tang, Z., Zhou, Y., Zhang, Y.: Multitier fog computing with large-scale IoT data analytics for smart cities. IEEE Internet Things J. **5**(2), 677–686 (2018)
2. Nan, Y., et al.: Adaptive energy-aware computation offloading for cloud of things systems. IEEE Access **5**, 23947–23957 (2017)
3. Sarkar, S., Chatterjee, S., Misra, S.: Assessment of the suitability of fog computing in the context of internet of things. IEEE Trans. Cloud Comput. **6**(1), 46–59 (2018)
4. Travelling team: en-USThe Changing Face of World Missions - The Strategic Context. http://www.thetravelingteam.org/articles/the-changing-face-of-world-missions-the-strategic-context
5. Yadav, P., Kar, S.: Evaluating the impact of region based content popularity of videos on the cost of CDN deployment. In: National Conference on Communications (NCC) 2020, pp. 1–6 (2020)
6. Fan, Q., Ansari, N.: Workload allocation in hierarchical cloudlet networks. IEEE Commun. Lett. **22**(4), 820–823 (2018)
7. Guo, S., Liu, J., Yang, Y., Xiao, B., Li, Z.: Energy-efficient dynamic computation offloading and cooperative task scheduling in mobile cloud computing. IEEE Trans. Mob. Comput. **18**(2), 319–333 (2019)
8. Deng, R., Lu, R., Lai, C., Luan, T.H., Liang, H.: Optimal workload allocation in fog-cloud computing toward balanced delay and power consumption. IEEE Internet Things J. **3**(6), 1171–1181 (2016)
9. Yousefpour, A., Ishigaki, G., Gour, R., Jue, J.P.: On reducing IoT service delay via fog offloading. IEEE Internet Things J. **5**(2), 998–1010 (2018). arXiv: 1804.07376
10. Rahbari, D., Nickray, M.: enTask offloading in mobile fog computing by classification and regression tree. enPeer-to-Peer Netw. Appl. **13**(1), 104–122 (2020). https://doi.org/10.1007/s12083-019-00721-7
11. Liu, Y., Yu, F.R., Li, X., Ji, H., Leung, V.C.: Hybrid computation offloading in fog and cloud networks with non-orthogonal multiple access. In: IEEE INFOCOM 2018 - IEEE Conference on Computer Communications Workshops (INFOCOM WKSHPS), pp. 154–159, April 2018
12. Majeed, A.A., Kilpatrick, P., Spence, I., Varghese, B.: Modelling Fog Offloading Performance, arXiv:2002.05531 [cs], February 2020
13. Zhang, G., Shen, F., Liu, Z., Yang, Y., Wang, K., Zhou, M.-T.: FEMTO: fair and energy-minimized task offloading for fog-enabled IoT networks. IEEE Internet Things J. **6**(3), 4388–4400 (2019)
14. Gu, L., Zeng, D., Guo, S., Barnawi, A., Xiang, Y.: enCost efficient resource management in fog computing supported medical cyber-physical system. enIEEE Trans. Emerg. Topics Comput. **5**(1), 108–119 (2017). http://ieeexplore.ieee.org/document/7359164/
15. Akbar, A., Ibrar, M., Jan, M.A., Bashir, A.K., Wang, L.: SDN-enabled adaptive and reliable communication in IoT-fog environment using machine learning and multiobjective optimization. IEEE Internet Things J. **8**(5), 3057–3065 (2021)

A Survey on DDoS Attacks on Network and Application Layer in IoT

Nimisha Pandey(✉) and Pramod Kumar Mishra

Banaras Hindu University, Varanasi 221005, UP, India
{nimisha.pandey17,mishra}@bhu.ac.in

Abstract. IoT makes the devices remotely accessible and these devices are used to collect data for analysis at a later stage. IoT was expected to bring revolutionary changes in the way people use technology. However, security vulnerabilities in IoT make it insecure towards the possible attacks like DoS/DDoS attacks. Detection of DDoS attacks is required to protect the IoT systems from attackers and evade financial losses. This paper presents a review of solutions proposed for the detection of DDoS attacks in the network layer and application layer of IoT. Application layer attacks have been increasing because they are sophisticated and are difficult to differentiate from real users. A huge number of papers have contributed to the network layer but issues are still faced in application layer in IoT. We have reviewed the issues in application layer protocols in IoT as well. The need of development of countering DDoS in application layer of IoT is also addressed.

Keywords: DDoS attack · Attack detection · IoT · Network layer · Application layer

1 Introduction

DoS attacks aim to achieve resource exhaustion at the target system by bombarding the target system with unnecessary redundant traffic. When this attack is done through distributed sources, it is known as Distributed Denial of Service (DDoS) attacks (Bhattacharyya and Kalita 2016). This is done to hamper the availability of target system and impact its catering of legitimate requests. The attackers launch this attack for political reasons, for levying huge financial losses on the corporations or are inspired from malicious intentions towards other users.

IoT is the network of heterogenous devices connected together with new protocols and standards that fulfill its requirements like IEEE 802.15.4. The network is resource-constrained since these devices are not necessarily computing devices with high storage and ample battery power. They usually have limited resources to perform small tasks for example, bulbs, camera, baby monitor, etc. This technology is developing and therefore, there is a threat of possible security vulnerabilities in the underlying protocols (Rahman and Shah 2016). Due to resource constrained nature of IoT, security concerns escalate since it will be easy to cause big impact even through low-intensity attacks. Therefore, the DoS/DDoS attacks must be detected and mitigated in an IoT system as soon as

I. Woungang et al. (Eds.): ANTIC 2021, CCIS 1534, pp. 240–250, 2022.
https://doi.org/10.1007/978-3-030-96040-7_19

possible. The basic techniques used for preventing DDoS attacks are resource accounting and resource multiplication. Resource accounting means that giving resources to users according to their behaviors i.e., if a user seems malicious then not allocating resources to that user. Resource multiplication is basically increasing the number of resources on al large scale so that attackers are not able to overwhelm the system by hijacking the resources (Asosheh and Ramezani 2008). Resource multiplication is not possible in most of IoT systems because of their resource-constrained nature and remote location. Resource accounting does not seem very robust in case of network-based attacks since attackers can utilize the loopholes in the protocols to defy this technique.

In modern times due the advent of botnets, the security concerns for IoT have increased. Botnets are a collection of compromised devices which were attacked by a malware created by the attacker (Kambourakis et al. 2017; Kolias et al. 2017). The users of the devices in a botnet, also known as bots, have no idea that their devices are compromised. The malware named Mirai attacked several smart home devices like baby monitors, doorbells, etc. to create a botnet. This botnet was leveraged to launch a huge attack on the DNS provider company Dyn in 2016 with a peak of 620 Gbps (Sam Egbo 2018). Until now, IoT systems have been used to be a part of botnets and are used to launch DDoS attacks on other systems. But when the IoT systems will be widely accepted across all walks of life then these systems will be an easy target for the attackers. The pictorial view of direct DDoS attack and reflector-based DDoS attack is defined in Figs. 1 and 2.

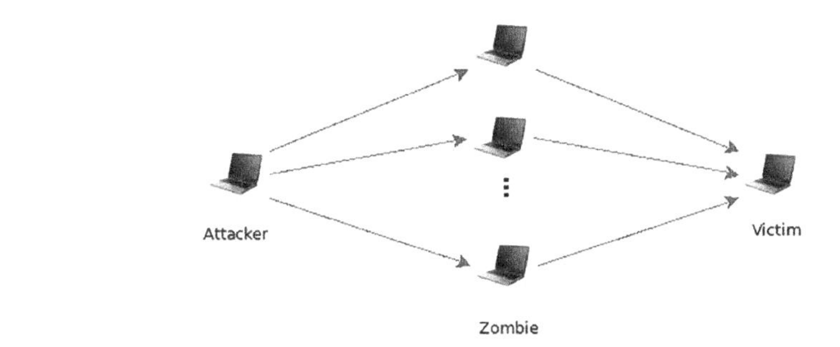

Fig. 1. Pictorial view of simplified DDoS attack

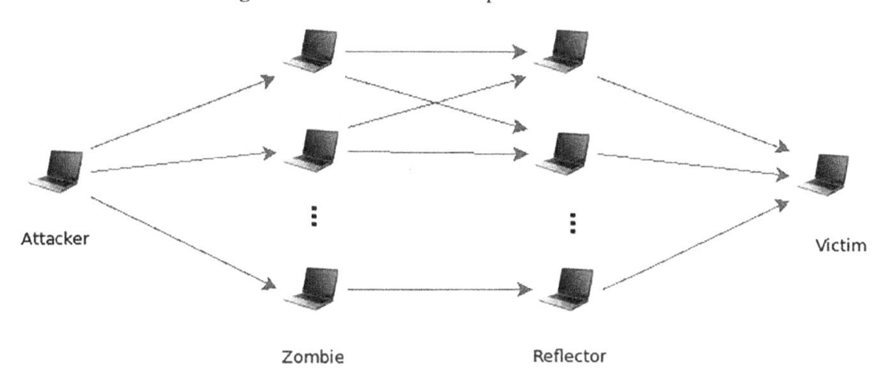

Fig. 2. Pictorial view of simplified Reflector based DDoS attack

DDoS normally attack in all the layers of IoT but most generally it affects the Network layer and Application layer. The review of these two layers is discussed in the next section.

2 DDoS in Network Layer of IoT

In a DDoS attack, multiple devices attack the victim system hence the characteristics of malicious networks can be used to detect the attack. These characteristics can be number of source IP addresses, number of source ports, number of destination IP address, number of destination ports used, protocol and many more. Researchers have proposed different techniques to detect these attacks in different layers of IoT. Many papers have used information theory metrics for the detection (Behal and Kumar 2017; Xiang et al. 2011; Mao et al. 2018; Idhammad et al. 2018; Bhuyan et al. 2016; Koay et al. 2018; Swami et al. 2019; Zhou et al. 2019). Some have used the machine learning and deep learning algorithms to differentiate between attack traffic (Tahsien et al. 2020; Lima Filho et al. 2019; Meidan et al. 2018; Doshi et al. 2018). Many researchers have used deep-learning techniques in DDoS detection in IoT (Doriguzzi-Corin et al. 2020; Farukee et al. 2020; Ma et al. 2020; Hussain et al. 2020; McDermott et al. 2018; Roopak et al. 2020). Sachdeva et al. (2016) have focused on high-rate DDoS attacks and differentiated between flash events and high-rate DDoS attacks. Doshi et al. (2021) have worked for the stealthier low-rate DDoS attacks and proposed techniques for differentiating them from normal traffic. Researchers have also created mathematical and computational models to discover the defense mechanisms (Liu and Qiu 2017; Maciá-Fernández et al. 2009; Maciá-Fernández et al. 2008). Furthermore, many papers have utilized of SDN for the detection and mitigation of the DDoS attacks in IoT and proposed SDN-based approaches for the detection (Bhayo et al. 2020; Sharma et al. 2019; Demetriou et al. 2017a, 2017b). Some works have also addressed the issues in SDN which make it vulnerable to DDoS attacks and have proposed novel approaches to protect the SDN based IoT networks from DDoS attacks (de Assis et al. 2020; Wang et al. 2021). Tushir et al. (2021) have investigated the impact in connectivity and consumption of power during the TCP-based and UDP-based DDoS attacks. Many researchers have proposed application-based solutions like for smart homes (Tushir et al. 2020; Paudel et al. 2019; Tanwar et al. 2017; Serror et al. 2018; Chitnis et al. 2016), smart grids, smart healthcare etc. Li et al. (2020) have worked for real-time detection of the attacks using network layer features.

Figure 3 shows the general method of detection of DDoS attacks in the network layer of IoT. When a system receives suspicious traffic, its IDS uses the network traffic details to create the features. It calculates the value of features and feeds them in the classifier. This classifier can be based on statistical methods, machine learning, deep learning, etc. The classifier, in general, differentiates the attack traffic from the legitimate traffic. Here, the detection of DDoS is done on the basis of network characteristics.

In the above study, it is clear that the increased propagation and ubiquity of IoT devices have also increased security issues in network layer. To properly secure IoT devices, many research and practical challenges are still to be investigated in this layer. In this layer, the goal of the attacker is to exhaust network layer resources like bandwidth etc.

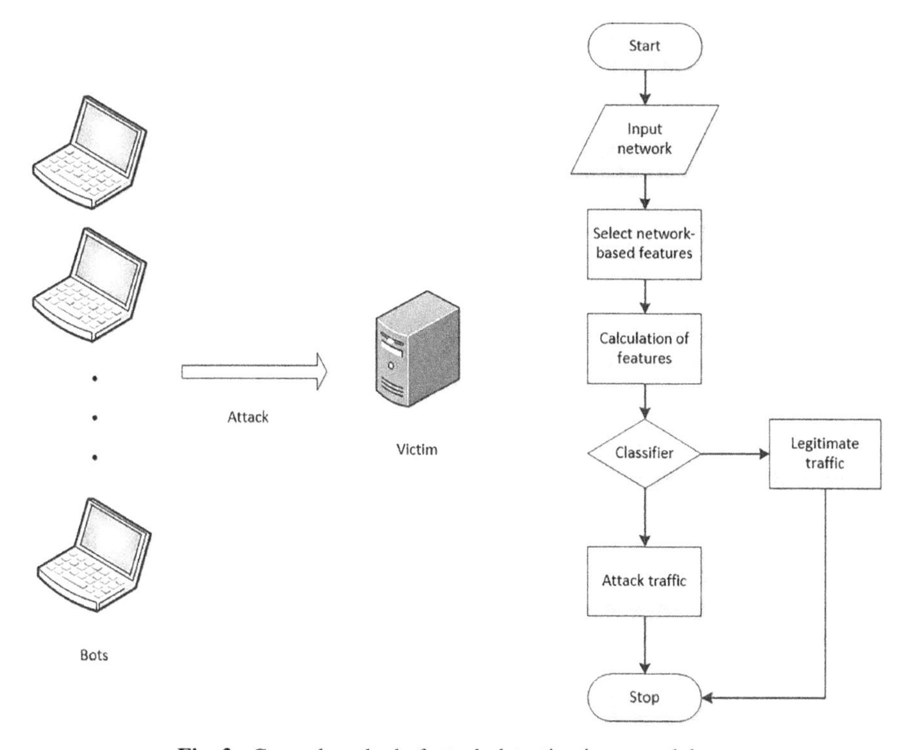

Fig. 3. General method of attack detection in network layer.

Possible Solution of the Problem: The machine learning algorithms along with information theory-based solutions and deep learning-based solutions generally give superior results in the detection of DDoS attacks in IoT.

3 DDoS in Application-Layer of IoT

Application layer attacks are same as the network layer attacks. Low-rate application layer attacks are more popular as they are very similar to the benign traffic. These attacks also exploit the resources by sending such highly computational requests for achieving resource exhaustion. Various protocols have been introduced for the IoT but they have DoS vulnerabilities. The CoAP protocol is an application layer protocol for IoT and is equivalent to HTTP protocol. DTLS protocol is equivalent to the TLS protocol which works with the TCP protocol. It works with UDP protocol which is used in the scenarios where power and time delay are constrained, making it favorable for IoT. DTLS provides security to CoAP and works between transport layer and application layer. Some of the protocols of application layer protocols are given in Fig. 4. There are various attacks possible in application layer in IoT. In slow send attacks, the attacker sends the packets very slowly in order to hold the connection for longer period. Similarly, in slow read attack, the attacker will consume the server's response very slowly. Low-rate attacks either compromise the system performance by stealthy approach or they request huge resources and hold them for long periods.

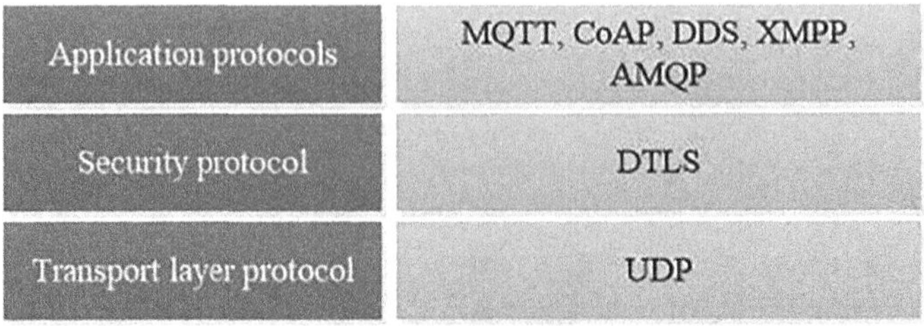

Application protocols	MQTT, CoAP, DDS, XMPP, AMQP
Security protocol	DTLS
Transport layer protocol	UDP

Fig. 4. Application layer protocols and corresponding lower layer protocols for CoAP in IoT.

Nebbione and Calzarossa (2020) have addressed the security problem from IoT perspective. They have studied the application layer protocols of IoT, viz. MQTT, AMQP, XMPP, SSDP, CoAP, DDS, mDNS. They have reviewed the security vulnerabilities of these protocols and have addressed that DoS attack is possible in all these protocols in IoT. In this study, they have concluded that limited research has been done in this field and further work is required to secure the application layer of IoT. Bhosale et al. (2017) have presented a taxonomy of DDoS attacks on application layer and their defense mechanisms. They have discussed about the various types of DDoS attacks, the tools that are used to launch these attacks and their prevention mechanisms. They have reviewed the various solutions proposed for application layer-based DDoS attacks. Rahman and Shah (2016) have reviewed the security problems in CoAP protocol. They have discussed various solutions proposed and further security challenges in CoAP.

Tiloca et al. (2017) argued that DTLS is vulnerable to DoS attacks which open multiple half-open sessions. DTLS uses a Cookie to determine whether the source address of the packet is spoofed or not, and thereby determining whether it is a legitimate packet or a DoS attack packet. Tiloca et al. (2017) argued that the cookie used in DTLS can be intercepted by an attacker and can also be used to make the DoS attack worse. They considered only the specific attack in which an attacker tries to open multiple half-open DTLS sessions. They also argued that while using the pre-shared keys, the scaling of the network is not possible. As the number of nodes increase in the network, the number of keys required for the communication between each pair of nodes also increase. They proposed the Derived Key Mode scheme which uses a Trust Anchor (TA) entity providing the keys for secure communication between client and server and is also used in detecting DDoS attack.

Moreover, since DTLS was designed for UDP (which is an unreliable protocol), it simply discards the duplicate messages and preserves the association (Rescorla and Modadugu 2017). Therefore, replay detection feature is not mandatory in DTLS. Wei and Du (2018) argued that the DTLS is vulnerable to replay message attacks which can eventually be a DDoS attack. They proposed that in order to mitigate such DDoS attacks, an upper limit must be set by the users on the number of replays of a message. Furthermore, Raza et al. (2012) proposed compression of DTLS header using compression mechanisms of 6LoWPAN. Raza et al. (2013) proposed a framework called Lithe which compresses the DTLS header with 6LoWPAN compression techniques and integrate the compressed DTLS with CoAP. Martins et al. (2016) proposed to amalgamate smart homes and smart grids for better monitoring of energy requirements and utilizing energy resources more efficiently. In order to assess the performance of DTLS and CoAP in this amalgamation, they argued that spurious CoAP requests are sent at border router to initiate a DoS attack. They send spurious CoAP request to the border router in attack intervals of 1000 ms, 500 ms and 700 ms. In case of 500 ms, 90% packet loss is witnessed in 2 min of the attack.

Gonzalez et al. (2014) analyzed the impact caused by DoS attacks on web servers. They proposed that the metrics based on packet loss and response time are not suitable to use in the detection of application-level attacks. This is because application layer attacks do not have an impact on the lower layers. So, they proposed a new metric named Loss for application server. They have created a testbed for four web servers viz. Apache on Windows and Linux, IIS, Tomcat on Windows, Linux-Default and Linux-Enhanced and Nginx on Linux and Windows. They launched slow send attack, slow read attack and low-rate attack on these servers. They found that Tomcat with Linux-default and Linux-enhanced was most affected by a loss of 100% under slow read attack while Tomcat with Windows showed a loss of 60% under low-rate and slow send attacks. Apache server on Windows suffers 90% loss in low-rate attack and slow send attack.

Singh et al. (2018) review the detection mechanisms for the HTTP GET Flood attacks. Based on user behavior, they have proposed four features to differentiate between these attacks, flash events and normal traffic. The four features are request index, response index, repetition index and popularity index. Request index, in general, deals with the number of requests for resources sent by a user in a duration of time. They used Chebyshev's inequality theorem (Amidan et al. 2005). Response index deals with the amount of data requested by a user in a duration of time. They used box plot to calculate response index. Popularity index assumes that legitimate users request for the popular web pages while the bots request for more random set for web pages. It deals with the randomness of the requests from the user. Another observation made by the author is that bots repeat their requests in a short duration of time more frequently in comparison to the legitimate user. This fact is used to create the repetition index. They have detected 12 different strategies used by the attacker in launching DDoS attack. They are random, flash, constant, main page, dominant page, repeated, replay flood, random page, Hot pages, high burst, continuous, and session flood. They have created a testbed and collected data. They have used ML classifiers like Naïve Bayes, JRip, Random Forests, J48, Support Vector Machines (SVM), and IBk (an implementation of kNN in Weka). Their approach has received a result of 97.46% accuracy.

Bravo and Mauricio (2018) have proposed an approach for the detection of application-layer DDoS attacks. They have stored all established connections and their invoked processes in a data bank. The data bank is analyzed by the interaction detector. From this data bank, feature extraction is done by Javascript programming in real time. Features extracted are mouse move, mouse highlight, mouse click, mouse drag and drop, key press and release event, and mouse wheel and scroll. On the basis of these features, eight characteristics are defined. Their classifier basically assumes any user a real one if one characteristic is true, otherwise it is considered an attack. In case of real user, further requests are allowed while in case of suspicious activity a captcha test is done. The tools used for creating attacking requests are LOIC, GoldenEye and OWASP Tool. The outcome of their approach was a 100% detection rate for attacks form all the three tools.

Johnson Singh et al. (2016) have proposed the detection of application layer-based DDoS attacks through multilayer perceptron (MLP) classification algorithm and learning algorithm being a genetic algorithm. They have setup a web server, Apache wamp server (64 bits and PHP 5.5) 2.5 hosted on Windows OS7. The attacks tools used are Slowhttptest on RedHat6.6, LOIC and HAVIJ on Windows 8, HOIC and Anonymous DDoSer on Windows 7, R.U.D.Y. on CentOS and BONESI on Fedora 24. The features used for classification between attack and normal traffic are HTTP GET request count, variance and entropy for every connection. For the validation, they have used the dataset named Environmental protection agency- Hypertext transfer protocol (EPA-HTTP). Their approach has given 98.31% accuracy, 0.9962 sensitivity and 0.0561 specificity. Imperva mitigates a massive HTTP flood: 690,000,000 DDoS requests from 180,000 botnets Ips.

Sreeram and Vuppala (2019) aimed for fast and early detection of HTTP flood attacks. They identified 6 features namely absolute time interval id, session time observed, page access count, packet observed for each type, max number of sessions and minimum time interval. They have used a swarm intelligence-based bat algorithm for classification. The approach is tested on CAIDA dataset and gives better accuracy, precision, f1-score, detection rate and lesser processing time in comparison to other existing approaches namely, ARTP and FCAAIS. Following Table 1 show the analysis of DDoS attack in application layer with non-IoT considerations.

Table 1. Analysis of detection of DDoS attack in application layer in non-IoT environment.

Papers	Dataset	Attacks considered	Attack tools	Features	Result
Singh et al. (2018)	Created	HTTP GET flood	Apache Jmeter	Request index, response index, repetition index and popularity index	97.46% accuracy
Bravo and Mauricio (2018)	Created	HTTP POST flood attack and DDoS	LOIC, OWASP DOS HTTP POST and GoldenEye HTTP Denial of Service Tool	Mouse move, mouse highlight, mouse click, mouse drag and drop, key press and release event, and mouse wheel and scroll	100% detection rate

(continued)

Table 1. (*continued*)

Papers	Dataset	Attacks considered	Attack tools	Features	Result
Johnson Singh et al. (2016)	EPA-HTTP	DDoS on HTTP	Slowhttptest, LOIC, HAVIJ, HOIC, Anonymous DDoSer, R.U.D.Y., and BONESI	HTTP GET request count, variance and entropy for every connection	98.31% accuracy, 0.9962 sensitivity and 0.0561 specificity
Sreeram and Vuppala (2019)	CAIDA	HTTP flood attack	–	Absolute time interval id, session time observed, page access count, packet observed for each type, max number of sessions and minimum time interval	Better Accuracy, f1-score, detection rates, precision and lesser processing time as compared to ARTP and FCAAIS

On the above study it is clear that the goal of the attacker in application layer is to drain target resources of the IoT. IoT devices are not to user friendly, and user generally neglect the security of these devices because of unawareness, making them more exposed to attacks.

Possible Solution of the Problem: These attack viz. HTTP Flood; DNS Flood can be overcome using the CAPTCHA and similar types of techniques.

4 Conclusion and Future Work

In this work, we have presented a brief review of the DDoS attacks in the network layer and application layer of IoT. A lot of work has been done in network layer but comparatively less work is done for DDoS in application layer in IoT. The application layer protocols of IoT are prone to these security attacks and the resource-constrained nature makes it more challenging. The application layer protocols need to be free from security loopholes and lightweight too. In this paper, we have highlighted the solutions provided by different researchers in several related areas and using different approaches. The objective of this study is to provide the reader with a brief knowledge of the latest research in this area, paving way for further research in this topic.

References

Amidan, B.G., Ferryman, T.A., Cooley, S.K.: Data outlier detection using the Chebyshev theorem. In: 2005 IEEE Aerospace Conference, March 2005, pp. 3814–3819. IEEE (2005)

Asosheh, A., Ramezani, N.: A comprehensive taxonomy of DDoS attacks and defense mechanism applying in a smart classification. WSEAS Trans. Comput. **7**(4), 281–290 (2008)

Behal, S., Kumar, K.: Detection of DDoS attacks and flash events using novel information theory metrics. Comput. Netw. **116**, 96–110 (2017)

Bhattacharyya, D.K., Kalita, J.K.: DDoS Attacks: Evolution, Detection, Prevention, Reaction, and Tolerance. CRC Press (2016)

Bhayo, J., Hameed, S., Shah, S.A.: An efficient counter-based DDoS attack detection framework leveraging software defined IoT (SD-IoT). IEEE Access **8**, 221612–221631 (2020)

Bhosale, K.S., Nenova, M., Iliev, G.: The distributed denial of service attacks (DDoS) prevention mechanisms on application layer. In: 2017 13th International Conference on Advanced Technologies, Systems and Services in Telecommunications (TELSIKS), pp. 136–139. IEEE, October 2017

Bhuyan, M.H., Bhattacharyya, D.K., Kalita, J.K.: E-LDAT: a lightweight system for DDoS flooding attack detection and IP traceback using extended entropy metric. Secur. Commun. Netw. **9**(16), 3251–3270 (2016)

Bravo, S., Mauricio, D.: DDoS attack detection mechanism in the application layer using user features. In: 2018 International Conference on Information and Computer Technologies (ICICT), pp. 97–100. IEEE, March 2018

Chitnis, S., Deshpande, N., Shaligram, A.: An investigative study for smart home security: issues, challenges and countermeasures. Wirel. Sens. Netw. **8**(04), 61 (2016)

de Assis, M.V., Carvalho, L.F., Rodrigues, J.J., Lloret, J., Proença Jr., M.L.: Near real-time security system applied to SDN environments in IoT networks using convolutional neural network. Comput. Electr. Eng. **86**, 106738 (2020)

Demetriou, S., et al.: HanGuard: SDN-driven protection of smart home WiFi devices from malicious mobile apps. In: Proceedings of the 10th ACM Conference on Security and Privacy in Wireless and Mobile Networks, July 2017, pp. 122–133 (2017)

Demetriou, S., et al.: Guardian of the HAN: thwarting mobile attacks on smart-home devices using OS-level situation awareness. arXiv preprint. arXiv:1703.01537 (2017)

Doriguzzi-Corin, R., Millar, S., Scott-Hayward, S., Martinez-del-Rincon, J., Siracusa, D.: LUCID: a practical, lightweight deep learning solution for DDoS attack detection. IEEE Trans. Netw. Serv. Manage. **17**(2), 876–889 (2020)

Doshi, R., Apthorpe, N., Feamster, N.: Machine learning DDoS detection for consumer internet of things devices. In: 2018 IEEE Security and Privacy Workshops (SPW), May 2018, pp. 29–35. IEEE (2018)

Doshi, K., Yilmaz, Y., Uludag, S.: Timely detection and mitigation of stealthy DDoS attacks via IoT networks. IEEE Trans. Depend. Secur. Comput. (2021)

Farukee, M.B., Zaman Shabit, M.S., Rakibul Haque, M., Sarowar Sattar, A.H.M.: DDoS attack detection in IoT networks using deep learning models combined with random forest as feature selector. In: Anbar, M., Abdullah, N., Manickam, S. (eds.) Advances in Cyber Security: Second International Conference, ACeS 2020, Penang, Malaysia, December 8-9, 2020, Revised Selected Papers, pp. 118–134. Springer, Singapore (2021). https://doi.org/10.1007/978-981-33-6835-4_8

Gonzalez, H., Gosselin-Lavigne, M.A., Stakhanova, N., Ghorbani, A.A.: The impact of application-layer denial-of-service attacks. Case Stud. Secur. Comput. Achiev. Trends **261** (2014)

Hussain, F., Abbas, S.G., Husnain, M., Fayyaz, U.U., Shahzad, F., Shah, G.A.: IoT DoS and DDoS attack detection using ResNet. In: 2020 IEEE 23rd International Multitopic Conference (INMIC), November 2020, pp. 1–6. IEEE (2020)

Idhammad, M., Afdel, K., Belouch, M.: Detection system of HTTP DDoS attacks in a cloud environment based on information theoretic entropy and random forest. Secur. Commun. Netw. (2018). https://doi.org/10.1155/2018/1263123

Johnson Singh, K., Thongam, K., De, T.: Entropy-based application layer DDoS attack detection using artificial neural networks. Entropy **18**(10), 350 (2016)

Kambourakis, G., Kolias, C., Stavrou, A.: The Mirai botnet and the IoT zombie armies. In: MILCOM 2017–2017 IEEE Military Communications Conference (MILCOM), October, 2017. pp. 267–272. IEEE (2017)

Koay, A., Chen, A., Welch, I., Seah, W.K.: A new multi classifier system using entropy-based features in DDoS attack detection. In: 2018 International Conference on Information Networking (ICOIN), January 2018, pp. 162–167. IEEE (2018)

Kolias, C., Kambourakis, G., Stavrou, A., Voas, J.: DDoS in the IoT: Mirai and other botnets. Computer **50**, 80–84 (2017). https://doi.org/10.1109/MC.2017.201

Li, J., Liu, M., Xue, Z., Fan, X., He, X.: Rtvd: a real-time volumetric detection scheme for DDoS in the internet of things. IEEE Access **8**, 36191–36201 (2020)

Lima Filho, F.S.D., Silveira, F.A., de Medeiros Brito Junior, A., Vargas-Solar, G., Silveira, L.F.: Smart detection: an online approach for DoS/DDoS attack detection using machine learning. In: Security and Communication Networks,2019 (2019)

Liu, C., Qiu, J.: Performance study of 802.11 w for preventing DoS attacks on wireless local area networks. Wirel. Person. Commun. **95**(2), 1031–1053 (2017)

Ma, L., Chai, Y., Cui, L., Ma, D., Fu, Y., Xiao, A.: A deep learning-based DDoS detection framework for Internet of Things. In: ICC 2020–2020 IEEE International Conference on Communications (ICC), June 2020, pp. 1–6. IEEE (2020)

Maciá-Fernández, G., Díaz-Verdejo, J.E., García-Teodoro, P.: Mathematical model for low-rate DoS attacks against application servers. IEEE Trans. Inf. Forens. Secur. **4**(3), 519–529 (2009)

Maciá-Fernández, G., Díaz-Verdejo, J.E., García-Teodoro, P.: Evaluation of a low-rate DoS attack against application servers. Comput. Secur. **27**(7-8), 335–354 (2008)

Mao, J., Deng, W., Shen, F.: DDoS flooding attack detection based on joint-entropy with multiple traffic features. In: 2018 17th IEEE International Conference on Trust, Security and Privacy in Computing and Communications/12th IEEE International Conference on Big Data Science and Engineering (TrustCom/BigDataSE), pp. 237–243. IEEE (2018). https://doi.org/10.1109/TrustCom/BigDataSE.2018.00045

Martins, R.D.J., et al.: Performance analysis of 6LoWPAN and CoAP for secure communications in smart homes. In: 2016 IEEE 30th International Conference on Advanced Information Networking and Applications (AINA), March 2016, pp. 1027–1034. IEEE (2016)

Meidan, Y., et al.: N-baiot—network-based detection of IoT botnet attacks using deep autoencoders. IEEE Pervas. Comput. **17**(3), 12–22 (2018)

McDermott, C.D., Majdani, F., Petrovski, A.V.: Botnet detection in the internet of things using deep learning approaches. In: 2018 International Joint Conference on Neural Networks (IJCNN), July 2018, pp. 1–8. IEEE (2018)

Nebbione, G., Calzarossa, M.C.: Security of IoT application layer protocols: challenges and findings. Future Internet **12**(3), 55 (2020)

Paudel, R., Muncy, T., Eberle, W.: Detecting DoS attack in smart home IoT devices using a graph-based approach. In: 2019 IEEE International Conference on Big Data (Big Data), December 2019, pp. 5249–5258. IEEE (2019)

Rahman, R.A., Shah, B.: Security analysis of IoT protocols: a focus in CoAP. In: 2016 3rd MEC International Conference on Big Data and Smart City (ICBDSC), March 2016, pp. 1–7. IEEE (2016)

Raza, S., Trabalza, D., Voigt, T.: 6LoWPAN compressed DTLS for CoAP. In: 2012 IEEE 8th International Conference on Distributed Computing in Sensor Systems, pp. 287–289 (2012)
Raza, S., Shafagh, H., Hewage, K., Hummen, R., Voigt, T.: Lithe: lightweight secure CoAP for the internet of things. IEEE Sens. J. **13**(10), 3711–3720 (2013)
Rescorla, E., Modadugu, N.: RFC 6347: datagram transport layer security version 1.2. IETF, Technical Report, January 2012 (2017)
Roopak, M., Tian, G.Y., Chambers, J.: An intrusion detection system against DDoS attacks in IoT networks. In: 2020 10th Annual Computing and Communication Workshop and Conference (CCWC), January 2020, pp. 562–567. IEEE (2020)
Sachdeva, M., Kumar, K., Singh, G.: A comprehensive approach to discriminate DDoS attacks from flash events. J. Inf. Secur. Appl. **26**, 8–22 (2016)
Egbo, S.: The 2016 Dyn DDoS Cyber Attack Analysis: The Attack that Broke the Internet for a Day. CreateSpace Independent Publishing Platform, North Charleston (2018)
Serror, M., Henze, M., Hack, S., Schuba, M., Wehrle, K.: Towards in-network security for smart homes. In: Proceedings of the 13th International Conference on Availability, Reliability and Security, August 2018, pp. 1–8 (2018)
Sharma, P.K., Park, J.H., Jeong, Y.S., Park, J.H.: SHSec: SDN based secure smart home network architecture for internet of things. Mob. Netw. Appl. **24**(3), 913–924 (2019)
Singh, K., Singh, P., Kumar, K.: User behavior analytics-based classification of application layer HTTP-GET flood attacks. J. Netw. Comput. Appl. **112**, 97–114 (2018)
Sreeram, I., Vuppala, V.P.K.: HTTP flood attack detection in application layer using machine learning metrics and bio inspired bat algorithm. Appl. Comput. Inf. **15**(1), 59–66 (2019)
Swami, R., Dave, M., Ranga, V.: Defending DDoS against software defined networks using entropy. In: Proceedings - 2019 4th International Conference on Internet of Things: Smart Innovation and Usages, IoT-SIU 2019, pp. 1–5. IEEE (2019). https://doi.org/10.1109/IoT-SIU.2019.8777688
Tahsien, S.M., Karimipour, H., Spachos, P.: Machine learning based solutions for security of Internet of Things (IoT): A survey. J. Netw. Comput. Appl. **161**, 102630 (2020)
Tanwar, S., Patel, P., Patel, K., Tyagi, S., Kumar, N., Obaidat, M.S.: An advanced internet of thing based security alert system for smart home. In: 2017 International Conference on Computer, Information and Telecommunication Systems (CITS), July 2017, pp. 25–29. IEEE (2017)
Tiloca, M., Gehrmann, C., Seitz, L.: On improving resistance to Denial of Service and key provisioning scalability of the DTLS handshake. Int. J. Inf. Secur. **16**(2), 173–193 (2017). https://doi.org/10.1007/s10207-016-0326-0
Tushir, B., Dalal, Y., Dezfouli, B., Liu, Y.: A quantitative study of DDoS and e-DDoS attacks on WIFI smart home devices. IEEE Internet Things J. **8**(8), 6282–6292 (2020)
Tushir, B., Sehgal, H., Nair, R., Dezfouli, B., Liu, Y.: The Impact of DoS Attacks on Resource-constrained IoT Devices: A Study on the Mirai Attack. arXiv preprint. arXiv:2104.09041 (2021)
Wang, S., Gomez, K., Sithamparanathan, K., Asghar, M.R., Russello, G., Zanna, P.: Mitigating DDoS attacks in SDN-based IoT networks leveraging secure control and data plane algorithm. Appl. Sci. **11**(3), 929 (2021)
Wei, Y., Du, J.: The defect of DTLS toward detected aged packets. In: Proceedings of the 2nd International Conference on Cryptography, Security and Privacy, pp. 34–39 (2018)
Xiang, Y., Li, K., Zhou, W.: Low-rate DDoS attacks detection and traceback by using new information metrics. IEEE Trans. Inf. Forens. Secur. **6**(2), 426–437 (2011)
Zhou, L., Sood, K., Xiang, Y.: ERM: an accurate approach to detect DDoS attacks using entropy rate measurement. IEEE Commun. Lett. **23**(10), 1700–1703 (2019)

Reconsidering RTTY in Telecommunications Education: Historicity, Simplicity, and Interoperability

Joshua D. Reichard[1,2(✉)]

[1] Omega Graduate School, Dayton, TN 37321, USA
jreichard@ogs.edu
[2] Forbes School of Business and Technology - University of Arizona Global Campus, Chandler, AZ 85225, USA

Abstract. While contemporary technologies and modulation techniques and protocols like LoRa (Low Power Long Range) have become popular telecommunications techniques in the Internet of Things (IoT), tried and true methods of digital telecommunications techniques like Radio Teletype (RTTY) may prove useful in telecommunications education and practical applications. In this paper and to that end, I recommend reconsidering RTTY for purposes of telecommunications education because of its historicity and durability, its simplicity and perceptibility, and its hardware interoperability.

Keywords: Radio teletype · Telecommunications · LoRa · IoT · HF · Baudot

1 Introduction

While contemporary technologies and modulation techniques and protocols like LoRa (Low Power Long Range) have become popular telecommunications techniques in the Internet of Things (IoT), tried and true methods of digital telecommunications techniques like Radio Teletype (RTTY) may prove useful in telecommunications education and practical applications. Roberts [1] argued that technology educators have a responsibility to "break the cycle of rapid obsolescence" by teaching toward technological stability. In this paper and to that end, I recommend reconsidering RTTY for purposes of telecommunications education because of its historicity and durability, its simplicity and perceptibility, and its hardware interoperability.

Not unlike the rich history of RTTY, Chirp Spread Spectrum modulation (CSS) has been used in military applications since the 1940s, but in recent years, small microcontroller devices have made CSS applicable to the burgeoning Internet of Things (IoT) movement by providing manageable data and symbol rates for transmission of small amounts of data over several kilometers of distance. LoRa operates in the non-licensed band below 1 GHz and uses a derivative of "chirp spread spectrum modulation" to achieve relatively long-range communications between devices. LoRaWAN allows remote sensors, monitored by inexpensive microcontrollers, to transmit data on battery or solar

I. Woungang et al. (Eds.): ANTIC 2021, CCIS 1534, pp. 251–260, 2022.
https://doi.org/10.1007/978-3-030-96040-7_20

power back to receiving stations connected to the Internet. "The LoRaWAN network applies an adaptive modulation technique with multichannel multi-modem transceiver in the base station to receive a multiple number of messages from the channels" [2]. Typical of LoRa are "with smart sensing applications working on the IoT non-authored spectrum"; RTTY, on the other hand, is "keyboard-to-keyboard" telecommunication [3].

LoRa transceivers are intended for long range and low power distributed applications [4]. Mesh networks comprised of LoRa gateways from local and remote devices provide more complex data transfer [5]. However, amateur radio-derived protocols such as RTTY can provide unconventional opportunities for long range data transmission using the characteristics of HF ionospheric propagation. Such applications could complement existing LoRa networks rather than supplant them. Protocols such as RTTY may contribute a holistic vision for the future of IoT, which employs traditional wired Internet infrastructure, mesh networks, VHF, UHF, microwave, and even HF frequencies.

While HF frequencies introduce confounding challenges to transmissions to long range communication such as ionospheric fluctuation and space weather disruptions, VHF, UF, and microwave frequencies typically employed with LoRa are not without their challenges as well. Zourmand et al. [6] noted that within and between buildings, the range and quality of LoRA networks depend on "the distance from the gateway" and also "the effect of path loss due to the structural element". However, Zourmand et al. also [6] demonstrated that "a single receiver in the LoRa network may be able to handle many nodes at multiple locations within the area, unlike Wi-Fi-based system which needs to have many access points to increase the coverage area". The most promising applications of both LoRa, RTTY, and WiFi are potentially integrative and complimentary rather than competing alternative technologies. Different circumstances and applications may warrant the implementation of LoRa transmissions, HF or VHF RTTY transmissions, packet radio, and WiFi or cellular transmissions for a rich integration of Internet of Things devices and ad hoc networks.

2 RTTY: A Retrospective

Although "LoRa has become popular in the communication industry due to its long-range capabilities with low power consumption" [7]. Wolf [8] notes that Radio Teletype Radioteletype (RTTY) "is also widely used in naval communications through the use of high-power VLF (because of its ability to penetrate salt water) and HF transmissions". RTTY saw "an explosion of growth in the 1930s and subsequently, World War II" and was "the first truly digital (binary) method of communications".

While advances in higher speed wireless telecommunications technologies have permitted higher speeds and better error correction than RTTY transmissions, RTTY remains the most widely used modulation technique for "low frequency commercial and military communications around the world even today" [8]. The United States Navy, for example, still uses RTTY for high-speed ship-to-ship and ship-to-shore telecommunications. Of course, Teletype machines are still used for the hearing impaired.

RTTY is a form of signal modulation used in amateur radio, military, and telemetry applications. Modulation is the "basic process in any communication system" [9]. RTTY can be used with both audio Frequency Shift Keying (FSK) and Frequency Shift Keying (FSK). AFSK is transmitted when audio from a TNC or computer sound card to the audio input of a radio transmitter either via the microphone input or accessory jack. FSK is transmitted when on and off keying is sent from a TNC or Serial COM port to the FSK input of a transmitter. Typically, on amateur radio transceivers, RTTY is transmitted in Lower Side Band (LSB) modulation.

There are advantages and disadvantages of both AFSK and FSK. Audio levels of AFSK transmissions must be monitored to ensure the transmitter is not overdriven, but permits relatively quick and easy operation with a modern computer and soundcard. FSK is preferable when the transmitter supports it because levels are automatically controlled and various transceiver settings are controlled by the computer.

RTTY is typically transmitted at 45.45 baud with a 170 Hz shift [8]. In other words, RTTY is a "45.45 baud binary FSK" with two audio frequencies representing a lower "mark" as a binary 1 and a higher "space" representing" a binary 0. The lower mark frequency is usually modulated at 1275 Hz, and the higher frequency space frequency is modulated at 1445 Hz. Specifically, the two binary frequencies are typically expressed as a mark at 1275 Hz with a shift of 170 Hz for the space. Each character is preceded and terminated by start and stop bits whereby the stop bit is 1.42 times the length of the start bit [10]. According to Ramola, 45.45 baud is a "relatively long symbol time" has "been favored as being resistant to HF multipath effects and thus attributed to its robustness" [9].

RTTY uses a "a two state, five-bit code" whereby each character can represent either a letter or a figure; while this permits a larger character set with five bit encoding, the scheme is prone to errors in decoding [10]. Thus, a RTTY waveform would yield an expected spectral occupancy of "the expected spectral occupancy is some 91 Hz for each of the two data tones" [9].

With its five-bit code, RTTY employs the Baudot telegraph code. The *Murray* code, similar to Baudot code, is used internationally. In the United States, Baudot is defined by the Federal Communications Commission (FCC) and is defined as the *Internation Telegraphic No. 2 Baudot Code* in part 97.69 in the FCC Rules and Regulations. The Baudot standard defines the encoding of letters, numbers. Internationally, there is variation in other symbols and punctuation. In the United States, radio amateurs have commonly accepted "Military Standard" Baudot code for punctuation largely due to widespread access to surplus military teletype machines following World War II. Amateur radio operators in Europe and other regions of the world have standardized on the *CCITT No. 2* code standard [11].

In Baudot code, there are two possible states per pulse. Because each of the five data pulses can be in either a mark or space condition, a total of 2^5 or 32 possible codes combinations are possible. The 32 Baudot code combinations are insufficient to permit transmission of the English alphabet (26), numbers (10), and punctuation. This limitation was addressed by applying the Baudot code in two modes or cases: Letters (LTRS) case and Figures (FIGS) case. Two special characters, LTRS and FIGS, are used to notify the printer when incoming characters are either letters or figures [11].

Using a latching mechanism, the printer remains in the last received case until is changed by reception of another special character. Specific control operations such as LTRS, FIGS, Carriage Return (CR), Line Feed (LF), Space Bar (SP), and Blank (BLNK = no carriage movement) are available in both LTRS and FIGS case so they can be sent when operating in either case. The remaining 26 code combinations have different letter or numeral/symbol meanings depending upon whether preceded by a LTRS or FIGS character. Needless to say, the Baudot code is restricted to upper-case characters only [11].

RTTY and the Baudot code has enjoyed extensive commercial use across the globe for nearly a century and is still actively employed for international wire communications, press exchanges, and weather reports. However, the lack of additional characters or control codes is a limitation which will restrict the applicability of RTTY in the future [11]. When teleprinters were exchanged for personal computers beginning in the 1970s, the limitations of Baudot code became apparent, as various control codes of terminal applications could not be accommodated. Modern software, such as the open-source fldigi, provide a better user experience, but output is still clumsy compared to more sophisticated terminal protocols. Nevertheless, the marks and spaces of Baudot code can be represented in binary form and with digital RTTY software like fldigi, the two frequencies, separated by 170 Hz, are visualized on a graphic waterfall.

The bits in Table 1, are arranged in ascending order (b1 to b5), contrary to typical computer binary representation. In typical binary configuration (such as ASCII), the letter "A" would be represented by the binary value "00011" (in Baudot) but in the Table 1, it is represented as "11000". Therefore, in Table 1 below, the leftmost bit is the bit which is sent first.

Table 1. Standard Baudot code

Binary	Decimal	Hex	Octal	Letter	Figure
00000	0	0	0	Blank	Blank
00001	1	1	1	T	5
00010	2	2	2	CR	CR
00011	3	3	3	O	9
00100	4	4	4	Space	Space
00101	5	5	5	H	
00110	6	6	6	N	,
00111	7	7	7	M	.
01000	8	8	10	Line feed	Line feed
01001	9	9	11	L	)
01010	10	A	12	R	4
01011	11	B	13	G	&
01100	12	C	14	I	8

(*continued*)

Table 1. (*continued*)

Binary	Decimal	Hex	Octal	Letter	Figure
01101	13	D	15	P	0
01110	14	E	16	C	:
01111	15	F	17	V	;
10000	16	10	20	E	3
10001	17	11	21	Z	"
10010	18	12	22	D	$
10011	19	13	23	B	?
10100	20	14	24	S	BEL
10101	21	15	25	Y	6
10110	22	16	26	F	!
10111	23	17	27	X	/
11000	24	18	30	A	-
11001	25	19	31	W	2
11010	26	1A	32	J	'
11011	27	1B	33	Figure shift	
11100	28	1C	34	U	7
11101	29	1D	35	Q	1
11110	30	1E	36	K	(
11111	31	1F	37	Letter shift	

3 Contemporary Applications of RTTY

Although LoRa spread spectrum is the favored technique for the IoT, it has also been used to transmit the RTTY protocol; thus, these technologies are not mutually exclusive. Shaarook and Kesan [7] have utilized LoRa modulation to transmit RTTY from a 100 mW beacon transmitter on the Satish Dhawan Satellite (SD Sat), a 3U Cubesat built by Space Kidz India. Multiple ground stations around the world have confirmed reception from SD Sat. However, "because of the effects of Doppler shift produced by high orbital velocity in lower earth orbit and wide bandwidth of LoRa modulation", many are "skeptical about the capabilities of using LoRa with low-cost ground equipment and antenna systems". Nevertheless, the satellite transmits in FSK RTTY at 100 mW. The transmission includes telemetry data and data concerning the health of the satellite's onboard systems. The satellite is designed to comply with the rules of the International Telecommunications Union [7]. The SD Sat is an example of how an otherwise outmoded protocol like RTTY can be used in contemporary educational applications.

Further, Brown [12] proposed a "low-cost low-power HF telemetry transmitter that can relay a balloon's position along with system telemetry over thousands of miles", including RTTY transmissions. Such a configuration permit "anyone with an HF radio and computer connected to the Internet" to "act as a remote reception site to enable a large world-wide distributed network of ground stations". RTTY may prove useful as another means by which telemetry data may be sent from beacons and satellites.

Focusing on low-cost, Lasagani, Iqbal, and Mann [13] concluded that LoRa was that most effective wireless communication method for "remote control of grid-tied converts" among three viable options. Measuring ranges and data rates, LoRa prevailed. However, LoRa spread spectrum emissions require specialized hardware, specific chips, and relatively demanding processing power for decoding. Wolf argues that when all other telecommunications systems fail, "open air and ionospheric reflection propagation a medium" can be used to establish "robust wireless data links" "over hundreds or even thousands of miles" [8]. While LoRa networks may permit shorter-range IoT connectivity for sensors and smart devices, protocols such as RTTY may be used for longer range communications using ionospheric propagation at high frequencies.

4 Three Reasons to Teach RTTY

I suggest RTTY can be reconsidered to play an important role in telecommunications education. First, its historical significance and nearly century-long durability make it relevant for laying a strong foundation in the practical and theoretical aspects of telecommunications theory. Second, the simplicity of the Baudot code makes RTTY encoding perceptible and accessible for engineering, computer science, and telecommunication students. Simple yet rich educational experiences and activities can be constructed around RTTY encoding. Third, RTTY provides a unique opportunity for interoperability between older and contemporary hardware; early personal computers from the 1970–1980s can still handily encode and decode RTTY signals and simple RTTY transceivers can be constructed with modern microcontrollers. Acknowledging the serious limitations of RTTY, including its vulnerability to noise and interference, its lack of sophisticated error correction, and its painfully slow baud rate, RTTY may still play an important role in the future of telecommunications education.

4.1 Historicity and Durability

RTTY has a rich history in telecommunications and has been a durable method transcending other technological advancements. From the teletypewriter, to teleprinters, to personal computers, to microcontrollers, RTTY has stood the test of time. Its historical significance alone warrants its place in telecommunications education, but its durability justifies its enduring practicality as technological architectures change and improve.

The mark and space language of the RTTY Baudot code refer to the marking pen and moving paper strip of early telegraph recording machines. The pen was driven by an electro-mechanical solenoid that lowered the pen when the sending key of the transmission was in a "down" state, thus "marking" the paper strip. When the telegraph key was in an "up" position, the paper advanced and left a space. Some operators of the time came to interpret the Morse code transmissions by listening to the engagement of the solenoid and the mechanical movement of the pen [11]. Later teleprinter machines also used solenoids called "Selector Magnets" that open and close according to the signal current called a "Loop Current".

However, instead of the dits and dahs of earlier Morse code transmissions, Teletype machines used timed pulses which could represent letters, numbers, and symbols based on combinations of spaces and marks. In early teletype machines, the normal "rest" condition of the machine was with the loop current flowing; an interruption of the loop current would release the selector magnet solenoid to allow a cam to rotate in the machine. Transmission of Teletype characters begin with a space pulse representing "current off" which is called the "start pulse". The start pulse notifies the receiving Teletype machine to start receiving a character. Immediately following the start pulse, a series of data pulses are transmitted as marks or spaces to indicate the encoding of the character; combinations of the five data pulses of Baudot code (whereas modern ASCII encoding uses 8-bits), followed by a stop pulse, which gives the Teletype machine time to rest [11].

While most RTTY is transmitted with computers today, the mechanical simplicity of early teletype machines is worthy of study in the contemporary context. The historical use of Baudot code and Teletype machines in early military and commercial telecommunications provides a rich foundation upon which contemporary students can build. Because of its proximity to early telegraphy and its importance in World War II, RTTY is a rare example of a historical telecommunications method which is still widely used in amateur radio and military applications. Even analog telephones are phasing out in favor of digital voice transmissions via cellular networks, but RTTY is a rare example of a telecommunications protocol which has endured nearly a century of technological change. Although there are clearly still applications for analog telecommunications, digital telecommunications are more "immune towards erroneous signals, noise and the other stray signals that have the potential to infect the useful information" [9]. RTTY is among the simplest and most durable of digital telecommunications protocols.

Although there is a rush to teach students current trends such as LoRa or other wireless protocols, there is not yet any indication such technologies will have meaningful staying power. Because of the simplicity of Baudot and uncomplicated nature of RTTY transmissions, they may prove to be valuable first steps in learning telecommunications theory for both computer science and engineering students. Naugler and Surendran [14] argued that complicated tools can confound the student's learning experience in technological fields and sometimes, less sophisticated tools are more effective at teaching fundamental concepts. Making a distinction between education and training, Naugler and Surendran contend that technology educators tend to prioritize tools over concepts and devolve technological education into training on how to use specific hardware or software rather than understanding the fundamental theoretical constructs which underly them. In like manner, an overemphasis on practical implementation of IoT, LoRa, microcontrollers, or complicated transceiver ICs may obscure fundamental aspects of "how" telecommunications work at their most fundamental levels. Reconsidering RTTY may be an antidote to such a problem.

At just five bits, the Baudot code remains simple and, from an educational perspective, perceptible. Although the modulation of either AFSK or FSK requires more sophisticated engineering knowledge, the notion of encoding five bits using the Baudot code is not outside the reach of secondary or even primary school students. Students can learn Baudot code, translate symbols and words, and even simulate RTTY transmissions without any involving any real equipment. More advanced undergraduate engineering students can examine waveforms with an oscilloscope to identify the marks and spaces of FSK RTTY transmissions or examine demodulated audio waveforms either at the receiving or sending ends of an AFSK transmission. Such practical and tangible exercises are not quite as perceptible when working with CSS at high frequencies. Learning about and from RTTY, in its simplicity, may help reduce the overall complexity often unnecessarily introduced in technology education [15].

Further, computer science students can code audio processing software to modulate and demodulate RTTY signals. Such signals need not be transmitted via radio frequencies; experimentation with signal process and prototypical error detection can be constructive exercises. Modern open-source software such as fldigi, which already accommodates RTTY encoding and decoding, may also play a role in such educational experimentation.

4.2 Hardware Interoperability

Not unlike the unnecessary complexities introduced in technology education, the telecommunications industry itself has been accelerating in unnecessarily complexity for decades [16]. While LoRa hardware and protocols provide some measure of promise for more open IoT telecommunications, the dependence on complicated integrated circuits, chip manufacturers, and relatively powerful microcontrollers and single board computers (like Arduinos and Raspberry Pis) limit interoperability. Ultimately, protocols such as TCP/IP, HTTPS, and UDTP are standardized protocols which make the Internet possible, not all telecommunications are Internet-dependent. Arnold and Lennhartz [17] propose that "communications services choices are conscious and reciprocal" and "people separate their social ties using the boundaries of communications services". While RTTY is not likely to displace Internet instant messaging services, SMS, or email, it does represent long-standing interoperability that transcends technological changes – even the Internet itself.

To demonstrate such interoperability, telecommunications students can build RTTY data modems from basic schematics. Bales proposed a scheme for hardware-based decoders of both RTTY and PSK31 for low-power amateur radio applications. Bales [10] notes that the "RTTY decoder is only mediocre, but the code should still serve as a very good start for developing low-milliwatt decoders in the future" [10]. The frequency shift in RTTY is small (170 Hz) compared to the mark and space frequencies (1275 Hz and 1445 Hz). "RTTY is defined by the speed at which the carrier is shifted (symbol rate), and how much the frequency of the carrier is shifted" [8]. In Bales' [10] demodulator, RTTY and PSK31 decode schemes use hardware timers to detect incoming symbols, which are accumulated until valid messages are received. A simple low-power RTTY demodulator can be constructed, tested, and integrated with other IoT transceivers.

Moreover, Lechowicz and Kokar [18] proposed an "ontology-based waveform reconfigurability that allows radios of different hardware or software architectures, using different software APIs and even non-uniform waveform description schemas, to interoperate". In fact, they developed a proof-of-concept system whereby they were able to evaluate "three different waveforms (BPSK31, QPSK31 and RTTY)", translate them into their interoperable ontology-based waveform, and successfully transfer them from one node to another, and reconstruct them on the receiving node into fully functional software modules implementing the waveforms. They note that both radios can then use the reconstructed waveforms for further communication.

5 Conclusion/Summary

In this paper, I recommended reconsidering RTTY for purposes of telecommunications education because of its historicity and durability, its simplicity and perceptibility, and its hardware interoperability. "Communication has occupied an important part of our lives without which information exchange cannot be imagined" [9]. And while RTTY may seem like a relic of a technological bygone era, its persistence in amateur radio and other applications make it a unique candidate for reconsidering how and what telecommunications students are taught concerning signal modulation and encoding/decoding digital protocols. Admittedly, RTTY is not likely to displace other more reliable error-correcting protocols in the cornucopia of technologies in the Internet of Things. In fact, Ramola [9] questions whether any "true coherent demodulation of Phase Shift Keying could ever be achieved non-cabled system since random phase changes would introduce uncontrolled phase ambiguities". But, because RTTY has survived this long there is no reason to count it out or to discount the possibility of creative and innovative uses in future telecommunications.

References

1. Roberts, E.: The dream of a common language: the search for simplicity and stability in computer science education. ACM SIGCSE Bull. **36**(1), 115–119 (2004)
2. Sinha, R.S., Wei, Y., Hwang, S.: A survey on LPWA technology: LoRa and NB-IoT. ICT Express **3**(1), 14–21 (2017). https://doi.org/10.1016/j.icte.2017.03.004
3. Nurelmadina, N., et al.: A systematic review on cognitive radio in low power wide area network for industrial IoT applications. Sustainability **13**(1), 338 (2021). https://doi.org/10.3390/su13010338
4. Khutsoane, O.C., Isong, B., Abu-Mahfouz, A.: IoT devices and applications based on LoRa/LoRaWAN. In: IECON 2017 - 43rd Annual Conference of the IEEE Industrial Electronics Society, pp. 6107–6112 (2017)
5. Lee, H., Ke, K.: Monitoring of large-area IoT sensors using a LoRa wireless mesh network system: design and evaluation. IEEE Trans. Instrum. Meas. **67**(9), 2177–2187 (2018)
6. Zourmand, A., Kun Hing, L., Wai Hung, C., AbdulRehman, M.: Internet of Things (IoT) using LoRa technology. In: 2019 IEEE International Conference on Automatic Control and Intelligent Systems (I2CACIS), pp. 324–330 (2019)
7. Shaarook, R., Kesasn, S.: Low Poer LoRa transmission in low earth orbiting satellites. In: Luštrek, M. (ed). Intelligent Environments 2021: Workshop Proceedings of the 17th International Conference on Intelligent Environments. ISO Press (2021)

8. Wolf, J.: Long Haul VLF/LF/HF Data Networks. INFSCI 1072. School of Information Science – University of Pittsburgh (2016). https://www.liltechdude.com/portfolio/Data_Networks.pdf. Accessed 03 Nov 2021
9. Ramola, S.: Digital communication-technology and advancements. Adv. Electron. Electr. Eng. **4**(4), 367–374 (2014)
10. Bales, B.: Low Power RTTY and PSK31 Decoder for Ham Radio Applications. Chancellor's Honors Program Projects (2011). https://trace.tennessee.edu/utk_chanhonoproj/1350. Accessed 03 Nov 2021
11. Henry, G.: ASCII, Baudot, and the Radio Amateur (n.d.). https://www.digigrup.org/ccdd/rtty.htm. Accessed 03 Nov 2021
12. Brown, B.: Multi-mode transmitter for high-altitude balloon telemetry. In: Academic High Altitude Conference, no. 1 (2011). https://doi.org/10.31274/ahac.5607
13. Lasagani, K., Iqbal, T., Mann, G.: A comparison of low cost wireless communication methods for remote control of grid-tied converters. In: 2017 IEEE 30th Canadian Conference on Electrical and Computer Engineering (CCECE), pp. 1–4 (2017)
14. Naugler, D., Surendran, K.: Simplicity first: use of tools in undergraduate computer science and information systems teaching. Inf. Syst. Educ. J. **2**(5) (2004)
15. Roberts, E.: The dream of a common language: the search for simplicity and stability in computer science education. ACM SIGCSE Bull. **36**(1), 115–119 (2004)
16. Fisher, M.A., Rana, S., Egelhaaf, C.: Interoperability in a multiple-provider telecommunications environment. In: Proceedings Second International Enterprise Distributed Object Computing (Cat. No. 98EX244), pp. 296–303 (1998)
17. Arnold, R., Schneider, A., Lennartz, J.: Interoperability of interpersonal communications cervices – a consumer perspective. Telecommun. Policy **44**(3), 101927 (2020)
18. Lechowicz, L., Kokar, M.M.: Waveform reconstruction from ontological description. Analog. Integr. Circ. Sig. Process. **78**(3), 753–769 (2013). https://doi.org/10.1007/s10470-013-0228-2

Intelligent Computing

Deep Learning Identifies Tomato Leaf Disease by Comparing Four Architectures Using Two Types of Optimizers

Mohamed Bouni[1](✉), Badr Hssina[1], Khadija Douzi[1], and Samira Douzi[2]

[1] Laboratory LIM, IT Department FST Mohammedia, Hassan II University, Casablanca, Morocco
mohamed.bouni1-etu@etu.univh2c.ma, badr.hssina@fstm.ac.ma
[2] FMPR, Mohammed V University in Rabat, Rabat, Morocco
s.douzi@um5r.ac.ma

Abstract. The agricultural industry is vital to a country's economic development. It has already made a significant contribution to industrialized nations' economic prosperity, and its role in the development of less developed countries is crucial. Plant diseases are a significant source of concern in agriculture, affecting food supply and public health. There are no early detection methods for these diseases. Detecting diseases in plants with the naked eye is a tedious task that takes time and accuracy. This study uses deep learning architectures such as AlexNet, VGG16Net, ResNet, and Dense Net to detect the tomato leaf diseases. The plant village dataset is used to evaluate the model and realized the Dense Net model with the Rmsprop optimization technique obtained the greatest result with the best accuracy 99.99.

Keywords: Agriculture · Big Data · Intelligence artificial · Machine learning · Deep learning · Plant village · AlexNet · VGG · ResNet · DensNet · Adam · Rmsprop · CNN

1 Introduction

Agriculture and farming are considered as one of the earliest and highly essential occupations in the world. They take a significant role in the prosperity of the economic field. According to the Food and Agriculture Organization of the United Nations (FAO), the global population will reach 9.2 billion by 2050 (Santos et al. [1]). More than two billion mouths to feed by then will necessitate an upsurge in agricultural production by 70% to satisfy the need. Therefore, the urge to find a more innovative approach to become more practical about farming and being highly productive is fundamental. Artificial intelligence is the most effective technological solution to deal with the escalating population and global climatic fluctuations. It is intensifying its substantial impacts on agriculture. Artificial intelligence in agriculture has been demonstrated to be a revolutionary technology in securing high productivity and good quality crop yield. With the employment of artificial in- diligence, farmers can now evaluate climate conditions, water utilization, and the environmental state of the soil amassed from their planting and sowing to

I. Woungang et al. (Eds.): ANTIC 2021, CCIS 1534, pp. 263–273, 2022.
https://doi.org/10.1007/978-3-030-96040-7_21

make critical and timely resolutions on business possibilities. Farmers can decide the most viable harvest varieties that specific year. Big Data assessment similarly enhances irrigation, lessens greenhouse gas emissions, and identifies the accurate soil, brightness, food, and water conditions essential for increased production. Plant diseases represent one of the important leading causes that result in the devastation of plants and crops, affecting the economy by suffering substantial economic losses. Detecting these diseases in the initial phases will empower farmers to overcome and treat them properly. With the proliferation of digitalization, artificial intelligence has provided numerous tools and advanced technologies such as computer vision and machine learning, image recognition that permit the farmer to detect areas infected and treat them appropriately at the early stages. While back, the farmer did not likely detect the plant disease until a broadly noticeable field area was inflicted and would have been required to spray the whole field compared to treating a small part of it. Zhao et al. [2] several deep handcrafted feature-based images identification techniques were widely used before deep learning became popular. It is called handcrafted methods because of the human expertise and many factors involved in the process. These methods were computationally and time demanding due to extensive preprocessing, feature extraction, and classification. Unlike the handcrafted-based methods that need separate procedures, Deep learning capabilities have allowed researchers to build systems that can be taught and evaluated end-to-end. Convolutional Neural Networks (CNNs) have been used in various sectors, including agriculture, automation, and robots. In agriculture, computer vision and CNNs are used to recognize plants and their associated diseases. As a contribution, we treated a simple example of the use of deep learning in agriculture concerning the detection of plant diseases in tomatoes. In this study, we used pre-trained deep neural networks: AlexNet, VGG16, ResNet, and DesnseNet to classify tomato diseases. The model is trained and fine-tuned according to the different changes of hyperparameters. The performance of the network was verified on the very famous disease prediction data set the "Plant Village Data Set", and very interesting findings were achieved.

2 Related Work

Over the last 50 years, there has been a sustainable development in artificial intelligence because of its robustness. It is omnipresent in all fields, one of these fields agriculture The applications of DL (Deep Learning) in agriculture are spread over several domains since 14 domains have been identified in recent papers. The most popular domains are: 1. Identification of the disease and 2. Plant recognition and identification of land cover and weeds. CNN has been the most popular architecture used, while the few alternatives are based on FCN (Fully Convolutional Network) or RNN (Recurrent neural network). Santos et al. [1] reviews research efforts based on deep learning applied to agricultural fields. It examines the agricultural field and describes the problem it focuses on, lists the technical details such as the architecture and model of deep learning, describes the data source, reports the overall accuracy of each work compared to alternative methods, verifies the equipment used the possible application in real-time. The indigent findings that in-depth learning achieved a high degree of accuracy in most of the work reviewed, marking a higher accuracy than other traditional techniques. Patel et al. [3] cover the study of different methods of detecting plant leaf and fruit diseases using a neural network. The authors present a review of the use of neural network models in plant disease detection. Adhikari et al. [4] aimed to classify and detect plant diseases, particularly in tomatoes, automatically. The critical process of the project includes Image processing, image acquisition, image investment adjustment, feature extraction, and classification based on the convolutional neural network (CNN). For hardware, Raspberry Pi is the central computing unit. Zhang et al. [5] apply the deep convolutional neural network (CNN) to identify tomato leaf diseases by learning by transfer. AlexNet, Google Net, and Reset have been used as the backbone of CNN. The best-combined model was used to modify the structure to explore the performance of CNN's complete training and development. The highest accuracy of 97.28% for tomato leaf identification the disease is achieved by the optimal Res Net model with stochastic gradient descent (SGD). Rangarajan et al. [6] employed images of tomato leave (6 diseases and one healthy class) obtained from the Plant Village dataset contribute to two deep learning architectures: AlexNet and VGG16 net. The role of the number of images and the significance of hyper parameters, including mini-batch size, weight, and bias learning rate, in classification accuracy and run-time, were analyzed. In another study, Stanojevic et al. [7] researchers proposed CNN-based systems that distinguish between 13 kinds of disease and healthy ones in five crops using a unique dataset comprised of pictures gathered from various online sources. Top-1 success rates of 96.3% were obtained, while top-5 success rates of 99.99% were achieved using the suggested method. Similarly, Lu et al. [8] have suggested a detection method for rice disease using deep convolutional neural networks an image dataset including 500 actual Images of damaged and healthy rice leaves and stems was used to train neural networks to detect ten prevalent rice diseases. The suggested model had an average accuracy of 95.48%, which was attained. Tan et al. [9] presented a CNN-based method for identifying apple pathology images, with a self-adaptive momentum strategy being utilized to change CNN parameters while the images were being processed. Using the proposed method, the recognition accuracy reached 96.08%, with relatively fast convergence. Adhikari et al. [10] tried to automatically categorize and detect plant illnesses, with a particular emphasis on tomato infections. Image Pre-processing, picture capture, and image investment adjustments were performed to prepare the input data. Later, a convolutional neural network was used to conduct feature extraction and classification, which resulted in a successful outcome (CNNs). The model was verified using a custom-generated dataset (a collection of internet pictures) and produced some promising results,

3 Methodology

The Plant Village dataset is used to collect images of 9 different diseases and healthy tomato crop samples, the original tomato leaf disease data set used in this study comes from the Plant Health Open Access Image Library [11], Table 1 and Fig. 1 list the health status of tomato crops and 8 other diseases. Table 1 and Fig. 1 list the health status of tomato crops and 8 other diseases.

In the segmented image, all background pixels in the three channels are set to 0 except those related to leaves. A total of 7,301 samples of 9 diseases and health levels were considered. For selected deep learning algorithms [AlexNet, ResNet, DenseNet, and VGG16], the original input image of 256 × 256 pixels is adjusted to 224 × 224 pixels. The following will briefly introduce the deep learning architecture.

Fig. 1. Sample images from Plant Village dataset

Table 1. Statistics of dataset

Label	Category	Number of samples
1	Bacterial spot	900
2	Early blight	500
3	Healthy	701
4	Late blight	800
5	Leaf mold	400
6	Septoria leaf spot	800
7	Spider mites	800
8	Target spot	700
9	Mosaic virus	160
10	Yellow leaf curl virus	1540

3.1 AlexNet

Compared with all classic machine learning and computer vision methods, AlexNet has obtained the most advanced recognition accuracy. This was a watershed in the history of machine learning and computer vision visual recognition and classification tasks, and it marked the beginning of a surge in interest in deep learning. The architecture of AlexNet [12] follows the number of convolutional layers and pooling layers. The first convolutional layer uses 96 different 11 × 11 size receive filters, and performs convolution and maximum pooling through local response normalization (LRN). A 3 × 3 filter with a step size of 2 is used to perform the maximum pooling operation. In the second layer, use the 5 × 5 filter to perform the same operation. 384, 384, and 296 feature maps are used for the third, fourth, and fifth convolutional layers. In AlexNet mode, several convolutional layers are followed by ReLU, maxpooling and normalization layers The output of all convolutional layers, including the last two fully linked layers, go through ReLU, a non-linear, non-saturating activation function. The final fully connected layer has been adjusted to include the total number of classes, namely 9 classes and 1 health class, for a total of 10. The architecture of AlexNet is shown in Fig. 2.

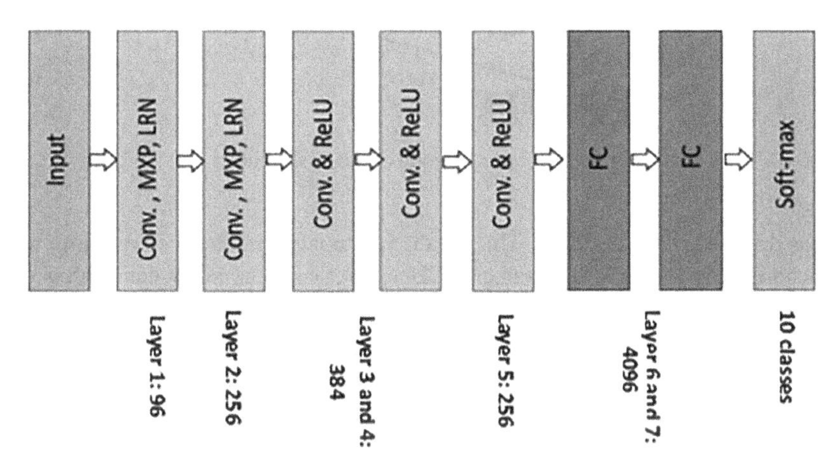

Fig. 2. The architecture of AlexNet

3.2 ResNet

The residual network design is the winner of ILSVRC 2015. Resnet was created by Kaiming. He set out to build an ultra-deep network that would not suffer from the vanishing gradient problem that plagued previous generations. ResNet uses a series of layers, including 34, 50, 101, 152, and even 1202 layers. The famous ResNet50 network has 49 convolutional layers, and finally a fully connected layer. We chose ResNet50 for this research, which is also a high-performance network. In this study, the last three layers of ResNet are enhanced by the full link layer, sofmax layer, and classification layer. The fully connected layer is replaced with 10 neurons, corresponding to the tomato leaf disease classification. As a result, the ResNet structure is modified. The size of the ResNet input image should be 224 × 224. The basic diagram of the ResNet block [13] is shown in Fig. 3.

3.3 DenseNet

Gao Huang and coworkers created DenseNet in 2017. The output of each layer is connected to all subsequent layers in a dense block [14], which consists of densely connected CNN layers. Therefore, it is constructed through dense connections across layers, and

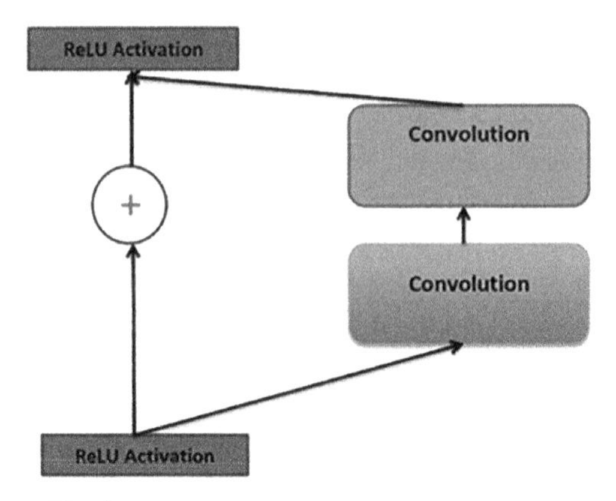

Fig. 3. The basic diagram of the ResNet block

thus has the label of "DenseNet". This idea is very convincing for feature reuse, which significantly reduces network parameters. DenseNet contains many dense blocks and transition blocks between two dense blocks. Each layer takes all the feature maps of the previous layer as input. For object recognition tasks, this novel model shows the most advanced accuracy with moderate network parameters. In this study, we use DensNet for transfer learning, where we use the value of pre-learned weights and modify the classification layer according to our categories (i.e., 10 categories).

3.4 VGG-16

The VGG16 pre-trained net is based totally on AlexNet's stacked layout, however with extra convolution layers (as supplied in Fig. 4). It has 13 convolution layers, every of that is observed by using a ReLU layer. just like AlexNet, a number of the convolution layers are followed with the aid of max-poolsing to decrease the dimension. In assessment to AlexNet, which uses filters of greater outsized dimensions, convolution layers use smaller filters of measurement 3×3. The wide variety of parameters is decreased through the usage of smaller filters, and a ReLU layer is introduced after every convolution layer to growth non-linearity. Basic building block of VGG [15] network Convolution is presented in the Fig. 4.

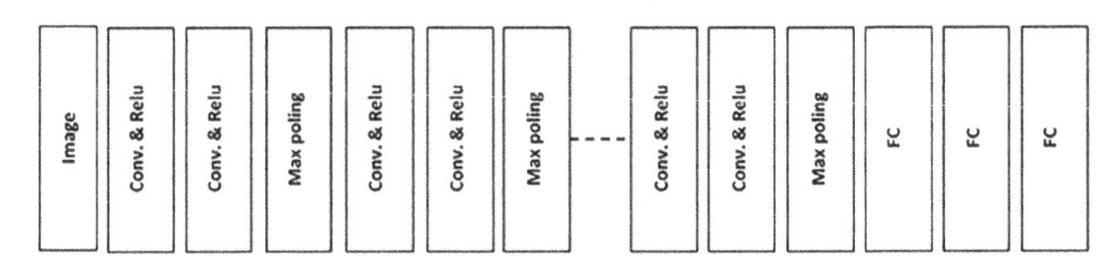

Fig. 4. Basic building block of VGG network: Convolution (Conv) and FC for fully connected layers

4 Experiments

The data improvement method was completed within the 1st part of the experiment. though the performance of deep neural network systems is best compared with typical machine learning or computer vision ways, they have over-fitting problems. Overfitting will be defined because the choice of hyperparameters, system regularization, or the utilization of an outsized range of images for training. we tend to adopted many enhancement methods to extend the quantity of images within the dataset. The techniques used include geometric transformation (resize, rotation, horizontal flip, cropping) and intensity transformation (contrast and brightness, color, noise). This work uses a transfer learning strategy, which requires the employment of a pre-trained deep learning model to classify new classes of objects. first initial a part of the study, the improved images were input into the pre-trained AlexNet, DenseNet, ResNet, and VGG16 networks. Since the model is trained to classify 1000 objects of various classes within the ImageNet dataset, the last layer has been replaced with associate output layer with a similar variety of classes. during this case, there are 10, with 9 diseases and one health level. A softmax layer and a fully connected layer are added to the architecture. Therefore, the last 3 layers of all models are adjusted. as a result of these are already pre-trained networks and are fine-tuned for recognizing ImageNet, the learning rate of the first layers is kept to a minimum. the number of epochs is set to 40[Max], and the batch size is set to 32.

5 Results

This section will reveal the experiment and discuss the results. All experiments are run on Windows 10, Python and Kaggle platform, TPU and 16 GB RAM the first experiment aims to find the best optimization strategy for detecting tomato leaf disease between Adam and RmsProp by integrating the pre-trained network for each network during this experiment, set the following hyperparameters: batch size is set to 32, initial learning rate is set to 0.001, and max epoch is ready to 40. The gradient decay rate is set to 0.9, the square gradient decay rate is 2 to 0.999, and the denominator of the Adam optimizer set is 10^{-8}. Table 2 and Fig. 5 show the accuracy of several networks using Adam optimizer. Table 3 and Fig. 6 show the accuracy of several networks using RmsProp optimizer. Table 4 and Fig. 7 show the accuracy of several networks using Adam and RmsProp optimizers.

Table 2. Accuracy, validation, test and time of execution of VGG-16, AlexNet, DenseNet and ResNet using Adam

Modele		VGG16		AlexNet		DenseNet		ResNet	
Number of Epochs		20	30	30	40	20	10	20	15
Accuracy	Training	0,858	0,997	0,855	0,900	0,999	0,996	0,999	0,998
	Validation	0,883	0,988	0,894	0,912	0,997	0,992	0,994	0,994
	Test	0,879	0,989	0,878	0,885	0,996	0,98	0,995	0,992
Loss	Training	0,396	0,006	0,327	0,291	0,002	0,013	0,002	0,006
	Validation	0,361	0,013	0,3	0,25	0,005	0,021	0,013	0,013
	Test	0,138	0,036	0,332	0,323	0,01	0,058	0,015	0,022
Time of execution		2260	3390	3360	4480	4280	2140	2420	1815

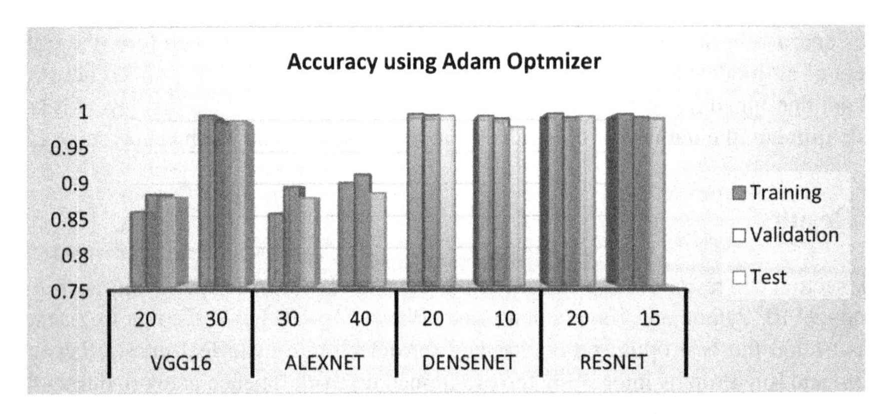

Fig. 5. Accuracy of different networks using Adam

Table 3. Accuracy, validation, test and time of execution of VGG-16, AlexNet, DenseNet and ResNet using RmsProp

Models		VGG16		AlexNet		DenseNet		ResNet	
Number of Epochs		20	30	30	40	20	15	20	15
Accuracy	Training	0,886	0,981	0,91	0,923	0,999	0,999	0,998	0,999
	Validation	0,912	0,967	0,927	0,929	0,998	0,996	0,996	0,996
	Test	0,916	0,954	0,954	0,592	0,996	0,995	0,994	0,984
Loss	Training	0,346	0,055	0,256	0,22	0,002	0,001	0,003	0,002
	Validation	0,273	0,088	0,22	0,194	0,01	0,01	0,011	0,02
	Test	0,238	0,123	0,123	0,15	0,013	0,014	0,019	0,038
Time of execution		2800	4200	4800	6400	3760	2070	2820	2115

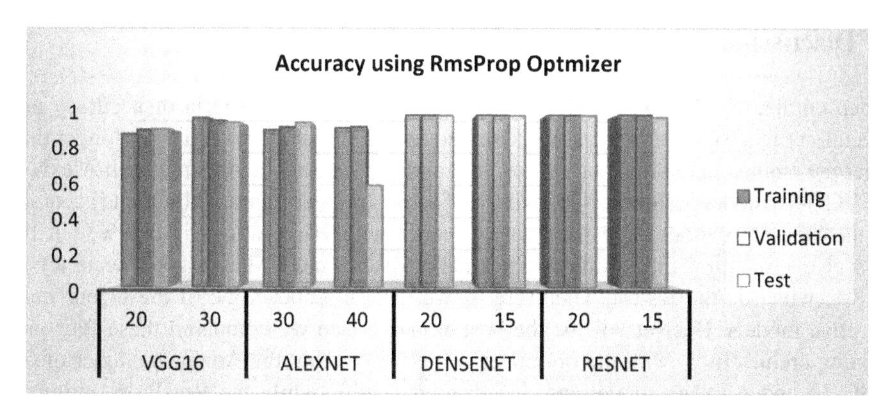

Fig. 6. Accuracy of different networks using RmsProp

Table 4. Accuracy, validation, test and time of execution of VGG-16, AlexNet, DenseNet and ResNet using Adam and RmsProp

Models		VGG16		AlexNet		DenseNet		ResNet	
Optimizer		Adam	Rms Prop	Adam	Rms Prop	Adam	Rms Prop	Adam	Rms Prop
Accuracy	Training	0,997	0,981	0,9	0,923	0,996	0,999	0,998	0,999
	Validation	0,988	0,967	0,912	0,929	0,992	0,996	0,994	0,996
	Test	0,989	0,954	0,885	0,952	0,98	0,995	0,992	0,984
Loss	Training	0,006	0,055	0,291	0,22	0,013	0,001	0,006	0,002
	Validation	0,013	0,088	0,25	0,194	0,021	0,01	0,013	0,02
	Test	0,036	0,123	0,232	0,15	0,058	0,014	0,022	0,038
Time of execution		3390	4200	4480	6400	2140	2070	1815	2115

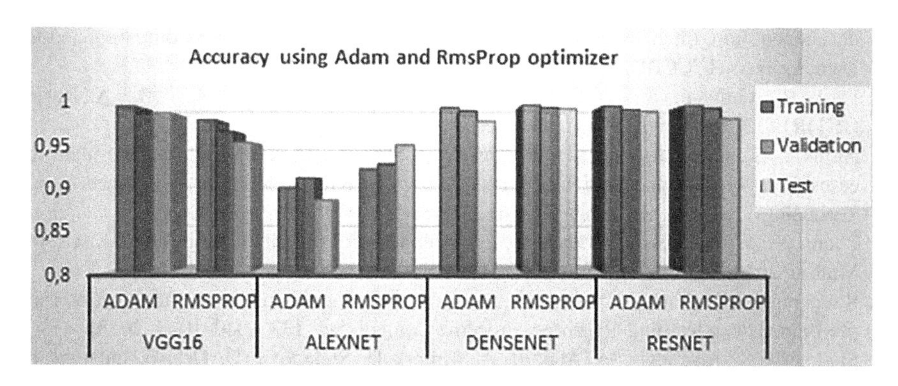

Fig. 7. Accuracy of different networks using Adam and RmsProp optimizer

6 Discussion

When comparing the four models using the two evaluation criteria of accuracy and execution time, we observed that AlexNet requires more iterations and takes longer than other pre-trained networks, but has lower accuracy. Similarly, compared with AlexNet, the VGG-16 model performs well in terms of accuracy and execution time. The accuracy of the DenseNet model and the ResNet model with 20 iterations is 0.99%, but the DenseNet execution time is the longest. The ResNet model has a good accuracy of 99.8%, which is the fastest. Therefore, if we have to choose one of these four most effective models, ResNet will be the best choice. Once we compared these four pre-training architectures with the optimizer, we observed that the Adam optimizer of the VGG-16 and AlexNet models produced good results, while the RmsProp optimizer of DenseNet and ResNet provided excellent results. As a general conclusion, DenseNet with Rmsprop is the most effective and suitable model for tomato leaf disease prediction.

7 Conclusion

This paper focuses on applying transfer learning to classify the tomato leaf disease using deep convolutional neural networks. The used networks are based on the VGG-16, DenseNet, ResNet, and AlexNet that are pre-trained. We analyzed the relative performance of these networks using Adam and RmsProp optimization methods, and discovered that the DenseNet with RmsProp optimizer method achieves the top outcome with the highest accuracy 99.8%. More research can be done to detect all forms of plant diseases, not only detecting them but also offering remedies.

References

1. Santos, L., Santos, F.N., Oliveira, P.M., Shinde, P.: Deep learning applications in agriculture: a short review. In: Silva, M.F., Lima, J.L., Reis, L.P., Sanfeliu, A., Tardioli, D. (eds.) Robot 2019: Fourth Iberian Robotics Conference: Advances in Robotics, Volume 1, pp. 139–151. Springer, Cham (2020). https://doi.org/10.1007/978-3-030-35990-4_12
2. Zhao, J.-C., Guo, J.-X.: Big data analysis technology application in agricultural intelligence decision system. In: 2018 IEEE 3rd International Conference on Cloud Computing and Big Data Analysis (ICCCBDA), pp. 209–212. IEEE (2018)
3. Patil, R., Gulvani, S.: Plant Disease Detection Using Neural Network: A Review (JETIR1902C27), vol. 6 (2019). www.jetir.org
4. Adhikari, S., Saban Kumar, K., Balkumari, L., Shrestha, B., Baiju, B.: Tomato plant diseases detection system using image processing. In: 1st KEC Conference on Engineering and Technology, Lalitpur, vol. 1, pp. 81–86 (2018)
5. Zhang, K., Wu, Q., Liu, A., Meng, X.: Can deep learning identify tomato leaf disease? Adv. Multim. **2018**, 1–10 (2018)
6. Rangarajan, A.K., Purushothaman, R., Ramesh, A.: Tomato crop disease classification using pre-trained deep learning algorithm. Procedia Comput. Sci. **133**, 1040–1047 (2018)
7. Sladojevic, S., Arsenovic, M., Anderla, A., Culibrk, D., Stefanovic, D.: Deep neural networks based recognition of plant diseases by leaf image classification. Comput. Intell. Neurosci. **2016**, 1–11 (2016)

8. Lu, Y., Yi, S., Zeng, N., Liu, Y., Zhang, Y.: Identification of rice diseases using deep convolutional neural networks. Neurocomputing **267**, 378–384 (2017)
9. Tan, W., Zhao, C., Wu, H.: CNN intelligent early warning for apple skin lesion image acquired by infrared video sensors. **22**, 67–74 (2016)
10. Adhikari, S., Shrestha, B., Baiju, B., Kumar, S.: Tomato plant diseases detection system using image processing. In: 1st KEC Conference on Engineering and Technology, Lalitpur, vol. 1, pp. 81–86 (2018)
11. Hughes, D., Salathé, M.: An open access repository of images on plant health to enable the development of mobile disease diagnostics. arXiv preprint. arXiv:1511.08060 (2015)
12. Pak, M., Kim, S.: A review of deep learning in image recognition. In: 4th International Conference on Computer Applications and Information Processing Technology (CAIPT), pp. 1–3 (2017)
13. Alom, Z., Taha, T.M., Yakopcic, C., Westberg, S., Sidike, P., Nasrin, M.S.: The History Began from AlexNet: A Comprehensive Survey on Deep Learning Approaches, p. 39 (2018)
14. Huang, G., Liu, Z., Van Der Maaten, L., Weinberger, K.Q.: Densely connected convolutional networks. In: Proceedings of the IEEE Conference on Computer Vision and Pattern Recognition, pp. 4700–4708 (2017)
15. Qassim, H., Verma, A., Feinzimer, D.: Compressed residual-VGG16 CNN model for big data places image recognition. In: 2018 IEEE 8th Annual Computing and Communication Workshop and Conference (CCWC), Las Vegas, January 2018, pp. 169–175 (2018). https://doi.org/10.1109/CCWC.2018.8301729

Object Detection and Face Recognition Using Yolo and Inception Model

Yatharth V. Kale, Ashish U. Shetty, Yogeshwar A. Patil, Rajeshwar A. Patil, and Darshan V. Medhane(✉)

Indian Institute of Information Technology, Pune, Maharashtra, India

Abstract. The task of facial recognition involves recognising faces in images, while object detection entails determining the location of objects in images. To accomplish this goal, we have developed a model capable of detecting objects as well as recognizing faces. The YOLO (You Only Look Once) model was used to detect objects in the image. If a person is detected by the model, then a cropped image of the person's face is passed to the pre-trained Inception model with additional custom hidden layers. The Inception model is trained on a facial dataset of size 1821 which consists of 5 classes. The Siamese network identifies the person by referring to the database of known people. By adding Siamese network, the framework becomes more scalable and adaptable. The testing accuracy of person recognition is 93.75%.

Keywords: YOLO · Inception · Siamese · Face recognition · Object detection · Convolutional neural network

1 Introduction

Facial recognition and object detection are an important field of study in modern deep learning. Face recognition is a classification problem that necessitates a separate image of each individual in order to execute the task of recognition. Considering the fact that most cameras in natural scenarios won't be capturing close shots of people, the goal of recognising faces in these images becomes a challenging task. To do this, we employed the YOLO model's object detection to obtain isolated images for each individual in the image. Other than faces, the YOLO model also helps in detecting objects in the surrounding.

Face recognition and object detection have a plethora of applications. The applications include identifying bankcards, controlling access, searching mug shots, monitoring security, and monitoring attendance. Object detection is a method for identifying objects in images and videos of a specific class (for example, humans, vehicles, and buildings). Autonomous vehicles, navigation for blind people are some of its applications. Siamese neural networks use a unique structure to make decisions about the similarity between inputs.

I. Woungang et al. (Eds.): ANTIC 2021, CCIS 1534, pp. 274–287, 2022.
https://doi.org/10.1007/978-3-030-96040-7_22

As we know that facial recognition is a highly computational task and requires a significant amount of time on lower-end devices, we have developed a novel idea to reduce the said computations. Most of our surrounding consists of objects rather than people so by using the YOLO model to filter out those frames that do require the more "heavier" inception model from those that don't, we have been able to save over (X)% amount of computational cost and time, where X is proportional to the ratio of objects which are not a person and the total number of objects.

2 Related Work

Multiple Object detection and identification from an image is hard for a machine. Although notable efforts have been made in the past in detecting an object [23].

There are several object detection methods namely (R-CNN, SPP-Net, Fast R-CNN, Faster R-CNN, Mask-R-CNN, SSD, YOLO). R-CNN and Fast R-CNN uses a selective search method to extract regions from images but it is a slow and time-consuming process that affects the performance of the network. In comparison to its predecessors, Faster R-CNN uses a separate network to determine region proposals instead of a selective search algorithm to identify them [17]. This paper [9] presents a way of using Faster R-CNN for face recognition. By adding a branch to Faster R-CNN that outputs a binary mask which would indicate whether a pixel is part of an object, Mask R-CNN takes a step further by locating exact pixels of each object instead of just bounding boxes [19]. Mask R-CNN improves results by combining the generated masks with classifications and bounding boxes.

Detection of multiple objects from an image is effectively done using YOLO (You look only Once) [15]. In this paper [1,16] the authors have proposed a modified network inspired from the YOLOv1 network model. The new model consists of an optimized loss function present in YOLOv1 model, inception model structure and a spatial pyramid pooling layer. This model effectively extracts features from images, which performs much better in object detection [16]. YOLO network is capable of making classification and bounding box regression at the same time unlike Faster R-CNN thus making it faster [5]. This article [2] provides a comparison of YOLO model and Faster R-CNN for Car Detection using Unmanned Aerial Vehicles. Usually, deep learning methods do not have a high detection rate when used under small datasets, so [11] proposes a novel image detection technique using YOLO to effectively increase its detection rate.

The Inceptionv3 model is widely used in image recognition which is trained on the imagenet dataset. The inception model gets deeper without increasing the computation exponentially. It is capable of performing several convolutions and max-pooling parallelly in a single step instead of lining them which increases the parameters, in turn, the computation to a large extent. It is capable of giving predictions from the intermediate layer instead of going through all the layers

[20]. This article [14] shows a comparative study of inception model with other models for facial expression detection.

This article [3] proposes a combination of the SSD model (where transfer learning is performed on the SSD_VGG16 model) and the Inceptionv3 model (trained using transfer learning utilizing ImageNet weights) [16] is used to classify 6 currency notes and 5 human faces using the SSD model. A performance evaluation of the above model is based on the intersection over union(IoU) method [3]. The classification accuracy of SSD is higher than YOLO but if precision is not much important, then YOLO will be the best choice to move forward since detecting the objects is very efficient and has a low processing time, allowing its use in real-time applications [18].

Siamese network is a modern architecture build to enhance the face identification technique in deep learning era [13]. Siamese network is a class of neural network architecture that contains two or more identical sub-network [7]. It is used to find the similarities between the input images by comparing the feature vector of the two sub-networks. The main benefit of using a Siamese network is that the database can be scalable [22]. These articles [4,5] shows the use of Siamese network for object tracking. A new method of finding matching and non-matching pairs of images with the Siamese network is discussed in [12]. Feature vectors based on neural networks can be used to represent the images, whose similarity can be measured with Euclidean distance.

In this paper, we are proposing a novel framework for objection detection and person recognition which uses the YOLOv3 model for object detection and the Inception model for face recognition. We are using YOLOv3 since it is much faster than other models with a nominal trade-off in accuracy, the Inception model which gets deeper without increasing the computation exponentially and using Siamese like network which makes this approach modular and scalable.

3 Proposed Design

The model consists of two main components YOLOv3 (You Only Look Once) model and the Inception model. The YOLO model outputs all the objects in the input image by drawing boxes around them. If an object is a person, then its cropped image is passed to the Inception model which is indeed responsible for recognizing the person. The method used to recognize the person's image is implemented using the concept of Siamese network. The Fig. 1 shows the control flow of diagram of the deep learning framework.

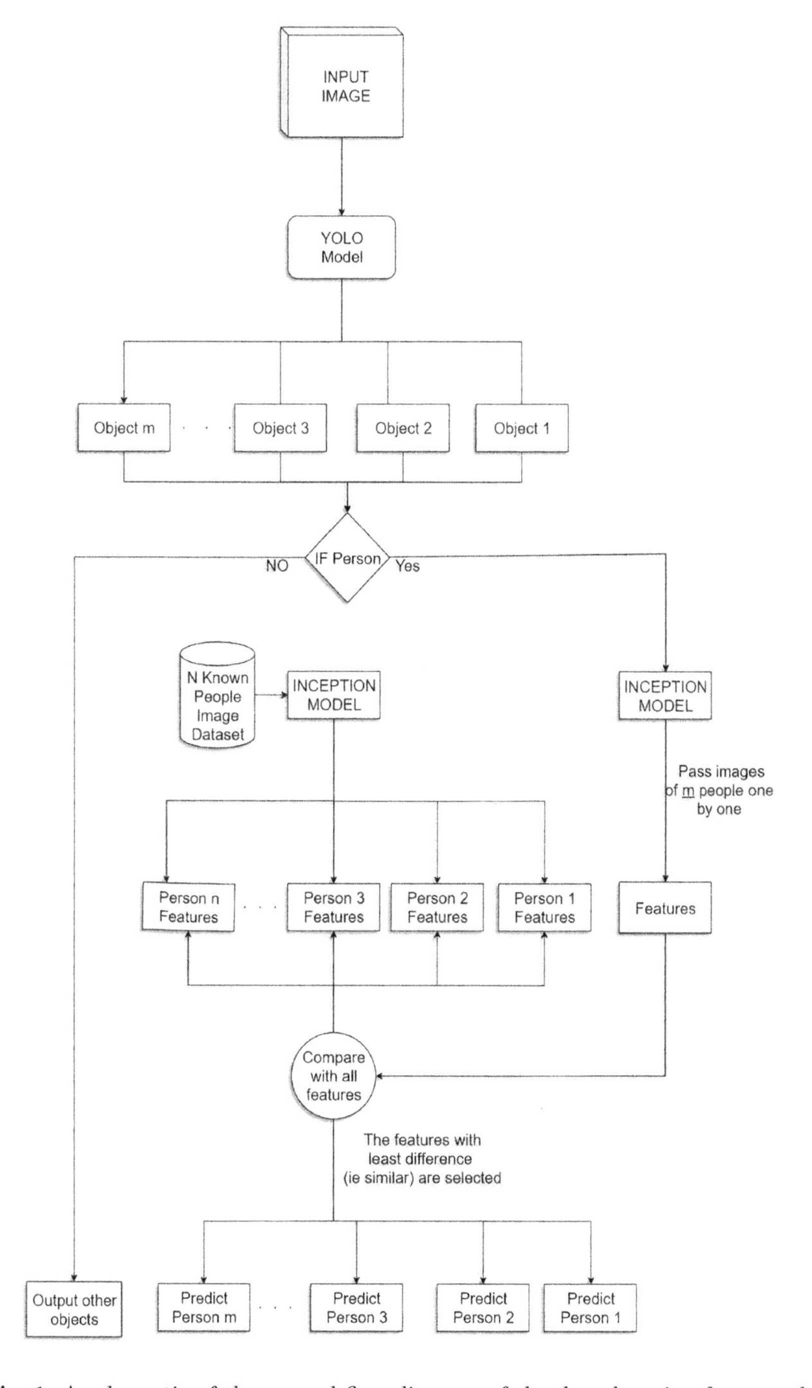

Fig. 1. A schematic of the control flow diagram of the deep learning framework

3.1 YOLO Model

YOLO ("You Only Look Once") is an effective real-time object recognition algorithm. YOLO is implemented in Darknet which is an open-source neural network framework. YOLOv3 uses a variant of Darknet. The Darknet model consists of a 53-layer network which is trained on Imagenet dataset and 53 more layers are stacked onto it, giving a 106 layer fully convolutional architecture for YOLOv3 [16]. Figure 2 depicts the YOLO model.

The YOLOv3 ("You Only Look Once") model is implemented using transfer learning (using pre-trained weights). There are 80 classes of objects in the model. When given an input image it passes through a series of steps. Firstly, the number of anchor boxes and their dimensions are fixed, the image is passed into the model in form of a grid. The center of the objects is assigned to some square of the grid. Intersection over Union (IoU) is used to determine which anchor box fits the object best. There are several boxes assigned for a single object to resolve this non-max suppression is applied which would only select the boxes with maximum probability and discard other boxes with high Intersection over Union (IoU) with respect to the selected box. The model outputs an image with bounding boxes around the detected objects along with their corresponding labels. The coordinates of these bounding boxes can be used to crop the objects from the image. The segments of the image which have their corresponding label as 'person' are fed to the inception model one by one for further classification and the other objects are output along with their labels.

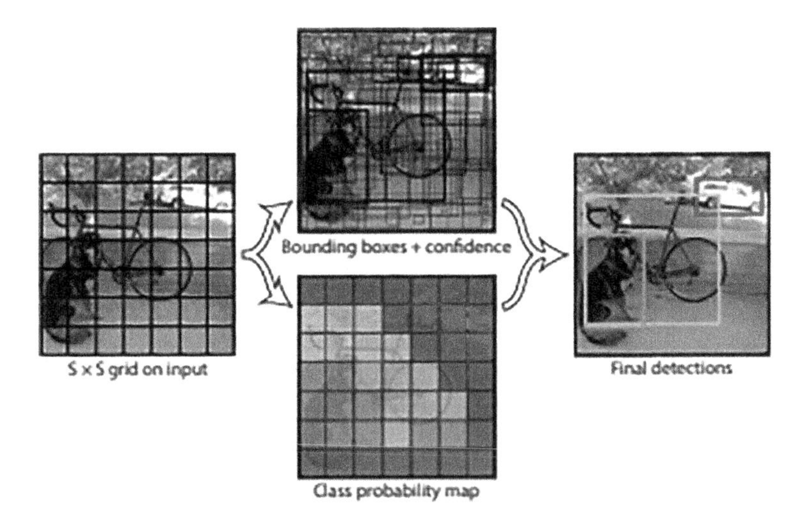

Fig. 2. YOLO model [21]

3.2 Inception Model

Inception is a convolutional neural network designed to reduce computing costs without reducing accuracy, and to make the architecture easy to extend or adapt without sacrificing performance or efficiency. Multiple convolutional layers and pooling layers can be applied simultaneously to the same input, the results are then concatenated which in turn reduces the computational cost [21]. The Fig. 3 shows the architecture of Inceptionv3 model.

The InceptionV3 model which is pre-trained on the ImageNet dataset is loaded along with its pre-trained weights. All the weights of the model are set to non-trainable and the last layer (softmax layer) is removed from the model. Custom hidden layers are added to the network namely (1 pooling layer, 5 dense layers with "relu" activation function, 4 dropout layers) and a softmax layer with 5 classes.

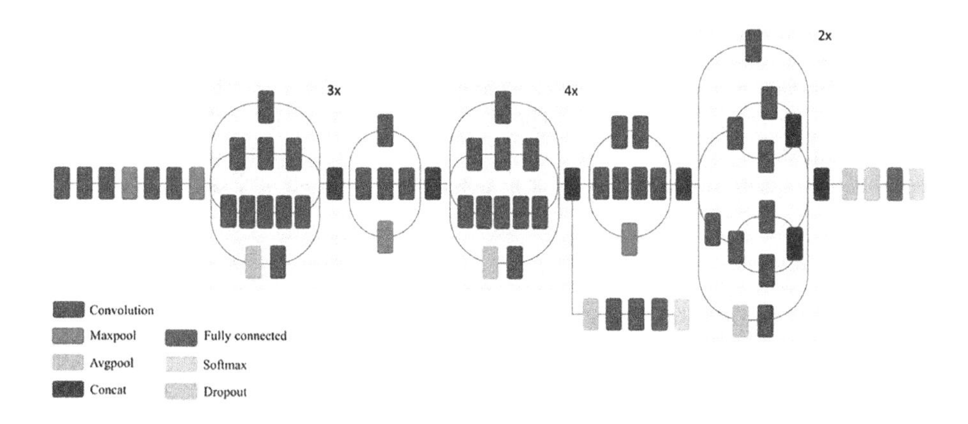

Fig. 3. Inceptionv3 model [20]

3.3 Calculation of Feature Array for Prediction

The facial images from our dataset are passed to the trained inception model. The model produces a feature array of size (1,64) for each image. Then take average over all feature arrays corresponding to each class. These averaged feature arrays will be used for prediction.

3.4 How to Perform Predictions

When we pass an image to this model it will give a new feature array. Now we take the element-wise absolute difference (Manhattan distance) between this array and the saved feature array of each class. This will give us a new array for each of the class. If this image is similar to any of the class, then the new array generated from it will have the least value.

3.5 Why We Need Threshold?

If the image does not match any of the classes the model will still give one of the classes as an answer (which has the minimum value). To avoid this, we introduce a threshold value for each class. If the calculated difference array is above the threshold of each class, then the image is classified as an unknown person.

3.6 Calculation of Threshold

The Manhattan distance [6], between the saved feature array of each class and the feature array of each image in the dataset is calculated.

The threshold is decided based on the maximum possible accuracy on the test dataset, i.e. (True positive + True negative/Total) should be maximum.

3.7 Prediction

After the Inception model has trained on the dataset, the last layer (softmax layer) is removed. When we predict the image on this model it gives an array of features corresponding to the input image. We pass images corresponding to each class in this model and calculate the average array of features for each class and save the feature arrays. Figure 4 shows the prediction for proposed model.

The cropped image of the person is passed to this model by the YOLO model. When we predict the image on this model it gives a new feature array. Now, this array is compared with each of the saved feature arrays corresponding to each class. They are compared using absolute difference. If the difference between the feature of the input image and a saved feature is below a certain threshold the label corresponding to the saved feature is predicted but in case of ties (if the difference between the input image and more than one features is less than the threshold) the difference closer to zero is selected.

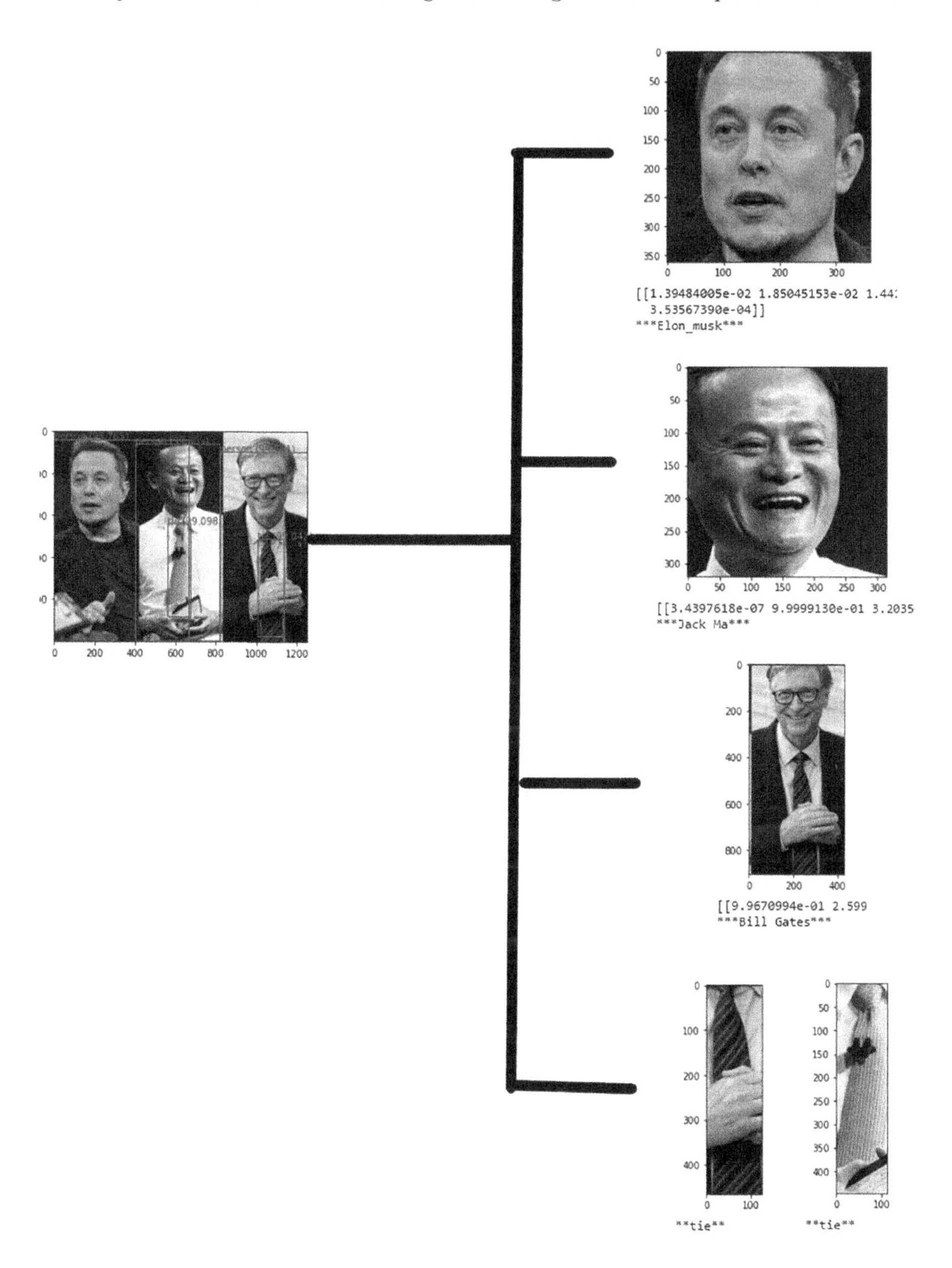

Fig. 4. Our model's prediction

3.8 Evaluation Criteria

In the prediction phase, six quantitative performance measures were computed to access the reliability of trained models using the validation data, including precision, recall, f1-score, accuracy, macro-avg and weighted-avg [8]. These metrics are computed based on True Negative (TN), True Positive (TP), False Negative (FN), False Positive (FP).

$$Precision = \frac{TP}{TP + FP} \quad (1)$$

$$Recall = \frac{TP}{TP + FN} \quad (2)$$

$$F1Score = 2 * \frac{Precision * Recall}{Precision + Recall} \quad (3)$$

$$Accuracy = \frac{TP}{TP + FN + TN + FP} \quad (4)$$

$$Weighted\_avg = F1class1 * W1 + F1class2 * W2 + F1class3 * W3 + \cdots + F1classn * Wn \quad (5)$$

F1classm : F1 score of class m

$$Macro\_avg = F1class1 + F1class2 + F1class3 + \cdots + F1classn \quad (6)$$

F1classm : F1 score of class m

4 Dataset

We have used a custom image dataset which has been gathered from google images. The collection contains five classes that correspond to the faces of five public figures. The total number of images are 1821 in a variety of settings. The images were further preprocessed by cropping and scaling in-order to remove irrelevant data. This dataset is used for training the proposed inception model.

5 Training Details

Tesla K80 GPU and 13 GB RAM used for training along with TensorFlow and Keras libraries in Google Colab, coded in Python.

The Inception model is trained on a dataset of 1821 face images of 5 people corresponding to the 5 classes of the softmax layer.

Data augmentation (rescaling, rotation, flipping) is applied to the dataset in order to expand the dataset and also make the model more robust to tilted or flipped images. The learning rate is set to 0.0001 and Adam optimizer [10] is used. The model is trained for 100 epochs. The model achieves an accuracy of 93.75% on the testing data. Figure 5 presents the accuracy vs epochs graph and Fig. 6 presents the loss vs epochs graph.

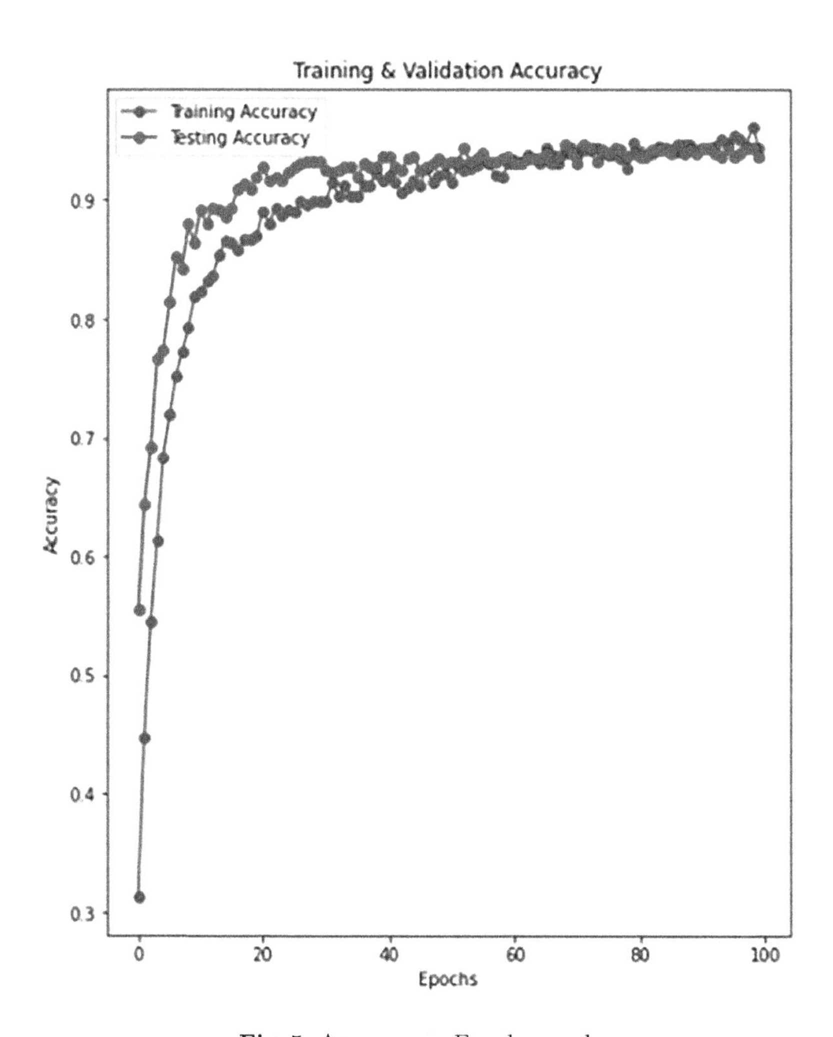

Fig. 5. Accuracy vs Epochs graph

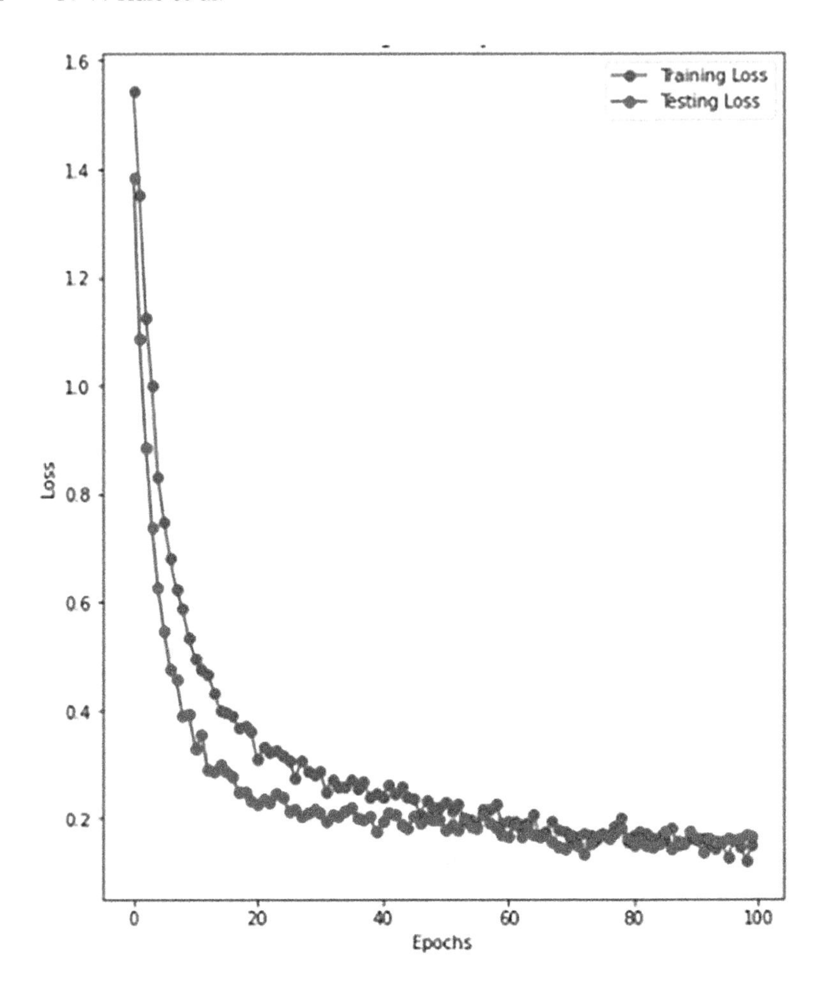

Fig. 6. Loss vs Epochs graph

6 Results

The pre-trained YOLO model detects objects and assigns them to one of the 80 classes of objects. If the label of the object is 'person' it is passed to the inception model which is trained on a face dataset with 5 classes. Inceptionv3 achieved an accuracy of 93.75% on the test dataset. The other objects are output along with their corresponding labels. The Fig. 7 shows the confusion matrix on the validation data and Table 1 shows the classification report on the validation data.

6.1 Comparitive Study

[3] in this paper, the author uses two models. The SSD model is used for detecting and classifying objects, while the Inceptionv3 model is used for recognizing

different human faces and currency notes. The SSD model has an accuracy of 67.8% and the Inceptionv3 model has an accuracy of 92.5% for human face recognition and 90.2% for currency recognition.

Our model detects 80 different classes of objects, the objects having labels as person are fed to Inceptionv3 model for recognition. Our model is recognizing five different human faces with an accuracy of 93.75%.

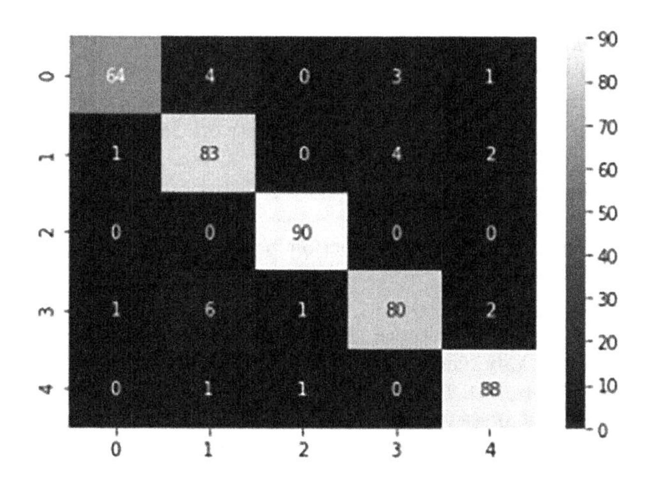

Fig. 7. Confusion matrix on validation set

Table 1. Classification report

Labels	Precision	Recall	f1-score	Support
gates(0)	0.97	0.89	0.93	72
jack(1)	0.88	0.92	0.90	90
modi(2)	0.98	1.00	0.99	90
musk(3)	0.92	0.89	0.90	90
trump(4)	0.95	0.98	0.96	90
Accuracy			0.94	432
Macro avg	0.94	0.94	0.94	432
Weighted avg	0.94	0.94	0.94	432

7 Conclusion Future Scope

It is possible to increase the number of known people by just adding few images without the need to train the model again on a large dataset. Using the YOLO model for Object Detection and Inception for face recognition makes it faster and effective by saving computation time, user-specific and adaptable.

Future work can be done to recognize faces and objects in real-time.

References

1. Ahmad, T., Ma, Y., Yahya, M., Ahmad, B., Nazir, S., et al.: Object detection through modified yolo neural network. Sci. Program. 2020 (2020)
2. Benjdira, B., Khursheed, T., Koubaa, A., Ammar, A., Ouni, K.: Car detection using unmanned aerial vehicles: comparison between faster R-CNN and YOLOv3. In: 2019 1st International Conference on Unmanned Vehicle Systems-Oman (UVS), pp. 1–6. IEEE (2019)
3. Bhole, S., Dhok, A.: Deep learning based object detection and recognition framework for the visually-impaired. In: 2020 Fourth International Conference on Computing Methodologies and Communication (ICCMC), pp. 725–728. IEEE (2020)
4. Dong, X., Shen, J.: Triplet loss in Siamese network for object tracking. In: Ferrari, V., Hebert, M., Sminchisescu, C., Weiss, Y. (eds.) ECCV 2018. LNCS, vol. 11217, pp. 472–488. Springer, Cham (2018). https://doi.org/10.1007/978-3-030-01261-8_28
5. Du, J.: Understanding of object detection based on CNN family and YOLO. J. Phys. Conf. Ser. **1004**, 012029 (2018)
6. Greche, L., Jazouli, M., Es-Sbai, N., Majda, A., Zarghili, A.: Comparison between Euclidean and Manhattan distance measure for facial expressions classification. In: 2017 International Conference on Wireless Technologies, Embedded and Intelligent Systems (WITS), pp. 1–4. IEEE (2017)
7. He, A., Luo, C., Tian, X., Zeng, W.: A twofold Siamese network for real-time object tracking. In: Proceedings of the IEEE Conference on Computer Vision and Pattern Recognition, pp. 4834–4843 (2018)
8. Hossin, M., Sulaiman, M.N.: A review on evaluation metrics for data classification evaluations. Int. J. Data Min. Knowl. Manag. Process **5**(2), 1 (2015)
9. Jiang, H., Learned-Miller, E.: Face detection with the faster R-CNN. In: 2017 12th IEEE International Conference on Automatic Face & Gesture Recognition (FG 2017), pp. 650–657. IEEE (2017)
10. Kingma, D.P., Ba, J.: Adam: a method for stochastic optimization. arXiv preprint arXiv:1412.6980 (2014)
11. Li, G., Song, Z., Fu, Q.: A new method of image detection for small datasets under the framework of yolo network. In: 2018 IEEE 3rd Advanced Information Technology, Electronic and Automation Control Conference (IAEAC), pp. 1031–1035. IEEE (2018)
12. Malkauthekar, M.: Analysis of Euclidean distance and Manhattan distance measure in face recognition. In: Third International Conference on Computational Intelligence and Information Technology (CIIT 2013), pp. 503–507. IET (2013)
13. Melekhov, I., Kannala, J., Rahtu, E.: Siamese network features for image matching. In: 2016 23rd International Conference on Pattern Recognition (ICPR), pp. 378–383. IEEE (2016)
14. Nivrito, A., Wahed, M., Bin, R., et al.: Comparative analysis between Inception-v3 and other learning systems using facial expressions detection. Ph.D. thesis, BRAC University (2016)
15. Redmon, J., Divvala, S., Girshick, R., Farhadi, A.: You only look once: unified, real-time object detection. In: Proceedings of the IEEE Conference on Computer Vision and Pattern Recognition, pp. 779–788 (2016)
16. Redmon, J., Farhadi, A.: YOLOv3: an incremental improvement. arXiv preprint arXiv:1804.02767 (2018)

17. Rohith Sri Sai, M., Rella, S., Veeravalli, S.: Object Detection and Identification a project report. Ph.D. thesis (2019)
18. Sanchez, S., Romero, H., Morales, A.: A review: comparison of performance metrics of pretrained models for object detection using the TensorFlow framework. IOP Conf. Ser. Mater. Sci. Eng. **844**, 012024 (2020)
19. Sumit, S.S., Watada, J., Roy, A., Rambli, D.: In object detection deep learning methods, YOLO shows supremum to mask R-CNN. J. Phys. Conf. Ser. **1529**, 042086 (2020)
20. Szegedy, C., et al.: Going deeper with convolutions. In: Proceedings of the IEEE Conference on Computer Vision and Pattern Recognition, pp. 1–9 (2015)
21. Szegedy, C., Vanhoucke, V., Ioffe, S., Shlens, J., Wojna, Z.: Rethinking the inception architecture for computer vision. In: Proceedings of the IEEE Conference on Computer Vision and Pattern Recognition, pp. 2818–2826 (2016)
22. Taigman, Y., Yang, M., Ranzato, M., Wolf, L.: DeepFace: closing the gap to human-level performance in face verification. In: Proceedings of the IEEE Conference on Computer Vision and Pattern Recognition, pp. 1701–1708 (2014)
23. Zou, Z., Shi, Z., Guo, Y., Ye, J.: Object detection in 20 years: a survey. arXiv preprint arXiv:1905.05055 (2019)

Research and Trends in COVID-19 Vaccines Using VOSviewer

Tayeb Brahimi[1](✉) and Hagar Abbas[2]

[1] Energy and Technology Research Center, Natural Science, Mathematics and Technology Unit, College of Engineering, Effat University, Jeddah, Saudi Arabia
tbrahimi@effatuniversity.edu.sa

[2] Computer Science Department, College of Engineering, Effat University, Jeddah, Saudi Arabia
haoabas@effat.edu.sa

Abstract. The World Health Organization (WHO) classified the latest coronavirus (COVID-19) as a pandemic on March 11, 2020. As a result, the pandemic has spread to practically every country on the planet. WHO's major goals for 2021 are to fight COVID-19, strengthen current health systems, increase access to COVID-19 treatment, and provide equitable and safe vaccines for all. As the number of scientific publications continues to expand, there is an increasing need to analyze factors and characteristics that contribute to highly published documents and highly cited articles. This study evaluates and identifies trends and studies in COVID-19 vaccines using the SCOPUS database and VOSviewer. The top five active countries on COVID-19 vaccines publication are the United States with 4168 documents, China with 2245 documents, Italy with 1512 documents, the United Kingdom with 1370 documents, and Spain with 663 documents. Results of network visualizations indicate that understanding the state-of-the-art COVID-19 pandemic is essential in planning future measures to fight COVID-19 and improve vaccination uptake.

Keywords: Covid-19 · Bibliometric analysis · VOSviewer · Vaccine

1 Introduction

A new coronavirus caused an outbreak of severe acute respiratory syndrome (SARS) in December 2019 in Wuhan, China. The COVID-19 pandemic has affected every segment of society [1]. There has been a global urgency to produce vaccines as a result [2]. Bibliometric is a statistical approach that may use mathematical ways to evaluate the research papers concerned with one subject quantitatively. Understanding bibliometric data can help you find research trends and elements that increase your article's citation rate. In particular, as journals and academics continue to use social media to distribute research, this is becoming increasingly crucial [3]. This evaluation is carried out using an online database Scopus, containing nearly all the major research papers published in the field. In fact, both the Web of Science (WoS) and Scopus databases (DBs), there is no consensus which one is better; however, in our case, our selection of Scopus DB is

I. Woungang et al. (Eds.): ANTIC 2021, CCIS 1534, pp. 288–296, 2022.
https://doi.org/10.1007/978-3-030-96040-7_23

purely a subscription-based decision. Since these databases are expensive data sources, and institutions often have to choose between them.

A vaccine for coronavirus disease 2019 (COVID-19) will help develop immunity to the virus that causes COVID-19. Although each type of the vaccine has a different mechanism, they all cause an immunological response to help the body fight viruses in the future [4]. Table 1 compares some of the COVID-19 vaccines [5, 6]. The efficacy (Relative Risk Reduction) [6] as of February 2021 shows Pfizer-BioNTech and Moderna among the highest. According to the WHO, the Pfizer/BioNTech vaccine was listed for Emergency Use Listing (EUL) on December 31, 2020; the AstraZeneca vaccine on February 16, Janssen vaacine on March 12, 2021, Moderna vaccine on April 30, 2021, Sinopharm vaccine on May 7, 2021, and Sinovac vaccine on June 1, 2021.

In a vaccinated population, the vaccine's efficacy represents the percentage of fewer people contracting the disease when in contact with the virus. As more people get vaccinated, it is expected to have fewer people getting the virus. The WHO stresses that individuals should continue to take precautions after getting vaccinated, such as physical distancing, wearing a mask, avoiding crowds, and cleaning hands. Because the COVID-19 pandemic has not been completely managed and further information from these references should be collected, it is urgently needed for bibliometric analysis. We conducted the present study to comprehensively understand COVID-19 vaccinations and prospective research avenues based on bibliometric analysis. The remainder of the paper is arranged in the following manner: Sect. 2 examines relevant literature; Sect. 3 outlines the research method and approach used in the study; Sect. 4 provides and discusses the findings of the current research, and Sect. 5 highlights the major conclusions of the study.

Table 1. Comparison between some types of COVID-19 vaccine [5, 6].

Company	Method	Efficacy	Origin
Pfizer/BioNTech	Uses Message RNA (mRNA) following SARSCov2 virus responsible for COVID-19	95.03%	UK/Sweden
Moderna/NIH	Uses Message RNA (mRNA)	94.08%	USA
AstraZeneca/Oxford	Uses genetically altered virus	66.84%	USA/Germany
Sputnik V	Uses adenoviral vectors, viruses responsible for human cold	90.97%	USA
Sinopharm	Uses inoculation technique	79%	Russia
Sinovac	Uses inoculation technique	78%	China
Johnson & Johnson	Uses genetically altered virus	66.62%	China

2 Literature Review

Many infectious diseases have occurred in the last two decades, including the COVID-19. The disease emerged in China and has since spread to over 200 countries and territories

[7]. The COVID-19 disease has rapidly spread at an exponential rate, with the number of cases increasing at a rapid pace. As reported by many COVID-19-related publications, the scientific community has responded quickly. The need to understand the factors that lead to highly impactful publications is growing. To that end, in May 2020, a study was carried out analyzing the characteristics of the top 50 cited COVID-19-related publications from the pandemic's early phases [8]. This disease's research is critical for determining pathogenic characteristics and developing therapeutic strategies. A study looked for COVID-19-related publications in the PubMed and WHO databases from December 2019 to March 18, 2020 [9]. The study focused on relevant observational and interventional studies. Just 564 publications met the inclusion criteria, according to the study. With 377 publications, China ranked first in total publications while Singapore ranked first in publications per million inhabitants [9]. Following the emergence of the COVID 19 global pandemic, there was an avalanche of science. In just a few months, over a thousand articles on this topic have emerged in peer-reviewed journals. On a macro level, the bibliometric aspects of these papers and those covering Coronaviruses, in general, were examined in a brief analysis [10].

Furthermore, a scoping review of the literature on COVID-19 findings revealed that, apart from medical and clinical aspects like vaccine and treatment safety, concerns about the COVID-19 pandemic have so far focused on patient transport and healthcare professionals safety [11]. The study's analysis also reveals several potentially serious safety issues that have arisen due to this worldwide health emergency, which has attracted only a limited amount of scientific attention. According to the report, having a full safety-scientific understanding of the COVID 19 issue will also aid in preparation for potential pandemics in the future [11]. In the subject of bibliometrics, bibliometric mapping is a major topic. Most bibliometrics studies rely on computer tools like SPSS and Pajek to generate basic graphical representations. However, it appears that there is a tendency toward bigger maps, for which such representations are insufficient. VOSviewer is software that allows you to study bibliometric maps in detail. Based on co-citation data, it may be used to create maps of authors or journals. For maps with at least 100 elements or more, viewing capabilities are very handy [14].

3 Methodology

Published documents about COVID-19 vaccines were obtained from the Scopus database published between 2019 to 2021. Citation information such as the author, author ID, year, source title, volume issue page, abstract and keywords, affiliation, language, and publishers were retrieved from the database. The open-source tool VOSviewer (version 1.6.16) (www.vosviewer.com) to obtain a deeper insight and evaluate current research trends on COVID-19 vaccines has been used. The nearest matching publication was conducted using the keywords (("covid-19" or "coronavirus" or "covid") and ("covid-19 vaccine" or "vaccine")), which was used as the title keyword. To access Scopus, we logged onto our university's website and entered the previously specified search keywords. Publication year, journal name, author, affiliation, keywords, document type, abstract, and the number of citations for each document that fit the requirements were all exported into CSV format. The retrieval date was on October 16, 2021.

VOSviewer uses a co-occurrence matrix to create a map. The process of making a map is broken down into three phases. The co-occurrence matrix is used to create a similarity matrix in the first phase. The similarity matrix is used to create a map using the VOS mapping approach in the second phase. Finally, the map is translated, rotated, and reflected in the third phase [14]. Step one and step two use the following equations with the similarity s_{ij} given by

$$s_{ij} = \frac{c_{ij}}{w_i w_j}, \quad (1)$$

where w_i and w_j represent either the total number of occurrences of items i and j or the total number of co-occurrences of these items, and C_{ij} represents the number of co-occurrences of items i and j. The number of objects to be mapped is denoted by n. The VOS mapping approach creates a two-dimensional map in which the items 1, …, n are placed so that the distance between any two items i and j properly represents their similarity s_{ij}.

$$V(\mathbf{x}_1, \ldots, \mathbf{x}_n) = \sum\nolimits_{i<j} s_{ij} \|\mathbf{x}_i - \mathbf{x}_j\|^2, \quad (2)$$

$$\frac{2}{n(n-1)} \sum\nolimits_{i<j} \|\mathbf{x}_i - \mathbf{x}_j\| = 1. \quad (3)$$

where $\|\bullet\|$ represents the Euclidean norm and $x_i = (x_{i1}, x_{i2})$ signifies the position of item i in a two-dimensional map. The objective function is minimized while keeping the limitation in mind.

4 Results and Discussion

In this study, we used the Scopus collection database for documents published in 2020/2021. We limited it to the English language using the search string (("covid19" OR "covid-19" OR "SARSCov2" OR "SARS-CoV-2" OR "mRNA" OR "2019-nCoV" OR "corona" OR "corona virus" OR "Coronavirus" OR "Novel-Coronavirus") AND ("pfizer" OR "Pfizer-Biotech" OR "Moderna" OR "AstraZeneca" OR "AstraZeneca-Oxford" OR "Johnson & Johnson" OR "janssen" OR "Sputnik" OR "Sinopharm" OR "Sinovac" OR "Alpha corona-virus" OR "αCoV" OR "Alphacoronavirus" OR "B.1.1.7" OR "Beta corona-virus" OR "βCoV" OR "betacoronavirus" OR "B.1.351" OR "Gamma corona virus" OR "γCoV" OR "Gamma corona-virus" OR "P.1" OR "Delta corona virus" OR "ΔCoV" OR "Delta coronavirus" OR "δCoV")) AND (limit-to (pubyear, 2021) OR limit-to (pubyear, 2020)) AND (limit-to (language, "English")). Using the Title-Abstract-Keyword option in Scopus and limiting the result to journals only, the initial search yielded 15696 documents distributed as shown in Fig. 1, with 11512 documents in Medicine alone. The top five active countries on COVID-19 vaccines publication, Fig. 2, are the United States with 4168 documents, China with 2245 documents, Italy with 1512 documents, the United Kingdom with 1370 documents, and Spain with 663 documents.

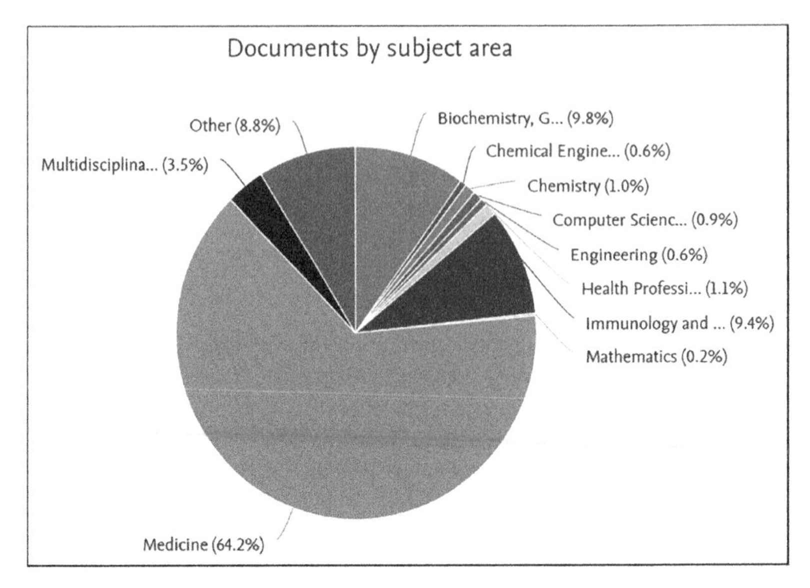

Fig. 1. Distribution of documents by subject area for COVID-19 and vaccines.

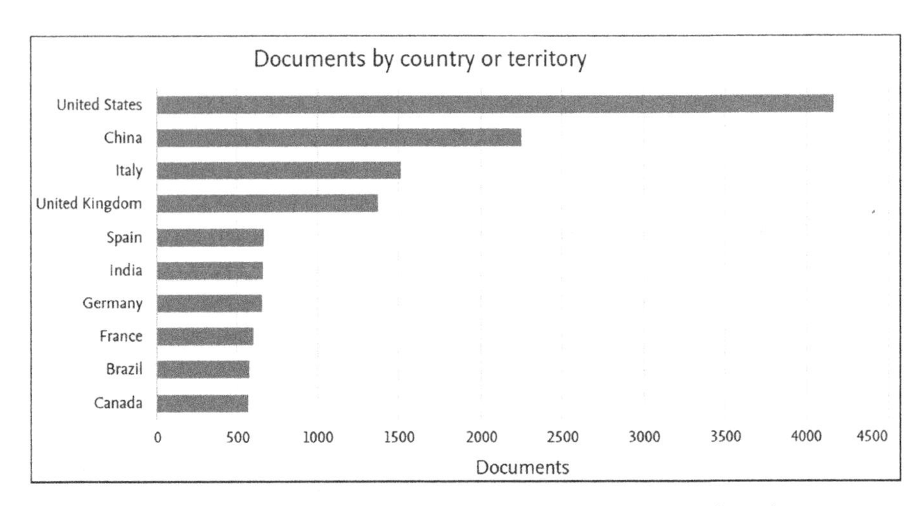

Fig. 2. Distribution of documents by country for COVID-19 and vaccines.

Limiting the output to TITLE yields 520 documents; these are the articles where the keywords in the string described above appear in the title. Compared to Fig. 1, the percentage of documents in Medicine is still high (53.4%). Using VOSviewer, the visualized maps based on a network of Scopus publications, authors, research organizations, countries, and keywords have been created. The document labels represent items in the network visualization using a circle. The size of the label and the circle surrounding it are determined by the weight of the item being labeled. The cluster to which it belongs determines the color of an object, and the lines connecting items represent the links between them. The final retrieved documents had all of the document's keywords, with a minimum occurrence of a keyword equal to five times; 337 of the 2824 keywords met this condition; the remaining papers did not.

The top dominant keywords occurrence with the highest total link strength (TLS) was COVID-19 (294, TLS = 4882), Sars-cov-2 (267, TLS = 4631), Coronavirus Disease (143, TLS = 3202), Covid-19 Vaccines (136, TLS = 2709), and Vaccination (105, TLS = 1938). Figure 3 shows the VOSviewer network visualization map for all keywords co-occurrence of COVID-19 and vaccine. In Fig. 4, we use keywords provided by the paper's authors and occurred more than five times in the Scopus core database. In this case, 36 of the 807 keywords fulfilled the requirement. The minimum number of occurrences of a keyword was set at 5. The top dominant keywords occurrence with the highest total link strength was Sars-cov-2 (127, TSL = 261), COVID-19 (134, TLS = 234), B.1.1.7 or Alpha variant (35, TLS = 101), Vaccine (39, TLS = 82), Variant (13, TLS = 47), and Vaccination (105, TLS = 1938).

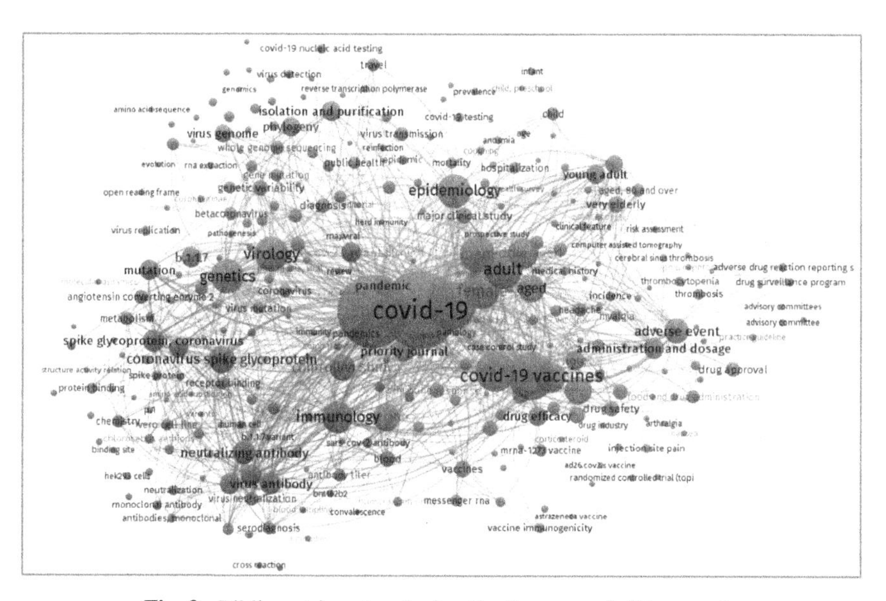

Fig. 3. Bibliometric network visualization map of all keywords.

Figure 5 depicts a bibliometric study of the co-authorship of COVID-19 and vaccine publications by different countries. The largest nodes are for the USA, and the UK, implying that these two countries produce most publications. Additionally, the distance between them is quite low, implying a high degree of co-authorship.

Fifteen thousand six hundred ninety-six publications on COVID-19 vaccines indexed in the Scopus core database have been analyzed. The main partners of the United States are the United Kingdom, Germany, Italy, Spain, Canada, and India. Saudi Arabia's main partners are the United States, the United Kingdom, and India; the total link strength of Saudi Arabia is 7, with a total of 71 citations. A version of the item density visualization is shown in Fig. 6.

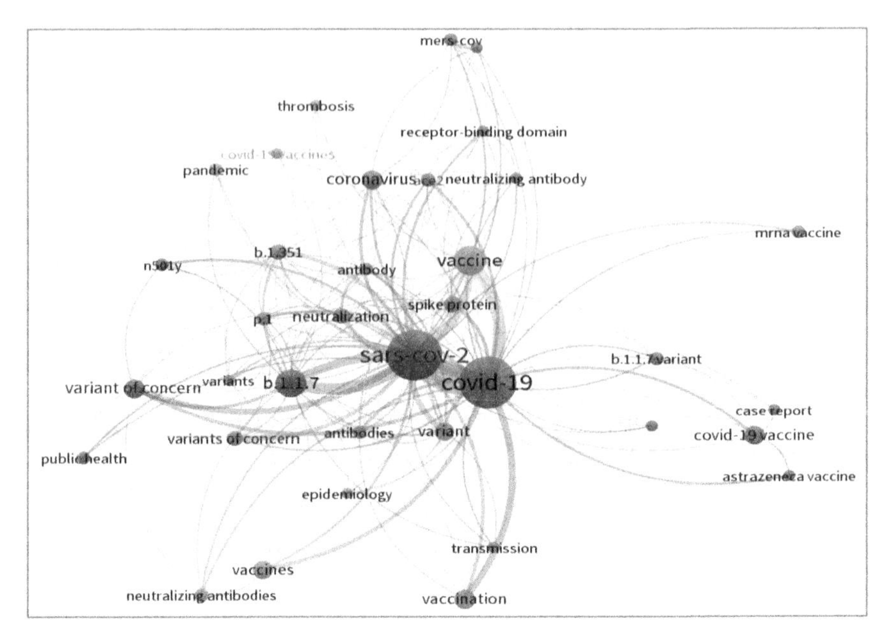

Fig. 4. Bibliometric network visualization map of author's keywords.

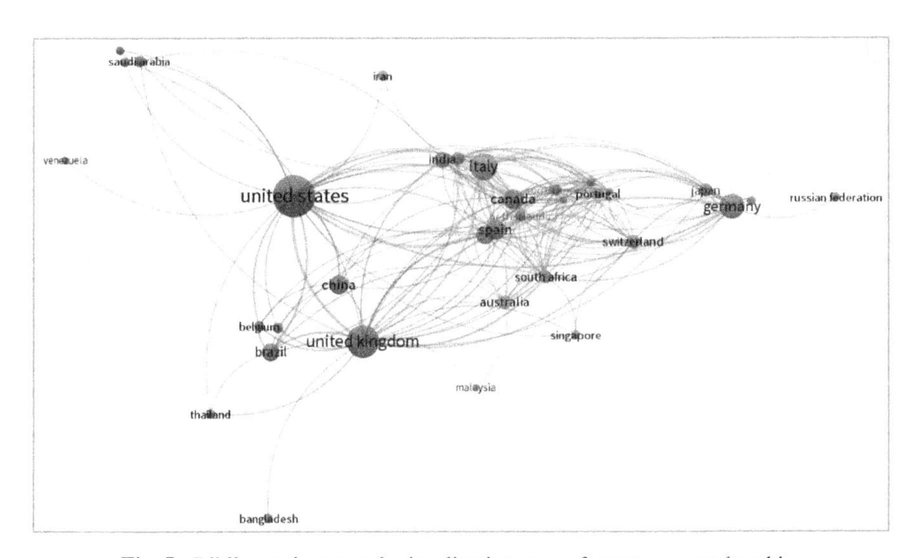

Fig. 5. Bibliometric network visualization map of country co-authorship.

Topics covered in the literature include coronavirus infection, clinical trials, and immunology. ChAdOx1 nCoV-19 is the AstraZeneca vaccine developed by the UK University in partnership with the biopharmaceutical company AstraZeneca. Pfizer's vaccine, BNT162b2, was developed in partnership with German biotech firm BioNTech

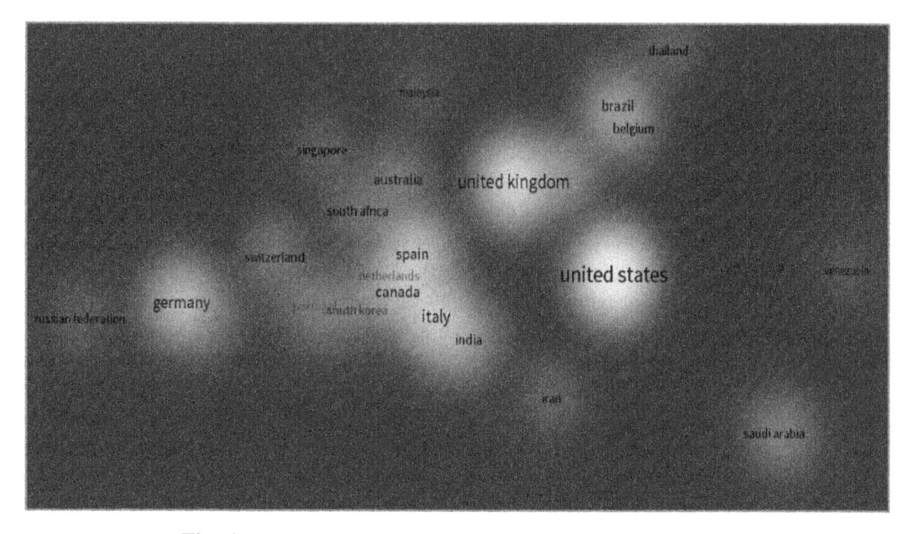

Fig. 6. Item density visualization for Covid-19 and vaccine

by one of the world's largest pharmaceutical firms located in New York. The Moderna vaccine, called mRNA-1273, was developed in collaboration with the US National Institutes of Health by a Massachusetts-based biotech company [12]. Pfizer BioNTech, Moderna, and Oxford-AstraZeneca have confirmed that their COVID-19 vaccine candidates are successful [13]. That explains having the United States and United Kingdom at the top of the most active countries on the publications of the COVID-19 vaccines.

5 Conclusion

The COVID-19 pandemic has had a huge influence on all facets of society. As a result, there has been a growing global demand for vaccines. The number of scientific publications on the COVID-19 and vaccine is too large to be assessed by traditional methods, such as narrative review, scoping review, or systematic literature review. In recent years, we are witnessing a great number of publications based on bibliometrics; some which may be attributed to the expansion of scientific research as a whole. Bibliometrics presents a statistical methodology that employs mathematical methods to assess research papers on a given subject quantitatively. It is becoming a standard and thorough method for analyzing and interpreting massive scientific data sets. Our study helps us dissect the evolution of Covid-19 and vaccine while also giving light on the new areas of research that are developing in healthcare and medicine. Results show that the United States has the most documents published in the Scopus core database about COVID-19 vaccines, followed by the United Kingdom. Topics on COVID-19 research involved many fields; understanding the state-of-the-art of the current events of the COVID-19 pandemic is crucial in planning future efforts towards fighting COVID-19 and improving vaccine acceptance.

References

1. Yang, L., Tian, D., Liu, W.: [Strategies for vaccine development of COVID-19]. Sheng Wu Gong Cheng Xue Bao Chin. J. Biotechnol. **36**(4), 593–604 (2020). https://doi.org/10.13345/j.cjb.200094
2. A novel coronavirus outbreak of global health concern. The Lancet. https://www.thelancet.com/journals/lancet/article/PIIS0140-6736(20)30185-9/fulltext. Accessed 17 Mar 2021
3. Yu, Y., et al.: A bibliometric analysis using VOSviewer of publications on COVID-19. Ann. Transl. Med. **8**(13), 816 (2020). https://doi.org/10.21037/atm-20-4235
4. Comparing the differences between COVID-19 vaccines. Mayo Clinic. https://www.mayoclinic.org/coronavirus-covid-19/vaccine/comparing-vaccines. Accessed 21 Mar 2021
5. Covishield and Covaxin: what we know about India's Covid-19 vaccines. BBC News, 9 March 2021. https://www.bbc.com/news/world-asia-india-55748124. Accessed 21 Mar 2021
6. Olliaro, P., Torreele, E., Vaillant, M.: COVID-19 vaccine efficacy and effectiveness—the elephant (not) in the room. Lancet Microbe **2**(7), e279–e280 (2021). https://doi.org/10.1016/S2666-5247(21)00069-0
7. ElHawary, H., Salimi, A., Diab, N., Smith, L.: Bibliometric analysis of early COVID-19 research: the top 50 cited papers. Infect. Dis. Res. Treat. **13**, 1178633720962935 (2020). https://doi.org/10.1177/1178633720962935
8. Chahrour, M., et al.: A bibliometric analysis of COVID-19 research activity: a call for increased output. Cureus **12**(3) (2020). https://doi.org/10.7759/cureus.7357
9. Becker, R.C.: Covid-19 treatment update: follow the scientific evidence. J. Thromb. Thrombolysis **50**(1), 43–53 (2020). https://doi.org/10.1007/s11239-020-02120-9
10. Haghani, M., Bliemer, M.C.J., Goerlandt, F., Li, J.: The scientific literature on Coronaviruses, COVID-19 and its associated safety-related research dimensions: a scientometric analysis and scoping review. Saf. Sci. **129**, 104806 (2020). https://doi.org/10.1016/j.ssci.2020.104806
11. Latest: more European countries suspend use of AstraZeneca's vaccine. Science, 16 March 2021. https://www.nationalgeographic.com/science/article/coronavirus-vaccine-tracker-how-they-work-latest-developments-cvd. Accessed 19 Mar 2021
12. Ask an expert: why are there so many COVID-19 vaccines — and is it better to have more? https://www.gavi.org/vaccineswork/ask-expert-why-are-there-so-many-covid-19-vaccines-and-it-better-have-more. Accessed 18 Mar 2021
13. De Soto, J.A.: Evaluation of the Moderna, Pfizer/Biotech, Astrazeneca/Oxford and Sputnik V Vaccines for Covid-19. J. Med. Clin. Sci. **7**(1), 1 (2021). https://doi.org/10.15520/arjmcs.v7i01.246
14. van Eck, N.J., Waltman, L.: Software survey: VOSviewer, a computer program for bibliometric mapping. Scientometrics **84**(2), 523–538 (2010). https://doi.org/10.1007/s11192-009-0146-3

Ensuring Security of User Data Using Federated Learning for the Sentiment Classification

Pravin Kumar[1,2](✉) and Manu Vardhan[1]

[1] Department of Computer Science and Engineering, Nit Raipur, Raipur, Chhattisgarh, India
pk.mmmec@gmail.com, mvardhan.cs@nitrr.ac.in

[2] Department of Electronics and Communication (Computer Science and Engineering), University of Allahabad, Prayagraj, India

Abstract. Federated Learning, which we can name as Secured Learning, is the latest entrant into the domain of Machine Learning. It is a secure way to train machine learning models on user's private data without compromising their privacy. This article demonstrates federated training of sentiment140 dataset with applications in sentiment classification. In this proposed approach, the model that is to be trained on a particular user's data is sent into that user's mobile device. This is different from the conventional approach because training takes place on the user's device rather than the cloud.

Moreover, the data of the user is completely encrypted before getting used for training. After the training is completed, essential statistics are uploaded back to the cloud, and the model is deleted from the user's device. The model trains on different devices under feasible conditions, i.e., the device is plugged in and connected to the internet and gets updated each time. Hence, this approach is novel because it preserves data privacy which has become a burning issue nowadays. To implement the proposed approach, we used different Neural Network architectures and compared their performances, including variants such as Feed Forward, Convolutional, Recurrent, etc. We used Tensor flow's Federated Core package to design these architectures, train models, and simulate Secured Learning. For visualizing the simulation of the entire backend, we developed a dashboard-like web interface. This dashboard displays how training on various devices occurs and how the model will constantly learn and get updated by user data.

Keywords: Federated learning · Machine learning · Sentiment analysis

1 Introduction

Sentiment analysis, or sentiment identification, has been increasingly important in business and market decisions in recent years. Sentiment classification is an important classification issue that can be solved using different lexicon-based and deep neural network-based classifiers. While the sentence is positive or negative, the extracted emotion represents the sentence information. Despite the promising findings, most existing sentiment analysis techniques rely on a single dataset and are trained directly on the

I. Woungang et al. (Eds.): ANTIC 2021, CCIS 1534, pp. 297–306, 2022.
https://doi.org/10.1007/978-3-030-96040-7_24

target dataset. The training dataset, i.e., user information, may be influenced by security issues, data-specific sentiment analysis may be dangerous. Furthermore, because the model's architecture and training are data-specific, and it's difficult for the model to directly transfer information from the training dataset to the testing dataset, especially if the datasets are leaked. It implies that a well-trained model on the training dataset may perform poorly on the testing dataset, implying that existing proposed models are not resilient.

In brief, we can claim that the previous sentiment analysis approaches, which were employed to model on a dataset, is not friendly to the model security. Using federated learning-based model training for datasets will, on the surface, enhance the model's performance and resilience. As a result, using federated learning for sentiment analysis modelling makes sense. Federated learning is a promising ideology to unite isolated datasets for machine learning problems. In the federated learning framework, no raw data are exchanged among participating entities. Instead, parameter gradients and aggregated communicated between servers during collective optimization. As a result, institutes can collaborate without revealing private information by contributing their data collection to the training of a unified model. This functionality is beneficial when dealing with sensitive data such as personal preferences, financial transactions, medical information, and so on. Furthermore, more business-to-client model training apps are gaining widespread attention. Aside from this business-to-client collaboration, there are other intriguing uses across institutions like Medical image analysis, smart retail, fraud detection etc.

Traditional Machine learning approaches require huge amount of data and training to obtain accurate results. Our approach is more flexible in the sense that training of a model happens without actually 'seeing' the data. This addresses issues namely data privacy by encrypting the data before training. Also since the training is decentralized, many devices mainly can simultaneously train the model and the changes from all devices are applied back to the model. Thus the model keeps updating as and when training happens in a new device. We apply this approach in 'Sentiment Analysis of Twitter Data'. What we aim to achieve is distributing the data which in this case are tweets according to users and training a 'Sentiment Analysis' model which is a Long Short Term Model over this distributed data. This trained model is distributed across a client pool and every time new data generated by the client is used to train the model and the updates metrics and statistics are updated in the original model. This updated model is then distributed to the client pool. This ensures the user's private data which in this case are user tweets, are securely trained at the client end and the results are however updated over the entire client pool.

Authors address the following three task in the article:

a. To implement 'Sentiment Analysis' of Sent140 data which is a collection of over 1.6 million tweets by using Federated Recurrent Neural Network architectures.
b. To use Tensorflow's Federated Core package and design architectures, train models and simulate Secured Learning.
c. To develop a dashboard like web interface to visualize the simulation of the entire backend.

2 Literature Review

Federated learning is a novel approach which allows mobile devices to collaboratively learn a shared prediction model while keeping all the data on the particular device, so that machine learning can be done without storing data in the cloud. Federated learning was first introduced by Google for the same purpose. In [1], researchers discussed a system as designed for Federated learning for mobile devices based on TensorFlow, Federated learning is the technique used to update the central model with the resulting metrics. In [2], they propose ways to decrease the uplink communication costs. They experimented both Convolutional and Recurrent Neural Networks and concluded that their approach decreased the cost by an order of two magnitudes. In [3] they showcase the efficiency of their approach simulated from two popular image classification datasets. They conclude that RNNs and CNNs are the most efficient approaches in their implementation. Federated learning was earlier experimented with various types of kernels in Support Vector Machines. Variations of federated learning like Horizontal Federated Learning and Vertical Federated Learning were also explored to facilitate efficient and secure sharing of data between firms as seen in [12–14], and [15].

Nowadays people express voraciously on social media platforms. One such platform is Twitter which is 'opinion rich' platform. Hence the tweets data from twitter is widely used for sentiment analysis. One such dataset used is 'Sent140 twitter data' which is a collection of over 1.6 million tweets. In [4], data from about 1.6 lakh tweets from Thinknook website concludes that LTSMs are best suited for 'Sentiment Analysis'. In [5], tweets related to the Typhoon Yolonda were scraped to survey the emotions of the residents of the Philippines towards the typhoon. The paper uses bidirectional Recurrent Neural Networks. They conclude that 51% of the tweets were positive. In Both the papers each tweet is considered as a sentence and each sentence is comprised of words. Each word is converted into vectors and is used for training. Sentiment analysis can also be applied to detect bots on social media platforms like [6, 7]. In [8] the text document was converted into image and then Convolutional Neural Network was used to detect sentiment. Recurrent Neural Networks are also used to identify sentiment like in [9]. An ensemble model of CNN and RNN was experimented in [10]. Most popularly RNN was used with Long Short Term Memory as demonstrated in [17]. A different approach integrating Deep Learning and Grammar rules can be seen in [11].

3 Proposed Methodology

We have developed a web based dashboard to simulate the working of the backend and for testing the model. The entire structure of our project is as follows:

Training and Testing

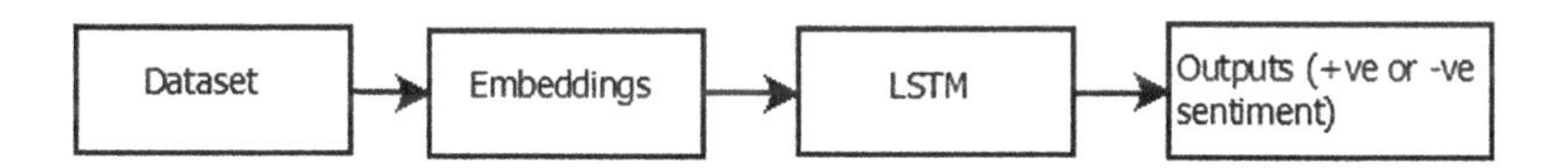

Prediction

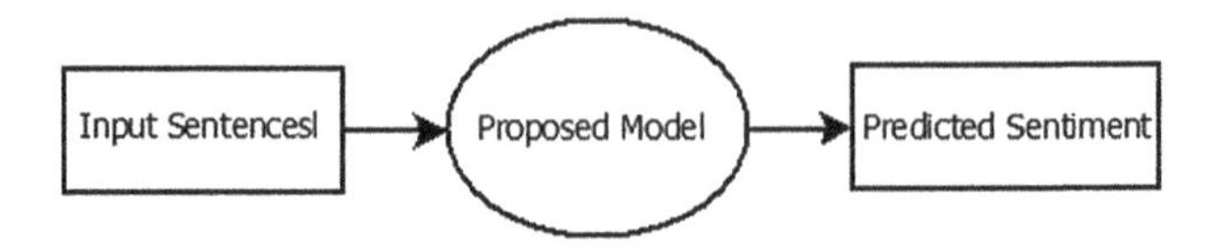

Proposed Algorithm

Algorithm 1: Ensuring security of user data using federated learning for the sentiment classification
Input: path of folder containing Twitter Dataset
Output: Accuracy of the proposed model

1. *Client pool is identified*
2. *select subset of clients satisfying parameters*
3. *select required clients*
4. *Send the model and train on device (client)*
5. *send stats and metrics to central model*
6. *delete models in the clients*
7. *update the central model*
8. *send the updated model to the clients*
9. *use this for next prediction*

4 Dataset Description and Data Pre-processing

The dataset we used in our project is the Sentiment140 dataset obtained from twitter API. This is a diverse collection of about 1,600,000 i.e., 1.6 million labelled tweets. The tweets are labelled as either 0 or 4 which are negative and positive tweets respectively. We extracted about 1008 for training from the train dataset of Sentiment140 dataset. The parameters considered for selecting the tweets are: We selected users having an average number of tweets i.e., 140–200 tweets per user. A total of 42 users were obtained who are considered as the client pool for simulation. From each of these 42 clients, we selected 24 tweets each with balanced number of positive and negative tweets. Thus we obtain 1008 tweets. For testing we selected 450 tweets which are the tweets of these 42 clients in the test dataset. The reason we chose a rather small number of tweets both for training and due to the limited computational resources we have. The accuracy and efficiency of our architecture can be fairly improved by increasing the number of tweets for training provided if we have better computational capabilities. We set the limit of 200 tweets per user because the dataset is crawling with a large number of 'Bot' users who have more than 500 identical tweets which are redundant to train. This dataset being in raw text form, however, can't be passed onto the model directly. It needs to undergo some pre-processing before it is training ready. We use the following pre-processing steps.

As it is common for the users to include URLs frequently in the tweets, we first identify these URLs from a tweet using regex and remove them. Then we remove the twitter handles which starts with '@'. We remove all the numbers. Then we tokenize the sentences using 'tweet tokenizer' function available in Natural Language Tool Kit (NLTK) library [16]. It also reduces the count of letters in a particular word if it occurs more than three times together in that word. For example, the word 'hellooooo' will be reduced to 'hellooo'. Next, all the words whose length is less than three are removed. Then we convert the data into Hierarchical Data Format (HDF) which is commonly used to store large amounts of scientific data in nested dictionaries. The tweets extracted were pre-processed to remove all numeric and special characters so that only text is remaining of the tweets. These tweets were converted to vectors for the prediction model. These steps are applied to both train set and test set.

5 Results and Discussions

In this section, we illustrate what the model can accomplish and how it works. A simulation which shows interaction between client models and server model and predictions is very necessary. The following simulation is shown in Fig. 1.

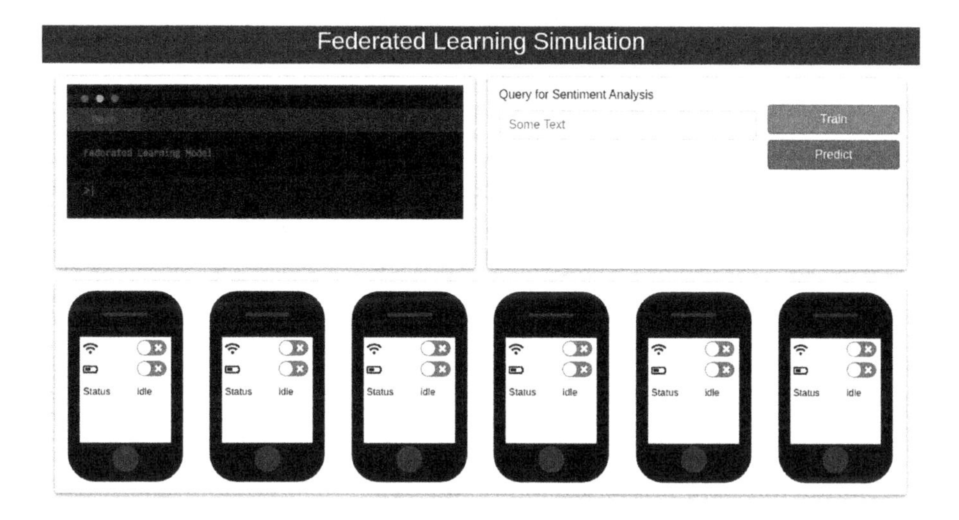

Fig. 1. Federated learning simulation dashboard

The simulation is divided into three parts as shown in the figure above.

1. Device Simulation.
2. Server Simulation.
3. Prediction Interface.

And also the above web application shows the interactions between these entities in a synchronized fashion.

5.1 Device Simulation

The device here can be anything like a mobile, IOT device, controller etc.., but for simplicity we are taking a mobile device. The following Fig. 2 shows the device simulation.

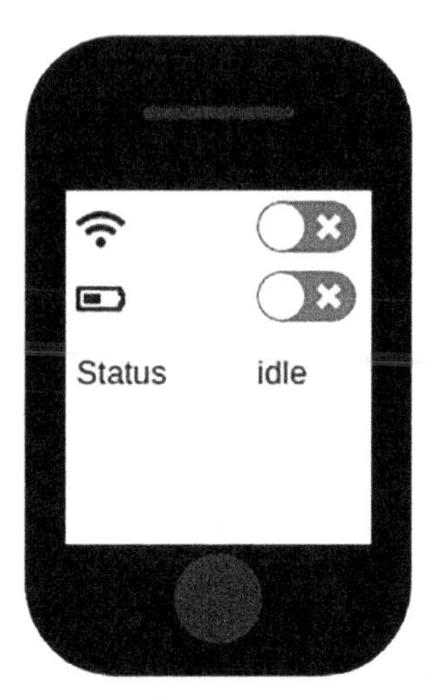

Fig. 2. Device simulation

There are three important requirements for federated learning to work. They are: internet should be turned on. The device should be plugged to a charging point. It has enough space for running the model. This is the initial state of device with internet and charging turned off. When all the above three conditions satisfies it goes into the state like this.

Fig. 3. Ready device

This Fig. 3 shows that the device is ready and can participate in federated learning. We are taking a default condition that every phone has enough memory to train a sentiment analysis model. Once the phone is in ready mode it can move in next mode on pressing train button. This mode lasts until the completion of training of model in local area of mobile and stops and get back to previous state on completion.

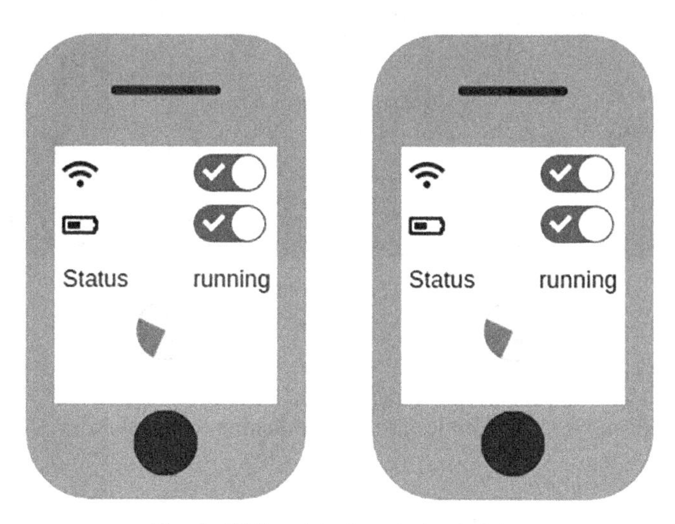

Fig. 4. Different devices during training

This training time depends on the size of the dataset and also processing power of the device and it varies from device to device like in Fig. 4. Once the training is done in the devices they send their updates to main model running in the server.

5.2 Server Simulation

The server shown in Fig. 5 is also simulated to get an idea of what's happening inside it. The main model applies federated averaging algorithm on all the updates received from its clients.

Once the instance of model is trained it sends the updates to all its clients for synchronizing and better performance of client models. It will take a good amount of time period to train the model from updates of all the clients.

5.3 Prediction Interface

The prediction of output can be done any time after the model is saved. After each iteration of the above processes the model accuracy will be increased to a small extent. For high output through federated learning we need a large set of clients. Below we project and describe our results and hyper parameters in detail in Table 1. We use categorical cross entropy loss and categorical accuracy as metrics to evaluate our model. Considering the first encounter between federated learning and sentiment analysis we got encouraging results which can be further improved by proper tuning of parameters and experimenting with various other Neural Network architectures.

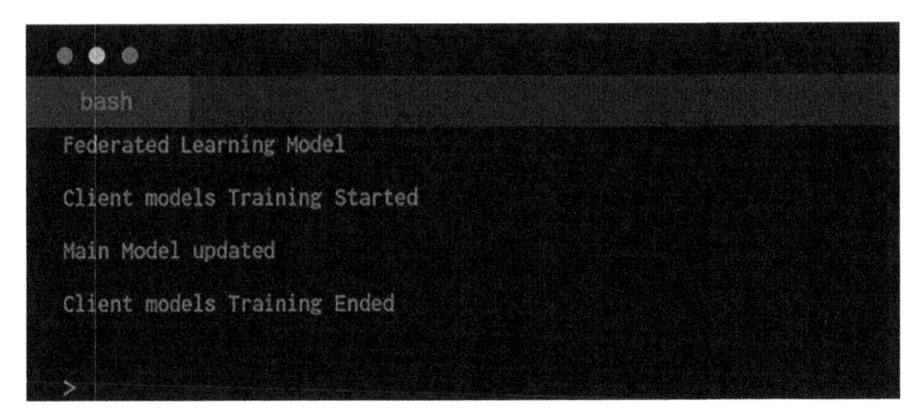

Fig. 5. Server simulation

5.4 Parameters

The parameters and the setup values taken in evaluating the results are:

Table 1. Hyper parameters

Epochs	10
Optimizer	SGD(stochastic gradient descent)
Rounds	10
Learning rate	0.02
Loss fn	Sparse categorical cross entropy loss
Metrics	Sparse categorical accuracy

The optimizer used is SGD with 10 epochs and with sparse categorical cross entropy loss function. The learning rate applied is 0.02. The final accuracy achieved is 74%. The test accuracy of 74% with minimal tuning of the parameters is good sign going forward for researchers. As it can be seen with this much of accuracy, the metrics from the evaluation are encouraging for training a Neural Network for sentiment analysis task in a relatively new federated setting to produced desired results. The loss shown in Fig. 6 can be further optimized and the accuracy can be further boosted by experimenting with different architectures and parameters. We used a limited amount of data because of our computational restrictions which happens to be a contributing factor for the less accuracy as Neural Networks are data hungry. Because with the increase in the training data, their capability to generalize increases. Hence using more data might boost the results further.

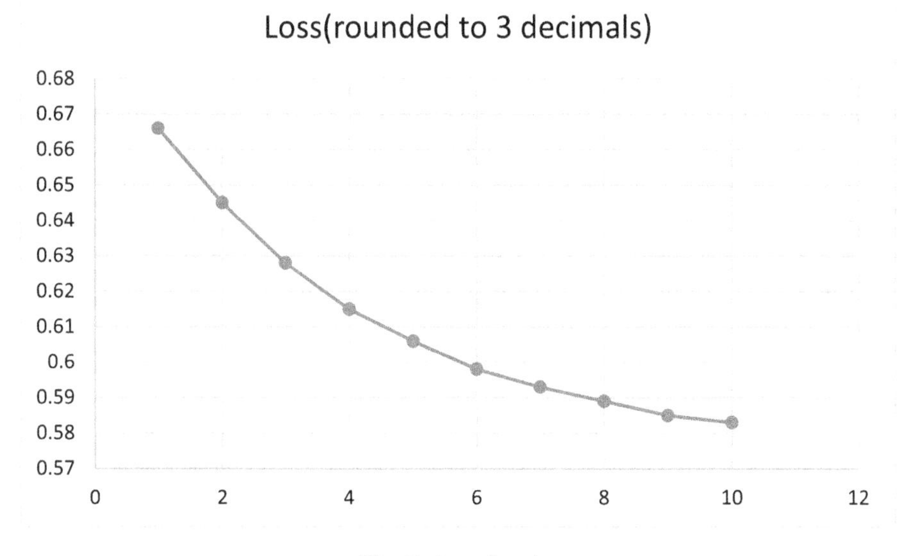

Fig. 6. Loss function

6 Conclusions

Federated learning offers researchers a viable framework for securing users' data during model training without exposing data. Unlike earlier research, which has focused on data security before model training, this study focuses on privacy problems during training, which is more suited for federated learning. Sentiment analysis, a traditional NLP job, is implemented on the Federated Learning platform. Extensive testing has revealed that the hyper parameters configuration significantly influences the Federated Learning models' performance. As the problem is relatively new and on a final note, we can say that federated learning is still under experimental stage and it has a bright future in various applications. Machine learning models on private data like photos, documents, etc. can be done in a federated fashion protecting user privacy. Federated learning should be applied in other areas of machine learning like Regression, Support Vector Machine etc. It should be experimented with light weight machine learning models and be supported on devices which have less memory and computation resources. Federated learning implementations like horizontal federated learning and vertical federated learning can be implemented in future for employing more secure learning across devices. Different architectures like the combination of RNN and CNN can be used to improve the accuracy of the model.

References

1. Bonawitz, K., et al.: Towards federated learning at scale: System design. arXiv preprint. arXiv: 1902.01046 (2019)
2. Ro, J., et al.: Communication-efficient agnostic federated averaging. arXiv preprint. arXiv: 2104.02748 (2021)

3. Yurochkin, M., et al.: Bayesian nonparametric federated learning of neural networks. In: International Conference on Machine Learning. PMLR (2019)
4. Reddy, D.M., Reddy, N.V.: Twitter sentiment analysis using distributed word and sentence representation. arXiv preprint. arXiv:1904.12580 (2019)
5. Imperial, J.M., et al.: Sentiment analysis of typhoon related tweets using standard and bidirectional recurrent neural networks. arXiv preprint. arXiv:1908.01765 (2019)
6. Kudugunta, S., Ferrara, E.: Deep neural networks for bot detection. Inf. Sci. **467**, 312–322 (2018)
7. Schmitt, M., et al.: Joint aspect and polarity classification for aspect-based sentiment analysis with end-to-end neural networks. arXiv preprint. arXiv:1808.09238 (2018)
8. Merdivan, E., et al. Image-based natural language understanding using 2d convolutional neural networks. arXiv preprint arXiv:1810.10401 (2018)
9. Shrestha, N., Nasoz, F.: Deep learning sentiment analysis of amazon.com reviews and ratings. arXiv preprint. arXiv:1904.04096 (2019)
10. Minaee, S., Azimi, E., Abdolrashidi, A.A.: Deep-sentiment: sentiment analysis using ensemble of CNN and bi-LSTM models. arXiv preprint. arXiv:1904.04206 (2019)
11. Dashtipour, K., et al.: A hybrid Persian sentiment analysis framework: integrating dependency grammar based rules and deep neural networks. Neurocomputing **380**, 1–10 (2020)
12. González-Serrano, F.-J., Navia-Vázquez, Á., Amor-Martín, A.: Training support vector machines with privacy-protected data. Pattern Recogn. **72**, 93–107 (2017)
13. Yu, H., Jiang, X., Vaidya, J.: Privacy-preserving SVM using nonlinear kernels on horizontally partitioned data. In: Proceedings of the 2006 ACM Symposium on Applied Computing (2006)
14. Rahulamathavan, Y., et al.: Privacy-preserving multi-class support vector machine for outsourcing the data classification in cloud. IEEE Trans. Depend. Secur. Comput. **11**(5), 467–479 (2013)
15. Vaidya, J., Hwanjo, Y., Jiang, X.: Privacy-preserving SVM classification. Knowl. Inf. Syst. **14**(2), 161–178 (2008)
16. Loper, E., Bird, S.: Nltk: The natural language toolkit. arXiv preprint. cs/0205028 (2002)
17. Sundermeyer, M., Schlüter, R., Ney, H.: LSTM neural networks for language modeling. In: Thirteenth Annual Conference of the International Speech Communication Association (2012)

Gender Recognition by Voice Using Machine Learning

Rohit Bhatia(✉) and Nagendra Pratap Singh(✉)

National Institute of Technology Hamirpur, Hamirpur, India
{cs16mi519,nps}@nith.ac.in

Abstract. Recently, gender recognition has become an important area of research. Gender recognition can be used in various fields like for security purposes, speaker identification and speaker recognition. Various techniques has been used to identify the gender of person like using facial analysis, voice identification, machine learning, deep learning, using features like LPC and MFCC. This paper deals with identifying the gender using Acoustic properties of voice using Machine learning and how the accuracy vary when dataset goes through different transformation. It shows that we achieve maximum accuracy when uniform transformation is applied on dataset in case of KNN and SVM both.

Keywords: Acoustic features · MFCC · LPC · KNN · SVM · Uniform distribution · Normalization · Gaussian distribution

1 Introduction

Voice signal has been using for different purposes like speech recognition, speaker identification and verification, gender classification. Different fields requires different types of properties of signals for evaluation. Voice signals are preprocessed to get accurate parameters for evaluation. Preprocessing may contains noise cancelling, windowing, framing, FFT, STFT. Different types of features are there like Acoustic properties, LPC, Mel-frequency cepstral coefficients (MFCC) now has been greater influence in the field of voice signals. Parameters extracted can be used integratedly as feature vector to analyse the voice signals. Acoustic properties of voice signal plays major role in identifying the person's gender.

Acoustic parameters of voice may vary according to sample characteristic's settings like intensity, time duration, frequency and filtering. These acoustic properties are keys that can be used to detect gender of speaker using voice.

The voice is considered as an effective communication method which consist of unique and para linguistic features like age, accent, gender, language and emotional state. They are of unique nature from all other sound waves because every wave comprises of a different frequency. Identification of human gender using voice is always been a tedious task for sound scientists who bring into play various softwares like customer relationship management (CRM) strategy systems which depends on gender. Now, to recognize gender using voice, a very

I. Woungang et al. (Eds.): ANTIC 2021, CCIS 1534, pp. 307–318, 2022.
https://doi.org/10.1007/978-3-030-96040-7_25

important step is ADC i.e. Analog digital conversion, which is required to get important features from a sound wave. An analog to digital converter (ADC), converts any analog signal into computable data, which helps processing, storing, and making it accurate and reliable by error minimization. Therefore, accuracy of voice features plays a important role in improving the classifier's efficiency because sound signals may contains background noise which will lead to inaccurate parameters. therefore, removal of non useful features from voice signal becomes an important task before constructing a classifier.

1.1 Types of Algorithms

Text Dependent: In Text Dependent Algorithms, the text used in testing phase is subset of text used during training. Models are trained on some specific text and the same text is used during testing phase.

Text Independent: In Text Independent Algorithms, the text used in training is independent of text used during testing. Models are trained on the basis of universal speech factors rather than predefined text.

1.2 Gender Recognition Basics

Any voice related model will have 3 basic requirements:

Vocabulary: Vocabulary plays an important role in speaker recognition models. Vocabulary is a collection of words that a computer can recognise efficiently from speech.

Utterance: Now, different words can have different accents of speaking which varies from speaker to speaker. Therefore, Utterance is the ability to recognise a word in a variety of contexts (Accent)

Accuracy: Accuracy can be defined as how accurately, the system is capable of recognising the particular word accurately.

The rest of the paper is arranged as: Sect. 2 comprises of related work review of the past studies. Section 3 describes the proposed scheme which consist of dataset and libraries and performance analysis. In Sect. 5, there is result simulation and analysis and Sect. 6 comprises of conclusion.

2 Related Work

An Pitch based algorithm is proposed in [6] for the detection of gender using analysis of the non stationary behavior of voice signal. Concept of peak detection is used to get the pitch and the frequencies which are dominant in sound wave and then fundamental frequency are selected from them.

In [3], Two algorithms are provided to automatically detect speaker's gender using audial features generated from the particular's voice. Speech data oriented here, to improve algorithms is deducted from a large dataset. Only audial vowel details and fricatives used for developing and testing algorithms.

In [2] various attributes such as energy, mfcc and pitch are extracted to identify the gender. SVM classifier has been used to classify the features of voice. Data set consist of total of 280 voice files comprised of 140 male and female samples each.

In [10], A binary classification algorithm based on decision tree is used to classify the gender. Random data is converted into systematic data and tree sorting is used to determine gender.

In [1], intensity of utterances is considered as the key factor which can be used to differentiate between male and female voices. To get the intensity, the Simpson's rule of numerical integration is used by utilizing area under the normalized curve.

Authors in [5], proposed an efficient gender detection (GD) algorithm based on time domain. Authors has used Autocorrelation K-Means as a (ACF) as classifier. An efficient frame selection technique is used to get the pitch of speech signal.

In [8], there is a novel method for feature extraction from audio signal in order to find out the gender as male or female. At initial stage, data pre-processing is performed for the noise-less plain data. Then, this processed data is fed to a multi-layer architecture model to get the parameters. First layer is related to calculation of fundamental frequency using autocorrelation function, spectral entropy and flatness and mode frequency. Another layer is related to mapping of pre-processed data into suitable range using linear interpolation function to and get MFCC from data. Different types of datasets are used: RAVDESS, TIMIT and BGC which is self created and the used machine learning classifiers are: KNN and SVM.

In [9], a system is designed to identify the gender of human, which is text independent, is developed. MFCC coefficients are extracted from voice signals using GMM. In this research, Gaussian mixture number, MFCC coefficients and their effects is analyzed.

In [4], Authors introduces a stacked machine learning algorithm to find the gender on the basis of the audial features of signal and then there is comparison of existing classifiers as CART, Random forest and Neural Network and their performances.

In [7], there is a comparison of performances of various pitch detection algorithms. In this paper, the author has discussed various Types of Pitch Detectors and problems associated with them and also various pitch detection algorithms.

3 Methodology

3.1 Overview

Our task is to record a voice, then extract features from it, then apply algorithm to find out if the voice belongs to a male or a female
Given: A voice recorded through a microphone
Predict: If the voice belongs to a male or a female

3.2 Data Preparation

On acoustic signals, Specan measures 29 acoustic parameters or characteristics for which the start and end times must be specified. Dataset used here comprises of 21 Acoustic Voice Parameters which are used to trained the model and feature Matching. Data set preparation comprises of Data collection, Data preprocessing Front-end processing: - this step related to the—signal processing part, which converts the continuous signals into discrete signals. removes the noise or background noise from signals using different techniques

3.3 Feature Extraction

The process of extracting different features from a voice signal, such as power, pitch, and vocal tract structure, is known as feature extraction. The process of transforming these features into signal parameters via differentiation and concatenation is known as parameter transformation. This step for extracting the acoustic features from signals of speakers to construct set of feature vectors for use in training and testing phases. The extracted set of features vectors will be stored in database which will be used further for decision making.

3.4 Decision-Making

This stage determines the speaker's identification by comparing unknown feature vectors to the whole pool of models and selecting the model that best matches the unknown feature vectors. Two models has been used in this study: KNN and SVM. Performance of classifiers is compared when dataset goes through different transformations.

3.5 Dataset and Libraries

Dataset used in this paper consist of 21 Acoustic parameters of voice signal and combining them to analyze the gender of person using different machine learning algorithms. Dataset is in the form of CSV file and contains 3168 rows and 21 columns where the last column is label as male or female. Libraries used for performing experiment are data analysis libraries like pandas, numpy and scikit (Fig. 1).

3.6 Components of a Typical Automatic Speaker Recognition System

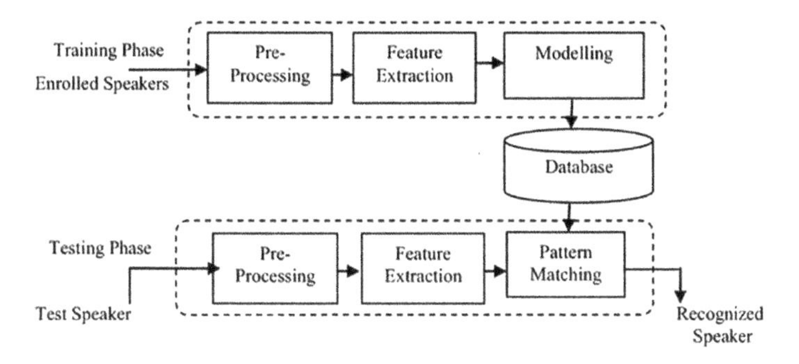

Fig. 1. Speaker recognition system

3.7 Features

Features used are 21 Acoustic Parameters which are used to identify the voice as male or female. Overview of dataset can be seen at https://www.kaggle.com/primaryobjects/voicegender.

4 Performance Analysis

4.1 KNN Analysis

Dataset is divided into 30 test set and 70 training set. When KNN is applied on Normal dataset, Figure below shows the performance of KNN on Dataset. The graph shows the training and testing accuracy at different values of k. According to graph it shows the maximum Testing accuracy of 70.76% at k = 8.

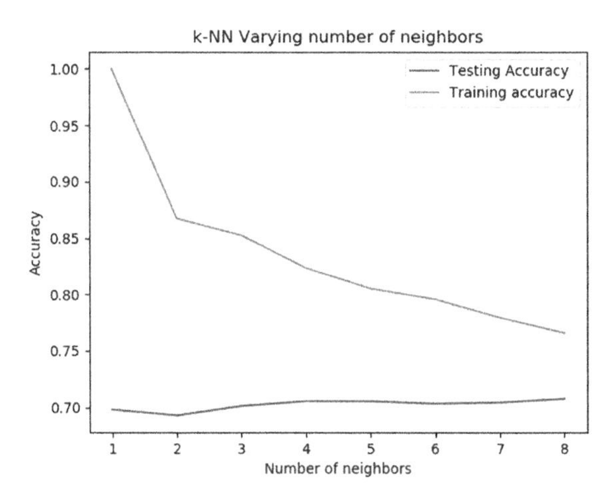

Fig. 2. Training and testing accuracy for different values of k on normal dataset.

4.2 Uniform Distribution

When KNN is applied on Uniformly distributed dataset, It shows maximum accuracy of 97.47% at k = 1.

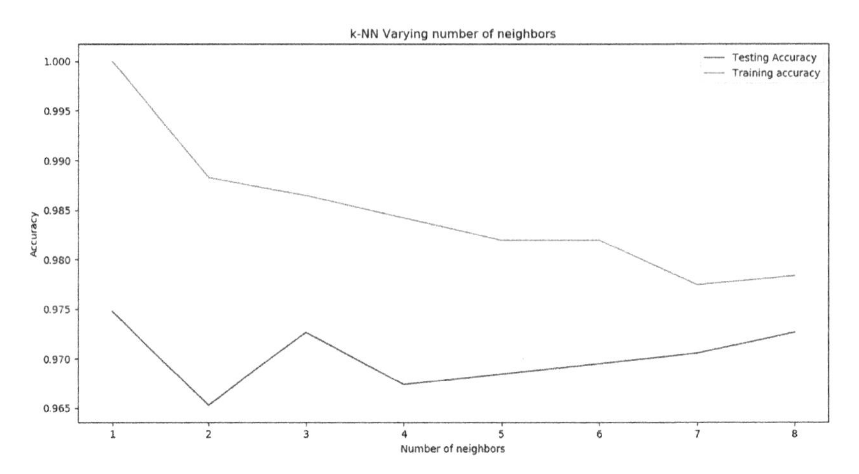

Fig. 3. Training and Testing accuracy for different values of k on uniformly distributed dataset.

4.3 Normalization

When Normalization is applied on dataset, it shows the maximum Testing accuracy of 78.75% at k = 3.

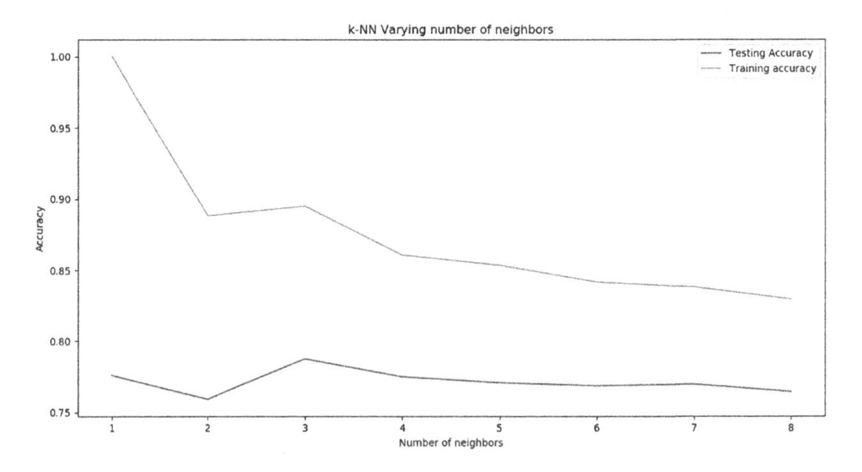

Fig. 4. Training and testing accuracy for different values of k on normalized dataset.

4.4 Gaussian Distribution

Performance of Gaussian distribution is compared below. It shows maximum accuracy of 77.91% at k = 8.

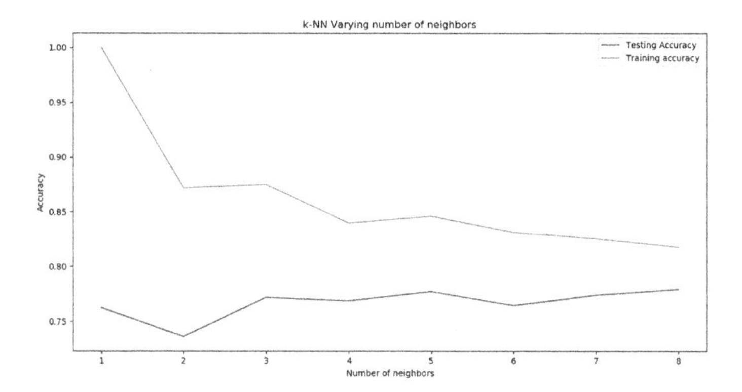

Fig. 5. Training and testing accuracy for different values of k on Gaussian dataset.

5 Simulation Result and Analysis

5.1 Results for Normal KNN Is Given Below

As according to the graph in Fig. 2, we can see that how the accuracy varies of model when KNN is applied on normal data set for different k. According to graph, it shows the maximum accuracy at k = 8. Table below shows the result at k = 8 (Table 1).

Table 1. Result for normal KNN at k = 8

.....	Precision	Recall	f1-score	Support
0	0.71	0.71	0.71	476
1	0.71	0.71	0.71	475
Micro avg.	0.71	0.71	0.71	951
Macro avg.	0.71	0.71	0.71	951
Weighted avg.	0.71	0.71	0.71	951

5.2 Results for Uniform Distribution Is Given Below

As according to the graph in Fig. 3, we can see that how the accuracy varies of model when KNN is applied on uniformly distributed dataset for different k. According to graph, it shows the maximum accuracy at k = 1. Table below shows the result at k = 1 (Table 2).

Table 2. Result for uniformly distributed dataset KNN at k = 1

.....	Precision	Recall	f1-score	Support
0	0.97	0.97	0.97	476
1	0.97	0.97	0.97	475
Micro avg.	0.97	0.97	0.97	951
Macro avg.	0.97	0.97	0.97	951
Weighted avg.	0.97	0.97	0.97	951

5.3 Results for Normal Distribution Is Given Below

As according to the graph in Fig. 4, we can see that how the accuracy varies of model when KNN is applied on normalize dataset for different k. According to graph, it shows the maximum accuracy at k = 3. Table below shows the result at k = 3 (Table 3).

Table 3. Result for normalized KNN at k = 3

.....	Precision	Recall	f1-score	Support
0	0.81	0.76	0.78	476
1	0.77	0.82	0.79	475
Micro avg.	0.79	0.79	0.79	951
Macro avg.	0.79	0.79	0.79	951
Weighted avg.	0.79	0.79	0.79	951

5.4 Results for Gaussian Distribution Is Given Below

As according to the graph in Fig. 5, we can see that how the accuracy varies of model when KNN is applied on normal dataset for different k. According to graph, it shows the maximum accuracy at k = 8. Table below shows the result at k = 8 (Table 4).

Table 4. Result for Gaussian distributed dataset KNN at k = 8

.....	Precision	Recall	f1-score	Support
0	0.77	0.79	0.78	476
1	0.78	0.77	0.78	475
Micro avg.	0.78	0.78	0.78	951
Macro avg.	0.78	0.78	0.78	951
Weighted avg.	0.78	0.78	0.78	951

5.5 Results for SVM on Normal Dataset

Dataset is divided into 30 test set and 70 training set. When SVM is applied on Normal dataset, Data below shows the performance of SVM on Dataset. Simulation has benndone to find out the best parameters i.e. C and g of SVM on which the model will perform best. Hence, all the results shown further are best in respective distribution (Table 5).

Table 5. Result for support vector machine on normal data set

.....	Precision	Recall	f1-score	Support
0	0.98	0.88	0.93	476
1	0.89	0.98	0.94	475
Macro avg.	0.94	0.93	0.93	951
Weighted avg.	0.94	0.93	0.93	951

Hence, Analyzing from the table it can be seen that SVM on dataset shows the maximum accuracy of 93.37% for c = 1.4 and gamma = 0.1.

5.6 Results for SVM on Normalized Dataset

Table 6. Result for support vector machine on normalized data set

.....	Precision	Recall	f1-score	Support
0	0.67	0.76	0.71	476
1	0.72	0.63	0.68	475
Macro avg.	0.70	0.70	0.69	951
Weighted avg.	0.70	0.70	0.69	951

Hence, Analyzing from the table it can be seen that SVM on Normally distributed dataset shows the maximum accuracy of 69.61% for c = 1.4 and gamma = 1.4 (Table 6).

5.7 Results for SVM on Uniformly Distributed Dataset

Table 7. Result for support vector machine on uniformly distributed data set

.....	Precision	Recall	f1-score	Support
0	0.99	0.99	0.99	476
1	0.99	0.99	0.99	475
Macro avg.	0.99	0.99	0.99	951
Weighted avg.	0.99	0.99	0.99	951

Hence, Analyzing from the table it can be seen that SVM on uniformly distributed dataset shows the maximum accuracy of 98.63% for c = 1.4 and gamma = 1.3 (Table 7).

5.8 Results for SVM on Gaussially Distributed Dataset

Table 8. Result for support vector machine on Gaussially distributed data set

.....	Precision	Recall	f1-score	Support
0	0.99	0.98	0.98	476
1	0.98	0.99	0.98	475
Macro avg.	0.98	0.98	0.98	951
Weighted avg.	0.98	0.98	0.98	951

Hence, Analyzing from the table it can be seen that SVM on Gaussially distributed dataset shows the maximum accuracy of 98.31% for c = 1.3 and gamma = 0.1 (Table 8).

5.9 Comparison

Distribution	Value of C	Value of G	Accuracy
No distribution	1.4	0.1	93.37%
Normal distribution	1.4	1.4	69.61%
Uniform distribution	1.4	1.3	98.63%
Gaussian distribution	1.3	0.1	98.31%

6 Conclusion and Future Scope

6.1 Introduction

The aim of dissertation was to find out how the Accuracy Varies for gender recognition through voice on different values of k when dataset goes through different operations using KNN. In Voice operations, gender recognition is a critical task Because Identifying gender through voice Parameters, which change according to environment factors makes it more vulnerable to wrong predictions. Because some of parameters may change when the environment condition changes, so, voice pre-processing becomes a very important task to extract the parameters accurately. There are plenty of techniques which has been developed for more accurate results. The dataset used in this dissertation which includes set of acoustic parameters can be used to detect the gender of a person using voice. moreover, it can be seen that the extraction of these parameters is fully dependent on the background noise. if there is less noise in background, parameters can be extracted efficiently and hence, accuracy will be increased. Gender recognition has become a popular area of research in voice due to its wide variety of applications like security purposes, Speaker identification, Military applications. This chapter summarises all of the work done during the dissertation and examines the field's future prospects.

6.2 Conclusion

A detailed introduction to Gender Recognition was discussed in Chap. 1. We discussed about how an Audio signal is understand by machine for machine learning applications of voice. In this chapter, we see a typical representation of sound wave and types of audio features for machine learning. In Chap. 2, we discussed about some techniques which already has been implemented. Then in Chap. 3, we discussed about the approach used and the steps we follow to reach a conclusion. It contains Data preparation, Feature extraction, speaker database, Decision making and the libraries. A typical Automatic speaker recognition system is shown. Pre processing Part of Voice contains feature engineering, Fourier transform and feature extraction. then there is set of acoustic features used for model and classifier is discussed and working of classifier is shown.

Then later, Types of Distribution is discussed which contains Normal Distribution, Uniform distribution and Gaussian distribution and why they are important.

In later chapter, we see the performance of all distributions and it comes out that in case of KNN, of all distributions, Uniform distribution shows maximum accuracy of 97.47% at $k = 1$ and for SVM, again Uniform distribution shows maximum accuracy of 98.63%.

6.3 Future Scope

In the dissertation, it is stated that, Gender detection has been the topic of a great deal of study. Different types of gender detection techniques are in focus now like using facial expressions and voice analysis. MFCC's are most popular features in gender recognition using voice. MFCC's can be considered as future of voice.

References

1. Alsulaiman, M., Ali, Z., Muhammad, G.: Gender classification with voice intensity. In: 2011 UKSim 5th European Symposium on Computer Modeling and Simulation, pp. 205–209 (2011)
2. Chaudhary, S., Sharma, D.K.: Gender identification based on voice signal characteristics. In: 2018 International Conference on Advances in Computing, Communication Control and Networking (ICACCCN), pp. 869–874 (2018)
3. Childers, D.G., Wu, K., Bae, K.S., Hicks, D.M.: Automatic recognition of gender by voice. In: ICASSP-88: International Conference on Acoustics, Speech, and Signal Processing, vol. 1, pp. 603–606 (1988)
4. Gupta, P., Goel, S., Purwar, A.: A stacked technique for gender recognition through voice. In: 2018 Eleventh International Conference on Contemporary Computing (IC3), pp. 1–3 (2018)
5. Kumari, M., Ali, I.: An efficient algorithm for gender detection using voice samples, pp. 221–226, November 2015
6. Kumari, M., Talukdar, N., Ali, I.: A new gender detection algorithm considering the non-stationarity of speech signa, pp. 141–146, November 2016
7. Rabiner, L., Cheng, M., Rosenberg, A., McGonegal, C.: A comparative performance study of several pitch detection algorithms. IEEE Trans. Acoust. Speech Signal Process. **24**(5), 399–418 (1976)
8. Uddin, M., Hossain, Md., Pathan, R., Biswas, M.: Gender recognition from human voice using multi-layer architecture, September 2020
9. Yücesoy, E., Nabiyev, V.V.: Gender identification of a speaker using MFCC and GMM. In: 2013 8th International Conference on Electrical and Electronics Engineering (ELECO), pp. 626–629, (2013)
10. Zhong, B., et al.: Gender recognition of speech based on decision tree model. In: Proceedings of the 3rd International Conference on Computer Engineering, Information Science Application Technology (ICCIA 2019), pp. 571–577. Atlantis Press, July 2019

An AI Driven Approach for Multiclass Hypothyroidism Classification

Riju Das, Sachin Saraswat, Divyansh Chandel, Somnath Karan, and Jyoti Singh Kirar(✉)

Banaras Hindu University, Varanasi, India

Abstract. Hypothyroidism is a condition when the thyroid gland produces less hormone than a normal range. As the symptoms of hypothyroidism are not clear, in past days many researchers have tried to work on fine-tuning hypothyroidism detection techniques. In this work, we have proposed a multiclass hypothyroidism detection technique using machine learning methods. Exploratory data analysis-based feature selection has proven helpful in eliminating irrelevant features for classification. Different classifiers have been experimented and compared to find the best-fitted classifier among the experimented ones for multiclass hypothyroidism classification. Experimental results on publicly available dataset shows the higher efficacy of the proposed method.

Keywords: Hypothyroidism · Decision tree · K nearest neighbor · Random forest · Support vector machine

1 Introduction

Thyroid diseases are increasing at an alarming rate nowadays. With the high rate of increase in thyroid patients, it is important to detect thyroid diseases accurately and fast. It is to be noted that nearly 33% of the Indian population suffers from various kinds of thyroid disorders. In addition to that, the female population [1] is 3–10 times more prone to different kinds of thyroid diseases than men. Females belonging to the age group [2] of 46–54 are most affected by hypothyroidism.

The thyroid gland is a butterfly-shaped endocrine gland situated on the front side of the neck that is responsible for the secretion of thyroid hormone. This hormone plays an important role in various important body functions like breathing, body weight control, heart rate, effective muscle strength, etc. Abnormal secretion of this hormone leads to various kinds of thyroid disorders. Thyroid disorders can cause high blood sugar, higher cholesterol level, anxiety, low fertility levels, and cardiovascular disorders. So, thyroid diseases are of grave concern. In India, about 50% of total cases of hypothyroidism remain undetected for years.

The seed of curing these disorders lies in the detection of these diseases at the right time. Infection trauma and stress are the most common factors that dictate the abnormal secretion of the thyroid gland. If the thyroid gland makes more hormone than ordinary levels, many bodily functions speed up, resulting in a condition named Hyperthyroidism.

I. Woungang et al. (Eds.): ANTIC 2021, CCIS 1534, pp. 319–327, 2022.
https://doi.org/10.1007/978-3-030-96040-7_26

On the other hand, if the thyroid gland produces less amount of hormone, many bodily functions slow down, resulting in a condition named Hypothyroidism. One of the main reasons for this hypothyroidism is an autoimmune disorder called Hashimoto's disease [3]. Some symptoms of thyroid dysfunction are increase in thyroid-stimulating Hormone (TSH), a decrease in free T4 (FT4; thyroxine), weight gain, a decrease in appetite, slow pulse, fatigue, and decreased metabolism.

Hypothyroidism can broadly be classified into Primary, Secondary, and Compensated Hypothyroidism. In primary hypothyroidism [4], the thyroid is being stimulated properly but isn't able to produce enough thyroid hormones for the body to function properly. Hence, thyroid itself is the source of the problem. In secondary hypothyroidism [5], the pituitary gland is not stimulating the thyroid gland to produce enough hormones. In other words, the problem is not actually with the thyroid gland. Compensated hypothyroidism [6] is the combination of elevated serum thyrotropin and normal serum thyroxine.

Even at the current scenario, the laboratory-based detection of thyroid disorders is more time-consuming, less effective, complex, and requires more extensive knowledge and experience for detecting perfectly. Even then, the accuracy of that detection is questionable. There are many constraints in the accuracy of these techniques, such as the machinery having some specific longevity, and with the increase in time, the accuracy of the detection degrades. At the same time, if the technician has little experience, the result can be faulty. So, to remove these redundancies, machine learning methods can be very useful as they never compromise with accuracy; at the same time, minimal data and time is required to detect the diseased state.

Hypothyroidism is the most common thyroid disorder, and at the same time, incurable. Despite the lack of a complete cure, the disease can be kept in control if detected earlier. The symptoms of hypothyroidism are not well-defined or exclusive, for example, tiredness can easily be associated with sleep apnea or narcolepsy, resulting in detecting the actual disorder late, and recovery may not be possible for late detection. In the current work, we have tried to find out a better and more accurate machine-learning process to detect hypothyroidism at its early stages, for a better prognosis of the disease.

The rest of the paper is organized as follows. Section 2 discusses the literature survey. Section 3 elaborates the proposed method for hypothyroidism detection. Section 4 focuses on experimental results and Sect. 5 concludes the article with some research directions.

2 Literature Survey

Technology is improving in leaps and bounds every day. With the advancement in artificial intelligence, the accuracy of many works have been drastically increased. Medical science is not an exception. There are several machine learning algorithms such as random forest, decision tree, naïve Bayes, SVM, and ANN that are used extensively in the detection and prognosis of many diseases. As our concern is about thyroid diseases only, many scientists have proposed many machine learning algorithms also to the specific problem of the detection of thyroid diseases. Some ground-breaking foundational studies have been laid down by the past workers, like, Chen et al. have proposed an expert system consisting of 3 stages using the support vector machines (SVM) model for the thyroid disease diagnosis [7]. Ammulu and Venugopal made an exploratory data analysis using the random forest algorithm to predict hypothyroid disorder by taking the data from the UCI repository [8]. Shankar and Lakshman used optimal feature selection and kernel-based support vector machine models to classify thyroid patients [9]. Ionita has used various classification models such as Naive Bayes, Radial Basis Function Network, and Decision Tree, and proposed a comparative study in thyroid prediction [10]. Ahmed and Soomrani have proposed a Thyroid Disease Type Diagnostics framework that helps in cleaning the medical data such that physicians can diagnose thyroid disorders properly [11]. Presently, we have tried to analyze different classification techniques proposed by other scientists and make a comparative study in those classification techniques directed towards investigating and ameliorating this specific problem, and conclude the best fitting tool to model the detection of hypothyroidism and fill in the gaps still present. For this purpose, we have used four main classification techniques - Decision tree, K Nearest Neighbor classifier (KNN), Random Forest classifier, and Support Vector Machine classifier.

3 Proposed Method

Supervised learning is the basic method that we have adopted for our comparative study on hypothyroid detection algorithms. It is a technique directed towards designing artificial intelligence (AI). Essentially, in this method, some computer algorithm is trained with some specific input data (also called a vector) which has been designated for some specific output (also called the 'supervisory flag'). The model is thus trained like this, till it can ascertain the underlying relationships and patterns between the input data and output labels [12], consequently enabling it to harvest fairly accurate labeling results, even when it is presented with data that has never been seen before.

This method has been seen to be fairly able at solving problems on classification and regression, for example, determining the category to which a news article belongs, or predicting sales volume for some given date in the future. Hence, its application widely varies from healthcare research to marketing to global finances. Precisely, supervised learning aims to make sense of acquired or fed data within the settings of a specific question.

Figure 1 shows the flow diagram of the proposed model.

- Step 1: Dataset description, and selecting features based on relevance of the attributes.
- Step 2: Data preprocessing and cleaning has been performed on the publicly available dataset used in this work to eliminate the redundant features
- Step 3: Hypothyroidism detection/classification using relevant subset of features from step 1 into four classes

Class 0	Class 1	Class 2	Class 3
Compensated hypothyroid	Negative	Primary hypothyroid	Secondary hypothyroid

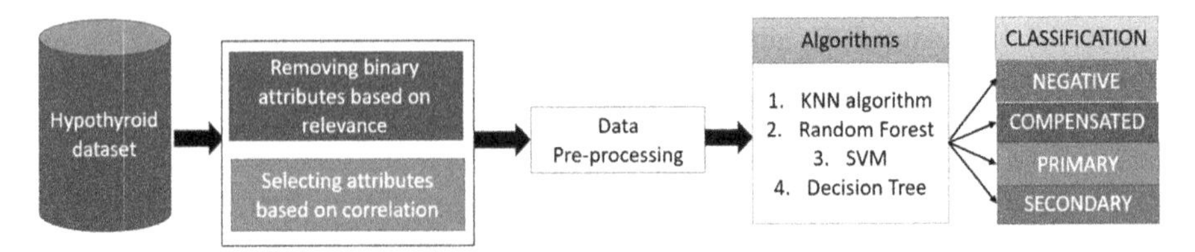

Fig. 1. Flow diagram of the proposed model

The detailed explanation of all the three steps are discussed in Sect. 4.

4 Experiments, Results and Discussion

4.1 Dataset Description

The data for this work on thyroid diseases have been taken from the UCI machine learning repository. The dataset consists of 3771 unique instances with 30 attributes, given by- age - Age of the person, sex - Male or Female, on_thyroxine - true or false, on_antithyroid_medication - true or false, sick - true or false,, thyroid_surgery - true or false, I131_treatment - true or false, query_hypothyroid - true or false, query_hyperthyroid -true or false, lithium - true or false, goiter - true or false, tumor - true or false, hypopituitary- true or false, psych - true or false, TSH_measured - true or false, TSH - thyroid stimulating hormone floating value, T3_measured - true or false, T3 - triiodothyronine value, TT4_measured- true or false, TT4 - Thyroxine value, T4U_measured- true or false, T4U - numerical value, FTI_measured- true or false, FTI -Free Thyroxine Index, TBG_measured- true or false, TBG - Thyroid-Binding Globulin value, referral_source - different sources of referrals, Class - different types of thyroid.

We have dropped the column of TBG as we have no known value in that column. Next, columns involving questionnaires about different kinds of tests performed or not. Such attributes are: for example, T4U_measured- true or false if T4U is measured, then the answer is true, and we will get the value of T4U for the corresponding individual, otherwise it will be a missing value. Hence, the T/F section won't create any differences in the analysis carried out. So, the column has been dropped. Now, we have merged the classes 'Primary_hypothyroidism', 'Secondary_hypothyroidism', and 'Compensated_hypothyroidism', and named it as Hypothyroidism class. Now, we have tried to find the correlation between the targeted classes, i.e. Hypothyroidism class and Negative class in order to check the dependency of various attribute. The attributes with correlation >0.03 have been listed below,

```
age        0.030603
sex        0.030154
TSH       0.454049
T3        0.132520
TT4       0.252506
T4U       0.053266
FTI        0.235387
```

So, only 7 attributes remain. We have worked on these attributes to make the comparative study more relevant. The attributes chosen are enlisted here as,

- Age
- Sex
- TSH- Thyroid Stimulating hormone
- T3 – Triiodothyronine
- TT4 - Total Thyroxine
- T4U – Thyroxine Utilization Rate
- FTI – Free Thyroxine Index

4.2 Data Cleaning

We can see from the Fig. 2 that some of our attributes are left-skewed. We have applied Log transformation to the data to solve this problem. After applying log transformation, we can see that our problem has been resolved (Fig. 3).

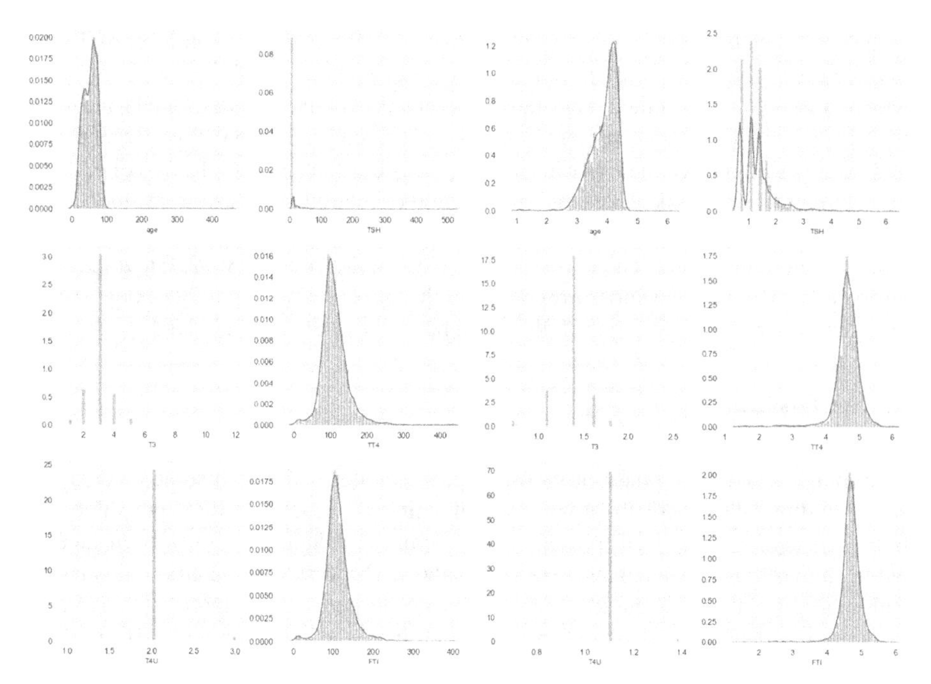

Fig. 2. The plot of skewed data

Fig. 3. Data after applying the logarithmic transformation

Also, by using count plot to our target column. We found that our data is highly imbalanced. For solving this problem, we have used Random Over Sampler. What it does is, duplicates the minority class in our data. As shown in Fig. 4 and Fig. 5.

The random forest classifier is proven to be the most efficient classifier among the classifiers we have worked on, in the given data set. The attributes we have worked on are – Age, Sex, T3, TT4, TSH, FTI, and T4U.

It is known that females belonging to the age group of 46–54 are most affected by hypothyroidism, and at the same time, females are 3–10 times more prone to hypothyroidism than men. So, age and sex are equally important in the study of hypothyroidism.

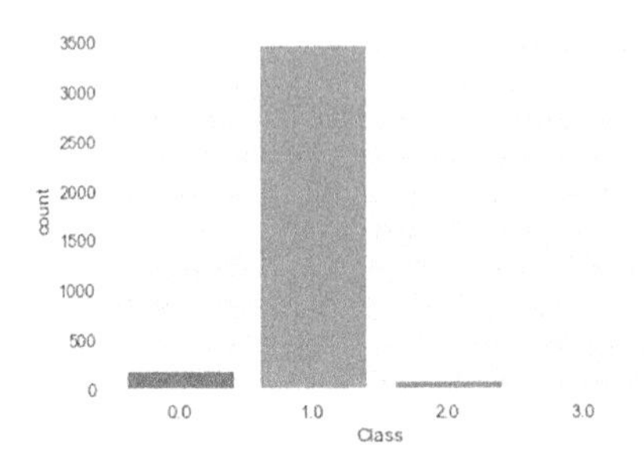

Fig. 4. The plot of imbalanced data

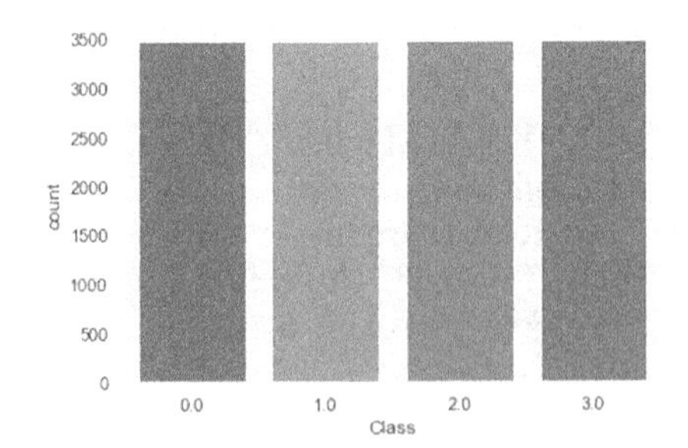

Fig. 5. Data after applying random over sampler

TSH is the Thyroid Stimulating Hormone. The over or under- secretion of TSH hormone can regulate the hypo or hyperthyroidism condition. The normal level of TSH is 0.4–4.0 mU/L. Now, T4 or thyroxine resides in our body in two forms - free thyroxine which is present in the blood, and thyroxine which is bound to protein. FTI measures the total free thyroxin that is present in the blood, that can enter the various target tissues and exerts its effect. On the other hand, T4U measures the amount of thyroxine, that is bound to protein. TT4 measures the total amount of thyroxine present in the blood. The normal range of T4 in the blood is 5.0–12.0 Finally, T3 is the amount of Triiodothyronine present in the blood. The normal range of T3 in the blood is 80–220 ng/dl. So, the attributes which we have chosen regulate the thyroid conditioning in our body and are also the primary manifestations of abnormal behavior of the thyroid gland.

4.2.1 Classification

As is the case for other machine learning algorithms, supervised learning too, is based on training. During the phase of its training, the model is fed with some labeled data sets, that direct the model about what output is linked to each specific input value. The trained system is henceforth presented with some test data, which has been labeled, but the algorithm remains unaware of these labels. The very specific aim of this testing data is to estimate the accuracy of the algorithm in performing upon unlabeled data. A supervised learning calculation probes the training information and then produces an indirect function, that can be employed for mapping new illustrations. An ideal improvement of such a model will take into account the reckoning of the algorithm to decide class names for unseen cases effectively. This in turn requires the ability of the model to estimate and sum up from the training information to be mapped on to hidden circumstances in a 'sensible' manner. We have used the following classifiers for multiclass classification:

1. *K-Nearest Neighbour classifier* [13]
2. *Support Vector Machine Classifier* [14, 15]
3. *Decision tree Classifier* [16]
4. *Random forest Classifier* [17]

Figure 6 shows the confusion matrix obtained using different classifiers.

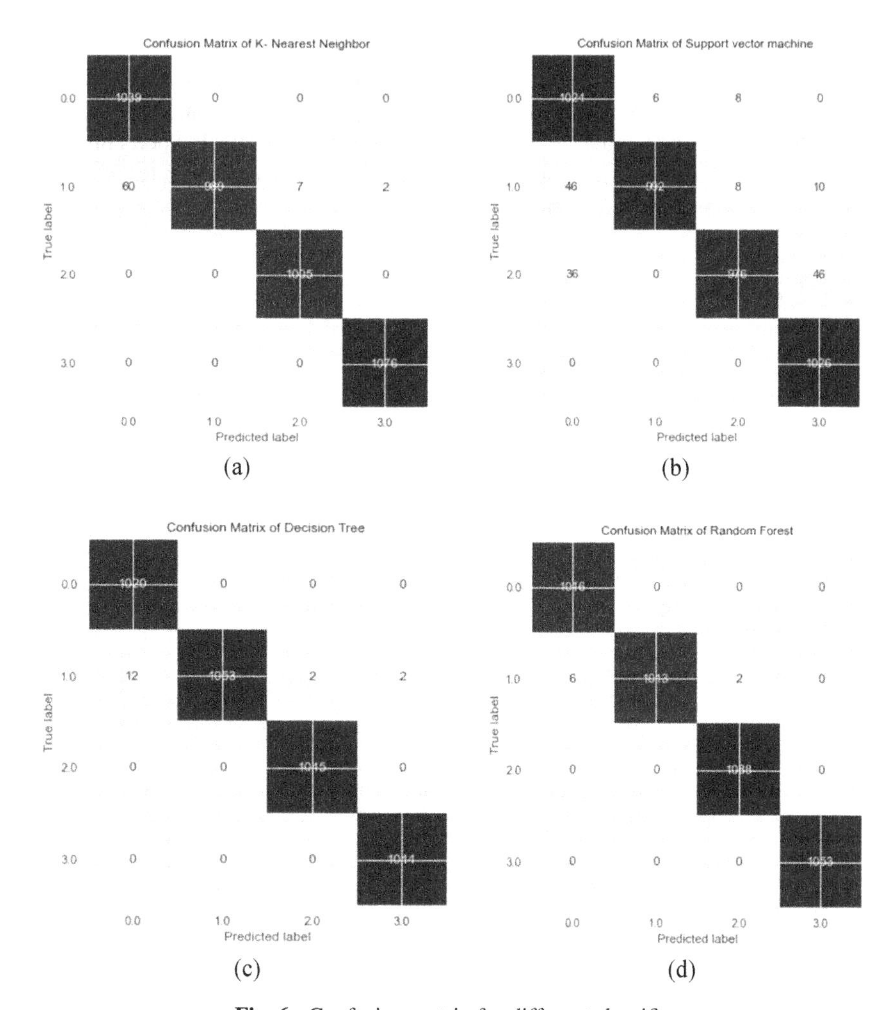

Fig. 6. Confusion matrix for different classifiers

The performance by each classifier is given in the Table 1 below in terms of classification accuracy, precision, recall and F1 score,

Table 1. Comparison of the different parameters for KNN, SVM, Decision tree, and Random Forest.

Classifier	Accuracy	Precision	Recall	F1_Score
KNN	98.3	98.4158	98.3695	98.3457
SVM	96.1	93.2990	96.2100	96.1693
Decision tree	99.5	99.6137	99.6174	99.6258
Random forest	**99.81**	**99.8073**	**99.8051**	**99.8041**

From Fig. 6 and Table 1, we can deduce that the random forest classifier is proven to be the most efficient classifier among the different classifiers on the given dataset. The Random forest algorithm detects hypothyroidism with 99.81% efficiency and performs best in terms of each parameter. So it can be proposed to the medical experts that, they can choose this classification technique along with the above given attributes to detect a hypothyroidism condition with high efficacy.

4.2.2 Comparison

Banu, Gulmohamed [18] have proposed a 6-fold based LDA algorithm in the same dataset. Tandan, S.R. [19] have proposed a 4-fold based Cart4.5 algorithm. Now, comparison of different measures of their algorithm with that of the Random-forest algorithm is given in the table below.

Table 2. Comparison of the different parameters for Cart4.5, LDA, and Random Forest

Classifier	Accuracy	Precision	Recall	F1_Score
C 4.5 (4-fold)	99.5	99.5	99.6	99.5
LDA(6-fold)	99.62	99.6	99.6	-
Random forest	**99.81**	**99.8073**	**99.8051**	**99.8041**

From Table 2 it is cleared that Random-forest algorithm with proposed data pre-processing techniques gives the best results among all.

5 Conclusion and Future Directions

In this work, we proposed a AI driven approach for multiclass classification of hypothyroidism detection using exploratory data analysis and supervised machine learning approach. We have used 7 features and four classifiers for a comparative study. It can be said that in the current study, we have achieved a fairly high accuracy of 98.6% in the hypothyroidism detection problem using random forest classifier. But, the number of attributes that have been dealt with is fairly large and involves many laboratory tests whose accuracy is questionable. At the same time, there are also financial and time constraints. More laboratory tests incur more cost and the time to get those results has a directly proportional incremental relationship with the number of tests being done. So, in the future, it can be a good realistic aim to keep the accuracy level uncompromised, even after limiting the attributes.

References

1. Rakov, H., et al.: Sex-specific phenotypes of hyperthyroidism and hypothyroidism in mice. Biol. Sex Differ. **7**(1), 1–13 (2016)

2. Samuel, E.J.V., et al.: Increasing prevalence of osteoporosis, hypothyroidism and endogenous estrogen with germ cells. Open J. Rheumatol. Autoimmun. Dis. **4**(3), 131–137 (2014)
3. Lorini, R., et al.: Hashimoto's thyroiditis. Pediatric endocrinology reviews: PER, vol. 1, pp. 205–211 (2003); discussion 211
4. Allahabadia, A., et al.: Diagnosis and treatment of primary hypothyroidism. Br. Med. J. **338**, b725 (2009)
5. Kostoglou-Athanassiou, I., Ntalles, K.: Hypothyroidism-new aspects of an old disease. Hippokratia **14**(2), 82 (2010)
6. Benediktsson, R., Toft, A.D.: Management of the unexpected result: compensated hypothyroidism. Postgrad. Med. J. **74**(878), 729–732 (1998)
7. Chen, L., et al.: Mining health examination records—a graph- based approach. IEEE Trans. Knowl. Data Eng. **28**(9), 2423–2437 (2016)
8. Ammulu, K., Venugopal, T.: Thyroid data prediction using data classification algorithm. Int. J. Innov. Res. Sci. Technol **4**(2), 208–212 (2017)
9. Shankar, K., et al.: Optimal feature- based multi-kernel SVM approach for thyroid disease classification. J. Supercomput. **76**(2), 1128–1143 (2020)
10. Ioniţă, I., Ioniţă, L.: Prediction of thyroid disease using data mining techniques. BRAIN. Broad Res. Artif. Intell. Neurosci. **7**(3), 115–124 (2016)
11. Ahmed, J., Soomrani, M.A.R.: TDTD: thyroid disease type diagnostics. In: 2016 International Conference on Intelligent Systems Engineering (ICISE). IEEE (2016)
12. Kotsiantis, S.B., Zaharakis, I., Pintelas, P.: Supervised machine learning: a review of classification techniques. Emerg. Artif. Intell. Appl. Comput. Eng. **160**(1), 3–24 (2007)
13. Deng, Z., et al.: Efficient kNN classification algorithm for big data. Neurocomputing **195**, 143–148 (2016)
14. Mavroforakis, M.E., Theodoridis, S.: A geometric approach to support vector machine (SVM) classification. IEEE Trans. Neural Netw. **17**(3), 671–682 (2006)
15. Kirar, J.S., Agrawal, R.K.: Composite kernel support vector machine based performance enhancement of brain computer interface in conjunction with spatial filter. Biomed. Signal Process. Control **33**, 151–160 (2017)
16. Freund, Y., Mason, L.: The alternating decision tree learning algorithm. in icml. Citeseer (1999)
17. Liaw, A., Wiener, M.: Classification and regression by randomForest. R News **2**(3), 18–22 (2002)
18. Banu, G.: Predicting thyroid disease using linear discriminant analysis (LDA) data mining technique. Commun. Appl. Electron. **4**, 4–6 (2016). https://doi.org/10.5120/cae2016651990
19. Tandan, S.R.: Diagnosis and classification of hypothyroid disease using data mining techniques. IJERT **6** (2013)

An Edge-Enabled Guidance Algorithm for Autonomous Robotic Navigation

Oladayo O. Olakanmi[1,2], Mbadiwe S. Benyeogor[2(✉)], Kosisochukwu P. Nnoli[3], and Olusegun I. Lawal[4]

[1] Department of Electrical and Electronics Engineering, University of Ibadan, Ibadan, Nigeria
olakanmi.oladayo@ui.edu.ng
[2] Cybersecurity and Cybertronics Research Centre (CCRC), OEMA Tools and Automation Ltd., Ibadan, Nigeria
[3] Department of Computer Science and Electrical Engineering, Jacobs University, Bremen, Germany
k.nnoli@jacobs-university.de
[4] Advanced Aerospace Engine Laboratory, National Space Research and Development Agency (NASRDA), Abuja, Nigeria

Abstract. In this paper, we develop an algorithm for a mobile robot to navigate from one point to another in the geodetic coordinate system, while avoiding obstacles, using edge computing devices and navigational sensors. This enables the robot to detect changes in its position, respond, and control its mechanical behavior towards following specified path points to reach the designated location on the earth's surface. Several experiments were performed in an open field to evaluate the efficiency of the algorithm and the accuracy of the sensors. Results show that the guidance control algorithm for autonomous robotic navigation using edge computing devices is practicable, cost-saving, and reliable.

Keywords: Automatic guidance · Autonomous robots · Edge computing · Intelligence system · Navigational sensors

1 Introduction

Autonomous mobile robots are robots that can perform a series of intelligent motions based on their control algorithms, sensors' data, and radioed instructions, with limited human involvement [6]. These robots are usually operated at standoffs to carry out automated self-decision and to correct their trajectory based on their sensor's data. This class of robot has stimulated the interest of many roboticists in the field of autonomous systems, thus leading to several applications in smart agriculture, sample-return mission, autonomous driving, etc. [7]. The amount of computing power required by an autonomous robot

Supported by the Nigerian Communications Commission (NCC) and the University of Ibadan.

I. Woungang et al. (Eds.): ANTIC 2021, CCIS 1534, pp. 328–341, 2022.
https://doi.org/10.1007/978-3-030-96040-7_27

depends on the type and complexity of the guidance algorithm adopted by the robot. Meanwhile, recent advances in microprocessor technology has availed the roboticists with robust and versatile edge computing devices, thereby allowing them to develop and use a powerful guidance algorithm for efficiency improvement in autonomous robots.

In this paper, we propose a novel guidance control algorithm for the guidance system of a mobile robot. We also demonstrate the effectiveness of the algorithm for point-to-point autonomous navigation of a mobile robot. Therefore, we proceed with the introduction of some relevant literature in Sect. 2. The experimental platform and the control strategies for the coordination of our robotic system are discussed in Sects. 3 and 4, respectively. In Sect. 5, we would discuss and visualize the results of field tests and evaluation of the deployed algorithms and control strategies and then conclude our findings in Sect. 6 with future recommendations.

2 Related Works

Most vehicular systems or autonomous guidance systems that involve a localization algorithm for computing the true location of a robot, depend on either satellite-based radio navigation systems, local beacons, landmark detection, or their combination. These could also be used to guide a robot along some predetermined path or signal it when it arrives at some target location [8]. For autonomous navigation, the System-On-Chip (SoC) computer and the microcontroller have proven to be effective edge computing devices. An exciting demonstration of this can be seen in the work of Efaz in [2], which involves the design of a speed-controlled path-finding obstacle avoidance robot, using Object Oriented Programming (OOP) techniques.

Recent edge-computing devices compete well with many desktop machines in terms of computing power, graphics processing and versatility. These are the aforementioned SoC computers. This is attributed to the continual shift from single-core to multi-core processors in embedded systems and edge computing [11]. With SoC platforms, the robot software developer can now create complex edge computing algorithms that were only possible with conventional desktop computers.

An autonomous robotic guidance system can also serve as a smart agricultural technology. In this wise, an electromechanical farming equipment can be integrated into the robotic system and operated purely as an automated subsystem, while the robot navigates autonomously as a multisensory system. One possible application is auto-mechanical seed planting, which is an application of interest in modern farming [12]. A typical development in this area was the autonomous navigation seed-sower by [4], which shows how a route tracking control algorithm could be used to navigate a robotic planter along the paths in which seeds are to be sown.

The present paper is an advancement on the methods contained in the reviewed literature, geared towards practical applications. Here, the focus is on multisensing, mechanical automation, and autonomous guidance using appropriate algorithms. In particular, this paper adopts some of the results of our previous works in [1,5,10], from where we mastered the rudiments of both wireless and autonomous motion control. The results obtained from the previous work serve as the technical background for this paper.

3 Experimental Platform

To test our computational hypothesis, we adopted the robotic system that was developed in [5], as the experimental platform. The Fig. 1 shows the physical composition of this system. The details of its underlying mechanics are given in [5]. Hence, the rest of this paper will focus on the formulation of edge computing algorithms to control the navigation of our robot as an autonomous system.

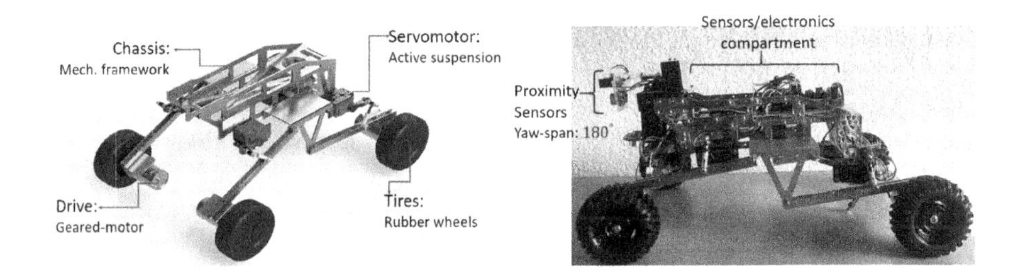

Fig. 1. Experimental platform (Left: 3D model, Right: Complete system)

3.1 Control Architecture

The control architecture of the robotic system is shown in Fig. 2. This consist of the command system (base station + edge computer) and the control system (companion computer + main controller), along with other peripherals (sensors, actuators, speed controller, etc.). The companion computer executes high-level computations in support of the main controller. The main controller performs all low-level computing and control functions. At the command of the system manager at the based station, the companion computer is signalled to activate the robot in either autonomous or remote control work mode. Any of the operating modes would trigger the selection of specific embedded functions in the main controller to drive the robot via the speed controller in the desired direction.

The main function of the companion computer is to maintain communication between the main controller and the base station via a remote access connection [10]. The companion computer also computes the localization algorithm on behalf of the main controller. The companion computer is necessary because its functions require high computing power and the scheduling function of an operating system. At the lower level, the algorithms that are implemented on the main controller include the obstacle avoidance algorithm (i.e., Algorithm 2) and automatic navigation algorithm (i.e., Algorithm 3). These are supported by 4 navigational sensors, of which types and functions are cataloged in Table 1. The ultrasonic and Infrared (IR) distance sensors are mounted on the frontal projection of the robot's chassis through a servo-controlled revolver with a yaw rotation span of 0° to 180°.

Table 1. List of navigational sensors on board the robot and their function

S/N	Sensor type	Function
1	Ultrasonic sensor	Obstacle proximity measurement
2	Infrared sensor	Obstacle proximity measurement
3	Global positioning system (GPS) sensor	Geo-spatial position measurement
4	Compass sensor	Measures the bearing of the robot

Using the Universal Asynchronous Reception and Transmission (UART) protocol, a serial communication channel is established between the companion computer and the main controller, to enable real-time transfer of information, control signals, and computational requests between the two devices.

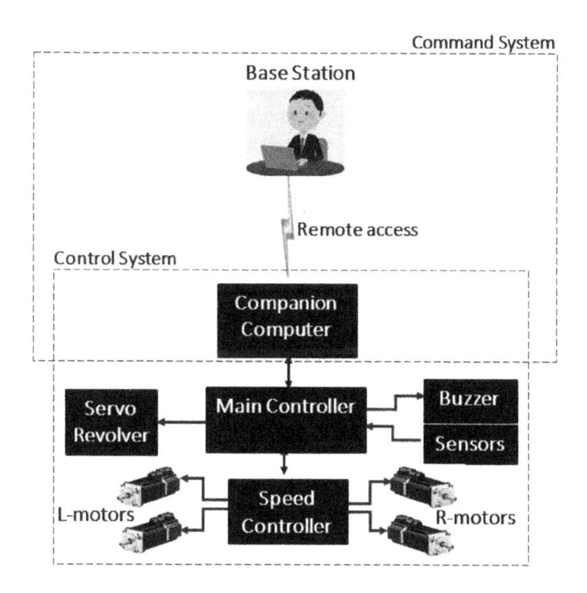

Fig. 2. The control architecture

4 Control Strategies

The control scheme (i.e., INITIAL_MOVE function in Algorithm 1) of our robot's control flow involves two basic functions for the direct control of its motion. These functions are the CHANGE_PATH in Algorithm 2, and AUTO_NAVIGATE in Algorithm 3. These could call upon one another and other embedded functions to make the robot act as an autonomous agent. Algorithm 1 starts up the robotic system once its power switch is turned on. It coordinates Algorithms 2 and 3, which are the actual autonomous control functions of the robot, of whose precision depends on the accuracy of sensors.

Algorithm 1. Motion initialization function

```
Require: distF
 1: function INITIAL_MOVE
 2:     repeat
 3:         Move FORWARD                                   ▷ Set robot in Motion
 4:         continue                                       ▷ To keep moving
 5:                                  ▷ ----------Decision----------
 6:         if distF ≤ 40cm then
 7:             call CHANGE_PATH                           ▷ Avoid obstacles
 8:         else if 40cm < distF ≤ 120cm then
 9:             Move FORWARD                               ▷ Move forward
10:         else if distF ≥ 120cm then
11:             call AUTO_NAVIGATE                         ▷ Move to target
12:         end if
13:     until P1[xi, yi] ≈ P2[xf ± δx, yf ± δy]
14: end function
```

4.1 Proximity Sensing and Obstacle Avoidance

To control the movement of our robot during obstacle avoidance, we developed a mean of detecting the proximity of obstacles to it as shown in Fig. 3. This involves the usage of ultrasonic and IR sensors to simultaneously measure this proximity, as a way to minimize the error associated with each of the sensors, while also harnessing their peculiar advantages. Therefore, we formulated a data fusion scheme that uses the Moving Average Filter (MAF) and the covariance formula to fuse the incoming data from the two sensors, thereby minimizing the chances of error in proximity measurement. According to [9], the mathematical derivation of the MAF is given by

$$\bar{d_k} = \frac{d_{k-n+1} + d_{k-n+2} + d_k}{n}, \tag{1}$$

where $\bar{d}_k$ is the average from $(k-n+1)^{\text{th}}$ to k^{th} measurement values and n is the total number of values. Hence, the moving average of the previous measurement is given as,

$$\bar{d}_{k-1} = \frac{d_{k-n} + d_{k-n+1} + d_{k-1}}{n}, \tag{2}$$

$$\therefore \quad \bar{d}_k = \bar{d}_{k-1} + \frac{d_k - d_{k-n}}{n}. \tag{3}$$

The Eq. (3) is the MAF formula in the form of a recursive function, which is used as a function to calculate the moving average of the streams of measurement data from each proximity sensor. Therefore, two moving averages are computed at every instant – one for the ultrasonic sensor and the other for the infrared sensor. Prior to fusing these two moving averages, we derive a covariance (Cov) formula which is applied to ensure that the raw measurements from the two sensors are consistent with each other (i.e., both sensors are ranging between the same obstacle). The Cov is given as

$$Cov = \sum_{index=1}^{n} \frac{(L_1[index] - Ave_1) \cdot (L_2[index] - Ave_2)}{n-1}, \tag{4}$$

where $L_1 \Leftarrow dist_1$ denote proximity measurements from the ultrasonic sensor, and $L_2 \Leftarrow dist_2$ denote proximity measurements from the infrared sensor. The $index$ is the sampling integer (where $index = 1, 2, ..., n$), and n is the total number of measurement samples.

At any instant, if the value of Cov in Eq. (4) is positive, the mean value of the two moving averages is calculated as $dist_F$ and returned as the measurement estimate of an object's proximity away from the robot. However, if the value of Cov is negative, the data fusion process is repeated. With this scheme, we significantly reduced the error in measurements by the robot's proximity sensors to a level that is acceptable and applicable for the obstacle avoidance motion control of our robot. Again, the algorithm for obstacle avoidance is given in Algorithm 2. This involves several calls to various embedded maneuvering functions in the main controller, to find the most obstacle-free direction, before returning control to the INTEL_SCHEMA in Algorithm 1. Therefore, our robot could reliably and autonomously avoid both static and moving obstacles along its navigational pathway to a given target location.

4.2 Automatic Navigation

For accurate maneuverability of the robot, an algorithm is required to iteratively control the movement of the robot from one point $\boldsymbol{P}_1[x_i, y_i]$ to another point $\boldsymbol{P}_2[x_f, y_f]$ in the geographic coordinate system, until the final destination is reached. To do this, the "Haversine formula" will be applied according to [3], for computing the bearing of $\boldsymbol{P}_2[x_f, y_f]$ from $\boldsymbol{P}_1[x_i, y_i]$ as ψ. The formula for calculating ψ is therefore given in Eq. (5) as,

Algorithm 2. Obstacle avoidance algorithm

```
Require: dist_F
 1: function CHANGE_PATH
 2:     call BRAKE
 3:     call REVERSE
 4:     continue for 1.5 s
 5:     call BRAKE
 6:                          ▷ —-Measure proximity of obstacle on the right—-
 7:     point proximity_sensor right
 8:     continue for 0.5 s
 9:     dist_Right ← PROX_ESTIMATE
10:     continue for 0.5 s
11:                          ▷ —-Measure proximity of obstacle on the left——
12:     point proximity_sensor left
13:     continue for 0.5 s
14:     dist_Left ← PROX_ESTIMATE
15:     continue for 0.5 s
16:
17:     point proximity_sensor forward                              ▷ Default
18:     continue for 0.5 s
19:                          ▷ ————————-Decision——————————
20:     if dist_Right > dist_Left then
21:         while φ ≠ 90° do
22:             Move RIGHT
23:             if φ ≈ 90° then
24:                 call INITIAL_MOVE
25:             end if
26:         end while
27:     else if dist_Left > dist_Right then
28:         while φ ≠ 270° do
29:             Move LEFT
30:             if φ ≈ 270° then
31:                 call INITIAL_MOVE
32:             end if
33:         end while
34:     else
35:         call INITIAL_MOVE
36:     end if
37: end function
```

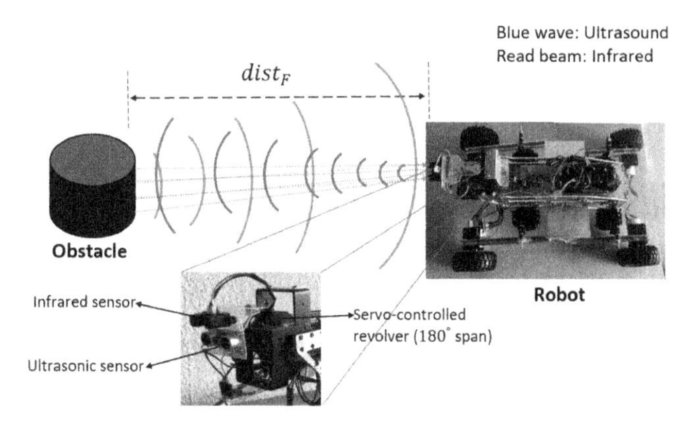

Fig. 3. Depiction of obstacle's proximity measurement using both the ultrasonic and infrared sensor (Note: $dist_F$ is the proximity of obstacle from the robot)

$$\psi = \arctan(X, Y), \tag{5}$$

where

$$X = \cos x_i \cdot \sin \Delta y \tag{6}$$

and

$$Y = \cos \theta_1 \cdot \sin x_f - \sin x_f \cdot \cos x_f \cdot \cos \Delta y. \tag{7}$$

The values x_i and x_f represents geographical latitudes, while y_i and y_f represent the longitude. For automatic navigation, the real-time bearing of the robot with respect to the magnetic north, φ is required. The vector $\boldsymbol{P}_1[x_i, y_i]$ and the bearing φ is iteratively read from a GPS sensor and a compass sensor, respectively. The vectors $\boldsymbol{P}_2[x_f, y_f]$ are supplied as way points along the desired route for the robot's navigation. The value ψ is calculated with Eq. (5) serves as the reference angle for steering the robot towards the next way point while depending on the value of $error = \varphi - \psi$. The goals of this path tracking algorithm are to:

1. calculate the bearing, ψ of $\boldsymbol{P}_2[x_f, y_f]$ from $\boldsymbol{P}_1[x_i, y_i]$.
2. orient the robot's motion along the bearing, ψ until $error = \varphi - \psi = 0$.
3. cause the robot to move in this direction until $\boldsymbol{P}_2[x_f, y_f]$.
4. take $\boldsymbol{P}_2[x_f, y_f]$ as $\boldsymbol{P}_1[x_i, y_i]$, and copy the next way-point to $\boldsymbol{P}_2[x_f, y_f]$.
5. repeat the process until the last way-point is reached.

In pseudocode, this task is more sufficiently described as Algorithm 3.

The boundary conditions for which Algorithm 3 is valid are as follows:

1. The earth is assumed to be a perfect sphere.
2. The point-wise range of navigation is limited to 500 m.

Algorithm 3. Automatic navigation algorithm

,

Require: $dist_F$, $\boldsymbol{P}_1[x_i, y_i]$, $\boldsymbol{P}_2[x_f, y_f]$, & φ
Ensure: ψ, $error = \varphi - \psi = 0$, & $\boldsymbol{P}_2[x_f, y_f] = \boldsymbol{P}_1[x_i, y_i]$

```
1: function AUTO_NAVIGATE
2:     while dist_F ≥ 120 cm do
3:         repeat
4:             P_1[] ← SENSORS[lat_1, lon_1, φ]
5:             x_i ← P_1[lat_1]
6:             y_i ← P_1[lon_1]
7:             φ ← P_1[φ]                                  ▷ Robot yaw angle
8:             P_2[] ← (Next   waypoint, [lat_2, lon_2])
9:             x_f ← P_2[lat_2]
10:            y_f ← P_2[lon_2]
11:            Δx ← (x_f − x_i)                            ▷ Compute target's bearing
12:            Δy ← (y_f − y_i)
13:            X ← cos x_f ∗ sin (Δy)
14:            Y ← cos x_i ∗ sin x_f − sin x_i ∗ cos x_f ∗ cos (Δy)
15:            ψ ← tan^−1 (X, Y)
16:            while φ ≠ ψ ± δψ do                          ▷ Steering control
17:                if ψ ≤ 180° then                         ▷ 3-figure bearing
18:                    Move RIGHT
19:                else
20:                    Move LEFT
21:                end if
22:            end while
23:            while φ = ψ ± δψ do                          ▷ Position control
24:                if P_1[x_i, y_i] ≠ P_2[x_f ± δx, y_f ± δy] then
25:                    Move FORWARD                         ▷ Move to target
26:                else
27:                    Call BRAKE                           ▷ Stop at target
28:                end if
29:            end while
30:            P_1[x_i, y_i] ← P_2[x_f, y_f]                ▷ Shift waypoint
31:            UpdateP_2[lat_2, lon_2]                      ▷ Update next waypoint
32:        until [P_2[lat_Final, lon_Final]
33:    end while
34: end function
```

3. The robot can not reverse its motion (i.e., it can only yaw and drive forward), except Algorithm 2 is called to perform obstacle avoidance.
4. Proximity of obstacles ahead must be greater than 120 cm.
5. Test navigation is performed in a controlled environment.

Fig. 4. Performance evaluations in field tests

5 Results and Discussion

Our navigational schemes and algorithms are implemented using the experimental platform in Sect. 3. We adopted the Raspberry Pi SoC and Arduino Mega Microcontroller as the companion and main controller, respectively. Experiments are conducted in an open field as shown in Fig. 4 to test the validity of our algorithms and the performance of the robot based on these algorithms. These involve the telemetry of navigational data to the base station for real-time analysis. Our experimental procedure, analytical techniques, and the results are discussed below in Subsects. 5.1 and 5.2.

5.1 Experimental Procedures and Results

As the navigational precision of our robotic system depends on the accuracy of its sensors, our field tests aim at evaluating the accuracy with which our robot could perform both obstacle avoidance and autonomous navigation, while maneuvering towards a target location.

Evaluation of Proximity Measurement Technique. To ensure proficiency in obstacle avoidance, we evaluate the accuracy of each of the two adopted proximity sensors and their fusion. To do this, we plot the real-time values from the ultrasonic sensor (i.e., $dist_{Ultrasonic}$), infrared sensor (i.e., $dist_{Infrared}$), their fusion (i.e., $dist_F$), and the true proximity of the obstacle from the robot (i.e., $dist_{Actual}$, as measured with a meter-rule) against time (in seconds). The result of this experiment is presented in Fig. 5. This visualizes the error present in the robot's sensitivity to the proximity of obstacles along its path.

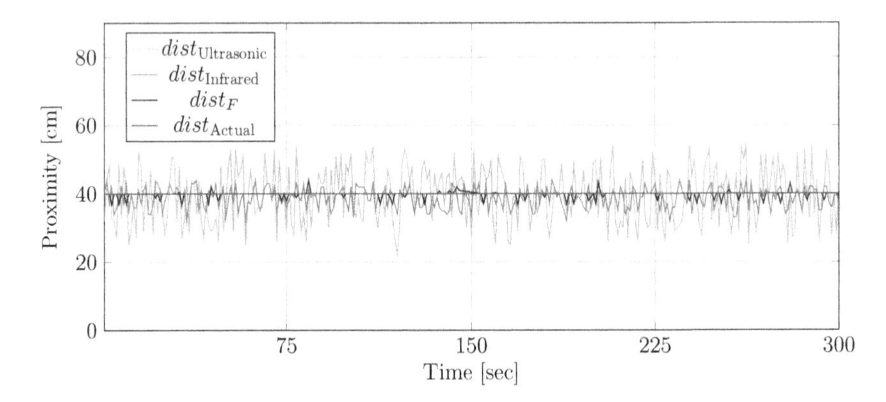

Fig. 5. Proximity sensors values (i.e., $dist_{\text{Ultrasonic}}$, $dist_{\text{Infrared}}$, $dist_F$, and $dist_{\text{Actual}}$) vs time

Graphical Visualization of the Robot's Routes. To evaluate how accurately our robot could autonomously move to a target location, we visualized the actual navigational routes of our robot in comparison to the desired path, using real-time position estimates from the GPS and compass sensors. For experimentation, variations in the navigational environment included the number of prestationed obstacles along the robot's travel path and the length of the path. The results of this experiment are presented in Figs. 6, 7 and 8.

5.2 Discussion

The plots in Fig. 5 shows the effectiveness of the fusion scheme we deployed in combining the moving average measurements from the ultrasonic and infrared sensors. We observe that this fusion process increases the accuracy in the measurement of an obstacle's proximity to the robot, unlike the direct measurements from the respective sensors (i.e., $dist_{Ultrasonic}$ and $dist_{Infrared}$), which as standalone devices, contains a higher level of noise.

According to our observations in Figs. 6, 7 and 8, Algorithm 3 is effective at homing the robot to a position near the target location. In particular, the first path of the plot in Fig. 6 (i.e., distance between the starting point and the first way point, which is obstacle free) shows that in the absence of obstacles, the robot will move along a near-straight line from its starting point to the immediate target location. This is made evident by the relatively less number of turning points. Comparatively, the plots in Figs. 7 and 8 show that the motion of the robot to a target point will exhibit more turning points, deviations from the expected path, and course correction with increase in the range of the navigation path and the number of obstacles along the path.

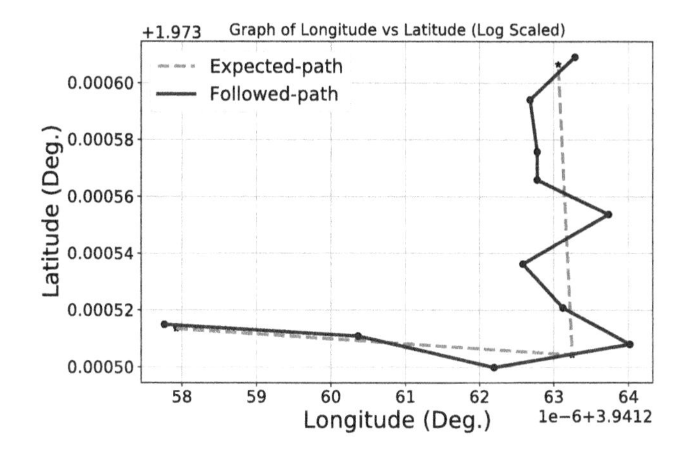

Fig. 6. Path-plot 1: Graph visualizing the robot's routes (number of way-points = 2, number of obstacles = 4)

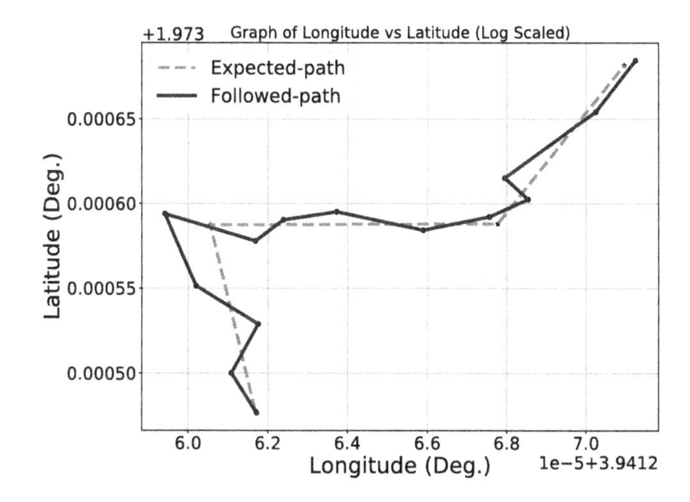

Fig. 7. Path-plot 2: Graph visualizing the robot's routes (number of way-points = 3, number of obstacles = 6)

Limitations. Based on our field observations, the technical possibility of flaws in sensor measurements limits the performance of our robot. Other limitations arise from the topography of the terrain and the mechanical constraints of the robot's drive system. We observe that unlike on paved paths, the robot experiences transitional difficulties and inadequate steering power when maneuvering over rough terrain and grassland; which is as a result of increased friction between the wheels and the ground. Apart from these constraints, the overall performance of our robot is satisfactory and meets the design objectives.

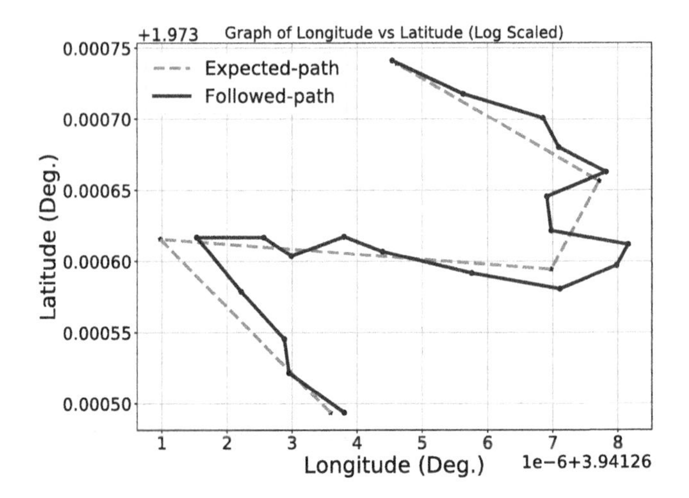

Fig. 8. Path-plot 3: Graph visualizing the robot's routes (number of way-points = 4, number of obstacles = 10)

6 Conclusion

In this paper, an edge-based autonomous robotic system was developed for point-to-point navigation using geospatial data. This robotic system is able to avoid obstacles on its path to the target location using the fusion measurements of how far the obstacles are from it. For the operation of this robot, the obstacle avoidance and automatic guidance algorithms are used. These use information about how far the obstacles are from the robot and the robot's location on the earth to autonomously navigate it from an initial point to a target location, without collision with any object along its path. The embedded hardware of the robot comprises two edge devices – the main controller and the companion computer, which work together as complimentary systems to implement the robot's control algorithms. Field tests were also conducted to evaluate the performance of the robot. These included the evaluation of how accurately the robot could detect obstacles along its travel path and maneuver them and how precisely it could get to its target location. Results show that our robot performs as expected, regardless of its operational constraints. In essence, our work proves that edge devices like microcontrollers and SoC computers are applicable to the development of intelligent and autonomous systems. Future developments in this area may focus on exploring the potential applications of our system; such as autonomous seed planting, environmental monitoring, logistical automation, etc. We therefore hope that our work stimulates interests geared toward the practical applications of our robot and its corresponding algorithms.

References

1. Benyeogor, M.S., Nnoli, K.P., et al.: An algorithmic approach to adapting edge-based devices for autonomous robotic navigation. EAI Endorsed Trans. Context-Aware Syst. Appl. (2021). Online First, https://doi.org/10.4108/eai.5-8-2021.170559
2. Efaz, E.T.: Pathfinder: design of an indicative featured and speed controlled obstacle avoiding robot (2018). https://create.arduino.cc/projecthub/maverick/pathfinder-229d5d
3. Faulkner, P.: Notes for applied mathematics in trigonometry and earth geometry/-navigation. Aust. Senior Math. J. **18**(1), 55–58 (2004)
4. Fernandez, B., Herrera, P.J., Cerrada, J.A.: A simplified optimal path following controller for an agricultural skid-steering robot. IEEE Access **7**, 95932–95940 (2019)
5. Gratton, E., Benyeogor, M., Nnoli, K., et al.: Multi-terrain quadrupedal-wheeled robot mechanism: Design, modeling, and analysis. Eur. J. Eng. Res. Sci. **5**, 24–33 (2020)
6. Herget, C., Grasz, E., Merrill, R.: Applications of intelligent telerobotic control. Annual Rev. Autom. Program. **16**, 149–152 (1991). Artificial Intelligence in Real-time Control 1991. https://doi.org/10.1016/0066-4138(91)90025-7
7. Hongo, T., Arakawa, H., Sugimoto, G., Tange, K., Yamamoto, Y.: An automatic guidance system of a self-controlled vehicle. IEEE Trans. Ind. Electron. IE **34**(1), 5–10 (1987). https://doi.org/10.1109/TIE.1987.350916
8. Li, Z., Huang, J.: Study on the use of Q-R codes as landmarks for indoor positioning: preliminary results. In: 2018 IEEE/ION Position, Location and Navigation Symposium (PLANS), pp. 1270–1276 (2018). https://doi.org/10.1109/PLANS.2018.8373516
9. Mohsin, O.Q.: Mobile robot localization based on Kalman filter: Dissertations and theses. Paper 1529. Portland State University Library (2000). https://doi.org/10.15760/etd.1528
10. Olakanmi, O., Benyeogor, M.: Internet based tele-autonomous vehicle system with beyond line-of-sight capability for remote sensing and monitoring. Internet Things **5**, 97–115 (2019)
11. Parviainen, O.: Software coven: OpenMP parallel computing in Raspberry-Pi (2014). https://www.softwarecoven.com/parallel-computing-in-embedded-mobile-devices/. Accessed 14 Apr 2021
12. Reid, J.F., Zhang, Q., Noguchi, N., Dickson, M.: Agricultural automatic guidance research in North America. Comput. Electron. Agric. **25**(1), 155–167 (2000). https://doi.org/10.1016/S0168-1699(99)00061-7

An Adaptive Algorithm for Emotion Quotient Extraction of Viral Information Over Twitter Data

Pawan Kumar[1], Reiben Eappen Reji[1], and Vikram Singh[2](✉)

[1] National Institute of Technology, Surathkal, India
{pawan.181ee133,reubeneappenreji.181ee136}@nitk.edu.in
[2] National Institute of Technology, Kurukshetra, India
viks@nitkkr.ac.in

Abstract. In social media platforms, a viral information or trending term draws attention, as it asserts the impact of user content towards topic/terms. In real-time sentiment analysis, these viral terms could deliver potential insights for the analysis and decision support. A traditional sentiment analysis tool generates the level of predefined sentiments over social media content for the defined duration and lacks in the extraction of emotional impact created by the same. In these settings, it is a multifaceted task to estimate precisely the emotional quotient viral information creates. A novel algorithm is proposed, to (i) *extract the sentiment and emotions quotient of current viral information over twitter*, (ii) *compare co-occurring trending/viral information*, (iii) *in-depth analysis of potential Twitter text data*. The generated emotion quotients and micro-sentiment reveals several valuable insight of a viral/trending topic and assists in decision support. A use-case analysis over real-time extracted data asserts significant insights, as generated sentiments and emotional effects reveals co-relations caused by viral/trending information. The algorithm delivers an efficient, robust, and adaptable solution for the sentiment analysis also.

Keywords: Big data · Emotion quotient · Sentiment analysis · Twitter

1 Introduction

The traditional social media platforms, e.g., Twitter, *Facebook*, etc. cater to the global users and list their personal information and media. The heterogeneous user data is often utilized for deriving common sentiments or trending information. The trending or viral information primarily harnesses the global content shared co-related to a particular topic and hash tag keywords [1].

A naive user or new user usually refers to this trending or viral list of information to see the most occurring or contributory piece of information [2, 3]. In this process, a user simply refers to the viral information and explores the related term over the Twitter API, without cognitive awareness of the emotional effect of viral information. A piece of viral information may have a list of information that may trigger the emotional effect

I. Woungang et al. (Eds.): ANTIC 2021, CCIS 1534, pp. 342–358, 2022.
https://doi.org/10.1007/978-3-030-96040-7_28

on the user and lead to emotional splits or swings on the choice of information. User assistance is pivotal for the user, which may assist the user to showcase the emotional effects that viral information may carry. Though social media platforms offer limited or no functions or aspect-related views on the API for the generic user.

For example: Viral information related to '*Covid-2019*' may cause significant emotional effects on the citizens in current time. The cause-effect analysis over twitter may assist in administration (*Healthcare* offices) to track the sources/persons to take precautionary measures proactively. Similarly, for spectrum of applications where these estimated statistics may play a significant role:

- Sentiment analysis on social media is extensively used in the *Stock market* and *crypto market* to observe current trends and potential of *Panic sell*.
- Several sentiment models adapted on political elections recently, e.g. for US elections and Indian Elections.
- Similarly, a government office could utilize to track civil riot origins before they become uncontrollable, etc.

Typically, the designed algorithm for the sentiment analysis and emotion quotients (*EQ*) statistics could serve several pivotal objectives, as asserted by the experimental analysis also [7, 8]. The sentiment and *EQ* statistics generated could be utilized in several application areas: decision-making, advertising, public administrations, etc. Though, generating these statistics for real-time published data from the twitter data is a complex and multifaceted computing task [13, 14].

1.1 Motivation and Research Questions

The sentiment analysis is a complex computing task, mainly due to the semantic correlation that exists between the user-generated data and targeted sentiment level and created emotional quotients [14]. The task becomes multifaceted, primarily, when it is aimed for deriving the '*emotion effect*' co-located to a sentiment, as a micro-level. In these settings, a strategy could be the need of the hours that acquire the real-time twitter data and deliver the insights.

The research questions (RQs) are formalized to assist design of proposed adaptive strategy for the estimation of *sentiment level (SL)* and *emotion quotient (EQs)* of viral information on the real-time basis:

RQI: What are the key twitter data elements/features to extract the *SL* and *EQs*?
RQII: How to estimate the *SLs* and *EQs* and co-located overlap on both estimates?
RQIII: What *SL* and *EQ* statistics asserted for spectrum of application domains?

The designed *RQs* assist in conducting overall work and validate its feasibility for analytics and just-in-time decision-making over real-time published twitter data.

The key contribution is a robust and adaptive algorithm for sentiment and emotion evaluation, on just-in-time estimation for an interactive data play. Other contributions are as follows:

(i) A portable and adaptive *UI*, to assist on generates the real-time statistics (*emotion and sentiment polarities*) for an emotion value 'as query' or viral information.
(ii) The strategy outlines pivotal features of text-based sentiment and emotion analysis on social media, e.g. *subjectivity, statement polarity, emotions expressed*, etc.
(iii) The experimental assessment asserts the overall accuracy upto 89% and 90%, respectively for *sentiment* and *EQs* estimation. The overall performance achieved is at significant-level in the view of real-time soft data analysis challenges.

The paper is organized as: *Sect.* 2 lists the relevant research efforts to the sentiment and emotional statistics. *Section* 3 elaborates the conceptual schema and internals of the designed strategy, with formulas and working example. *Section* 4 describes the experimental assessment on the traditional metrics and advanced measures. Conclusion listed at last.

2 Related Work

In recent years, developing novel algorithms for sentiment and threaded emotional analytics estimation on soft data, particularly at micro-level on viral or trending information is area of interest. The located research areas fall under two heads:

2.1 Sentiment Analysis Over Soft Data (*Reviews/Posts/Viral Information*)

In recent years, the research efforts made on the accurate estimation of sentiment statistics, with a listed core task (i) *an automatic identification of relevant and text with opinion or documents* [15–20], (ii) *preparation of sentiment and threaded sentiment analysis.* Existing strategies and methods employed mainly *rule-based* and *statistical machine learning* approaches for these inherent tasks, e.g. opinion mining and sentiment analysis [22, 23].

A comprehensive survey is presented in [34] with two broad set of strategies (*opinion mining and sentiment analysis*). Whereas, Turney [38] asserts that an unsupervised algorithm, could be more suitable for the *lexicon-based* determination of *sentiment phrases* using function of opinion words over the word/sentences or document corpus, same is supported in [4, 36]. Another work in [5] highlighted the use of *SentiWordNet* as lexicon-based sentiment classifier over document and text segregation, as it may contains opinion strength for each term [22]. A prototype in [9], used the *SentiWordNet* Lexicon to classify reviews and [6] build a dictionary enabled sentiment classification over reviews with embedded adjectives.

Further, several work used Naïve-byes and SVM for sentiment analysis of movie reviews supported by inherent features, unigrams, bigrams, etc. These experimentations reveal that with feature a greater accuracy could be achieved sentiment polarity and statistics generation [23, 25].

In the recent potential work, a subsequence *kernel-based* voted *perceptron* prototype is created, and it is observed that the increase in the number of false positives is strongly co-related with the number of true positives. The designed models reveal its resiliency over intermediate star rating reviews classification, though five-star rating reviews is not utilized while training the model. Similar model is used for the sentiment analysis over the *microblog posts* using two phases: first phase involves partition of subjective and objective documents based on created and further for the generation of sentiment statistics (as *positive* and *negative*) in the second phase [10, 11, 34].

2.2 Emotion Quotients (EQs) Over Soft Data

The accurate detection of inherent '*Emotion*' over a text data using natural language processing and text analytics to discover people's feelings located subarea of research work. The usage of it could be tracking of disasters and social media monitoring.

Tracking user's opinions and inherent emotional quotients using posted soft data reveals interesting insights, e.g. tracking and analyzing Twitter data for election campaigns [6, 21, 22, 39, 41, 43]. There are several research studies asserts that sentiment topics and emotion topic/terms delivers promising outcomes for the generation of both polarities, such as for the tracking and monitoring '*earthquake disasters*' using '*Weibo*' a Chinese social media content is used to see the sentiments generated and sensitization [44]. In this, the proposed framework detected disasters related sentiment over massive data from a micro-blogging stream and to filter the negative messages to derive co-located event discovery in a post-disaster situation [23, 24, 37, 38].

The emergence of spectrum of social media platforms justified the need of social analytics for decision-making [38]. A system for tracking of sentiment on news entities over time [35, 39, 42], the socio-politics issues are detected over real-time streams. In this, sentiment-spike detection has been generated in [25–29, 40], Twitter data and analyzed the sentiment towards 70 entities from different domains. Similarly, in [30–33, 41, 45] a system to tracking health trends using microblogs data for the identification of province of several health related sentiment and co-located emotions are used. The authors introduced an open platform that uses crowd-sourced labeling of public social media content.

The key challenges in the accurate estimation sentiment and co-located *EQs* is the scalability of soft data and its rate of change for each inherent levels. Though, building an information system on the top of social media platform content could offer several useful takeaways to government's official and decision-makers. An end-to-end adaptive system is a focus of the system to generate these statistics to the spectrum of user, ranging from naïve to policy maker.

3 Proposed Strategy

The traditional social media platform, e.g. *Twitter, Facebook*, etc. caters global user's personal intents using posted media. The posted heterogeneous user data is pre-processed for acquiring generic sentiments and *EQs* of a viral information. The trending or viral information primarily harnesses the global content shared to a particular topic and keyword (e.g., *hashtag*). In this setting, a user simply refers to the viral information/hashtag or manually explores the related term over the twitter API, without cognitive awareness of the emotional effect of viral information. Though, social media platforms offer limited or no functions or aspect related views on the API for the generic user. The design system assists a naive user to understand its sentiment impact and further emotional quotients (EQs).

3.1 Conceptual Framework

A novel strategy for the real-time generation of emotional quotients of viral/trending information on twitter is designed. Figure 1 illustrates the internal computing blocks and their interactions for the intended objectives. The proposed framework begins with a traditional data collection over twitter API. The data extraction is driven by the user inputs, e.g. *keyword/hashtags, number of tweets, and duration.* The retrieved tweets from the *API*, are now to be stored in a temporary storage for later text-processing and feature extraction.

The local twitter data storage is also connected to the computing clock 'text preprocessing', each tweet extracted must go through local text processing and further supplied to the 'feature-extraction'. Further, a small computing thread is kept within the 'feature extraction' computing block for the estimation '*sentiment score (SC)*' and '*emotional quotient (EQ)*' co-located twitter data objects.

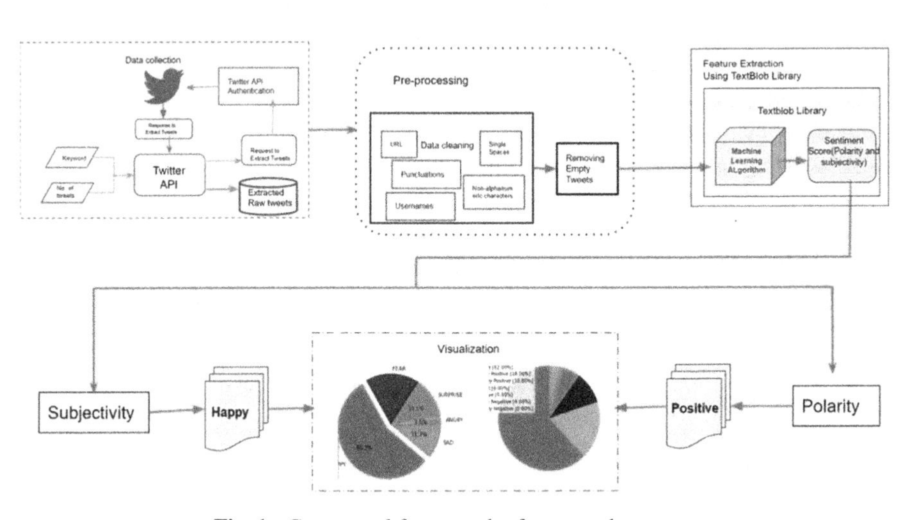

Fig. 1. Conceptual framework of proposed strategy

3.2 Generating Sentiment Level (SL) and Emotional Quotient (EQ)

The aim of the designed system is to generate the sentiments and emotion quotient. A prospective user (e.g., naïve, decision-maker, business analytics, admin, Govt. official, etc.) submit the data request over the user interface, using keywords, number of tweets of interest and name of emotional (optional). The designed system, evaluates the both statistics over a real-time.

The pre-processing stage, each extracted tweet is divided into tokens with estimated probability (T_{prob} = *Happy, Sad,* etc.). The python sentiment analysis is conducted using *NLTK* library [12]. The probability score ($WGT_{Prob}.$) is weighted a value, as to account of fewer negative tweets as there are positive and neutral ones. Additionally, token below a *threshold count* is truncated, since it is not significant and often little contributory. The latter is determined using the H_{10} entropy, formalized as *Eq. 1* as

$$\mathrm{H}_{10}(\mathrm{token}) = -\sum\nolimits_{\mathrm{s}\in\mathrm{sentiment}} (\mathrm{p}_{(\mathrm{s|token})} \log_{10} \mathrm{p}_{(\mathrm{s|token})}). \tag{1}$$

The measures for *positive, neutral* and *negative* emojis are found. Finally, as given that a tweet is composed of several words; all the different features are aggregated /summed for each word, as to obtain an overall *tweet_value* ($\mathrm{T_v}$), normalized by the *tweet_length* (T_{ct}). The *positive score* (s^+) and *negative scores*(s^-) for each tweet are determined as average of the both scores using *Eq. 2* and *Eq. 3* respectively. The overall *Sentiment Score* (SC) is estimated for the locating the topic proportion.

$$\mathrm{s}^+ = \frac{\sum_{\mathrm{i}\in\mathrm{t}} \mathrm{pos\_score_i}}{n} \tag{2}$$

$$\mathrm{s}^- = \frac{\sum_{\mathrm{i}\in\mathrm{t}} \mathrm{neg\_score_i}}{n} \tag{3}$$

$$\text{Overall Sentiment Score } (SC) = (s^+ - s^-) \tag{4}$$

To extract emotion from a tweet, the topical words (bigram) are taken from tweet content, based on '*item response theory*' [12] and further categorized using its unsupervised features. The proposed algorithm is based on '*Topic proportion*' that helps to identify related sentiment terms located to a topic sentiment lexicon.

Algorithm: Extracting *SL* and EQ of *Tweet_objects*

Input: Topic name (T_{kw}), No. of Tweets (*ToI*), and Emotion _name (*Emo*)

Output: *Tweet_list, EQ, SL*

```
Step 1: Cleaning up the tweets                     /*denoising the ttweets such as links,@,#,etc.*/
      df['Tweet'] = df['Tweet'].apply(cleanUpTweets)
      all_tweets=df['Tweet'].tolist()
      df = df.drop(df[df['Tweet'] == ' '].index)
Step 2: Gathering and Storing Emotions             /*text2emotion EQ value generated for each tweet*/
      dict = te.get_emotion(text)
      for key in dict:
        if (key == "Happy"):
          Happy.append(dict[key])/*Similartly also for "Sad","Fear" etc*/
Step 3: Finding the dominating Emotion     /*For each tweet, store the dominating emotion values */
      for i in range(0,NoOfTweets):
      maxel=max(Happy[i],Sad[i],Angry[i],Fear[i])
           if((maxel==Happy[i])and(maxel!=0)):
             Happy2.append(all_tweets[i])                /*Similarly also for Sad2, Fear2 etc*/
Step 4: Finding the overall percentage of each emotion after analyzing each tweet on the topic
                    /*All the values in each emotion are added and for each emotion percent is calculat-
                          ed by dividing summation of values for that emotion by the total sum*/
      happysum = sum(Happy)
      totalsum = zip(Happy, Sad, Angry, Surprise, Fear)
      for x in totalsum:
         for y in x:
           total = total + y;
        happypercent = (happysum / total)                         * 100/*Similarly the percentage
                                                                  of other emotions is calculated*/
 Step5: Polarity analysis                  /*For each tweet, polarity is analyzed& values are clas-
                                                                          sified into three groups */
      for tweet in self.tweets:
        analysis = TextBlob(tweet.text)
        polarity += analysis.sentiment.polarity
          if (analysis.sentiment.polarity== 0 neutral += 1;
          elif (analysis.sentiment.polarity> 0 and analysis.sentiment.polarity<0.3):wpositive += 1;
          elif (analysis.sentiment.polarity>0.3 and analysis.sentiment.polarity<=0.6): positive += 1;
          elif (analysis.sentiment.polarity> 0.6 and analysis.sentiment.polarity<= 1):spositive += 1;
         elif (analysis.sentiment.polarity> -0.3 and analysis.sentiment.polarity<= 0):wnegative += 1;
         elif (analysis.sentiment.polarity> -0.6 and analysis.sentiment.polarity<= -0.3): negative += 1;
        elif (analysis.sentiment.polarity> -1 and analysis.sentiment.polarity<= -0.6):snegative += 1;
```

The proposed algorithm is driven on the topic model, as extracted *tweet-object* is a mixture of one or more topics. A lexicon approach measures the sentiment of a group of documents '*corpus*', with the help of dictionary of words with associated polarity scores. A dictionary of lexicons elements is added externally to the corpus for the purpose of enhancing the embedded lexicons. The proposed *lexicon-based* sentiment estimation for *positive* and *negative* sentiment using *Eq. 5* and *Eq. 6* respectively, and formalized as following,

$$P(w|+) = \frac{M_w}{|N^+|} \tag{5}$$

$$P(w|-) = \frac{M_w}{|N^-|}. \tag{6}$$

Here, M_w is the set of messages containing lexical token '*w*'. The *positive* and *negative* sentiments are coded as N^+, N^-, for each message (message) '*m*', the log likelihood ratio is calculated using the *Eq. 7*:

$$S_m = \sum\nolimits_{i=1}^{n} \log\left(\frac{P(w_i|-)}{P(w_i|+)}\right) \tag{7}$$

Here, *lexical unit* of the dictionary is presented by *w* and *n* is the number of words and collocations included in the dictionary, existences with the sentence *m*.

3.3 System Use-Cases of Sentiment and Emotional Analytics

The designed system is plugged with an interactive user interface (UI) for several types of users. The UI aimed is to offer estimated statistics and processed data for the different decision making and analytics purposes. There are several use-cases of designed system listed during the design phase, e.g. related viral information to an emotional value, comparing the emotional causes of more than one viral/trending information, etc.

The *first use-case* is coming from a naive user, as '*basic search on social media data for a sentiment and emotion value*', illustrated in Fig. 2. Here, two parts of UI steers user cognitive tasks, e.g. search, exploration, browse, etc., on the real-time analytics. For a user input '*keyword/hashtag/emotion name*', the system extracts the related tweets and prepares intermediate statistics to be shown in the graphical scheme.

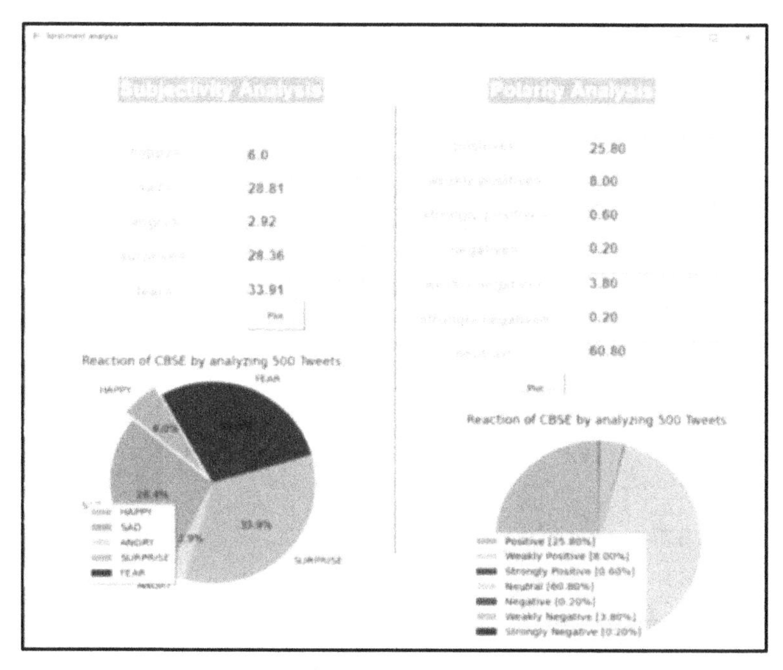

Fig. 2. User interface (UI) for 1st use-case, for the sentiment and EQs explorations.

Here, *Part 1* illustrates both information using pie chart with scores 'as % values', and within it tweets lists of the user interest is available. In *Part 2*, the feature of compares and explores offers a new dimension of the designed prototype. Here, a user may be interested to compare effects of pair of viral and trending information, for playing with the relevant social media data, further two trending or viral information or may delve into the deeper view of these statistics and values and related tweets.

The *second use-case* is robust exploration into the relevant viral/trending information for a user query. The designed system may be adapted for the social data exploration within emotion quotient, illustrated in Fig. 3. The matching viral information easily adapted for the analytics.

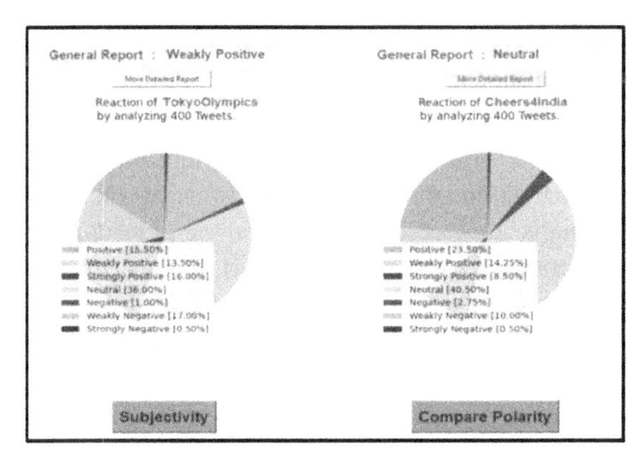

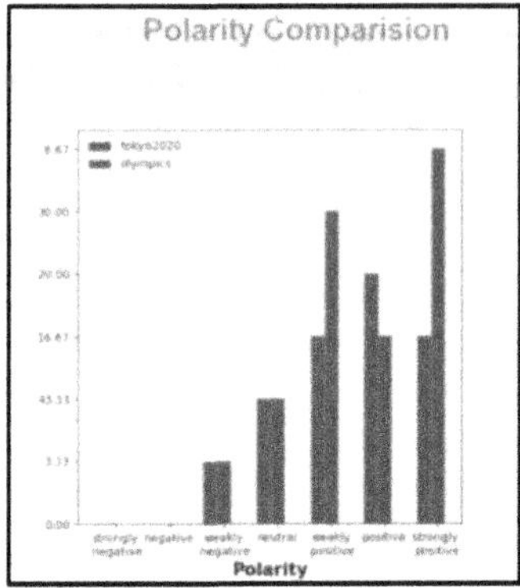

Fig. 3. User interface for comparing *SL* and *EQs* with *polarity* (*'Tokyo2020' & 'Olympic'*).

Further, *third use-case* is a capacity, to deliver a matching list of trending/viral information's for and matching viral the system for the exploration within the tweets data for an input, '*emotion quotient*'. The viral/trending topics may be extracted with the presence of the same emotional quotient values.

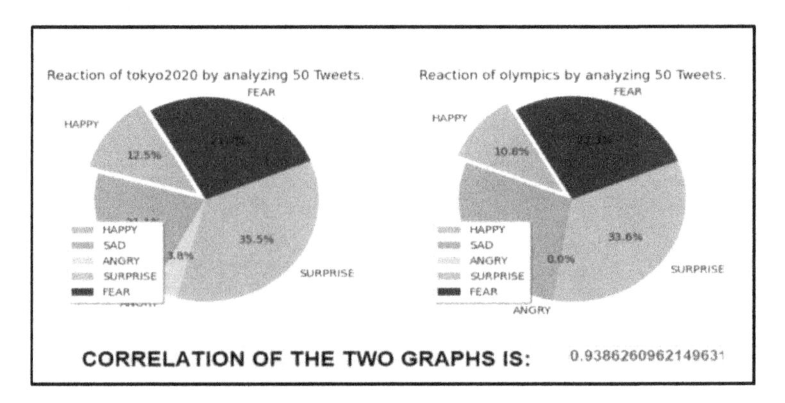

Fig. 4. User interface for (*'tokyo2020'* & *'olympic'* for 100 *IoT) sentiment* and *EQ* polarity.

The *fourth use-case* is an inherent capability for systematic comparative view between more than one user inputs (*trending and viral information*) and its detailed view on the emotion quotients (EQs) and further exploration on the generated tweet text, also illustrated in Fig. 4.

3.4 Working Example

The *Olympic* is the world's biggest sporting event, as plethora of sentiments and emotion affect attached, as abundant amount of social media data is contributed globally on social networking platforms. Similarly, the designed algorithm may be adapted for event, as it caters diverse user and their relevant content and Meta data, e.g. comments, share, tagging, repost, etc.

The designed prototype is configured over the twitter API and its settings. The data relevant data may extract on real-time basis and subsequently pre-processed, using traditional text processing. The interactive UI is designed to capture user cognitive actions support visualizations of derived analytics for the provided social media data.

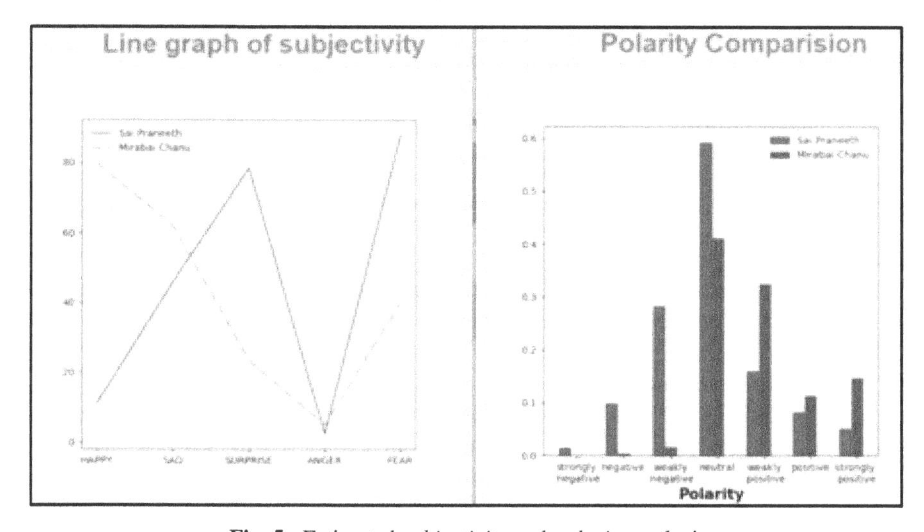

Fig. 5. Estimated *subjectivity* and *polarity* analysis

For instance, the men's single badminton player '*Sai Praneeth*', who lost both his initial matches and is out of the tournament, results to polarity positives (22.8%) is less than negative's (30.8%), here '*Fear*' and '*Surprise*' are dominating. Further, between '*Sai Praneeth*' and '*MirabaiChanu*' as trending pattern, polarity graph reveals interesting patterns, despite difference on sport zones, in Fig. 5 under Subjectivity graph. In subjectivity *Happiness* surrounds *Mirabai Chanu* with some sadness, for *Sai Praneeth surprise*, *Fear*, *Sadness* is mostly visible.

Next, *Sai Praneeth* and *KentoMomota* are compared to evolving polarity patterns. Subjectivity analysis appears similar, except to '*fear*' emotion, as higher fear is surrounded to '*KentoMomota*', illustrated in Fig. 5, under Polarity graph, while polarity graph reveals '*KentoMomota*' with higher positivity, while '*Sai Praneeth*' has trace of '*weak negativity*'.

4 Performance Assessment and Evaluation

4.1 Data Settings

The experimental setup includes software used *Jypter notebook*, *Visual Studio* and *PyCharm*. Twitter API isused for the real-time data extraction, at instance to be extracted 1 to 800 tweets for comparing 1 to 1600 tweets for analytics. There are several libraries, e.g. *tweepy, re, text2emotion, textblob, pandas, numpy, matplotlib, sys, csv* and *Tkinter* (for user interface). The hardware components includes, 2 PCs with specification as: *AMD Ryzen 5 2500U processer with Radeon Gfx 2.00 GHz, RAM 8 GB* and another with processor of *Intel Core i5(8250U)CPU @ 1.60 GHz, Intel UHD* graphics 620, 12 GB RAM).

The user interface (UI) is designed using *Python library Tkinter* and statistics are using *Python library Matplotlib* for extracted tweet objects for a user request. The real-time extraction of *tweets* objects using *API* and further cleaned and stored into *Pandas Data Frame*. The number of tweets for extraction is related to a user input, as for each user input it is related. The experiments are conducted using the several input values, 1 to 1600 numbers of tweets extracted on real-time basis on the prototype.

4.2 Performance on Sentiments and Emotion Quotients (EQs) Estimation

The performance evaluation of the designed system for the real-time data processing to estimate the underlying statistics outlines several insights, specific to system's feasibility and its viability for just-in-time decision making and analysis. The *Query Length (QL)*, *Number of tweet of interests (ToI)*, and *Performance (processing time)* are employed as key performance assessment parameters.

Here, *QL* defined as the dimension of the *keywords/hashtasg*, i.e. *No. of characters* in the user input *keywords or hashtags*. For a user input '*Tokyo2020*', *QL is* 9. Similarly, *ToI* value directs system to extract at least these many recent tweets from API on in real-time basis, e.g. '*Tokyo2020*' with *ToI* value 20 fetches recent 20 tweets at the time.

Figure 6 illustrates the *overall processing time*, formalized as '*total time required for the preparation of both statistics on real-time*', real-time appearing viral topics dated on *30 July 2021*. For each topic, relevant tweet set is extracted and subsequently ranked for implicit pre-processing. In this, at least *1000* tweets are fetched on co-located viral topics, based on priority based preference. A generic estimation of processing time usually increased with the higher no. of tweets, as higher *no. of tweets* harness increased coverage of sentiments and emotions.

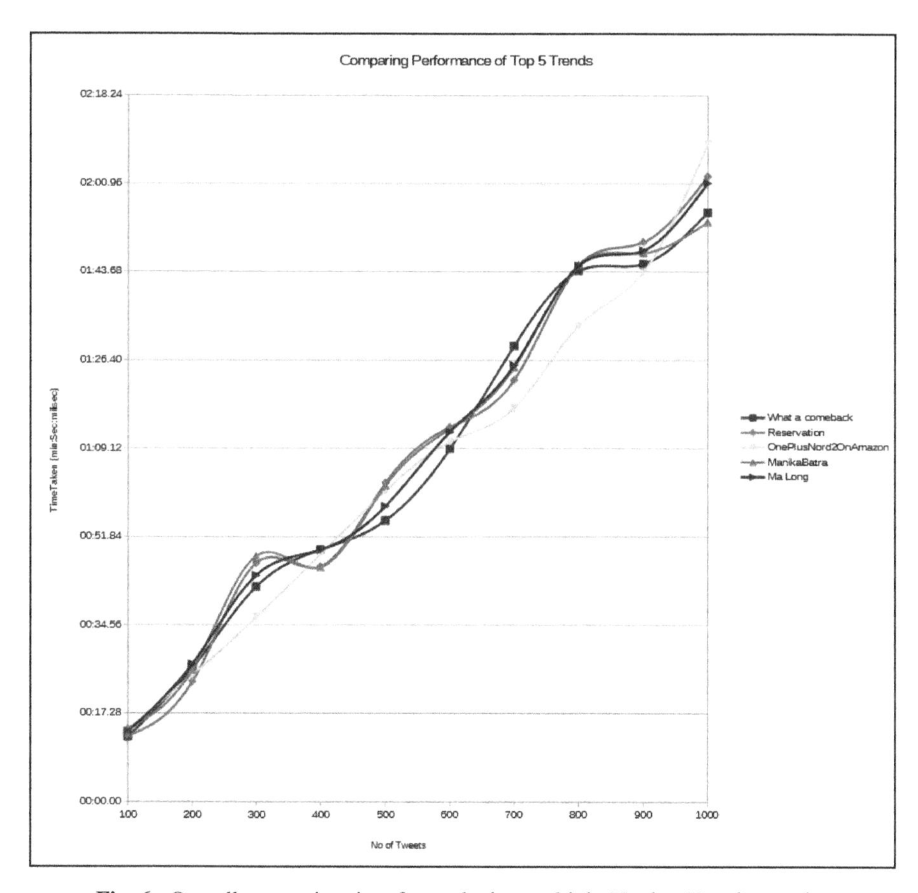

Fig. 6. Overall processing time for analyzing multiple *Viral or Trending* topics

Figure 7 depicts the *overall processing time* for the user submitted query or selected viral topics on real-time data. The different size of user input, as query length (QL) is adapted for the assessment, with an aim to highlight the fact that as *QL* increased to a level affects the overall computational time. In a generic settings, to different viral/trending topics appearing, as to ensure the *variable QL* and observed the processing time patterns for at least recent 1000 tweets on each viral or trending topics.

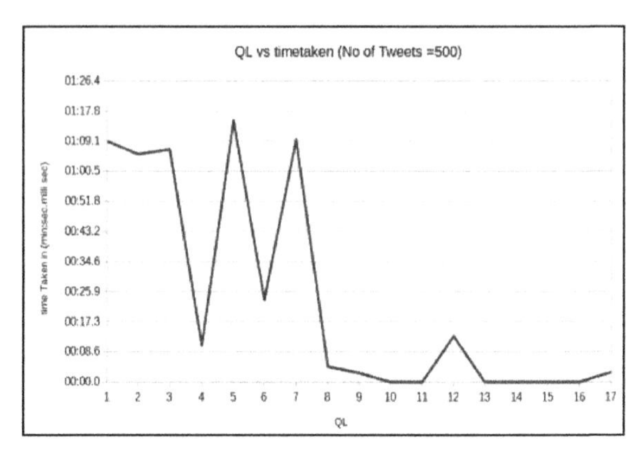

Fig. 7. Overall query response time (QRT) for different size of user query, as '*QL*'

4.3 Accuracy on Sentiment and Emotional Quotient Generation

The evaluation of accuracy in the detection of accurate sentiment levels (SLs), as sentiment levels are formalized as '*the quantum of a sentiment contributes to viral information, w.r.t user contributed data*', and each user specified inputs is pivotal for the analysis. The performance statistics for the estimation of accuracy on specified *sentiment-levels* are listed in Table 1. A brief comparison over *500 ToIs* asserts overall accuracy 85% on the prediction of sentiment-levels (e.g. *positive, negative, and neutral* as 89%, *88.58%, and 77.62%*).

Table 1. Accuracy on estimation of each specified *sentiment -levels.*

			Predicted		
			Positive	Negative	Neutral
			234	***108***	***155***
Actual	Positive	***181***	143	14	24
	Negative	***177***	54	86	37
	Neutral	***139***	37	8	94

Table 2 lists the overall accuracy statistics for the generation of emotion quotients (EQs), here EQ is formalized as '*the emotion influx created by the viral information on different fundamental emotions*'. The designed algorithm predicts EQs accuracy of 87%, with 89%, 85.88%, 85.66%, 89.88% and 88% for spectrum of the emotions, e.g. *Happy, Sad, Fear, Surprise and Anger* respectively.

Table 2. Accuracy on estimation of each specified emotion.

			Predicted				
			Happy	Sad	Fear	Surprise	Anger
			56	*24*	*22*	*12*	*16*
Actual	Happy	*33*	22	5	2	2	2
	Sad	*40*	15	6	8	6	5
	Fear	*18*	6	2	7	0	3
	Surprise	*15*	5	5	1	4	0
	Anger	*24*	8	6	4	0	6

4.4 Overall Retrieval Performance

The performance of a designed sentiment and EQs estimation on the traditional retrieval metrics are key indicators for the feasibility for decision making and related functions. The traditional metrics are adapted for the evaluation of the system performance, e.g. *Precision, recall and f-measure.* The precision is adapted in its fundamental notion, as a measure of 'precisely matched results to the user input', and recall as a measure '*closely relevant result to the user query*'. *F-measure* is a geometric mean of precision and recall.

Table 3 lists these indicators, when experimented with varying degree of user input (*Query length* and *ToI*).

Table 3. Traditional retrieval metrics for *sentiment level* and *emotion quotient estimation*

Sentiment type	Metric (scores)	
Positive	Precision	0.611
	Recall	0.389
	F-measure	0.475
Negative	Precision	0.796
	Recall	0.224
	F-measure	0.351
Neutral	Precision	0.606
	Recall	0.240
	F-measure	0.344

Emotion type	Metric (scores)	
Happy	Precision	0.393
	Recall	0.259
	F-measure	0.312
Sad	Precision	0.250
	Recall	0.113
	F-measure	0.156
Fear	Precision	0.318
	Recall	0.108
	F-measure	0.161
Surprise	Precision	0.333
	Recall	0.048
	F-measure	0.084
Anger	Precision	0.375
	Recall	0.063
	F-measure	0.107

5 Conclusion

The data generated over the various Social media platforms trigger significant changes on the public sentiment and emotional flux. A novel algorithm is designed to estimate the sentimental and emotional quotient of viral or trending information in real-time. The work is in line with current need to textual emoticons mining in several real-life application scenarios. The algorithm estimates sentiment and EQs of a user requested viral/trending information over the twitter on just-in-time basis over real-time twitter data. The approach builds a corpus of tweets and related fields where each tweet is classified with respective emotion based on lexicon approach. The systematic evaluation asserts the significance of delivered statistics for the user input '*viral information*', and its usability. The feasibility analysis of statistics for real-time analysis and decision-making is uncovered using 04 use-cases, also outlines the key features of social media data for the purpose.

The embedding interactive user interface is one of future scope of the current algorithm. An intent model for the estimating the user interest and its correlation with current trending or viral information in the social media platforms is another tentative direction.

References

1. Bikel, D.M., Sorensen, J.: If we want your opinion. In: International conference on semantic computing (ICSC 2007), pp. 493–500 (2007). https://doi.org/10.1109/ICSC.2007.81
2. Cambria, E., Schuller, B., Xia, Y., Havasi, C.: New avenues in opinion mining and sentiment analysis. IEEE Intell. Syst. **28**(2), 15–21 (2013). https://doi.org/10.1109/MIS.2013.30
3. Chen, R., Xu, W.: The determinants of online customer ratings: a combined domain ontology and topic text analytics approach. Electron. Comm. Res.(2016). https://doi.org/10.1007/s10660-016-9243-6
4. Ding, X., Liu, B., Yu, P.S.: A holistic lexicon-based approach to opinion mining. In: Proceedings of the 2008 International Conference on Web Search and Data Mining, pp. 231–240 (2008). https://doi.org/10.1145/1341531.1341561
5. Esuli, A., Sebastiani, F.: Sentiwordnet: A publicly available lexical resource for opinion mining. In: Proceedings of 5th Language Resources and Evaluation, Vol. 6, pp. 417–422 (2006)
6. Fei, G., Liu, B., Hsu, M., Castellanos, M., Ghosh, R.: A dictionary-based approach to identifying aspects implied by adjectives for opinion mining. In: Proceedings of 24th International Conference on Computational Linguistics, p. 309 (2012)
7. Feldman, R., Fresco, M., Goldenberg, J., Netzer, O., Ungar, L.: Extracting product comparisons from discussion boards. In: Seventh IEEE International Conference on Data Mining (ICDM 2007) (pp. 469–474) (2007). https://doi.org/10.1109/ICDM.2007.27. A model for sentiment and emotion analysis of unstructured 197 123
8. Godbole, N., Srinivasaiah, M., Skiena, S.: Large-scale sentiment analysis for news and blogs. Proc. Int. Conf. Weblogs Soc. Media (ICWSM) **7**(21), 219–222 (2007)
9. Hamouda, A., Rohaim, M.: Reviews classification using sentiwordnet lexicon. In: World Congress on Computer Science and Information Technology (2011)
10. Jindal, N., Liu, B.: Mining comparative sentences and relations. In: Proceedings of the 21st National Conference on Artificial Intelligence, Vol. 2, pp. 1331–1336 (2006)

11. Van de Kauter, M., Breesch, D., Hoste, V.: Fine-grained analysis of explicit and implicit sentiment in financial news articles. Expert Syst. Appl. **42**(11), 4999–5010 (2015). https://doi.org/10.1016/j.eswa.2015.02.007
12. Loper, E., Bird, S.: Nltk: the natural language toolkit. *arXiv preprint cs/0205028* (2002)
13. Li, Y., Qin, Z., Xu, W., Guo, J.: A holistic model of mining product aspects and associated sentiments from online reviews. Multimed. Tools Appl. **74**(23), 10177–10194 (2015). https://doi.org/10.1007/s11042-014-2158-0
14. Liu, B.: Sentiment analysis and subjectivity. Handbook Nat. Lang. Proc. **2**, 627–666 (2010)
15. Liu, B.: Opinion mining and sentiment analysis. In: Web data Mining: Exploring Hyperlinks, Contents, and Usage Data, pp. 459–526 (2011). https://doi.org/10.1007/978-3-642-19460-3_11
16. Liu, B.: Sentiment analysis and opinion mining. Synth. Lect. Human Lang. Technol. **5**(1), 1–167 (2012). https://doi.org/10.2200/S00416ED1V01Y201204HLT016
17. Liu, P., Gulla, J.A., Zhang, L.: Dynamic topic-based sentiment analysis of large-scale online news. In: Cellary, W., Mokbel, M.F., Wang, J., Wang, H., Zhou, R., Zhang, Y. (eds.) WISE 2016. LNCS, vol. 10042, pp. 3–18. Springer, Cham (2016). https://doi.org/10.1007/978-3-319-48743-4_1
18. Ma, Y., Chen, G., Wei, Q.: Finding users preferences from large-scale online reviews for personalized recommendation. Electron. Commer. Res. **17**(1), 3–29 (2017). https://doi.org/10.1007/s10660-016-9240-9
19. Mo, S.Y.K., Liu, A., Yang, S.Y.: News sentiment to market impact and its feedback effect. Environ. Syst. Dec. **36**(2), 158–166 (2016). https://doi.org/10.1007/s10669-016-9590-9
20. Montoyo, A., Martı́Nez-Barco, P., Balahur, A.: Subjectivity and sentiment analysis: an overview of the current state of the area and envisaged developments. Dec. Supp. Syst. **53**(4), 675–679(2012). https://doi.org/10.1016/j.dss.2012.05.022
21. Nassirtoussi, A.K., Aghabozorgi, S., Wah, T.Y., Ngo, D.C.L.: Text mining of newsheadlines for forex market prediction: a multi-layer dimension reduction algorithm with semantics and sentiment. Expert Syst. Appl. **42**(1), 306–324 (2015). https://doi.org/10.1016/j.eswa.2014.08.004
22. Ohana, B.: Opinion mining with the sentwordnet lexical resource. M.Sc. dissertation, Dublin Institute of Technology (2009)
23. Pang, B., Lee, L.: A sentimental education: Sentiment analysis using subjectivity summarization based on minimum cuts. In: Proceedings of the 42nd Annual Meeting on Association for Computational Linguistics, p. 271 (2004). https://doi.org/10.3115/1218955.1218990
24. Pang, B., Lee, L.: Opinion mining and sentiment analysis. Found. Trends Inf. Retr. **2**(1–2), 1–135 (2008). https://doi.org/10.1561/1500000011
25. Pang, B., Lee, L., Vaithyanathan, S.: Thumbs up?: sentiment classification using machine learning techniques. In: Proceedings of the ACL-02 Conference on Empirical Methods in Natural Language Processing, Vol. 10, pp. 79–86 (2002). https://doi.org/10.3115/1118693.1118704
26. Parkhe, V., Biswas, B.: Sentiment analysis of movie reviews: finding most important movie aspects using driving factors. Soft. Comput. **20**(9), 3373–3379 (2016). https://doi.org/10.1007/s00500-015-1779-1
27. Peng, J., Choo, K.K.R., Ashman, H.: Astroturfing detection in social media: using binary n-gram analysis for authorship attribution. In: Proceedings of the 15th IEEE International Conference on Trust, Security and Privacy in Computing and Communications (TrustCom 2016), pp. 121–1286 (2016)
28. Peng, J., Choo, K.K.R., Ashman, H.: Bit-level n-gram based forensic authorship analysis on social media: Identifying individuals from linguistic profiles. J. Netw. Comput. Appl. **70**, 171–182 (2016). https://doi.org/10.1016/j.jnca.2016.04.001

29. Peng, J., Detchon, S., Choo, K.K.R., Ashman, H.: Astroturfing detection in social media: a binary n-gram-based approach. Concurr. Comput. Pract. Exp. (2016). https://doi.org/10.1002/cpe.4013.198 J. K. Rout et al. 123
30. Pro¨llochs, N., Feuerriegel, S., Neumann, D.: Enhancing sentiment analysis of financial news by detecting negation scopes. In: Proceedings of the 48th Hawaii International Conference on System Sciences (HICSS), pp. 959–968 (2015). https://doi.org/10.1109/HICSS.2015.119
31. Robinson, R., Goh, T.T., Zhang, R.: Textual factors in online product reviews: a foundation for a more influential approach to opinion mining. Electron. Commer. Res. **12**(3), 301–330 (2012). https://doi.org/10.1007/s10660-012-9095-7
32. Rout, J., Dalmia, A., Choo, K.K.R., Bakshi, S., Jena, S.: Revisiting semi-supervised learning for online deceptive review detection. IEEE Access **5**(1), 1319–1327 (2017). https://doi.org/10.1109/ACCESS.2017.2655032
33. Rout, J., Singh, S., Jena, S., Bakshi, S.: Deceptive review detection using labeled and unlabeled data. Multimed. Tools Appl. **76**(3), 3187–3211 (2017). https://doi.org/10.1007/s11042-016-3819-y
34. Sadegh, M., Ibrahim, R., Othman, Z.A.: Opinion mining and sentiment analysis: a survey. Int. J. Comput. Technol. **2**(3), 171–178 (2012)
35. Song, L., Lau, R.Y.K., Kwok, R.-W., Mirkovski, K., Dou, W.: Who are the spoilers in social media marketing? Incremental learning of latent semantics for social spam detection. Electron. Commer. Res. **17**(1), 51–81 (2016). https://doi.org/10.1007/s10660-016-9244-5
36. Taboada, M., Brooke, J., Tofiloski, M., Voll, K., Stede, M.: Lexicon-based methods for sentiment analysis. Comput. Linguist. **37**(2), 267–307 (2011). https://doi.org/10.1162/COLI_a_00049
37. Tang, H., Tan, S., Cheng, X.: A survey on sentiment detection of reviews. Expert Syst. Appl. **36**(7), 10760–10773 (2009). https://doi.org/10.1016/j.eswa.2009.02.063
38. Turney, P.D.: Thumbs up or thumbs down?: semantic orientation applied to unsupervised classification of reviews. In: Proceedings of the 40th Annual Meeting on Association for Computational Linguistics, pp. 417–424 (2002). https://doi.org/10.3115/1073083.1073153
39. Wang, D., Li, J., Xu, K., Wu, Y.: Sentiment community detection: exploring sentiments and relationships in social networks. Electron. Commer. Res. **17**(1), 103–132 (2017). https://doi.org/10.1007/s10660-016-9233-8
40. Zheng, L., Wang, H., Gao, S.: Sentimental feature selection for sentiment analysis of chinese online reviews. Int. J. Mach. Learn. Cybern. **6**. https://doi.org/10.1007/s13042-015-0347-4
41. Alves, A.L.F.: A spatial and temporal sentiment analysis approach applied to Twitter microtexts. J. Inf. Data Manag. **6**, 118 (2015)
42. Chaabani, Y., Toujani, R., Akaichi, J.: Sentiment analysis method for tracking touristics reviews in social media network. In: Proceedings of the International Conference on Intelligent Interactive Multimedia Systems and Services, Gold Coast, Australia, 20–22 May 2018 (2018)
43. Contractor, D.: Tracking political elections on social media: applications and experience. In: Proceedings of the Twenty-Fourth International Joint Conference on Artificial Intelligence, Buenos Aires, Argentina, 25–31 July 2015 (2015)
44. Bai, H., Yu, G.: A Weibo-based approach to disaster informatics: incidents monitor in post-disaster situation via Weibo text negative sentiment analysis. Nat. Hazards **83**, 1177–1196 (2016)
45. Brynielsson, J., Johansson, F., Jonsson, C., Westling, A.: Emotion classification of social media posts for estimating people's reactions to communicated alert messages during crises. Secur. Inf. **3**(1), 1–11 (2014). https://doi.org/10.1186/s13388-014-0007-3

Comparative Analysis of Edge Detectors Applying on the Noisy Image Using Edge-Preserving Filter

Tapendra Kumar(✉) and Shawli Bardhan

School of Computing, Indian Institute of Information Technology Una, Una 177220, Himachal Pradesh, India

Abstract. In computer vision, edge detection is a fundamental technique. It is used as a pre-processing technique to make image segmentation, pattern recognition, and feature extraction more comfortable. Digital images are often corrupted by the noise that causes the detection of spurious edges during edge detection. Thus, we'd like to suppress the maximum amount of noise as potential while retaining important image features such as edges, corners, and other sharp structures. This research compares multiple edge detection methods applied to a filtered image by adding speckle noise. In this paper, four edge detection operators have been applied to an image denoised by various edge-preserving filters, and their performance is evaluated based on the performance metrics peak signal-to-noise ratio (PSNR) and mean squared error (MSE). Images from the Barcelona Images for Perceptual Edge Detection Dataset (BIPED) are used for performance evaluation of filters and edge detection techniques. The experimental results show that a bilateral filter with a Canny edge detection operator is the most optimized method for edge detection of speckle-noise-affected images.

Keywords: Image processing · Edge detection · Noise · Edge-preserving filters · PSNR · MSE · BIPED

1 Introduction

Edge detection is one of the most prevalent and widely used computer vision algorithms, and it is responsible for detecting the most important features and properties of the objects in a digital image. These properties of the object could include anomalies in the photometrical, geometrical shape, and other bodily appearance of the object. Dissimilarities in the image's grey level are caused by these characteristics.

An edge is a set of connected pixels where an abrupt change of brightness or intensity occurs. Noisy images will cause worse performance in edge detection since it prompts variability in the local contrast along an edge. Furthermore, at low signal-to-noise ratios (SNR), known as edge contrast, divided by the noise level, local contrast reversals can occur [1]. So, to effectively extract edges from a noisy image, we have to denoise it first. The most common approach for removing noise is to smooth the images with a Gaussian filter. Although smoothing reduces noise, it can potentially blur and weaken contrast

I. Woungang et al. (Eds.): ANTIC 2021, CCIS 1534, pp. 359–373, 2022.
https://doi.org/10.1007/978-3-030-96040-7_29

around edges or even blend adjacent edges. It has been a challenging task to minimize noise as much as feasible while retaining the important structural properties of the image. Over the years, several non-linear filtering techniques have been proposed that reduce the unwanted effects of linear filtering. They can smooth away noise while retaining the image details and local geometries. Therefore, this paper proposes an optimized approach for edge detection of noisy images based on edge-preserving denoising. This work familiarizes with multiple traditional edge detection techniques and improves edge detection with edge-preserving filters. Then a comparative analysis is conducted in this research to see the optimized approach for edge detection using various metrics such as PSNR and MSE.

This paper is organized into seven sections. Section 2 summarizes the literature review on edge detection of noisy images. Section 3 discusses some of the edge-preserving filters that will be used to filter noisy pictures before edge detection. A brief description of the frequently applied edge detection methods is presented in Sect. 4. Section 5 presents methods of performance evaluation for filters and edge detectors. Experimental methodology is presented in Sect. 6, and finally, in Sect. 7, results and concluding remarks are given.

2 Review of Previous Work

Over the years, lots of work has been done towards analyzing the effect of noise on several edge detection techniques and the importance of filtration before edge detection. D Poobathyand and RM Chezian [2] compared the performance of several edge detection operators using the MSE and PSNR matrices and found that canny operator to be the best amongst others in edge detection accuracy and also the most time-consuming. Normally, Images are convolved with a filter for smoothing out noise before edge detection. Fawwaz et al. [3] has suggested a Gaussian bilateral filter for this purpose and by comparing its performance with Gaussian filter using PSNR value, concluded that bilateral filter performs better at preserving edges during filtering and using canny detector after Bilateral Filter is the most efficient approach for edge detection of Satellite imagery. A gaussian filter is also used in the standard canny edge detection technique for smoothing, but it performs very poorly in removing salt and pepper noise. Sekehravani et al. [4] used of a canny edge detection algorithm based on the median filter for edge detection of salt and pepper noise affected image and showed that using the suggested method, edges in noisy images can be effectively located and superior to the traditional Canny algorithm in terms of edge and detail identification. R Maini and JS Sohal [5] studied the Prewitt operator's performance for noisy images and observed that it detects edges of images affected by Poisson noise effectively, but that it performs poorly for other types of noise. Ruslau et al. [6] performed edge detection in noisy image and concluded that the LoG operator was able to perform better in detecting edges in the noisy image and reducing noise than Canny. However, for noisy images with step edges, the Canny operator's edge detection results can be improved by choosing the appropriate threshold values. Z Hussain and D Agarwal [7] compared various techniques used for edge detection of flame images. They observed that Local Binary Pattern (LBP) yields better results than the Canny operator for edge detection of flame images. Edge detection by LBP produces clear and continuous edges, which are required for estimating different flame characteristics. R. Swarnalakshmi [8] evaluated the performance of the some widely used edge detection methods, namely Sobel, Prewitt, and Canny. He used salt & pepper noise affected images that have been filtered with a median filter for the experiment and found that in case of detecting edges of medical image and Satellite image, canny edge detection exhibits the best performance. However, in the case of detecting edges of the Lena image, the Sobel edge detection technique outperforms other methods. M Roushdy [9] presented a comparative study of edge detection operators under noise conditions and showed that Boie-Cox, Shen-Castan, Canny, LoG, Prewitt, and Sobel perform better, respectively. It also concludes that using morphological filters before applying edge detection operators increases their efficiency. RK Singh and D Shaw [10] presented a comparative study of the effect of various types of noise namely Speckle noise, Salt and Pepper noise, and Gaussian noise of different variances on images and their edge maps computed by various edge detection operators. Aishwarya et al. [11] presented an evaluation of filter performance and found that edge detection methods combined with the BM3D filter have done a better job of detecting edges compared to other filters like median, Gaussian, and bilateral filters. For identifying digits from noisy SVHN pictures, Fawwaz et al. [12] recommended using Bilateral filters before applying an edge detector. They found that the edge map of the image affected with speckle noise is better than that of Salt & Pepper, and Gaussian noise and digits still can be recognized in the edge map. Canny operator provides the best performance for identifying edges of SVHN images affected by Speckle noise.

3 Edge Preserving Denoising Techniques

Digital images can be affected by noise during the acquisition, transmission, and compression processes, resulting in deteriorated visual image quality. Image denoising is to approximate the original image by eliminating noise from an image. However, because noise, edge, and texture all represent variations in pixel intensity values, it is also possible that in the denoising process, noise, as well as the edge pixels, are smoothed out due to the inability to differentiate them. As a result, denoised images would certainly lose some features. A primary problem is to recover important information from a noisy image in the denoising process to obtain a high-quality image. In recent decades, several non-linear filtering techniques have been proposed that preserve edges and other sharp structures of the image while reducing noise. In this section, we will be focusing on some of the most promising edge-preserving denoising techniques.

3.1 Adaptive Median Filter

The Adaptive Median Filter (AMF) uses spatial processing to identify pixels of an image that have been affected by impulsive noise. AMF compares each pixel with its surrounding neighboring pixels to label pixels as noise. The fundamental difference between AMF and median filter is that the size of the neighborhood and the comparison threshold are both adjustable. If a pixel differs from the majority of its neighbors while also not being structurally aligned with those that are similar to it, it is classified as impulsive noise. Pixels labeled as noise have their values replaced by the median values of their neighbors that passed the noise labeling test. The figure below depicts a flow chart of the AMF process (Fig. 1):

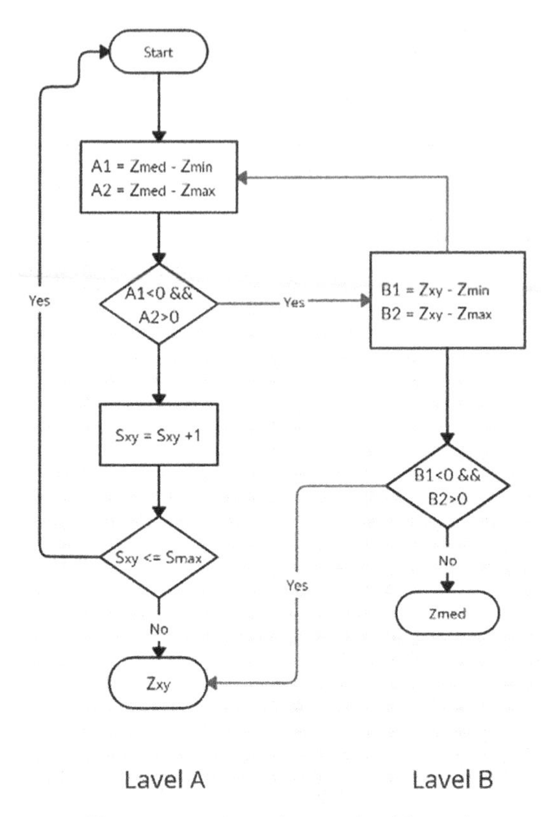

Fig. 1. Flow chart of AMF algorithm [13]

Here Z_{xy} represents the gray level of the pixel at location (x, y) of the corrupted image, S_{xy} is the size of surrounding neighbourhood. And Z_{min}, Z_{med} and Z_{max} minimum, median and maximum grey levels of the pixels in the neighbourhood respectively. And S_{max} is the maximum allowed size of S_{xy} [14].

3.2 Bilateral Filter

As its name indicates, the bilateral filter incorporates two types of filtering, namely domain, and range filtering. By averaging pixel intensity values with weights that decrease with distance, domain filtering enforces geometric locality between pixels based on their spatial proximity. On the other hand, range filtering enforces photometric locality between pixels by weighing image intensity values with weights that increase with the increasing similarity of neighboring pixels to the given reference pixel

[15]. The main principle bilateral filter is that for one pixel to impact another in the denoising process, it must not only be nearby but also have a similar intensity value. It replaces each pixel's intensity values with the weighted average of intensity values from its neighbors. This weight may well be calculated using a Gaussian distribution. Weights are determined not just by the Euclidean distance but also by the photometric differences between pixels. The bilateral filter is:

$$P^{filtered} = \frac{1}{W_p} \sum\nolimits_{x_i \int \Omega} P(x_i) k_r(||P(x_i) - P(x)||) k_s(||x_i - x||) \tag{1}$$

Here W_p is normalization factor which is defined as:

$$W_p = \sum\nolimits_{x_i \in \Omega} k_r(||P(x_i) - P(x)||) k_s(||x_i - x||) \tag{2}$$

Here, k_s is the spatial kernel and k_r is the range kernel (defined over the set of intensities the image pixels can take). For computational simplicity both kernels are defined using Gaussians.

3.3 Anisotropic Diffusion

Anisotropic diffusion is an extension of the diffusion process that means it generates a family of parameterized images, but each one is a mix of the original image and a filter based on the original image's local content. As a result, this transforms the original image in a non-linear and space-variant manner. The space-variant filter is isotropic in its original formulation by Perona and Malik [16], but it depends on the image content to approximate an impulse function near to edges and other structures that should be retained in the image throughout the many levels of the resultant scale space. A more general formulation permits the locally adapted filter to be really anisotropic near to the edges: it has an orientation determined by the structure that is extended along with the structure and narrows across the structure [17]. As a result, the images generated preserve linear structures while smoothing along with these structures. Both of these instances can be represented by a generalization of the diffusion equation, in which the diffusion coefficient is a function of image position and assumes a matrix value rather than being a constant scalar.

3.4 Guided Filter

Simple explicit linear translation-invariant (LTI) filters like the box filter are frequently employed for image denoising, edge detection, and other applications. The spatial kernels of LTI filters are unaffected by image features. However, in many circumstances, extra image information from a guidance image may be required in the denoising process for the filtering process. To solve this issue, a Guided filter is used, the filtering output of which is locally a linear transformation of the guidance image. The statistics of an area in the appropriate spatial neighborhood in the guidance image, which can be the same as the original image or a different one, are used to generate the denoised image using a guided filter. This filter preserves edges while smoothing out the noise similar to the bilateral filter, except it is free of gradient reversal artifacts and has a linear computational complexity [18]. The filtering output of a general linear transform invariant filtering process at pixel i, O_i can be given by weighted average:

$$O_i = \sum_j W_{ij}(G)I_j \tag{3}$$

Where W_{ij} represents filter kernel. Using the assumption of linearity between guidance image and output image we can also write output image as:

$$O_i = a_k G_i + b_k, \forall_G \in \omega_k \tag{4}$$

Where a_k and b_k are linear coefficients that are expected to be constant in window ω_k centered at pixel k. Using linear regression to minimize the cost function:

$$a_k = \frac{\frac{1}{|\omega|}\sum_{i\in\omega_k} G_i I_i - \mu_k \overline{I_k}}{\sigma_k^2 + \int} \tag{5}$$

$$b_k = \overline{I_k} - a_k \mu_k \tag{6}$$

Here, $\in$ represents a regularization parameter, μ_k and σ_k^2 represent the mean and variance of G in ω_k, $|\omega|$ represents the total number of pixels present in ω_k, and $\overline{I_k} = \frac{1}{|\omega|}\sum_{i\in\omega_k} I_i$ is the mean of input pixel values in ω_k. After obtaining the linear coefficients a_k and b_k, filtration output O_i can be calculated by:

$$O_i = \frac{1}{|\omega|}\sum_{k:i\in\omega_k}(a_k G_i + b_k) \tag{7}$$

Using Eq. (3), (4), (5), (6), and (7) kernel weights can be given by [18]:

$$W_{ij} = \frac{1}{|\omega|^2}\sum_{k:i\in\omega_k}\left(1 + \frac{(G_i - \mu_k)(G_j - \mu_k)}{\sigma_k^2 + \epsilon}\right) \tag{8}$$

4 Edge Detection Operators

The main objective of edge detection is to detect the object's shape information, as well as the image's reflectance. Edge detection minimizes the data that must be processed for object detection but preserves the characteristics and features of the objects present in an image. The outcome of implementing edge detection operators to an image is a binary image called an edge map, in which a group of connected pixels denote discontinuity in surface orientation or edge and designate the borderlines of objects denoted as white. Different types of edge-detection approaches can be broadly classified into:

- First derivative/Gradient-based edge detectors
- Second derivative/Zero-crossing edge detectors
- Optimal Edge Detector

4.1 Gradient-Based Edge Detectors

The directional change of the image function is expressed by its gradient, a two-dimensional vector of its partial derivatives [19]:

$$\nabla f = \begin{bmatrix} G_x \; G_y \end{bmatrix} = \begin{bmatrix} \frac{\partial f}{\partial x} \; \frac{\partial f}{\partial y} \end{bmatrix} \tag{9}$$

Gradient magnitude $|\nabla f|$ and its direction θ can be calculated by [19]:

$$|\nabla f| = \sqrt{G_x^2 + G_y^2} \tag{10}$$

$$\theta = \tan^{-1}\left(\frac{G_x}{G_y}\right) \tag{11}$$

However, to approximate the gradient magnitude by absolute value:

$$|\nabla f| \approx |G_x| + |G_y| \tag{12}$$

Because the digital image intensity function is discrete in nature, so to compute its derivatives, we have to consider it as the result of sampling some underlying differentiable intensity function at the image pixels. So the derivative can be estimated as a function of the sampled intensity function. Approximations of these derivative functions, on the other hand, can be specified with varying degrees of accuracy. For example, derivatives can be approximated by finite differences. The first-order differences of the image g in the vertical and horizontal direction are given by the forward or backward difference.

Robert Operator. The Roberts cross-operator is a differential operator that offers a simple approximation to the image's gradient using discrete differentiation. It is accomplished by calculating the sum of the squares of the differences of intensity values of diagonally adjacent pixels.

$$G_x = \begin{bmatrix} 1 & 0 \\ 0 & -1 \end{bmatrix} \text{ and } G_y = \begin{bmatrix} 0 & -1 \\ 1 & 0 \end{bmatrix}$$

The intensity values of output pixels are the estimated spatial gradient magnitude of the original image at that pixel. It measures the spatial gradient of an image. Thus, it highlights those areas that have high spatial frequency and are often considered to be the edge. The main advantage of this operator is that it can detect edges at angles of 45° and 135° from the horizontal, and are very quick to compute [20]. However, its accuracy is low because it produces a weak response to genuine edges that have less gradient magnitude. Also, since its uses are relatively small kernels, it is highly susceptible to noise.

Sobel Operator. The Sobel operator is based on computing the central difference in the direction of derivatives. However, unlike the Prewitt filter, it gives more weight to the central pixel for averaging. This method computes partial derivatives of the image intensity function by convolving the inputted image with a pair of 3x3 convolution kernels in the horizontal and vertical directions.

$$G_x = \begin{bmatrix} 1 & 0 & -1 \\ 2 & 0 & -2 \\ 1 & 0 & -1 \end{bmatrix} \text{ and } G_y = \begin{bmatrix} 1 & 2 & 1 \\ 0 & 0 & 0 \\ -1 & -2 & -1 \end{bmatrix}$$

These kernels are most suited for detecting edges running horizontally and vertically [20]. The computational cost of this operator is low. It also has better noise suppression characteristics compared to Robert and Prewitt operators. However, in the edge detection process, the genuine edges in the input image will produce lines in the output edge map, but due to the smoothing performed by the Sobel operator, they are several pixels wide, resulting in thick and inaccurate edges.

There are various gradient-based operators for edge detection, such as Prewitt, Robinson, and Krrish. However, only Robert and Sobel are mentioned in this work.

4.2 Second Derivative Based Edge Detectors

The edge detection operators discussed in the previous section compute the gradient magnitude at each pixel and, if it exceeds a threshold, it is considered an edge point. Because of this, too many edge points are detected, in result creating thick edges. A better methodology for getting sharper and more accurate edges would be to consider only points of locally maxima gradient magnitude or local maxima or minima in the first derivative as edge points. So, these methods report pixels as edge points where zero crossing is found in the second order derivative of the image function. In 2D, the second derivative can be computed by two operators: the Laplacian and the second directional derivative.

Laplace Operator. The Laplace operator, often known as the Laplacian, is a differential operator that is defined by the divergence of a function's gradient. It is the 2D isotropic equivalent of the second-order spatial derivative of the image intensity function. Laplacian of a function f (x, y) is defined by:

$$\nabla^2 f = \frac{\partial^2 f}{\partial x^2} + \frac{\partial^2 f}{\partial y^2} \quad (13)$$

Approximating second derivatives along the x and y directions [21]:

$$\frac{\partial^2 f}{\partial x^2} = f(p, q-1) - 2f(p, q) + f(p, q+1) \quad (14)$$

$$\frac{\partial^2 f}{\partial y^2} = f(p-1, q) - 2f(p, q) + f(p+1, q) \quad (15)$$

This is the appropriate approximation of the second-order partial derivatives centered around (p, q).

By combining these partial derivative approximations, convolution with the following kernel can be used to approximate the Laplacian [21]:

$\nabla^2 \approx$

0	1	0
1	-4	1
0	1	0

When the operator's output transitions through zero, the Laplacian identifies the existence of an edge point. Uniform zero areas (trivial zeros) are disregarded. Using linear interpolation, the zero-crossing points can be approximated to subpixel resolution theoretically, although the result may be incorrect due to noise [21]. Therefore, filtering out noise before edge enhancement using Laplacian is recommended to reduce the influence of noise.

4.3 Optimal Edge Detector

The Canny operator was designed as an optimal edge detection operator. It is a multi-step algorithm, which can be divided into mainly five different steps. In the first stage, a gaussian filter is used to reduce the noise in the image. The smoothed picture is then subjected to a second stage in which a simple two-dimensional first derivative is applied to compute the local intensity gradients and edge direction at every point of the image and highlight those regions that have high gradient magnitude. The output image should ideally have thin edges. So in the third stage, we use non-maximum suppression to thin out edges and eliminate spurious edge detection responses. The method iterates over all of the points on the gradient image, preserving the pixel if it is a local maximum in its neighborhood in the edge directions, otherwise setting pixel value to 0 to have a thin line in the resultant edge map. The remaining edge pixels after non-maximum suppression give a more realistic depiction of genuine edges. However, there are still some edge pixels left that are not part of an edge, but a result of color variation or noise. To accommodate for these erroneous responses, edge pixels with low gradient magnitude must be filtered away, whereas edge pixels with high gradient magnitude must be preserved. So, to achieve this in the fourth stage, we select low and high threshold values and mark edge pixels as strong pixels if their gradient magnitude is higher than the high threshold that is sure to contribute to the final edge. We mark an edge pixel as a weak pixel if its gradient magnitude lies between low and high threshold values. The edge pixel will be discarded if its gradient magnitude is less than the low threshold. In the final stage hysteresis algorithm transforms weak pixels into strong ones based on the double threshold results, if at least one of the pixels in the 8-connected neighborhood of the one being processed is a strong pixel [22].

5 Performance Evaluation

There are mainly two ways for evaluating edge detector performance: subjective and objective methods. Subjective approaches entail showing a group of people a sequence of edge maps and asking them to rate them on a scale. Although subjective methods can be implemented quite easily, they also have some limitations. The human eye can only distinguish a limited number of characteristics. That's why these methods should only be preferred over objective methods if we want high-quality pictures as determined by human perception. On the other hand, objective methods require a ground-truth image (GTI) as a reference and compare the amount of error in the processed edge map by comparing it with the GTI. To compare the difference between original and processed images, the two most commonly used methods are statistical methods, including PSNR and MSE. Even though objective methods are broadly utilized, however, they depend stringently on numeric comparison and do not consider any degree of biological factors of the human visual system. For example, an image whose error metric is high may look better than an image whose error metric is low. The MSE between two images f_{ij} and g_{ij} is defined as:

$$MSE = \frac{1}{mn} \sum_{i=0}^{m-1} \sum_{j=0}^{n-1} (f_{ij} - g_{ij})^2 \tag{16}$$

One drawback of MSE is that it depends highly on image intensity scaling. To overcome this drawback, PSNR scales the MSE according to the image range.

$$PSNR = 10 \log_{10} \left(\frac{\max_i^2}{MSE} \right) \tag{17}$$

6 Experimentation

This experiment was done to analyze and compare the performance of filters and edge detection operators to extract a clean edge map of a noisy image. For this experiment image from the BIPED dataset has been used. This dataset is publicly available as a benchmark for evaluating edge detection techniques. The ground truth (GT) image of this dataset includes both the segmentation and the boundary. Speckle noise of 0.02 variance is applied to the image. Then, various edge-preserving filters like Bilateral filter, Anisotropic Diffusion, Guided filter, and AMF are applied to the noisy image to filter out the noise from the image before edge detection (Fig. 2).

The subjective assessment of the outputs of selected edge detectors can be achieved by comparing the resulting images visually. Figure 3 shows the results of applying several edge detection operators to images affected by speckle noise. Figures 4, 5, 6, and 7 show the comparison of the edge maps of the image filtered by the Bilateral filter, Guided filter, and Anisotropic diffusion, respectively, generated by different edge detectors. For performance comparison, objective methods were utilized by computing PSNR and MSE of the edge map with respect to the original GTI.

Fig. 2. Original image, noisy image, and GTI respectively

Fig. 3. Edge maps noisy image

Here, Table 1 illustrates the PSNR and MSE value of GTI and the edge maps produced by edge preserving filter with edge detection operators. PSNR values ranged from 14.24 dB to 61.02. It is quite low for edge maps produced by gradient-based operators and the highest for edge maps produced by canny operators. MSE values follow the exact opposite trend. The MSE values are lowest for edge maps filtered by a canny operator and the highest for edge maps produced by gradient-based operators. PSNR value is highest for edge map produced by bilateral filter with canny edge detection operator.

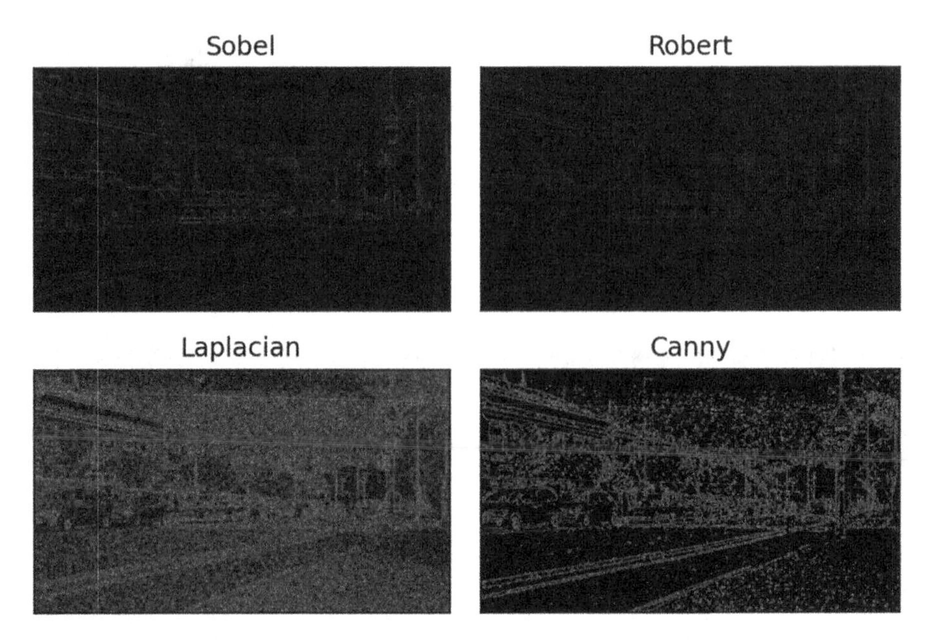

Fig. 4. Edge maps image filtered by AMF

Fig. 5. Edge maps image filtered by bilateral filter

Fig. 6. Edge Maps Image Filtered by guided filter

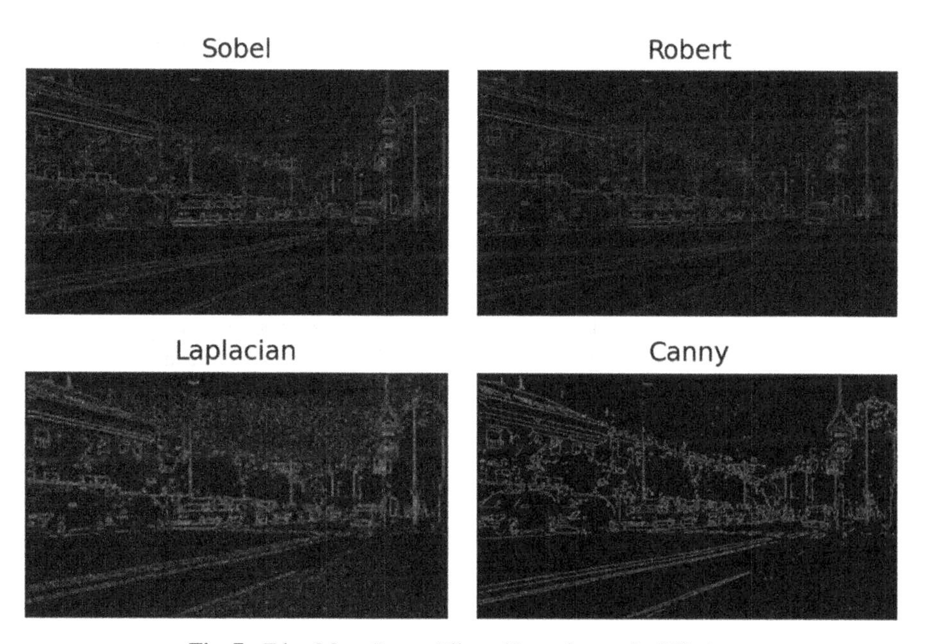

Fig. 7. Edge Maps Image Filtered by anisotropic diffusion

Table 1. PSNR b/w GTI and detected edge maps

	Sobel		Robert		Laplacian		Canny	
	PSNR	MSE	PSNR	MSE	PSNR	MSE	PSNR	MSE
Guided Filter	14.242019	2448.378	14.241635	2448.594	28.841729	84.899	59.912974	0.066
Adaptive Median Filter	14.242562	2448.072	14.242304	2448.217	28.721634	87.280	57.246559	0.122
Anisotropic Diffusion	14.242212	2448.269	14.241620	2448.603	29.302751	76.349	60.012849	0.0648
Bilateral Filter	14.242118	2448.488	14.241823	2448.488	28.822524	85.276	60.018668	0.0647

7 Results and Conclusion

The primary objective of this study was to discuss and compare the different approaches to denoise and extract a clear edge map of a noisy image. Under noisy conditions, subjective evaluation of the resultant edge maps suggests that Canny, Laplacian, Sobel, and Robert exhibit better performance, respectively. By visualizing Figs. 4–7, it can be shown that, in comparison to Sobel, Laplacian produces low-quality edge maps. But while first-order derivatives operators such as Sobel can easily compute edges along with their orientation, edge detection is unreliable, more susceptible to noise, and results in thicker edges. However, according to Table 1, the Sobel operator is showing less sensitivity to noise than Robert. Even though the Laplacian operator provides more accurate edges and its PSNR values are greater than those of Sobel and Robert, it is clear from the figures that it is more susceptible to noise even compared to gradient operators, because when the noise level is high, the grey level varies and may result in many zero crossings. Canny edge detection in comparison to Sobel, Robert, and Laplacian is complex and computation is time-consuming. But detection of edges by the canny operator is less susceptible to noise. It provides better quality edge maps relative to other edge detectors. In addition, it has been observed that in the removal of speckle noise, AMF is less effective than other filters. By visually comparing edge maps in Fig. 3 to other edge maps, it can also be observed that filtering is necessary as a preliminary procedure in edge detection for a noisy image. Finally, after analysis of different types of filtering and edge detection techniques, it is concluded that the best approach to finding a clear edge map of an image affected by speckle noise is to first denoise it with a bilateral filter and then compute the edge map using a canny edge detector.

In future work, we can improve the quality of the image by designing a new filter over the limitations so that the image can be enhanced by its resolution by reducing the noise.

References

1. Ofir, N., Galun, M., Alpert, S., Brandt, A., Nadler, B., Basri, R.: On detection of faint edges in noisy images. IEEE Trans. Pattern Anal. Mach. Intell. **42**(4), 894–908 (2019)

2. Poobathy, D., Chezian, R.M.: Edge detection operators: peak signal to noise ratio based comparison. IJ Image Graph. Signal Proc. **10**, 55–61 (2014)
3. Fawwaz, I., Zarlis, M., Rahmat, R.F.: The edge detection enhancement on satellite image using bilateral filter. IOP Conf. Ser. Mater. Sci. Eng. **308**, 012052 (2018)
4. Sekehravani, E.A., Babulak, E., Masoodi, M.: Implementing canny edge detection algorithm for noisy image. Bulletin of Electrical Engineering and Informatics **9**(4), 1404–1410 (2020). https://doi.org/10.11591/eei.v9i4.1837
5. Maini, R., Sohal, J.S.: Performance evaluation of Prewitt edge detector for noisy images. GVIP J. **6**(3), 39–46 (2006)
6. Ruslau, M.F.V., Pratama, R.A., Nurhayati, S.A.: Edge detection in noisy images with different edge types. IOP Conf. Ser. Earth Environ. Sci. **343**(1), 012198 (2019)
7. Hussain, Z., Agarwal, D.: A comparative analysis of edge detection techniques used in flame image processing. Int. J. Adv. Res. Sci. Eng. IJARSE **4**, 1335–1343 (2015)
8. Swarnalakshmi, R.: A survey on edge detection techniques using different types of digital images. Int. J. Comput. Sci. Mob. Comput. **3**(7), 694–699 (2014)
9. Roushdy, M.: Comparative study of edge detection algorithms applying on the grayscale noisy image using morphological filter. GVIP J. **6**(4), 17–22 (2006)
10. Singh, R.K., Shaw, D.: Experimental analysis of impact of noise on various edge detection techniques. In: Proceedings of the World Congress on Engineering, Vol. 1 (2016)
11. Aishwarya, K.M., Rao, A.A., Singh, V.: A Comparative study of edge detection in noisy images using BM3D filter. Int. J. Eng. Res. Technol. (IJERT) **5**(9), 142–147 (2016)
12. Insidini Fawwaz, N.P., Dharshinni, F.A.: Noise effect analysis on edge detection in detecting digits with bilateral filter. J. Phys. Conf. Ser. **1230**(1), 012095 (2019)
13. Sarker, S., Chowdhury, S., Laha, S., Dey, D.: Use of non-local means filter to denoise image corrupted by salt and pepper noise. Sign. Image Proc. **3**(2), 223 (2012)
14. Lei, P.E.N.G.: Adaptive median filtering. In: Seminar Report, Machine Vision, Vol. 140 (2004)
15. Tomasi, C., Manduchi, R.: Bilateral filtering for gray and color images. In: Sixth International Conference on Computer Vision (IEEE Cat. No. 98CH36271), pp. 839–846. IEEE (1998)
16. Perona, P., Malik, J.: Scale-space and edge detection using anisotropic diffusion. IEEE Trans. Pattern Anal. Mach. Intell. **12**(7), 629–639 (1990)
17. Weickert, J.: Anisotropic diffusion in image processing, Vol. 1, pp. 59–60. Stuttgart, Teubner (1998)
18. He, K., Sun, J., Tang, X.: Guided image filtering. IEEE Trans. Pattern Anal. Mach. Intell. **35**(6), 1397–1409 (2012)
19. Gonzalez, R.C., Woods, R.E., Masters, B.R.: Digital image processing, Third Edition. J. Biomed. Opt. **14**(2), 029901 (2009)
20. Kaur, J., Kumar, A.: Evaluating the shortcomings of edge detection operators. Int. J. Adv. Res. Comput. Sci. Softw. Eng. (2015)
21. Jain, R., Kasturi, R., Schunck, B.G.: Machine vision, Vol. 5, pp. 309–364. McGraw-HillNew York. https://doi.org/10.1007/978-3-662-47794-6
22. Moeslund, T.: Canny Edge Detection. Retrieved December 3, 2014 (2009)

Noise-Resilient Ensemble Learning Using Evidence Accumulation

Gaëlle Candel[1,2(✉)] and David Naccache[1]

[1] Département d'informatique de l'ENS ENS, CNRS, PSL University, Paris, France
{gaelle.candel,david.naccache}@ens.fr
[2] Wordline TSS Labs, Paris, France
gaelle.candel@worldline.com

Abstract. *Ensemble Learning* methods combine multiple algorithms performing the same task to build a group with superior quality. These systems are well adapted to the distributed setup, where each peer or machine of the network hosts one algorithm and communicate its results to its peers. Ensemble learning methods are naturally resilient to the absence of several peers thanks to the ensemble redundancy. However, the network can be corrupted, altering the prediction accuracy of a peer, which has a deleterious effect on the ensemble quality. In this paper, we propose a noise-resilient ensemble classification method, which helps to improve accuracy and correct random errors. The approach is inspired by *Evidence Accumulation Clustering*, adapted to classification ensembles. We compared it to the naive voter model over four multi-class datasets. Our model showed a greater resilience, allowing us to recover prediction under a very high noise level. In addition as the method is based on the evidence accumulation clustering, our method is highly flexible as it can combines classifiers with different label definitions.

Keywords: Classification · Distributed systems · Ensemble learning · Evidence accumulation clustering · Label corruption

1 Introduction

Ensemble Learning [23] methods combine several algorithms performing the same task to obtain a better-quality group. Ensemble learning methods play on diverse group aspects: the number of algorithms [2,10], their weighting based on their contribution [11,18,19], and their selection based on their diversity [12].

Ensemble learning methods are well adapted to the distributed setup, where several machines host each a single algorithm and send their results to a central node aggregating the results [1,6,14]. They can be adapted to decentralized peer-to-peer networks [14], where a dynamic group collaborate to improve its accuracy by electing a leader or by aggregating the group's results. Distributed systems are prone to network failures, where communications are broken between some nodes, or corrupted with noise [17]. In addition, nodes can change of behavior if controlled by malicious entities.

I. Woungang et al. (Eds.): ANTIC 2021, CCIS 1534, pp. 374–388, 2022.
https://doi.org/10.1007/978-3-030-96040-7_30

Ensemble methods are resilient to the absence of one or more weak learners thanks to group redundancy [12,16]. However, the corruption of a learner's predictions is equivalent to a negative change of accuracy, which has a deleterious effect on the group quality. Thus, there are two ways to deal with corrupted computers: detecting inaccurate peers to avoid data pollution or resilience to error. The detection can be done using network monitoring methods, or exploiting trust to weigh peers based on their past contributions. However, this approach is not adapted to a dynamic environment such as a peer-to-peer network where a peer lifetime is very short and may change temporary behavior. In contrast, being resilient to error is more suitable as all inputs are accepted but more challenging to design as it requires smart correction algorithms.

In this article, we propose a noise-resilient ensemble classification method, correcting errors while improving accuracy. The method uses the Evidence Accumulation Clustering approach to rectify class boundaries and correct corrupted labels by performing a local weighted vote. The approach was tested under several noise condition over four datasets and tolerated high noise levels without accuracy degradation.

This paper is structured as follows. The first section presents the related works regarding ensemble learning methods and resilience to error. The second section details the proposed ensemble classification method. The datasets and the classifiers' setup are detailed in the experimental section, followed by the results. Finally, the paper ends with a discussion and a conclusion.

2 Related Works

Ensemble classification methods adopt different strategies to increase accuracy. The simplest one is *bagging* [2] – for bootstrap aggregating, learners are trained over randomly sampled subsets. A related approach exploits *random projections* [10], which creates a different view of the data, making it less dependent on the pre-processing step. Rather than sampling items at random, selection can be made on features [15], reducing the computational complexity of each classifier while allowing the ensemble to classify incomplete items with missing features. A classifier often makes very few errors in very dense areas where a single class is represented, because there is no ambiguity. Near a boundary, the classifier is less exact because this area is less dense; therefore difficult to learn correctly. Ensemble methods help mostly to correct these areas. A good example is decision trees with rough decision boundaries, while a *Random Forest* [3] has smooth boundaries with various shapes. The *bagging* approaches exploit the fact that random initialization would lead to diverse classifiers, compensating errors within the group. However, it requires a large number of classifiers and fails to improve when classifiers are correlated.

Boosting [19] solves the correlation problem by weighting classifiers based on the improvement they can lead to. On the other hand, *ensemble pruning* [12] removes correlated classifiers that lead to no improvement, decreasing the overall complexity while preserving the accuracy. The ensemble size is smaller for these methods but is still resilient to the absence of several classifiers during the inference stage [16]. *Boosting* and *ensemble pruning* target the problem of *how to construct* a full ensemble using very few classifiers. As they require a setup phase to evaluate the classifiers, these approaches are not adapted to dynamic environments, such as P2P networks, as peers can join and leave at any time.

Most ensemble classification methods assume that no issue can occur within the ensemble, i.e. all predictions are transmitted without errors. The centralizing node is always available, but some classification peers can be down. In the case of Internet-of-Things networks, like Wireless Sensor Network (WSN) or Vehicular Ad Hoc Network (VANET), the devices are prone to fault, failures and attacks [17,20,21]. Several techniques exist to detect deceptive nodes, but it requires some time and often needs centralized monitoring capabilities. To our knowledge, no work assumed transmission of corrupted predictions to the aggregator node, nor solution in case of unavailability. It is equivalent to a completely decentralized P2P network where there is no leader and corrupted peers can participate to the classification task.

The other branch of ensemble learning is *ensemble clustering* [22] that combines clustering algorithms. There is no way to know if a partitioning is *good* because there is no ground truth. Therefore, ensemble clustering algorithms must simultaneously deal with the good and bad clustering without any other help. Therefore, this class of algorithms will be helpful to deal with corrupted predictions. While clustering is analogous to classification as it assigns labels to items, the approaches shared by ensemble classification and clustering are limited to bagging and weighting [11,18], where weights are assigned at *inference time* allowing to handle corrupted output more carefully.

Evidence Accumulation Clustering [7,8,11,13,18] is one of the main Ensemble clustering methods, which has the advantage of not requiring to match labels between the different clustering. The approach starts by gathering the co-clustering frequencies of all possible pairs of items into a co-association (CA) matrix of size $n \times n$ for n items. The similarity matrix obtained is then re-clustered to obtain the final partitioning. A hierarchical algorithm is often used [7,8] as it allows the final user to decide on the clustering granularity. The major drawback of evidence accumulation clustering is the complexity and the memory footprint because of the $n \times n$ matrix. Using *Single-Linkage* (SL) hierarchical clustering [7], the authors proposed to limit the CA matrix to the $k = 20$ nearest neighbors, as SL does not consider distant neighbors. Depending on the dataset, the approach led to better results than other clustering methods exploiting the full matrix while reducing the overall computational complexity.

We inspired ourselves from the evidence accumulation clustering methods and the nearest-neighbors trick to deal with noise and improve accuracy. Nonetheless, our approach differs significantly from it as the CA weights are exploited to refine labels rather than clustered to obtain the final classification. The approach will be detailed in the following section.

3 Label Refinement with Implicit Boundary Learning

In this section, we will detail our proposed ensemble classification method exploiting the evidence accumulation clustering method. The task is to classify an unlabeled dataset $X = \{\mathbf{x}_1, \mathbf{x}_2, ..., \mathbf{x}_n\}$ with $n = |X|$ elements. The ensemble is a group $\mathcal{A} = \left\{A^{(1)}, A^{(2)}, ..., A^{(k_p)}\right\}$ of k_p classifiers, called sometimes *peers*. The way they are obtained impacts the ensemble accuracy – as in any other ensemble classification methods – but does not impact the overall process. The prediction made by the peer p is denoted $\hat{Y}^{(p)} = A^{(p)}(X) \in \left(\mathcal{L}^{(p)}\right)^n$, where $\mathcal{L}^{(p)}$ is the set of classes that p can distinguish, possibly different from the other classifiers. We will discuss later about this particularity.

The goal of the proposed approach is to rectify the label of an item based on its neighborhood. In general, a classifier makes errors on items close to a class boundary. A peer will collect the peers' opinion to know if two items are on the same side of the class boundary. These weights will be used to rectify uncorrect labels using a weighted voter model. The following paragraphs describe the process and explain the motivation of the different choices.

3.1 Gathering Co-association Matrices

The initial prediction $\hat{Y}^{(p)}$ of the peer p is transformed into the local CA matrix $M^{(p)}$, where $M^{(p)}(\mathbf{x}, \mathbf{x}') = 1$ if $A^{(p)}(\mathbf{x}) = A^{(p)}(\mathbf{x}')$ else 0. Only the pairs $M(\mathbf{x}, \mathbf{x}')$ concerning the k-nearest neighbors (k-NN) of $\mathbf{x} \in X$ (denoted $\mathcal{N}_k(\mathbf{x})$, with $\mathbf{x} \notin \mathcal{N}_k(\mathbf{x})$) are computed. As mentioned in [7], this trick reduces the memory footprint and computational cost from $\mathcal{O}(n^2)$ to $\mathcal{O}(kn)$.

After prediction, peers exchange their results, allowing them to compute locally the average CA matrix $\mathcal{M}$:

$$\mathcal{M}(\mathbf{x}, \mathbf{x}') = \frac{1}{k_p} \sum_{p=1}^{k_p} M^{(p)}(\mathbf{x}, \mathbf{x}') \tag{1}$$

Peers may get different results of $\mathcal{M}$ as a perfect communication model is not assumed here. Some peers may not receive the results from all of their peers, or may get truncated messages depending on the network stability. They replace in that case $\frac{1}{k_p}$ by the number of message received for a particular pair $(\mathbf{x}, \mathbf{x}')$. Unless very few messages are received by a peer, this would not impact the outcome of the process.

Nearest Pairs: The computation of $\mathcal{M}$ for the closest pairs only is motivated by the idea that NN are likely to belong to the same class unless near a class boundary. In a multi-class classification problem, negative information – indicating that two items belong to different classes – is less informative than positive information. For example, for d classes and two items, there are $d \times (d-1)$ assignment possibilities for negative evidence while only d for positive ones. With that many possibilities, it is unlikely to find the correct classes by chance using negative information. Additionally, for an item in a class representing $100 \times \alpha\%$ of the dataset, $100 \times (1-\alpha)\%$ of the pairs would be 0, leading to many unnecessary data as α shrinks with the number of classes and the class imbalance.

Class Matches: The use of pairs of co-association makes it possible to dispense with the definition of peer classes $\mathcal{L}^{(p)}$. $\mathcal{M}$ gather the number of times two items are classified together, regardless of the possible labeling errors or granularity level. The evidence accumulation clustering approach allows combining classification to clustering algorithms as they both produce labeling, one following a ground-truth definition, the other defined from scratch. Nevertheless, the combination of algorithms with different labels definitions assumes that most of the boundaries are common to the different classes/clusters

As an example, suppose we have three classifiers with different goals, one classifier focusing on *animals* [dog, wolf, cat, lion, frog], another on *sizes* [small, medium, large], and the last on *colors* [white, gray, black, green]. The *animal* and *size* classifiers are compatible as one size corresponds to a unique set of animals (small $=$ {frog}, medium $=$ {dog, cat}, large $=$ {wolf, lion}). Therefore, all the *size* boundaries are preserved by the *animal* classifiers, with some additional. However, the *animal* and *color* classifiers share only the boundary between frog and the other animals which corresponds to the boundary between green and the other colors. Therefore, these two classifiers are weakly compatible.

To obtain compatible classifiers, classes must be derived from a primary set of classes, as in *Error Correcting Output Code* (ECOC) [4] where classes are grouped into two groups to train a binary classifier on it. Another option is available if classes organized into a hierarchical ontology. In this case, high and low-level classes could be combined as several low-level classes derived from a single high-level class. Therefore, the high-level class boundary is preserved within the low-level classes. However, the combination of algorithms with different class definitions impacts the strength of $\mathcal{M}$ as the ensemble boundaries are less pronounced. This would have almost no impact on the high-level classifiers as low-level ones share their boundaries. Still, low-level classifiers would be as not all their boundaries are not matched by higher-level classifiers.

Exchanging Pairs: Depending on the network quality and the data privacy preferences, a peer may prefer to exchange its CA matrix or its raw predictions.

The raw predicitions require $8 \times n$ bits per message if a label is encoded over an octet in terms of *bandwidth requirements*. When sending the binary CA matrix, it requires $k \times n$ bits per message. In both cases, we assume that all peers have the same set of items with the same indexing; therefore the features and index do not need to be exchanged. The value of k in our setups is relatively small (≈ 10); therefore one or the other option is equivalent in size.

In an *insecure environment*, peers may prefer not to exchange their raw prediction as an eavesdropper would get roughly labeled data for free. With a CA matrix, the eavesdropper can only recover a partial clustering, limiting the amount of information leakage. In addition, the encryption scheme [9] can be used in binary form, allowing everyone to get the total number of votes for each item without knowing what peers are voting for.

3.2 Label Refinement Phase

The refinement phase exploits $\mathcal{M}$ to adjust the initial predictions $\{\hat{Y}^{(p)}\}_{p=1:k_p}$. This step is done locally where each peer will refine its own prediction $\hat{Y}^{(p)}$, without any communication. The initial classification $y_0^{(p)}(\mathbf{x}) = A^{(p)}(\mathbf{x})$ is updated using the following equation:

$$y_{t+1}^{(p)}(\mathbf{x}) = \arg\max_{\ell \in \mathcal{L}^{(p)}} \sum_{\mathbf{x}' \in \mathcal{N}_k(\mathbf{x})} \mathcal{M}(\mathbf{x}, \mathbf{x}')\delta\left(y_t^{(p)}(\mathbf{x}'), \ell\right) \tag{2}$$

In other words, $y^{(p)}(\mathbf{x})$ is updated with a weighted voter model, using $\mathcal{M}$ to adjust the neighbors' label contribution. The update process is repeated several times until labels do not change significantly from their previous estimation.

The process can be seen as a horizontal voter model, where instead of looking at all superposed predictions $\{A^{(p)}(\mathbf{x})\}_{p=1:k_p}$, item's nearest neighbors' label $\{A^{(p)}(\mathbf{x}')\}_{\mathbf{x}' \in \mathcal{N}(\mathbf{x})}$ are taken into account. By re-using its prediction, the peer prevents itself from label contamination from noisy or malicious peers as only the weights can be altered but not the labels.

The weights in $\mathcal{M}$ reflect the probability of two items of being in the same class, which indirectly encodes the ensemble's boundary location as it concerns the nearest neighbors. Near a boundary, items on the same side have high weights, whereas items from the other side have low weights. Therefore, only the most relevant items on the same boundary side would contribute positively.

If a boundary in $\mathcal{M}$ does not pre-exist in $M^{(p)}$, nothing would change p's prediction near the boundary as an item is surrounded by items with the same label. This case occurs when a classifier has a label definition different from the group, which explains why we can combine classifiers with different objectives without issue.

The results of this weighting scheme differ from the k-NN outcome, where all items contribute equally or differently as in the wk-NN version. When density is not homogeneous, one class may contribute more than another because items are easier to find in the neighborhood. Therefore, items would be misclassified, whereas in our approach, relevant items are identified, allowing to rectify labels independently of the density.

3.3 Algorithm Complexity

The complexity of the proposed approach is decomposed as follow. The prediction step is at least in $\mathcal{O}(n)$ per peer. Next, the matrices $\{M^{(p)}\}_{p=1:k_p}$ are each obtained in $\mathcal{O}(kn \log n)$ to search for the k-NN and issue the $n \times k$ matrix. The averaging of the matrices $\{M^{(p)}\}_{p=1:k_p}$ into $\mathcal{M}$ is made in $\mathcal{O}(knk_p)$. Last, the refinement step costs $\mathcal{O}(knT)$ for T iterations. In total, each peer has a complexity of $\mathcal{O}((\log n + k_p + T)kn)$, which is larger than other ensemble methods (like boosting with $\mathcal{O}(k_p n)$), but the cost is close to be linear in n. The total cost is quadratic with the number of peers as they each need to average k_p matrices into $\mathcal{M}$. This cost can be significantly reduced using *Gossip* approaches by performing local averaging.

4 Experimental Setup

4.1 Datasets

We selected four multi-class datasets from the UCI machine learning repository [5] with a sufficient number of instances ($\sim$ 1.000 per class).

A large number of instances allows us to train many classifiers on disjoint item sets while preserving a large part for testing. Some of these datasets were already split into the *training* and *testing* part. We did not keep this split and merged the two sets to obtain a larger experiment testing set.

We tested our approach over multi-class datasets because binary problems are easier to solve than multi-class problems. Multi-class problems have a lower accuracy baseline, therefore a larger margin for improvement. In our proposed approach, we used only positive evidences. In a binary problem, negative evidence is equivalently informative as there are only two class possibilities. For this particular case, $\mathcal{M}$ could be used differently to improve accuracy.

Table 1. Datasets description.

Dataset	n	$\lvert\mathcal{L}\rvert$	MinClass	MaxClass
DryBean	13.611	7	3.83%	26.05%
PenDigit	10.992	10	9.59%	10.40%
Statlog	6.435	6	9.73%	23.82%
USPS	9.298	10	7.61%	16.70%

Table 1 summarizes the characteristics of the different datasets. The table gathers the number of samples n, the number of classes $|\mathcal{L}|$, and the proportion of the less and most represented classes MinClass and MaxClass respectively. The datasets are correctly balanced, with enough samples in the lowest class, so we do not need to care about class unbalance. We pre-processed all datasets the same way, using z-normalization to give an equal contribution to each feature.

4.2 Weak Learner Setup

In our experiments, we will test the influence of the number of classification peers in the system. We want to train them on a disjoint subset of data to ensure they all are different. As the datasets are limited in size, we selected the k-NN neighbors classifier for simplicity and the low training data requirements. To avoid class imbalance, each classifier's training was composed of d items from each class. d was set to 3 leading to 30 items per classifiers for a dataset with 10 classes like *USPS* and *PenDigit*.

The *USPS* has the lowest ratio $\frac{n}{|\mathcal{L}|}$. In an ensemble with 20 classifiers, it uses 600 items for training, representing less than 7% of the total dataset size. For the *DryBean* dataset with the largest ratio, 3% of the dataset would be used in the largest training condition. This places the experiment under weakly supervised learning conditions, letting more space for accuracy improvement.

5 Results

In this section, we present the results of the experiments comparing our ensemble method – denoted LR for *Label Refinement* – to the simple voter model – denoted VM – using the same ensemble of classifiers. As the LR method is performed locally, not all the different peers obtain the same results. The different predictions can be centralized and aggregated using a voter model. This third possibility is denoted $LR + VM$ in the experiments.

5.1 Accuracy Improvement Under Stable Conditions

The first experiment explores the impact of the ensemble's size on accuracy. In this setup, the peers are assumed to be trustful and communication perfect, such as the matrices $M^{(p)}$ are exchanged without errors. Figure 1 displays the results of the three different setups and the baseline corresponding to a classifier alone.

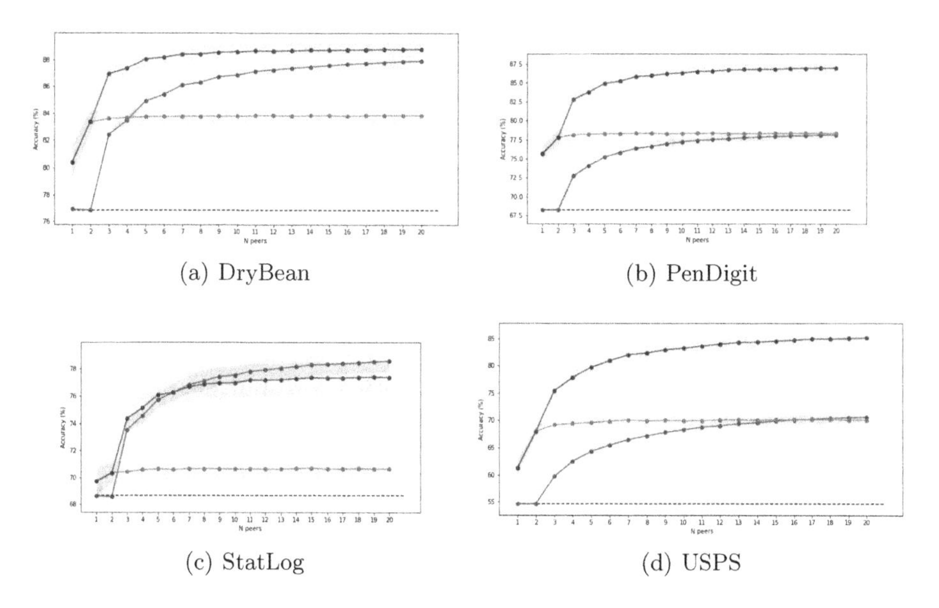

(a) DryBean (b) PenDigit

(c) StatLog (d) USPS

Fig. 1. Accuracy evolution varying with the number of peers. Dashed line: average accuracy of a classifier; orange: LR; green line: $LR + VM$; blue line: VM. Shaded areas correspond to the standard deviation of the mean accuracy over 50 trials. (Color figure online)

As expected, all ensemble methods are more accurate than the average classifier. The accuracy gain of each method varies from one dataset to another, regardless of the number of classes, number of test samples or the base accuracy.

For all approaches, there is a point for 1 and 2 peers. In the case of the voter model, the results correspond to the average learner accuracy as the scheme cannot help in this configuration. For LR and $LR+VM$, the first point corresponds to the mean accuracy of a learner p refining its prediction with a binary matrix $M^{(p')}$ from another peer p' without combining it with its own. For 2 peers, the internal matrix $M^{(p)}$ is combined with an external one $M^{(p')}$. The LR approach led to accuracy gain for all datasets in this configuration.

Compared to the voter model, the label refinement approach is quickly beaten over the *Drybean* and *StatLog* datasets. However, it is more competitive over the *PenDigit* and *USPS* datasets where the voter model can only reach the same accuracy after gathering 15 learners in the ensemble. The accuracy gain obtained with LR is quite stable after gathering just a few peers. 5 peers are enough to reach the maximal accuracy gain with LR. In comparison, the accuracy of an ensemble using 20 peers can still improve by adding peers under the voter model.

When applying a voter model *after* the label refinement step, an additional gain of accuracy is observed. The gain is non-negligible and allows better accuracy with $LR + VM$ over three of the four datasets with a large margin. While the LR accuracy is stable after gathering 5 peers, the combination of LR to VM is beneficial as $LR+VM$ can still improve accuracy by adding more peers. Nonetheless, it requires another communication phase to collect all the refined predictions.

5.2 Resilience to Output Corruption

This second experiment simulates an environment where *output* labels are corrupted at random. The corruption could happen in the peer memory or during transmission. In the real world, it could be materialized by a truncation of the results (one part is unreadable or non-received), or by the addition of noise (some labels are flipped). A noise level α corresponds to the replacement of $\alpha\%$ of the labels by a random label from $\mathcal{L}$. A deletion of $\alpha\%$ of the input could be considered the same way, as missing data are inputted with random labels, to the difference that the position of the incorrect label is known.

In this configuration, we compare the model's ability to recover the true labels $\mathcal{Y}$ using only the corrupted labels $\{\hat{Y}_c^{(p)}\}_{p=1:k_p}$. The $\{M_c^{(p)}\}_{p=1:k_p}$ matrices are also derived from the *corrupted* labels and the refinement step starts with the peer's corrupted labels $\hat{Y}_c^{(p)}$. The results of the three different methods over the four datasets are displayed in Fig. 2.

The line corresponding to an ensemble of size 1 corresponds – as in the first experiment – to the average accuracy of corrupted peers in the case of the VM, and the accuracy of a peer refining its corrupted output $\hat{\mathcal{Y}}_c^{(p)}$ with a received binary matrix $M_c^{(p')}$. When increasing the noise level, the VM accuracy decreases linearly. It is almost equal to $\text{acc}(\alpha) = (\text{acc}_0 - |\mathcal{L}|^{-1}).(1-\alpha) + |\mathcal{L}|^{-1}$, where acc_0 is the accuracy without noise, and $|\mathcal{L}|^{-1}$ the probability to select a correct label at random. The dashed line corresponds to this equation to allow an easier comparison of the different methods.

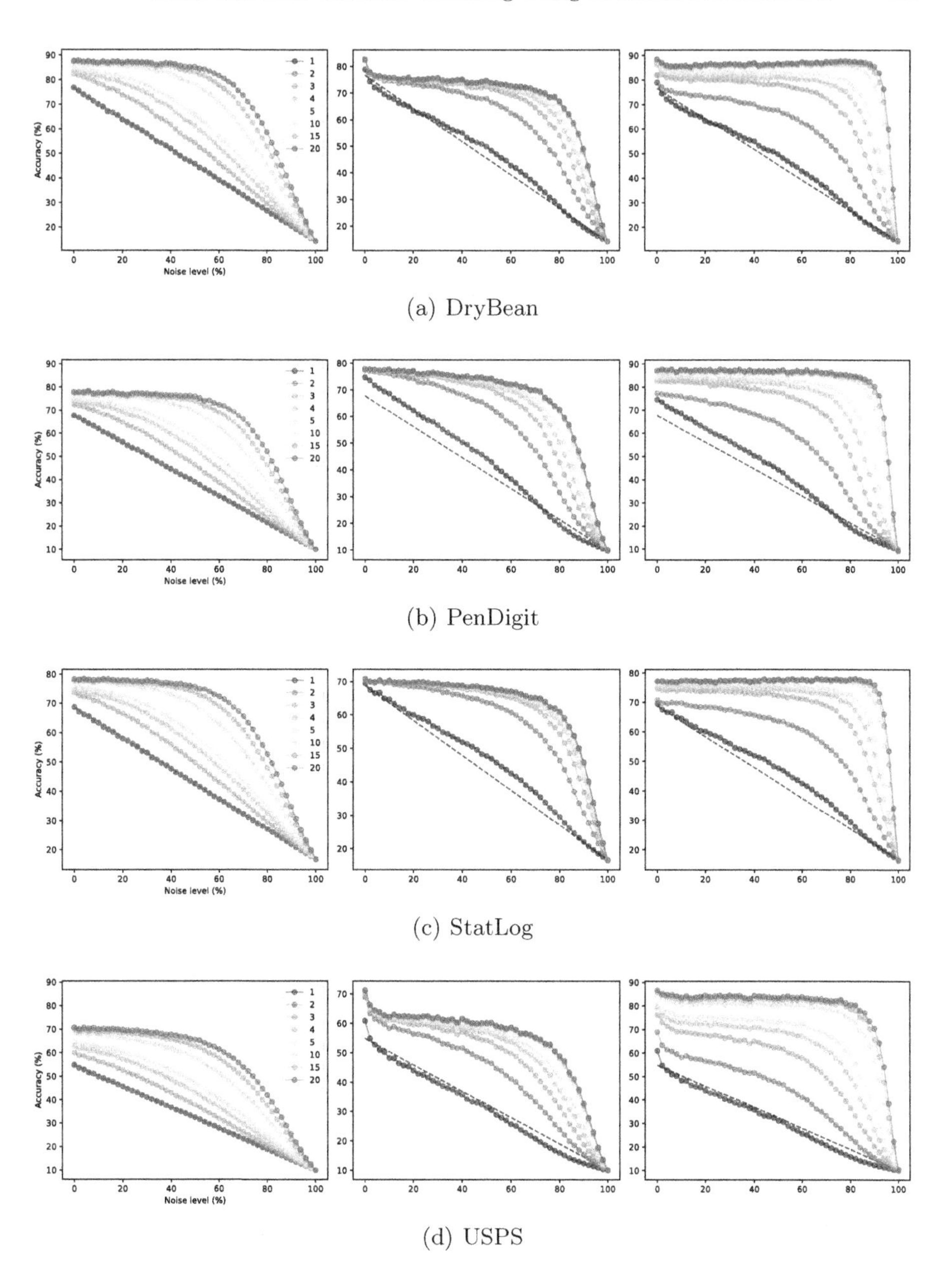

(a) DryBean

(b) PenDigit

(c) StatLog

(d) USPS

Fig. 2. Resilience to output corruption under increasing level of noise, with boundary parameter $k = 10$. For each dataset, the ensemble models are ordered as follow: left: VM, middle: LR, right: $LR + VM$. Each curve corresponds to a particular ensemble size; the number of peers is indicated in the first figure's legend. The dashed line corresponds to the average corrupted learner accuracy. For VM, the curves corresponding to the ensemble of sizes 1 and 2 overlap with the dashed line. The noise is increased by step of 2%, from 0% to 100%. Each experiment has been repeated 25 times.

The methods do not have the same response to an increase in noise. VM requires much more peers to obtain a resilience equivalent to LR. A change from 1 to 5 learners offers limited improvement for VM, while in the case of LR, an ensemble with 5 learners is almost as stable as an ensemble of 20 learners. The accuracy of LR using one external matrix M_c for refinement does not offer resilience, as depending on the dataset and the noise level the accuracy is a little bit greater or lower than the baseline accuracy. However, moving to the use of 2 matrices M_c, the resilience of LR surpasses the voter model. $LR + VM$ is even more resilient than the two other methods with very flat horizontal curves

When looking at the red curves corresponding to ensembles of size 20, there are two observable domains: *resilience* and *fragility*. In the *resilience* domain, an increase of noise leads to a small decrease of accuracy, while in the *fragility* domain this increase leads to an important change of accuracy. The transition between the two domains depends on the method used and the ensemble size. The transition occurs around 65% of noise for VM, nearby 80% for LR, and 90% for $LR + VM$. Therefore, the label refinement approach and its variant lead to more resilience than the voter model with the same ensemble.

5.3 Boundary Size Influence

In the previous experiments, the boundary parameter k controlling the number of nearest neighbors was set to $k = 10$. This parameter is important as it controls the number of items that can induce a label change. Therefore, the third experiment studies the impact of k over the resilience behavior for a fixed number of peers $k_p = 10$. The results are displayed in Fig. 3.

Under the absence of noise, the increase of k has a positive effect on the accuracy of the LR method but is limited in amplitude. In the presence of noise, a larger value of k prevents a premature loss of accuracy.

The configuration with $k = 1$ does not seem to have any impact on the accuracy. In this setup, the only possibility for an item is to inherit from its neighbor's label unless they already have the same label. If $\mathbf{x}$ and $\mathbf{x}'$ are reciprocal neighbors, they will exchange their labels forever. If they are not (i.e., $\mathbf{x} \in \mathcal{N}(\mathbf{x}')$ but $\mathbf{x}' \notin \mathcal{N}(\mathbf{x})$), then $\mathbf{x}$ would take $\mathbf{x}'$'s label. The number of possible changes in this setup is limited, which might explain the small difference from the average classifier accuracy.

The accuracy under noisy condition increase is better recovered with a larger k. The switch from *resilience* to *fragility* also changes from 40–50% for $k = 3$ to 80–90 for $k = 20$. The resilience behavior is easily obtained, with $k = 10$ providing almost the same resilience to noise as $k = 20$. Compared to the curves presented in Fig. 2, an increase of the boundary k in LR leads to greater resilience in accuracy than the addition of more peers in the VM. A sufficient k needs to be combined to a sufficient k_p to benefit from the ensemble size and items' neighborhood stability.

The greater resilience to noise with a larger k can be simply explained. In a dense area where all items belong to the same class (before corruption), all nearest items would receive the same weight. Therefore the weighted vote will be equivalent to a normal vote. As k increases, more intact labels would be included in the vote, making the decision more stable. Consequently, items in this area will easily recover their initial labels. The items near the boundary benefit from the same effect because they are surrounded by boundary items that are possibly misclassified. Therefore, k needs to be larger for these area to recover.

When k is small, the density is almost invariant; therefore, all classes might be represented in equal proportion. However, the density can vary from one location to another within the radius when extending the radius. If located in a very dense area, a class will be more numerically represented. The vote in (2) is biased by the weights, but also by the number of items that contribute. As an example, if there is one item $\mathbf{x}'$ in the neighborhood of $\mathbf{x}$ belonging to the correct class ℓ with weight $\mathcal{M}(\mathbf{x}, \mathbf{x}') = 1$, but 10 other items of the same incorrect class ℓ_w with weight $\mathcal{M}(\mathbf{x}, \mathbf{x}') = 0.11$, then ℓ_w would win the vote. Therefore, the parameter boundary k must be chosen appropriately to avoid this type of issue. In addition, the process becomes computationally expensive.

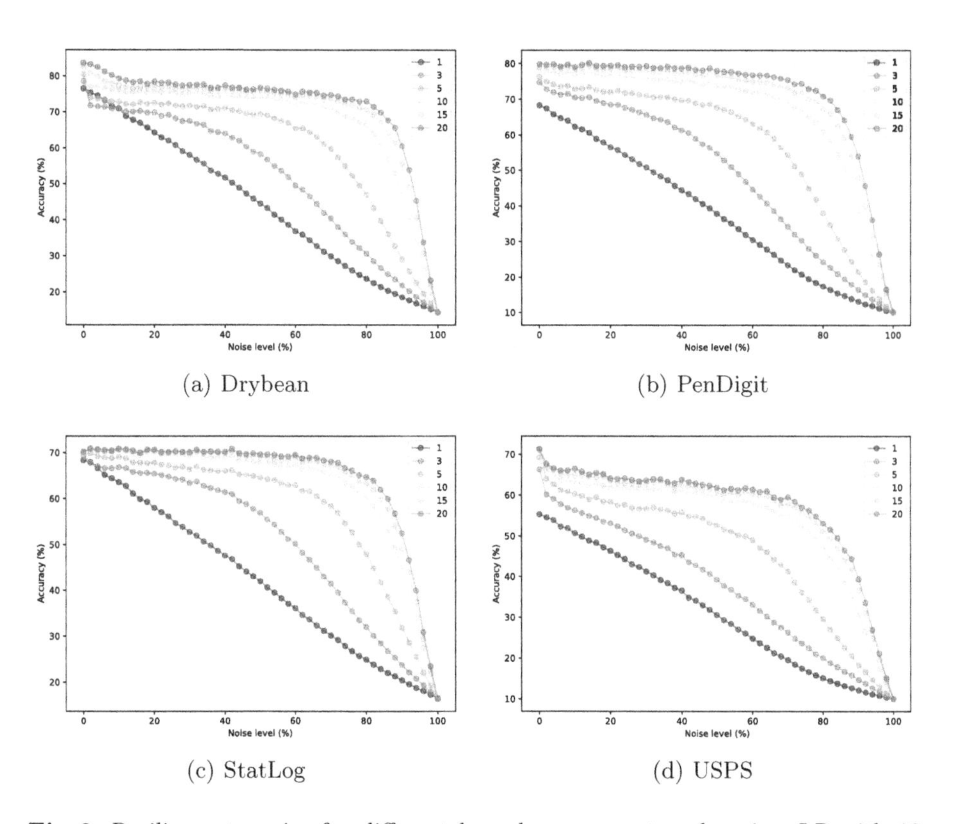

(a) Drybean (b) PenDigit

(c) StatLog (d) USPS

Fig. 3. Resilience to noise for different boundary parameters k, using LR with 10 classifiers in the ensemble.

6 Discussion

6.1 Instance vs Batch Classification

One major difference between our ensemble method and classical ensemble methods is the input size. Traditionally, the ensemble process one sample at a time, allowing us to predict labels of any sample size without constraint. Our approach is inspired by *Evidence Accumulation Clustering*, where samples are processed by batch. Therefore, it limits the usability of use-cases that do not require real-time answers. Nonetheless, for applications with no real-time constraint, data can be stored into a buffer and classified when the collected data is sufficient.

In the different experiments, we classified all the items left in the test set together. When reducing the test size, the system can become unstable if too few samples are available. In a batch, items can be classified into *core* items that are easily classified by all peers, with most of their nearest neighbors belonging to the same class, and *boundary* items where errors are more frequent. Misclassified core items are easily rectified thanks to their strong neighborhood. However, boundary items rely on core items to get good reference labels, as other boundary items may not be reliable enough. If the batch size is too small, core areas would be restricted to very few points, and would not be large enough to offer stable references to other items.

We tested over the *DryBean* dataset to reduce the batch size. Before reaching the system instability for $k = 10$ and its 7 classes, the critical size limit was 150 items. Under this size, the mean accuracy decreased while the variance increased.

One way to work on smaller batches is to adapt the boundary parameter k. Decreasing its value may help to keep core areas stable over small batch sizes, while a too large k might destroy core areas due to misclassified boundary items. Another strategy to work on small batches is to classify the batch X together with X' obtain from a database.

The full process is performed on $X_0 \cup X_1$, which would not suffer from possible instabilities.

6.2 Combining Clustering with Classification

The use of co-association matrices $\{M^{(p)}\}_{p=1:k_p}$ rather than raw predictions $\{\hat{Y}^{(p)}\}_{p=1:k_p}$ gives the possibility to adapt our approach to clustering or mix clustering with classification results. The combination of the two types of algorithms may help to improve classification accuracy. However, the clustering's benefits might be more limited.

One of the main clustering problems is the selection of the number of clusters. Unfortunately, ensemble clustering algorithms often cluster $\mathcal{M}$ using a hierarchical algorithm that circumvents the cluster number problem. The label refinement scheme will help adjust the boundary but will neither help find the best number of clusters.

A direction to explore is how to fuse (or split) clusters using ensemble information. A possibility would be to search for clusters boundaries looking at the local CA matrix $M^{(p)}$ that does not exist in the ensemble matrix $\mathcal{M}$. By detecting weak cluster boundaries, the clusters separated by these boundary could be merged. The same reasoning could be applied for splitting clusters by searching for a strong boundary in $\mathcal{M}$ that does not exist in $M^{(p)}$. This direction can be explored and compared to a more direct clustering of $\mathcal{M}$.

6.3 Ensemble of n-ary Classifiers

Multi-class classifiers can be built by combining binary classifiers, as in the ECOC [4]. Each classifier is trained over 2 classes obtained by splitting the d_0 initial classes at random. This class grouping exploits *class synergies* as two similar classes are easier to distinguish from the other classes when grouped than each alone. The original multi-class would be identified by looking at the ensemble classification binary code obtained.

More generally, d_0 classes can be fused into d_1 classes. However, the transformation of the multi-class prediction to fewer classes can obfuscate the results for privacy reasons. Another motivation is the reduction of the transmitted bit when exchanging the raw predictions.

If $d_0 \gg d_1$, the amount of preserved boundaries is $1 - \frac{1}{d_1}$ (the boundary between two classes merged into the same group is no more identifiable, but still observable for two classes in a distinct group). It is at worse 50% for a binary grouping, impacting $\mathcal{M}$ in these proportion.

Looking at particular class boundaries, a proportion α of the classifiers still see this boundary and would vote something different for each pair of items. On the other hand, a proportion $(1-\alpha)$ do not see it and would vote 1 for any pairs. Therefore, we could rewrite $\mathcal{M}$ as $\mathcal{M}(\mathbf{x}, \mathbf{x}') = \alpha m(\mathbf{x}, \mathbf{x}') + (1-\alpha)$ where $m(\mathbf{x}, \mathbf{x}')$ is the average co-association score of the classifiers seeing the boundary. When performing the vote using Eq. (2), the part $(1 - \alpha)$ contributes equally for all neighbors leading to a very limited effect on the vote outcome.

7 Conclusion

In this article, we proposed a noise-resilient ensemble classification method exploiting the co-classification of nearest items. The system works in two steps: an *exchange phase* where peers communicate their prediction, and a local *label refinement phase* where labels are adjusted using a weighted voter model.

This ensemble approach was tested on four different datasets, where it led to accuracy improvement. Moreover, under the noisy condition, the proposed ensemble method was highly resilient to noise. It allowed it to preserve accuracy with much fewer peers than an ensemble combining its prediction using a voter model. The approach is flexible, as clustering peers can participate in helping classification peers. Additionally, classifiers with different objectives can be combined if some of their class shares the same boundaries.

In future works, we will analyze the data factors impacting the accuracy to quantify the possible gain, and investigate how to estimate the boundary parameter k to better adapt to the dataset to classify.

References

1. Abualkibash, M., ElSayed, A., Mahmood, A.: Highly scalable, parallel and distributed adaboost algorithm using light weight threads and web services on a network of multi-core machines. arXiv abs/1306.1467 (2013)
2. Breiman, L.: Bagging predictors. Mach. Learn. **24**, 123–140 (2004)
3. Breiman, L.: Random forests. Mach. Learn. **45**, 5–32 (2001). https://doi.org/10.1023/A:1010933404324
4. Dietterich, T.G., Bakiri, G.: Solving multiclass learning problems via error-correcting output codes. arXiv cs.AI/9501101 (1995)
5. Dua, D., Graff, C.: UCI machine learning repository (2017). http://archive.ics.uci.edu/ml
6. Fan, W., Stolfo, S., Zhang, J.: The application of adaboost for distributed, scalable and on-line learning. In: KDD 1999 (1999)
7. Fred, A., Jain, A.K.: Combining multiple clusterings using evidence accumulation. IEEE Trans. Pattern Anal. Mach. Intell. **27**, 835–850 (2005)
8. Galdi, P., Napolitano, F., Tagliaferri, R.: Consensus clustering in gene expression. In: di Serio, C., Liò, P., Nonis, A., Tagliaferri, R. (eds.) CIBB 2014. LNCS, vol. 8623, pp. 57–67. Springer, Cham (2015). https://doi.org/10.1007/978-3-319-24462-4_5
9. Hao, F., Ryan, P., Zielinski, P.: Anonymous voting by two-round public discussion. IET Inf. Secur. **4**, 62–67 (2010)
10. Khan, Z., et al.: Ensemble of optimal trees, random forest and random projection ensemble classification. Adv. Data Anal. Classif. **14**(1), 97–116 (2019). https://doi.org/10.1007/s11634-019-00364-9
11. Li, T., Ding, C.: Weighted consensus clustering. In: SDM (2008)
12. Margineantu, D., Dietterich, T.G.: Pruning adaptive boosting. In: ICML (1997)
13. Monti, S., Tamayo, P., Mesirov, J., Golub, T.: Consensus clustering: a resampling-based method for class discovery and visualization of gene expression microarray data. Mach. Learn. **52**, 91–118 (2004)
14. Ormándi, R., Hegedüs, I., Jelasity, M.: Gossip learning with linear models on fully distributed data. Concurr. Comput. Pract. Experience **25**, 556–571 (2013)
15. Pes, B.: Ensemble feature selection for high-dimensional data: a stability analysis across multiple domains. Neural Comput. Appl. **32**(10), 5951–5973 (2019). https://doi.org/10.1007/s00521-019-04082-3
16. Probst, P., Boulesteix, A.: To tune or not to tune the number of trees in random forest? J. Mach. Learn. Res. **18**, 181:1–181:18 (2017)
17. Ratasich, D., Khalid, F., Geissler, F., Grosu, R., Shafique, M., Bartocci, E.: A roadmap toward the resilient internet of things for cyber-physical systems. IEEE Access **7**, 13260–13283 (2019)
18. Ren, Y.Z., Domeniconi, C., Zhang, G., Yu, G.X.: Weighted-object ensemble clustering. In: IEEE 13th International Conference on Data Mining, pp. 627–636 (2013)
19. Schapire, R., Freund, Y.: Boosting: foundations and algorithms. Kybernetes **42**(1), 164–166 (2012)
20. Sen, J.: A survey on wireless sensor network security. arXiv abs/1011.1529 (2009)
21. Sheikh, M.S., Liang, J.: A comprehensive survey on VANET security services in traffic management system. Wirel. Commun. Mob. Comput. **2019**, 2423915:1–2423915:23 (2019)
22. Strehl, A., Ghosh, J.: Cluster ensembles – a knowledge reuse framework for combining multiple partitions. J. Mach. Learn. Res. **3**, 583–617 (2002)
23. Zhou, Z.: Ensemble Methods: Foundations and Algorithms. Chapman and Hall/CRC, New York (2012)

SIITR: A Semantic Infused Intelligent Approach for Tag Recommendation

M. Anirudh[1,2], Gerard Deepak[1,2](✉), and A. Santhanavijayan[1,2]

[1] Department of Computer Science and Engineering, National Institute of Technology, Tiruchirappalli, Tiruchirappalli, India
[2] SRM Institute of Science and Technology, Ramapuram, Chennai, India
gerard.deepak.christuni@gmail.com

Abstract. In the present-day time, Tag recommendation is of utmost importance as it heads into annotations and labeling of several entities or elements, be it images or data over the World Wide Web. Thereby, in recent times of Semantic Web, tagging is one of the crucial aspects to incorporate meanings to the entities or elements on the World Wide Web. As a result, socially relevant tag recommendation is of extreme value. This paper puts forth a Semantic infused intelligent approach is proposed to recommend tags for socially relevant images using Jaccard and Cosine similarity and SVM to classify the documents. The above experiments have been conducted for Heritage CHIC dataset and New Image Corpus and the accuracy percentage of 94.48% is achieved.

Keywords: Cosine · Image-tagging · Jaccard · SVM · Semantic similarity

1 Introduction

Image tagging is a process that enables users to identify text in images and make them more searchable. It is done by creating a template that automatically tags each image whereas social tagging is a process that enables people to organize and identify online content by its various keywords. The need of social tagging is to organize content on a website or social bookmark and with respect to other similar approaches. The problem arises due to the increased data in the World Wide Web which comprises of all the multimedia data which incites time complexity throughout the Web. So, tagging of these data or naming them accordingly, especially images are vital. Search engines and image retrieval systems retrieve images that are requested but it makes the process hefty if there are no proper tagging of the images and social awareness in required in order to bridge the cognitive gap between Semantic and Social Web. An image's tag provides an identity to that particular image like an identification which makes tagging of images a vital process [1]. An Untagged image can lead to high complexity and it even become non retrievable when an image search engine is acting on it or any misannotated image will never get retrieved by any intelligent information retrievable system. As a result, there is a need for tagging especially for the contents in the Semantic Web which is an extension of the existing World Wide Web that was modified to make the data present into

I. Woungang et al. (Eds.): ANTIC 2021, CCIS 1534, pp. 389–399, 2022.
https://doi.org/10.1007/978-3-030-96040-7_31

machine-readable. These modifications were made to represent the metadata that possess ontologies that can brief concept used, describe the relationship between entities and their categories. Socially relevant tag recommendation is an important factor regarding images, it marks its respective genre based on its relevance and that makes the whole process effortless and swift.

Motivation: The present data on the World Wide Web is exploding and retrieval of items from the World Wide Web is a cumbersome task. To facilitate this information retrieval tagging items on the World Wide Web is of foremost importance. So, tagging of every form of contents throughout the Web like multimedia, images, videos, text etc. is required which makes its retrieval effortless. In present times of Semantic Web, a need for semantic induced approach for recommendation for tags apart from this a socially aware approach is required because the Social Web must be given an equal importance in bridging the gap between cognitive, semantic and the real-world web knowledge. As a result, socially aware semantically driven tag recommendation approach is required.

Contribution: A semantically infused approach has been proposed for tag recommendation. Tag preprocessing is done using tokenization, lemmatization, stop word removal and NER. Relevant documents are yielded using the classification algorithm called Support Vector Machine (SVM). Datasets are obtained from Heritage CHIC and New Image Corpus to enrich the tag space. Semantic similarities such as Jaccard and Cosine to further diversify and obtain distinct results. An Accuracy of 95.49%, nDCG of 0.96 and FDR of 0.07 has been achieved by the proposed approach.

Organization: The remaining paper is structured as follows. Section 2 elaborates the Related Works. Section 3 explains the Proposed System Architecture in detail. Section 4 consolidates the Implementation and the results of the proposed architecture. In Sect. 5, depicts the conclusion.

2 Related Works

Pushpa et al. [1], have proposed a strategic tag recommendation system which possess an automatic relevant tag assigning feature that prioritizes user's choice. Suman Banerjee et al. [2], have put forth a Multistep recommendation system where the neighbor user item set as well as user feature matrix, user priority and item priority are taken into consideration and recommendation are done mainly based on item priority. A. K Singh et al. [3], developed using a collaborative filtering method. Topical Folksonomy definitely ensures the amount of diversity in the results which is highly preferrable but since collaborative filtering is required, rating based on profile matrices of the topic and user topic have to be computed for which the individual rating of item set is required. As a result, a lag in this approach can be identified that can be improved when collaborative filtering is replaced by some other classification or semantic infused techniques. Ralf Krestel et al. [4], created a Latent Dirichlet Allocation for tag recommendation or LDA model is a topic modelling algorithm to recommend tags of tag set created and addressed by many users which makes the resources easier to search and it helps in increasing the probability of diversity in the final results. Junjie Zhang et al. [5] proposes heritage image annotation using collective knowledge by uncovering the links between images and their corresponding meta data and conducts a metric learning to extract their contents.

Chandramani Chaudhary et al. [6], This framework proposes a fresh method for tag assignment that considers the properties of various markup features, such as Naturalness, Newness, and Rarity that helps in enriching the semantic depiction of the images. Shafin Rahman et al. [7], In Zero-shot image tagging is a challenging problem where multiple labels are attached to an image. The goal of this paper is to provide an automatic approach to discover and identify the scene concepts associated with an image using Multiple instance learning framework. Jianlong Fu et al. [8], The proposed work introduces a deep learning system which is weakly-supervised that is trained from images taken from websites to reduce manpower that was tuned to be levelled with the arbitrary tags based in a feature map of its deep layer. G Chen et al. [9], proposed a new ranking function that takes into account the global context of labels. introduce the concept of multi-labelling problem and develop it as a ranking solution for structured output prediction using Convoluted Neural Network (CNN). Hebatallah A. Mohamed Hassan et al. [10] had proposed a Tag recommendation system which is content-based that uses deep neural networks to semantic vectors that encodes abstract and titles of scientific articles and bidirectional fated recurrent units were used to carry out their experiment on datasets from CiteULike. Benjamin Adrian et al. [11] had designed a customized approach in constructing a semantically driven tag recommendation system that are completely centered on Web 2.0 services and Semantic Web ontologies. Bo Jiang et al. [12] has put forth a model in recommending tags using K-Nearest Neighbor (KNN) that are completely based on social comment network. In [13–25] several semantic frameworks in support of the literature of the proposed frameworks have been depicted.

3 Proposed System Architecture

Figure 1 depicts the architecture of the Tagging of socially relevant photos with tags of heritage value and some public events. The process proceeds from tag pre-processing, document classification that yields the relevant documents and At First, Tag pre-processing is done by identifying the tags or annotations present with the images based on tokenization, lemmatization, stop word removal and Named Entity Recognition (NER). The Tag aware approach begins by pre-processing the tags, labels, annotations, or descriptions are first Tokenized by ensuring the descriptions or multi-word labels into individual terms. Lemmatization secures the morphological or inflectional term from its Tokenized base form. Then the process enters stop word removal, Here the elimination of stop words like of, the, an etc. take place. NER is the final pre-processing step in which the Entities are recognized. After the pre-processing steps, it yields a set of pre-processed set of documents instead of tag terms. The Document classification is executed with SVM using the dataset, SVM are prioritized in this process which requires an Indistinct and computationally inexpensive classification algorithm. At each of the tag terms, the Term frequency and the Inverse Document Frequency is applied from the classified document set to obtain much more relevant documents based on the uniqueness of a term within the document and its frequency within that document. The relevant documents are yielded and from the set of documents, The Tag K or Top relevant

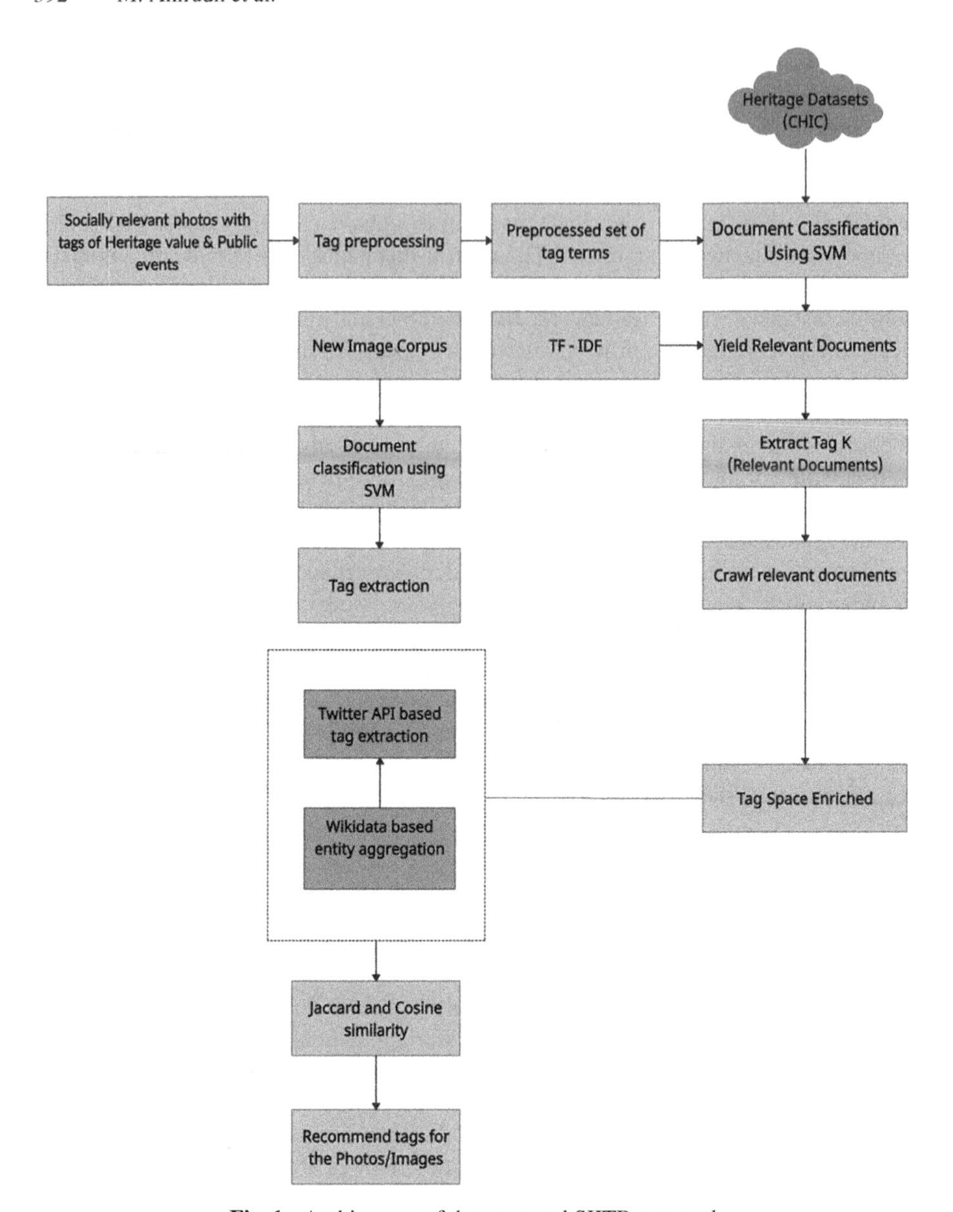

Fig. 1. Architecture of the proposed SIITR approach

documents are extracted. The keywords from these documents are obtained and are fed into the tag space. The new image corpus houses lots of relevant heritage images with some specifications of tags, where the document and the image classification are done by SVM.

SVM is a supervised machine learning model that analyses the given data using Regression and classification. It is a discriminative classifier that is parted by a hyperplane that distinguishes based on the given datasets. From the given datasets, it categorizes the input based on its suitable support vectors. Support vectors are the data points which are closest to its opposing data class. These data points are the key distinguishing factor of each dataset. SVMs are best for working with linear and non-linear models; specially to transform any non-linear model to a linear model. From the existing images, the tags will be further extracted and are loaded into the tag space which is then enriched by Wikidata. Wikidata is used for entity aggregation which is accessed by an API and all the socially relevant tags or element from the tag space and from Wikidata API are identified. Finally, the Images will be recommended with their corresponding tags. *The semantic similarities are computed using Jaccard and Cosine.* The New image corpus is classified using SVM and the relevant tags are extracted. Its anti-semantic computation is done with Jaccard and as well as with Cosine to be loaded onto the Tag space. Wikidata is a one of the vast sources of open data that contain information about various locations, art and literature. It is a knowledge graph of multilingual semantics that powers some projects like Wikipedia. It's QID (a unique label) is used to identify any given data based on its attribute and significance even if it is known for multiple aliases and statements. Terms specific to a document are identified and counted based on its singularity and the repetitions of these terms are called as Term Frequency (TF). While Inverse Document Frequency (IDF) determines the rarely used important terms from a document rather than its repeated words of less importance. So, to incorporate these specific important ones among the rest IDF is used. *Jaccard similarity gets the information whether the given documents are similar and distinct based on their uniqueness and repetition of the terms provided. cosine similarity from its name can be defined as the cosine parameter of the angle between a pair of vectors and their space.* It can be defined as the distance between the vectors and their space.

$$\mathrm{tf}(\mathrm{t},\mathrm{d}) = \frac{\text{terms in document}}{\text{words in document}} \tag{1}$$

$$\text{tf-idf}(\mathrm{t},\mathrm{d}) = \mathrm{tf}(\mathrm{t},\mathrm{d}).\log\left(\frac{\mathrm{N}}{\mathrm{df}+1}\right) \tag{2}$$

$$\text{jaccard similarity} = \frac{|\mathrm{A}\cap\mathrm{B}|}{|\mathrm{A}\cup\mathrm{B}|} = \frac{|\mathrm{A}\cap\mathrm{B}|}{|\mathrm{A}|+|\mathrm{B}|-|\mathrm{A}\cap\mathrm{B}|} \tag{3}$$

$$\text{cosine similarity} = \frac{\mathbf{B}\cdot\mathbf{G}}{\|\mathbf{B}\|\,\|\mathbf{G}\|} = \frac{\sum_{i=1}^{n}\mathrm{B_i}\mathrm{G_i}}{\sqrt{\sum_{i=1}^{n}\mathrm{B_i^2}}\sqrt{\sum_{i=1}^{n}\mathrm{G_i^2}}} \tag{4}$$

Here Eqs. (1) and (2) depicts the tf and idf. Equations (3) and (4) represents Jaccard Similarity and cosine similarity respectively.

4 Implementation and Result Analysis

Python with anaconda and Intel i5 processor with Nvidia Graphic card and 16 GB RAM is used to carry out the implementation. MySQL lite is used as the back end. Hash table and hash map are incorporated from collection frame. The NLP tasks were performed using python's Natural Language Toolkit Library. Datasets were obtained from Cultural Heritage in CLEF (CHIC) which is an open-source dataset that comprises of over 170,000 documents from 13 languages and New Image Corpus that houses lot of heritage images with some specification of tags. The experiment was conducted for 5871 Queries whose ground truth has been collected for the proposed model.

The proposed SIITR approach for tag recommendation is evaluated offer performance using Precision, Recall, Accuracy, F-Measure, nDCG and FDR as the potential matrix. The reason for using Recall, Accuracy, Precision and F-Measure is to compute the relevance on the results yielded, while the nDCG or the Normalized Discounted Cumulative Gain measures the diversity of the results and False Discovery Rate (FDR) is a method of estimating the number of null hypotheses that are false which are furnished by the proposed SIITR. In order to evaluate the performance of the SIITR, it is baselined with SpatioTag, MPBSTI, CFTR and LDATR. Table 1 Depicts the algorithm for the proposed SIITR approach for tag recommendation.

$$\text{Precision, P} = \frac{|(\text{relevant docs}) \cap (\text{retrieved docs})|}{|(\text{retrieved docs})|} \tag{5}$$

$$\text{Recall, R} = \frac{|(\text{relevant}) \cap (\text{retrieved})|}{|(\text{relevant})|} \tag{6}$$

$$\text{Accuracy} = \left(\frac{P+R}{2}\right) \tag{7}$$

$$\text{F-Measure} = \left(\frac{2.P.R}{P+R}\right) \tag{8}$$

$$\text{DCG}_\text{p} = \sum_{i=1}^{p} \frac{2^{\text{rel}_i} - 1}{\log_2(i+1)} \tag{9}$$

$$\text{nDCG}_\text{p} = \frac{DCG_p}{IDCG_p} \tag{10}$$

$$FDR = 1 - R \tag{11}$$

Equations (5), (6), (7) and (8) depicts Precision, Recall, Accuracy, F-Measure and Eq. (9) measures the Discounted Cumulative Gain (DCG). Equation (10) measures Normalized Discounted Cumulative Gain (nDCG). Equation (11) displays the False Discovery Rate (FDR).

From Table 2, It is found that SpatioTag yields 85.69% of Precision, 87.36% of Recall, 86.52% of Accuracy, 86.51% of F-Measure, 0.88 of nDCG and 0.15 of FDR. MPBSTI yields 87.18% of Precision, 89.66% of Recall, 88.42% of Accuracy, 88.40% of

Table 1. Proposed SIITR algorithm

```
Input: Photos extracted from social media
Output: Tags recommended
Begin
    Step 1: Tags are pre-processed using tokenization, lemmatization,
  stop word removal and NER and save as a set of tags Tg from the
  socially relevant photos
    Step 2: while (Tg. next()! = NULL)
                 for each tag Tg
                      for dataset from CHIC and Image corpus
                          yield relevant documents based on TF-
  IDF and SVM
                      end inner for
                 end outer for
            end while
    Step 3: for the relevant documents
               Obtain keywords ks
               for keywords ks
                  enrich ks using Tag space
               end inner for
             end outer for
    Step 4: From the new image corpus the relevant documents are
  classified using SVM and the tags are extracted
    Step 5: for tagspace enriched keywords ks
               for tags extracted from new image corpus
                    Perform entity aggregation based on Wikidata
                    Perform TWITTER API based tag extraction
               end inner for
            end outer for
    Step 6: Semantic similarity is computed using Jaccard and Cosine
  similarity.
    Step 7: if (ssm>0.75)
               Recommend tags for the photo/images
            end if
    Step 8: The tags are recommended
End
```

Table 2. Comparison of Performance of the proposed SIITR with other approaches

Search technique	Average precision %	Average recall %	Average accuracy %	Average F-Measure %	nDCG	FDR
SpatioTag [1]	85.69	87.36	86.52	86.51	0.88	0.15
MPBSTI [2]	87.18	89.66	88.42	88.40	0.79	0.13
Collaborative Filtering for Tag Recommendation (CFTR) [3]	84.12	86.77	85.44	85.42	0.81	0.16
LDATR [4]	86.33	89.69	88.01	87.97	0.90	0.14
Proposed SIITR	**93.79**	**95.18**	**94.49**	**94.48**	**0.96**	**0.07**

F-Measure, 0.79 of nDCG and 0.13 of FDR. CFTR yields 84.12% of Precision, 86.77% of Recall, 85.44% of Accuracy, 85.42% of F-Measure, 0.81 of nDCG and 0.16 of FDR. LDATR yields 86.33% of Precision, 89.69% of Recall, 88.01% of Accuracy, 87.97% of F-Measure, 0.90 of nDCG and 0.14 of FDR. Proposed SIITR yields an average Precision of 93.79%, average Recall of 95.18%, average Accuracy of 94.49%, Average F-Measure of 94.48%, nDCG of 0.96 and FDR of 0.07.

The SpatioTag recommends tags in tags for photograph using the spatial information using a flickr dataset. Content based similarity and Text based similarity is computed using the normalized pointwise mutual information strategy and a clustering is performed based on the npmi value. It yields fairly high Precision, Recall, Accuracy, F-Measure and nDCG but there is always a scope for improvement by incorporating much better artificially intelligent schemes and paradigms for recommending images by giving importance to the heritage.

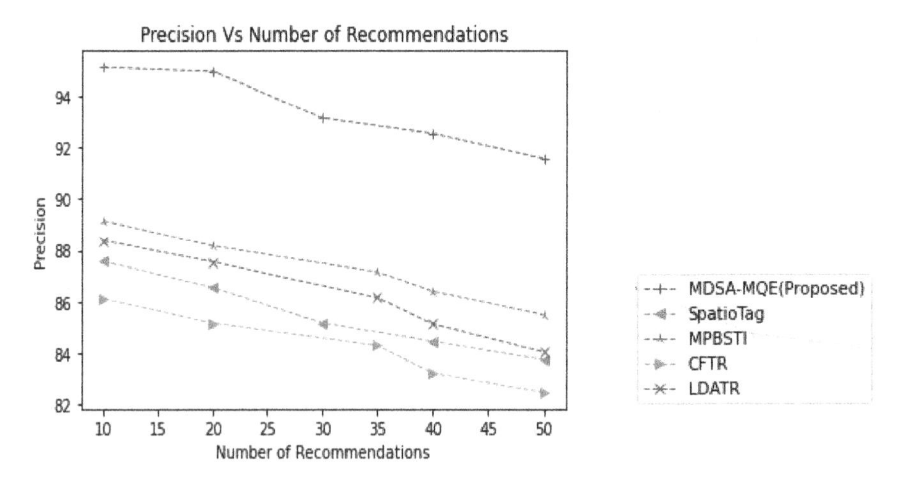

Fig. 2. Precision Vs number of recommendations

From Fig. 2. the Precision-Recall curve is depicted makes it clear that the proposed approach has the better Precision vs no of recommendations curve compared to other approaches with respect to the no of recommendations. Spatio Tag is specifically designed to ensure the tag validation and increase the elements in suggested tags based on its content and text-based similarity which is them computed using the Normalized Pointwise Mutual information strategy [1]. In MPBSTI where the neighbor user item set as well as user feature matrix, user priority and item priority are taken into consideration and recommendation in mainly based on item priority. The relation between user priority and item priority is incorporated in this approach. This is completely a non-semantic approach which is unsuitable for data set with high data density. The user item set, user priority and item priority are computed to obtain the ratings, which can be challenging because rating based on social trust item tags and finding implicit values and baselining these values become a very tough task including the cold-start problem which are solved using this method. Where there is a definite scope for making a query centric approach for recommendation of tags. Diversity of tags is quite low because there is no diversified knowledge fed into the system [2]. CFTR is used. Topical Folksonomy ensures the amount of diversity in the results which is highly preferrable but since collaborative filtering is required, rating based on the profile matrices of topic and user topic have to be computed for which the individual rating of item set is required. As a result, a lag in this approach can be identified that can be improved when collaborative filtering is replaced by some other classification or semantic infused techniques [3]. Latent Dirichlet Allocation for tag recommendation or LDA model is a topic modelling algorithm but when LDA is singularly used to generate a controlled vocabulary, or a good annotation item set it becomes insufficient. There should be a much better scheme to ensure its relevance enhancement, but LDA does help in increasing the probability of diversity in the final results but it's still less yielding if not integrated with any other schemes [4].

The Proposed SIITR is a semantically intelligent integrative approach which incorporates several models namely tf, idf, svm based image document classification, wiki data and twitter API. The incorporation of idf, tf ensures that the inverse document frequency and term frequency are computed considering the occurrence of terms within the matrix and across the document corpus and frequency of terms within the matrix and across the document corpus. Therefore, the terms are given the priority within a corpus, document and between documents over a Web document corpus. SVM is used for initial classification of documents because it is a linear binary classifier which is quite light weight in nature and since the approach only demands a basic classification. As a result, the usage of SVM; a good choice, it makes the document light weight, computationally light and inexpensive. Using Wikidata ensures the background knowledge is implicated into the approach where the API is used to access the Wikidata and Wikidata yields the cognitive gap between the existing World Wide Web knowledge and the knowledge incorporated to the framework, it induces the cognitive gap, it increases the density of background knowledge and there by using Wikidata makes it quite dense. So, this enhances the knowledge density, background knowledge is incorporated and thereby it yields a better approach based on Precision, Recall, Accuracy, F-Measure and nDCG which are also increased. Twitter API makes sure that twitter tags are also incorporated which ensures that social awareness is also integrated into the framework and then the Jaccard and Cosine semantic similarities are used together to make sure that the relevance computation is also much better than the other traditional approaches.

5 Conclusion

A Socially relevant Tag recommendation system is developed that uses a semantic infused approach. The proposed system focuses on pre-processing the tags from the input images from which the pre-processed tags yield the documents that are classified using SVM into its relevance. Jaccard and Cosine semantic similarities are integrated to obtain the Top relevant documents or Tag-K documents which are further incorporated into the tag space. Finally, the experiments are carried out for Heritage CHIC dataset and new image corpus that yields 94.48% F-Measure with very high nDCG and with low FDR are achieved for the proposed method.

References

1. Pushpa, C.N., Deepak, G., Pradeep, T.J., Venugopal, K.R.: SpatioTag: a strategic approach for recommending image tags for socially relevant photos using spatial information. Int. J. Comput. Sci. Eng. **6**(7), 261–267 (2018)
2. Banerjee, S., Banjare, P., Pal, B., Jenamani, M.: A multistep priority-based ranking for top-N recommendation using social and tag information. J. Ambient. Intell. Humaniz. Comput. **12**(2), 2509–2525 (2020). https://doi.org/10.1007/s12652-020-02388-y
3. Singh, A.K., Nagwani, N.K., Pandey, S.: TAGme: a topical folksonomy based collaborative filtering for tag recommendation in community sites. In: Proceedings of the 4th Multidisciplinary International Social Networks Conference, pp. 1–7, July 2017
4. Krestel, R., Fankhauser, P., Nejdl, W.: Latent dirichlet allocation for tag recommendation. In: Proceedings of the third ACM Conference on Recommender Systems, pp. 61–68, October 2009
5. Zhang, J., Wu, Q., Zhang, J., Shen, C., Lu, J., Wu, Q.: Heritage image annotation via collective knowledge. Pattern Recogn. **93**, 204–214 (2019)
6. Chaudhary, C., Goyal, P., Prasad, D.N., Chen, Y.P.P.: Enhancing the quality of image tagging using a visio-textual knowledge base. IEEE Trans. Multimedia **22**(4), 897–911 (2019)
7. Rahman, S., Khan, S., Barnes, N.: Deep0tag: Deep multiple instance learning for zero-shot image tagging. IEEE Trans. Multimedia **22**(1), 242–255 (2019)
8. Fu, J., Wu, Y., Mei, T., Wang, J., Lu, H., Rui, Y.: Relaxing from vocabulary: robust weakly-supervised deep learning for vocabulary-free image tagging. In: Proceedings of the IEEE International Conference on Computer Vision, pp. 1985–1993 (2015)
9. Chen, G., Xu, R., Yang, Z.: Deep ranking structural support vector machine for image tagging. Pattern Recogn. Lett. **105**, 30–38 (2018)
10. Hassan, H.A.M., Sansonetti, G., Gasparetti, F., Micarelli, A.: Semantic-based tag recommendation in scientific bookmarking systems. In: Proceedings of the 12th ACM Conference on Recommender Systems, 465–469, September 2018
11. Adrian, B., Sauermann, L., Roth-Berghofer, T.: Contag: a semantic tag recommendation system. Proc. I-Semant. **7**, 297–304 (2007)
12. Jiang, B., Ling, Y., Wang, J.: Tag recommendation based on social comment network. Int. J. Digit. Content Technol. Appl. **4**(8), 110–117 (2010)

13. Deepak, G., Ahmed, A., Skanda, B.: An intelligent inventive system for personalised webpage recommendation based on ontology semantics. Int. J. Intell. Syst. Technol. Appl. **18**(1–2), 115–132 (2019)
14. Kumar, A., Deepak, G., Santhanavijayan, A.: HeTOnto: a novel approach for conceptualization, modeling, visualization, and formalization of domain centric ontologies for heat transfer. In: 2020 IEEE International Conference on Electronics, Computing and Communication Technologies (CONECCT), pp. 1–6. IEEE, July 2020
15. Kaushik, I.S., Deepak, G., Santhanavijayan, A.: QuantQueryEXP: a novel strategic approach for query expansion based on quantum computing principles. J. Discrete Math. Sci. Cryptogr. **23**(2), 573–584 (2020)
16. Deepak, G., Priyadarshini, J.S.: Personalized and enhanced hybridized semantic algorithm for web image retrieval incorporating ontology classification, strategic query expansion, and content-based analysis. Comput. Electr. Eng. **72**, 14–25 (2018)
17. Leena Giri, G., Deepak, G., Manjula, S., Venugopal, K.: OntoYield: a semantic approach for context-based ontology recommendation based on structure preservation. In: Chaki, N., Cortesi, A., Devarakonda, N. (eds.) Proceedings of International Conference on Computational Intelligence and Data Engineering. LNDECT, vol. 9, pp. 265–275. Springer, Singapore. https://doi.org/10.1007/978-981-10-6319-0_22
18. Kumar, N., Deepak, G., Santhanavijayan, A.: A novel semantic approach for intelligent response generation using emotion detection incorporating NPMI measure. Procedia Comput. Sci, **167**, 571–579 (2020)
19. Deepak, G., Teja, V., Santhanavijayan, A.: A novel firefly driven scheme for resume parsing and matching based on entity linking paradigm. J. Discrete Math. Sci. Cryptogr. **23**(1), 157–165 (2020)
20. Surya, D., Deepak, G., Santhanavijayan, A.: KSTAR: a knowledge based approach for socially relevant term aggregation for web page recommendation. in: Motahhir, S., Bossoufi, B. (eds.) ICDTA 2021. LNNS, vol. 211, pp. 555–564. Springer, Cham (2021). https://doi.org/10.1007/978-3-030-73882-2_50
21. Deepak, G., Kumar, N., Bharadwaj, G.V.S.Y., Santhanavijayan, A.: OntoQuest: an ontological strategy for automatic question generation for e-assessment using static and dynamic knowledge. In: 2019 Fifteenth international conference on information processing (ICINPRO), pp. 1–6. IEEE, December 2019
22. Srivastava, R.A., Deepak, G.: PIREN: prediction of intermediary readers' emotion from news-articles. In: Shukla, S., Unal, A., Varghese Kureethara, J., Mishra, D.K., Han, D.S. (eds.) Data Science and Security. LNNS, vol. 290, pp. 122–130 Springer, Singapore (2021). https://doi.org/10.1007/978-981-16-4486-3_13
23. Rithish, H., Deepak, G., Santhanavijayan, A.: Automated Assessment of question quality on online community forums. In: Motahhir, S., Bossoufi, B. (eds.) ICDTA 2021. LNNS, vol. 211, pp. 791–800. Springer, Cham. https://doi.org/10.1007/978-3-030-73882-2_72
24. Deepak, G., Rooban, S., Santhanavijayan, A.: A knowledge centric hybridized approach for crime classification incorporating deep bi-LSTM neural network. Multimedia Tools Appl. **80**(18), 28061–28085 (2021). https://doi.org/10.1007/s11042-021-11050-4
25. Krishnan, N., Deepak, G.: KnowSum: knowledge inclusive approach for text summarization using semantic allignment. In: 2021 7th International Conference on Web Research (ICWR), pp. 227–231. IEEE, May 2021

OntoHDClass: Ontology Driven Approach for High Dimensional Data Classification Integrating Semantic Measures and Recurrent Neural Network

N. Ramanathan[1], Gerard Deepak[2(✉)], and A. Santhanavijayan[2]

[1] Department of Computer Science and Engineering, SRM Institute of Science and Technology, Ramapuram, Chennai, India

[2] Department of Computer Science and Engineering, National Institute of Technology Tiruchirappalli, Tiruchirappalli, India

gerard.deepak.christuni@gmail.com

Abstract. Clustering the high-dimensional data has remained a burdensome job. Existing clustering algorithms developed are significantly inept. Frequently, in high dimensional data many noises and disturbances are formed and data is masked in existing clusters. Subsequently, we have presented an Ontology Driven Approach for High Dimensional Data Classification integrating Semantic Measures and Recurrent Neural Network to solve the above problem in this current Paper. The first phase OntoHDClass identifies the problem domain and formulate domain ontologies, accomplishes subspace relevance analysis by computing concept similarities. The second phase, after detection of dense regions elimination of outliers takes place and inliers are extracted. The extracted inliers then yield elite group by Cross entropy and Pearson's correlation. The third phase involves classification of the space using Recurrent Neural Network. The Result are found to be high that the OntoHDClass system reaches an overall Precision, Accuracy, and False Negative Rate of 97.32%, 96.37%, and 0.05% respectively. Elimination of outliers advances the subspace clustering with more accuracy and the elite subspace that are produced is indicated by our result. OntoHDClass can learn the clusters in very subspaces with Lower fault rate besides higher accuracy.

Keywords: Concept similarity · Cross entropy · High dimensional data · Ontology · Subspace clustering

1 Introduction

Through the advancement of extensive and global computing generation of data stood in an increased progress. As a greater number of devices are interconnected and involving storage of growing size of information and are difficult to maintain. In a dataset the dimensions correspond to number of attributes/features that are present in a dataset. A dataset that consists of huge sum of features is called High Dimensional Data. The sum of features can surpass total no. of observations. For example, Data of Health care patient can

I. Woungang et al. (Eds.): ANTIC 2021, CCIS 1534, pp. 400–409, 2022.
https://doi.org/10.1007/978-3-030-96040-7_32

be multi-dimensional. In an instance there would be 50+ measured/documented parameters from body fluid pressure, genetic background, nutrition, operations, treatments, diagnosed diseases, blood analysis. These data set extant innovative challenges meant for unsupervised learning. Subspace clustering stand one response to those encounters. They outclass in circumstances comparable to those pronounced overhead. Therefore, it is highly difficult for classification of the data. So, a best-in-class classification approach has been put forwarded in this paper with existing data via subspace exploration.

Motivation: Information mining, data sets, artificial intelligence and examining are instances of Dimensionally cursed phenomena. The overall setting of these issues is that the accessible information become meager when the volume of the space increments so quick and when the dimensionality increments. To take the state of sparsity the distance and areas of the datapoint focuses are more scatter and dissipated. Numerous dimensions might be of no importance and can veil existing groups in noisy data, data sparsity is one of the features of the menace of dimensionality. The firmness of the procedure can be regularized via subspace exploration namely the small-dimension structures.

Contribution: High-variance or overfitting condition are due to training a model with sparse information. Feature selection also known as elimination of unconnected dimensions is one such technique to perform. In addition, features can be difficult to define as some correlation may happen between diverse dimensions. A clustering technique which discovers clusters within diverse subspaces is called Subspace clustering. This paper proposes a methodology to identify subspace in the given data. The First stage of OntoHDClass identifies the problem domain and formulate domain ontologies, completes relevance analysis intended for the subspace by computing concept similarities. The second phase starts with the exploration, finding of dense sections, it banishes outliers and inliers are extracted. The third phase involves classification of the space using Recurrent Neural Network. The extracted inliers then yield elite group by Cross entropy and Pearson's correlation. Ontology Driven Approach for High Dimensional Data Classification integrating Semantic Measures and RNN when compared to some traditional system show that a higher accuracy can be attained. The OntoHDClass system achieves an overall Precision, Recall, Accuracy, F-measure, and FNR of 97.32%, 95.18%, 96.37%, 96.23%, and 0.05% respectively.

Organization: The remaining paper shows the flow of the process involved. The Sect. 2 reports briefly about the related research done in OntoHDClass. Section 3 describes the strategic system architecture titled as OntoHDClass. Section 4 comprises largely of the implementation method, the results attained from it and the evaluation. Lastly Sect. 5 consists of conclusion.

2 Related Works

This segment confers about the subspace clustering's past studies towards high dimensional data. Elankavi et al. [1], have put forward an algorithm which is Fast-clustering for picking the subset of features and this Fast-clustering algorithm exhibits high efficiency and effectiveness in discovering the subset of features. Radhika et al. [2], have put forwarded two methods, one for computing outliers in addition another for mining elite subspace from the given Multidimensional data. RMSC main goal is to use analytical research procedure and establish a lower error rate besides higher accuracy in all the ways with noise and without noise. Lei et al. [3], have put forwarded a fuzzy c-means clustering algorithm. Chen et al. [4], have put forward a fast-clustering algorithm in DBSCAN built on pruning distance calculations that are gratuitous. Shakya et al. [5], have put forwarded anomaly detection using a support vector machine and Naive Bayes.

Chakraborty et al. [6], have proposed a modest Lasso Weighted k-means algorithm intended for efficient sparse clustering technique for multi-dimensional data. Arias-Castro et al. [7], have proposed a method with a lasso penalty term which is an adaptive method framed as a hill-climbing tactic. The main idea is the minimization of clusters in a regularized sum of square cluster. Sun et al. [8], have projected a high-dimensional cluster analysis which is used to remove redundant variables using k-mean clustering and regularization which can perform cluster observation that are similar simultaneously. McWilliams et al. [9], have proposed a knowledge of Predictive subspace clustering. The proposed algorithm works on estimation of cluster wise PCA parameters, whereas simultaneously splits the data into clusters. Kadir et al. [10], have put forwarded masked EM process for clustering High dimensional data. They have applied the masking strategy for classification of unsupervised learning application for hard EM algorithm and the gaussian fitting mixture.

Wang et al. [11], introduced a way which is proposed for high-dimensional regression problem novel outlier detection. Gan et al. [12], have proposed an iterative practice that is designed to develop and optimize the functional objective for the planned algorithm and begin the convergence. Chormunge et al. [13], have proposed an innovative way to solve the problems of high and multi-dimensional data where good feature subset can be found by clustering and integrating with correlation measure. In [14–24], several ontological frameworks in support of the literature of the proposed framework has been depicted.

3 Proposed System Architecture

The architecture for the proposed OntoHDClass for High Dimensional Data is shown in the Fig. 1. The put forward architecture has three phases. The 1st phase is identifying the problem domain and formulating domain ontologies using which we compute concept similarity and eliminate outliers. The 2nd phase is extracting inliers and yielding the elite group by using Pearson's correlation and Cross entropy. The 3rd phase is classifying the remaining space using RNN and clustering the neighborhood using the extracted inliers. The core purpose of this OntoHDClass system is the extended increase designed for a high-dimensional data in clustering accuracy.

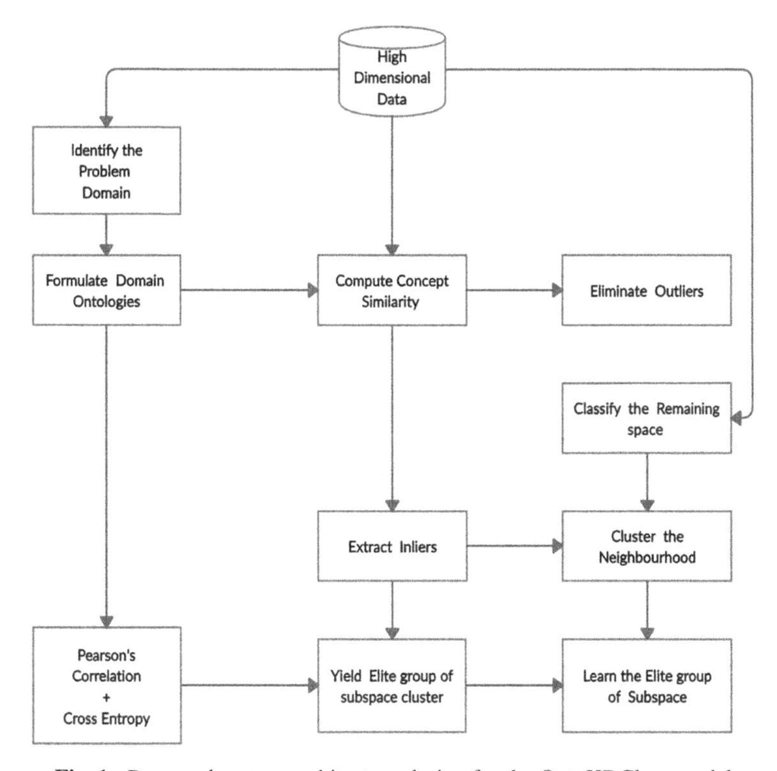

Fig. 1. Proposed system architecture design for the OntoHDClass model

The first phase entails, Identifying the problem domain and formulating domain ontologies. Domain Ontologies are modelled using OntoCollab and Web Protégé. Web Protégé is based on user modelling/problem domain. OntoCollab is for automatic and dynamic modelling of ontologies. Identify Domain based on the domain terms, crawl index words from the World Wide Web and dynamically formulate Ontologies. Also, Static Modelling of Ontologies by users is achieved, then using the formulated domain ontologies. concept similarity is computed to identify similar concept names with the related similarity degrees. Concept Similarity ranges from 0 and 1. 0->dissimilar, 1->similar and 0.5–1 ->similar, using these outliers are eliminated. Eliminating Concept Similarity from OntoHDClass. The second phase involves extracting inliers and yielding the elite group by using Pearson's correlation and Cross entropy formulated from the domain ontologies.

Here Eqs. (1) and (2) depicts the Pearson's correlation and Cross-entropy respectively.

$$(r = X\,\Sigma ab - (\Sigma a)(\Sigma b))/\sqrt{}([X\,\Sigma a^2 - (\Sigma a^2)][X\,\Sigma b^2 - (\Sigma b^2)]) \tag{1}$$

To discover how stout a relationship is between the data is described as Pearson's correlation. Information about the magnitude of the corelation is given together with the relationship direction of the corelation. A value between negative one and positive one is given as output for the formulas, where: 1 directs a solid positive relationship. −1 directs a solid negative relationship. Where X = sum of pairs of scores Σab = summation of the products of paired scores Σa = summation of × scores Σb = summation of y scores Σa2 = summation of squared × scores Σb2 = summation of squared y scores.

$$l(a, b) = -a(y)\,\log b(y) \tag{2}$$

The Cross-entropy Eq. (2) assumes a(y) is the probability of the event y in a, b(y) is the probability of event y in b and log is the base-2 logarithm, denotation that the outcomes are in bits. The difference between dual probability distribution in amount is called cross-entropy for a random variable given. Values of zero and one are used to encode the class labels while preparing the data for classification.

Eliminating Pearson's Correlation and Cross Entropy from OntoHDClass. The third phase involves classification of the remaining space using RNN, clustering the neighborhood with the inliers and learning the elite group of subspaces. RNN is used for classification, it is where the previous steps outputs are inserted or fed into the current step as input to produce the following output, the hidden states recall the info about a sequence. Pitfall of the RNN approach is training is an exhausting and very hard job and gradient vanishing and exploding complications. The advance is that, without having to perform multipart transformations, you can capture a much more nuanced relationship between your data points. Thus, this reduces the complexity of parameters.

4 Implementation and Result Analysis

We have presented an Ontology Driven Approach for High Dimensional Data Classification integrating Semantic Measures and RNN to solve the above problem. The main idea is formulating domain ontologies using which we compute concept similarity and eliminate outliers and extracting inliers and yielding the elite group by using Pearson's correlation and cross entropy. The key study goal is to device a method called OntoHDClass. Using clustering procedure to make it robust against contrary form of noise, data sparsity and to extract the elite group. The Dataset used in OntoHDClass is obtained from UCI Machine Learning Repository where dataset of Gas sensor array exposed to turbulent gas mixtures is taken which is an open-source dataset that comprises of over 1500 attributes. The Algorithm of the proposed OntoHDClass for High Dimensional Data is made known in Table 1.

It is inferable that the performance of the put forward strategy is computed using F-Measure, Recall, Precision, Accuracy, and FNR as the possible measure. Precision is represented as the proportion of the retrieved and significant ontologies towards the overall sum of recovered ontologies. Recall is the share of ontologies recovered and applicable towards the total sum of ontologies that are relevant. For precision and recall measures, accuracy is specified as the average. Equations (3), (4), (5), (6) and (7) depicts the Recall, Precision, Accuracy, FNR of the system and F-Measure.

$$\text{Recall\%} = \frac{\text{True number of Positives}}{\text{True number of Positives} + \text{False number of Negatives}} \tag{3}$$

$$\text{Precision\%} = \frac{\text{True number of Positives}}{\text{True number of Positives} + \text{False number of Positives}} \tag{4}$$

Table 1. Proposed OntoHDClass algorithm

Input: High Dimensional Data (Data points), Formulated Domain Ontologies **Output:** Elite group of subspaces.

Begin
 	
Step 1: The input data is pre-processed in such a way it encounters and identifying the problem domain and formulating domain ontologies.
 	
Step 2: From the data set, concept similarity Cs is being calculated
If (Cs<0.5)
Dissimilar, remove the outliers
If (Cs>0.5)
Similar, extract inliers
 	
Step 3: Eliminating Concept Similarity from OntoHDClass Eliminating Pearson's Correlation and Cross Entropy from OntoHDClass.
 	
Step 4: Extracting inliers and yielding the elite group by using Pearson's correlation and cross entropy.
 	
Step 5: Classifying the remaining space using RNN and clustering the neighborhood using the extracted inliers.
 	
End

$$\text{Accuracy\%} = \frac{\text{Precision} + \text{Recall}}{2} \tag{5}$$

$$\text{False Negative Rate} = 1 - \text{Recall} \tag{6}$$

$$\text{F-Measure\%}\frac{2(\text{Precision} \times \text{Recall})}{(\text{Precision} + \text{Recall})} \tag{7}$$

The effectuation of the put forward approach is estimated by comparing it by means of RMSC [2], Fuzzy c-means Clustering [3], DBScan [4], and SVM+Naïve Bayes [5]. Also, since the proposed approach is by integrating Semantic measures and RNN it can be deduced that the Proposed OntoHDClass has better Precision and Recall, 97.32% and 95.18% respective in comparison to Clustering technique Fuzzy c-means. DBScan, SVM+Naïve Bayes, RMSC,88.62% and 84.18%,84.23% and 81.79%,80.23% and 73.12%,91.61% and 87.32% respectively. Fuzzy c-means it is best used in pattern recognitions and a clustering technique which allows two or more clusters for one data. In contrast k-means clustering allows data points completely belong to one cluster. DBscan is called as Density based spatial clustering it is used to find different shapes of clusters and density. SVM+Naïve Bayes where Naïve Bayes considers when we have a single feature that all the attributes in a data are distinct to each other calculate the posterior probability. It is difficult to calculate the posterior probability as sometimes we can get a zero-probability problem. In support vector machine information focuses are plotted in n-dimensional space. where n is the quantity of highlights then the grouping is finished by choosing a reasonable hyper-plane that separates two classes.

Intended for this implementation. Four present methods are measured as baseline, made known in Table 2. The foremost uses RMSC [2] and the put forwarded OntoHD-Class model's performance rise is related to it has precision of 5.71%, recall of 7.86, accuracy of 7.10%, F-measure of 6.82% and False Negative of 0.07. The following next method uses Fuzzy c-means Clustering [3] to accomplish the identical task and the projected model OntoHDClass model's performance rise related to the problem has precision of 8.7%, recall of 11%, accuracy of 10.05%, F-measure of 9.89% and False Negative of 0.09. The 3rd method is DBScan [4] and the projected OntoHDClass model's performance-growth related to the problem has precision of 13.09%, recall of 13.39%, accuracy of 13.59%, F-measure of 12.86% and False Negative of 0.13. The 4th approach is SVM+Naïve Bayes [5] and the projected OntoHDClass model's performance-upsurge related to it has precision of 17.09%, recall of 22.06%, accuracy of 18.56%, F-measure of 19.72% and False Negative of 0.21%. Thus, the put forward system is a lot better than the prevailing approaches as it dynamically models ontology from the data. Also, Static Modeling of Ontologies by users is achieved and the dynamically versioned Ontology is more used for knowledge combined using the Pearson's correlation and

Table 2. Evaluation of Performance of the proposed OntoHDClass

Search technique	Average precision %	Average recall %	Accuracy %	F-Measure %	FNR 1-recall
RMSC [2]	91.61	87.32	89.27	89.41	0.12
Fuzzy c-means Clustering [3]	88.62	84.18	86.32	86.34	0.14
DBScan [4]	84.23	81.79	82.78	83.37	0.18
SVM+Naïve Bayes [5]	80.23	73.12	77.81	76.51	0.26
Proposed OntoHDClass	**97.32**	**95.18**	**96.37**	**96.23**	**0.05**
Eliminating Ontologies from OntoHDClass	89.18	84.12	87.79	86.57	0.14
Eliminating Pearson's Correlation and Cross Entropy	87.21	83.78	86.18	85.46	0.14
Eliminating Concept Similarity from OntoHDClass	85.72	80.18	83.39	82.85	0.20

cross entropy Measure. The research was trained and arranged, quite a lot of number of recommendations. Intended for this research purpose 5 different recommendations are measured.

The Table 2 shows the performance assessment intended for the 5 recommendations and their performance evaluation for RMSC [2], Fuzzy c-means Clustering [3], DBScan [4], and SVM+Naïve Bayes [5], OntoHDClass. The graph Fig. 2 presents the results analysis of the Proposed OntoHDClass's and Fuzzy c-means Clustering, DBScan, SVM+Naïve Bayes, RMSC system Accuracy Percentage Vs No. of Recommendations of the OntoHDClass model and the graph tells us that the OntoHDClass has higher accuracy. Hence, A technique meant for high dimensional data is offered by OntoHDClass which is also cost-effective subspace clustering process.

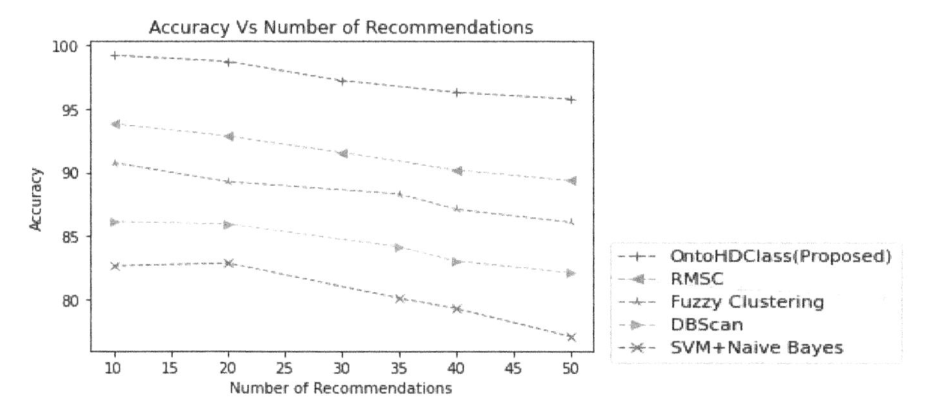

Fig. 2. Proposed system Accuracy Percentage Vs No. of Recommendations of the OntoHDClass model

From the line graph shown in the Fig. 2. we can identify that the put forwarded OntoHDClass has higher precision percentage when related to all other baseline models. Our approach uses an ontology base model using Concept similarity, Pearson's corelation and elimination of outliers increases the model accuracy and helps to perform subspace clustering. Even with the increase in number of recommendations, the put forwarded approach has much higher precision.

5 Conclusion

There is constantly a requirement for subspace clustering and still there exists issues where dimensions might be unrelated in addition veil the clusters to have noisy data, data sparsity. Accordingly, we outlined an exceptionally straightforward strategy where we define computing the utmost efficient subspaces or the elite group from the high dimensional data. This study has put forwarded an OntoHDClass model for subspace clustering and retrieving the elite group. Subsequent to contrasting the current frameworks for subspace clustering, data containing outliers when handled by our OntoHDClass yields precise results. Our method can be used to extract valuable information as it displays the exciting behavior of produced data sets with low dimensionality. So, we have presented an Ontology Driven Approach for High Dimensional Data Classification integrating Semantic Measures and RNN to solve the above problem. An overall F-Measure of 96.23% has been reached with a very low FDR of 0.05 which makes the projected OntoHDClass the best-in-class approach for finding elite group of subspaces.

References

1. Elankavi, R., Kalaiprasath, R., Udayakumar, D.R.: A fast-clustering algorithm for high-dimensional data. Int. J. Civ. Eng. Technol. (IJCIET) **8**(5), 1220–1227 (2017)
2. Radhika, K.R., Pushpa, C.N., Thriveni, J., Venugopal, K.R.: RMSC: robust modelling of subspace clustering for high dimensional data. In: 2017 International Conference on Advances in Computing, Communications and Informatics (ICACCI), pp. 1535–1539. IEEE, September 2017
3. Lei, T., Jia, X., Zhang, Y., He, L., Meng, H., Nandi, A.K.: Significantly fast and robust fuzzy c-means clustering algorithm based on morphological reconstruction and membership filtering. IEEE Trans. Fuzzy Syst. **26**(5), 3027–3041 (2018)
4. Chen, Y., Tang, S., Bouguila, N., Wang, C., Du, J., Li, H.: A fast-clustering algorithm based on pruning unnecessary distance computations in DBSCAN for high-dimensional data. Pattern Recogn. **83**, 375–387 (2018)
5. Shakya, S., Sigdel, S.: An approach to develop a hybrid algorithm based on support vector machine and Naive Bayes for anomaly detection. In: International Conference on Computing, Communication and Automation (ICCCA), pp. 323–327. IEEE, May 2017
6. Chakraborty, S., Das, S.: Detecting meaningful clusters from high-dimensional data: a strongly consistent sparse center-based clustering approach. IEEE Trans. Pattern Anal. Mach. Intell. (01), 1 (2020)
7. Arias-Castro, E., Pu, X.: A simple approach to sparse clustering. Comput. Stat. Data Anal. **105**, 217–228 (2017)
8. Sun, W., Wang, J., Fang, Y.: Regularized k-means clustering of high-dimensional data and its asymptotic consistency. Electron. J. Stat. **6**, 148–167 (2012)
9. McWilliams, B., Montana, G.: Subspace clustering of high-dimensional data: a predictive approach. Data Min. Knowl. Disc. **28**(3), 736–772 (2014)
10. Kadir, S.N., Goodman, D.F., Harris, K.D.: High-dimensional cluster analysis with the masked EM algorithm. Neural Comput. **26**(11), 2379–2394 (2014)
11. Wang, T., Li, Z.: Outlier detection in high-dimensional regression model. Commun. Stat. Theory Methods **46**(14), 6947–6958 (2017)
12. Gan, G., Ng, M.K.P.: Subspace clustering with automatic feature grouping. Pattern Recogn. **48**(11), 3703–3713 (2015)
13. Chormunge, S., Jena, S.: Correlation based feature selection with clustering for high dimensional data. J. Electr. Syst. Inf. Technol. **5**(3), 542–549 (2018)
14. Deepak, G., Teja, V., Santhanavijayan, A.: A novel firefly driven scheme for resume parsing and matching based on entity linking paradigm. J. Discrete Math. Sci. Cryptogr. **23**(1), 157–165 (2020)
15. Kumar, A., Deepak, G., Santhanavijayan, A.: HeTOnto: a novel approach for conceptualization, modeling, visualization, and formalization of domain centric ontologies for heat transfer. In: 2020 IEEE International Conference on Electronics, Computing and Communication Technologies (CONECCT), pp. 1–6. IEEE, July 2020
16. Deepak, G., Ahmed, A., Skanda, B.: An intelligent inventive system for personalised webpage recommendation based on ontology semantics. Int. J. Intell. Syst. Technol. Appl. **18**(1–2), 115–132 (2019)

17. Deepak, G., Priyadarshini, J.S.: Personalized and enhanced hybridized semantic algorithm for web image retrieval incorporating ontology classification, strategic query expansion, and content-based analysis. Comput. Electr. Eng. **72**, 14–25 (2018)
18. Kumar, N., Deepak, G., Santhanavijayan, A.: A novel semantic approach for intelligent response generation using emotion detection incorporating NPMI measure. Procedia Comput. Sci. **167**, 571–579 (2020)
19. Giri, G.L., Deepak, G., Manjula, S., Venugopal, K.: OntoYield: a semantic approach for context-based ontology recommendation based on structure preservation. In: Chaki, N., Cortesi, A., Devarakonda, N. (eds.) Proceedings of International Conference on Computational Intelligence and Data Engineering. LNDECT, vol. 9, pp. 265–275. Springer, Singapore (2018). https://doi.org/10.1007/978-981-10-6319-0_22
20. Kaushik, I.S., Deepak, G., Santhanavijayan, A.: QuantQueryEXP: a novel strategic approach for query expansion based on quantum computing principles. J. Discrete Math. Sci. Cryptogr. **23**(2), 573–584 (2020)
21. Rithish, H., Deepak, G., Santhanavijayan, A.: Automated assessment of question quality on online community forums. In: Motahhir, S., Bossoufi, B. (eds.) ICDTA 2021. LNNS, vol. 211, pp. 791–800. Springer, Cham (2021). https://doi.org/10.1007/978-3-030-73882-2_72
22. Surya, D., Deepak, G., Santhanavijayan, A.: KSTAR: a knowledge based approach for socially relevant term aggregation for web page recommendation. In: Motahhir, S., Bossoufi, B. (eds.) ICDTA 2021. LNNS, vol. 211, pp. 555–564. Springer, Cham (2021). https://doi.org/10.1007/978-3-030-73882-2_50
23. Deepak, G., Kumar, N., Bharadwaj, G.V.S.Y., Santhanavijayan, A.: OntoQuest: an ontological strategy for automatic question generation for e-assessment using static and dynamic knowledge. In: 2019 Fifteenth International Conference on Information Processing (ICINPRO), pp. 1–6. IEEE, December 2019
24. Srivastava, R.A., Deepak, G.: PIREN: prediction of intermediary readers' emotion from news-articles. In: Shukla, S., Unal, A., Varghese Kureethara, J., Mishra, D.K., Han, D.S. (eds.) Data Science and Security. LNNS, vol. 290, pp. 122–130. Springer, Singapore (2021). https://doi.org/10.1007/978-981-16-4486-3_13

M-polynomial and Degree-Based Topological Indices of Subdivided Chain Hex-Derived Network of Type 3

Shikha Rai and Shibsankar Das(✉)

Department of Mathematics, Institute of Science, Banaras Hindu University, Varanasi 221005, Uttar Pradesh, India
shib.iitm@gmail.com

Abstract. The evaluation of degree-based topological indices with the help of M-polynomial is a recent concept. The Hex-derived network is derived from the hexagonal network and has different applications in the area of telecommunications networks, pharmaceutics and electronics. In the current work, we obtain a subdivided chain Hex-derived network of type three of dimension n. We compute the degree-dependent topological indices of the subdivided network by using direct method and M-polynomial method. Additionally, we pictorially illustrate the M-polynomial and the corresponding topological descriptors of the said network to understand their geometrical behavior. The results achieved can lay a base for further research into subdivided chain Hex-derived networks, their properties and applications.

Keywords: Degree-dependent topological descriptors · Subdivided chain Hex-derived network of third type · M-polynomial · Graph polynomial

1 Introduction

Chemical Graph Theory (CGT) is a discipline of mathematical chemistry which makes use of graph theory to the study and theoretical analysis of molecular structures. Let $G = (V(G), E(G))$ be an undirected, unweighted, connected and simple graph with sets of vertices $V(G)$ and edges $E(G)$. The degree of vertex $v \in V(G)$, counts the number of edges that are incident on v and it is denoted by $d(v)$ [29].

A *molecular graph* G is a pictorial demonstration of the structural formula of a molecular compound, where a vertex corresponds to the atom and an edge corresponds to the chemical bond of the compound. In CGT, a *topological index* (usually termed as a *connectivity index*) is a type of molecular descriptor that is evaluated from a chemical compound's molecular graph. In particular, it converts

The present investigation has been carried out under the financial supports from BHU-fellowship to Shikha Rai from Banaras Hindu University (BHU), Varanasi.

I. Woungang et al. (Eds.): ANTIC 2021, CCIS 1534, pp. 410–424, 2022.
https://doi.org/10.1007/978-3-030-96040-7_33

a molecular structure's chemical information into a valuable number that determines its topology and has also seen a significant role in the study of Quantitative Structure-Property Relationship (QSPR)/Quantitative Structure-Activity Relationship (QSAR). In the area of Chemical Informatics (i.e., Cheminformatics), the QSPR/QSAR is employed to analyze the physical and chemical characteristics and biological and pharmacological activities of chemical compounds with no participation in laboratory trials [10]. Using the statistical analysis and modeling of the topological indices of a molecular structure one can identify the structural features used in the study of QSAR and QSPR [11].

Therefore, it is vital to explore the topological indices of molecules, in order to utilize the suitable indices to acquire an appropriate relationship between the chemical properties and their underlying molecular structures.

There is a wide collection of topological descriptors of a molecular structure that exhibits a strong relationship with its chemical, physical, and biological properties such as heat of reaction, melting point, boiling point, enthalpies of formation, gas-chromatographic retention, vapor pressure, toxicity, Kovat's retention index, viscosity, observed bioactivities, aromatic stability, the radius of gyration, π-electron energy, resonance energy, etc.

Some important classes of topological descriptors are degree-dependent topological indices [17], counting related topological indices [22] and distance-based topological indices [2]. Here, we focus on some degree-dependent topological descriptors derived from the expression of the M-polynomial for the network we have considered. The Zagreb indices (first and second) have been proposed by Gutman and Trinjastić [15] in 1972. Instead of focusing on the external edges and vertices, the Zagreb indices give priority to the internal edges and vertices of a molecular graph. These are prescribed as

$$M_1(G) = \sum_{v_1v_2 \in E(G)} (d(v_1) + d(v_2)) \quad \text{and} \quad M_2(G) = \sum_{v_1v_2 \in E(G)} (d(v_1)d(v_2)).$$

On the other hand, the modified Zagreb indices are presented in [24] and mathematically described as

$$^{m}M_2(G) = \sum_{v_1v_2 \in E(G)} \frac{1}{d(v_1)d(v_2)}.$$

The Randić index is a very well-known degree-based topological descriptor, proposed in 1975 by Milan Randić [27]. In the area of drug design, the Randić index has widespread applicability. In 1998, Amić et al. [1] and Bollobás et al. [3] introduced the general Randić index which is the generalized form of the Randić index and is given by the formula

$$R_\alpha(G) = \sum_{v_1v_2 \in E(G)} (d(v_1)d(v_2))^\alpha,$$

where $\alpha \in \mathbb{R}$, and R_α turns out to be Randić index for $\alpha = -1/2$. Also, the inverse Randić index [1] is prescribed as

$$RR_\alpha(G) = \sum_{v_1v_2 \in E(G)} \frac{1}{(d(v_1)d(v_2))^\alpha}.$$

The harmonic index [13] is an alternative of the Randić index. It is represented by

$$H(G) = \sum_{v_1v_2 \in E(G)} \frac{2}{d(v_1)+d(v_2)}.$$

Another one is the inverse sum (indeg) index [28] that forecasts the complete surface area of octane isomers. It is mathematically specified as

$$ISI(G) = \sum_{v_1v_2 \in E(G)} \frac{d(v_1)d(v_2)}{d(v_1)+d(v_2)}.$$

The symmetric division (deg) index is introduced in [28], which helps assess the entire surface area of polychlorobiphenyls. It is mathematically formulated as

$$SDD(G) = \sum_{v_1v_2 \in E(G)} \left\{ \frac{\min(d(v_1),d(v_2))}{\max(d(v_1),d(v_2))} + \frac{\max(d(v_1),d(v_2))}{\min(d(v_1),d(v_2))} \right\}.$$

For the analysis of heat of formation of alkanes, the augmented Zagreb index is proposed by Furtula et al. [14] and it is formulated as

$$AZ(G) = \sum_{v_1v_2 \in E(G)} \left\{ \frac{d(v_1)d(v_2)}{d(v_1)+d(v_2)-2} \right\}^3.$$

Usually, to compute the numeric values of topological indices, formal definitions are used. Also, we have another way of finding values for these indices by considering the derivative or integral or both integral and derivative of the graph polynomial at a particular point. A number of graph polynomials have been introduced in past literature; a few of them are the Hosoya polynomial [19], the M-polynomial [26], the Tutte polynomial [21], the matching polynomial [12], the Schultz polynomial [16], and the Clar covering polynomial [30]. The M-polynomial is one of these polynomials that is used to estimate the degree-dependent topological descriptors of a molecular structure.

Definition 1 ([9]). *The expression of M-polynomial for a graph G is*

$$M(G;x,y) = \sum_{\delta \le i \le j \le \Delta} m_{i,j}(G)\, x^i y^j,$$

where $\Delta = max\{d(v)|v \in V(G)\}$, $\delta = min\{d(v)|v \in V(G)\}$ *and* $m_{i,j}(G)$ *counts the total number of edges* $v_1v_2 \in E(G)$ *in such a manner that for* $i,j \ge 1$, *we have* $d(v_1) = i$ *and* $d(v_2) = j$.

For a graph G, a degree-dependent topological descriptor is described, in [8], as a graph invariant, denoted by $I(G)$ and represented by $I(G) = \sum_{i \leq j} m_{i,j}(G) f(i,j)$.

Theorem 1 **([9], Theorems 2.1, 2.2).** *Let $G = (V, E)$ be an undirected, connected and simple graph. Then*

(i) If $I(G) = \sum_{e=v_1v_2 \in E(G)} f(d(v_1), d(v_2))$, *where* $f(x,y)$ *is a polynomial in* x *and* y, *then*

$$I(G) = f(D_x, D_y)(M(G;x,y))|_{x=y=1}.$$

(ii) If $I(G) = \sum_{e=v_1v_2 \in E(G)} f(d(v_1), d(v_2))$, *where* $f(x,y) = \sum_{i,j \in \mathbb{Z}} \alpha_{i,j} x^i y^j$, *then* $I(G)$ *can be obtained from* $M(G;x,y)$ *using the operators* D_x, D_y, S_x, *and* S_y.

(iii) If $I(G) = \sum_{e=v_1v_2 \in E(G)} f(d(v_1), d(v_2))$, *where* $f(x,y) = \frac{x^r y^s}{(x+y+\alpha)^t}$, *where* $r, s \geq 0$, $t \geq 1$ *and* $\alpha \in \mathbb{Z}$, *then*

$$I(G) = S_x^t Q_\alpha J D_x^r D_y^s (M(G;x,y))|_{x=1}.$$

where,

$$D_x(\mathrm{f}(x,y)) = x\frac{\partial(f(x,y))}{\partial x}, \qquad D_y(f(x,y)) = y\frac{\partial(f(x,y))}{\partial y}$$

$$S_x(f(x,y)) = \int_0^x \frac{f(t,y)}{t}dt, \qquad S_y(f(x,y)) = \int_0^y \frac{f(x,t)}{t}dt,$$

$$\mathrm{J}(f(x,y)) = f(x,x), \qquad Q_\alpha(f(x,y)) = x^\alpha f(x,y), \alpha \neq 0.$$

In the year 2015, Deutsch and Klavžar [9] proposed the idea of the M-polynomial which has a significant role in the determination of degree-dependent topological descriptors. There are several papers [4–7,20,23] for different structures wherein the M-polynomial and their concerned degree-dependent topological indices are determined.

In this work, we calculate the same for the subdivided chain Hex-derived network of type three of dimension n (*SCHDN*3[n]). For this, we construct the *SCHDN*3[n] network using the subdivision operation on chain Hex-derived network of type three of dimension n (*CHDN*3[n]) in Sect. 2. Section 3 deals with the direct enumeration of degree-dependent topological indices of the *SCHDN*3[n] network. In Sect. 4, we obtain a closed algebraic form of M-polynomial for the network and use it to evaluate these degree-based topological indices for the *SCHDN*3[n] network. We also plot the M-polynomial and associated indices for various dimensions in Sect. 5 to understand their geometrical behavior.

2 Construction of Subdivided Chain Hex-Derived Network of Type Three of Dimension n

Silicates are the most common rock-forming minerals and also the biggest, complicated, and most interesting minerals. Silicates' basic unit is the tetrahedron (SiO_4). Tetrahedra (SiO_4) is found in almost all silicates. A silicate sheet consists of tetrahedron rings connected from one ring to another by oxygen ions in a 2-dimensional plane, forming a sheet-like structure. As a chemical entity, the oxygen atoms of the (SiO_4) tetrahedron are represented by the corner vertices, whereas the central vertex represents the silicon atom. Graphically, these corner atoms are called oxygen vertices, the central atom as the silicon vertex, and the bonds between them are called an edge. Figure 1 depicts a simplified two-dimensional image of a (SiO_4) tetrahedron. Different types of silicate structures can be created by linking or combining the (SiO_4) tetrahedron. Based on the method of linking tetrahedral silicate units, the silicates can be divided into the following categories: Cyclic silicates, Chain silicates, Pyrosilicates, Sheet silicates, Orthosilicates, Three-dimensional silicates. A chain silicate network is generated by connecting 'n' tetrahedral (SiO_4) units in a linear pattern. Figure 2 depicts a 5-dimensional chain silicate network.

The construction of the $CHDN3[n]$ network is based on the structure of third type of Hex-derived network [25] and chain silicate network [18]. For more details, please see [5].

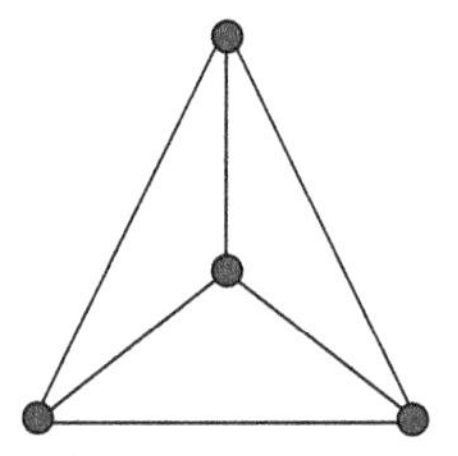

Fig. 1. Graphical illustration of SiO_4 tetrahedron in two dimensions.

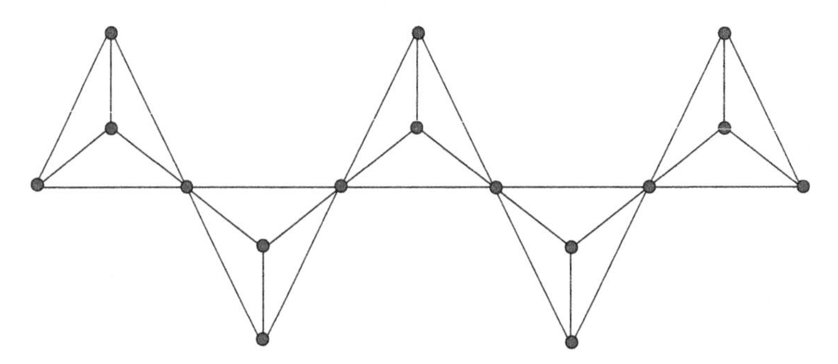

Fig. 2. 5-dimensional chain silicate network.

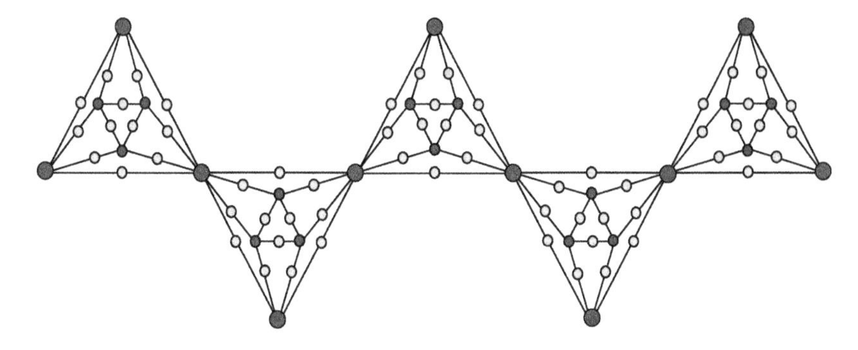

Fig. 3. Subdivided chain Hex-derived network of third type of dimension 5 (*SCHDN*3[5]).

Elementary operations, also known as graph edit operations, generate a new graph from an existing one by making a simple change, such as adding or removing a vertex or an edge, edge contraction, merging and splitting vertices, and so on. The graph formed by subdividing each edge of graph G is called the *subdivision graph*. As a result, with this basic subdivision operation on *CHDN*3[n] network, we can create a new chemical network of our interest that we call a subdivided chain Hex-derived network of third type of dimension n (*SCHDN*3[n]), where $n \geq 2$. The graphical view of the *SCHDN*3[5] network is illustrated in Fig. 3.

3 Computation of Topological Indices of *SCHDN*3[*n*]

This section focuses on the direct assessment of the degree-dependent topological descriptors of the *SCHDN*3[n] network by using their explicit formulas discussed in Sect. 1.

Theorem 2. *Suppose SCHDN*3[n] *denotes the subdivided chain Hex-derived network of type three of dimension n, where $n \geq 2$. Then*

(i) $M_1(SCHDN3[n]) = 16(11n - 2)$.
(ii) $M_2(SCHDN3[n]) = 64(4n - 1)$.
(iii) ${}^{m}M_2(SCHDN3[n]) = \frac{1}{2}(5n + 1)$.
(iv) $R_\alpha(SCHDN3[n]) = 8^\alpha(16n + 8) + 16^\alpha(8n - 8)$.
(v) $RR_\alpha(SCHDN3[n]) = \frac{1}{8^\alpha}(16n + 8) + \frac{1}{16^\alpha}(8n - 8)$.
(vi) $SDD(SCHDN3[n]) = 2(37n - 7)$.
(vii) $H(SCHDN3[n]) = \frac{8}{15}(13n + 2)$.
(viii) $ISI(SCHDN3[n]) = \frac{32}{15}(16n - 1)$.
(ix) $AZ(SCHDN3[n]) = 192n$.

Proof. Underneath, we calculate the degree-dependent topological indices of the *SCHDN*3[n] network by using their respective explicit formulas as indicated in Sect. 1.

(i) **First Zagreb Index:**

$$\begin{aligned} M_1(SCHDN3[n]) &= \sum_{v_1v_2 \in E(SCHDN3[n])} (d(v_1) + d(v_2)) \\ &= (16n+8) \times (2+4) + (8n-8) \times (2+8) \\ &= 6(16n+8) + 10(8n-8) \\ &= 16(11n-2). \end{aligned}$$

(ii) **Second Zagreb Index:**

$$\begin{aligned} M_2(SCHDN3[n]) &= \sum_{v_1v_2 \in E(SCHDN3[n])} (d(v_1)d(v_2)) \\ &= (16n+8) \times (2 \times 4) + (8n-8) \times (2 \times 8) \\ &= 8(16n+8) + 16(8n-8) \\ &= 64(4n-1). \end{aligned}$$

(iii) **Modified Second Zagreb Index:**

$$\begin{aligned} {}^mM_2(SCHDN3[n])) &= \sum_{v_1v_2 \in E(SCHDN3[n])} \frac{1}{d(v_1)d(v_2)} \\ &= (16n+8)\frac{1}{2 \times 4} + (8n-8)\frac{1}{2 \times 8} \\ &= \frac{1}{2}(5n+1). \end{aligned}$$

(iv) **General Randić Index:**

$$\begin{aligned} R_\alpha(SCHDN3[n]) &= \sum_{v_1v_2 \in E(SCHDN3[n])} (d(v_1)d(v_2))^\alpha \\ &= 8^\alpha(16n+8) + 16^\alpha(8n-8). \end{aligned}$$

(v) **Inverse Randić Index:**

$$\begin{aligned} RR_\alpha(SCHDN3[n]) &= \sum_{v_1v_2 \in E(SCHDN3[n])} \frac{1}{(d(v_1)d(v_2))^\alpha} \\ &= \frac{1}{8^\alpha}(16n+8) + \frac{1}{16^\alpha}(8n-8). \end{aligned}$$

(vi) **Symmetric Division (Deg) Index:**

$$\begin{aligned} SDD(SHDN3[n]) &= \sum_{v_1v_2 \in E(SHDN3[n])} \left\{ \frac{\min(d(v_1), d(v_2))}{\max(d(v_1), d(v_2))} + \frac{\max(d(v_1), d(v_2))}{\min(d(v_1), d(v_2))} \right\} \\ &= (16n+8)\left\{ \frac{\min(2,4)}{\max(2,4)} + \frac{\max(2,4)}{\min(2,4)} \right\} + (8n-8)\left\{ \frac{\min(2,8)}{\max(2,8)} + \frac{\max(2,8)}{\min(2,8)} \right\} \\ &= (16n+8)\left\{ \frac{2}{4} + \frac{4}{2} \right\} + (8n-8)\left\{ \frac{2}{8} + \frac{8}{2} \right\} \\ &= \frac{5}{2}(16n+8) + \frac{17}{4}(8n-8) \\ &= 2(37n-7). \end{aligned}$$

(vii) **Harmonic Index:**

$$
\begin{aligned}
H(SCHDN3[n]) &= \sum_{v_1v_2 \in E(SCHDN3[n])} \frac{2}{d(v_1)+d(v_2)} \\
&= (16n+8)\frac{2}{2+4} + (8n-8)\frac{2}{2+8} \\
&= \frac{1}{3}(16n+8) + \frac{1}{5}(8n-8) \\
&= \frac{8}{15}(13n+2).
\end{aligned}
$$

(viii) **Inverse Sum (Indeg) Index:**

$$
\begin{aligned}
ISI(SCHDN3[n]) &= \sum_{v_1v_2 \in E(SCHDN3[n])} \frac{d(v_1)d(v_2)}{d(v_1)+d(v_2)} \\
&= (16n+8)\frac{2\times 4}{2+4} + (8n-8)\frac{2\times 8}{2+8} \\
&= \frac{4}{3}(16n+8) + \frac{8}{5}(8n-8) \\
&= \frac{32}{15}(16n-1).
\end{aligned}
$$

(ix) **Augmented Zagreb Index:**

$$
\begin{aligned}
AZ(SCHDN3[n]) &= \sum_{v_1v_2 \in E(SCHDN3[n])} \left\{\frac{d(v_1)d(v_2)}{d(v_1)+d(v_2)-2}\right\}^3 \\
&= (16n+8)\left\{\frac{2\times 4}{2+4-2}\right\}^3 + (8n-8)\left\{\frac{2\times 8}{2+8-2}\right\}^3 \\
&= 8(16n+8) + 8(8n-8) \\
&= 192n.
\end{aligned}
$$

4 M-polynomial for $SCHDN3[n]$ Network

In the present section, we compute the degree-dependent topological indices of the $SCHDN3[n]$ network using the most recent M-polynomial methodology.

Theorem 3. *Suppose that $SCHDN3[n]$ network represents the subdivision of the $CHDN3[n]$ network, where $n \geq 2$. Hence, the M-polynomial of $SCHDN3[n]$ network is given by*

$$M(SCHDN3[n]; x, y) = (16n+8)x^2y^4 + (8n-8)x^2y^8.$$

Proof. See that, the number of edges and vertices of the $CHDN3[n]$ network are

$$|E(CHDN3[n])| = 12n, \quad \text{and} \quad |V(CHDN3[n])| = 5n+1.$$

Knowing the cardinalities of edges and vertices of the $CHDN3[n]$ network, one can simply estimate the cardinalities of edges and vertices of the $SCHDN3[n]$ network. Because the $SCHDN3[n]$ network is a subdivision of $CHDN3[n]$ network, hence

$$|E(SCHDN3[n])| = 2|E(CHDN3[n])| = 24n,$$
$$\text{and } |V(SCHDN3[n])| = |V(CHDN3[n])| + |E(CHDN3[n])| = 17n + 1.$$

Let us now consider three disjoint partitions of the vertex set $V(SCHDN3[n])$, based on the degree of vertices of the $SCHDN3[n]$ network. They are as follows.

$$V_1(SCHDN3[n]) = \{v \in V(SCHDN3[n]) : d(v) = 2\},$$
$$V_2(SCHDN3[n]) = \{v \in V(SCHDN3[n]) : d(v) = 4\},$$
$$V_3(SCHDN3[n]) = \{v \in V(SCHDN3[n]) : d(v) = 8\}.$$

And the cardinalities of such vertices are given by $|V_1(SCHDN3[n])| = 12n$, $|V_2(SCHDN3[n])| = 4n + 2$, $|V_3(SCHDN3[n])| = n - 1$. Next, the edge set $E(SCHDN3[n])$ of the $SCHDN3[n]$ network is partitioned into two distinct sub-classes as shown below.

$$E_1(SCHDN3[n]) = \{e = v_1v_2 \in E(SCHDN3[n]) : d(v_1) = 2, d(v_2) = 4\},$$
$$\text{and } E_2(SCHDN3[n]) = \{e = v_1v_2 \in E(SCHDN3[n]) : d(v_1) = 2, d(v_2) = 8\},$$

and their cardinalities are $|E_1(SCHDN3[n])| = 16n + 8$, $|E_2(SCHDN3[n])| = 8n - 8$. Thus, according to the basic definition of M-polynomial, we have

$$\begin{aligned} M(SCHDN3[n]; x, y) &= \sum_{i \le j} m_{i,j} x^i y^j \\ &= \sum_{2 \le 4} m_{2,4} x^2 y^4 + \sum_{2 \le 8} m_{2,8} x^2 y^8 \\ &= \sum_{uv \in E_1(SCHDN3[n])} m_{2,4} x^2 y^4 + \sum_{uv \in E_2(SCHDN3[n])} m_{2,8} x^2 y^8 \\ &= (16n + 8) x^2 y^4 + (8n - 8) x^2 y^8. \end{aligned}$$

Now, from the derived expression of $M(SCHDN3[n]; x, y)$ (as introduced in Theorem 3), we obtain the values of the correlated degree-dependent topological indices of the $SCHDN3[n]$ network in terms of dimension n. In fact, we present an alternative proof (based on M-polynomial) of Theorem 2 as follows.

*Proof (**M-polynomial-based proof of Theorem** 2).* For notational simplicity, let us denote $\Gamma(x, y) = M(SCHDN3[n]; x, y)$. Therefore, the M-polynomial for the $SHDN3[n]$ network is

$$\Gamma(x, y) = (16n + 8) x^2 y^4 + (8n - 8) x^2 y^8.$$

Table 1. Formulas to derive the degree-dependent topological indices in context of M-polynomial.

Sl. No.	Topological index	f(x,y)	Derived from $(M(G;x,y))$
(i)	First Zagreb Index	$x+y$	$M_1(G)=(D_x+D_y)(M(G;x,y))\|_{x=y=1}$
(ii)	Second Zagreb Index	xy	$M_2(G)=(D_xD_y)(M(G;x,y))\|_{x=y=1}$
(iii)	Modified Second Zagreb Index	$\frac{1}{xy}$	${}^mM_2(G)=(S_xS_y)(M(G;x,y))\|_{x=y=1}$
(iv)	General Randić Index	$(xy)^\alpha$	$R_\alpha(G)=(D_x^\alpha D_y^\alpha)(M(G;x,y))\|_{x=y=1}$
(v)	Inverse Randić Index	$\frac{1}{(xy)^\alpha}$	$RR_\alpha(G)=(S_x^\alpha S_y^\alpha)(M(G;x,y))\|_{x=y=1}$
(vi)	Symmetric Division (Deg) Index	$\frac{x^2+y^2}{xy}$	$SDD(G)=(D_xS_y+D_yS_x)(M(G;x,y))\|_{x=y=1}$
(vii)	Harmonic Index	$\frac{2}{x+y}$	$H(G)=2S_xJ(M(G;x,y))\|_{x=1}$
$(viii)$	Inverse Sum (Indeg) Index	$\frac{xy}{x+y}$	$ISI(G)=S_xJD_xD_y(M(G;x,y))\|_{x=1}$
(ix)	Augmented Zagreb Index	$(\frac{xy}{x+y-2})^3$	$AZ(G)=S_x^3Q_{-2}JD_x^3D_y^3(M(G;x,y))\|_{x=1}$

Note that, in Table 1, $D_x(f(x,y))=x\frac{\partial(f(x,y))}{\partial x}$, $D_y(f(x,y))=y\frac{\partial(f(x,y))}{\partial y}$,
$S_x(f(x,y))=\int_0^x\frac{f(t,y)}{t}dt$, $S_y(f(x,y))=\int_0^y\frac{f(x,t)}{t}dt$,
$J(f(x,y))=f(x,x)$, $Q_\alpha(f(x,y))=x^\alpha f(x,y)$, $\alpha\neq 0$.

In order to evaluate various degree-dependent topological indices, we determine some necessary terms which are calculated below.

$$
\begin{aligned}
D_x(\Gamma(x,y)) &= 2(16n+8)x^2y^4+2(8n-8)x^2y^8,\\
D_y(\Gamma(x,y)) &= 4(16n+8)x^2y^4+8(8n-8)x^2y^8,\\
D_yD_x((\Gamma(x,y)) &= 8(16n+8)x^2y^4+16(8n-8)x^2y^8,\\
S_x(\Gamma(x,y)) &= (8n+4)x^2y^4+(4n-4)x^2y^8,\\
S_y(\Gamma(x,y)) &= (4n+2)x^2y^4+(n-1)x^2y^8,\\
S_xS_y(\Gamma(x,y)) &= (2n+1)x^2y^4+\frac{1}{2}(n-1)x^2y^8,\\
D_x^\alpha D_y^\alpha(\Gamma(x,y)) &= 8^\alpha(16n+8)x^2y^4+16^\alpha(8n-8)x^2y^8,\\
S_yD_x(\Gamma(x,y)) &= (8n+4)x^2y^4+(2n-2)x^2y^8,\\
S_xD_y(\Gamma(x,y)) &= 2(16n+8)x^2y^4+4(8n-8)x^2y^8,\\
S_x^\alpha S_y^\alpha(\Gamma(x,y)) &= \frac{1}{8^\alpha}(16n+8)x^2y^4+\frac{1}{16^\alpha}(8n-8)x^2y^8,\\
S_xJ(\Gamma(x,y)) &= \frac{1}{3}(8n+4)x^6\frac{1}{5}(4n-4)x^{10},\\
S_xJD_xD_y(\Gamma(x,y)) &= \frac{4}{3}(16n+8)x^6+\frac{8}{5}(8n-8)x^{10},\\
S_x^3Q_{-2}JD_x^3D_y^3(\Gamma(x,y)) &= 8(16n+8)x^4+8(8n-8)x^8.
\end{aligned}
$$

Thus, according to the derivation formulas listed in Table 1, the requisite degree-dependent topological descriptors of the $SCHDN3[n]$ network are calculated as follows.

(i) **First Zagreb Index:**
$M_1(SCHDN3[n]) = (D_x + D_y)(\Gamma(x,y))|_{x=y=1} = 16(11n-2)$.

(ii) **Second Zagreb Index:**
$M_2(SCHDN3[n]) = D_x D_y(\Gamma(x,y))|_{x=y=1} = 64(4n-1)$.

(iii) **Modified Second Zagreb Index:**
${}^mM_2(SCHDN3[n]) = S_x S_y(\Gamma(x,y))|_{x=y=1} = \frac{1}{2}(5n+1)$.

(iv) **General Randić Index:**
$R_\alpha(SCHDN3[n]) = D_x^\alpha D_y^\alpha(\Gamma(x,y))|_{x=y=1} = 8^\alpha(16n+8) + 16^\alpha(8n-8)$.

(v) **Inverse Randić Index:**
$RR_\alpha(SCHDN3[n]) = S_x^\alpha S_y^\alpha(\Gamma(x,y))|_{x=y=1} = \frac{1}{8^\alpha}(16n+8) + \frac{1}{16^\alpha}(8n-8)$.

(vi) **Symmetric Division (Deg) Index:**
$SDD(SCHDN3[n]) = (S_y D_x + S_x D_y)(\Gamma(x,y))|_{x=y=1} = 2(37n-7)$.

(vii) **Harmonic Index:**
$H(SCHDN3[n]) = 2S_x J(\Gamma(x,y))|_{x=1} = \frac{8}{15}(13n+2)$.

(viii) **Inverse Sum (Indeg) Index:**
$ISI(SCHDN3[n]) = S_x J D_x D_y(\Gamma(x,y))|_{x=1} = \frac{32}{15}(16n-1)$.

(ix) **Augmented Zagreb Index:**
$AZ(SCHDN3[n]) = S_x^3 Q_{-2} J D_x^3 D_y^3(\Gamma(x,y))|_{x=1} = 192n$.

5 Graphical Overview

Here, we discuss the nature of M-polynomial and associated indices of the $SCHDN3[n]$ network graphically. In Table 2, we have calculated the numerical values of topological indices for varied dimensions $(2 \leq n \leq 10)$. Observe that, as we increase the value of dimension n, the value of indices increases as well, and their geometric structures are shown in Figs. 4, 5, and 6. Moreover, we plot the M-polynomial of the $SCHDN3[2]$ network in the domain $-0.5 \leq x, y \leq 0.5$. See Fig. 7.

Table 2. Numerical evaluation of the M-polynomial and topological indices of the $SCHDN3[n]$ network for various n.

Sl.	Dimension	$n=2$	$n=3$	$n=4$	$n=5$	$n=6$	$n=7$	$n=8$	$n=9$	$n=10$
	$\frac{M-polynomial}{TopologicalIndex}$	$40x^2y^4+80x^2y^8$	$56x^2y^4+16x^2y^8$	$72x^2y^4+24x^2y^8$	$88x^2y^4+32x^2y^8$	$104x^2y^4+40x^2y^8$	$120x^2y^4+48x^2y^8$	$136x^2y^4+56x^2y^8$	$152x^2y^4+64x^2y^8$	$168x^2y^4+72x^2y^8$
1	First Zagreb Index	320	496	672	848	1024	1200	1376	1552	1728
2	Second Zagreb Index	448	704	960	1216	1472	1728	1984	2240	2496
3	Modified Second Zagreb Index	5.50	8	10.50	13	15.50	18	20.50	23	25.5
4	General Randić Index ($\alpha=1/2$)	145.1371	222.3919	299.6468	376.9016	454.1564	531.4113	608.6661	685.9209	763.1757
5	Inverse Randić Index ($\alpha=1/2$)	16.1421	23.7990	31.4558	39.1127	46.7696	54.4264	62.0833	69.7401	77.3970
6	Symmetric Division (Deg) Index	134	208	282	356	430	504	578	652	726
7	Harmonic Index	14.9333	21.8667	28.80	35.7333	42.6667	49.6000	56.5333	63.4667	70.4000
8	Inverse Sum (Indeg) Index	66.1333	100.2667	134.40	168.5333	202.6667	236.80	270.9333	305.0667	339.2000
9	Augmented Zagreb Index	384	576	768	960	1152	1344	1536	1728	1920

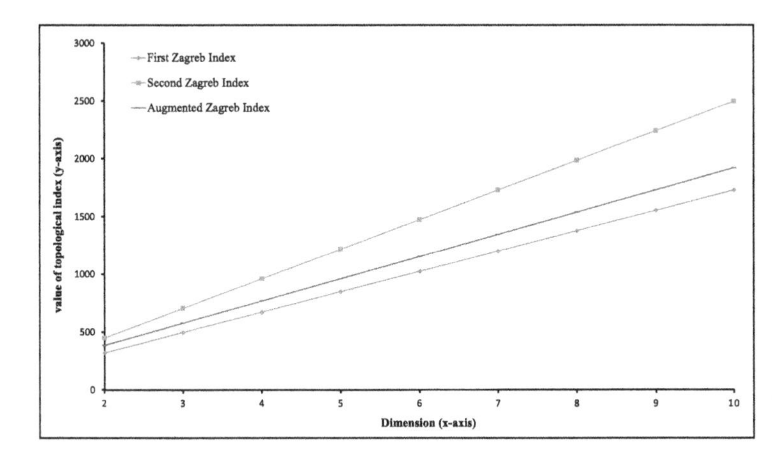

Fig. 4. Geometrical behavior of first Zagreb, augmented Zagreb, and second Zagreb indices of $SCHDN3[n]$ network for various n ($2 \leq n \leq 10$).

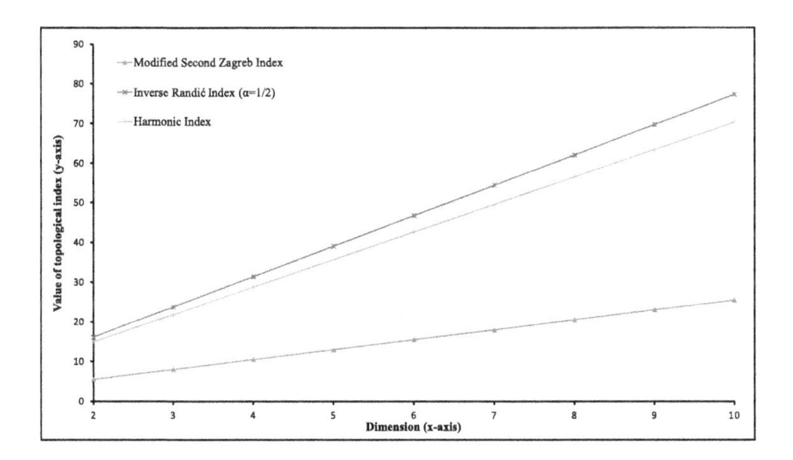

Fig. 5. Geometrical behavior of inverse Randić ($\alpha = 1/2$), harmonic, and modified second Zagreb indices of $SCHDN3[n]$ network for various n ($2 \leq n \leq 10$).

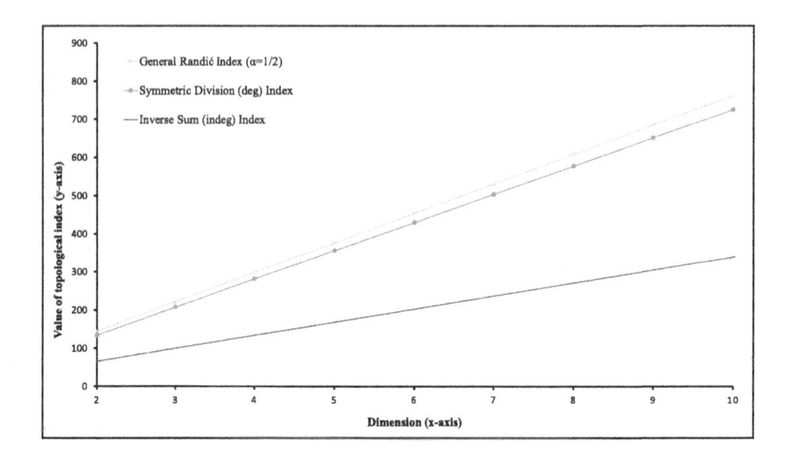

Fig. 6. Geometrical behavior of general Randić ($\alpha = 1/2$), inverse sum (indeg), and symmetric division (deg) indices of $SCHDN3[n]$ network for various n ($2 \leq n \leq 10$).

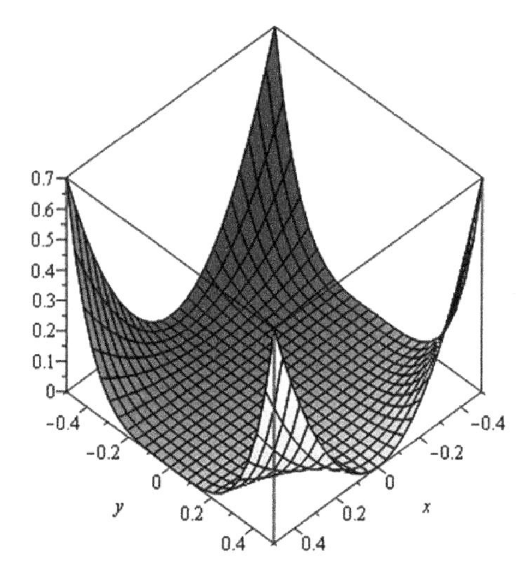

Fig. 7. Geometrical behavior of the M-polynomial of $SCHDN3[2]$ network, where $-0.5 \leq x, y \leq 0.5$.

6 Conclusions

In the present work, we have introduced the n-dimensional $SCHDN3[n]$ network which is acquired from the $CHDN3[n]$ network. Note that, the $SCHDN3[n]$ network is a more dense structure in comparison to the $CHDN3[n]$ network. We have directly calculated the degree-dependent topological indices by using their explicit formulas and seen that the calculation takes lots of time and steps. To overcome this, we have used the M-polynomial method, which is more simple, fast and compact. In this method, for the $SCHDN3[n]$ network, we have established a closed algebraic expression of M-polynomial and derived the associated degree-dependent topological indices correctly. Furthermore, we have figured the indices and M-polynomial of the $SCHDN3[n]$ network to better understand its geometrical properties. The results attained could play a significant role in broadening our perspective of the behavior and nature of such Hex-derived networks further.

References

1. Amić, D., Bešlo, D., Lučić, B., Nikolić, S., Trinajstić, N.: The vertex-connectivity index revisited. J. Chem. Inf. Comput. Sci. **38**(5), 819–822 (1998)
2. Balaban, A.T.: Highly discriminating distance-based topological index. Chem. Phys. Lett. **89**(5), 399–404 (1982)
3. Bollobás, B., Erdős, P.: Graphs of extremal weights. ARS Combin. **50**, 225–233 (1998)
4. Cancan, M., Afzal, D., Hussain, S., Maqbool, A., Afzal, F.: Some new topological indices of silicate network via m-polynomial. J. Discrete Math. Sci. Crypt. **23**(6), 1157–1171 (2020)
5. Das, S., Rai, S.: M-polynomial and related degree-based topological indices of the third type of chain hex-derived network. Malaya J. Matematik (MJM) **8**(4), 1842–1850 (2020)
6. Das, S., Rai, S.: M-polynomial and related degree-based topological indices of the third type of hex-derived network. Nanosystems Phys. Chem. Math. **11**(3), 267–274 (2020)
7. Das, S., Rai, S.: Topological characterization of the third type of triangular hex-derived networks. Sci. Ann. Comput. Sci. **31**(2), 145–161 (2021). https://doi.org/10.7561/SACS.2021.2.145
8. Deng, H., Yang, J., Xia, F.: A general modeling of some vertex-degree based topological indices in benzenoid systems and phenylenes. Comput. Math. Appl. **61**(10), 3017–3023 (2011)
9. Deutsch, E., Klavžar, S.: M-polynomial and degree-based topological indices. Iran. J. Math. Chem. **6**(2), 93–102 (2015)
10. Devillers, J., Balaban, A.T.: Topological Indices and Related Descriptors in QSAR and QSPAR. CRC Press, London (2000)
11. Emmert-Streib, F.: Statistical Modelling of Molecular Descriptors in QSAR/QSPR. Wiley, Weinheim (2012)
12. Farrell, E.J.: An introduction to matching polynomials. J. Comb. Theory Ser. B **27**(1), 75–86 (1979)
13. Favaron, O., Mahéo, M., Saclé, J.F.: Some eigenvalue properties in graphs (conjectures of Graffiti-II). Discret. Math. **111**(1–3), 197–220 (1993)
14. Furtula, B., Graovac, A., Vukičević, D.: Augmented Zagreb index. J. Math. Chem. **48**(2), 370–380 (2010)
15. Gutman, I., Trinajstić, N.: Graph theory and molecular orbitals. Total π-electron energy of alternant hydrocarbons. Chem. Phys. Lett. **17**(4), 535–538 (1972)
16. Gutman, I.: Some relations between distance-based polynomials of trees. Bulletin (Académie serbe des sciences et des arts. Classe des sciences mathématiques et naturelles. Sciences mathématiques) **131**(30), 1–7 (2005)
17. Gutman, I.: Degree-based topological indices. Croatica Chemica Acta **86**(4), 351–361 (2013)

18. Hayat, S., Imran, M.: Computation of topological indices of certain networks. Appl. Math. Comput. **240**, 213–228 (2014)
19. Hosoya, H.: On some counting polynomials in chemistry. Discret. Appl. Math. **19**(1–3), 239–257 (1988)
20. Kang, S.M., Nazeer, W., Zahid, M.A., Nizami, A.R., Aslam, A., Munir, M.: M-polynomials and topological indices of hex-derived networks. Open Phys. **16**(1), 394–403 (2018)
21. Kauffman, L.H.: A Tutte polynomial for signed graphs. Discret. Appl. Math. **25**(1–2), 105–127 (1989)
22. Khadikar, P.V., Deshpande, N.V., Kale, P.P., Dobrynin, A., Gutman, I., Domotor, G.: The Szeged index and an analogy with the Wiener index. J. Chem. Inf. Comput. Sci. **35**(3), 547–550 (1995)
23. Kwun, Y.C., Munir, M., Nazeer, W., Rafique, S., Kang, S.M.: M-polynomials and topological indices of V-Phenylenic nanotubes and Nanotori. Sci. Rep. **7**(1), 1–9 (2017)
24. Miličević, A., Nikolić, S., Trinajstić, N.: On reformulated Zagreb indices. Mol. Diversity **8**, 393–399 (2004)
25. Raj, F.S., George, A.: On the metric dimension of HDN 3 and PHDN 3. In: 2017 IEEE International Conference on Power, Control, Signals and Instrumentation Engineering (ICPCSI), pp. 1333–1336, September 2017
26. Raj, F.S., George, A.: Network embedding on planar octahedron networks. In: 2015 IEEE International Conference on Electrical, Computer and Communication Technologies (ICECCT), pp. 1–6. IEEE (2015)
27. Randić, M.: Characterization of molecular branching. J. Am. Chem. Soc. **97**(23), 6609–6615 (1975)
28. Vukičević, D., Gašperov, M.: Bond additive modeling 1. Adriatic indices. Croatica Chemica Acta **83**(3), 243–260 (2010)
29. West, D.B.: Introduction to Graph Theory, 2nd edn. Prentice Hall, Englewood Cliffs (2000)
30. Zhang, H., Zhang, F.: The Clar covering polynomial of hexagonal systems I. Discret. Appl. Math. **69**(1–2), 147–167 (1996)

Improved Detection of Coronary Artery Disease Using DT-RFE Based Feature Selection and Ensemble Learning

Ashima Tyagi(✉), Vibhav Prakash Singh, and Manoj Madhava Gore

Department of Computer Science and Engineering, Motilal Nehru National Institute of Technology Allahabad, Prayagraj, India
{ashima.2020rcs01,vibhav,gore}@mnnit.ac.in

Abstract. Coronary Artery Disease (CAD) is a sub-type of cardiovascular disease, causing a large number of deaths all over the world. It is increasingly important to diagnose this disease in an early stage for timely treatment. To detect this disease, it becomes crucial to recognize which feature contributes as a risk factor, which feature is irrelevant and provides inaccurate results. Machine learning techniques not only help in detecting or diagnosing the disease but also in identifying the pertinent risk factors responsible for causing heart disease. This paper proposes an ensemble model based on the Decision Tree-Recursive Feature Elimination (DT-RFE) feature selection technique to predict the risk of coronary artery disease. The introduced feature selection approach eliminates the irrelevant features based on the scores generated by the decision tree. Further, the remaining high-ranked features are used for building a model using an ensemble of SVM, k-NN, and Random Forest, and their prediction scores are combined using majority voting. The performance measures of the proposed work have been evaluated on a benchmark Z-Alizadeh Sani dataset and fifteen clinically correlated features are selected as significant risk factors. The experimental results demonstrate that the obtained relevant feature set with the ensemble model gives significantly encouraging detection performance compared to other methods.

Keywords: Coronary artery disease · Cardiovascular disease · Ensemble model · Feature selection techniques · Classification · Machine learning algorithms

1 Introduction

One of the most pernicious diseases in the world is heart disease. Based on the recent statistics of the World Health Organization (WHO) [1], 26 million of the world's adult population is suffering from heart diseases. Among many heart diseases, the most lethal is coronary artery disease. Coronary Artery Disease (CAD) arises when plaque gets accumulated in the coronary arteries [2]. A condition called atherosclerosis narrows the artery lumen by building plaques inside

I. Woungang et al. (Eds.): ANTIC 2021, CCIS 1534, pp. 425–440, 2022.
https://doi.org/10.1007/978-3-030-96040-7_34

it, which restricts the oxygenated blood flow to the heart [2]. The blood then becomes less oxygenated and is not suitable for heart muscles, resulting in chest pain, neck pain, shoulder and arms pain called Angina. This blockage of the less or no oxygenated blood causes a heart attack. The factors that pose a danger to the coronary artery are stress, smoking, high cholesterol, hypertension, diabetes, physical inactivity, unhealthy food intake, a person's genetic history, and so on [3]. The high risk of coronary artery disease is causing extensive research in this area. Early diagnosis of CAD is an important aspect and also a challenge in the medical industry. The primary reasons for the high death rate due to heart attack are lack of health awareness, knowledge, and insufficient diagnostic devices among patients and medical experts [4]. Early diagnosis will not only help in identifying the risk of CAD but will also decrease the deaths associated with the lethal disease.

Machine Learning (ML) is a branch of Artificial Intelligence (AI) that is increasingly utilized within the field of cardiovascular medicine, disease prediction, drug discovery, and in the measurement of the efficacy of treatments provided to patients [5,6]. Heart disease data available is huge, unstructured, and also consists of missing values, redundant, and irrelevant features that would diminish the prediction model's quality [7]. Due to the high dimensionality of data, learning models give biased and inaccurate results. Not all sets of features are significantly contributing to the prediction of the risk of the disease. It becomes difficult for the model to identify a set of important and relevant features that will contribute to the prediction of CAD. Including unnecessary features to build the model will give inaccurate results as the data will be overfitted. Therefore, research on feature selection techniques is implemented in recent times to control these issues by recognizing the selection of only a relevant set of features used for building the classifier. This will save effort, time and diminish the complexity of learning models.

There are several studies reported in the literature for the diagnosis and classification of CAD. A recent study [2], used the heterogeneous ensemble method for CAD prediction. They employed a random forest-based Boruta wrapper feature selection algorithm and feature importance of SVM for the prediction of CAD and selected five features [2]. After applying the 10-fold cross-validation method, a heterogeneous ensemble technique was built integrating three classifiers viz., k-NN, RF, and SVM. The study done in [8], used a computer-aided diagnosis system for diagnosing coronary artery disease on the Z-Alizadeh Sani dataset [9]. They have used SVM, C5.0, CHAID, and random forest to model with a 10-fold cross-validation method and determined the accuracy using some performance measures. According to their study, random trees performed best with an accuracy of 91.47% when used with 40 relevant features. Alizadeh Sani et al. have come up with an application of data mining methods that were established on the ECG symptoms and characteristics under the diagnosis of CAD [10]. They used the SMO-Naive Bayes hybrid algorithm that showed 88.52% accuracy after using the cross-validation method. A fuzzy expert system developed in [11] used the clinical parameters for the CAD problem. The accuracy achieved by the proposed algorithm was 84.20% [11]. The research done in [12] used cost-sensitive algorithms, SMO, SVM, C4.5, Naive Bayes, and k-NN after applying cross-validation on the sample. SMO achieved the highest accuracy of 92.09% as compared to other algorithms.

The research discussed till now is based on selecting a feature selection algorithm and the models are trained on the selected features to predict the risk of CAD. But the models in the recent research on the dataset have not shown remarkable performance and also showed low accuracy. A model has to be created that is built on a relevant set of features and gives accurate results. In this paper, we have used the DT-RFE feature selection technique to select the necessary set of features that are contributing as a pertinent risk factor for coronary artery disease. DT-RFE selected fifteen attributes that were used to build an ensemble of Support Vector Machines (SVM), k-Nearest Neighbor (k-NN), and Random Forest (RF) that performed better with high accuracy in comparison with the other feature selection methods.

The first section explains the methods and models used for the prediction. It explains feature selection methods and learning classifiers used for the CAD disease prediction. It also depicts the proposed framework used in the study. The next section provides result analysis which discusses the performance metrics achieved by using feature selection methods on the dataset. The paper concludes with the ensemble of three base classifiers using the DT-RFE technique and its associated performance.

2 Methods and Models

The proposed framework of this work is depicted in Fig. 1, in which first the data preprocessing task is performed followed by the proposed DT-RFE feature selection technique for selecting relevant features. Then the ensemble model is trained with the selected set of features for the detection of CAD. A detailed description of the proposed framework is given in Subsects. 2.1, 2.2, 2.3 and 2.4.

2.1 Data Preprocessing

Preprocessing of data is the foremost step to be followed for analyzing the dataset and creating learning models from it. Following preprocessing tasks are carried out on the dataset.

- *Scaling of Numerical features:* The features are in different ranges and hence they are not normalized or standardized. Working on the data without normalizing it, will lead to inaccurate results. Hence, scaling is done to all the numerical independent features in the dataset to normalize the data to be in a particular range. For scaling purposes "Standard Scaler" is used to convert the data into a form such that distribution has 0 as the mean and standard deviation of 1.

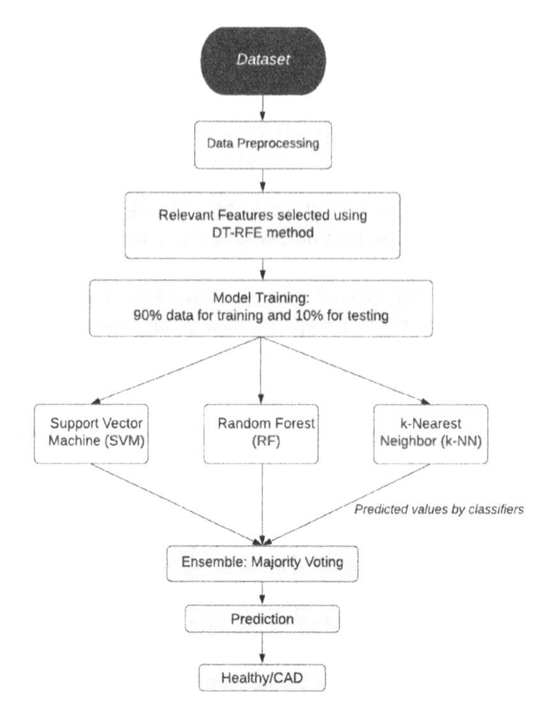

Fig. 1. Proposed framework

- *Label Encoding:* The categorical features have to be converted into numerical form so that the data will be treated as numerical data by the learning model. To achieve this we have applied label encoding and one-hot encoding. To perform one-hot encoding the categorical features need to have labels. Hence, label encoding is performed that provides labels from values of 0 and m_classes-1 where m is the no. of labels (distinct). It gives the same value to a feature if the label repeats itself. But doing label encoding isn't enough. The model will assume that the data lies between 0 and $n - 1$ which is not the real case. To eliminate this confusion one-hot encoding is applied.
- *One-hot Encoding:* The next step after the categories are transformed into 0,1,2,.... values, is to one-hot encode them. This method splits a categorical column into multiple columns where 0s and 1s are used to replace the numerical values. The number of columns divided is the total number of categories that are there in the data. This will lead the data to a form that can be directly given to the machine learning model for training purposes.

2.2 Decision Tree-Recursive Feature Elimination Method

The goal of feature selection is to decrease the dimensions of the dataset and reduce the dataset complexity [13]. To select some relevant set of features from a large set, we have applied the DT-RFE method to the dataset. RFE is a wrapper feature selection method that wraps another learning classifier to select features. It works with all the features in the dataset and then removes them until the relevant features are left. This can be done by first building a machine learning model and providing ranks to the features. The unimportant features are eliminated and then the model is built again on the remaining features [14]. These steps are performed iteratively till the important features are left [14]. In our study, we have used decision trees as the base classifier which is wrapped by the recursive feature elimination method to select features and set the total number of features to fifteen. The algorithm for DT-RFE is explained below:

Input: Sample with their known targets

1. Initialize original feature set as R = 1,2,...,n and selected features set as K = [].
2. for i = 1,2,....,n do
3. Create RFE model for the decision tree classifier and select features
4. Calculate ranking scores for all the features
5. Select features with higher ranking
6. Update the feature set K
7. Remove left over features
8. until R = []
9. end for

Output: 'K' feature set with higher ranking.

DT-RFE first takes all the features and eventually eliminates the irrelevant features till the relevant features persist according to scores provided to features. The RFE ranking returns an array as an output which depicts the positive integer values representing the ranking of the feature. A feature with a lower score indicates a higher ranking and vice versa. This approach eliminates the low rank or irrelevant features and selects high-ranked features. DT-RFE selected fifteen relevant features that were treated as a great risk factor for coronary artery disease.

2.3 Classifiers

In this paper, we have used three base classifiers for the prediction of CAD and the details are given below:

- Support Vector Machines: SVM is a supervised learning classifier that works for both regression and classification problems. If you give any labeled training data to the SVM algorithm, the data will be divided into different classes [15]. SVM creates a hyperplane in an N-dimensional space [16]. This is done to classify the data points distinctly. The kernel used in this research is 'Linear' SVM.

- k-Nearest Neighbors: One of the widely used classification algorithms is k-NN. It classifies new objects according to the training data. The objects are classified on the training data according to the nearest distance of the data point to a new object based on a distance formula like euclidean distance [17]. In this study k-NN used cv = 20 meaning 20 experiments will be done. Or in other words, train test split will be done 20 times to get better results. After checking the accuracy rate Vs the k value of the graph, no. of neighbors used in k-NN is 6.
- Random Forest Classifier: Random forest is an ensemble model that creates an ensemble of various decision trees. For discrimination, the information entropy function is used [18]. Like SVM, it can solve classification problems and regression problems. Prediction is higher if the number of trees is higher. Random forest first collects random samples of data. The decision trees are created for each sample and the tree that shows better prediction results is taken [19]. The number of estimators used here is 200, i.e. 200 decision trees are used for training the model.

2.4 Ensemble Framework

We have applied the DT-RFE feature selection technique to identify the relevant set of features for training the model. The DT-RFE method selected 15 features and these features are used for creating an ensemble of SVM, k-NN, and Random Forest. 90% of the data is given for model training and 10% is given for model testing. SVM, k-NN, and RF are used as base classifiers for creating an ensemble model. The outcomes of these base classifiers are combined using majority voting as described in Eq. (1) to detect whether a patient is healthy or is suffering from CAD.

$$\sum_{t=1}^{T} d_{t,j} = \max_{j=1}^{c} \sum_{t=1}^{T} d_{t,j} \tag{1}$$

where, $d_{t,j}$ is the decision of which classifier to choose from the tth classifiers, t = 1....., T. Here, the number of classifiers are denoted by T and j = 1....c where c is the total classes. If t^{th} classifier decides to choose a class w_j, then $d_{t,j}$ =1, else it is 0.

3 Analysis and Discussion of Results

3.1 Dataset and Preprocessing

The dataset used for testing the performance of the model is - Z-Alizadeh Sani dataset [9] created by Dr. Zahra Alizadeh Sani, Associate Professor of cardiology, Iran University, Tehran, Iran. The total number of instances is 303 with 56 attributes that mean the dataset consists of the records of 303 patients based on 56 features. There are no null values or missing values in the dataset. The features are distinguished into four areas: demographic, symptoms & examination,

ECG, and laboratory & echo features. They are of integer type as well as real. The demographic features are Age, Sex, BMI, HTN, Weight, DM, Length, Current Smoker, FH, EX-Smoker, CVA, Obesity, CRF, Airway Disease, Thyroid Disease, CHF, DLP. Symptoms & Examination includes BP, Weak Peripheral Pulse, PR, Systolic Murmur, Edema, Lung rales, Diastolic Murmur, Typical Chest Pain, Function Class, Atypical, Dyspnea, Nonanginal, Exertional CP, LowTH Ang. The attributes under ECG are Q Wave, St Elevation, St Depression, Poor R Progression, Tinversion, LVH, BBB. Laboratory & echo attributes are FBS, HDL, CR, TG, BUN, LDL, ESR, HB, Na, K, WBC, Neut, Lymph, PLT, Region RWMA, EF-TTE, VHD. The feature 'Cath' is the categorical feature that denotes whether the patient is normal or having CAD.

There are a total of 55 independent features, and one i.e. 'Cath' is the dependent feature or target feature. Cath determines whether a person is suffering from Coronary Artery Disease or not. It contains 2 labels 'CAD' and 'Normal'. CAD means a patient has the disease and normal means the patient does not have the disease. 72.0% of patients have coronary artery disease while 28.9% of patients are healthy. Figure 2 signifies that 216 patients are having the disease while 87 are normal patients.

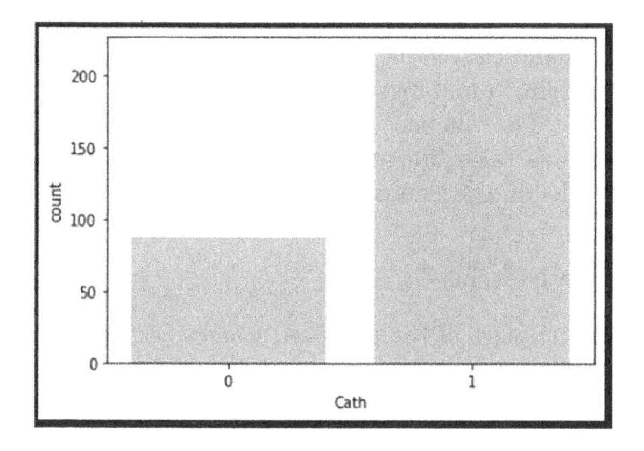

Fig. 2. Cath distribution

There are 22 categorical features in the dataset which are having a value in categories while there are 34 numerical features. The correlation among data is represented by a correlation matrix where a value near 1 represents a positive correlation and a value near to 0 or is negative represents a negative correlation. The correlation matrix is depicted in Fig. 3.

Data preprocessing is done after analyzing the data. The dataset is having a mix of categorical and numerical features. A model cannot be built with this kind of dataset. So, there are some feature processing techniques to convert the features into the form that a machine learning model can work upon. The dataset is first scaled using Standard Scaler to normalize the dataset. Scaling is applied

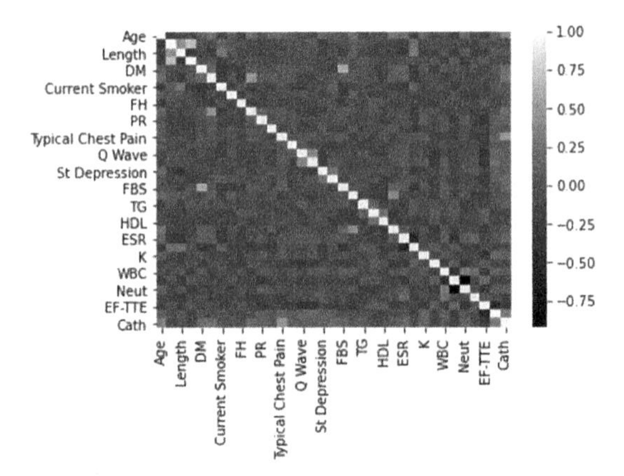

Fig. 3. Correlation matrix

to all 35 numerical features. Then label encoding is done to provide labels to the 21 categorical features except 'Cath' as it is the target feature. After label encoding, one-hot encoding is applied to the dataset. After applying one-hot encoding the dimensions of the data have become 303 * 78. The model is trained on the training data and to test the model we provide testing data. So, the split of the dataset into training and testing data is done using "train-test-split". The test size taken is 0.1. That means, 90% of the data is taken for model training and 10% of the data is taken for model testing. Hence, 272 samples are taken for training the model and 31 samples are taken for testing the model.

3.2 Performance Metrics

The detection performances of the proposed framework are evaluated using the following metrics.

Confusion Matrix: A confusion matrix is an n*n matrix that determines the model performance, where n is the number of classes [20]. A 2*2 confusion matrix is represented, as shown in Fig. 4, where:

True positive (TP): Reflects the predicted positive class by the model is the same as the actual positive class.
True negative (TN): Reflects the predicted negative class by the model is the same as the actual negative class.
False positive (FP): Reflects the predicted positive class by the model is not the same as the actual negative class.
False negative (FP): Reflects the predicted negative class by the model is not the same as the actual positive class.

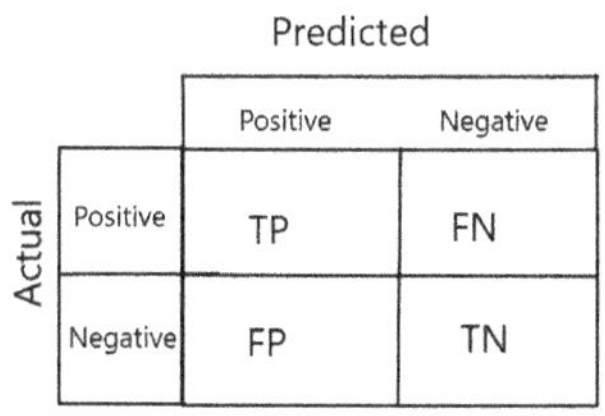

		Predicted	
		Positive	Negative
Actual	Positive	TP	FN
	Negative	FP	TN

Fig. 4. Confusion matrix

Accuracy: Accuracy is defined as the ratio of accurate predictions to the total number of predictions [21]. It is defined as:

$$Accuracy = \frac{TP + TN}{TP + TN + FP + FN} \tag{2}$$

Precision: Precision determines what proportion of positively identified values were true or correct. It is defined as:

$$Precision = \frac{TP}{TP + FP} \tag{3}$$

Recall: Recall determines what proportion of actual positives were identified correctly by the model. It is defined as:

$$Recall = \frac{TP}{TP + FN} \tag{4}$$

3.3 Discussion

Performance Measures Without Feature Selection. The learning models are first created using all 56 features without the feature selection method. As shown in Table 1, SVM, k-NN, and RF each showed an accuracy of approximately 90% when modeled individually. Table 2 shows that SVM achieved the highest precision of 91% as compared to k-NN and RF. Whereas, SVM achieved a recall of 95% lower than k-NN and RF that showed a recall of 100% as shown in Table 3. But all these results are not accurate as the data was overfitted. That means, all the features were used to build the model. Hence, it is important to apply feature selection techniques to select only a relevant set of features to build models with accurate results.

Performance Measures Using Mutual Information Method. Mutual Information Feature selection method is applied to the dataset to discover only a set of features that are relevant to the prediction of the CAD. The feature selection technique is first applied with k = all (all features) which gives the scores. The scores which are greater than 0.04 are taken for training the model. So, after applying this method only 15 features are taken into consideration for building the model. Selected Features are: 'Age', 'DM', 'HTN', 'EF-TTE', 'Typical Chest Pain', 'St Elevation', 'Tinversion', 'FBS', 'CR', 'Lung rales_2', 'Diastolic Murmur_2', 'Atypical_1', 'Atypical_2', 'Nonanginal_2', 'Region RWMA'. The mutual information feature selection method is found to be satisfactory in predicting the accuracy of the base classifiers. From Table 1, k-NN achieved an increase in the accuracy to 96% using the features selected by the mutual information method. Although SVM and Random Forest showed 90.3% accuracy that is the same as before. Tables 2 and 3 depict the precision and recall for the classifiers with the mutual information method. SVM and RF showed a precision of 88% while k-NN achieved 95%. The recall is always 100% for all classifiers.

Table 1. Accuracy (%) using different feature selection methods

Methods	W/O feature selection method	Mutual_Info	Boruta	DT-RFE
SVM	90.3	90.3	90.3	93.5
k-NN	90.3	96	93.5	93.5
Random Forest	90.3	90.3	87	90.3

Table 2. Precision (%) using different feature selection methods

Methods	W/O feature selection method	Mutual_Info	Boruta	DT-RFE
SVM	91	88	88	92
k-NN	88	95	95	100
Random Forest	88	88	85	92

Table 3. Recall (%) using different feature selection methods

Methods	W/O feature selection Method	Mutual_Info	Boruta	DT-RFE
SVM	95	100	100	100
k-NN	100	100	95	91
Random Forest	100	100	100	100

Performance Measures Using Boruta Feature Selection Method. Boruta Feature selection method is a wrapper feature selection method that is built on random forest. The number of estimators is set to auto i.e. determined automatically based on the size of the dataset. The maximum iterations are 50. After applying this method only 15 features are found relevant and are given to the base classifiers. Selected Features are 'Age', 'DM', 'HTN', 'BP', 'FBS', 'Typical Chest Pain', 'BMI', 'TG', 'ESR', 'K', 'EF-TTE', 'Nonanginal_1', 'Region RWMA', 'Atypical_1', 'Atypical_2'. Observing Table 1, k-NN accuracy got improved to 93.5% as compared to accuracy when all features were used to build the model. Here, no. of neighbors selected for k-NN are 4. Random Forest showed a decline in the accuracy to 87%. The accuracy of SVM is still 90.3% same as before. From Table 2, we can see that the precision achieved by SVM is 88%, k-NN is 95%, and RF is 85%. Table 3 shows that SVM and RF showed a recall of 100% while k-NN showed a recall of 95%.

Performance Measures Using Recursive Feature Elimination Method. RFE method is a backward feature selection method. It is using a Decision Tree classifier as the base learner for feature selection. After applying this method only 15 features are selected for training the model. Selected Features are 'Age', 'Length', 'Weight', 'BMI', 'HTN', 'DM', 'Current Smoker', 'Typical Chest Pain', 'St Elevation', 'Tinversion', 'CR', 'TG', 'ESR', 'Region RWMA', 'Nonanginal'. Table 1 depicts that the RFE feature selection method was proved out to be considerably better for the k-NN classifier and SVM with an accuracy of 93.5%. Whereas, random forest showed an accuracy of 90.3%. Here, k-NN showed accuracy when no. of neighbors selected is 5. We can notice that precision, in Table 2, is 92% for SVM and RF while it is 100% for k-NN. Recall given in Table 3 signifies that SVM and RF achieved a recall of 100% while k-NN achieved 91% recall. After the detailed analysis of performance in Tables 1, 2 and 3, the detection performance of DT-RFE showed significantly encouraging results in comparison with the other two feature selection methods. Hence, the DT-RFE method is used to create an ensemble of SVM, k-NN, and RF.

Performance of Proposed Ensemble Model with DT-RFE. To improve the accuracy for detecting CAD, we used an ensemble of three classifiers, SVM, k-NN, and RF. DT-RFE method showed significantly encouraging accuracy for predicting CAD. Hence, this method is used for selecting a relevant set of features for building the ensemble model. DT-RFE is first applied to all the features to find out the important set of features for which a total of 56 iterations are performed. Figure 5 illustrates the accuracy for the ensemble model when a different number of features are used. The classifier performed better with fifteen features with an accuracy of 96.7%.

The features selected by the DT-RFE method are Age, Length, Weight, BMI, HTN, DM, Current Smoker, Typical Chest Pain, St Elevation, Tinversion, CR, TG, ESR, Region RWMA, Nonanginal. New training data is created using these selected features and given to the ensemble model. We have used 90% data for training and 10% data for model testing.

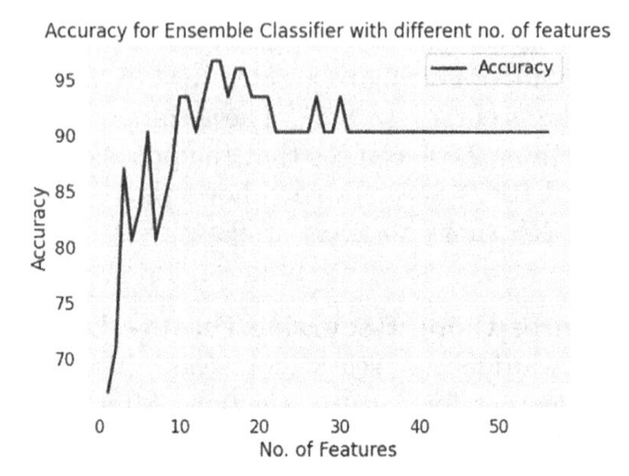

Fig. 5. Accuracy for ensemble model with different set of features

Voting Classifier is used for majority voting to select the accuracy among three base classifiers- SVM, k-NN, and RF. An ensemble model is a learning classifier that combines several models and trains on them. It predicts the output according to the highest probability of the classes used in the prediction. Here we have used hard voting. We have given 90% of the training data to SVM, k-NN, and RF classifiers respectively and 10% of the testing data is then given to each of these models for the detection of the disease. Majority voting counts the votes of the three classifiers and then selects the majority class as the output to detect whether a patient has the disease or not. The accuracy achieved by the ensemble model is 96.7% which is better than the individual algorithms. The proposed ensemble model has increased the confidence of the result for the prediction of CAD.

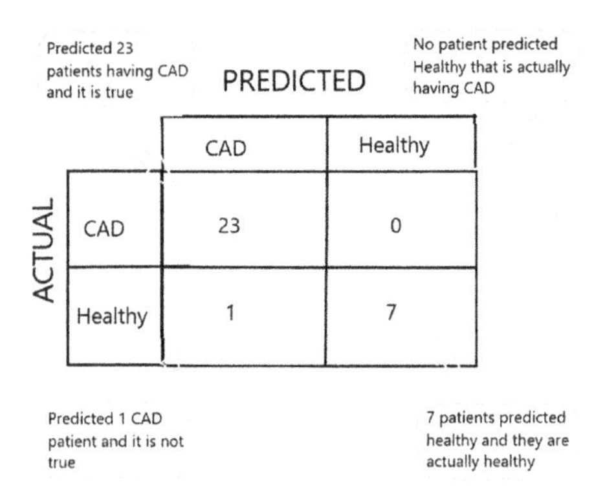

Fig. 6. Confusion Matrix for ensemble model

Figure 6 represents the confusion matrix for the ensemble model with recursively eliminated features. It depicts that the model predicted 30 (23 + 7) predictions correctly while only one prediction is false. Corresponding to the confusion matrix; accuracy, precision, and recall have been calculated for the ensemble model.

Table 4. Performance of ensemble classification using different feature selection methods (%)

Methods	Accuracy	Precision	Recall
W/O feature selection method	90.3	88	100
Mutual information	93.5	92	100
Boruta	90.3	88	100
DT-RFE	96.7	95	100

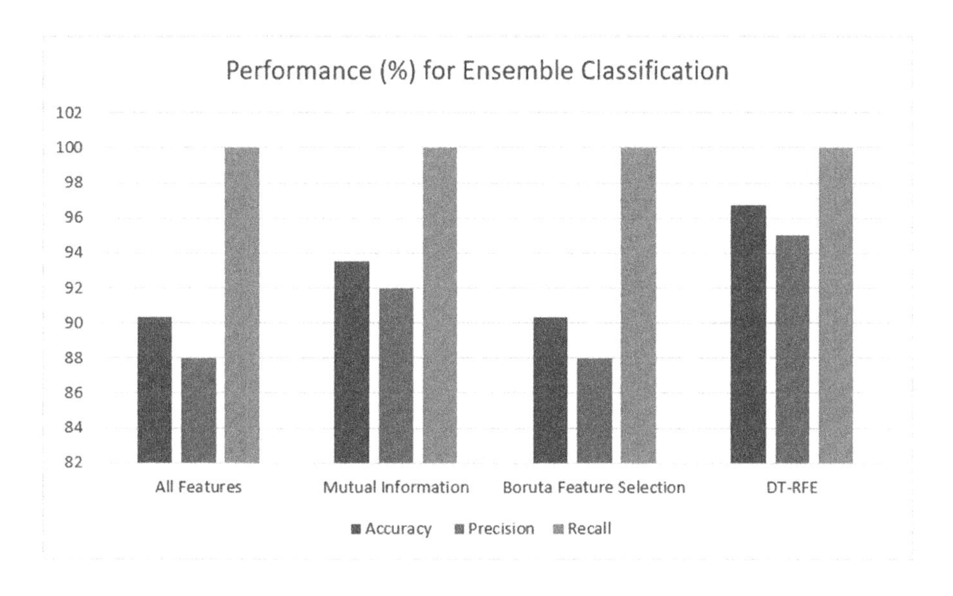

Fig. 7. Performance (%) for ensemble classification

Table 4 outlines the accuracy, precision, and recall of the Ensemble model using different feature selection techniques. It can be perceived that the accuracy is 90.3% when no feature selection method is used. Boruta also gave an accuracy of 90.3%. The accuracy for ensemble when mutual information used is 93.5%. The ensemble model achieved an accuracy of 96.7% when DT-RFE features are used. The precision of boruta method is 88% while the precision for the mutual information method is 92%. Also, the precision is 88% when no feature selection method is used. Ensemble with recursive eliminated features showed a precision of 95%. The recall of the ensemble classification is always 100%. Accuracy, precision, and recall for all the ensemble models built using different feature selection methods are depicted in the bar graph shown in Fig. 7.

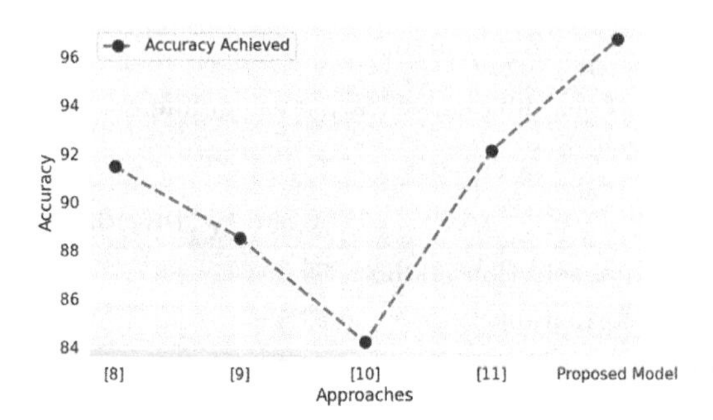

Fig. 8. Accuracy achieved in different approaches

Comparative Analysis with Other Approaches: Figure 8 shows accuracy achieved by the recent research done in [8–11] which is compared in the line graph along with the accuracy achieved in this research using an ensemble model with DT-RFE method. The introduced ensemble model with DT-RFE gives significantly encouraging detection performance in terms of accuracy as compared to the other methods.

4 Conclusion

In this paper, an ensemble classification model was developed with relevant features using the DT-RFE technique to predict the risk of coronary artery disease. The data was processed by doing scaling, label encoding, and one-hot encoding. DT-RFE technique was used to select a subset of features that are relevant to build learning models. Three machine learning algorithms viz; SVM, k-NN, and RF were used to predict whether a patient is having the risk of coronary artery disease or not.

DT-RFE method proved to be a better feature selection method as compared to other methods to select a relevant set of features. It selected fifteen clinically correlated features that were important in the prediction. Hence, the features selected by this method were used to create an ensemble of three base classifiers - SVM, k-NN, and RF. We used majority voting that selected the majority of high probability class and it gave an accuracy of 96.7%. Hence, the ensemble model gave remarkably better results as compared to the individual machine learning classifiers. The majority voting performed a reliable detection of CAD with the fifteen most relevant features and also increased the confidence of the result.

References

1. Zomorodimoghadam, M., Abdar, M., Davarzani, Z., Zhou, X., Plawiak, P., Acharya, U.R.: Hybrid particle swarm optimization for rule discovery in the diagnosis of coronary artery disease. Expert. Syst. **38**, e12485 (2019)
2. Velusamy, D., Ramasamy, K.: Ensemble of heterogeneous classifiers for diagnosis and prediction of coronary artery disease with reduced feature subset. Comput. Methods Programs Biomed. **198**, 105770 (2021)
3. Prabhakaran, D., Jeemon, P., Roy, A.: Current epidemiology and future directions. Ann. Phys. **133**(16), 1605–20 (2016)
4. Prabhakaran, D., Singh, K., Roth, G.A., Banerjee, A., Pagidipati, N.J., Huffman, M.D.: Cardiovascular diseases in India compared with the United States. J. Am. Coll. Cardiol. **72**(1), 79–95 (1905)
5. Md Idris, N., Chiam, Y.K., Varathan, K.D., Wan Ahmad, W.A., Chee, K.H., Liew, Y.M.: Feature selection and risk prediction for patients with coronary artery disease using data mining. Med. Biol. Eng. Comput. **58**(12), 3123–3140 (2020). https://doi.org/10.1007/s11517-020-02268-9
6. Krittanawong, C., et al.: Machine learning prediction in cardiovascular diseases: a meta-analysis. Sci. Rep. **10**(1), 16057 (2020)
7. Sun, S.: An innovative intelligent system based on automatic diagnostic feature extraction for diagnosing heart diseases. Knowl.-Based Syst. **75**, 224–238 (2015)
8. Joloudari, J.H., et al.: Ranking the significant features using a random trees model. Int. J. Environ. Res. Public Health **17**(3), 731 (2020)
9. Sani, Z.A.: Z-Alizadeh Sani dataset. UCI Machine Learning Repository (Online) (2017)
10. Alizadehsani, R., et al.: Diagnosis of coronary artery disease using data mining techniques based on symptoms and ECG features. Eur. J. Sci. Res. **82**(4), 542–553 (2010)
11. Pal, D., Mandana, K., Pal, S., Sarkar, D., Chakraborty, C.: Fuzzy expert system approach for coronary artery disease screening using clinical parameters. Knowl.-Based Syst. **36**(13), 162–174 (2012)
12. Alizadehsani, R., Hosseini, M.J., Sani, Z.A., Ghandeharioun, A., Boghrati, R.: Diagnosis of coronary artery disease using cost-sensitive algorithms, pp. 9–16 (2012)
13. Tadist, K., Najah, S., Nikolov, N.S., Mrabti, F., Zahi, A.: Feature selection methods and genomic big data: a systematic review. J. Big Data **6**, 1–24 (2019)
14. Brownlee, J.: Recursive Feature Elimination (RFE) for feature selection in Python. Machine Learning Mastery, 25 May 2020 (Online)
15. Katarya, R., Srinivas, P.: Predicting heart disease at early stages using machine learning: a survey. In: International Conference on Electronics and Sustainable Communication Systems (ICESC), pp. 302–305 (2020)
16. Gandhi, R.: Support Vector Machine - Introduction to Machine Learning Algorithms. Towards Data Science, 7 June 2018 (Online)
17. Larose, D.T.: Discovering Knowledge in Data an Introduction to Data Mining, 2nd edn. Wiley Interscience, Hoboken (2005)
18. Fan, R., Zhang, N., Yang, L., Ke, J., Zhao, D., Cui, Q.: AI-based prediction for the risk of coronary heart disease among patients with Type 2 Diabetes Mellitus. Sci. Rep. **10**(1), 14457 (2020)
19. Karaca, Y., Cattani, C.: Naive Bayesian classifier. Comput. Methods Data Anal. 229–250 (2018)

20. Bhandari, A.: Everything you Should Know about Confusion Matrix for Machine Learning. Analytics Vidhya, 17 April 2020 (Online)
21. Wenxin, X.: Heart disease prediction model based on model ensemble. In: International Conference on Artificial Intelligence and Big Data (ICAIBD), pp. 195–199 (2020)

Comparative Study of Marathi Text Classification Using Monolingual and Multilingual Embeddings

Femida Eranpurwala[1,2](✉), Priyanka Ramane[1], and Bharath Kumar Bolla[3]

[1] UpGrad Education Private Limited, Nishuvi, Ground Floor, Worli, Mumbai 400018, India
[2] Liverpool John Moores University, Rodney House, Liverpool L3 5UX, UK
[3] University of Arizona, Tucson, AZ 85721, USA

Abstract. Text classification, a common task in Natural Language Processing, is used to classify sentiments, food reviews and many more. With the increased adaption of regional languages in the digital era, the necessity of text classification in regional languages has also increased. As one of the 22 constitutional languages, Marathi is spoken by approximately 83 million people. This widespread use of Marathi has led researchers to perform Marathi text classification; however, these research works pre-dominantly were based on language specific manual features based on grammar rules, morphological clues and syntactic dependencies between the words. This work compares and evaluates various pre-trained monolingual and multilingual embeddings with machine learning models to perform Marathi text classification. Our experiments show that the transformer based multilingual embeddings with Long Short Term Memory (LSTM) architecture outperform monolingual embeddings with better recall, precision, accuracy and F1-Score with lesser training time.

Keywords: Marathi · Text classification · Embedding · Monolingual · Multilingual · IndicFT · fastText · IndicBERT · MuRIL · XLM-R · LSTM · biLSTM · NLP

1 Introduction

Text classification has been one of the essential tasks in NLP with widespread applications like Movie review analyzing, Sentiment analysis, Intent detection, etc. In recent days, text classification in regional language has gained much attention and companies have started adapting to it to serve their customers better. As a result, text classification in regional languages has gained focus and is an open research area.

Marathi is one of the most spoken regional languages of the world. With a low digital footprint performing text classification is always challenging. In addition to the data challenges, grammatical and syntactic rules of Marathi make text classification further complex. Marathi is a free order language, which implies that the sentence's meaning remains the same even if the sequence of the word changes. Example, रामाने रावणाला मारले is equivalent to रावणाला रामाने मारले. With 52 alphabets, 16 vowels and 36 consonants, Marathi

I. Woungang et al. (Eds.): ANTIC 2021, CCIS 1534, pp. 441–452, 2022.
https://doi.org/10.1007/978-3-030-96040-7_35

is one of the few morphologically rich languages, allowing for the creation of numerous morphological variants by inflections on a single root word. Word sense disambiguation, a challenging task in natural language processing, becomes even more difficult in Marathi due to the language's free order sequencing and inflectional variants.

The emergence of transfer learning and the subsequent use of pre-trained word embeddings has reduced the reliance on the size of a dataset. These deep learning-based pre-trained word embeddings capture the relationships within the various words, showing the word analogy in vector space, represented in an equation form. When words are plotted using these word embeddings, similar word vectors fall close together, indicating a higher degree of word similarity. In the last two-three years, embeddings have evolved from context free embedding to modern days contextual embeddings like CoVe, ELMo and BERT. Context-free embeddings generate a global vocabulary by ignoring the meaning of words in different contexts; consequently, they fail to handle polysemy effectively.

The purpose of this study is to investigate and compare the performance of multiple pre-trained context-free monolingual and contextual multilingual word embeddings in conjunction with supervised machine learning algorithms in order to determine which combination produces the best results for Marathi text classification. This study intends to cater as a baseline experiment to automate text classification for Marathi. Also, this study further contributes to various text related tasks such as Name entity recognition, question answering system and information retrieval by capturing several aspects of the pre-trained embeddings.

2 Related Work

Various researchers and linguistics have studied Indian regional languages to perform text classification. [1] conducted their work in 2009 to perform text classification using the Vector space model and Artificial Neural Network for Tamil. [2] performed text classification on Gujarati text documents. This work involved experiments with and without feature selection in conjunction with the Naïve Bayes algorithm. One study [3] used various classifiers such as KNN, Decision tree and Naïve Bayes to classify Tamil, Kannada and Telugu text. In the pre deep learning era, one study [4] used Principal Component Analysis and modified Label Induction Grouping algorithm for document classification in Marathi. Further, [5] proposed classification based on ontology and Modified KNN, SVM and Naïve Bayes.

Due to less digital presence, auto capturing the language specific complex relationships is a difficult task hence the feature selection and feature reduction were primarily dependent on manual rules which identifies language specific inflectional, morphological and grammatical patterns. This rule based curated data was an input for the machine learning algorithms. Hence, most of the regional language studies, including Marathi text classification, focus on finding various rules and feeding the curated data in machine learning algorithms. The process of rule-based feature generation often comes with the limitation to perform the automatic text classification. However, with various NLP trends, many researchers have started using the advanced techniques to perform the text classification. [6] have used the Hierarchical Attention (HAN)-Based Encoder-Decoder Model for Marathi text classification. [7] have used the KNN along with term-weighting for Marathi text classification in conjunction with parallel algorithm on GPU.

As a result of the recent advancements in word embeddings, numerous researchers have proposed reusable pre-trained monolingual and multilingual embeddings. [8] and [9] proposed multiple pre-trained monolingual embeddings in various languages using the Skip-gram [10] and CBOW [10] models, respectively. BERT [11] based multilingual embeddings and [12, 13, 14] are the current state-of-the-art embeddings that have outperformed traditional algorithms. These embeddings can capture the complex language specific relationships when used with machine learning algorithms with less manual intervention and can be used for many downstream NLP tasks.

We drew inspiration for this work from the research [15] and [16]. BERT-based embeddings were used and experimented with in these studies for text classification in a variety of Indian languages in order to determine the best performing embedding.

3 Data Sourcing

In this study, the experimental dataset is obtained from Kaggle. This dataset is the Marathi news headline dataset and was contributed by [17]. This dataset is available in CSV format.

There are two attributes in the data set:

1. *headline:* - This column contains free text Marathi news headlines from news articles that should be classified according to their subject matter.
2. *label:* - This column contains the categories of news that correspond to the class variable.

A raw input dataset sample is given below:

Table 1. Input data sample

headline	***label***
बहुजन वंचित आघाडीची पहिली यादी जाहीर, प्रकाश आंबेडकरांच्या मतदारसंघावरुन सस्पेंन्स कायम	state
गृहमंत्रिपद हे पार्ट टाइम नाही, राजीनामा द्या, तृप्ती देसाईंची मुख्यमंत्र्यांवर टीका	state
शरद पवार यांनी मिलिंद नार्वेकर यांच्या कानात काय सांगितलं असेल?	state
ग्रेट भेटमध्ये अंजली भागवत - भाग 3	state
'तारक मेहता..'चा कलाकार विशाल ठक्कर 11 दिवसांपासून बेपत्ता	state
'दंगल'मुळे स्मॉल स्क्रीन थिएटर्सला फायदा	entertainment
चाहत्याने खांद्यावर हात ठेवताच सोनू निगमने केलं असं काही की सारेच झाले अवाक्!	entertainment
जळगावातल्या 'त्या' नरभक्षक बिबट्याने आतापर्यंत घेतलेत 6 बळी	state
मराठवाड्यात पावसाचं कमबॅक ; मुंबईसह राज्यातही संततधार	state

There are three categories for the class variable: entertainment, state and sport; hence this problem falls under multiclass classification. Out of 9763 data rows, a validation dataset of size 1000 was created using stratified random sampling method. The below table (Table 2) lists the training data with the respective class-wise distribution.

As indicated in the below table (Table 2), the dataset appears to be somewhat unbalanced; however, this was addressed by evaluating the outcome in terms of F1-Score and accuracy.

Table 2. Input data class distribution

Label (class)	Total count	Proportion (percent)
Sports	895	10
Entertainment	2379	27
State	5431	62

Type-Token ratio was calculated to aid in the comprehension of data variability. Type-Token ratio was used instead of standardized Type-Token since the dataset is comparatively small and was able to capture the variability as expected. The dataset contains 58941 tokens and 22429 types, resulting in a Type-Token ratio of approximately 38%, indicating that the dataset contains relatively lesser variation. The character distributions in each class are shown below in Fig. 1. The overall distribution shows a similar pattern except for the class sports, which has scattered data. The average character count ranges between 40–70 characters.

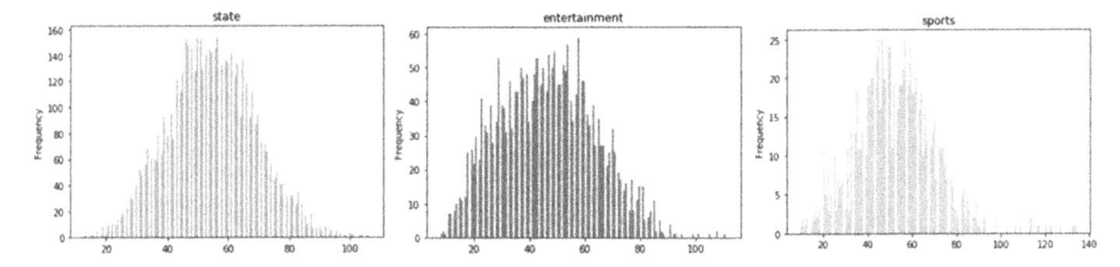

Fig. 1. Character distribution in each data row

Figure 2 illustrates the distribution of words within each class. These distributions follow a similar pattern to character distributions (Fig. 1), illustrating the dispersed nature of sports data. The typical word count is between 5–7.

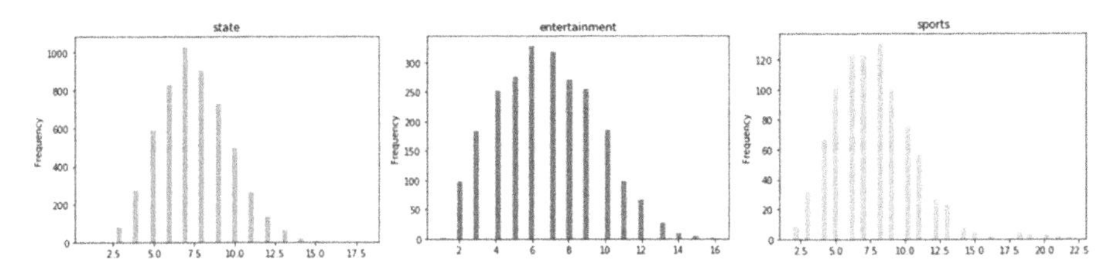

Fig. 2. Word distribution in each data row

Observatory Pointers

- Dataset used in this study is well structured and well-formatted with formal news language with significantly little junk data.
- Character and Word count distribution suggest that sports category data is more scattered than the other two categories.
- TTR with 38% suggests the controlled variability of the dataset and required no further data cleaning.

4 Research Methodology

There are four primary components of the methodology, i.e., data preparation, word embedding generation, classification and evaluation. The methodology follows the following path shown in the flowchart (Fig. 3):

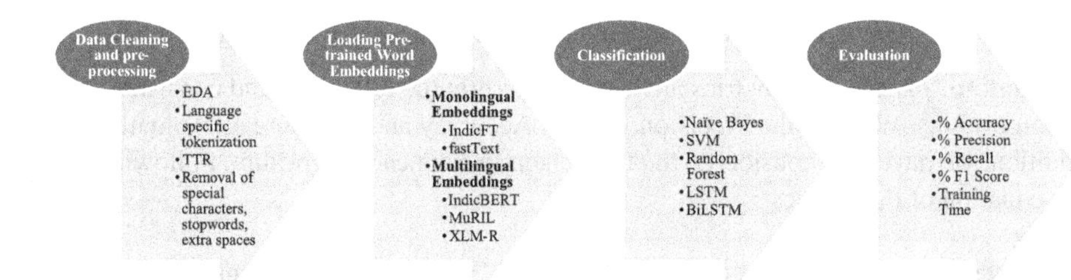

Fig. 3. Flowchart of the methodology followed

Data Cleaning and Pre-processing: - This is the first and most critical step in any machine learning project. Consequently, data cleaning is subdivided into numerous subtasks. The scope and complexity of these subtasks vary according to the data and nature of the project. In this study, we removed special characters, English alphabets, extra spaces, single letter words, junk characters and language-specific stop words from text data. Additionally, we used a language-specific tokenizer, which assists in tokenizing words according to their inflection rules to handle the suffix or prefix of a word. E.g. word ' आघाडीची' will be tokenized into ' आघाडी, ' ची'. The data cleaning activities were limited to the experiments involving traditional machine learning models such as Naïve Bayes, SVM and Random Forest and not for deep learning-based models such as LSTM and biLSTM. After removing the stop words, SVM, Random Forest and Naïve Bayes performed better; however, removing stop words decreased the accuracy of LSTM and biLSTM; hence while modeling these two algorithms, we provided raw data as input.

Loading Pre-trained Word Embeddings: - In any NLP project, text data needs to be converted into corresponding numerical vector representation to be consumed in the machine learning model. We used Marathi specific pre-trained monolingual and multilingual word embeddings in this work instead of creating embeddings from inception.

We have used the monolingual embeddings IndicFT [8] and fastText [9] and multilingual embeddings IndicBERT [12], MuRIL [14] and XLM-R [13]. IndicFT was trained on the IndicCorp corpus using the skip-gram model on a vocabulary size of 258413. fastText learns representations for character n-grams in order to represent words as the sum of n-gram vectors. It was trained on Common Crawl and Wikipedia dataset using the CBOW model with position-weighted vocabulary of size 99882. IndicBERT was trained on ALBERT [18] using standard masked language model with vocabulary 200K and total parameter of 33M. XLM-R was trained on mBERT with vocabulary 250K and total parameter of 270M. MuRIL was trained on BERT base encoder model using Masked Language Modeling and Translation Language Modeling with a vocabulary of 197K and a total of 236M parameters.

Classification: - Classification is a supervised machine learning task in which the model needs to predict a class or label variable for input data by using set of machine learning algorithms. In this study, we used traditional SVM, Naïve Bayes and Random Forest algorithms and advanced fairly complex LSTM and biLSTM algorithms. Naïve Bayes algorithm is generally used as a baseline performer in text classification. SVM and Random Forest are comparatively high in complexity as compared to Naïve Bayes.

Evaluation: - Evaluation metrics are used to interpret the reliability and consistency of the algorithms. We used the Precision, Recall, Accuracy and F1-Score as evaluation. In addition, we have also considered the time taken to train each algorithm. Following are the equation of the metrics:

- **Accuracy:** Accuracy is the most frequently used and intuitive performance indicator. It is defined as the ratio of accurately predicted to observed data. Accuracy is a valid measure when dealing with balanced datasets.

$$Accuracy = \frac{True\,Positive + True\,Negative}{True\,Positive + False\,Positive + True\,Negative + False\,Negative}$$

- **Recall:** The ratio of correctly predicted observations to all observations in the actual class is defined as recall.

$$Recall = \frac{True\,Positive}{True\,Positive + False\,Negative}$$

- **Precision:** Precision is the ratio of correctly predicted observations by a model to the total predicted positive observations.

$$Precision = \frac{True\,Positive}{True\,Positive + False\,Positive}$$

- **F1-Score:** The F1-Score is defined as the weighted average of precision and recall; as such, it takes into account both false positives and negatives. The F1-Score is beneficial in the imbalance dataset.

$$F1 - Score = 2 * \frac{Precision * Recall}{Precision + Recall}$$

5 Results

We created word vectors using two monolingual embeddings, IndicFT and fastText, and three multilingual embeddings, XLM-R, MuRIL and IndicBERT. These five embeddings were evaluated on five classification algorithms; Naïve Bayes, Support Vector Machine, Random Forest, LSTM and biLSTM.

Table 3. Model Result Comparison and Evaluation

Type of embedding	Word embedding	Classifier	Accuracy (percent)	Precision (percent)	Recall (percent)	F1-score (percent)	Model training time
Monolingual embeddings	IndicFT	Naïve Bayes	60.0	59.0	60.0	60.0	00:00:00.06
		Random Forest	66.0	71.0	66.0	57.0	00:02:17.17
		SVM	75.0	75.0	76.0	75.0	00:07:54.83
		LSTM	**80.0**	**80.0**	**80.0**	**80.0**	**00:02:11.87**
		biLSTM	80.0	80.0	80.0	80.0	00:04:14.11
	fastText	Naïve Bayes	56.0	58.0	56.0	56.0	00:00:00.07
		Random Forest	66.0	70.0	66.0	58.0	00:02:18.06
		SVM	74.0	74.0	75.0	74.0	00:01:17.26
		LSTM	74.0	74.0	75.0	74.0	00:02:12.06
		biLSTM	74.0	74.0	75.0	75.0	00:04:17:33
Multilingual embeddings	IndicBERT	Naïve Bayes	53.0	60.0	53.0	56.0	00:00:00.08
		Random Forest	66.0	68.0	66.0	59.0	00:16:48.44
		SVM	71.0	70.0	71.0	68.0	00:01:22.93
		LSTM	80.0	80.0	80.0	80.0	02:35:06.21
		biLSTM	80.0	80.0	80.0	80.0	02:36:37.44
	MuRIL	Naïve Bayes	83.0	84.0	83.0	84.0	00:00:00.06
		Random Forest	82.0	83.0	82.0	80.0	00:02:55.68
		SVM	88.0	89.0	89.0	89.0	00:00:24.15
		LSTM	**93.0**	**93.0**	**93.0**	**93.0**	**03:07:29.17**
		biLSTM	**93.0**	**94.0**	**94.0**	**94.0**	**03:15:30.77**
	XLM-R	Naïve Bayes	83.0	83.0	83.0	83.0	00:00:00.07

(*continued*)

Table 3. (*continued*)

Type of embedding	Word embedding	Classifier	Accuracy (percent)	Precision (percent)	Recall (percent)	F1-score (percent)	Model training time
		Random Forest	82.0	84.0	83.0	81.0	00:03:10.47
		SVM	86.0	86.0	86.0	86.0	00:01:31.04
		LSTM	91.0	91.0	91.0	91.0	03:13:50.75
		biLSTM	90.0	91.0	91.0	91.0	03:10:57.83

We conducted a total of twenty-five experiments combining the aforementioned embeddings and classification algorithms. The comparisons and evaluations in Table 3 suggests; those multilingual embeddings outperform monolingual embeddings. Within the multilingual embeddings, MuRIL with LSTM and MuRIL with biLSTM combination give the most optimal performance followed by XLM-R. Within the monolingual embeddings, IndicFT along with LSTM and IndicFT with biLSTM top the chart.

The results table (Table 3) also shows that monolingual embeddings take less time to train than multilingual embeddings. Training time for multilingual embeddings is higher due to the complex architecture of multilingual embeddings than monolingual embeddings.

The below graph (Fig. 4) depicts the accuracy of the best model combinations MuRIL with biLSTM and LSTM. After ten epochs, the train and validation curves converge and reach a plateau.

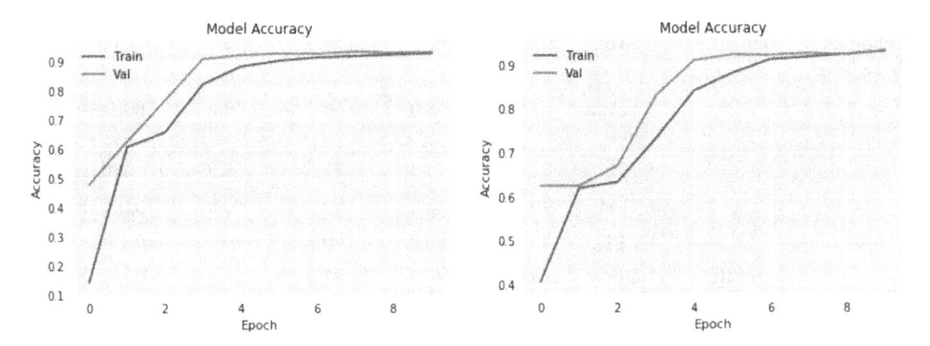

Fig. 4. Accuracy of MuRIL trained with biLSTM (Left) and MuRIL trained with LSTM (Right)

The hyperparameters finalized for the two best model combinations of MuRIL are in Table 4. Except for the units, other hyperparameters are the same.

Table 4. Hyperparameters for experiments involving MuRIL

Hyperparameter	biLSTM	LSTM
Units	4	16
Dropout	0.1	0.1
return_sequences	True	True
Optimizer	Adam optimizer	Adam optimizer
Learning rate	0.000001	0.000001

The graph (Fig. 5) gives the comparison of the average metrics from monolingual and multilingual embeddings. These metrics are obtained by averaging the metrics obtained with the monolingual and multilingual embeddings across five machine learning algorithms. We can deduce that MuRIL's multilingual embedding performed the best, followed by XLM-R, which lags fractionally behind. IndicFT outperforms fastText in the monolingual category. MuRIL achieves an average performance gain of 15–19% over traditional monolingual embeddings IndicFT and fastText across all evaluation criteria. The superior performance of multilingual embeddings over monolingual embeddings marks a significant advancement in the field of NLP.

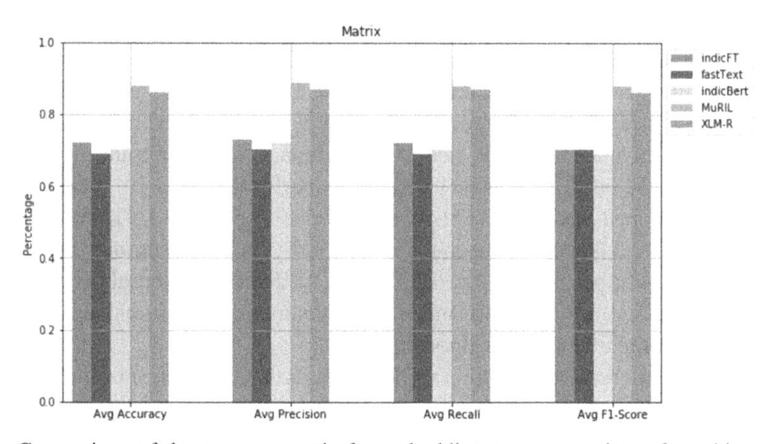

Fig. 5. Comparison of the average metric for embedding across a variety of machine learning algorithms

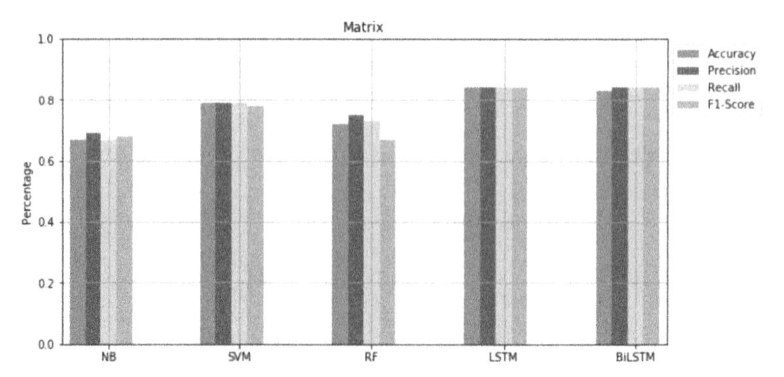

Fig. 6. Comparison of the average performance metric for classifiers across a variety of multilingual and monolingual embeddings

The above graph (Fig. 6) compares the average metrics obtained from the classification algorithms across monolingual and multilingual embeddings vectorized text. Here, we can observe biLSTM and LSTM have consistently outperformed the other traditional algorithms. Among the traditional algorithm, SVM has performed consistently as compared to Naïve Bayes and Random Forest.

6 Conclusion

As seen in the assessment table, multilingual embeddings have shown ~15% performance gain compared to traditional monolingual embeddings. Multilingual embeddings, when coupled with LSTM and biLSTM, give the state-of-the-art performance. The superior performance of multilingual embeddings and LSTMs results from the high-quality word embeddings generated by BERT-based multilingual embeddings and the LSTM's sequence handling capability. Moreover, if coupled with LSTM and biLSTM, monolingual embeddings give a reasonably decent performance with minimum training time. LSTM and biLSTM have accounted for superior performance as compared to Naïve Bayes, Random Forest and SVM. This is mainly due to the way LSTM and biLSTM captures the long term dependencies in the sentences by handling complex relationships between the words impeccably. Also, LSTM and biLSTM are Neural Network based techniques with access to various advanced hyperparameters which enables the superior performance as compared to other algorithms.

The experiments performed in this work, reassure the importance of the pre-trained embeddings and ability to perform even with limited data. One interesting finding from the study is that even weak classifiers such as Naïve Bayes improve metrics by 15–20% when used with multilingual embeddings such as MuRIL and XLM-R and outperform monolingual embeddings with LSTM and biLSTM. We can infer that high-quality embedding is sufficient enough to bring performance improvement even with weak classifiers. As GPU is not available for model inference in all the systems, MuRIL and SVM can be considered the optimal combination to set the benchmark with the high performance in less time.

7 Contribution to Knowledge

This work sets the baseline performance for the Marathi text classification using the various pre-trained monolingual (IndicFT, fastText) and multilingual embeddings (MuRIL, IndicBERT, XLM-R). These embeddings along with results and experiments can further be explored and extended to various NLP task such as information extraction, information retrieval and topic modelling. Also, this comparative study helped in identifying the infrastructure friendly embeddings by suggesting IndicFT embedding as optimal for projects with low infrastructure whereas MuRIL to achieve state of the art in projects with high availability of infrastructure resources by minimal algorithmic hyperparameter updates.

8 Future Recommendation

Future work can include data augmentation to handle the class imbalance, further enhance the model's performance. Also, implementing knowledge distillation of the various work embeddings can reduce the training and inference time and dependency on high infrastructural resources. Exploring the complex neural network architecture, which combines CNN, LSTM, dense layer and batch normalisation can help capture the complex relationships between Marathi sentence formation and improve the model's accuracy.

References

1. Rajan, K., Ramalingam, V., Ganesan, M., Palanivel, S., Palaniappan, B.: Automatic classification of Tamil documents using vector space model and artificial neural network. Expert Syst. Appl. **36**, 10914–10918 (2009). https://doi.org/10.1016/j.eswa.2009.02.010
2. Rajnish, M., Saini, J.: Classification of Gujarati documents using Naïve Bayes classifier. Indian J. Sci. Technol. **10**, 1–9 (2017). https://doi.org/10.17485/ijst/2017/v10i5/103233. https://indjst.org/articles/classification-of-gujarati-documents-using-nave-bayes-classifier
3. Swamy, M., Hanumanthappa, M., Mohan, J.: Indian language text representation and categorization using supervised learning algorithm, pp. 406–410 (2014). https://doi.org/10.1109/ICICA.2014.89. https://ieeexplore.ieee.org/document/6965081
4. Narhari, S.A., Shedge, R.: Text categorization of Marathi documents using modified LINGO. In: International Conference on Advances in Computing, Communication and Control (ICAC3), Mumbai, pp. 1–5 (2017). https://doi.org/10.1109/ICAC3.2017.8318771. https://ieeexplore.ieee.org/document/8318771
5. Bolaj, P., Govilkar, S.: Text classification for Marathi documents using supervised learning method. Int. J. Comput. Appl. **155**, 6–10 (2016). https://doi.org/10.5120/ijca2016912374. https://www.ijcaonline.org/archives/volume155/number8/bolaj-2016-ijca-912374.pdf
6. Deshmukh, R.D., Kiwelekar, A.: Classification of Marathi text using hierarchical attention (HAN)-based encoder-decoder model. In: Pandian, A.P., Palanisamy, R., Ntalianis, K. (eds.) Proceedings of International Conference on Intelligent Computing, Information and Control Systems. AISC, vol. 1272, pp. 721–736. Springer, Singapore (2021). https://doi.org/10.1007/978-981-15-8443-5_62

7. Lade, S., Bhosale, G., Sonavane, A., Gaikwad, T.: Parallel implementation of Marathi text news categorization using GPU. In: Swain, D., Pattnaik, P.K., Athawale, T. (eds.) Machine Learning and Information Processing. AISC, vol. 1311, pp. 365–372. Springer, Singapore (2021). https://doi.org/10.1007/978-981-33-4859-2_36
8. Kunchukuttan, A., Kakwani, D., Golla, S., Gokul, C., Bhattacharyya, A., Khapra, M.: AI4Bharat-IndicNLP corpus: monolingual corpora and word embeddings for Indic languages. https://arxiv.org/pdf/2005.00085.pdf (2020)
9. Grave, E., Bojanowski, P., Gupta, P., Joulin, A., Mikolov, T.: Learning word vectors for 157 languages. arXiv:1802.06893 (2018)
10. Mikolov, T., Chen, K., Corrado, G., Dean, J.: Efficient estimation of word representations in vector space. ICLR. https://arxiv.org/abs/1301.3781 (2013)
11. Devlin, J., Chang, M., Lee, K., Toutanova, K.: BERT: pre-training of deep bidirectional transformers for language understanding. NAACL-HLT. https://arxiv.org/pdf/1810.04805.pdf (2019)
12. Kakwani, D., Kunchukuttan, A., Golla, S., Gokul, N.C., Bhattacharyya, A., Khapra, M.: IndicNLPSuite: monolingual corpora, evaluation benchmarks and pre-trained multilingual language models for Indian languages, pp. 4948–4961 (2020). https://doi.org/10.18653/v1/2020.findings-emnlp.445. https://indicnlp.ai4bharat.org/papers/arxiv2020_indicnlp_corpus.pdf
13. Conneau, A., et al.: Unsupervised cross-lingual representation learning at scale. ACL. https://arxiv.org/pdf/1911.02116.pdf (2020)
14. Khanuja, S., et al.: MuRIL: multilingual representations for indian languages. arXiv:2103.10730 (2021)
15. Saha, D., Paharia, N., Chakraborty, D., Saha, P., Mukherjee, A.: Hate-Alert@DravidianLangTech-EACL2021: ensembling strategies for transformer-based offensive language detection. arXiv:2102.10084 (2021)
16. Kulkarni, A., Mandhane, M., Likhitkar, M., Kshirsagar, G., Jagdale, J., Joshi, R.: Experimental evaluation of deep learning models for Marathi text classification. arXiv:2101.04899 (2021)
17. Gaurav: Marathi News dataset (2020). https://www.kaggle.com/disisbig/marathi-news-dataset
18. Lan, Z., Chen, M., Goodman, S., Gimpel, K., Sharma, P., Soricut, R.: ALBERT: a lite BERT for self-supervised learning of language representations. arXiv:1909.11942 (2020)

Deep Learning Approaches for Automated Diagnosis of COVID-19 Using Imbalanced Training CXR Data

Ajay Sharma(✉) and Pramod Kumar Mishra

Department of Computer Science, Institute of Science, Banaras Hindu University, Varanasi 221005, India
{ajay.sharma17,mishra}@bhu.ac.in

Abstract. Due to the exponential rise of COVID-19 worldwide, it is important that the artificial intelligence community address to analyze CXR images for early classification of COVID-19 patients. Unfortunately, it is very difficult to collect data in such epidemic situations, which is essential for better training of deep convolutional neural networks. To address the limited dataset challenge, the author makes use of a deep transfer learning approach. The presence of limited number of COVID-19 samples may lead to biased learning due to class imbalance. To resolve class imbalance, we propose a new class weighted loss function that reduces biasness and improves COVID-19 sensitivity. Classification and preprocessing are two concrete components of this study. For classification, we compare five pre-trained deep neural networks architectures i.e. DenseNet169, InceptionResNetV2, MobileNet, Vgg19 and NASNetMobile as a baseline to achieve transfer learning. This study is conducted using two fused datasets where samples are collected from four heterogeneous data resources. Based on number of classes we make four different classification scenarios to compare five baseline architectures in two stages. These scenarios are COVID-19 vs non- COVID-19, COVID-19 vs Pneumonia vs Normal, COVID-19 pneumonia vs Viral vs Bacterial pneumonia vs Normal and COVID19 vs Normal vs Virus + Bacterial pneumonia. The primary goal of this study is to improve COVID-19 sensitivity. Experimental outcomes show that DenseNet169 achieves the highest accuracy and sensitivity for COVID-19 detection with score of 95.04% and 100% for 4-class classification and 99.17% and 100% for 3 class-classification.

Keywords: Deep learning · Chest X-ray · COVID-19 classification · Pneumonia classification · Image processing · Imbalanced data

1 Introduction

Coronavirus disease 2019 (COVID-19) [1], was declared a global pandemic by WHO just in less than four months when 3.3 million confirmed and 238,000 deaths were reported as of 2nd May 2020. COVID-19 disease was originated by SARSCoV-2, till February 7 2021, 105,394,301 confirmed cases and with 2,302,302 deaths were reported

I. Woungang et al. (Eds.): ANTIC 2021, CCIS 1534, pp. 453–472, 2022.
https://doi.org/10.1007/978-3-030-96040-7_36

globally to WHO [2]. Due to lack of sufficient treatment and its extremely contagious nature, it is essential to prevent its spreading and to develop different methodologies that can early classify findings of COVID-19. Worldwide, many scientists of medicine, clinical study, Artificial intelligence and others are trying to build up different approaches to easily categorize and to prevent the spreading of COVID-19 and any such pandemic in future. Initial testing indicates that reverse-transcriptase polymerase chain reaction (RT-PCR) has very low sensitivity for COVID-19 [3] but radiological examinations are found more useful in assessment and diagnosis of disease evolution. According to most of clinical studies worldwide, exposed that lung infection has been seen in COVID-19 patients [28]. For lung-related infection CT (chest-computed tomography) is utmost effective technique for COVID-19 diagnosis but it is expensive. Screening done by CT on COVID-19 patients showed more sensitivity [4] as compared to initial testing technique [3, 10] RT-PCR. But, due to unexpected surge in COVID-19 pervasiveness it becomes difficult to make routine use of CT because of portability issue and higher cost than CXR. Currently, CXR modality is considered a standard tool to detect infection attacks of COVID-19. In past CXR images are widely used to analyze the lungs to diagnose abscesses, pneumonia, tuberculosis [21], lung inflammation and enlarged lymph nodes [9]. Also, radiological examinations disclosed that because of similarity in viral pneumonia and COVID-19 [28], some patients of COVID-19 infection were identified as pneumonia. Thus, early detection and isolation of COVID-19 cases is significant to prevent it from [33] spreading.

In order to make more proficient future healthcare systems, more research should be carried out on medical images and radiological examinations to develop better diagnostic systems that assist doctors in early diagnosis. In regard of COVID-19, radiological examinations have undoubtedly shown certain abnormal patches in CXR [28] of patients having COVID-19 and pneumonia. Accordingly, numerous CNN models [6] have been explored actively for the classification of CXR [13–15, 50] and CT images [7, 12, 28] having COVID-19 symptoms. There exist several popular deep learning models that have been proven very useful for numerous radiological applications for classification of pneumonia and tuberculosis [16, 22, 27]. However, CNN models have successfully analyzed CT and video endoscopy [7] more efficiently compared to radiologists. Several studies [5, 8] have shown that deep learning models developed for image analysis tasks has surpassed performance label of radiologists. Initially, Wang et al. [7] proposed VGG16 model especially for classification of pneumonia in lungs. Following this, Rajpurkar et al. [8] developed a system based on deep CNN model named DenseNet201 to distinguish pneumonia among various pathologies using CXR images. CheXNet methodology was later proposed using 121 layered CNN architecture by making the largest ChestX-ray14 dataset [11] dataset holding more than 100,000 frontal-view X-ray images. Comparative study has successfully shown that proposed model gains superior label of performance compared to radiologists based on F1-score metric. Another CheXpert [9] model proposed for diagnosing 14 different pathologies from CXR radiographs outperforms the results shown by three radiologists [9, 10]. These studies motivate researchers to develop new models or use existing models for the diagnosis of COVID-19 infection in CXR radiographs.

The research contribution by Wang and Wong [28] proposed deep learning-based model named COVID-Net for diagnosis of COVID-19 in CXR with 80% sensitivity. Following this, numerous deep learning applications including AlexNet, Inception, GoogleNet, VGG16, MobileNetV2, VGG19, ResNet50, InceptionResNetV2, and many more have been actively explored for drawing successful [22] conclusions for COVID-19 from [52, 53] CXR images. But, due to limited availability of COVID-19 Chest X-ray samples as compared to other similar characteristics pathology like pneumonia may resulted in biased learning [29]. Although, motivated by success of these models in the early classification of COVID-19 cases by finding certain abnormalities in CXR images enforce us to explore further deep learning methodologies handle such epidemic data in future [50, 51]. Major contribution of this paper is to develop a CNN architecture appropriate for handling limited imbalanced training dataset based on better image processing methods.

The whole structure of this article is defined as follows: In Sect. 2, we discuss the existing contributions of research studies. In Sect. 3 we discuss dataset collection following this in Sect. 4 we discuss current methodology that has been used to develop reliable model using image preprocessing and weighted loss approach. In Sect. 5 we discuss results obtained by the proposed methodology and attention visualization using Grad-CAM. Finally, Sect. 6 ends with a conclusion.

2 Literature Review

With the advancement in artificial intelligence and deep learning, it has been widely applicable for various radiological examinations on CXR images for analyzing respiratory diseases in last decade. Inspired by early success for interpretation of radiological examinations successfully using deep learning, various research studies have been conducted for identification of COVID infected patients by CXR images [12–14]. Majority of these studies use similar COVID-19 dataset collected by Cohen et al. in combination with subset of various pneumonia datasets. For instance, the research study proposed in [15] author used three CNN-based models for detection of COVID-19, these baseline architectures are InceptionResNetV2, InceptionV3 and ResNet50. Experiments had been performed on dataset containing only 50 COVID-19 and 50 normal patients. Result highlights that ResNet50 achieves good accuracy of 98%. Another deep learning pre-trained architecture VGG16 [17] was used to categorize COVID-19 against control and pneumonia classes. The study was carried out on balanced dataset containing 132 COVID-19, 132 pneumonia and 132 controls images. This achieves 100% sensitivity for identifying COVID-19 whereas less for other two. In [18] by Abdullahi et al. employed a pretrained deep learning model for classification of COVID-19 pneumonia, against BP, VP and control CXR examples with different classification strategies. This study experimented AlexNet model on various classification strategies for comparison of COVID-19 class, these are normal vs COVID-19, normal vs bacterial pneumonia, normal vs viral pneumonia and bacterial pneumonia vs COVID-19 whereas in 3- way (normal vs COVID-19 vs. bacterial pneumonia) is compared and in 4-way classification all are considered a different label.

In another study [19], Xception is employed as baseline architecture for detection of COVID-19. This transfer learning approach make use of 500 normal, 127 COVID-19, and

500 pneumonia samples collected from different heterogeneous sources. The proposed model achieves an accuracy value of 97% for covid-19 detection. In [20], the author tested seven CNN architectures using imbalanced dataset containing 25 COVID infected and 50 normal patients. DenseNet121 and VGG19 pretrained models present best results with F1-score of 0.91 and 0.89 for COVID-19 and normal samples. In [23], four different CNN architectures (DenseNet-121, SqueezeNet, ResNet50 and ResNet18) were used to facilitate transfer learning. For the experimental study 5, 000 no-finding and pneumonia images and 184 COVID-19 samples has been collected. The study claims 98% recall and 93% specificity. In [24], five deep learning models (DenseNet169, ResNet, ResNet-v2, Inception-v3, Inception and NASNetLarge) selected as baseline. Resampling and entropy-based approach is used to handle data imbalance. NesNetLarge achieves an accuracy of 98% and 96% and recall of 90% and 91% in 2 and 3 class problem.

Another approach was employed in [25] to mitigate data imbalance by generating duplicate samples for better training. With the help of generative adversarial network (GAN) similar samples of CXR images were produced. This study uses dataset with a corpus of 1, 124 normal and 403 COVID-19 images. With augmentation author presents that accuracy improved from 85% to 95% using VGG16 network as backbone. Another study in [26] was conducted for diagnosing COVID-19 in binary and multiclass problem. The comparative study used similar dataset as in [19] but make comparison between covid-19, control and covid-19, pneumonia and control samples. Modified DarkNet model employed as baseline model with cross-validation ($k = 5$) achieves 98% and 87% accuracy score in 2-way and multi-way classification. In [30], GAN model has been employed for increasing the corpus of data with data augmentation. Dataset used to study contains 307 images having 4 different classes: COVID-19, normal, VP and BP. AlexNet, Restnet18, GoogleNet were the CNN architectures experimented for classification. GoogleNet achieved improved accuracy score of 99%.

In [1], found new way to train models based on patch-wise approach. Analogous, concept of transfer learning with ResNet18 has experimented as classification network model and to achieve segmentation FC-DenseNet is used. The study was carried out using a corpus of COVID-19 (180), BP (54) and VP (20), normal (191) and tuberculosis (57). This patch-wise study attained an accuracy score 89%. Specifically, wong and wang et al. introduced DL based COVID-Net [31] framework in early times. Experiments comprised on standard dataset comprising of 13975 CXR images, including classes pneumonia, normal and 266 COVID examples. COVID-Net model achieves better recall and accuracy score of 91% and 93.3% for COVID-19 cases as compared to VGG19 and ResNet50. Another COVID-19 detection model named COVID-AID [32] built on 121 layered DenseNet model investigated on covid-chestxray-dataset [37] for covid cases. It distinguishes COVID-19 infection cases with 100% sensitivity among VP, BP and control cases. In [33], another DeTrac (Decompose, Transfer and Compose) architecture based on deep and shallow transfer learning (VGG19, ResNet, AlexNet, GoogLeNet and squeezeNet) to classify COVID-19 (105) against normal (80) and SARS (11) cases. DeTrac achieves an accuracy of 93.1% for covid-19 detection.

3 Dataset Description

For this research study, we COVID-19 CXR dataset used in above studies but with few updated COVID-19 samples. In this study, we make two separate datasets COVIDx and COVIDy by utilizing data from different heterogeneous CXR datasets. COVIDx dataset makes by fusion of three popular publicly available datasets i.e. COVID-19 chest Xray-dataset [37], RSNA (Radiological Society of North America [34]) and USNLM (U.S. national library of medicine [36]). Covid chest X-ray dataset is a publicly available database of CXR images collected by Cohen et al. [37] related to MERS, ARDS, SARS, COVID-19 pneumonia, viral pneumonia etc. from various resources accessible at different public domains [37]. Another NIH CXR14 [36] dataset is used for collecting pneumonia and normal CXR images. For more robust training and testing another NLM(MC) dataset is used for collecting more normal images. The complete description of the number of images sampled for making fused dataset COVIDx is shown in Table 1. Further in the second stage, we prepare another dataset COVIDy where we use the same subset of samples from COVIDx frontal-view chest X-ray images for Covid-19 and normal but new chest-Xray-pneumonia dataset [35] is used for extracting images of CXR viral /bacterial pneumonia labeled differently for more robust model development. The aim for making this dataset is to identify how better developed deep learning model can identify Covid-19 pneumonia when bacterial pneumonia (BP) and viral pneumonia (VP) are included differently. Table 1 represents a detailed description about the datasets used for collecting CXR images, their respective class, number of samples present in a particular class, and total number of samples.

Table 1. Class summary and data resources related to fused datasets.

	Dataset used	Class	#	Total
COVIDx	Covid CXR [37]	COVID-19	193	**193**
		Pneumonia	38	**520**
	RSNA [34]	Pneumonia	482	
		Normal	443	**523**
	NLM(MC) [36]	Normal	80	
COVIDy	COVIDx	Covid-19	193	**193**
		Normal	173	**173**
	CXR Pneumonia [35]	VP	122	**122**
		BP	113	**113**

4 Proposed Contributions

In the following subsections, we are going to discuss methodology and proposed contributions for better identification of COVID-19 CXR samples. In the first subsection, we discuss basic initial preprocessing of an image and the class imbalance approach used. In Further subsections, the two-stage classification Process, classification architecture and evaluation metrics have been discussed.

4.1 Image Preprocessing

Computer-aided diagnosis of medical images needs identification of an essential region of interest whereas image preprocessing helps to reduce time and space complexity of an algorithm as well as error rate. Basic preprocessing steps involved are as follows:

CXR images for this study are collected from different heterogeneous data sources. These sources may represent heterogeneity in shape, size, datatype, scanning condition, acquisition condition, range and postprocessing etc. CXR images are normalized first to ensure mean value to zero and standard deviation to one [38]. Further, their data types from uint8/uint16 converted to uniform format float32. Also, images are reshaped to 320 * 320 * 3 so that the classification process can be achieved rapidly.

Histogram equalization (HE) is used to enhance contrast by adjusting intensity values. Initially, the images from RGB are converted to L*a*b* color space. L*a*b* model not only represents colors of CMYK and RGB but also expresses color alleged by human eye. Medical image analysis prefers L*a*b* because it compensates inequality in color distribution of the RGB color model. L* is luminance component that is taken into consideration by HE [39]. Conversion from RGB to L*a*b* directly not possible so it is processed as i.e., RGB-XYZ-L*a*b* where XYZ is another color space. This resulted preprocessed image helps classification network to focus on more relevant features present within lungs. Due to implementation of above algorithm, there may be a chance of uncertainty present at pixel level in filtered image [48]. So, the adaptive total variation filter method [49] is used for denoising of uncertainty present at pixel level. In Fig. 1 we have shown raw covid image, histogram Equalised and denoised image, normalized image with distribution of pixel intensities corresponding to the image.

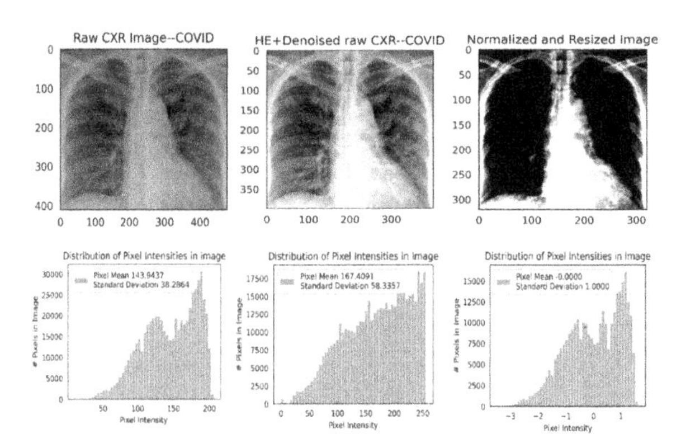

Fig. 1. Corresponding image and distribution of pixel intensities at different stages of preprocessing.

Let I represent grayscale image over given bounded set Ω subset of R^2 then denoised image d approximately matches an observed image y = (y1, y2)∈ Ω, is represented by Eq. (1) as:

$$d = \arg\min_{d}\left(\int_{\Omega}(d - I.\ln d)dy + \left(\int_{\Omega}(w(y)|\nabla d|dy\right)\right) \tag{1}$$

Where $w(y) = \frac{1}{1+K \, mod \, G_\sigma * \nabla d'}$, $GaussianKernel(G_\sigma)$ with variance(σ), contrast parameter (K > 0) and a convolution operator (*). After initial preprocessing, images are converted to grayscale for thresholding process [26]. In order to prevent symbolic and textual noise in images, binary thresholding is applied on initial preprocessed image by thresholding process using Eq. (2).

$$M(x, y) = \begin{cases} Max_{th}, & I(x, y) \geq Min\_th \\ 0, & otherwise \end{cases} \tag{2}$$

Where I (x, y) be input image, *Min_th* and *Max_th* are the minimum and maximum threshold [27] values. Also, morphological analysis of binary threshold image helps to correctly classify some of the dark misclassified [41] areas of target image that are below threshold label. Erosion operation is performed on the threshold image followed by a dilation operation that helps to generate better binary mask [43]. Figure 2 shows the results obtained at various stages of preprocessing related to COVID, normal, virus and bacteria image. First row represents original image, second row shows results after standardization and resizing of image whereas third row represents image resulted after denoising an image and final row shows image obtained by binary thresholding process.

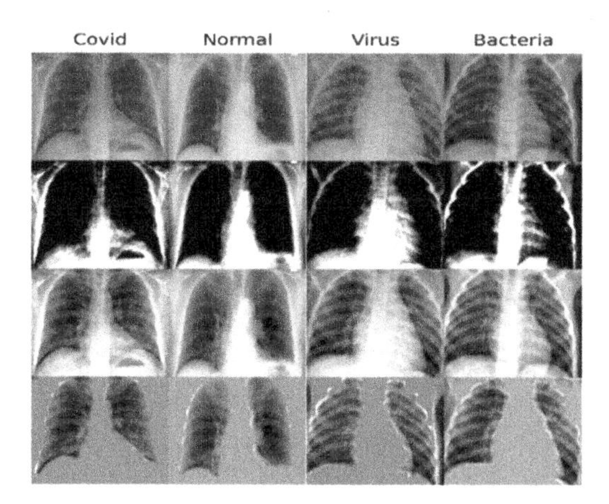

Fig. 2. Results of various stages of CXR preprocessing

The fused CXR datasets COVIDx and COVIDy contain a corpus of 146 training COVID samples with proportion to positive training cases of normal and pneumonia pathology. Similarly, COVIDy virus and bacteria samples are less in number than other two pathologies. Such imbalanced dataset does not ensure smooth learning of models [42]. So, we used weighted loss-based approach to handle class imbalance challenge. We assign new weights to each class to achieve equal contribution to loss for positive as well as negative cases for each class. For this we multiply each example in training set by a class specific weighted factor (Wpos and Wneg) corresponding to each class computed using Eq. (5) and (6), so that each class contributes equally to the loss computed using Eq. (6) and (7).

$$freq_{pos} = \frac{total\ positive\ examples}{N} \tag{3}$$

$$freq_{neg} = \frac{total\ negative\ examples}{N} \tag{4}$$

$$W_{pos} = freq_{neg}\ and\ W_{neg} = freq_{pos} \tag{5}$$

$$L(X, y_{covid}) = \begin{cases} -w_{pos,covid} logP(Y = 1|X) ifY = 1 \\ -w_{neg,covid} logP(Y = 0|X) ifY = 0 \end{cases} \tag{6}$$

$$L(X, y) = L(X, y_{covid}) + L(X, y_{normal}) + L(X, y_{bacteria}) + L(X, y_{virus}) \tag{7}$$

4.2 Classification Process

Based on the number of classes present in COVIDx and COVIDy CXR image datasets we make binary and multiclass classification strategies for COVID-19 classification as shown in Table 2 with features represented as classes, labels and dataset used in this study. Whereas, viral pneumonia (VP) and bacterial pneumonia (BP) are taken from CXR pneumonia dataset [35] are also combined as other pneumonia class to compare against COVID and normal cases.

Table 2. Summary of classification scenarios and corresponding labels

Classes	Labels	Dataset
COVID-19 vs non- COVID-19	{0, 1}	COVIDx, Binary
COVID-19 vs normal vs pneumonia	{0, 1, 2}	COVIDx, Multi-class
COVID-19 vs normal vs viral pneumonia vs bacterial pneumonia	{0, 1, 2, 3}	COVIDy Multi-class
COVID-19 vs normal vs other pneumonia	{0, 1, 2}	COVIDy Multi-class

All the classification strategies are managed using different one hot encoded data frame. Table 3 depicted clearly that how many CXR images are distributed into each training, testing and validation for different classification scenarios. Since, covid CXR dataset used for the proposed study does not previously hold any data for training and testing purpose. So, we randomly distribute CXR fused datasets into training, testing and validation set in proportion as shown in Table 3. Data splitting has been performed on patient level to ensure no data leakage between the training, validation and testing dataset.

Table 3. Distribution summary of data into training, testing and validation

Pathology	COVIDx dataset (stage 1)						COVIDy dataset (stage 2)					
	2-class classification			3-class classification			3-class classification			4-class classification		
	Train	Val	Test	Train	Val	Test	Train	Val	Test	Train	Val	Test
COVID -19	146	16	32	146	16	32	137	24	32	137	24	32
Pneumonia	852	76	114	426	37	56	–	–	–	–	–	–
Normal				426	39	58	145	27	41	145	27	41
VP	–	–	–	–	–	–	153	34	48	78	15	23
BP	–	–	–	–	–	–				75	19	25
Total	**998**	**92**	**146**	**998**	**92**	**146**	**435**	**84**	**121**	**435**	**84**	**121**

4.3 Classification Network Architecture

Our classification network aims to classify given image into one of pathology class by extracting features using state-of-the-art deep learning models. We are intended to prevent overfitting by using limited number of data points as CXR images. Deep learning models are data hungry require big volume of data. To work with limited training CXR examinations we adopted deep transfer learning-based approaches using pretrained ImageNet weights. Instead of using categorical cross-entropy loss [44] function we trained SOTA models with improved class weighted loss function to reduce biasness due to class imbalance. All these strategies support in making training process stable even when the size of dataset is small.

Regarding classification we choose five state-of-the-art DL architectures, DenseNet169, InceptionResNetV2, MobileNet, Vgg19 and NASNetMobile [46–48] as baseline models. In Fig. 3 we have shown the classification network architecture that classifies input samples taken from two datasets COVIDx and COVIDy based on different pathological characteristics. These baseline architectures use class weighted loss function to handle class imbalance over different classification scenarios. We added new layers at top of these baseline model. The output of these base models is taken as input followed by flatten layer, Dense layer having 512 neurons, Dropout layer with value of 0.5 to avoid overfitting accompanied with ReLU activation function. Also, L1 regularization is used in dense layer with low learning rate to prevent overfitting. Most popular

ReLU (Rectified linear unit) activation function used by internal layers whereas SoftMax is used by final dense layer as Eq. (12) and Eq. (13). In addition, finally, an output layer consists of 2, 3 or 4 neurons based on the number of classes with SoftMax activation function.

Once models training is over, these models classify input test image into one of pathology class based on maximum probability value given using softmax function. To know the interpretability of models we visualize attention maps by extracting gradients from last convolutional layer to interpret why that sample is classified to particular class. Models takes input processed in 3 channel and resized to (224, 224) before training various SOTA deep learning approaches. Minibatch gradient descent with batch size 32, optimization approach as Adam with initial learning rate 0.00001 and early stopping strategy with patience value 5 is used for training and testing. On early stopping, learning rate is reduced by 10. Also, model finetuning is performed for last 10 epochs with similar parameters by unfreezing top 20 layers. The entire process is accomplished in two stages. In stage 1, the classification network is trained and tested using COVIDx dataset to make one of these possible predictions based on maximum probability value i.e., a) normal, b) non-COVID19 and c) COVID-19 infection or simply a) covid-19 infection or non-covid-19. In stage 2, the classification network is trained and tested using COVIDy dataset. Here pneumonia cases are separated as viral and bacterial pneumonia for more robust and comparative analysis using SOTA approaches mentioned above. This entire work has been carried out using Tensorflow and Keras library via N-Vidia Titan GPU.

$$ReLu = \max(0, x) \tag{8}$$

$$Softmax = \frac{1}{1 + e^{-x}} \tag{9}$$

4.4 Performance Measures

Deep learning models has been evaluated using performance metrices accuracy, sensitivity (Recall), positive predictive value (PPV) and F1-score computed using Eq. (10–13) [16, 45]. COVID-19 sensitivity and COVID-19 PPV are also recorded to find best model for COVID-19 identification. For any input image I_X, class having maximum confidence score as output by softmax function is considered as final prediction for calculating confusion matrix.

$$\text{Accuracy} = \frac{(TN + TP)}{(TN + TP + FN + FP)} \tag{10}$$

$$\text{Sensitivity} = \frac{(TP)}{(TP + FN)} \times 100 \tag{11}$$

$$\text{Postive Prediction(PPV)} = \frac{(TP)}{(TP + FP)} \tag{12}$$

$$F1\,score = 2 \times \frac{PPV \times \text{Sensitivity}}{PPV + \text{Sensitivity}} \tag{13}$$

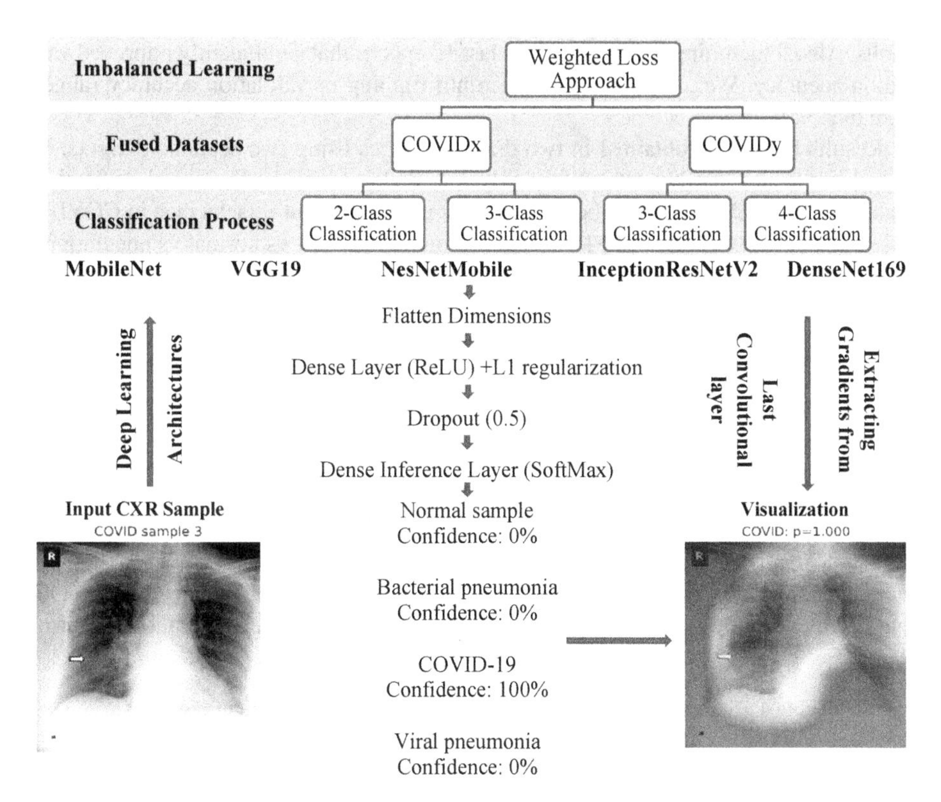

Fig. 3. Classification network architecture

5 Results and Discussion

In this section, we discuss the results achieved after experimentally analyzing CXR images using five transfer learning architectures as baseline models i.e., DenseNet169, InceptionResNetV2, MobileNet,Vgg19 and NASNetMobile. These models are quantitatively evaluated using sensitivity, Positive predictive value, accuracy, F1-score, COVID-19 sensitivity, and COVID-19 positive predictive value. Sample images used for COVID-19 are taken from similar dataset used in majority of studies, So COVID -19 sensitivity and COVID 19 PPV are also taken into consideration. Independent testing set is used for performance analysis whereas model development has been carried out using train and validation sets listed in Table 3. In Table 4 we summarized the best results achieved by CNN architectures on 4 different binary and multiclass classification strategies shown in Table 2. Classification strategy in Table 4 is represented as CX (m, n) and CY (m, n) Where CX and CY refers to COVIDx and COVIDy dataset and "m" is stage number and "n" is the number of classes considered among different classification strategies. All these experiments are performed using improved class weighted loss function. After extensive number of trials, we observed that no particular model performs well on all classification scenarios. But various significant observations have been drawn from these results. Also, Finetuning is performed for last 10 epochs that significantly improved validation accuracy. We aim to train models until training or validation accuracy ranges more than 95%.

Results have been obtained in two different stages using two separate datasets. In stage 1, we used COVIDx dataset for study with 3 classes resulted in two classification strategies. MobileNet achieves best overall accuracy score but it lacks race in COVID-19 sensitivity and COVID -19 PPV in classifying COVID-19 vs normal vs pneumonia. DenseNet169 claimed to be best for classification of COVID-19 with sensitivity score of 93.75% and PPV of 100%. Similarly, in stage 2 we used COVIDy dataset for more robust analysis of pneumonia into viral and bacterial pneumonia as in other studies. Here, also DenseNet169 achieves best sensitivity value of 100% in both 3 class and 4 class scenarios. Accuracy score is also considerably better than other SOTA models. Also, MobileNet achieves the same but less accurate in distinguishing viral and bacterial pneumonia. DenseNet169 also achieves best PPV value of 96.77% and 100% respectively. However, identification of viral and bacterial pneumonia separately helps more wisely in the identification of COVID-19 cases correctly. Also, VGG19 and NASNetMobile gives very low confidence score as output probability and DenseNet169 gives highest confidence score on given covid-19 sample as input. In Table 4 we have shown the results obtained by five models in stage 1 and stage 2 separately for each classification scenario.

Table 4. Summary of classification results obtained with different scenarios

Stage no.	Classification model	Classification process	Accuracy (%)	sensitivity (%)	Precision (%)	F1-score (%)	COVID19 sensitivity	COVID-19 (PPV) (%)
Results of Stage 1 over (COVIDx Dataset)	MobileNet	CX (1,3)	86.99	86.99	87.11	86.98	90.62	93.55
		CX (1,2)	95.89	95.89	96.10	95.73	81.25	100
	VGG19	CX (1,3)	76.71	76.71	77.58	76.34	56.25	85.71
		CX (1,2)	95.89	95.89	96.10	95.73	81.25	100
	NASNetMobile	CX (1,3)	78.77	78.77	79.89	78.70	65.62	91.30
		CX (1,2)	80.82	80.82	80.82	80.82	56.25	56.25
	InspectionResNetv2	CX (1,3)	78.77	78.77	80.50	78.89	68.75	100
		CX (1,2)	95.21	95.21	95.25	95.05	81.25	96.30
	DenseNet169	CX (1,3)	**89.04**	**89.04**	**89.25**	**89.12**	**93.75**	**100**
		CX (1,2)	**96.58**	**96.58**	**96.79**	**96.63**	**96.88**	**88.57**
Results of Stage 2 over (COVIDy Dataset)	MobileNet	CY (2,4)	90.91	90.91	90.82	90.00	100	96.97
		CY (2,3)	96.69	96.69	96.86	96.70	100	91.43
	VGG19	CY (2,4)	88.43	88.43	88.66	88.25	93.75	93.75
		CY (2,3)	95.04	95.04	95.18	95.01	87.50	96.55
	NASNetMobile	CY (2,4)	76.03	76.03	77.18	76.30	87.50	73.68
		CY (2,3)	89.26	89.26	89.21	89.10	78.12	89.29
	InspectionResNetv2	CY (2,4)	85.95	85.95	86.54	86.24	93.75	90.91
		CY (2,3)	95.04	95.04	95.15	95.09	93.75	88.24
	DenseNet169	CY (2,4)	**95.04**	**95.04**	**95.15**	**94.98**	**100**	**96.97**
		CY (2,3)	**99.17**	**99.17**	**99.19**	**99.17**	**100**	**100**

* *Bold represents the best model*

Furthermore, majority of studies have taken subset of pneumonia and normal samples from benchmark datasets but COVID-19 images are similar in majority of studies. Also, we need to distinguish COVID-19 sample more accurately. Results show that DenseNet169 is best in detecting COVID-19 cases. Outcome values achieved by proposed DenseNet169 shows significant improvements in the identification of COVID-19 cases among others. Likewise, in Fig. 5 we have shown that how training and validation accuracy of DenseNet169 varies with respect to number of epochs. Subplots a) and b) in first row showed graph of training and validation accuracy at y-axis with respect to epoch number at x-axis in 3-class and 2-class respectively in stage 1 using dataset COVIDx. Similarly, subplot c) and d) shows the comparison of accuracy graph in stage 2 using dataset COVIDy in 4-class and 3-class respectively. In initial epochs validation accuracy seems to be more than training accuracy this is because of strong regularization.

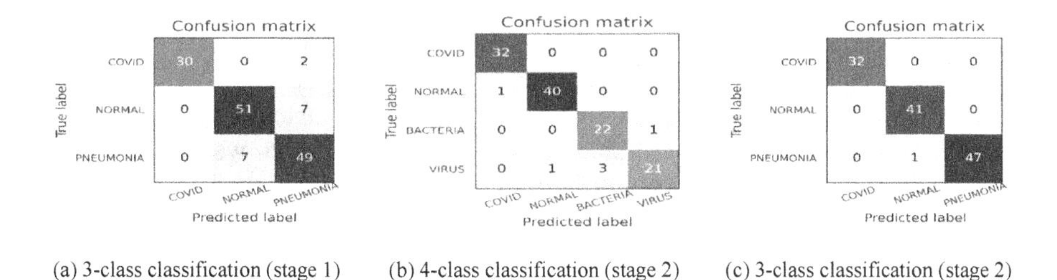

(a) 3-class classification (stage 1) (b) 4-class classification (stage 2) (c) 3-class classification (stage 2)

Fig. 4. Confusion matrices corresponding to 3-class and 4-class configurations by DenseNet169.

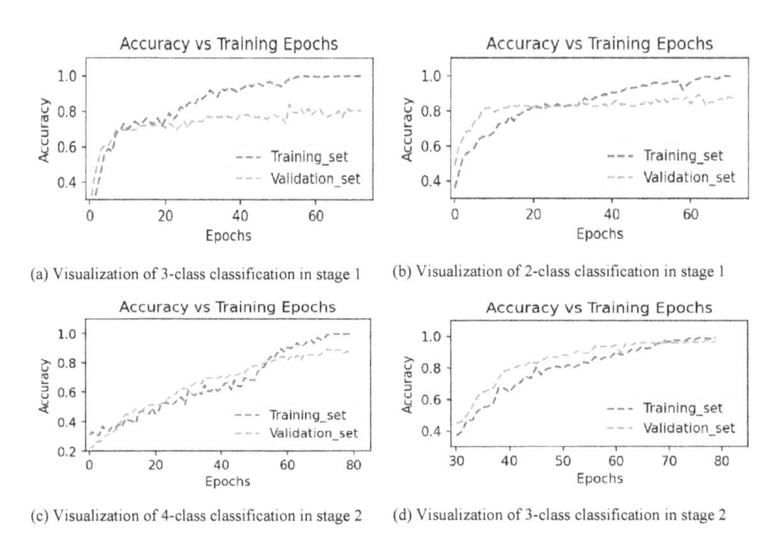

(a) Visualization of 3-class classification in stage 1 (b) Visualization of 2-class classification in stage 1

(c) Visualization of 4-class classification in stage 2 (d) Visualization of 3-class classification in stage 2

Fig. 5. Visualization of training and validation accuracy vs epoch number for DenseNet169

Based on the principle of class having maximum confidence score as a final predicted class, we have shown resulted confusion matrices for 3-class and 4-class classification scenario on test data in Fig. 4. From results, we can perceive that in stage 1 COVID-19 class gain sensitivity and PPV value of 93.75% and 100% is confused with pneumonia samples because of some overlapping characteristics. But further distinguishing of pneumonia into separate viral and bacterial pneumonia classes in stage 2 resolves the confusion with pneumonia sample. It achieves COVID-19 sensitivity of 100% in stage 2. Moreover, further observations sight that viral class gain sensitivity score of 84% and miserably confused with bacterial patients.

This study is supported by class weighted loss function to handle class imbalance. Further analysis of the proposed class weighted loss function while using proposed model has seen enormous improvements in results. In Table 5 we compared sensitivity values achieved by various classes using categorical cross-entropy loss and proposed weighted loss function on COVIDx dataset. Sensitivity value is observed after every 10 epochs as depicted in Table 5 under similar parameter value of learning rate. Due to a lesser number of COVID-19 samples in this dataset we observed clearly from table that categorical cross-entropy loss prioritizes learning of normal and pneumonia which is against our desired goal. But with weighted loss, improvements can be noticed clearly due to equal importance given to positive and negative samples of each class. We compared the results obtained by proposed DenseNet169 with various state-of-the-art proposed approaches as shown in Table 6, Table 7 and Table 8. In Table 6, we compared our proposed DenseNet169 with very first COVID CXR classification model COVID-Net [28] model that works for three classes as COVID-19, pneumonia and normal. In Table 7 further we compare the proposed model with a popular 4-class COVID-19 classification model named COVID-AID [33] based on DenseNet121 model with ChexNet weights. In Table 8 we compared the results of proposed model with other recently proposed state-of-the-art techniques. Majority of techniques are based on transfer learning and finetuning due to limited availability of COVID-19 CXR images. Table 8 contains features as the reference considered for comparison, best architecture, number of classes, technique used and performance measure in terms of accuracy, COVID-19 sensitivity and COVID-19 positive predictive value as per considerations taken in reference. Table 8 various state-of-the-art approaches are compared with proposed model.

Table 5. Sensitivity comparison with and without weighted loss over COVIDx dataset

Epochs	Classes	1–10	11–20	21–30	31–40	41–50	51–60	61–70	71–80
Categorical Cross Entropy Loss (%)	COVID-19	**6.45**	**12.90**	**16.13**	**19.35**	**58.06**	**61.29**	**64.52**	**74.19**
	Normal	65.52	75.86	81.03	82.76	81.03	82.76	79.31	84.48
	Pneumonia	64.29	69.64	71.43	73.21	75.00	78.57	78.57	75.00
Weighted loss approach (%)	COVID-19	**41.94**	**54.84**	**58.06**	**54.84**	**77.42**	**80.65**	**80.65**	**83.87**
	Normal	53.45	62.07	74.14	77.59	81.03	84.48	81.03	86.21
	Pneumonia	64.29	75.00	76.79	76.79	78.57	78.57	80.36	78.57

* *Bold represents difference in minority class sensitivity before and after data imbalance approach.*

Table 6. Class wise comparison of proposed model with COVID-Net

Pathology	COVID-Net		Proposed	
	PPV	Sensitivity	PPV	Sensitivity
COVID-19	98.9	91	**100**	**93.75**
Normal	90.5	95	87.93	87.93
Pneumonia	91.3	94	84.48	87.50

Table 7. Class wise comparison of proposed model with Covid-AID

Pathology		CovidAID		Proposed	
		PPV	Sensitivity	PPV	Sensitivity
4-class	COVID-19	93.8	100	**96.97**	**100**
	Normal	98.9	76.1	97.56	97.56
	VP	72.1	87.2	95.45	86
	BP	88.1	96.1	88	95.65
3-class	COVID-19	96.8	100	**100**	**100**
	Normal	98.9	74.4	97.62	100
	Pneumonia	86.8	99.5	100	97.92

Table 8. Comparison of proposed model with other state-of-the-art models

Ref.	Best architecture	Classes	Technique	Performance metrics (%)
[17]	VGG16	3	Fine tuning	$Se_{cov} = 100\%$
[19]	Xception	3	Fine tuning	$A_{cov} = 97\%$
[23]	ResNet50	3	Fine tuning	$Se_{cov} = 98\%$, $Sp_{cov} = 93\%$
[24]	NASNetLarge	3	Fine tuning	$A = 96\%$, $Se = 91\%$
[31]	COVID-Net	3	–	$A = 93.3\%$, $Se_{cov} = 91\%$
Proposed DenseNet169		**3**	**Fine tuning**	$\mathbf{A = 99.17\%}$, $\mathbf{Se_{cov} = 100\%}$ $\mathbf{Sp_{cov} = 100\%}$
[30]	GoogleNet	4	Fine tuning	$A = 99\%$
[1]	ResNet18	4	Fine tuning	$A = 89\%$, $Se_{cov} = 100\%$
[32]	DenseNet121(COVID-AID)	4	Fine tuning	$Se_{cov} = 100\%$
Proposed DenseNet169		**4**	**Fine tuning**	$\mathbf{A = 95.04\%}$, $\mathbf{Se_{cov} = 100\%}$ $\mathbf{PPV_{cov} = 96.97\%}$

* *Bold represents best model, Se represents sensitivity, A as Accuracy and Sp as specificity.*

5.1 Visualizing Learning with GradCAM

To demonstrate output qualitatively, we visualize attention maps focused by proposed model with Gradient weighted Class Activation Mapping (GradCAM). CAM (Class activation maps) aimed at increasing interpretability by looking at where model has focused while classifying given input image to a particular class. In Fig. 6, by taking given COVID-19 test input image we use the Grad-CAM technique to build heatmap showing significant regions of consideration for predicting certain pathological condition. We accomplish this by extracting gradients flowing into last convolutional layer of our proposed DenseNet169 corresponding each predicted class. This is more important to understand whether the interpretation of model focuses on right regions while classifying given input image. From row 1st to 3rd of Fig. 6, with given input positive COVID-19 sample, corresponding saliency map is shown next to the original image with correct predicted class and output probability at top of saliency map image. In saliency map, red signifies region of more importance. Whereas in row 4th and 5th indicate Grad-CAM heatmap on giving correctly classified COVID positive sample as input.

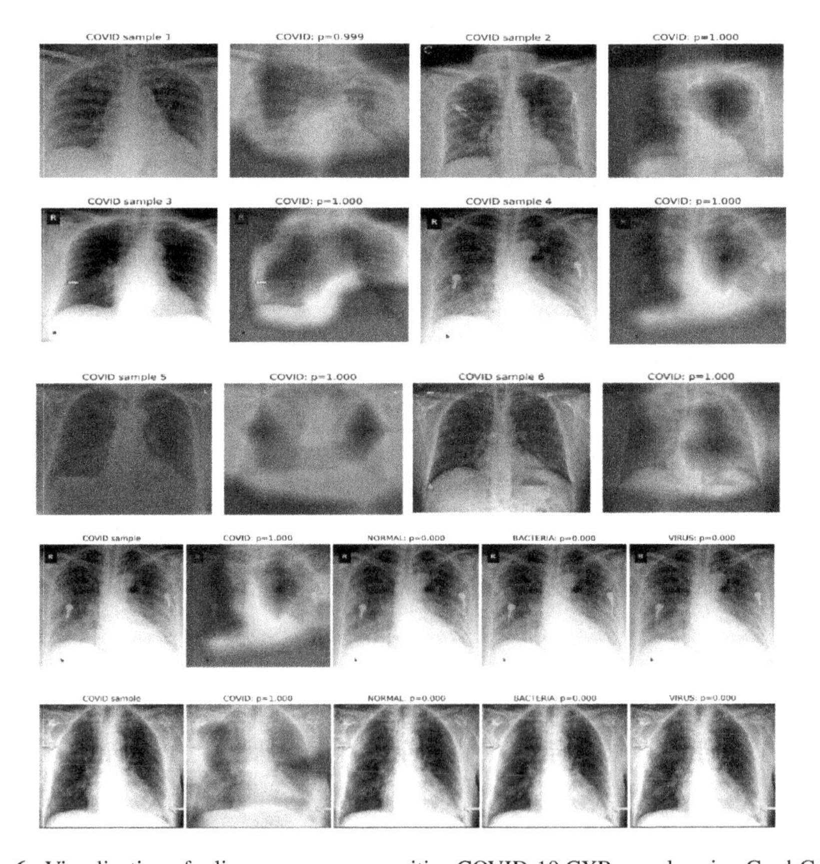

Fig. 6. Visualization of saliency maps over positive COVID-19 CXR sample using Grad-CAM

6 Conclusions

This article presents a comparative study of deep CNN models for automated detection of COVID-19 using limited CXR samples. Five different state-of-the-art transfer learning models are tested to aid the early classification of COVID-19. We observed that the emergency situation of COVID-19 completely shattered the healthcare sector worldwide due to its exponential behaviour and unidentified nature of the disease in its early stages. We believe that timely inference and correct diagnosis of type of disease can save millions of lives every year. Thus, computer-aided diagnosis significantly assists radiologists to capture more better images and real-time identification of COVID-19 or pneumonia just after the acquisition. The idea of early classification will open many application areas of CAD tools. It will be useful for the screening process at airports for early diagnosis of COVID-19, pneumonia and other types of disease. Now, it is the responsibility of artificial intelligence community to develop, explore and test new models to deal with current situation and any such epidemic situation in future. In this study, after number of trials we observed that DenseNet169 outperforms other four comparison models in terms of COVID-19 sensitivity in 3-class and 4-class classification. Also, we are in a view that proposed class weighted loss function performs better than categorical class entropy loss. Although, size of the dataset is small but outcomes look promising. Finally, in global perspectives after critically analyzing the related literature and the results obtained from DenseNet169, we can say that deep learning models has enormous potential in the classification of COVID-19 patients against others. Also, the series of DenseNet models have high sensitivity value and is likely to be more supportive in the early classification of COVID-19 cases among others.

References

1. Oh, Y., Park, S., Ye, J.C.: Deep learning COVID-19 features on CXR using limited training data sets. IEEE Trans. Med. Imaging **39**(8), 2688–2700 (2020)
2. WHO: Coronavirus disease 2019 (COVID-19) Dashboard (2021). https://covid19.who.int/. Accessed 7 Feb 2021
3. Waller, J.V., et al.: Diagnostic tools for coronavirus disease (COVID-19): comparing CT and RT-PCR viral nucleic acid testing. Am. J. Roentgenol. **215**(4), 834–838 (2020)
4. Khatami, F., et al.: A meta-analysis of accuracy and sensitivity of chest CT and RT-PCR in COVID-19 diagnosis. Sci. Rep. **10**(1), 1–12 (2020)
5. Wang, X., Peng, Y., Lu, L., Lu, Z., Bagheri, M., Summers, R.M.: Chestx-ray8: hospital-scale chest X-ray database and benchmarks on weakly-supervised classification and localization of common thorax diseases. In: Proceedings of the IEEE Conference on Computer Vision and Pattern Recognition, pp. 2097–2106 (2017)
6. Afshar, P., Heidarian, S., Naderkhani, F., Oikonomou, A., Plataniotis, K.N., Mohammadi, A.: COVID-CAPS: a capsule network-based framework for identification of COVID-19 cases from X-ray images. Pattern Recogn. Lett. **138**, 638–643 (2020)
7. Zreik, M., Van Hamersvelt, R.W., Wolterink, J.M., Leiner, T., Viergever, M.A., Išgum, I.: A recurrent CNN for automatic detection and classification of coronary artery plaque and stenosis in coronary CT angiography. IEEE Trans. Med. Imaging **38**(7), 1588–1598 (2018)
8. Rajpurkar, P., et al.: Deep learning for chest radiograph diagnosis: a retrospective comparison of the CheXNeXt algorithm to practicing radiologists. PLoS Med. **15**(11), e1002686 (2018)

9. Irvin, J., et al.: CheXpert: a large chest radiograph dataset with uncertainty labels and expert comparison. In: Proceedings of the AAAI Conference on Artificial Intelligence, vol. 33, no. 01, pp. 590–597, July 2019
10. Ketu, S., Mishra, P.K.: Enhanced Gaussian process regression-based forecasting model for COVID-19 outbreak and significance of IoT for its detection. Appl. Intell. **51**(3), 1492–1512 (2021). https://doi.org/10.1007/s10489-020-01889-9
11. Rajpurkar, P., et al.: CheXNet: radiologist-level pneumonia detection on chest X-rays with deep learning. arXiv preprint arXiv:1711.05225 (2017)
12. Perumal, V., Narayanan, V., Rajasekar, S.J.S.: Detection of COVID-19 using CXR and CT images using transfer learning and Haralick features. Appl. Intell. **51**(1), 341–358 (2020). https://doi.org/10.1007/s10489-020-01831-z
13. Chakraborty, M., Dhavale, S.V., Ingole, J.: Corona-Nidaan: lightweight deep convolutional neural network for chest X-Ray based COVID-19 infection detection. Appl. Intell. **51**(5), 3026–3043 (2021). https://doi.org/10.1007/s10489-020-01978-9
14. Sedik, A., Hammad, M., Abd El-Samie, F.E., Gupta, B.B., Abd El-Latif, A.A.: Efficient deep learning approach for augmented detection of Coronavirus disease. Neural Comput. Appl., 1–18 (2021)
15. Narin, A., Kaya, C., Pamuk, Z.: Automatic detection of coronavirus disease (COVID-19) using X-ray images and deep convolutional neural networks. arXiv preprint arXiv:2003.10849 (2020)
16. Sharma, A., Mishra, P.K.: Performance analysis of machine learning based optimized feature selection approaches for breast cancer diagnosis. Int. J. Inf. Technol., 1–12 (2021)
17. Civit-Masot, J., Luna-Perejón, F., Domínguez Morales, M., Civit, A.: Deep learning system for COVID-19 diagnosis aid using X-ray pulmonary images. Appl. Sci. **10**(13), 4640 (2020)
18. Ibrahim, A.U., Ozsoz, M., Serte, S., Al-Turjman, F., Yakoi, P.S.: Pneumonia classification using deep learning from chest X-ray images during COVID-19. Cognit. Comput., 1–13 (2021)
19. Das, N.N., Kumar, N., Kaur, M., Kumar, V., Singh, D.: Automated deep transfer learning-based approach for detection of COVID-19 infection in chest X-rays. Irbm (2020)
20. Hemdan, E.E.D., Shouman, M.A., Karar, M.E.: COVIDX-Net: a framework of deep learning classifiers to diagnose COVID-19 in X-ray images. arXiv preprint arXiv:2003.11055 (2020)
21. Ayaz, M., Shaukat, F., Raja, G.: Ensemble learning based automatic detection of tuberculosis in chest X-ray images using hybrid feature descriptors. Phys. Eng. Sci. Med. **44**(1), 183–194 (2021). https://doi.org/10.1007/s13246-020-00966-0
22. Srivastava, A., Mishra, P.K.: A survey on WSN issues with its heuristics and meta-heuristics solutions. Wireless Pers. Commun. **121**(1), 745–814 (2021). https://doi.org/10.1007/s11277-021-08659-x
23. Minaee, S., Kafieh, R., Sonka, M., Yazdani, S., Soufi, G.J.: Deep-COVID: predicting COVID-19 from chest X-ray images using deep transfer learning. Med. Image Anal. **65**, 101794 (2020)
24. Punn, N.S., Agarwal, S.: Automated diagnosis of COVID-19 with limited posteroanterior chest X-ray images using fine-tuned deep neural networks. Appl. Intell. **51**(5), 2689–2702 (2020). https://doi.org/10.1007/s10489-020-01900-3
25. Waheed, A., Goyal, M., Gupta, D., Khanna, A., Al-Turjman, F., Pinheiro, P.R.: CovidGAN: data augmentation using auxiliary classifier GAN for improved COVID-19 detection. IEEE Access **8**, 91916–91923 (2020)
26. Ozturk, T., Talo, M., Yildirim, E.A., Baloglu, U.B., Yildirim, O., Acharya, U.R.: Automated detection of COVID-19 cases using deep neural networks with X-ray images. Comput. Biol. Med. **121**, 103792 (2020)
27. Mishra, S.: Financial management and forecasting using business intelligence and big data analytic tools. Int. J. Financ. Eng. **5**(02), 1850011 (2018)

28. Zhou, T., Lu, H., Yang, Z., Qiu, S., Huo, B., Dong, Y.: The ensemble deep learning model for novel COVID-19 on CT images. Appl. Soft Comput. **98**, 106885 (2021)
29. Srivastava, A., Mishra, P.K.: State-of-the-art prototypes and future propensity stem on internet of things. Int. J. Recent Technol. Eng. (IJRTE) **8**(4), 2672– 2683 (2019). https://doi.org/10.35940/ijrte.D7291.118419
30. Loey, M., Smarandache, F., M Khalifa, N.E.: Within the lack of chest COVID-19 X-ray dataset: a novel detection model based on GAN and deep transfer learning. Symmetry **12**(4), 651 (2020)
31. Wang, L., Lin, Z.Q., Wong, A.: COVID-Net: a tailored deep convolutional neural network design for detection of COVID-19 cases from chest X-ray images. Sci. Rep. **10**(1), 1–12 (2020)
32. Mangal, A., et al.: CovidAID: COVID-19 detection using chest X-ray. arXiv preprint arXiv: 2004.09803 (2020)
33. Abbas, A., Abdelsamea, M.M., Gaber, M.M.: DeTrac: transfer learning of class decomposed medical images in convolutional neural networks. IEEE Access **8**, 74901–74913 (2020)
34. Stein, A.: Pneumonia dataset annotation methods. RSNA pneumonia detection challenge discussion, 2018 (2020). https://www.kaggle.com/c/rsna-pneumonia-detection-challenge/discussion/. Accessed 5 Dec 2020
35. Mooney, P.: Kaggle chest X-ray images (pneumonia) dataset (2018). https://www.kaggle.com/paultimothymooney/chest-xray-pneumonia. Accessed 5 Dec 2020
36. Jaeger, S., Candemir, S., Antani, S., Wáng, Y.X.J., Lu, P.X., Thoma, G.: Two public chest X-ray datasets for computer-aided screening of pulmonary diseases. Quant. Imaging Med. Surg. **4**(6), 475 (2014)
37. Cohen, J.P., Morrison, P., Dao, L.: COVID-19 image data collection. arXiv preprint arXiv: 2003.11597 (2020)
38. Mishra, S., Tripathi, A.R.: IoT platform business model for innovative management systems. Int. J. Financ. Eng. (IJFE) **7**(03), 1–31 (2020)
39. Abdullah-Al-Wadud, M., Kabir, M.H., Dewan, M.A.A., Chae, O.: A dynamic histogram equalization for image contrast enhancement. IEEE Trans. Consum. Electron. **53**(2), 593–600 (2007)
40. Mishra, S., Tripathi, A.R.: AI business model: an integrative business approach. J. Innov. Entrep. **10**(1), 1–21 (2021). https://doi.org/10.1186/s13731-021-00157-5
41. Mishra, S., Triptahi, A.R.: Platforms oriented business and data analytics in digital ecosystem. Int. J. Financ. Eng. **6**(04), 1950036 (2019)
42. Anand, A., Pugalenthi, G., Fogel, G.B., Suganthan, P.N.: An approach for classification of highly imbalanced data using weighting and undersampling. Amino Acids **39**(5), 1385–1391 (2010)
43. Mishra, S., Tripathi, A.R.: Literature review on business prototypes for digital platform. J. Innov. Entrep. **9**(1), 1–19 (2020). https://doi.org/10.1186/s13731-020-00126-4
44. Huan, E.-Y., Wen, G.-H.: Transfer learning with deep convolutional neural network for constitution classification with face image. Multimedia Tools Appl. **79**(17–18), 11905–11919 (2020). https://doi.org/10.1007/s11042-019-08376-5
45. Sharma, A., Mishra, P.K.: State-of-the-art in performance metrics and future directions for data science algorithms. J. Sci. Res. **64**(2) (2020)
46. Rajinikanth, V., Joseph Raj, A.N., Thanaraj, K.P., Naik, G.R.: A customized VGG19 network with concatenation of deep and handcrafted features for brain tumor detection. Appl. Sci. **10**(10), 3429 (2020)
47. Khan, S., Islam, N., Jan, Z., Din, I.U., Rodrigues, J.J.C.: A novel deep learning based framework for the detection and classification of breast cancer using transfer learning. Pattern Recogn. Lett. **125**, 1–6 (2019)

48. Chaturvedi, S.S., Tembhurne, J.V., Diwan, T.: A multi-class skin cancer classification using deep convolutional neural networks. Multimedia Tools Appl. **79**(39–40), 28477–28498 (2020). https://doi.org/10.1007/s11042-020-09388-2
49. Ketu, S., Mishra, P.K.: A hybrid deep learning model for COVID-19 prediction and current status of clinical trials worldwide. Comput. Mater. Continua **66**(2) (2020)
50. Arias-Londoño, J.D., Gomez-Garcia, J.A., Moro-Velázquez, L., Godino-Llorente, J.I.: Artificial Intelligence applied to chest X-ray images for the automatic detection of COVID-19. A thoughtful evaluation approach. IEEE Access (2020)
51. Chaurasia, B., Verma, A.: A comprehensive study on failure detectors of distributed systems. J. Sci. Res. **64**(2) (2020)
52. Mishra, S., Tripathi, A.R.: Platform business model on state-of-the-art business learning use case. Int. J. Finan. Eng. **7**(02), 2050015 (2020)
53. Mishra, S., Tripathi, A.R.: IoT platform business model for innovative management systems. Int. J. Finan. Eng. **7**(03), 2050030 (2020)

Load Forecasting for EV Charging Stations Based on Artificial Neural Network and Long Short Term Memory

Naval Kumar, Dinesh Kumar(✉), and Pragya Dwivedi

Department of Computer Science and Engineering, Motilal Nehru National Institute of Technology Allahabad, Prayagraj, India
{dinesh.kumar,pragyadwi86}@mnnit.ac.in

Abstract. Due to the increasing prices of nonrenewable resources and their fast depletion, Electric Vehicles (EVs) are proposed to be a viable and eco-friendly option for transportation. However, the increasing number of EVs poses a challenge for EV charging stations to fulfil their charging demands. EVs can manage and derive their charging strategies if the future charging load is known. This work addresses the EV load forecasting problem at a charging station and tries to predict charging requirements for the next hour, day and weekly basis through Artificial Neural Network (ANN) and Long Short-Term Memory Network (LSTM) approaches. In this work, the real dataset of the Adaptive Charging Network stationed at California Institute of Technology, USA is used. Experiments results show that LSTM proves to have more accuracy in terms of RMSE compared to ANN for all types of forecasting.

Keywords: Electric vehicles · Load forecasting · LSTM · ANN · EV charging

1 Introduction

Climate change has become a growing concern in recent years. As a result, the United Nations (UN) has included combating climate change as one of the Sustainable Development Goals (SDGs), with plans to raise $100 billion by 2020 to combat the crisis [19]. The transportation sector consumes more than a quarter of the world's energy [7]. According to the UN, two-thirds of the world's population is expected to live in cities by 2050, increasing the demand for vehicles to provide urban mobility and, as a result, increasing fossil fuel consumption and greenhouse gas emissions [18].

Due to the increasing prices of nonrenewable resources and their fast depletion, Electric Vehicles (EVs) are proposed to be a viable, eco-friendly, and viable option for transportation. This has resulted in the increasing demand for electric vehicles and hence their charging facilities. Consequently, the demand of electric vehicles and their charging facilities is increasing day by day. The process of charging an EV is different from fueling petrol/diesel in vehicles as it takes significant time although it depends upon the charging facility type to charge an EV. Therefore, there is always a need for a

I. Woungang et al. (Eds.): ANTIC 2021, CCIS 1534, pp. 473–485, 2022.
https://doi.org/10.1007/978-3-030-96040-7_37

coordination mechanism between the charging stations and EVs as there is a possibility of long queues at charging stations and large waiting times for EVs if the number of EVs significantly increases [13].

The rapid advancements in battery technology can be attributed to the massive increase in EV popularity in the context of smart cities. The latest EVs can travel between 300–500 Kms on a single charge, as opposed to older models, which would frequently travel less than 100 km on a single charge. As a result, the number of EV charging stations has increased, giving drivers more flexibility in planning their trips. Furthermore, EV reliability has improved significantly since the early days, resulting in increased consumer trust and satisfaction.

Despite the encouraging signs, there are a few obstacles to overcome. EV owners feel very inconvenience due to the long charging time which is significantly high as compared to fueling petrol/diesel. Furthermore, many EV owners rely only on public charging stations due to the unavailability of charging facilities at home. Because EVs require a lot of power, integrating them on a large scale will put a lot of strain on the grid [14]. Uncoordinated EV charging behavior is likely to exacerbate power distribution network degradation and instability [1]. As power generations resources are limited, the increasing number of charging stations is impractical due to power generation constraints. There are several other constraints such as inadequate physical space in urban areas [11]. Therefore, in order to increase EV user satisfaction, there is a need for a coordination mechanism that schedules EVs on a charging station. To provide charging services in a better way, demand forecasting plays a greater role in which a charging station predicts the estimated required energy in a given time duration and can plan/manage its charging resources accordingly [3].

In recent, emphasis has been given to data-driven approaches such as big data analytics and machine learning to solve the demand forecasting for an EV charging station [16]. An ML-enabled system learns from the historical data of charging load and EV owner behavior. Accurate predictions can be obtained after the training phase. Such predictions can then be used to improve EV charging scheduling strategies, predicting future demands, and finding the best locations for installing charging stations. These algorithms work very well if applied individually or in conjunction with other algorithms. ML algorithms have been shown in studies to be capable of providing good forecasting of EV charging demands, especially for time-series data, and can also be used to predict charging behavior [4].

We intend to come up with an idea of load forecasting at charging stations so that electric vehicle charging can be done smoothly and efficiently to the nearest charging station and the availability of energy at charging stations became easy. The main challenge is to predict the amount of energy/load charging stations require for the next hour/day/week. Forecasting can be short-term, mid-term, or long-term, depending upon the time duration for which forecasting is done. The three types of forecasting are described below [2].

1. Short-term load forecasting: It is done to predict load for the next few hours to few weeks. It has several applications in the power sector such as optimizing a power system by predicting load for the next hour or day, maintaining the stability of

the power system, etc. Short-term forecasting helps in minimizing the energy cost incurred during EV charging.
2. Medium-term load forecasting: It is done to predict load for a period extending from about three days to seven days in advance. It can be used for power system planning over a week or few weeks.
3. Long-term load forecasting: It is done to predict load for a period ranging from six months to five years. This forecasting helps in the decision making regarding infrastructure investment, charging policies, and charging pricing strategies

Predicting load at the charging station gives the precise estimate of charging demand which ultimately helps in deriving the optimal strategy for planning and operation. However, there are certain issues and challenges such as data security, processing cost, and data acquisition. Generally, a good forecasting model requires large data sets for forecasting.

Our work focuses on using Adaptive Charging Network data at Caltech and pre-processing the data to make the model ready which includes completion of missing data, splitting the data into train and test sets, setting the baseline for our work. We address the issue of EV load forecasting at a charging station using machine learning methods. We have considered the real data set and applied Artificial Neural Network (ANN) and Long Short Term Memory (LSTM) for forecasting the demand of energy at a particular station.

2 Related Work

In literature, some research works have used the Machine Learning and Deep Learning techniques to predict user behavior for EVs and electricity demand forecasting [3, 4, 10, 16]. The time series prediction methods have been used for predicting the future demand or requirements given its past history data. One of the well-known model among the time series prediction methods is the ARIMA model with Box-Jenkins approach. It is extensively used in research and real-world scenarios. However, choosing optimal parameter values in ARIMA is non-trivial and time-taking process. Recent advancement in hardware and software technologies bring a significant advancement in developing advanced machine learning algorithms such as deep learning, LSTM, reinforcement learning, Deep Q networks etc. The study in [17] shows that average reduction in error rates obtained by LSTM was between 84–87% when compared to ARIMA model. S. Hippert et al. [5] claimed that ANN is effective for load forecasting in terms of efficiency and accuracy. Recently, deep learning has been applied in image processing, Natural Language Processing, classification etc. Due to complex network structure and large number of hidden layers, Recurrent Neural Networks (RNN) have stronger and self-adaptive ability than ANN. But gradient vanishing problem is a big issue while using RNN. To remove this problem, LSTM was proposed. Zhu Juncheng et al. [20] applied the novel LSTM model to predict charging load of PEVs at two different time scale. In their experiments, lowest prediction error was observed in case of LSTM as compared to other models. Authors also applied deep learning approaches for super-short-term and minute-level short-term EV charging load forecasting. Deep learning has the advantages of using lesser number of features in the model training. However, it is also capable of capturing the potential load change features using only historical load data such as the nonlinear feature and temporal correlations. Authors applied different models like ANN, RNN, LSTM, Bi-LSTM in three different scenarios. They also investigated Fine-tuning and proper hyper-parameters for improving the performance. The experiment results shown in [20] conclude that LSTM models have higher accuracy in forecasting the short-term EV charging load.

3 Preliminaries

In this section, basic background details of ANN and LSTM are given.

3.1 Artificial Neural Network (ANN)

Artificial intelligence is the simulation of human neurological behavior, to enhance the speed, precision, and effectiveness of human efforts. The basis of artificial intelligence is artificial neurons. Several features like voice analysis, face recognition, forecasting of weather, and forecasting of future of transportation, electric vehicle charging, are available for use to human mankind [10]. The fundamental unit of an artificial neural network is a neuron. These artificial neurons are also called processing units. These processing units consist of input and output units, various forms of information are received through these input units, and the neural network learns patterns from this information and produces an output. There are various concepts of weightage of input, bias, outliers, to increase the efficiency of neural networks. Before passing through activation functions, inputs are given some weightage. There are several types of artificial neural networks. Some of them are Modular Neural Networks, Feedforward Neural Networks, Recurrent Neural Networks, Convolutional Neural Networks, Long/Short Term Memory [12]. Figure 1 shows the structure of an artificial neuron.

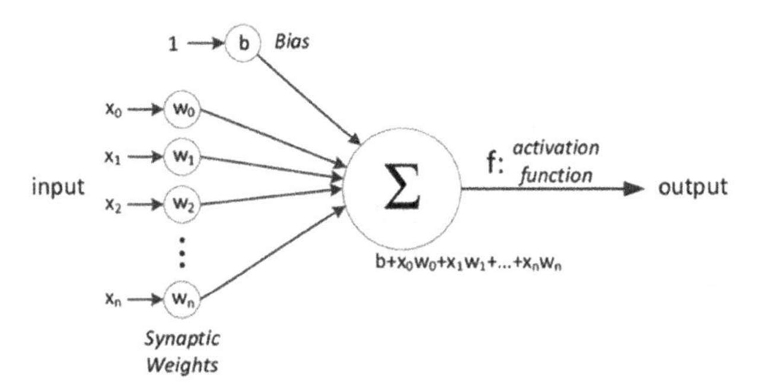

Fig. 1. Structure of an artificial neuron [12]

Layers of neurons are interconnected in a neural network, Number of neurons, type of neurons, activation functions of these neurons vary for different uses. Layers other than first and last are called hidden layers, Activation functions are mathematical functions.

Some of them are sigmoid, relu, softmax, softsign etc. [12]. Figure 2 shows the structure of an artificial neural network.

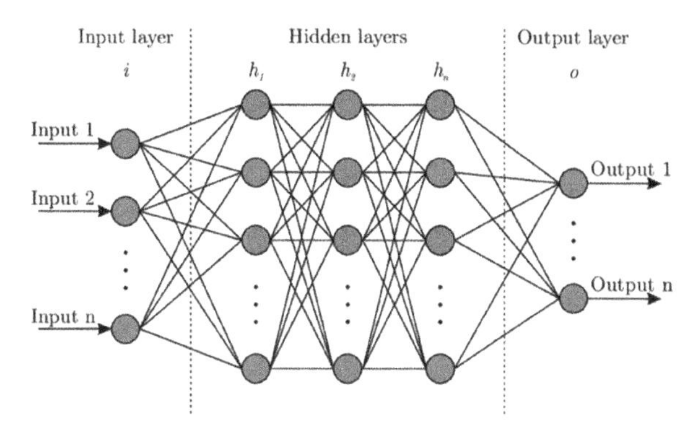

Fig. 2. Structure of an artificial neural network [12]

3.2 Long Short Term Memory (LSTM)

Long Short Term Memory was introduced by Schmidhuber Hochreiter in 1997 [6]. It is a special kind of Recurrent Neural Network and can handle long-term dependencies. When output has a long-term dependency of input points, then LSTM is required. Standard ANN will not remember this dependency. Long-term dependency comes into the picture when working with time-varied data, also in Natural Language Processing. Electric vehicle data is time-varied, with the dataset carrying starting and ending point of charging

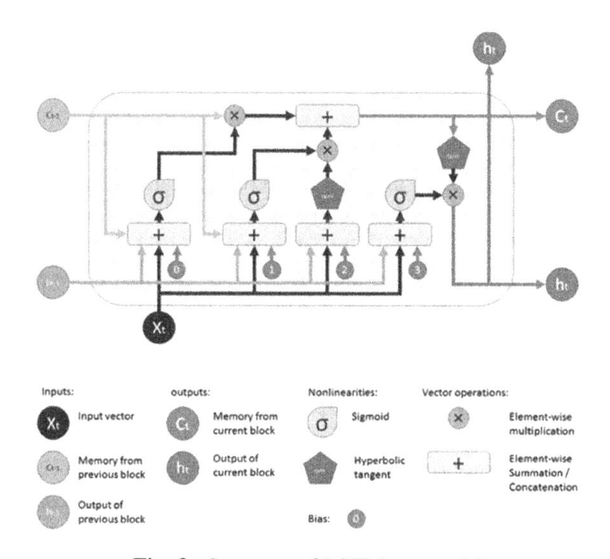

Fig. 3. Structure of LSTM neuron [6]

and the total power delivered to the vehicle [9]. Figure 3 shows the structure of an LSTM neuron.

3.3 Python Libraries

In this work, Keras, Pandas, Numpy, Scikit-learn and Matplotlib python libraries have been used for implementing our work [15].

4 Data Analysis and Forecasting Process

In this section, the Data analysis and forecasting process is described. Figure 4 shows the complete flow of the activities carried out in our proposed work.

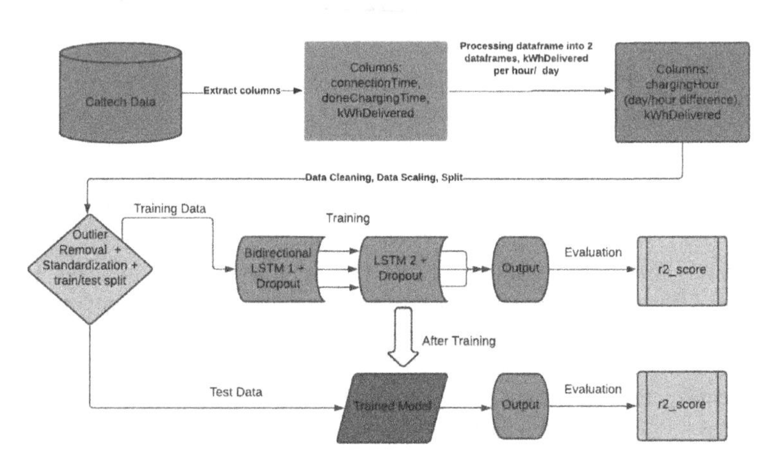

Fig. 4. Complete flow of the model

4.1 Introduction to the Dataset

We have used the Adaptive Charging Network (ACN) Dataset for training and testing [9]. ACN is an EV charging framework proposed in Caltech university that supports large-scale EV charging. This ACN network consists of 54 Electric Vehicles Supply Equipments (EVSEs) or charging stations. A 50 KW DC fast charger is used for EV charging. This ACN charging facility owned by PowerFlex is open to the public and is usually used by Non-Caltech persons also for charging their vehicles. As different users arrive and charge their vehicles, the dataset generated by ACN depicts irregular patterns.

The dataset is publicly available at https://ev.caltech.edu/dataset. This dataset consists of 30000 charging sessions and more sessions added daily. The data in this dataset is collected from two charging facilities located at different workplaces in California. These charging facilities are owned by PowerFlex, a startup based in California. This dataset is the most recent dataset related to EV charging and it is continually updated by adding sessions daily. Figure 5 shows the sample rows of the ACN data.

	connectionTime	doneChargingTime	kWhDelivered
3082	2020-02-29 19:52:17+00:00	2020-03-01 00:21:41+00:00	15.569
3083	2020-02-29 20:09:02+00:00	2020-02-29 21:31:04+00:00	7.942
3084	2020-02-29 21:38:42+00:00	2020-02-29 22:09:33+00:00	0.856
3085	2020-02-29 23:11:35+00:00	2020-03-01 01:15:51+00:00	6.790
3086	2020-02-29 23:35:14+00:00	2020-03-01 02:16:48+00:00	10.080

Fig. 5. Sample rows of caltech dataset [9]

Figure 6, 7 and 8 show the energy delivered Hour wise, Day wise and Week wise.

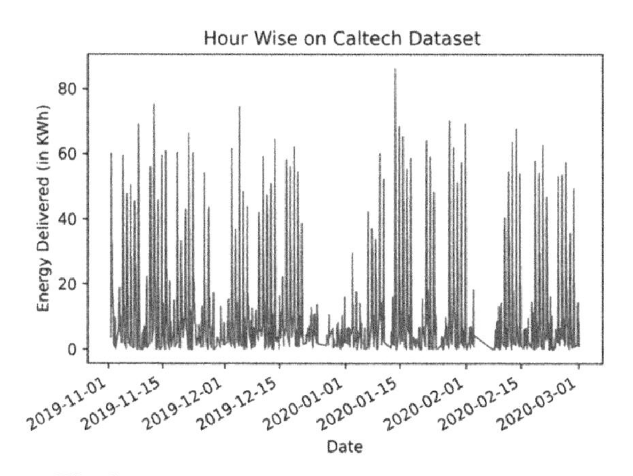

Fig. 6. Energy delivered vs date on hourly basis [9]

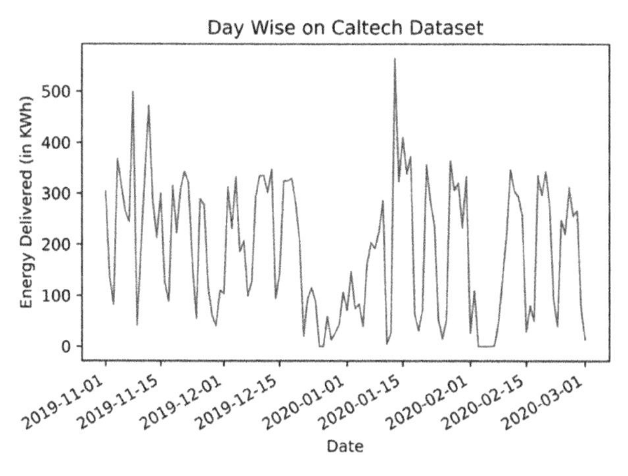

Fig. 7. Energy delivered day wise on caltech dataset [9]

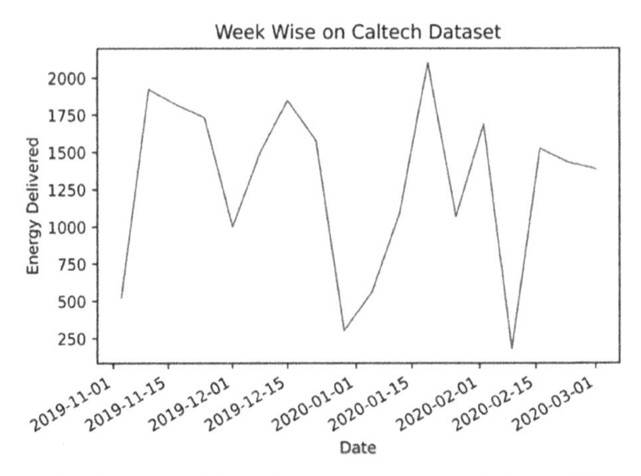

Fig. 8. Energy delivered week wise on caltech dataset [9]

4.2 Data Preprocessing

Raw data in a dataset must be converted to useful data for training and testing purposes. Three different models are used for day-wise forecasting, hour-wise, and week-wise forecasting. Then total power delivered to the vehicle is averaged per hour as well as per day. Removing outliers from the dataset is also essential. Outliers are sudden peaks and drops in power delivered. The whole dataset is divided into training and testing data in 80:20 ratios. The window size is also varied for both days wise and hour wise. Scikit learn is also used to normalize data.

1. **Removing Outliners**
   Removing outliers from a dataset is essential. To identify outliers in a dataset, boxplot is used. In boxplot, both the values below the lower bound and values above the upper bound are treated as outliers. In the next step, all such outliers' samples are collected and replaced. The replacement policy is based on the mean values of all outliers. All the outliers' values are replaced with the mean of all outliers' samples.
2. **Resampling of Data**
   Data of approximately 6 months are taken and resampled by converting in 1-hour multi-scale data every one hour for hour wise, 1-day multi-scale data every 1 day for day wise and similarly for week wise. The whole dataset is divided into training and testing data in 80:20 ratios.
3. **Normalized processing**
   Data normalization is also necessary in order to remove the influence of the data dimension on the final result. For normalization, maximum and minimum normalization showing in Eq. (1) below.

$$y = \frac{x - x_{min}}{x_{max} - x_{min}} \tag{1}$$

5 Training Model

Three separate models for both hours-wise, day-wise, and week-wise distribution are prepared. To predict current energy, we have taken the previous 7 days as input for day-wise forecasting and 24 h for hourly forecasting, and similarly 6 days for weekly forecasting. One Bidirectional LSTM layer and one unidirectional LSTM layer is used with different number of nodes. Some dropout is also used to avoid overfitting in the model. The input nodes and hidden nodes have been found using the Hit and Trial approach. Epoch values are also determined using hit and trial and observing error in the model.

The configurations and hyperparameter tuning details are as follows.

1. Number of hidden layers: There is no restriction on the number of layers to use so it can be used by hit and trial basis on individual problems. Generally, one hidden layer can be used for simple problem and two for complex. In our case we have used two layers.
2. Number of units in a dense layer and dropout: Number of units has also been decided using hit and trial by subsequently increasing number of units from 10–50 and 20% of dropout is used because it is best value to compromise between preventing model overfitting and retaining model accuracy.
3. Number of epochs and batch size: Epoch value represents the number of iterations to be run on the dataset. One of the most favorable way is to set a large epoch value. Whereas Batch size tell number of samples to work on before internal parameters of the model are updated. Generally, a good value is near to 32 or we can check with its multiples also.

6 Experimental Results and Discussion

We have trained and tested our model by doing simulations on a System having Intel i7 with 3.0 GHz processor, 16 GB RAM, and GeForce NVidia GTX-1650 GPU. All experiments are coded using Python in the Keras library. TensorFlow is used as a backend in our experiments. The source code of our proposed work is available at [8].

6.1 Model Evaluation

We have used the Root Mean Square Error (RMSE), one of the widely used metrics employed in EV charging load forecasting models, for measuring the effectiveness of our proposed work. RMSE is calculated using Eq. 2 given below.

$$RMSE = \sqrt{\frac{1}{N}\sum\nolimits_{i=1}^{n}\left(\hat{y}_i - y_i\right)^2} \tag{2}$$

6.2 Experimental Results

For the exhaustive evaluation, the whole study is done in different settings.

Figure 9 shows the comparison of ANN and LSTM on the prediction of energy required vs time on an hourly basis. According to the graph, both ANN and LSTM seem to follow the curve of the actual value of energy required because the interval of data nearly matches the hour interval but LSTM seems to have more accuracy compared to ANN.

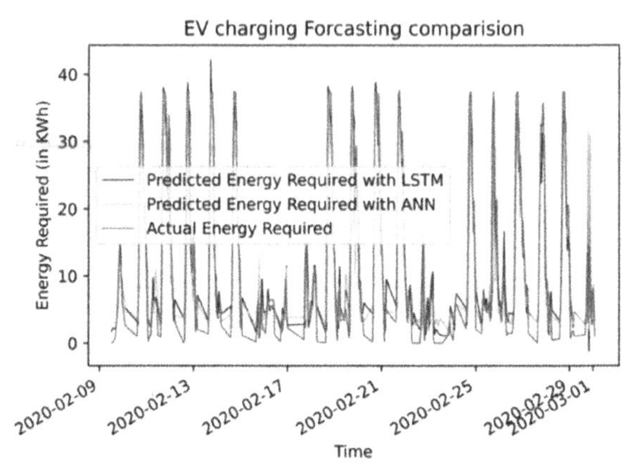

Fig. 9. Comparison of ANN and LSTM on prediction of energy required vs time on an hourly basis

Figure 10 shows the comparison of ANN and LSTM on the prediction of energy required vs time on daily basis. From Fig. 10, it can be seen that accuracy is reduced as compared to the Hourly-basis model but the model is following the curve and here also accuracy is more in LSTM as compared to ANN.

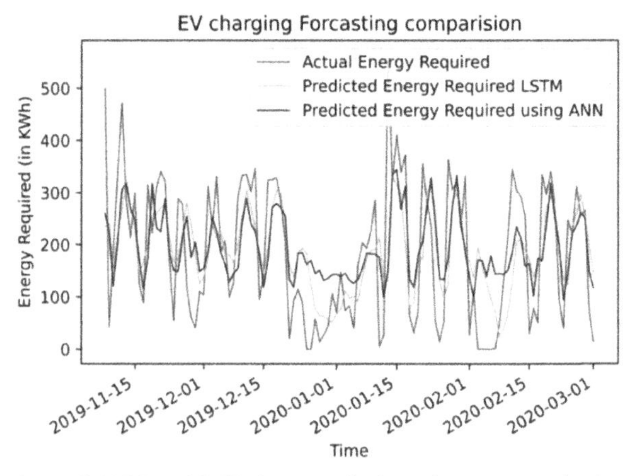

Fig. 10. Comparison of ANN and LSTM on prediction of energy required vs time on a daily basis.

Figure 11 shows the comparison of ANN and LSTM on the prediction of energy required vs time on a weekly basis. From Fig. 11, it can be noted that it doesn't follow the curve as expected because of training done through a small dataset as well as some data were missing as well but LSTM somewhat seems to follow the curve of actual energy required i.e. it achieves higher accuracy.

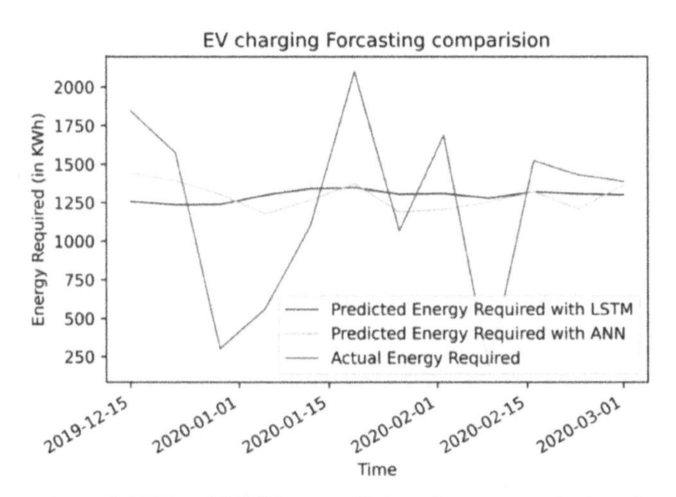

Fig. 11. Comparison of ANN and LSTM on prediction of energy required vs time on a weekly basis.

We trained RNN and ANN model with 80:20 training testing division and obtained the following accuracy on our dataset for Hourly, Daily Basis, and Weekly Basis. Table 1 shows the RMSE values for both ANN and LSTM based approaches.

Table 1. Accuracy of ANN and LSTM on various model in terms of RMSE values

Model	ANN	LSTM
Hour wise	0.7934518403188806	0.814473714353031
Day wise	0.299676795751963	0.356936411557772
Week wise	–0.106029677767993	0.0103159944429557

7 Conclusion

In this work, we addressed the EV load forecasting problem at a charging station and implemented the ANN and LSTM based approaches to solve it. Our work focuses on using Adaptive Charging Network data at Caltech and pre-processing the data to make it ready for analysis, including completing missing data and splitting the data into train and test sets. From experiments, it can be seen that LSTM proves to have more accuracy than ANN for all three models. For future work, other EV charging data can be explored. Also, there is a need to explore more attributes while making forecasting decisions.

References

1. Al-Awami, A.T., Sortomme, E.: Coordinating vehicle-to-grid services with energy trading. IEEE Trans. Smart Grid **3**(1), 453–462 (2012). https://doi.org/10.1109/TSG.2011.2167992
2. Al-Ogaili, A.S., et al.: Review on scheduling, clustering, and forecasting strategies for controlling electric vehicle charging: challenges and recommendations. IEEE Access **7**, 128353–128371 (2019). https://doi.org/10.1109/ACCESS.2019.2939595
3. Chiş, A., et al.: Reinforcement learning-based plug-in electric vehicle charging with forecasted price (2017). https://doi.org/10.1109/TVT.2016.2603536
4. Debnath, K.B., Mourshed, M.: Forecasting methods in energy planning models (2018). https://doi.org/10.1016/j.rser.2018.02.002
5. Hippert, H.S., et al.: Neural networks for short-term load forecasting: a review and evaluation. IEEE Trans. Power Syst. **16**, 1 (2001). https://doi.org/10.1109/59.910780
6. Hochreiter, S., Schmidhuber, J.: Long short-term memory. Neural Comput. **9**(8), 1735–1780 (1997). https://doi.org/10.1162/neco.1997.9.8.1735
7. IEA: Key World Energy Statistics 2020. https://www.iea.org/reports/key-world-energy-statistics-2020. Accessed 28 Aug 2021
8. Kumar, N.: EV Load Forecasting (2021). https://colab.research.google.com/drive/15Lx1puwm23RJoxsqEiM0tmKO8ITyLUtX
9. Lee, Z.J., et al.: ACN-data: analysis and applications of an open EV charging dataset. In: e-Energy 2019 - Proceedings of the 10th ACM International Conference on Future Energy Systems, pp. 139–149 (2019). https://doi.org/10.1145/3307772.3328313
10. Lopez-Garcia, T.B., et al.: Artificial neural networks in microgrids: a review. Eng. Appl. Artif. Intell. **95** (2020). https://doi.org/10.1016/j.engappai.2020.103894
11. Majhi, R.C., et al.: A systematic review of charging infrastructure location problem for electric vehicles. Transp. Rev. **41**(4), 432–455 (2021). https://doi.org/10.1080/01441647.2020.1854365
12. Majumder, M.: Artificial neural network. In: Interdisciplinary Computing in Java Programming, pp. 49–54 Springer, Boston (2015). https://doi.org/10.1007/978-981-4560-73-3_3
13. Mukherjee, J.C., Gupta, A.: A review of charge scheduling of electric vehicles in smart grid. IEEE Syst. J. **9**(4), 1541–1553 (2015). https://doi.org/10.1109/JSYST.2014.2356559
14. Patil, H., Kalkhambkar, V.N.: Grid Integration of electric vehicles for economic benefits: a review (2021). https://doi.org/10.35833/MPCE.2019.000326
15. Python Software Foundation: Welcome to Python.org. https://www.python.org/about/. Accessed 28 Aug 2021
16. Shahriar, S., et al.: Machine learning approaches for EV charging behavior: a review (2020). https://doi.org/10.1109/ACCESS.2020.3023388
17. Siami-Namini, S., et al.: A comparison of ARIMA and LSTM in forecasting time series. In: Proceedings - 17th IEEE International Conference on Machine Learning and Applications, ICMLA 2018 (2019). https://doi.org/10.1109/ICMLA.2018.00227
18. United Nations: 68% of the world population projected to live in urban areas by 2050, says UN | UN DESA | United Nations Department of Economic and Social Affairs (2018). https://www.un.org/development/desa/en/news/population/2018-revision-of-world-urbanization-prospects.html

19. United Nations General Assembly: Global indicator framework for the Sustainable Development Goals and targets of the 2030 Agenda for Sustainable Development. Work Stat. Comm. Pertain. to 2030 Agenda Sustain. Dev., pp. 1–21 (2020)
20. Zhu, J., et al.: Electric vehicle charging load forecasting: a comparative study of deep learning approaches. Energies **12**, 14 (2019). https://doi.org/10.3390/en12142692

Analysis of Long Term Trends of Rainfall in Jaipur

Vikas Bajpai(✉) and Anukriti Bansal

The LNM Institute of Information Technology, Jaipur, Rajasthan, India

Abstract. The importance of rainfall cannot be ignored. Fluctuations in rainfall can majorly impact people and especially the economy of a country that is majorly dependent on agriculture. Availability of water has a direct relation with the rainfall received over that region. This paper is on the trend analysis of rainfall in Jaipur, the capital city of the Indian state of Rajasthan. Temporal variation for monthly, seasonal (pre-monsoon, monsoon, and non-monsoon period), and annual rainfall trend analysis is done for the period of 61 years, ranging from 1957 to 2017. The major contribution of this paper is to identify whether there is an uptrend or a downtrend in the amount of precipitation received in the area under study. Daily rainfall data is collected from eight rain gauge stations installed by Water Resource Department, Govt of Rajasthan, fixed at various locations in the Jaipur district. It can be observed from the analysis that major rainfall occurs during the monsoon season (June-July-August-September).

Keywords: Trend analysis · Rainfall · Jaipur · Monsoon

1 Introduction

Rainfall is important for livelihood in each and every aspect. Observing the past and present trends of rainfall and its' analysis is equally important for taking decisions related to crops management [2], water distribution and management [1], agriculture planning [10], animal husbandry, flood management and control [5], disaster management [8], infrastructural decisions [21] etc. Research on rainfall analysis using meteorological data is going on since several decades. The present study focuses on analyzing rainfall data in Jaipur, the capital city of the state of Rajasthan, India for the duration of 1957 to 2017. The paper also identifies the extreme rainfall events[1], where the amount of rainfall observed is more than 244.5 mm.

[1] https://imdpune.gov.in/Weather/Reports/forecaster_guide.pdf.

I. Woungang et al. (Eds.): ANTIC 2021, CCIS 1534, pp. 486–501, 2022.
https://doi.org/10.1007/978-3-030-96040-7_38

Rainfall analysis for Jaipur, Rajasthan also becomes important as it helps in analyzing the water level of various lakes like Mansagar, Ramgarh, Chandlai, Ramgarh lakes present in and around the city. Long term trend analysis of rainfall is important for seasonal crops planning and specially for the ones where either very less or more rain is required. Kharif crop in this region mainly depends on rainfall and it's frequency. It is observed that with increasing temperature, the extreme rainfall events are increasing over India [7]. This kind of variation is not only observed in India but in several parts of globe as well [3,17,18,22].

In past some amount of rainfall trend analysis is done for some of the cities of Rajasthan like Chakshu, Udaipur, Kota and Jodhpur [12,13,20]

Ganesan et al. [6] made an attempt to find the trend of heavy rainfall in Jaipur for the period of 1901 to 1970 where the researchers identified that the majority of heavy rainfall days were observed in the month of July and August.

Section 2 describes the available data and the area under study. Various statistical approaches used for analysis are described in Sect. 3. Results and analysis of results are discussed in Sect. 4. Finally concluding remarks are made in Sect. 5.

2 Study Area and Data Availability

Pink city Jaipur, the capital city of Indian state of Rajasthan is located at 26° 55' north latitude and 75° 49' east longitude. It's boundary extends from 26° 46' north latitude to 27° 01' north latitude and 75° 37' east longitude to 76° 57' east longitude and covers an area of 484.6 square kilometres. The geographical coordinates and the location of Jaipur city on the map of Rajasthan is show in Fig. 1. Majorly this are is an urban populated area with a minute area of agricultural land. What makes taking Jaipur city for this study is the area of city and it's location in a state where we can observe plain area, hill ranges and desert. Largely Rajasthan comes under arid to semi- arid area and hence analysing the rainfall trend becomes crucial. Major supply of water for the entire city is the Bisalpur dam on Banas river. The refilling of this dam largely depends on monsoon rainfall only. Tourism is major revenue generation source not only in Jaipur but for entire Rajasthan and hence importance of this trend analysis increases. The historical time-series data of daily rainfall from eight rain gauge stations (catchments) for a period of 61 years was obtained from Special Project Monitoring Unit (SPMU), Hydrology Department, Water Resources, Government of Rajasthan. These rain gauge stations are installed by water resources, revenue department Rajasthan and Indian Meteorology Department (IMD). Daily rainfall values of all the catchments were summed to get monthly, seasonal and annual values. These summed up values were further analysed for trends and characterizing the rainfall as discussed in the subsequent sections.

3 Methodologies

While selecting the methodologies or tests for our trend analysis, we assumed to have no hypotheses and wanted to investigate, what is hidden in the data.

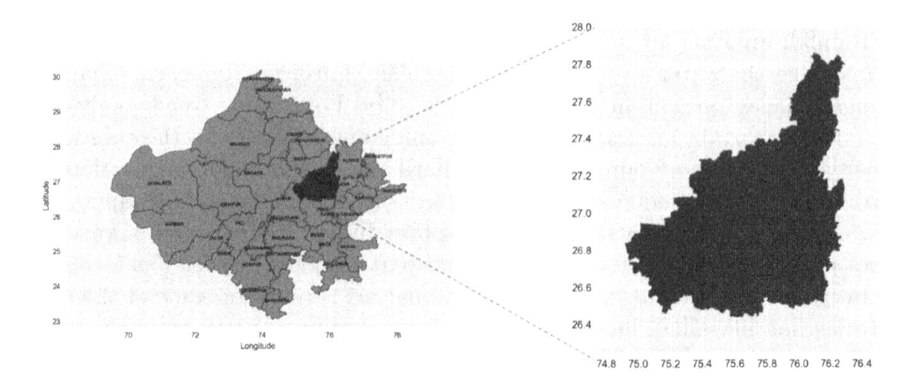

Fig. 1. Map and geographical coordinates of study area Jaipur

Considering this we applied the following tests for this study For non-parametric analysis, Mann-Kendall test was conducted on monthly rainfall values. Linear regression trend lines were investigated for Parametric tests. Apart from these parametric and non-parametric analysis, some other indices like average rainfall, normal average rainfall, Precipitation Concentration Index (PCI), Rainfall Anomaly Index (RAI), Index of wetness, Coefficient of Variation, Rainy and non rainy days (wet and dry days) are analyzed to further enhance this study.

3.1 Mann and Kendall Test

Mann and Kendall test [11,14] is a popular non-parametric test used for detecting the trends [4]. In this approach our idea is to identify whether there is an uptrend or downward trend in the amount of monthly rainfall received by the area under study. By assuming the null hypothesis that there isn't any trend in the time series rainfall data we applied Mann and Kendall test [16]. The other substitute hypothesis includes not-null, and positive or negative trend. In Mann and Kendall Test, we either reject the null hypothesis denoted by H_0 or accept H_α, which is the alternative hypothesis. In this, H_0 represents "absence of any monotonic trend" and H_α represents "presence of monotonic trend".

$$S = \sum_{k-1}^{n-1} \sum_{j-k+1}^{n} sign(x_j - x_k) \tag{1}$$

The variance for S is calculated as:

$$Variance(S) = \frac{1}{18}\left[n(n-1)(2n+5) - \sum_{p-1}^{g} t_p(t_p-1)(2t_p+5) \right] \tag{2}$$

The Mann Kendall test static, Z_{MK} is calculated as

$$\begin{aligned} Z_{MK} &= \frac{S-1}{Variance(S)}\ if\ S > 0 \\ &= 0\ if\ S = 0 \\ &= \frac{S+1}{\sqrt{Variance(S)}}\ if\ S < 0 \end{aligned} \tag{3}$$

3.2 Parametric Test

Linear regression is used as parametric test which helps in identifying a trend in a historical time series data. Unlike non-parametric tests, the assumption made in parametric test is that there exist a linear trend in rainfall data. The slope identified from the Eq. 4 helps in identifying the uptrend or down trend in daily, monthly and seasonal rainfall.

$$y = mx + i \tag{4}$$

where, m is the rainfall trend and y is the rainfall which is a dependent variable and x is time which is independent and i is the intercept.

3.3 Annual Rainfall and Normal Annual Rainfall

For annual rainfall, daily rainfall was summed up and for normal annual rainfall [9] average rainfall over a period of 30 consecutive years was calculated.

$$Normal\ rainfall = \frac{\sum_{i=1}^{30}(R_i)}{30} \tag{5}$$

where R_i is the amount of rainfall in the ith year.

3.4 Precipitation Concentration Index (PCI)

Precipitation Concentration Index (PCI) [15,19] is a very popular indicator to identify the rainfall fluctuations, that means whether the area under study received a uniform or non-uniform rainfall over a period of time. Table 1 gives the range of PCI and it's significance. If the PCI value is greater than 20 then it reflects that thee is a severe n on-uniformity in the rainfall distribution on monthly, yearly or seasonal basis. Similarly, if the PCI value is less than equal to 10 then there is a uniform distribution of rainfall. PCI can be calculated as

$$PCI_{annual} = \frac{\sum_{i=1}^{12} p_i^2}{(\sum_{i=1}^{12} p_i)^2} \times 100 \tag{6}$$

where p_i is the amount of rainfall in the ith month of the year.

Table 1. PCI range and it's significance

PCI range	Significance
PCI $\leq$ 10	Uniform rainfall distribution
10 $<$ PCI $\leq$ 15	Moderate rainfall distribution
15 $<$ PCI $\leq$ 20	Irregular distribution of rainfall
PCI $>$ 20	Strong irregularity of rainfall distribution

3.5 Rainfall Anomaly Index (RAI)

Rainfall Anomaly Index (RAI) is an important index to identify the intensity of humid and dry months, years or decades. In our analysis, annual RAI is calculated to get an idea about dry and wet years and the intensity of dryness or humidity from the period of 1957 to 2017. The positive and negative anomalies are calculated as follows

$$RAI_{positive} = +3 \times \frac{N - \bar{N}}{\bar{M} - \bar{N}} \tag{7}$$

$$RAI_{negative} = -3 \times \frac{N - \bar{N}}{\bar{X} - \bar{N}} \tag{8}$$

where, $RAI_{positive}$ is the positive rainfall anomaly index, $RAI_{negative}$ is the negative rainfall anomaly index, N is current year's rainfall, $\bar{N}$ is yearly average rainfall, $\bar{M}$ is average of ten maximum yearly rainfall and $\bar{X}$ is the average of ten minimum yearly rainfall from the historical time series data.

The range of RAI and the significance can be obtained from the Table 2.

Table 2. Annual Rainfall Anomaly Index (RAI) and it's significance

Annual Rainfall Anomaly Index (RAI)	Significance
RAI $>$ 4	Extremely humid
4 $\leq$ RAI $>$ 2	Very humid
2 $\leq$ RAI $>$ 0	Humid
0 $\leq$ RAI $>$ -2	Dry
-2 $\leq$ RAI $>$ -4	Very dry
-4 $\leq$ RAI	Extremely dry

3.6 Index of Wetness (IoW)

In order to find the wetness in an entire year, ratio of annual rainfall and normal annual rainfall is calculated.

$$IoW = \frac{Annual\ Rainfall}{Normal\ Rainfall} \times 100 \tag{9}$$

Index of wetness can give the measure of Rain deficiency and it can be calculated as

$$Percentage\ rain\ deficiency = 100 - Percentage\ of\ Index\ of\ Wetness \tag{10}$$

3.7 Coefficient of Variation (CoV)

Monthly rainfall variability for a period of 61 years was calculated by

$$CoV = \frac{Mean\ Monthly\ Rainfall}{Standard\ Deviation\ of\ Monthly\ Rainfall} \times 100 \tag{11}$$

Coefficient of variation gives the extent to which the rainfall varies and can be categorized as given in Table 3.

Table 3. Coefficient of variation and it's significance

CoV range	Significance
Coefficient of variation ≤ 20	Less variability in rainfall
$20 <$ coefficient of variation ≤ 30	Moderate variability in rainfall
Coefficient of variation > 30	High variability in rainfall

3.8 Rainy Days vs Non-rainy Days

Identification of number of rainy (wet days) and non-rainy (dry days) days is also important as it gives the trend of shrinking or expanding rainfall or monsoon rainfall. Agricultural practices in major parts of India can be impacted by this dry or wet spells. Sudden change in these numbers can actually impact agricultural, industrial and disaster management activities.

4 Results and Discussion

Yearly rainfall for the period of 61 years is plotted to show the variation and it can be observed from Fig. 2. The month-wise rainfall distribution and its variation can be observed from Fig. 3, where each month is depicted in a different color.

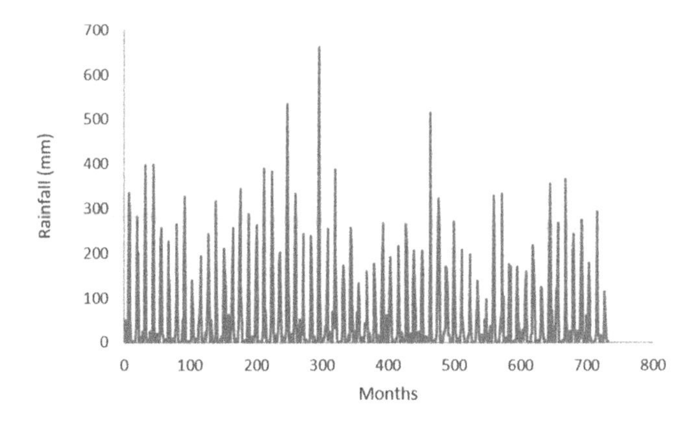

Fig. 2. Monthly rainfall over a period of 61 years (from 1957 to 2017)

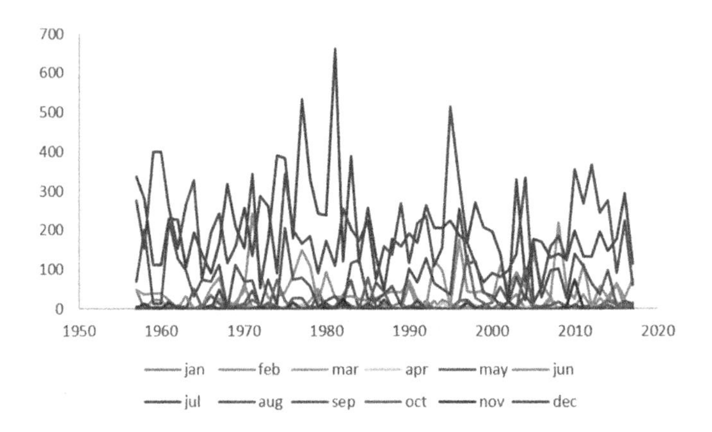

Fig. 3. Month-wise rainfall over a period of 61 years (from 1957 to 2017)

Table 4. Decadal monthly rainfall trend analysis using parametric approach

Month	1957–1966	1967–1976	1977–1986	1987–1996	1997–2006	2007–2017
January	Downtrend	Downtrend	Downtrend	Uptrend	Uptrend	Uptrend
February	Uptrend	Downtrend	Downtrend	Downtrend	Uptrend	Downtrend
March	Downtrend	Downtrend	Downtrend	Uptrend	Uptrend	Uptrend
April	Uptrend	Downtrend	Uptrend	Uptrend	Downtrend	Downtrend
May	Uptrend	Uptrend	Uptrend	Downtrend	Downtrend	Downtrend
June	Downtrend	Uptrend	Downtrend	Uptrend	Uptrend	Downtrend
July	Downtrend	Uptrend	Downtrend	Uptrend	Downtrend	Uptrend
August	Downtrend	Uptrend	Uptrend	Uptrend	Downtrend	Downtrend
September	Downtrend	Uptrend	Uptrend	Uptrend	Downtrend	Downtrend
October	Downtrend	Uptrend	Uptrend	Uptrend	Downtrend	Uptrend
November	Downtrend	Uptrend	Downtrend	Downtrend	Downtrend	Downtrend
December	Downtrend	Downtrend	Uptrend	Downtrend	Downtrend	Downtrend

Monthly and seasonal rainfall trend is depicted in Fig. 4 and Fig. 5 where we can observe the trend lines to identify the month-wise uptrend and downtrend. There is a marginal downtrend in the monsoon rainfall and a minor uptrend in the pre-monsoon rainfall. This observation can be very helpful for the Rabi and Kharif crops planning. Figure 6 shows the trend of maximum and total rainfall received over the period of 61 years and we can find that there is a noticeable downtrend in the yearly and maximum rainfall received. For the deeper analysis of the rainfall trend, decadal monthly trend analysis is done and it can be observed from Table 4 that there is a downtrend in the monthly decadal rainfall for the months of August and September for the last two decades whereas there is an uptrend in the July rainfall for the last decade.

Minimum, maximum and average monthly rainfall ranges and coefficient of variation of all the months for the entire duration is presented in Table 5. The data shows the fluctuation with the minimum monthly rainfall ranges from 0.00 to 15.041, whereas maximum monthly rainfall varies from 48.30 in December to 661.09 mm in July. The data depicts that there is less to high variability in the monthly rainfall patterns. The monthly normal annual rainfall for the periods, 1957–1986, 1967–1996, 1977–2006, and 1987–2016 was calculated and presented in Table 6. Some minor fluctuations can be observed in the normal annual rainfall for several months from the Table 6.

Mann Kendall test results are given in Table 7 which shows that we cannot say that there isn't any trend in the rainfall data. On majority of the occasions, upward monotonic trend is observed and on some occasions downward monotonic trend is also observed. In the duration 1987–1996, only the month of January, February and December observed downward trend whereas rest of the months show an uptrend which indicates a good sign as far as rainfall in considered. The observations for the duration of 2007–2017 shows that majority of the months are showing a downtrend which isn't a good sign as far as rainfall is considered.

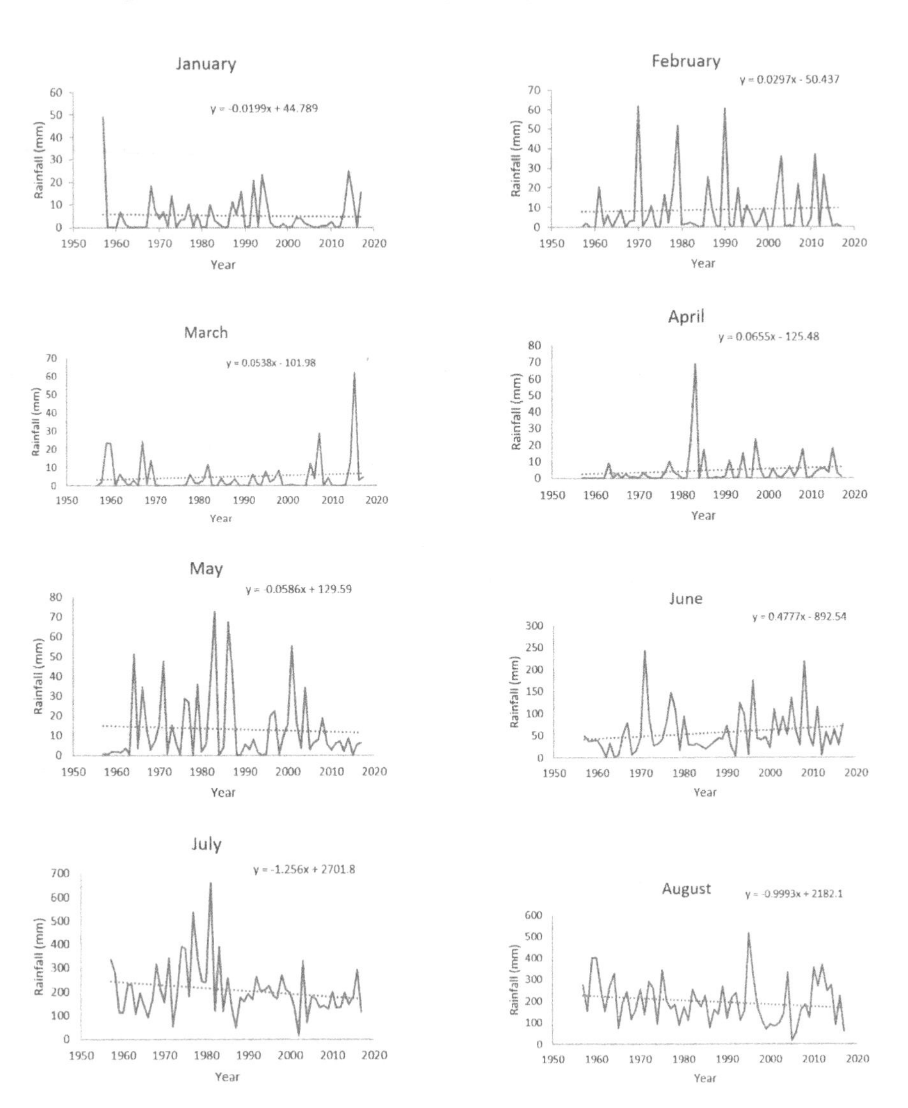

Fig. 4. Monthly rainfall trend analysis from the year 1957 to 2017

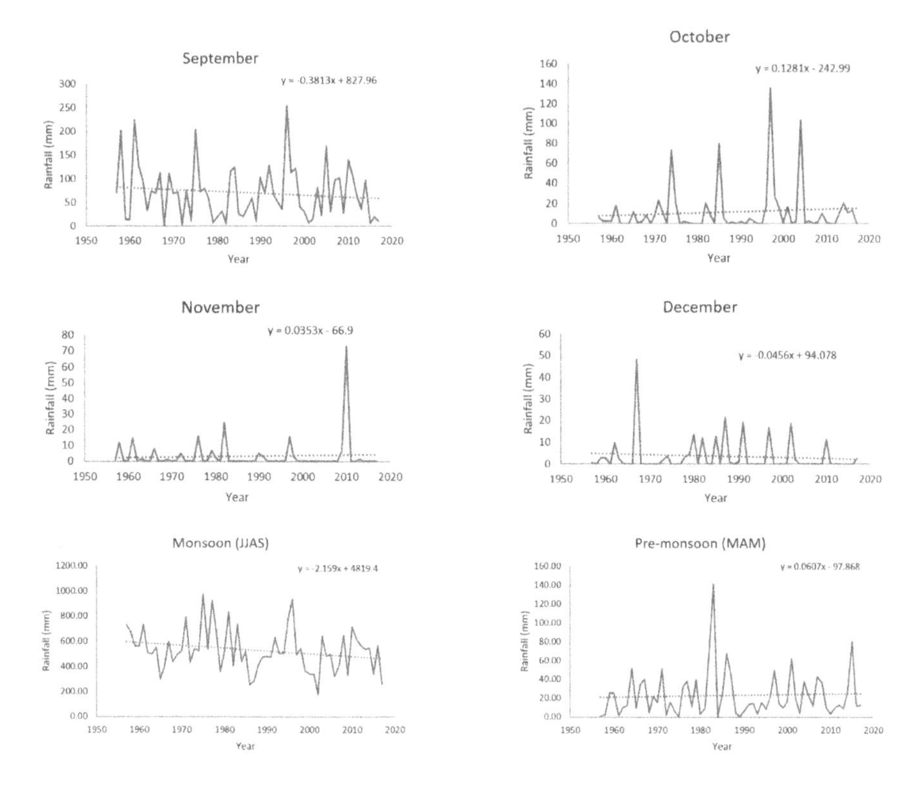

Fig. 5. Monthly and seasonal rainfall trend analysis from the year 1957 to 2017

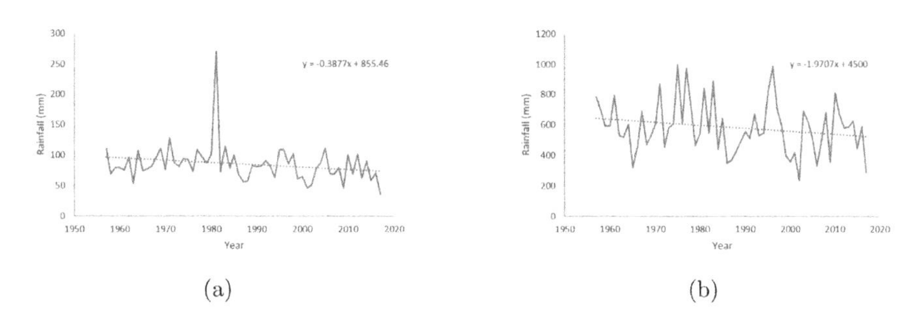

Fig. 6. Trend analysis of maximum and total rainfall received from the year 1957 to 2017

Table 5. Rainfall range (minimum, maximum, and average), associated coefficient of variation, and the rainfall variability

Month	Minimum	Maximum	Average	CV	Rainfall variability
January	0.00	48.86	5.19	16.60	Less variable
February	0.00	61.96	8.63	16.85	Less variable
March	0.00	62.00	4.89	20.39	Moderate variable
April	0.00	68.70	4.68	21.70	Moderate variable
May	0.00	72.588	13.14	13.40	Less variable
June	0.48	243.21	56.67	8.83	Less variable
July	15.041	661.09	206.07	5.49	Less variable
August	16.6	254.302	196.49	5.68	Less variable
September	0.00	135.96	70.27	8.35	Less variable
October	0.00	135.96	11.46	21.71	Moderate variable
November	0.00	73.200	3.25	31.58	High variable
December	0.00	48.30	3.50	22.91	Moderate variable

Table 6. Monthly normal annual rainfall during different time periods

Month	1957–1986	1967–1996	1977–2006	1987–2016
January	4.84	6.01	4.53	5.19
February	8.33	10.53	9.42	9.21
March	4.31	3.06	2.70	5.47
April	4.97	5.46	6.60	4.51
May	16.10	15.51	16.58	10.41
June	49.21	60.82	60.48	63.54
July	240.42	240.52	219.67	174.79
August	207.92	199.34	168.31	189.65
September	71.21	67.35	64.15	71.32
October	10.11	9.58	14.89	13.20
November	3.04	2.10	1.96	3.57
December	3.95	4.77	4.24	3.08
Total	624.40	625.04	573.52	553.93
JJAS	568.76	568.02	512.60	499.29

Table 7. Mann-Kendall test results of various decades

Month	1957–1966	1967–1976	1977–1986	1987–1996	1997–2006	2007–2017
January	−0.389	−0.337	0.181	−0.187	−0.185	0.296
February	0.181	0.091	−0.309	−0.019	−0.060	−0.072
March	0.078	−0.060	−0.147	0.330	0.173	0.414
April	0.416	0.289	−0.141	0.173	0.337	0.022
May	0.477	−0.037	0.112	0.283	0.127	−0.156
June	−0.037	0.455	−0.055	0.018	0.382	−0.022
July	−0.440	0.273	−0.382	0.309	−0.382	0.156
August	−0.220	0.164	−0.127	0.018	0.236	−0.244
September	0	0.200	0.236	0.273	0.164	−0.511
October	−0.208	−0.075	0.141	0.443	−0.250	0.163
November	−0.060	−0.092	−0.389	0.128	−0.426	−0.365
December	−0.057	0.337	0.019	−0.185	−0.093	0.108

Annual Precipitation Index (PCI) was calculated over a period of 61 years as shown in Fig. 7. It can be observed that the area under study has a very strong irregularity in the rainfall distribution. PCI values are ranging from 18 to 63. On very few occasions, the irregularity is moderate or else the rainfall distribution is heavily irregular. Decadal PCI also indicates that apart from the decade of 1967–1976 where there was a moderate irregularity, in rest other decades, Jaipur observed very strong irregularity in rainfall trends. For the period of 1997–2006 the total decadal rainfall decreased along with strong irregularities in rainfall distribution, as shown in Table 8 and Fig. 8.

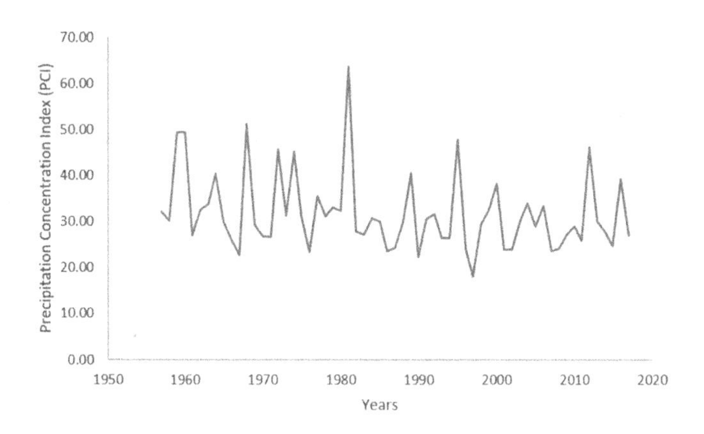

Fig. 7. Annual Precipitation Concentration Index (PCI) over a period of 61 years (from 1957 to 2017)

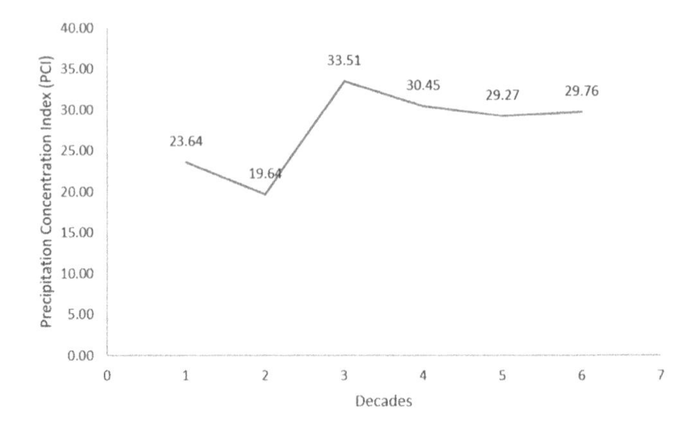

Fig. 8. Decadal Precipitation Concentration Index (PCI) (from 1957 to 2017)

Table 8. Decadal Analysis of rainfall received, associated Precipitation Concentration Index (PCI) and it's significance

Decades	PCI Range	Decadal rainfall (mm)	Significance
1957–1966	23.64	5509.84	Strong irregularity
1967–1976	19.64	6890.92	Moderate rainfall distribution
1977–1986	33.51	5551.89	Strong irregularity
1987–1996	30.45	6319.82	Strong irregularity
1997–2006	29.27	4693.63	Strong irregularity
2007–2016	29.76	5533.93	Strong irregularity

It can be observed from the Fig. 9 that throughout the duration of 61 year the rainy days and non-rainy days (dry days) are in a range and not much variation is observed. In the year 2002, least amount of rainy days were there with nearly 20 days of rainfall and in the year 1997 Jaipur observed rain for around 55 days. Similarly for the non-rainy days, in the same year of 1997, for around 310 days there was no rain and in the year 2002 nearly 345 days were dry days.

Rainfall Anomaly Index for the aforementioned duration of 61 years indicates that nearly 60% of the time duration was dry to extremely dry and remaining 40% of the duration was humid to extremely humid as shown in Fig. 10. There is also low to very high fluctuation in the Rainfall Anomaly Index as well. On some occasion like from the year 1960 to 1961 and 2001 to 2002 an abrupt change in RAI is observed.

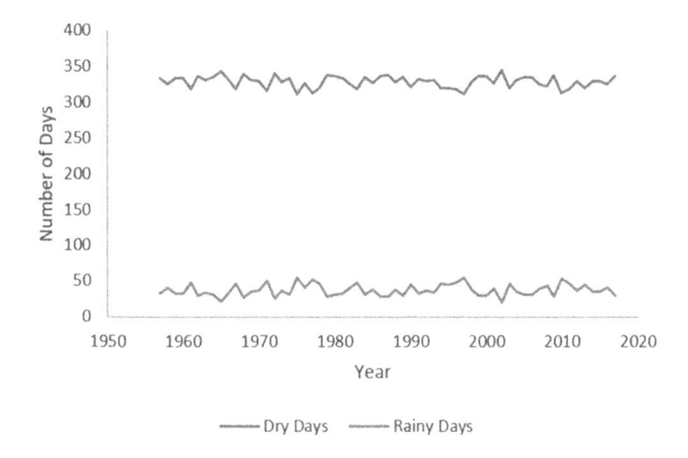

Fig. 9. Rainy days and non-rainy days (dry days) over a period of 61 years (from 1957 to 2017)

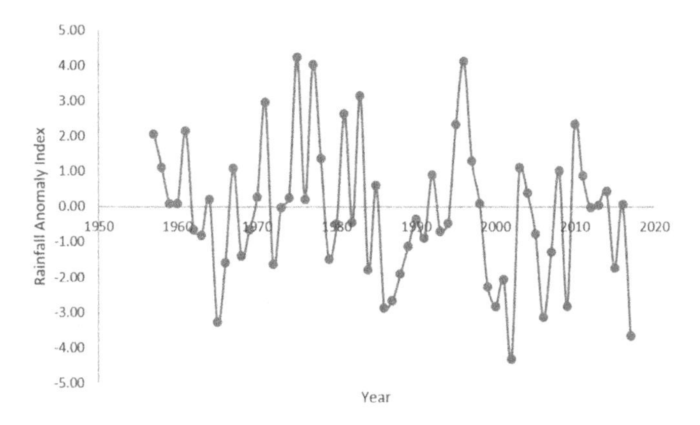

Fig. 10. Rainfall Anomaly Index (RAI) over a period of 61 years (from 1957 to 2017)

5 Conclusion

Amount of rainfall received is an important indicator for various purposes ranging agriculture to industrial. The major objective/purpose of this research article is to do the long term trend analysis of rainfall over Jaipur. Non-parametric Mann Kendall test, parametric linear regression analysis and evaluation of other indicators like Normal annual rainfall, Precipitation Concentration Index (PCI), Rainfall Anomaly Index (RAI), Index of Wetness, Coefficient of Variation, dry days vs wet days were calculated and analyzed for the period of 1957 to 2017. We can observe that there is a gradual change in the amount of rainfall received by the area under study. On rare occasions this change is abrupt or unexpected. With the availability of continuous large data record of rain the trend analysis is done successfully and the authors feel that this analysis will enable the water resource department to plan and develop optional water harvesting, allocation and usage policies and will improve the difference between the water requirements and it's availability for the city of Jaipur.

Acknowledgments. This work on rainfall trend analysis for the Jaipur city is in collaboration with Water Resources, Government of Rajasthan. We are thankful to Special Project Monitoring Unit, National Hydrology Project, Water Resources Rajasthan Jaipur, India for providing us the Rainfall data and valuable inputs for this study.

References

1. Brière, F.G.: Drinking-Water Distribution, Sewage, and Rainfall Collection. Presses inter Polytechnique (2014)
2. Calviño, P., Sadras, V.: Interannual variation in soybean yield: interaction among rainfall, soil depth and crop management. Field Crop Res. **63**(3), 237–246 (1999)
3. De Luis, M., Raventós, J., González-Hidalgo, J., Sánchez, J., Cortina, J.: Spatial analysis of rainfall trends in the region of Valencia (East Spain). Int. J. Climatol. J. Roy. Meteorol. Soc. **20**(12), 1451–1469 (2000)
4. Esterby, S.R.: Review of methods for the detection and estimation of trends with emphasis on water quality applications. Hydrol. Process. **10**(2), 127–149 (1996)
5. Esteves, L.S.: Consequences to flood management of using different probability distributions to estimate extreme rainfall. J. Environ. Manage. **115**, 98–105 (2013)
6. Ganesan, G., MC, P.: A study of heavy rainfall in and around Jaipur (1981)
7. Goswami, B.N., Venugopal, V., Sengupta, D., Madhusoodanan, M., Xavier, P.K.: Increasing trend of extreme rain events over India in a warming environment. Science **314**(5804), 1442–1445 (2006)
8. Goswami, P., Ramesh, K.: Extreme rainfall events: vulnerability analysis for disaster management and observation system design. Current Sci. **94**, 1037–1044 (2008)
9. Goyal, M.K.: Engineering Hydrology. PHI Learning Pvt., Ltd. (2016)
10. Halder, D., Panda, R., Srivastava, R., Kheroar, S., Singh, S.: Stochastic analysis of rainfall and its application in appropriate planning and management for eastern India agriculture. Water Policy **18**(5), 1155–1173 (2016)
11. Kendall, M.G., et al.: Multivariate Analysis, vol. 2. Griffin London (1975)
12. Kumar, Y., Kumar, A.: Spatiotemporal analysis of trend using nonparametric tests for rainfall and rainy days in jodhpur and Kota zones of Rajasthan (India). Arab. J. Geosci. **13**(15), 1–18 (2020)
13. Lal Meena, A., Bisht, P., et al.: Study of rainfall pattern in Chaksu Tehsil, Jaipur, Rajasthan, India (2020)
14. Mann, H.B.: Nonparametric tests against trend. Econometrica J. Econ. Soc., 245–259 (1945)
15. Martin-Vide, J.: Spatial distribution of a daily precipitation concentration index in peninsular Spain. Int. J. Climatol. J. Roy. Meteorol. Soc. **24**(8), 959–971 (2004)
16. McLeod, A.I.: Kendall Rank Correlation and Mann-Kendall Trend Test. R Package Kendall (2005)
17. Nyatuame, M., Owusu-Gyimah, V., Ampiaw, F.: Statistical analysis of rainfall trend for Volta region in Ghana. Int. J. Atmos. Sci. **2014** (2014)
18. Odjugo, P.A.: An analysis of rainfall patterns in Nigeria. Global J. Environ. Sci. **4**(2), 139–145 (2005)

19. Oliver, J.E.: Monthly precipitation distribution: a comparative index. Prof. Geogr. **32**(3), 300–309 (1980)
20. Rana, S., Deoli, V., Kashyap, P.: Temporal analysis of rainfall trend for Udaipur district of Rajasthan. Indian J. Ecol. **46**(2), 306–310 (2019)
21. Schlögl, M., Matulla, C.: Potential future exposure of European land transport infrastructure to rainfall-induced landslides throughout the 21st century. Nat. Hazard. **18**(4), 1121–1132 (2018)
22. Taschetto, A.S., England, M.H.: An analysis of late twentieth century trends in Australian rainfall. Int. J. Climatol. J. Roy. Meteorol. Soc. **29**(6), 791–807 (2009)

Experimentally Proven Bilateral Blur on SRCNN for Optimal Convergence

Divya Mishra[1](✉), Tayyibah Khanam[2], and Ivanshu Kaushik[3]

[1] Jamia Millia Islamia, New Delhi, India
dvmshr946@gmail.com
[2] Aligarh Muslim University, Aligarh, India
tayyibahkhanam@zhcet.ac.in
[3] Birla Institute of Technology and Science, Pilani, India
ivanshukaushik918@gmail.com

Abstract. Image super-resolution is a contentious issue in developing computer vision applications such as satellite imaging, graphics industry, medical diagnostics, and real-time scene monitoring for security and safety. CNN-based image super-resolution algorithms are the simplest and most resource-efficient in deep learning. The original SRCNN framework, on the other hand, incorporated Gaussian blur, which takes a long time to converge. This is not only computationally expensive, but it also extends the training time. In light of this, we retrained original SRCNN utilizing widely available blurring techniques and observed that the bilateral filter beats the Gaussian filter at optimal convergence. Our goal is to preserve the original SRCNN features while reducing training time and computational resources and hence, speed up the architecture with optimum number of epochs. This is, to the best of our knowledge, the first study of its kind, and no other study has implemented this alternative blurring training on SRCNN and ranked the best of them for image super-resolution.

Keywords: Blurring techniques · Image quality metrics · Super-resolution · SRCNN

1 Introduction

Super-resolution is the approach of creating a high-resolution(HR) image from a low-resolution(LR) image. Multiple steps towards super resolution have been included in sparse-coding based methods [27,28], such as overlapping densely cropped patches followed by mean subtraction and normalisation as a pre-processing step for further encoding image patches by Low resolution dictionary to recover high resolution patches. Furthermore, these existing approaches lack optimization. The above-mentioned sequential methodologies are analogous to the behavior of a deep convolution neural network [12]. This motivation prompted the authors to propose the SRCNN [4], the first convolution neural network-based super-resolution model. SRCNN is a multi-layer, end-to-end

I. Woungang et al. (Eds.): ANTIC 2021, CCIS 1534, pp. 502–516, 2022.
https://doi.org/10.1007/978-3-030-96040-7_39

deep convolutional network that takes a low-resolution image and produces a high-resolution image. The network is lightweight, with only three CNN layers and a small number of filters. It can execute on a CPU as well. Aside from that, the network is quicker than prior sparse example-based approaches, which are hampered by a long chain of encoded dictionaries.

Bicubic interpolation is used to first upscale the low resolution (LR) input picture. Then, between high resolution and low resolution samples, a mapping function is learned that consists of three operations: generating feature maps through patch extraction and representation, non-linear mapping among high-dimensional vectors that represent a high resolution image, and finally aggregation of all high-dimensional vectors to reconstruct a super resolution image. They discovered that perceptual quality may be enhanced much more if:

- For training purposes, a large number of datasets are available.
- When a dense network is utilised with more layers than three, and the number of filters is increased, the Peak signal to noise ratio(PSNR) improves considerably, as shown in Table 1, where F1, F2 and F3 are filter size in consecutive network layers of SRCNN [4].
- To achieve more super-resolved results, the network could be able to control all three channels instead of only the Y channel.

Table 1. PSNR values on increasing filter size in SRCNN

Filter size in 3 layers	PSNR
F1 = 9, F2 = 1, F3 = 5	32.52
F1 = 11, F2 = 1, F3 = 7	32.57
F1 = 9, F2 = 3, F3 = 5	32.66
F1 = 9, F2 = 5, F3 = 5	32.75

In Image pre-processing, most of the models utilise a Gaussian filter. In 2019, authors [1] tried to improve results for same SRCNN model by introducing a change in existing Relu activation function as modified and bilateral ReLU. This helps them in overall improvement in image quality. Another approach, is used by presenting a resilient loss function merged with MSE used in SRCNN based on the Canny operator's preservation of edges [21] which shows better results on both PSNR and SSIM. The authors [14] modified SRCNN using color feature based image super-resolution algorithm for underwater image applications. The author proposes a bilateral up-sampling network [30] which consists of a bilateral up-sampling filter used for single image super-resolution with arbitrary scaling factors. Our plan is to put three additional blurring methods: the average blur, median blur and bilateral blur to test the original SRCNN [4] model and retrain them thrice on each blur filter. For all five pictures evaluated, we discovered that bilateral outperforms than Gaussian and rest blurs for a smaller number of epochs (nearly 25). This also implies that if the number of epochs is kept around 25, we may utilise bilateral for better outcomes.

The following is a breakdown of the paper's structure: Sect. 2 briefly introduce widely used blurring techniques image processing techniques. Section 3 introduced by common image quality metrics like Mean Squared Error (MSE), Peak Signal to Noise Ratio (PSNR) and Structural Similarity Index Measure (SSIM). Section 4 accumulates conventional interpolation based methods for image super-resolution. Section 5 describe about the first deep learning based image SR model and how different existing blurs can be used in that SRCNN. Section 6 summarize the dataset used and evaluation result on each blur for all 5 test images. Section 7 finally discusses the conclusion and the potential of future work.

2 Blurring Techniques Widely Used in Image Processing Techniques

Blurring is a word that is commonly used to describe something that is smeared or unclear. This term is quite frequently used by the Image Processing and Computer Vision fraternity, with the goal to refer smoothing or de-sharpening of images.

A crucial step in image processing and computer vision tasks is to identify distinct objects [25], measure their sizes, to find influence on face Recognition Performance [10], in image restoration task [9,19,24] obtain the shape matrices etc. In such a situation, the earliest phases of object identification and localisation need 'edge detection. Edges define a strong divide between two visually distinct sets of pixels, resulting in abrupt variations in pixel intensity all along the edge.

The objective of blurring an image is to smooth out these abrupt variations in pixel intensities, which can be simply translated to smoothing out the image. The Computer Vision community refers to blurring or smoothing as applying a low pass filter to an image, which basically eliminates the "noise" while preserving other features of the image intact.

2.1 Median Blur

Median filtering [2] is a class of non-linear edge preserving filtering that is frequently employed in digital signal processing. The median filter works by traversing pixel by pixel through each image and replacing each pixel with the median of adjacent pixels. The pattern of neighbours is referred to as the "window," and it slides across the entire image, pixel by pixel. While the window for one-dimensional data, such as electronic signals, must encompass all entries over a certain radius or ellipsoidal region, the window for higher-dimensional data, such as images, must include all entries within a given radius. The median is generated by sorting all of the pixel values in the window into numerical order, then replacing the pixel in concern with the middle (median) pixel value.

2.2 Average Blur

The average filter [23] operates similarly to the median blur in that it moves across the image pixel by pixel, replacing each value with the average value of neighbouring pixels, including itself. Because the filter is the simplest of all existing filters, it contributes equally to the formation of the final image, as illustrated in Eq. 1.

$$K = \frac{1}{9} * \begin{bmatrix} 1 & 1 & 1 \\ 1 & 1 & 1 \\ 1 & 1 & 1 \end{bmatrix} \tag{1}$$

2.3 Gaussian Blur

A Gaussian filter can be considered as a non-uniform linear low-pass filter that preserves low spatial frequency and reduces image noise and speckles [5,18]. To avoid erroneous edge detection, such a denoising operation is necessary to eliminate the particularly high frequency components according to a given threshold. It's usually done by using a Gaussian kernel to convolve an image. As a result, Gaussian blur calculates a local average of intensities at each position. Gaussian filtering, as a consequence, is a weighted average of the intensity of neighbouring locations, with the weight decreasing as the distance from the centre (p) rises, expressed by Eq. (2). The Gaussian kernel in 2-D form is expressed in Eq. (3), where sigma is the standard deviation of the distribution and controls the variance along a mean value of a Gaussian distribution. The degree of blurring may thus be adjusted by varying the standard deviation and threshold values to detect the most general edges.

$$Gb[I]_p = \sum_{q \in S} G_\sigma(||p - q||) I_q \tag{2}$$

$$G_{2D}(x, y, \sigma) = \frac{1}{2\pi\sigma^2} e^{-\frac{x^2+y^2}{2\sigma^2}} \tag{3}$$

Consider this visualisation of a two-dimensional Gaussian function in Fig. 1 to better understand how the equations above operate.

While the standard deviation or sigma determines the width or 'spread' of the curve (larger the deviation, more the spread and flatter the shape), the height is determined by the mean of the normal distribution, or the weight given to the underlying pixel in the kernel.

The basic operation in linear image filtering is convolution by a positive kernel such as in Gaussian blur. Thus, The linear filter Gaussian blur [7] is an example of one that can be constructed quickly and efficiently.

2.4 Bilateral Blur

The bilateral filter [11], like the Gaussian convolution, is defined as a weighted average of pixels. While the Gaussian filter smoothes away noise or textures

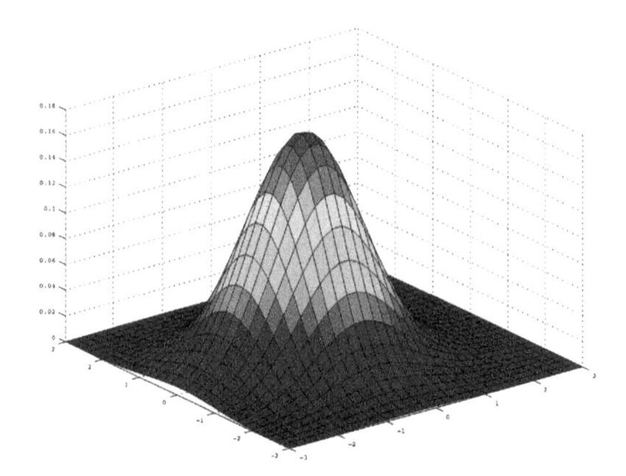

Fig. 1. Two dimensional Gaussian density function

without maintaining edges, the bilateral filter does so while taking intensity fluctuations into account. Bilateral filtering is based on the idea that two pixels are close to one another not only if their spatial locations are similar, but also if their photometric ranges are also similar. Equation (4) and Eq. (5) define the Bilateral filter and its normalization factor respectively.

$$BF[I]_p = \frac{1}{W_p} \sum_{q \in S} G_{\sigma_s}(||p - q||) G_{\sigma_r}(I_p - I_q) I_q \tag{4}$$

$$W_p = \sum_{q \in S} G_{\sigma_s}(||p - q||) G_{\sigma_r}(I_p - I_q) I_q \tag{5}$$

The filtering intensity is thus controlled by two weight parameters - σ_s and σ_r. As σ_r increases, the bilateral filter approaches purely gaussian characteristics and increasing σ_s smoothens larger features. The weights are multiplied in bilateral filtering, which implies that no smoothing happens as soon as one of the weights approaches zero. G_s is a spatial Gaussian that minimizes the influence of distant pixels, whereas G_r is a range Gaussian that minimizes the influence of pixels q with an intensity value differing from I_p as shown in Fig. 2.

3 Image Quality Metrics

Image enhancement, sometimes referred to as enhancing a digital image's visual quality, is a subjective practice. It is possible that the assertion that one approach creates a higher-quality image varies from person to person. As a result, quantitative and empirical measurements for comparing the impact of image enhancement algorithms on image quality must be established.

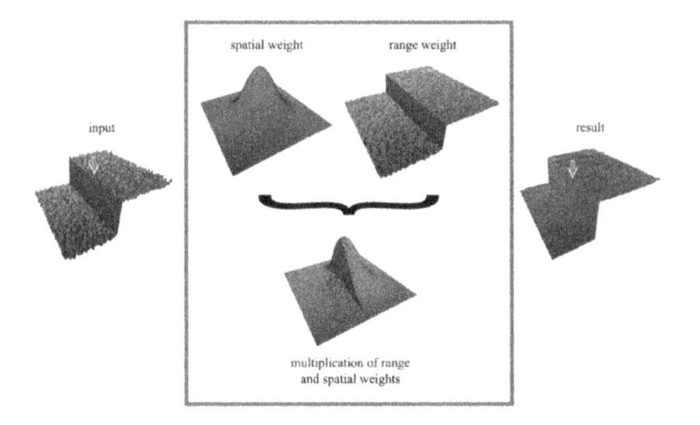

Fig. 2. Filtering process through bilateral blur

3.1 Mean Squared Error

In the recent past, the mean squared error (MSE) has been widely utilised for most practical uses. It allows us to compare our original image's "true" pixel values to our degraded image. The MSE is the square root of the "errors" in our actual and noisy images. The error is the difference in between values of the original image and the values of the degraded image. The MSE between an original image matrix f, and the degraded image matrix g is represented by Eq. (6):

$$MSE = \frac{1}{mn} \sum_{0}^{m-1} \sum_{0}^{n-1} ||f(i,j) - g(i,j)||^2 \tag{6}$$

3.2 Peak Signal to Noise Ratio

PSNR (peak signal-to-noise ratio) is a term for the ratio of a signal's greatest potential value (power) to the strength of distorting noise that influences its representation quality. The PSNR is generally represented in logarithmic dB because many signals have a high dynamic range (the difference between the greatest and lowest possible values of a changing quantity). When dealing with images instead of signals, PSNR may be thought of as a more advanced version of MSE. The MSE is a measure of the cumulative squared error between the original and degraded image, whereas the PSNR is a measure of the peak error between the original and degraded image. Equation 7 represents the PSNR value of the pair of images (f, g) in terms of the MSE described in Eq. (6).

$$PSNR = 20log_{10}(\frac{MAX_f}{\sqrt{MSE}}) \tag{7}$$

where MAX_f is the maximum signal value that exists in our original known to be "good image".

Although it is self-evident that a greater PSNR (equivalent to a lower MSE) indicates a better quality of the degraded picture g and hence a superior reconstruction technique [6]. As a result, the metric's range of validity must be utilised with extreme caution. PSNR has difficulties in assessing the quality of two virtually similar pictures with a near-zero MSE. Because division by zero is undefined, PSNR can't be quantified as a meaningful quality evaluation metric in all circumstances because it relies only on numerical comparisons. PSNR has lately been found to perform badly in contrast to other quality evaluation methods [8,26] such as the Structural Similarity Index Metric (SSIM) which also considers biological factors of human visual system along with numeric comparisons.

3.3 Structural Similarity Index Metric

As mentioned in the previous section, SSIM [26] is a perception-based metric that takes into consideration not only the difference in structural information between two pictures, but also fundamental perceptual notions like luminance masking and contrast masking. The concept of structural information relates to the assumption that pixels, especially when they are spatially comparable, exhibit substantial interdependencies. These dependencies hold vital information about the structure of the visual scene's entities. For example, Contrast masking lowers the visibility of image distortions in areas with substantial activity or "texture" in the image, whereas luminance masking reduces the visibility of image distortions in bright areas (in this context). SSIM focuses on measuring the perceptual difference between the two available image matrices rather than determining which image is superior in terms of visual appearance.

Calculation of the SSIM index is done on various image windows. Between two windows x and y of common size N * N, the measure between them is represented by Eq. (8).

$$SSIM(x,y) = [l(x,y)^{\alpha} \cdot c(x,y)^{\beta} \cdot s(x,y)^{\gamma}] \tag{8}$$

where the SSIM formula is a weighted combination of three comparison measurements between the samples of x and y: luminance (l), contrast (c) and structure (s), where α, β and γ represent the individual weights of all three components respectively.

4 Image Super Resolution

The computer vision community has conducted substantial research on the technique of upscaling and enhancing an image, popularly known as super resolution. The goal is to convert a lower-resolution original image to a higher-resolution image that is perceptually pleasing and realistic.

In the coming sections, we will discuss some of the initial upscaling algorithm [15], all based on linear methods of interpolation. Interpolation is the problem of approximating the value of a function for a non-given point in some space given the value of that function in points around (neighbouring) that point.

4.1 Nearest Neighbors Interpolation

The simplest method of interpolation is the Nearest Neighbour interpolation [13, 20, 22], also commonly known as the 'box filter'. The closest neighbour method selects the value of the nearest point while neglecting the values of neighbouring points, yielding a piecewise constant interpolant. Rather of utilising weighting criteria to determine an average value or complicated procedures to provide an intermediate value, this method simply identifies the "nearest" neighbouring pixel and assumes its intensity value.

4.2 Bilinear Interpolation

Also known as the 'tent' function [15] is a re-sampling method that estimates a new pixel value by taking the distance weighted average of the four nearest pixel values. It is possible to interpolate functions of two variables (such as x and y) on a two-dimensional grid using a bilinear interpolation technique. By first conducting linear interpolation in one direction and then in the opposite direction, bilinear interpolation is achieved. This interpolation is exclusively linear along lines parallel to the X or Y axis, and quadratic along all other straight lines.

4.3 Bicubic Interpolation

Because the interpolated surface obtained from bicubic interpolation [3] is smoother than similar surfaces derived from bilinear or nearest-neighbor interpolation, bicubic interpolation is usually favoured over bilinear or nearest-neighbor interpolation in image resampling. This is because Bi-linear determines the output using four nearest neighbours, whereas Bi-cubic employs sixteen (4 * 4) neighbours, slowing down the process but providing a better upsampled rendition of the original image. To fill up the gaps, the techniques indicated above gather data from neighbouring pixels. But why don't these tried-and-true ways work? A well-known truth in data processing is that data can't be further processed to deliver any information that doesn't already exist, which simply translates to - data can't be further processed to provide any information that doesn't already exist. We can't upgrade images using data processing methodologies since they aren't predictive.

This is where a neural network comes into play perfectly. Based on the weights it learns throughout the training process on a wide set of images, a neural network learns to hallucinate features. The first one in this direction was SRCNN [4] which simply learns a mapping between a low resolution and a high-resolution image. The aim is to train a neural network using a dataset made up of high-resolution images that have been downscaled and then input into the neural network. Because this is a kind of self-supervised learning, we may compare the neural network generated image to the ground truth image by downscaling high resolution samples to low resolution samples before feeding them into the network.

The SRCNN paper thus simply minimizes the squared difference (Mean Squared Error/MSE) of the pixel values while training to calculate the loss. Mean Squared Error, on the other hand, cannot be labeled a strong loss function since it only evaluates pixel-wise variations rather than structural information. As a result, the Structural Similarity Index (SSIM), a superior gauge of perceptual quality, works better in this situation. SSIM has been utilised as a loss function for image restorations and other reasons by a few academics, despite its origins as a quality evaluation tool.

5 SRCNN

Prior to SRCNN, image restoration was accomplished using a technique called Sparse Coding. Sparse coding, built on a complex pipeline and mathematical algorithms, extracted overlapping patches from an image, projected those patches to a higher resolution space, and then aggregated these high resolution vectors to reconstruct the image. Due to the complexity of sparse coding based image restoration, the authors of SRCNN attempted at recreating an identiacal Convolutional Neural Network pipeline for ease and simplicity.

Non-predictive techniques are unable to predict details in an image, resulting in upscaling losses; here is when the positives of using a neural network come into play. Based on the weights it learns throughout the training process on a wide set of images, a neural network learns to hallucinate features. The first one in this direction was SRCNN [4] which simply learns a mapping between a low resolution and a high-resolution image. The aim is to train a neural network using a dataset made up of high-resolution images that have been downscaled and then input into the neural network. Because this is a kind of self-supervised learning, we may compare the neural network generated picture to the ground truth image by downscaling high resolution samples to low resolution samples before feeding them into the network. To compute the loss, the SRCNN study simply minimises the squared difference (Mean Squared Error/MSE) of the pixel values while training.

5.1 Architecture Details

The SRCNN network is a short depth three-convolutional layer network. Each convolutional layer is responsible for a distinct function. The first convolutional layer serves as a low-level feature extractor, responsible for patch extraction and representation, while the second convolutional layer performs non-linear mapping and the third reconstructs the up-sampled image as shown in the Fig. 3.

The model receives as input a standard low resolution image created by adding Gaussian blur to the original high resolution image. When training, the output is set to the appropriate high-resolution image. In this case, the error is the difference between the ground truth high resolution image and the generated high resolution image from neural network.

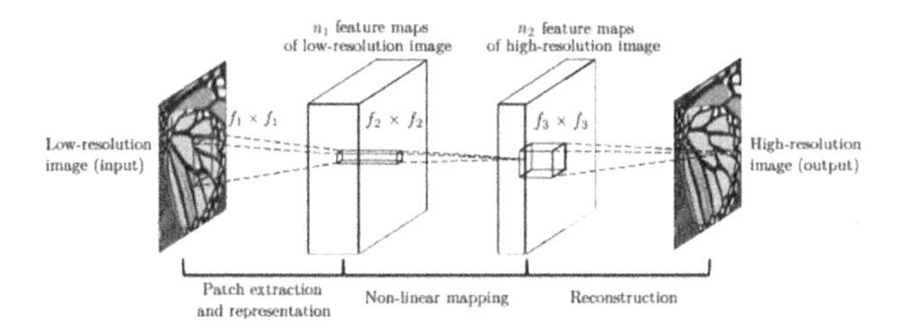

Fig. 3. Basic architecture of SRCNN [4]

The first convolutional layer separates or filters different locations from the low-resolution input image in an overlapping manner and turns them into a high-dimensional vector for further processing. The recovered high dimensional vector output of layer 1 is then nonlinearly mapped by the second convolutional layer to another high dimensional vector. Finally, the third convolutional layer reconstructs the original image by combining the aforementioned vectors into a high-quality image.

5.2 Modified-SRCNN

Advantages of Using Less Epochs: In machine learning, epochs represent the number of passes an algorithm makes across the whole training set. One epoch denotes a single run through the neural network algorithm of the whole training data forward and backward. Because gradient descent is an iterative process, updating the weights through a single epoch is unsatisfactory; hence, a sufficient number of epochs is required to make sure that our model's loss is properly mitigated. Although it is evident that more rounds of optimization would minimize the error on training data, there may come a point where the network becomes over-fit to the training data and begins to lose performance in terms of generalization to non-training (unseen) data. Furthermore, increasing the number of epochs not only increases training time but also becomes computationally expensive. As a result, a neural network is expected to be computationally efficient, converge as quickly as feasible in as few epochs as possible.

Experimenting with Various Blurring: In addition to the Gaussian blur utilized by the authors of SRCNN to create a low-dimensional, blurred image from a high-dimensional ground truth image, we performed our tests with three additional types of blurring filters: average blur, median blur, and bilateral blur. The objective of our research was to see how well a super-resolution neural network performed on differently blurred input images. While upsampling a low-resolution image, we were able to evaluate the relevance of linear and nonlinear filters, edges, neighbouring pixels, and other factors.

Reason of Why Bilateral? The Gaussian filter is an isotropic filter that smoothens out the whole image, including edges and high-frequency features. The bilateral filter, on the other hand, gives more human-like results even after blurring. It avoids smoothing off edges, curves, and other structural features that don't need to be. The reason for this is that a bilateral filter is a more modified version of a Gaussian filter. If two nearby pixel values are numerically close to each other and represent a comparable picture, the Gaussian coefficient is multiplied by a quantity close to 1 and the outcome is similar to the Gaussian blur. On the other hand, if two nearby pixel values are significantly different from one another, reflecting a sudden dissimilar scene in an image, a number close to 0 is multiplied by the Gaussian coefficient, turning off the Gaussian filter and resulting in bilateral filtering. Furthermore, the author [17] has already analysed these four frequently used blurring techniques and evaluated them on ten test pictures, revealing that bilateral blurring is the best approach on potential image quality measures PSNR, SSIM, and FSIM [29]. We are looking into the SRCNN model architecture leveraging that result.

6 Evaluation

6.1 Dataset

We ran our experiments on the Berkeley Segmentation Data Set 500, BSDS500 [16]. Originally developed for segmentation tasks, BSDS500 contains 200 natural images in the training set, 100 images in validation set and 200 images in the test set. The training and validation sets were used for training the modified SRCNN in our experiments, whereas the test set was used as the validation set. For testing purposes, we randomly picked 5 high resolution images from the internet as shown in Fig. 4.

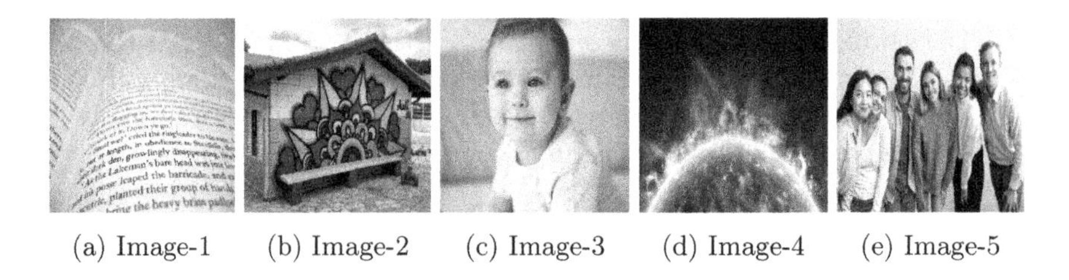

(a) Image-1 (b) Image-2 (c) Image-3 (d) Image-4 (e) Image-5

Fig. 4. Randomly selected 5 test images

6.2 Evaluation Results on Test Images

Our modified SRCNN architectures is trained for 15×10^6 back-propogations till 400 epochs using Stochastic Gradient Descent (SGD) optimizer and MSE loss. Additionally, to preserve the input image's original dimensions, we used padding that was not used previously in the original SRCNN paper.

From the observation through Table 2, Table 3, Table 4, Table 5 and Table 6, the PSNR values for bilateral blur among all blurs experimented are best till 25 epochs. The result is showing same trend for all 5 test-images. The re-implemented SRCNN is assessed using three distinct LR blurring techniques: average, median, and bilateral, and a potential image quality metric such as PSNR. The greater the value of the PSNR measure, the better the image quality.

Table 2. PSNR values for Image-1

Epochs	Gaussian (original)	Average blur	Median blur	Bilateral blur (Our)
5	13.6546	31.2339	30.8291	**31.5442**
10	32.0823	32.3148	31.5506	**34.7294**
15	32.5153	33.7281	32.8810	**38.2987**
20	33.4163	35.4244	34.7412	**38.6823**
25	34.3733	37.6884	36.5833	**38.7579**

Table 3. PSNR values for Image-2

Epochs	Gaussian (original)	Average blur	Median blur	Bilateral blur (Our)
5	12.3813	23.1625	23.8009	**26.2090**
10	25.0341	26.1183	26.2064	**27.6182**
15	26.6046	26.6944	26.8331	**28.7722**
20	26.9624	27.2860	27.4676	**28.8853**
25	27.4290	27.8657	27.8763	**28.8717**

Table 4. PSNR values for Image-3

Epochs	Gaussian (original)	Average blur	Median blur	Bilateral blur (Our)
5	12.8546	22.5428	22.4944	**24.0124**
10	23.1272	24.0977	24.0283	**24.5656**
15	24.4383	24.3235	24.3757	**25.0244**
20	24.6514	24.5969	24.7149	**25.1123**
25	24.8785	24.8410	24.9768	**25.1350**

Table 5. PSNR values for Image-4

Epochs	Gaussian (original)	Average blur	Median blur	Bilateral blur (Our)
5	21.7155	23.9870	23.8865	**25.4093**
10	24.7007	25.5559	25.6138	**26.3961**
15	25.8845	26.0202	26.2040	**27.1997**
20	26.2099	26.4581	26.7967	**27.3426**
25	26.5939	26.9021	27.2382	**27.3803**

Table 6. PSNR values for Image-5

Epochs	Gaussian (original)	Average blur	Median blur	Bilateral blur (Our)
5	12.4835	29.0525	29.1541	**30.9659**
10	29.8682	31.2727	31.1806	**32.5813**
15	32.0604	32.0230	32.1268	**34.8543**
20	32.1702	33.0057	33.2220	**35.0118**
25	32.7468	34.6621	34.6660	**35.0745**

7 Conclusion and Future Work

Image super-resolution is a popular and in-demand application in the field of image processing, but resource management is a significant difficulty in deep learning-based research. To resolve the convergence time, we conducted an experiment by training original SRCNN model thrice by using three different widely used blurring techniques. Those three blurring filters are average or box blur, median and bilateral blur. The goal of this experiment is to obtain the best outcomes with the minimal number of running epochs possible in order to conserve computational resources and time without compromising on image quality. On five randomly selected images, we utilised PSNR as a conventional image quality statistic to evaluate the findings. The least effective blur is as expected the average blur and the best is bilateral blur. This research will be useful not just for convolution neural network-based designs, but also for generative models such as Generative Adversarial Networks (GAN)-based super-resolution models in the future. New filters with features equivalent to bilateral filters can be designed as a result of this experiment, allowing them to converge even quicker than bilateral filters. In this way, we can actively contribute to the growing body of knowledge that aids practitioners, academics, and developers in choosing optimal blurring algorithms for image super-resolution with the minimum number of epochs.

References

1. Ahn, H., Chung, B., Yim, C.: Super-resolution convolutional neural networks using modified and bilateral ReLU. In: 2019 International Conference on Electronics, Information, and Communication (ICEIC), pp. 1–4 (2019). https://doi.org/10.23919/ELINFOCOM.2019.8706394
2. Chang, C., Hsiao, J., Hsieh, C.: An adaptive median filter for image denoising. In: 2008 Second International Symposium on Intelligent Information Technology Application, vol. 2, pp. 346–350 (2008). https://doi.org/10.1109/IITA.2008.259
3. Dengwen, Z.: An edge-directed bicubic interpolation algorithm. In: 2010 3rd International Congress on Image and Signal Processing, vol. 3, pp. 1186–1189, October 2010
4. Dong, C., Loy, C.C., He, K., Tang, X.: Image super-resolution using deep convolutional networks (2015)
5. Gedraite, E.S., Hadad, M.: Investigation on the effect of a Gaussian blur in image filtering and segmentation. In: Proceedings ELMAR-2011, pp. 393–396 (2011)
6. Gupta, P., Srivastava, P., Bhardwaj, S., Bhateja, V.: A modified PSNR metric based on HVS for quality assessment of color images. In: 2011 International Conference on Communication and Industrial Application, pp. 1–4, December 2011. https://doi.org/10.1109/ICCIndA.2011.6146669
7. He, H., Siu, W.C.: Single image super-resolution using gaussian process regression. In: CVPR 2011, pp. 449–456, June 2011. https://doi.org/10.1109/CVPR.2011.5995713
8. Horé, A., Ziou, D.: Image quality metrics: PSNR vs. SSIM. In: 2010 20th International Conference on Pattern Recognition, pp. 2366–2369 (2010). https://doi.org/10.1109/ICPR.2010.579
9. Huang, H.Y., Tsai, W.C.: Blurred image restoration using fast blur-kernel estimation. In: 2014 Tenth International Conference on Intelligent Information Hiding and Multimedia Signal Processing, pp. 435–438, August 2014
10. Knežević, K., Mandić, E., Petrović, R., Stojanović, B.: Blur and motion blur influence on face recognition performance. In: 2018 14th Symposium on Neural Networks and Applications (NEUREL), pp. 1–5, November 2018. https://doi.org/10.1109/NEUREL.2018.8587028
11. Kornprobst, P., Tumblin, J., Durand, F.: Bilateral filtering: theory and applications. Found. Trends Comput. Graph. Vis. **4**, 1–74 (01 2009). https://doi.org/10.1561/0600000020
12. LeCun, Y., et al.: Backpropagation applied to handwritten zip code recognition. Neural Comput. **1**(4), 541–551 (1989). https://doi.org/10.1162/neco.1989.1.4.541
13. Li, F., Shang, C., Li, Y., Yang, J., Shen, Q.: Interpolation with just two nearest neighboring weighted fuzzy rules. IEEE Trans. Fuzzy Syst. **28**(9), 2255–2262 (2020). https://doi.org/10.1109/TFUZZ.2019.2928496
14. Li, Y., et al.: Underwater image high definition display using the multilayer perceptron and color feature-based SRCNN. IEEE Access **7**, 83721–83728 (2019). https://doi.org/10.1109/ACCESS.2019.2925209
15. Lukin, A., Krylov, A., Nasonov, A.: Image interpolation by super-resolution (2010)
16. Martin, D., Fowlkes, C., Tal, D., Malik, J.: A database of human segmented natural images and its application to evaluating segmentation algorithms and measuring ecological statistics. In: Proceedings 8th International Conference Computer Vision, vol. 2, pp. 416–423, July 2001

17. Mishra, D., Mishra, A.: Comparison of blurring techniques for generative adversarial network-based superresolution models: an empirical study. Int. J. Res. Eng. Appl. Manag. **07** (2021)
18. Misra, S., Wu, Y.: Chapter 10 - machine learning assisted segmentation of scanning electron microscopy images of organic-rich shales with feature extraction and feature ranking. In: Misra, S., Li, H., He, J. (eds.) Machine Learning for Subsurface Characterization, pp. 289–314. Gulf Professional Publishing (2020). https://doi.org/10.1016/B978-0-12-817736-5.00010-7. https://www.sciencedirect.com/science/article/pii/B9780128177365000107
19. Motohashi, S., Nagata, T., Goto, T., Aoki, R., Chen, H.: A study on blind image restoration of blurred images using R-MAP. In: 2018 International Workshop on Advanced Image Technology (IWAIT), pp. 1–4, January 2018. https://doi.org/10.1109/IWAIT.2018.8369650
20. Ni, K.S., Nguyen, T.Q.: Adaptable k-nearest neighbor for image interpolation. In: 2008 IEEE International Conference on Acoustics, Speech and Signal Processing, pp. 1297–1300, March 2008. https://doi.org/10.1109/ICASSP.2008.4517855
21. Pandey, R.K., Saha, N., Karmakar, S., Ramakrishnan, A.G.: MSCE: an edge preserving robust loss function for improving super-resolution algorithms (2018)
22. Rukundo, O., Maharaj, B.T.: Optimization of image interpolation based on nearest neighbour algorithm. In: 2014 International Conference on Computer Vision Theory and Applications (VISAPP), vol. 1, pp. 641–647, January 2014
23. Singh, T.: Comparative analysis of image deblurring techniques. Int. J. Comput. Appl. **153**, 39–44 (2016). https://doi.org/10.5120/ijca2016912068
24. Veeramani, T., Rajagopalan, A.N., Seetharaman, G.: Restoration of foggy and motion-blurred road scenes. In: 2013 IEEE International Conference on Image Processing, pp. 928–932, September 2013. https://doi.org/10.1109/ICIP.2013.6738192
25. Wang, R., Li, W., Qin, R., Wu, J.: Blur image classification based on deep learning. In: 2017 IEEE International Conference on Imaging Systems and Techniques (IST), pp. 1–6, October 2017. https://doi.org/10.1109/IST.2017.8261503
26. Wang, Z., Bovik, A., Sheikh, H., Simoncelli, E.: Image quality assessment: from error visibility to structural similarity. IEEE Trans. Image Process. **13**(4), 600–612 (2004). https://doi.org/10.1109/TIP.2003.819861
27. Yang, J., Wright, J., Huang, T., Ma, Y.: Image super-resolution as sparse representation of raw image patches. In: 2008 IEEE Conference on Computer Vision and Pattern Recognition, pp. 1–8, June 2008. https://doi.org/10.1109/CVPR.2008.4587647
28. Yang, J., Wright, J., Huang, T.S., Ma, Y.: Image super-resolution via sparse representation. IEEE Trans. Image Process. **19**(11), 2861–2873 (2010). https://doi.org/10.1109/TIP.2010.2050625
29. Zhang, L., Zhang, L., Mou, X., Zhang, D.: FSIM: a feature similarity index for image quality assessment. IEEE Trans. Image Process. **20**(8), 2378–2386 (2011)
30. Zhang, M., Ling, Q.: Bilateral upsampling network for single image super-resolution with arbitrary scaling factors. IEEE Trans. Image Process. **30**, 4395–4408 (2021). https://doi.org/10.1109/TIP.2021.3071708

A Siamese Neural Network-Based Face Recognition from Masked Faces

Rajdeep Chatterjee[1], Soham Roy[1], and Satyabrata Roy[2](✉)

[1] School of Computer Engineering, KIIT University, Bhubaneswar 751024, India
[2] Manipal University Jaipur, Jaipur-Ajmer Expressway, Rajasthan 303007, India
satyabrata.roy@jaipur.manipal.edu

Abstract. In modern days, face recognition is a critical aspect of security and surveillance. Face recognition techniques are widely used for mobile devices and public surveillance. Occlusion is a challenge while designing face recognition applications. In the COVID19 pandemic, we are advised to wear a face mask in public places. It helps us prevent the droplets from entering our body from a potential COVID19 positive person's nose or mouth. However, it brings difficulty for the security personnel to identify the human face by seeing the partially exposed face. Most of the existing models are built based on the entire human face. It could either fail or perform poorly in the scenario as mentioned above. In this paper, a solution has been proposed by leveraging Siamese neural network for human face recognition from the partial human face. The prototype has been developed on the celebrity faces and validated with the state-of-the-art VGGFace2 (Resnet50) model. Our proposed model has performed well and provides very competitive results of 93% and $84.80 \pm 4.71\%$ best-of-five and mean accuracy for partial face-images, respectively.

Keywords: Deep learning · Face recognition · One-shot learning · Siamese network · VGGface2

1 Introduction

Since late 2019, the World has witnessed a new virus called COVID-19. It started in Wuhan city of China and spread to the rest of the World. Not only does it claim millions of human lives, but also the consequences are catastrophic economically. It brings the entire world to a standstill and disrupting every regular human activity. One way to prevent contact; is the safety precaution and preventive measures given by World Health Organization (WHO). Among these safety precautions, one of the most effective measures is to wear a face mask to prevent the virus from spreading. According to a WHO statement, it says "Masks should be used as part of a comprehensive strategy of measures to suppress transmission and save lives" [1–3]. The face is the most important means of identifying a person in various public places such as ATMs, railway stations, airports, etc. It becomes challenging to recognize a person wearing a mask. Most of the face recognition models have been

I. Woungang et al. (Eds.): ANTIC 2021, CCIS 1534, pp. 517–529, 2022.
https://doi.org/10.1007/978-3-030-96040-7_40

developed based on features extracted from the whole human face. Due to this, many of such algorithms perform poorly with half faces. It is imperative to effectively improve the existing face recognition approaches to learn and distinguish from the in-exposed face. The learning needs to be done only from the upper half of the face, visible without the mask.

A Siamese neural network-based face recognition model from the half-face images has been introduced to overcome the issue. The selection of a Siamese network over a typical Convolutional network for face recognition has been made because of its low computational requirement for training.

The paper has been organized into a total of six sections. Section 2 discusses the related research works. The background concepts have been explained in Sect. 3. The proposed model has been described in Sect. 4. It is followed by Sect. 5 which contains details of the datasets, experimental set-up, and results. Finally, the paper has been concluded in Sect. 6.

2 Related Work

A reasonable number of research articles have been available in recent years on the domain of face recognition. The primary focus is to apply learning generalized face recognition to work across domains. In the paper by Guo et al. [4], they use a 28-layer ResNet as the backbone, but with a channel-number multiplier of 0.5. This model has been implemented on a pre-trained model and trained on the Full face dataset. Accuracy of over 99% has been achieved. In the paper by Ling et al. [5], they have introduced an attention-based mechanism that uses standard convolutional neural networks (CNNs), such as ResNet-50, ResNet-101. Their work has enabled more discriminative power for deep face recognition.

In the paper by Song et al. [6], they propose to discard feature elements that occlusions have corrupted. They identify face occlusions using PDSN (Pairwise difference Siamese Network) and created an algorithm to recognize the face. However, the occlusions are randomly spread out all over the face. It is not necessarily restricted to the face portion from the nose. They used the Facescrub dataset with training data of 0.5 million images and gained an accuracy of 99.20%.

In the paper by Guo and Zhang [7], they study face recognition of imbalanced train image classes under different lightning, age pose, and variations. They choose a 34 layer standard residual network (ResNet-34) as their feature extractor. They introduced a concept called Classification vector-centered Cosine Similarity to train a better face feature extractor. A concept is introduced called underrepresented-classes promotion, which effectively addresses the data imbalance problem in one-shot learning. The recognition coverage rate has been increased from 25.65% to 94.89% at the precision of 99% for one-shot classes, while still keep an overall accuracy of 99.8% for regular classes.

Most of the work in this domain has been done using deep learning. These models are trained on multitudes of layers and millions of parameters. Many of them are already using pre-trained models and improving upon them. Needless to say that they are very hardware intensive and time-consuming. Although most of them have used full faces to extract facial features, few studies have been done with occlusions, a wide array of lightning, ambiance, and age to improve the model's applicability.

Currently, in the trying times as everyone is wearing a mask in public places, it is almost difficult for the existing models to recognize the face effectively. The problem in hand requires the usage of the face where the portion below the nose is in-exposed while taking into account only the top half of the face. Therefore, the new approach to dealing with the problem requires focusing on the face region, precisely above the tip of the nose. To the best of our knowledge, such an approach is not available in the literature. Here, a Siamese neural network-based face recognition model has been proposed to address the issue.

3 Background Concepts

"Siamese network" is a type of neural network architecture that has been first introduced by Bromley et al. [8] for signature verification work purposes. They have trained two similar *Siamese* neural networks having shared the same parameters and weights, which gave an output of two feature vectors when two input signatures are fed to the network. The outcome of two signatures is two vectors. These vectors are compared based on some distance measure that has been used as a loss function during learning. Later, Siamese network has been incorporated into other aspects in computer vision tasks including face verification [9], one-shot image recognition [10]. The crux of the Siamese neural network is to learn general feature representations with a distance (or similarity) metric calculated from the feature vectors from two similar inputs (images in our case). Siamese neural network is a class of network architectures that usually contains two identical networks. The two networks have the same number of layers and configurations with the same parameters and shared weights. The parameter updating in one network is reflected across the other network since the configuration is shared. This framework has been successfully used for dimensionality reduction in weakly supervised metric learning and face verification in [11]. The top layer of these networks consists of a loss function, which computes the similarity or dissimilarity score using the Euclidean distance or the cosine similarity between the feature vector representation on both networks. Two such popular loss functions used with the Siamese network are the contrastive loss [12] and the Triplet Loss [11]. In our work, we have used the Contrastive Loss function (see Eq. 1), which is defined as follows:

$$L(i_1, i_2, Label) = \alpha \times (1 - Label)D_w^2 + \beta \times Label \times max(0, m - D_w)^2 \quad (1)$$

Where i_1 and i_2 are two samples (image of the portion of the face above the tip of the nose), we refer to them as images in the rest of the paper for simplicity. The label is a binary value showing whether the two face images belong to the same class or not; α and β are two constants, both are set to 1 in our study, and m is the margin equal to 2. In Eq. 2, f is a function that maps an image to a vector feature space, i indicates an image, l denotes the index of an image (from image-bank), and w is the learnt weight of the underlying network.

$$\vec{x}_l = f(i_l, w) \tag{2}$$

Equation 3, is the pair-wise Euclidean distance computed from the feature vector representation from the two images through the network. Here, n is the number of embedding vector lengths (256), the model output.

$$D_w = \left(\sum_{j=1}^{n} |x_{1j} - x_{2j}|^p \right)^{\frac{1}{p}} \tag{3}$$

A Siamese network's objective is to make the feature vector representation of the input images that have the same class label closer and push away the feature vector representation of the input images are different class labels. Because of the Contrastive loss function (Eq. 1), after the learning/training stage of the model, the feature vector output has the characteristic that the Euclidean distance of the images of the same class is closer to the images of different classes. To decide whether two images belong to a similar class (label as 0) or a different class (label as 1), we need to determine a threshold value on the cosine dissimilarity of the distance between the embedded vector representations. This step is typically determined by training the network and studying similarity scores from genuine-genuine and genuine-fake images. A top K match is considered as qualification criteria from the image bank based on the threshold value.

VGGFace2 model [13] developed by researchers at the Visual Geometry Group at Oxford. The dataset has been prepared before modeling by the VGG team itself. Models are trained on the dataset, specifically a ResNet-50 and a SqueezeNet-ResNet-50 model (SE-ResNet-50 or SENet). Given an image as an input, the model gives the output of a 2048 vector embedding. The vector's length is normalized using the $L2$ vector norm (Euclidean distance from the origin). They obtain a vector for each image called a face descriptor. The way they get the face descriptor is that the extended bounding box of the face is resized to make the shorter side of the image 256 pixels; then the center 224×224 dimension crop of the face image is used as input to the network. The face descriptor is extracted from the layer adjacent to the classifier layer, the last layer of the model. It leads to a 2048 length vector representation of the image, which is then L2 normalized. The similarity between the two images is calculated by the cosine similarity function from the two face descriptors. The deep models (ResNet-50 and SENet) trained on VGGFace2 achieve state-of-the-art performance on the IJB-A, IJB-B, and IJB-C benchmarks.

4 Proposed Model

A research workflow diagram has been given in Fig. 1. It shows that the used pipeline takes an image or video as input and detects faces using Multi-task Cascaded Convolutional Neural Network (MTCNN) [14] algorithm. Subsequently,

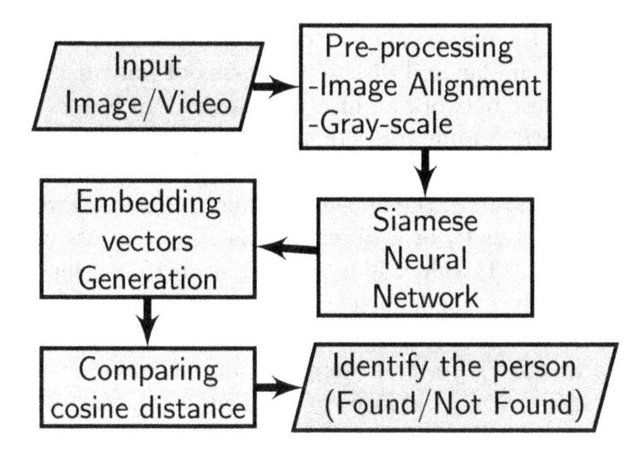

Fig. 1. Algorithm workflow of this paper

input image pre-processing is performed, such as gray scaling, face alignment, etc. Then, the proposed model has been used to match partial face image input with the existing face image-bank. Cosine distance metric has been employed to find out the face similarity.

In our proposed model, a Siamese neural network (SNN) is to be prepared for lightweight face image recognition. The model has been used to get the feature representation of an image when it is passed through the network. The model has two sub-networks, as with a Siamese network having shared weights and biases. An extra padding layer is required all along the layers to keep the image's dimension constant. The model has six layers in total. The first layer is a convolution $2d$ layer with an input channel of 1 as the image is gray-scale and the output has 4 channels. We use ReLu (Rectified Linear Unit) activation layer and batch normalization for the four output channels. We are using a kernel size of 3 and padding up with 1 to keep the image dimension the same. The second layer of the image is almost similar to the first one. In this layer, the only difference is that the input is of four channels and the output is eight channels, and the batch normalization is happening across all eight channels. The third layer is also familiar; here, the number of input channels is eight, and no output channel is also eight, followed by batch normalization. At the end of the third layer, we connect it to a fully connected sequential layer with the input dimension of $8 \times 100 \times 100$ and 512 nodes. In the fully connected layers also we are using the ReLu activation function. The second layer in the Linear space or the fifth layer in the model is a linear layer with input and output dimensions as 512 and 512. We are using ReLu activation here. Our model's final layer is linear with a 512 input dimension and a 256 output dimension. This 256 output dimension is our model's final output, which is the feature embedding of an input image when passed through the network. The Contrastive loss function is the objective function to learn the similar and dissimilar classes of images. A block diagram of the proposed Siamese network architecture is given in Fig. 2.

After the learning/training stage of the model, the last layer output is a vector of length 256. It has the characteristic that the Euclidean distance of the same class's images is closer to the images of different classes. Two images belong to a similar class (label as 0) or a different class (label as 1), which is decided on a threshold (Th. = 0.1) value. It is calculated on the cosine distance of the dissimilarity between the embedded vector representations. This step is typically determined by a rigorous study of the similarity scores from the genuine-genuine and genuine-fake images. After this step, we can determine a top K match for an anchor image(cropped masked face) in an image bank (consisting of the individuals' full faces). A bird's-eye view of the implementation has been given in Fig. 3.

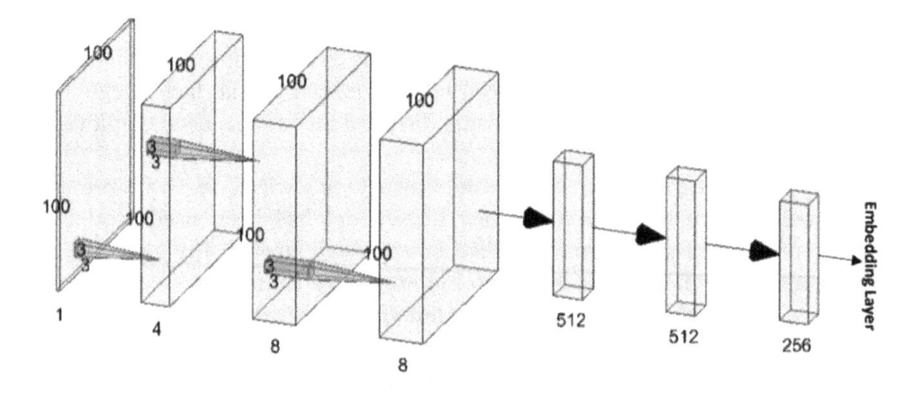

Fig. 2. Proposed Siamese Neural Network Architecture

5 Experimental Results and Analysis

5.1 Datasets

The dataset[1] contains 20 celebrity faces (*Blake Lively, Brad Pitt, Brendan Gleeson, Brian Cox, Brie Larson, Britney Spears, Brittany Snow, Bruce Lee, Bruno Mars, Cameron Diaz, Cate Blanchett, Chadwick Boseman, Chris Hemsworth, Chris Martin, Christian Bale, Emma Roberts, Emma Stone, Emma Watson, Eric Bana, Ethan Hawke*), with the images in each class varying in age, luminosity, brightness, expression and face alignment, and visual quality. The purpose of this is to generalize the data as much as possible for each class. Each folder contains around 100 images for a celebrity. Since a mask can be worn in varying

[1] Dataset: https://github.com/prateekmehta59/Celebrity-Face-Recognition-Dataset/blob/master/README.md.

degrees, it can be worn too high or low. The used dataset has been prepared in a way to resemble the realistic situation. A test dataset is also created from the same source data containing only 30 images per person.

5.2 System Configuration

The paper is implemented using Python 3.7 and PyTorch (GPU) 1.5.1 on an Intel(R) Core(TM) $i7 - 9750H$ CPU (9th Gen.) 2.60 GHz, 16 GB RAM and 6 GB NVIDIA GeForce RTX 2060 with 64 bits Windows 10 Home operating system. Again, the same codes have been validated on the Online Google Colab environment with more learning parameters. For implementing the VGGFace2 model on the same datasets for comparison, the Tensorflow-GPU 1.14 has been used simultaneously. The dimension of the output embedding is 256.

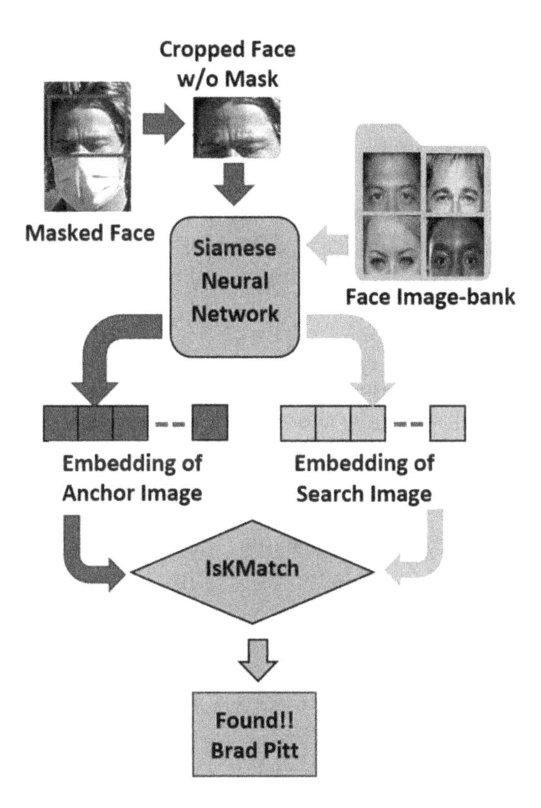

Fig. 3. Bird's-eye view of the proposed Siamese Neural Network pipeline

Three types of optimization techniques are used to explore the best possible model. These are AdaBelief (lr = $1e{-}3$, eps = $1e{-}16$ and betas = $(0.9, 0.999)$), Adam (lr = $1e{-}3$, weight_decay = 0.0005) and Stochastic Gradient Descent (SGD) (lr = $1e{-}3$, momentum = 0.9) [15,16].

5.3 Results

The proposed model has been implemented with different configurations. The first experiment (test-case-I): the configurations vary based on embedding vector length, threshold, optimization algorithms, and the number of learning epochs, etc. Three distinct types of optimization techniques have been employed to observe their effects. It is noticed that AdaBelief outperforms others in early epochs in terms of the total loss. However, SGD performs better than Adam and AdaBelief algorithms in longer iterations. The obtained results are given in Tables 1 and 2.

The best model from our experiments has been used on a randomly generated face test-image-bank (containing 20 persons × 35 half face-images) for evaluating the model accuracy (%). The second experiment (test-case-II): The anchor image is a half-face, and the test-image-bank contains half faces. It is very satisfactory that the proposed model built from scratch provides very efficient results in our experiments. The experiment has been executed 5 times independently to examine its robustness. The empirical data has been given in Table 3. The best of five (#Bo5) accuracy, the mean accuracy, and the standard deviation are 93%, 84.80%, and ±4.71, respectively.

Table 1. Results obtained from different optimization algorithms and half-face images using the proposed model where the threshold (Th.) = 0.1 and the embedding vector length = 64

Embedding length	Optim	Epoch	Total loss	Time (min.)
64	AdaBelief	100	0.0567	1.7320
	Adam		0.1595	1.7560
	SGD		0.0751	1.6506
	AdaBelief	500	0.0035	9.1326
	Adam		0.0493	8.9267
	SGD		0.0098	8.5420

Table 2. Results obtained from different optimization algorithms and half-face images using the proposed model where the threshold (Th.) = 0.1 and the embedding vector length = 256

Embedding length	Optim	Epoch	Total loss	Time (min.)
256	AdaBelief	100	0.2110	1.5600
	Adam		0.3450	1.6500
	SGD		0.2025	1.6300
	AdaBelief	500	0.0219	9.7975
	Adam		0.0508	9.4183
	SGD		0.0170	8.4834

Table 3. Results obtained from the proposed model for 5 independent runs with the threshold (Th.) = 0.1 and embedding vector length = 256

#Run	Accuracy (%)	Time (s)	Mean Acc.	Std.
1	82	3.1524	84.80	±04.71
2	79	2.9113		
3	**93**	3.0011		
4	84	2.8905		
5	86	3.2021		

In the third experiment (test-case-III): four images (*68bradpitt.png, 4blake-lively.png, 22chadwikboseman.png, and 45emmastone.png*) have been randomly picked from the test set, and a comparison between the proposed best model and the VGGFace2 have been made. The four images are of celebrities Brad Pitt, Blake Lively, Chadwick Boseman, and Emma Stone. A K-shot learning [17] approach has been used to recognize the actual person from a partial face image (anchor). Here, $K \geq 10$ is used. It suggests that the similarity score between 10 or more images and the anchor image is always less than or equals 0.1 (this threshold is achieved through an elbow technique). In literature, it is instructed to use 0.5 as the threshold value for VGGFace2 (if the similarity score falls below 0.5, it is considered a match). Even for the VGGFace2 implementation, $K \geq 10$ matches have been used to qualify a match.

In Table 4, the performance of our proposed Siamese neural network-based model outperforms the state-of-the-art VGGFace2 in terms of average time taken for searching and accuracy. It must be noted that the complete image bank has 20 different personalities, and the test-case-III has been implemented for the whole test dataset without the first match-only match (FMOM) policy.

Table 4. Results obtained from different celebrity half-face images using ours and VGGFace2 models, respectively. K = 10 indicates the dissimilarity values between anchor image and at least 10 images fall below the threshold (Th.)

Celebs (Anchor Image)	Models	Th.	Total Time	Avg. Time	Searched Images	Top Searched Names (K=10)	Scores
	Ours	0.1	0.0515	0.000098	531	Brad Pitt	0.0015
						Brendan	0.7730
						Chris martin	0.0136
	VGGFace2	0.5	0.5585	0.000112	549	Brad Pitt	0.2081
						Chris Hemsworth	0.4953
	Ours	0.1	0.4991	0.000092	541	Blake Lively	0.0011
	VGGFace2	0.5	0.5651	0.000128	442	Blake Lively	0.3154
						Cate Blanchett	0.3963
						Brad Pitt	0.4197
						Emma Stone	0.4252
						Brittany Snow	0.4263
	Ours	0.1	0.5286	0.000096	551	Chadwik Boseman	0.0122
	VGGFace2	0.5	0.5494	0.000104	529	Chadwik Boseman	0.2577
						Bruno Mars	0.4347
	Ours	0.1	0.5183	0.000094	553	Emma Stone	0.0842
	VGGFace2	0.5	0.5589	0.000124	452	Emma Stone	0.2769
						Cate Blanchett	0.3639
						Blake Lively	0.4284
						Chris Hemsworth	0.4350
						Brittany Snow	0.4476

All test cases (I, II, and III) are tested on the known faces (suggests model built using the same persons' images). Here, the model has been tested on few unknown faces (not used in learning). Two prominent examples are shown in Figs. 4, 5 and 6. In Figs. 4 and 5, comparisons have been given for Tom Cruise and Keanu Reeves based on both the full and half face similarity. Similarly, a dissimilarity comparison has also been shown in Fig. 6. Both Tom Cruise and Keanu Reeves face images have not been used while the model building; even then, the obtained results are accurate using the same embedding scheme and adopted threshold.

Similarity Score: 0.0000 (Th.=0.1)
They are same person, Tom Cruise

(a) Proposed model on unkonwn full face

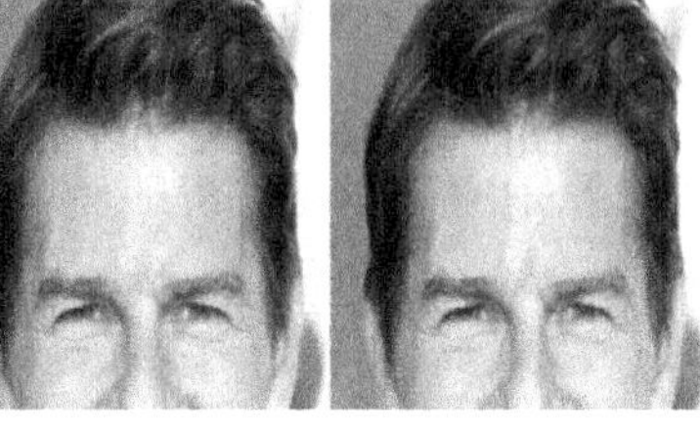

Similarity Score: 0.0000 (Th.=0.1)
They are same person, Tom Cruise

(b) Proposed model on unkonwn half face

Fig. 4. Example of Tom Cruise face similarity using the proposed model, where the embedding vector length = 256

Similarity Score: 0.0000 (Th.=0.1)
They are same person, Keanu Reeves

(a) Proposed model on unkonwn full face

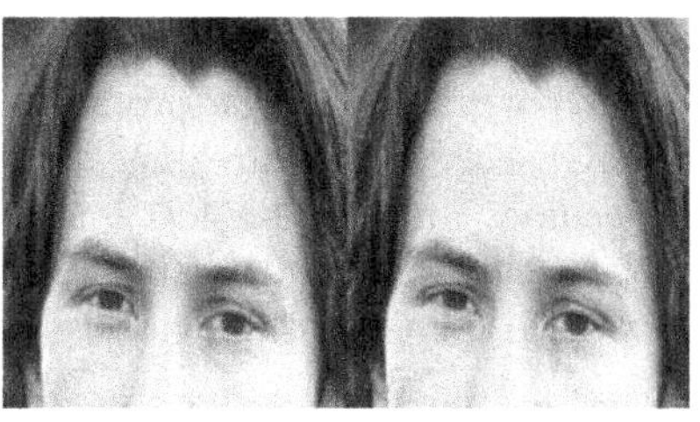

Similarity Score: 0.0000 (Th.=0.1)
They are same person, Keanu Reeves

(b) Proposed model on unkonwn half face

Fig. 5. Example of Keanu Reeves face similarity using the proposed model, where the embedding vector length = 256

Similarity Score: 0.4456 (Th.=0.1)
They are different persons.

(a) Proposed model on unkonwn full face

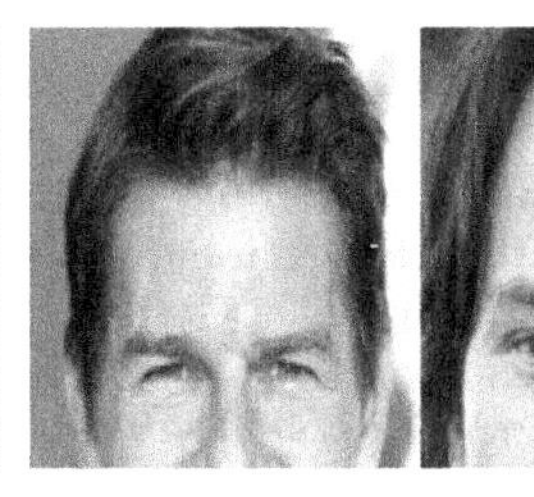

Similarity Score: 0.3062 (Th.=0.1)
They are different persons.

(b) Proposed model on unkonwn half face

Fig. 6. Example of face dissimilarity using the proposed model, where the embedding vector length = 256

The first adaption of the model, which we have built from scratch, gives encouraging results compared with the state-of-the-art. However, as the model is constructed from a moderate-sized dataset and constrained resources (hardware), we believe that if the architecture is tested in a broader dataset run on a high-performing machine, the robustness of the model would undoubtedly improve.

6 Conclusion

Deep learning-based face detection and recognition are typical computer vision applications. However, there is still room for improvement in human face recognition. It is already established that CNN-based deep learning requires many input images and takes a more prolonged training time. On the other hand, a Siamese neural network can work with fewer input data and provides similarity or dissimilarity scores between any two faces. The proposed SNN model developed based on half of the human face gives very competitive results in identifying a person from only the exposed upper portion of the face (mainly above the tip of the nose or nostrils). The proposed model takes less search-time/per image than VGGFace2 while searching a face. The proposed SNN model gives 93% #Bo5 accuracy and $84.80 \pm 4.71\%$ mean accuracy for the partial face matching.

In the future, the model can be extended to train based on a larger dataset and examined on a real-time video.

References

1. Eikenberry, S.E., et al.: To mask or not to mask: modeling the potential for face mask use by the general public to curtail the COVID-19 pandemic. Infect. Dis. Model. **5**, 293–308 (2020)
2. Li, T., Liu, Y., Li, M., Qian, X., Dai, S.Y.: Mask or no mask for COVID-19: a public health and market study. PLoS ONE **15**(8), e0237691 (2020)
3. Loey, M., Manogaran, G., Taha, M.H.N., Khalifa, N.E.M.: A hybrid deep transfer learning model with machine learning methods for face mask detection in the era of the COVID-19 pandemic. Measurement **167**, 108288 (2021)
4. Guo, J., Zhu, X., Zhao, C., Cao, D., Lei, Z., Li, S.Z.: Learning meta face recognition in unseen domains. In: Proceedings of the IEEE/CVF Conference on Computer Vision and Pattern Recognition, pp. 6163–6172 (2020)
5. Ling, H., Wu, J., Huang, J., Chen, J., Li, P.: Attention-based convolutional neural network for deep face recognition. Multimedia Tools Appl. **79**, 5595–5616 (2019). https://doi.org/10.1007/s11042-019-08422-2
6. Song, L., Gong, D., Li, Z., Liu, C., Liu, W.: Occlusion robust face recognition based on mask learning with pairwise differential Siamese network. In: Proceedings of the IEEE/CVF International Conference on Computer Vision, pp. 773–782 (2019)
7. Guo, Y., Zhang, L.: One-shot face recognition by promoting underrepresented classes. arXiv preprint arXiv:1707.05574 (2017)
8. Bromley, J., Guyon, I., LeCun, Y., Säckinger, E., Shah, R.: Signature verification using a "Siamese" time delay neural network. In: Advances in Neural Information Processing Systems, pp. 737–744 (1994)

9. Taigman, Y., Yang, M., Ranzato, M.A., Wolf, L.: DeepFace: closing the gap to human-level performance in face verification. In: Proceedings of the IEEE Conference on Computer Vision and Pattern Recognition, pp. 1701–1708 (2014)
10. Koch, G., Zemel, R., Salakhutdinov, R.: Siamese neural networks for one-shot image recognition. In: ICML Deep Learning Workshop, Lille, vol. 2 (2015)
11. Schroff, F., Kalenichenko, D., Philbin, J.: FaceNet: a unified embedding for face recognition and clustering. In: Proceedings of the IEEE Conference on Computer Vision and Pattern Recognition, pp. 815–823 (2015)
12. Chopra, S., Hadsell, R., LeCun, Y.: Learning a similarity metric discriminatively, with application to face verification. In: 2005 IEEE Computer Society Conference on Computer Vision and Pattern Recognition (CVPR'05), vol. 1, pp. 539–546. IEEE (2005)
13. Cao, Q., Shen, L., Xie, W., Parkhi, O.M., Zisserman, A.: VGGFace2: a dataset for recognising faces across pose and age. In: 2018 13th IEEE International Conference on Automatic Face & Gesture Recognition (FG 2018), pp. 67–74. IEEE (2018)
14. Zhang, K., Zhang, Z., Li, Z., Qiao, Yu.: Joint face detection and alignment using multitask cascaded convolutional networks. IEEE Signal Process. Lett. **23**(10), 1499–1503 (2016)
15. Zhuang, J., et al.: AdaBelief optimizer: adapting stepsizes by the belief in observed gradients. arXiv preprint arXiv:2010.07468 (2020)
16. Ketkar, N.: Introduction to PyTorch. In: Deep Learning with Python, pp. 195–208. Springer, Cham (2017). https://doi.org/10.1007/978-1-4842-2766-4_12
17. Sung, F., Yang, Y., Zhang, L., Xiang, T., Torr, P.H.S., Hospedales, T.M.: Learning to compare: relation network for few-shot learning. In: Proceedings of the IEEE Conference on Computer Vision and Pattern Recognition, pp. 1199–1208 (2018)

MCDA Based Swimmers Performance Measurement System

Jakub Więckowski[1], Aleksandra Bączkiewicz[2], Bartłomiej Kizielewicz[1], Andrii Shekhovtsov[1], and Wojciech Sałabun[1](✉)

[1] West Pomeranian University of Technology in Szczecin, ul. Żołnierska 49, 71-210 Szczecin, Poland
wojciech.salabun@zut.edu.pl

[2] University of Szczecin, ul. Cukrowa 8, 71-004 Szczecin, Poland

Abstract. MCDM methods are effectively used in many practical areas and scientific disciplines. These methods' primary tasks are to evaluate alternatives from the point of view of multiple criteria and build rankings results according to the decision maker's preferences. On the other hand, MCDM methodology and current research's versatility confirm the usefulness of these methods both in the area of building synthetic indicators of assessing the achievement of specific goals and their broader use in the area of performance measurement. Due to the existing research gap involving the lack of usage of MCDM methods in the broad area of education performance measurement, this article attempts to present an MCDM-based approach for swimmers' performance measurement. In addition to the indicated practical contribution, the article also addresses methodological challenges. Using the VIKOR and TOPSIS methods, the influence of input data preprocessing form (techniques of normalization of input data) and uncertainty of measurement data (interval representation of data) on the resultant ranking.

Keywords: Sport management · Uncertain data · MCDM · Decision support

1 Introduction

Decision making is a problem that arises in every aspect of life today. Problems related to decision making are found often in sports [44], e-commerce systems [2], energy [3], supplier selection [50,56], medicine [4] or construction [8]. The offered options are often similar to each other, making it difficult to choose the optimal solutions. The problems occurring during the decision-making process are influenced by many criteria and the definition of a hierarchy of criteria concerning each other [60]. Thus, the determination of optimal choices depends, to a large extent, on the expert knowledge that has been used to solve the problem. To support the expert in making decisions, many methods are used as benchmarks to assess the quality of the offered alternative in the context of the whole

I. Woungang et al. (Eds.): ANTIC 2021, CCIS 1534, pp. 530–545, 2022.
https://doi.org/10.1007/978-3-030-96040-7_41

set of alternatives under consideration [46]. Over time, more and more of such models are being developed, and their quality can be seen in the fact that they are willing to use them for problems requiring the selection of optimal solutions.

The issue of taking a multi-criteria approach into account in the problem-solving process also arises in sport. Recently, the sports environment, coaches and players have significantly increased usage of intelligent methods in the training process [16]. Their application allows the introduction of training methods that ensure better results while maintaining a sustainable exploitation level. Besides, subjects such as, neural networks or machine learning are used in models of prediction of athletes' performance and models assessing the physical conditions of athletes for a given sport and competition [5]. Increased interest in intelligent training support systems on the part of sports clubs results in specialists' increased attention to the development of such systems. The consequence of such actions is the possibility of improving the results achieved and maximising athletes' potential using these systems.

The methods belonging to the MCDM (Multi-Criteria Decision Making) group of methods can be used to create a multi-criteria model to assess the set of options [24,43]. They are of high interest to experts trying to solve problems concerning selecting optimal decisions with numerous criteria that determine alternatives. Two of the chosen methods, namely VIKOR (Vlsekriterijumska Optimizacija I KOmpromisno Resenje) and TOPSIS (Technique for Order Preference by Similarity to an Ideal Solution), are based on a distance-based approach [68]. These methods have been used to solve risk assessment problems [27], selection of material suppliers [70], resource management planning [34] and sports [11]. Due to their high practical potential, MCDM methods are widely used in the area of performance measurement. The literature analysis shows many successful attempts to adapt this methodology for the needs of performance evaluation, e.g. banks [13], logistics and supply chain [40] or financial performance [1]. The current state of the art is available, for example, in the works [57] and [12]. However, the analysis of the area of MCDM in education reveals a clear gap. The available works are either not directly oriented to assessing educational effects of a given block of knowledge or course, but the assessment of the performance of whole units [69] or elements of their strategies [9].

It is also worth pointing out the motivation for using MCDA (Multi-Criteria Decision Analysis) methods, which have already been used to create assessment models in many sports. Various multi-criteria methods are selected to design such models, with attempts to ensure that the performance characteristics of the methods are suited to the assessment problem at hand. For example, the MCDA-based approach in evaluating basketball players in [23] considered incomplete knowledge, and the model itself was based on the COMET (Characteristic Objects Method) method. In the football theme, an attempt was made to identify the winner of the FIFA 2014 Golden Ball Award, and the evaluation of the players was made based on the AHP method [31]. E-sports ranking identification presented in [64], was based on the COMET method and Triangular Fuzzy Numbers (TFNs), and the results obtained by players of a popular e-sports game were evaluated to create a ranking of these players. The system presented in [10] for assessing basketball players' performance and teams were developed based on the TOPSIS method. MCDA methods are also used in issues directly related to sport, including sports institutions' management or the construction of sports facilities. Problems such as the evaluation of factors delaying the creation of sports facilities [17], the assessment of sports management journals [53] or the study of the quality of performance of sports organizations in the management of these organizations [32] are addressed. The analysis of the field of sport has shown that MCDA methods are frequently used for diverse problems within this topic.

The literature studies clearly indicate, that the results obtained with the different MCDM methods are inconsistent [55]. It is due to many reasons. The algorithms of the various MCDM methods differ significantly [41,42], the modelling of weights [61] or uncertainties in the models are also different. According to many researchers [15,19,38], in solving problems using MCDM methods, an important aspect is determining the connection between model's input data and the results obtained. Many MCDM methods require the input data to be in a normalized form [39] using one of the available normalization techniques [7]. As pointed in work [37,46,66], this creates a natural background to study the influence of the form of the model input data (including the normalization technique used) on the form of the resulting model [47,72]. The need for objectified comparisons of the resulting rankings shows that the similarity coefficients can be used to quantify the similarity of two rankings [45,51]. It is an effective way of checking whether the results obtained from different methods guarantee a meaningful solution to the problem [49].

It was therefore decided to use the TOPSIS and VIKOR methods in this study. Both methods are based on the same principles of so-called "Reference Points" [25]. They differ only in the adopted aggregation function of assessments and in the way of input data normalization [36]. In the TOPSIS method, the aggregation function maximizes the distance to the anti-ideal solution and minimizes the distance to the ideal solution [52]. In the case of the VIKOR method, the aggregation function only minimizes the distance to the ideal solution [71]. As far as normalization is concerned, in the VIKOR method data normalization is linear, while in the TOPSIS method it is vectorial [35].

In this paper, swimmers performance measurement system is proposed. The influence of different normalization techniques on the form of the final ranking of alternatives was investigated. The VIKOR method and the expansion on interval arithmetic of the TOPSIS method were used for comparison purposes. In the TOPSIS method, five normalization methods were used, when in the VIKOR method, raw data were taken into consideration while solving the problem. Spearman correlation coefficient, Pearson correlation coefficient and WS similarity coefficient were used to compare the resulting rankings.

The rest of the paper is organised as follows. Section 2 provides an introduction to the basic assumptions of the VIKOR method, the TOPSIS method and similarity coefficients. Section 3 contains a study case in which the MCDM methods were used to solve a complex multi-criteria problem. Section 4 is a comparison of rankings obtained from VIKOR and TOPSIS methods using selected similarity coefficients. A summary of the results and conclusions from the conducted research are presented in Sect. 5.

2 Preliminaries

2.1 VIKOR Method

Opricovic developed the basic principles of the VIKOR (Vlsekriterijumska Optimizacija I KOmpromisno Resenje) method in 1998 [35,36]. The method covers issues such as the necessity to define the vector of weights for criteria based on expert knowledge, modifications of the decision matrix, and calculating the distance to the ideal solution based on the collected set of alternatives [29,59]. The final ranking of alternatives is calculated based on the obtained preferences [48]. The received ranking depends largely on the initial vector of weights for the criteria using expert knowledge. This vector should meet the following condition (1):

$$\sum_{i=1}^{N} w_i = 1, \quad \text{, where } N \text{ is a number of criteria} \tag{1}$$

Each criterion is defined as one of two types of criterion (2). If it is a cost type, its value should be as low as possible. Where a criterion is of the profit type, it should be of the highest possible value. Each value of a criterion is then multiplied by the predefined weighting for that criterion (3).

$$\begin{array}{lll} f_i^* = max_j f_{ij}, & f_i^- = min_j f_{ij} & \text{if the } i-th \text{ criteria is a profit type;} \\ f_i^* = min_j f_{ij}, & f_i^- = max_j f_{ij} & \text{if the } i-th \text{ criteria is a cost type} \end{array} \tag{2}$$

$$w_j \cdot \frac{f_j^* - f_{ij}}{f_j^* - f_i^-} \tag{3}$$

The distance to the ideal solution is then calculated based on previously obtained transformations of the decision matrix. The formulas used result in three final rankings S (4), R (5) and Q (6).

$$S_i = \sum_{j=1}^{N} w_j \cdot \frac{f_j^* - f_{ij}}{f_j^* - f_i^-} \tag{4}$$

$$R_i = max_j \left[w_j \cdot \frac{f_j^* - f_{ij}}{f_j^* - f_i^-} \right] \tag{5}$$

$$Q_i = v \cdot \frac{S_i - S^*}{S^- - S^*} + (1 - v) \cdot \frac{R_i - R^*}{R^- - R^*} \tag{6}$$

2.2 TOPSIS Method

The TOPSIS method (Technique for Order Preference by Similarity to an Ideal Solution) was developed in 1992 [25,65]. This technique is combined with other MCDM methods to acquire new features such as robustness to the ranking reversal paradox [20]. Chen and Hwang proposed an approach where the main idea was to calculate the distance to the ideal solution [18,58]. Based on these distances, a ranking of preferences for alternatives was then calculated [22]. The defined decision matrix should be normalized to obtain correct final results [6,33]. After that, a weighted normalized decision matrix should be calculated using the formula presented below (7):

$$\begin{array}{ll} v_{ij} = w_i \cdot r_{ij} & \text{, where } r_{ij} \text{ is a value from decision matrix} \\ j = 1, \ldots, J & \text{, where } J \text{ is a number of alternatives} \\ i = 1, \ldots, N & \text{, where } N \text{ is a number of criteria} \end{array} \tag{7}$$

Positive and negative ideal solutions for a defined decision-making problem should also be identified (8):

$$\begin{aligned} A^* &= \{v_1^*, \ldots, v_n^*\} = \left\{\left(\max_j v_{ij} | i \in I^P\right), \left(\min_j v_{ij} | i \in I^C\right)\right\} \\ A^- &= \left\{v_1^-, \ldots, v_n^-\right\} = \left\{\left(\min_j v_{ij} | i \in I^P\right), \left(\max_j v_{ij} | i \in I^C\right)\right\} \end{aligned} \tag{8}$$

where I^C stands for cost type criteria and I^P for profit type.

Negative and positive distance from an ideal solution should be calculated using the n-dimensional Euclidean distance. To apply such calculations, formula presented below should be used (9):

$$\begin{aligned} D_j^* &= \sqrt{\sum_{i=1}^n \left(v_{ij} - v_i^*\right)^2}, \quad j = 1, \ldots, J \\ D_j^- &= \sqrt{\sum_{i=1}^n \left(v_{ij} - v_i^-\right)^2}, \quad j = 1, \ldots, J \end{aligned} \tag{9}$$

The last step is to calculate the relative closeness to the ideal solution (10):

$$C_j^* = \frac{D_j^-}{\left(D_j^* + D_j^-\right)}, \quad j = 1, \ldots, J \tag{10}$$

2.3 Interval TOPSIS Method

Main assumption of this extension of the conventional TOPSIS method is that each element of the decision matrix is represented by an interval value [14,62]. The subsequent steps to calculate the final alternatives preferences are presented below.

Normalization of the interval-valued decision matrix performed for the lower (x_{ij}^L) and upper (x_{ij}^U) bound of the interval (11):

$$\begin{aligned} r_{ij}^L &= \frac{x_{ij}^L}{\left(\sum_{k=1}^m \left(\left(x_{kj}^L\right)^2 + \left(x_{kj}^U\right)^2\right)\right)^{\frac{1}{2}}}, \quad i = 1, \ldots, m, \quad j = 1, \ldots, n \\ r_{ij}^U &= \frac{x_{ij}^U}{\left(\sum_{k=1}^m \left(\left(x_{kj}^L\right)^2 + \left(x_{kj}^U\right)^2\right)\right)^{\frac{1}{2}}}, \quad i = 1, \ldots, m, \quad j = 1, \ldots, n \end{aligned} \tag{11}$$

Calculation of the weighted values for each interval value (12):

$$\begin{aligned} \nu_{ij}^{L} &= w_j \cdot r_{ij}^{L}, \quad i = 1, \ldots, m; \quad j = 1, \ldots, n \\ \nu_{ij}^{U} &= w_j \cdot r_{ij}^{U}, \quad i = 1, \ldots, m; \quad j = 1, \ldots, n \end{aligned} \tag{12}$$

Identification of the Positive and Negative Ideal Solution as follows (13):

$$\begin{aligned} A^{+} &= \left\{v_1^{+}, v_2^{+}, \ldots, v_n^{+}\right\} = \left\{\left(\max_i v_{ij}^{U} | j \in I^{P}\right), \left(\min_i v_{ij}^{L} | j \in I^{C}\right)\right\} \\ A^{-} &= \left\{v_1^{-}, v_2^{-}, \ldots, v_n^{-}\right\} = \left\{\left(\min_i v_{ij}^{L} | j_i \in I^{P}\right), \left(\max_i v_{ij}^{U} | j \in I^{C}\right)\right\} \end{aligned} \tag{13}$$

where I^P stands for profit type criteria and I^C for cost type.

Calculation of the distance from the Ideal Solution is described as (14):

$$\begin{aligned} S_i^{+} &= \left\{\sum_{j \in K_b} \left(v_{ij}^{L} - v_j^{+}\right)^2 + \sum_{j \in K_c} \left(v_{ij}^{U} - v_j^{+}\right)^2\right\}^{\frac{1}{2}}, \quad i = 1, \ldots, m \\ S_i^{-} &= \left\{\sum_{j \in K_b} \left(v_{ij}^{U} - v_j^{-}\right)^2 + \sum_{j \in K_c} \left(v_{ij}^{L} - v_j^{-}\right)^2\right\}^{\frac{1}{2}}, \quad i = 1, \ldots, m \end{aligned} \tag{14}$$

Calculation the relative closeness to the Ideal Solution (15):

$$RC_i = \frac{S_i^{-}}{S_i^{+} + S_i^{-}}, \quad i = 1, 2, \ldots, m \quad 0 \leq RC_i \leq 1 \tag{15}$$

2.4 Similarity Coefficients

Similarity coefficients make it possible to compare the two rankings in terms of the positions obtained or the value of preferences by successive alternatives [21]. The higher the coefficient value achieved, the greater the similarity of the compared rankings [67]. To compare the rankings obtained with the use of MCDM methods, three coefficients were selected: Spearman correlation coefficient (16), Pearson correlation coefficient (17) and WS similarity coefficient (18).

$$r_s = 1 - \frac{6 \cdot \sum d_i^2}{N \cdot (N^2 - 1)} \tag{16}$$

where d_i - distance between ranked alternatives in both rankings, N - amount of alternatives.

$$\rho(a, b) = \frac{E(ab)}{\sigma_a \sigma_b} \tag{17}$$

where σ_a means preference for alternative i from first method, σ_b is the preference for alternative i from second method and E is a cross-correlation between a and b.

$$WS = 1 - \sum \left(2^{-x_i} \frac{|x_i - y_i|}{max\{|x_i - 1|, |x_i - N|\}}\right) \tag{18}$$

where x_i means position in the reference ranking, y_i is the position in the second ranking and N is a number of ranked elements.

3 Study Case

Swimming is a sport in which many factors influence the final result. The results achieved are affected by the way athlete's training, their diet and physical conditions. Also, each swimming style is characterised by special properties that make the athletes' physical build more suited to a particular swimming style than another. A group of 6 swimmers was collected to determine the athletes' ranking and investigate the influence of physical conditions on the predisposition to compete in the backstroke style. Moreover, the aim was to measure the progress of student's practicing swimming. The data were obtained in an analogue way with specialised equipment and measurements taken on the tested group of swimmers. Based on the data, a ranking of the swimmers' predisposition to participate in backstroke competitions was determined using MCDA methods, and then the obtained results were compared.

The problem under consideration requires assessing the physical conditions of professional swimmers to participate in backstroke style competitions. Each swimmer has different body parameters, which makes them better or worse able to compete with the best in a given swimming style. By measuring the selected body parameters shown in Table 1, it can be determined which competitor has the best predisposition for swimming in backstroke style. The 14 most important characteristics that are most important when moving in the water and therefore affect the quality and pace of swimming are selected. Body parameters which were taken into consideration were height, height of body in sitting position, arm span, palm width, foot length, pelvic width, shoulders width, inhalation chest circumference, exhalation chest circumference, waist circumference, hips circumference, thigh circumference, tension biceps circumference and extension biceps circumference. The selected criteria are reliable and can be used to qualitatively assess the athletes [26,28,30,54,63].

Table 1. Criteria C_1–C_{14} defined for the determined problem, where p - profit and c - cost.

C_i	C_1	C_2	C_3	C_4	C_5	C_6	C_7	C_8	C_9	C_{10}	C_{11}	C_{12}	C_{13}	C_{14}
C_{type}	p	p	p	p	p	c	p	p	c	c	c	p	p	c

The collected data were obtained from measurements of selected body parameters among 6 swimmers who have been ranked high in the Polish Swimming Championship in recent years. The decision matrix, based on which comparison of predispositions of swimmers will be made, is presented in Table 2. For the correct performance of selected MCDM methods, it was also necessary to define the weight vector for the proposed criteria based on expert knowledge. It is presented in Table 3.

Two methods belonging to the MCDM methods were chosen to solve the presented problem. Both are based on a distance-based approach, which makes

Table 2. Decision matrix for defined alternatives A_i and criteria C_j.

A_j	C_1	C_2	C_3	C_4	C_5	C_6	C_7	C_8	C_9	C_{10}	C_{11}	C_{12}	C_{13}	C_{14}
A_1	180	93	190	11.0	27.0	34	50	103	96	80	96	56	34.0	32
A_2	180	90	191	8.0	27.0	32	50	106	93	85	90	59	33.0	28
A_3	186	96	191	10.0	28.0	34	50	110	101	84	98	59	39.5	37
A_4	190	107	191	11.5	28.0	34	48	102	93	84	98	61	39.0	35
A_5	191	100	196	13.0	27.5	36	50	102	92	79	93	54	35.0	30
A_6	196	100	206	12.0	30.0	41	54	119	107	93	100	63	39.5	35

the principle of operation of both similar and allows for a meaningful comparison of the received rankings. Below are presented the results obtained using the VIKOR and TOPSIS methods, respectively. Then, to check the result for the decision matrix using intervals, the Interval TOPSIS method was used, which is a variation of the classic TOPSIS method. It allows the results to be simulated if the parameters' measurement values are changed to both larger or smaller.

Table 3. Weights for specified criteria C_1–C_{14}.

C_i	C_1	C_2	C_3	C_4	C_5	C_6	C_7	C_8	C_9	C_{10}	C_{11}	C_{12}	C_{13}	C_{14}
w_i	0.12	0.07	0.11	0.11	0.11	0.05	0.08	0.07	0.07	0.05	0.05	0.05	0.03	0.03

3.1 VIKOR Method Assessment

For the presented decision matrix, further actions were performed according to the procedure presented above for the VIKOR method. During the calculation, the defined weights for the appropriate criteria were used. After execution all transformations, the preferences of alternatives for the three rankings S, R and Q were obtained. These preferences, together with the positions on which the alternatives were classified, are shown in Table 4.

Table 4. Preference ranking obtained with the VIKOR method.

A_j	S	Position	R	Position	Q	Position
A_1	0.702	6	0.120	5.5	1.000	6
A_2	0.698	5	0.120	5.5	0.995	5
A_3	0.616	4	0.103	3.5	0.725	4
A_4	0.515	3	0.103	3.5	0.601	3
A_5	0.465	2	0.092	2	0.430	2
A_6	0.294	1	0.070	1	0.000	1

3.2 TOPSIS Method Assessment

The conducted analysis for the TOPSIS method took into account the normalization of decision matrix using five selected normalization methods. For this purpose, minmax, max, sum, vector and logarithmic normalization methods were used. It allowed for assessing the obtained rankings and the impact of the normalization used on the results obtained using this method for the same input data. The results of alternatives preferences and positional ranking for each performed normalization are presented in Table 5.

Table 5. Preference ranking of TOPSIS method using different types of normalization.

A_j	MinMax	Pos.	Max	Pos.	Sum	Pos.	Vector	Pos.	Log	Pos.
A_1	0.298	6	0.405	5	0.411	5	0.407	5	0.536	5
A_2	0.301	5	0.308	6	0.283	6	0.297	6	0.232	6
A_3	0.383	4	0.415	4	0.424	4	0.422	4	0.607	4
A_4	0.484	3	0.556	3	0.553	3	0.559	3	0.661	3
A_5	0.537	2	0.631	2	0.645	2	0.642	2	0.857	1
A_6	0.706	1	0.656	1	0.648	1	0.656	1	0.804	2

3.3 Interval TOPSIS Method Assessment

Another objective was to examine the impact of the use of intervals in the designated range [0.95, 1.05] of the nominal value on the achieved preference values of the alternatives and the positions taken in the general ranking. For this purpose, the Interval TOPSIS method was used, which allows for the conversion of value representation in the decision matrix from sharp values to interval values. The results obtained are presented in Table 6.

Table 6. Preference rankings of Interval TOPSIS method with range [0.95, 1.05].

A_j	Preference	Position
A_1	0.5608	4
A_2	0.5586	6
A_3	0.5646	2
A_4	0.5640	3
A_5	0.5599	5
A_6	0.5672	1

3.4 Rankings Comparison

The obtained rankings with preferences and positions in the considered set were evaluated through three similarity coefficients. The similarities were examined separately within the results obtained using the TOPSIS method, the VIKOR method. Additionally, the rankings of the Interval TOPSIS method and the traditional TOPSIS method were compared.

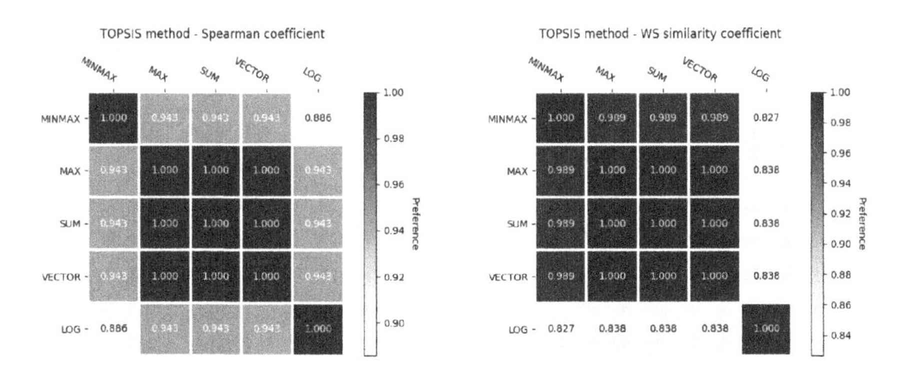

Fig. 1. Comparison of rankings correlation for TOPSIS method for Spearman coefficient and WS similarity coefficient.

The obtained similarity coefficients are presented in Table 7. The highest similarity was found in the rankings analyzed using Spearman correlation coefficient and WS similarity coefficient, where four pairs of compared rankings obtained the similarity of 1.00, meaning their equality. On the other hand, the least similar pair of rankings were those obtained using the Interval TOPSIS method and the classic TOPSIS method using logarithmic normalization, with the result of 0.371. The visualizations of the rankings correlations for the TOPSIS method, VIKOR method are presented in Fig. 1 and 2 respectively.

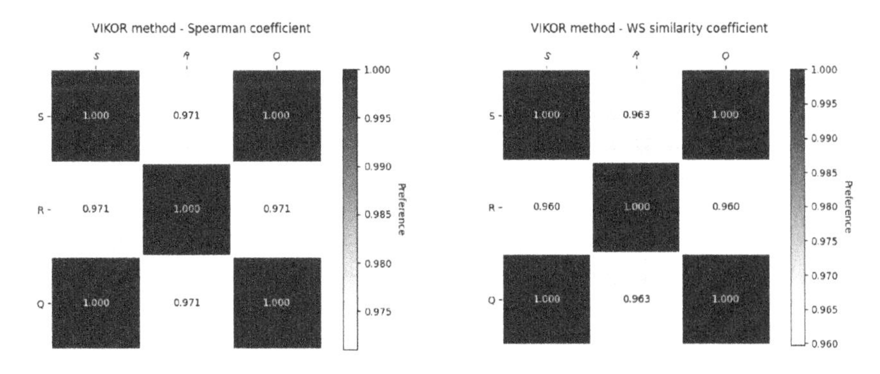

Fig. 2. Comparison of rankings correlation for VIKOR method for Spearman coefficient and WS similarity coefficient.

Table 7. Rankings comparison with Spearman, Pearson and WS coefficients.

Rankings		Spearman	Pearson	WS
R_1	R_2	Value	Value	Value
TOPSIS MINMAX	TOPSIS MAX	0.943	0.927	0.989
TOPSIS MINMAX	TOPSIS SUM	0.943	0.899	0.989
TOPSIS MINMAX	TOPSIS VECTOR	0.943	0.912	0.989
TOPSIS MINMAX	TOPSIS LOG	0.886	0.799	0.826
TOPSIS MAX	TOPSIS SUM	1.000	0.996	1.000
TOPSIS MAX	TOPSIS VECTOR	1.000	0.999	1.000
TOPSIS MAX	TOPSIS LOG	0.943	0.932	0.837
TOPSIS SUM	TOPSIS VECTOR	1.000	0.998	1.000
TOPSIS SUM	TOPSIS LOG	0.943	0.959	0.837
TOPSIS VECTOR	TOPSIS LOG	0.943	0.946	0.837
VIKOR S	VIKOR R	0.971	0.978	0.963
VIKOR S	VIKOR Q	1.000	0.994	1.000
VIKOR R	VIKOR Q	0.971	0.994	0.959
INTERVAL TOPSIS	TOPSIS MAX	0.600	0.544	0.830
INTERVAL TOPSIS	TOPSIS MINMAX	0.486	0.686	0.806
INTERVAL TOPSIS	TOPSIS SUM	0.600	0.533	0.830
INTERVAL TOPSIS	TOPSIS VECTOR	0.600	0.541	0.830
INTERVAL TOPSIS	TOPSIS LOG	0.371	0.542	0.723

4 Results and Discussion

According to the performed research, the alternative A_6 was the most willingly chosen and gained first place in 6 out of 7 analyzed rankings. The only case in which it did not score the best result was using TOPSIS method and logarithmic normalization, where the alternative A_5 scored better. The alternative A_4 in each of the case was ranked third.

Comparing the results obtained with the VIKOR and TOPSIS methods, it can be noticed that the positional rankings are highly similar. On the other hand, when applying the Interval TOPSIS method, whose task was to examine the impact of a change in the representation of sharp values on the interval values, it can be observed that the order of allocated positions is different than in the case of the other two methods. The interval method's use guaranteed second place for the alternative A_3, which was ranked fourth in the rankings of the other methods. The alternative A_5, which was at the top of the rankings of the VIKOR and TOPSIS methods, in this case, was only ranked fifth.

It is also worth noting the differences in the values of preferences obtained using the MCDM methods used. In the case of the VIKOR and TOPSIS methods, the differences between the assessments of preferences of alternatives were significant, showing the disproportions between better and worse choices. In the case of Interval TOPSIS, the differences between all assessed alternatives were not so significant. The difference between the best and worst evaluated alternatives was only 0.0086. Similar preference ratings of alternatives make it difficult to assess the quality of the alternatives unambiguously, as they are at a very similar level.

5 Conclusions

Making decisions follows a daily routine, and it requires choosing the optimal options. This process can be based on models that support decision making based on expert knowledge. To create such models, multi-criteria decision making methods can be used, which allow comparing alternatives from the set based on the selected set of criteria.

As shown in this paper, MCDA methods can provide a strong methodological background in constructing a performance measurement system in swimming. Selected MCDM methods, namely VIKOR and TOPSIS, which are based on a distance-based approach, were used to support the decision-making process. The research was conducted on a selected group of professional swimmers from whom measurement data of specific body parameters were collected. Attempts were made to answer the question of which of the swimmers have the best predispositions for swimming in backstroke style. The results obtained indicated that by using the VIKOR and TOPSIS method, a nearly unambiguous result was obtained. It was confirmed by using similarity coefficients and showing high similarity of the received rankings. Moreover, it was decided to use the Interval TOPSIS method to simulate situations in which there is uncertainty about the change of measurement parameters' values. This procedure has shown that with the introduction of the uncertainty, the received ranking has changed significantly.

Future directions are worth considering to apply a greater amount of MCDM methods to determine the problem to receive more benchmarkable results. Additionally, it is worth considering the set values of the weights for criteria, which are selected based on the expert knowledge, and to a large extent influence the final results. The model presented can be used to manage and assign athletes to events for which they are more suitable to compete.

References

1. Abdel-Basset, M., Ding, W., Mohamed, R., Metawa, N.: An integrated plithogenic MCDM approach for financial performance evaluation of manufacturing industries. Risk Manag. **22**(3), 192–218 (2020). https://doi.org/10.1057/s41283-020-00061-4
2. Bączkiewicz, A., Kizielewicz, B., Shekhovtsov, A., Wątróbski, J., Sałabun, W.: Methodical aspects of MCDM based e-commerce recommender system. J. Theor. Appl. Electron. Commer. Res. **16**(6), 2192–2229 (2021)

3. Bączkiewicz, A., Kizielewicz, B., Shekhovtsov, A., Yelmikheiev, M., Kozlov, V., Sałabun, W.: Comparative analysis of solar panels with determination of local significance levels of criteria using the MCDM methods resistant to the rank reversal phenomenon. Energies **14**(18), 5727 (2021)
4. Balubaid, M.A., Basheikh, M.A.: Using the analytic hierarchy process to prioritize alternative medicine: selecting the most suitable medicine for patients with diabetes. Int. J. Basic Appl. Sci. **5**(1), 67 (2016)
5. Bauer, H.H., Stokburger-Sauer, N.E., Exler, S.: Brand image and fan loyalty in professional team sport: a refined model and empirical assessment. J. Sport Manag. **22**(2), 205–226 (2008)
6. Behzadian, M., Otaghsara, S.K., Yazdani, M., Ignatius, J.: A state-of the-art survey of TOPSIS applications. Expert Syst. Appl. **39**(17), 13051–13069 (2012)
7. Bolstad, B.M., Irizarry, R.A., Åstrand, M., Speed, T.P.: A comparison of normalization methods for high density oligonucleotide array data based on variance and bias. Bioinformatics **19**(2), 185–193 (2003)
8. Božanić, D., Milić, A., Tešić, D., Salabun, W., Pamučar, D.: D numbers - FUCOM - fuzzy RAFASI model for selecting the group of constructions machines for enabling mobility. Facta Universitatis, Series: Mechanical Engineering (2021)
9. Chen, J.K., Chen, I.S.: Using a novel conjunctive MCDM approach based on DEMATEL, fuzzy ANP, and TOPSIS as an innovation support system for Taiwanese higher education. Expert Syst. Appl. **37**(3), 1981–1990 (2010)
10. Dadelo, S., Turskis, Z., Zavadskas, E.K., Dadeliene, R.: Multi-criteria assessment and ranking system of sport team formation based on objective-measured values of criteria set. Expert Syst. Appl. **41**(14), 6106–6113 (2014)
11. Dey, P.K., Ghosh, D.N., Mondal, A.C.: A MCDM approach for evaluating bowlers performance in IPL. J. Emerg. Trends Comput. Inf. Sci. **2**(11), 563–73 (2011)
12. Ferreira, F.A., Ilander, G.O.P.B., Ferreira, J.J.: MCDM/A in practice: methodological developments and real-world applications. Manag. Decis. **57**, 295–299 (2019)
13. Ferreira, F.A., Santos, S.P., Rodrigues, P.M.: Adding value to bank branch performance evaluation using cognitive maps and MCDA: a case study. J. Oper. Res. Soc. **62**(7), 1320–1333 (2011)
14. Giove, S.: Interval TOPSIS for multicriteria decision making. In: Marinaro, M., Tagliaferri, R. (eds.) WIRN 2002. LNCS, vol. 2486, pp. 56–63. Springer, Heidelberg (2002). https://doi.org/10.1007/3-540-45808-5_5
15. Gouveia, M.C., Dias, L.C., Antunes, C.H.: Additive DEA based on MCDA with imprecise information. J. Oper. Res. Soc. **59**(1), 54–63 (2008)
16. Güllich, A., Emrich, E.: Individualistic and collectivistic approach in athlete support programmes in the German high-performance sport system. Eur. J. Sport Soc. **9**(4), 243–268 (2012)
17. Gunduz, M., Tehemar, S.R.: Assessment of delay factors in construction of sport facilities through multi criteria decision making. Prod. Plann. Control **31**(15), 1291–1302 (2020)
18. Huang, Y.S., Li, W.H.: A study on aggregation of TOPSIS ideal solutions for group decision-making. Group Decis. Negot. **21**(4), 461–473 (2012)
19. Hyde, K., Maier, H.R., Colby, C.: Incorporating uncertainty in the PROMETHEE MCDA method. J. Multi-Criteria Decis. Anal. **12**(4–5), 245–259 (2003)
20. Kizielewicz, B., Shekhovtsov, A., Sałabun, W.: A new approach to eliminate rank reversal in the MCDA problems. In: Paszynski, M., Kranzlmüller, D., Krzhizhanovskaya, V.V., Dongarra, J.J., Sloot, P.M.A. (eds.) ICCS 2021. LNCS, vol. 12742, pp. 338–351. Springer, Cham (2021). https://doi.org/10.1007/978-3-030-77961-0_29

21. Kizielewicz, B., Wątróbski, J., Sałabun, W.: Identification of relevant criteria set in the MCDA process-wind farm location case study. Energies **13**(24), 6548 (2020)
22. Kizielewicz, B., Więckowski, J., Wątrobski, J.: A study of different distance metrics in the TOPSIS method. In: Czarnowski, I., Howlett, R.J., Jain, L.C. (eds.) Intelligent Decision Technologies. SIST, vol. 238, pp. 275–284. Springer, Singapore (2021). https://doi.org/10.1007/978-981-16-2765-1_23
23. Kizielewicz, B., Dobryakova, L.: MCDA based approach to sports players' evaluation under incomplete knowledge. Procedia Comput. Sci. **176**, 3524–3535 (2020)
24. Kou, G., Peng, Y., Wang, G.: Evaluation of clustering algorithms for financial risk analysis using MCDM methods. Inf. Sci. **275**, 1–12 (2014)
25. Lai, Y.J., Liu, T.Y., Hwang, C.L.: TOPSIS for MODM. Eur. J. Oper. Res. **76**(3), 486–500 (1994)
26. Lätt, E., et al.: Physiological, biomechanical and anthropometrical predictors of sprint swimming performance in adolescent swimmers. J. Sports Sci. Med. **9**(3), 398 (2010)
27. Liu, H.-C.: FMEA using uncertainty theories and MCDM methods. In: FMEA Using Uncertainty Theories and MCDM Methods, pp. 13–27. Springer, Singapore (2016). https://doi.org/10.1007/978-981-10-1466-6_2
28. Lowensteyn, I., Signorile, J.F., Giltz, K.: The effect of varying body composition on swimming performance. J. Strength Cond. Res. **8**(3), 149–154 (1994)
29. Mardani, A., Zavadskas, E.K., Govindan, K., Amat Senin, A., Jusoh, A.: VIKOR technique: a systematic review of the state of the art literature on methodologies and applications. Sustainability **8**(1), 37 (2016)
30. Moura, T., Costa, M., Oliveira, S., Júnior, M.B., Ritti-Dias, R., Santos, M.: Height and body composition determine arm propulsive force in youth swimmers independent of a maturation stage. J. Hum. Kinet. **42**(1), 277–284 (2014)
31. Mu, E.: Who really won the FIFA 2014 Golden Ball Award?: what sports can learn from multi-criteria decision analysis. Int. J. Sport Manag. Mark. **16**(3–6), 239–258 (2016)
32. O'Boyle, I., Hassan, D.: Performance management and measurement in national-level non-profit sport organisations. Eur. Sport Manag. Q. **14**(3), 299–314 (2014)
33. Olson, D.L.: Comparison of weights in TOPSIS models. Math. Comput. Model. **40**(7–8), 721–727 (2004)
34. Opricovic, S.: Fuzzy VIKOR with an application to water resources planning. Expert Syst. Appl. **38**(10), 12983–12990 (2011)
35. Opricovic, S., Tzeng, G.H.: Compromise solution by MCDM methods: a comparative analysis of VIKOR and TOPSIS. Eur. J. Oper. Res. **156**(2), 445–455 (2004)
36. Opricovic, S., Tzeng, G.H.: Extended VIKOR method in comparison with outranking methods. Eur. J. Oper. Res. **178**(2), 514–529 (2007)
37. Palczewski, K., Sałabun, W.: Influence of various normalization methods in PROMETHEE II: an empirical study on the selection of the airport location. Procedia Comput. Sci. **159**, 2051–2060 (2019)
38. Pelissari, R., Oliveira, M.C., Abackerli, A.J., Ben-Amor, S., Assumpção, M.R.P.: Techniques to model uncertain input data of multi-criteria decision-making problems: a literature review. Int. Trans. Oper. Res. **28**(2), 523–559 (2021)
39. Podvezko, V.: The comparative analysis of MCDA methods SAW and COPRAS. Eng. Econ. **22**(2), 134–146 (2011)
40. Radović, D., et al.: Measuring performance in transportation companies in developing countries: a novel rough ARAS model. Symmetry **10**(10), 434 (2018)

41. Roszkowska, E., Wachowicz, T.: Analyzing the applicability of selected MCDA methods for determining the reliable scoring systems. In: Bajwa, D.S., Koeszegi, S., Vetschera, R. (eds.) Proceedings of the 16th International Conference on Group Decision and Negotiation Bellingham, pp. 180–187. Western Washington University (2016)
42. Roy, B., Vanderpooten, D.: The European school of MCDA: emergence, basic features and current works. J. Multi-Criteria Decis. Anal. **5**(1), 22–38 (1996)
43. Sałabun, W., Piegat, A.: Comparative analysis of MCDM methods for the assessment of mortality in patients with acute coronary syndrome. Artif. Intelli. Rev. **48**, 557–571 (2017). https://doi.org/10.1007/s10462-016-9511-9
44. Sałabun, W., et al.: A fuzzy inference system for players evaluation in multi-player sports: the football study case. Symmetry **12**(12), 2029 (2020)
45. Sałabun, W., Urbaniak, K.: A new coefficient of rankings similarity in decision-making problems. In: Krzhizhanovskaya, V.V., et al. (eds.) ICCS 2020. LNCS, vol. 12138, pp. 632–645. Springer, Cham (2020). https://doi.org/10.1007/978-3-030-50417-5_47
46. Sałabun, W., Wątróbski, J., Shekhovtsov, A.: Are MCDA methods benchmarkable? A comparative study of TOPSIS, VIKOR, COPRAS, and PROMETHEE II methods. Symmetry **12**(9), 1549 (2020)
47. Sałabun, W., et al.: How the normalization of the decision matrix influences the results in the VIKOR method? Procedia Comput. Sci. **176**, 2222–2231 (2020)
48. Sanayei, A., Mousavi, S.F., Yazdankhah, A.: Group decision making process for supplier selection with VIKOR under fuzzy environment. Expert Syst. Appl. **37**(1), 24–30 (2010)
49. Shekhovtsov, A., Kołodziejczyk, J.: Do distance-based multi-criteria decision analysis methods create similar rankings? Procedia Comput. Sci. **176**, 3718–3729 (2020)
50. Shekhovtsov, A., Kozlov, V., Nosov, V., Sałabun, W.: Efficiency of methods for determining the relevance of criteria in sustainable transport problems: a comparative case study. Sustainability **12**(19), 7915 (2020)
51. Shekhovtsov, A., Sałabun, W.: A comparative case study of the VIKOR and TOPSIS rankings similarity. Procedia Comput. Sci. **176**, 3730–3740 (2020)
52. Shih, H.S., Shyur, H.J., Lee, E.S.: An extension of TOPSIS for group decision making. Math. Comput. Model. **45**(7–8), 801–813 (2007)
53. Shilbury, D., Rentschler, R.: Assessing sport management journals: a multidimensional examination. Sport Manag. Rev. **10**(1), 31–44 (2007)
54. Siders, W.A., Lukaski, H.C., Bolonchuk, W.W.: Relationships among swimming performance, body composition and somatotype in competitive collegiate swimmers (1993)
55. Sironen, S., Leskinen, P., Kangas, A., Hujala, T.: Variation of preference inconsistency when applying ratio and interval scale pairwise comparisons. J. Multi-Criteria Decis. Anal. **21**(3–4), 183–195 (2014)
56. Stević, Ž, Pamučar, D., Puška, A., Chatterjee, P.: Sustainable supplier selection in healthcare industries using a new MCDM method: measurement of alternatives and ranking according to COmpromise solution (MARCOS). Comput. Ind. Eng. **140**, 106231 (2020)
57. Stojčić, M., Zavadskas, E.K., Pamučar, D., Stević, Ž, Mardani, A.: Application of MCDM methods in sustainability engineering: a literature review 2008–2018. Symmetry **11**(3), 350 (2019)

58. Syamsudin, S., Rahim, R.: Study approach Technique for Order of Preference by Similarity to Ideal Solution (TOPSIS). Int. J. Recent Trends Eng. Res **3**(3), 268–285 (2017)
59. Tong, L.I., Chen, C.C., Wang, C.H.: Optimization of multi-response processes using the VIKOR method. Int. J. Adv. Manuf. Technol. **31**(11–12), 1049–1057 (2007)
60. Triantaphyllou, E.: Multi-criteria decision making methods. In: Multi-Criteria Decision Making Methods: A Comparative Study, pp. 5–21. Springer, Cham (2000). https://doi.org/10.1007/978-1-4757-3157-6_2
61. Triantaphyllou, E., Baig, K.: The impact of aggregating benefit and cost criteria in four MCDA methods. IEEE Trans. Eng. Manage. **52**(2), 213–226 (2005)
62. Tsaur, R.C.: Decision risk analysis for an interval TOPSIS method. Appl. Math. Comput. **218**(8), 4295–4304 (2011)
63. Tuuri, G., Loftin, M., Oescher, J.: Association of swim distance and age with body composition in adult female swimmers. Med. Sci. Sports Exerc. **34**(12), 2110–2114 (2002)
64. Urbaniak, K., Wątróbski, J., Sałabun, W.: Identification of players ranking in e-sport. Appl. Sci. **10**(19), 6768 (2020)
65. Wang, T.C., Lee, H.D.: Developing a fuzzy TOPSIS approach based on subjective weights and objective weights. Expert Syst. Appl. **36**(5), 8980–8985 (2009)
66. Wang, Y.M., Elhag, T.M.: On the normalization of interval and fuzzy weights. Fuzzy Sets Syst. **157**(18), 2456–2471 (2006)
67. Warrens, M.J.: Similarity coefficients for binary data: properties of coefficients, coefficient matrices, multi-way metrics and multivariate coefficients (2008)
68. Wei, J., Lin, X.: The multiple attribute decision-making VIKOR method and its application. In: 2008 4th International Conference on Wireless Communications, Networking and Mobile Computing, pp. 1–4. IEEE (2008)
69. Wu, H.Y., Chen, J.K., Chen, I.S., Zhuo, H.H.: Ranking universities based on performance evaluation by a hybrid MCDM model. Measurement **45**(5), 856–880 (2012)
70. Yazdani, M., Chatterjee, P., Zavadskas, E.K., Zolfani, S.H.: Integrated QFD-MCDM framework for green supplier selection. J. Clean. Prod. **142**, 3728–3740 (2017)
71. Zhang, N., Wei, G.: Extension of VIKOR method for decision making problem based on hesitant fuzzy set. Appl. Math. Model. **37**(7), 4938–4947 (2013)
72. Zolfani, S., Yazdani, M., Pamucar, D., Zarate, P.: A VIKOR and TOPSIS focused reanalysis of the MADM methods based on logarithmic normalization. arXiv preprint arXiv:2006.08150 (2020)

Can MCDA Methods Be Useful in E-commerce Systems? Comparative Study Case

Bartłomiej Kizielewicz[1], Aleksandra Bączkiewicz[2,3], Andrii Shekhovtsov[1], Jakub Więckowski[1], and Wojciech Sałabun[1](✉)

[1] Research Team on Intelligent Decision Support Systems, Department of Artificial Intelligence and Applied Mathematics, Faculty of Computer Science and Information Technology, West Pomeranian University of Technology in Szczecin, ul. Żołnierska 49, 71-210 Szczecin, Poland
wojciech.salabun@zut.edu.pl

[2] Institute of Management, University of Szczecin, ul. Cukrowa 8, 71-004 Szczecin, Poland

[3] Doctoral School of University of Szczecin, ul. Mickiewicza 16, 70-383 Szczecin, Poland

Abstract. Shopping via e-commerce sites is becoming increasingly popular among customers. More and more such sites are being created, and more marketing activities and innovative solutions are needed to attract customers' attention to increase competitiveness and stand out on the market. An effective tactic is to take the consumer's needs into account as much as possible and keep them satisfied to become regular customers and recommend the place to their family and friends. For responding to the customers' needs, it is essential to recognise and understand them. The ever-increasing variety of products on the market and the need to consider an expanding number of technical parameters of equipment and devices make the selection of purchased products and goods by consumers more and more challenging. The problem of selecting purchased products is, therefore, a multi-criteria problem. An intuitive approach and consideration of only the main selection criteria may result in inappropriate choices. Multi-criteria decision-analysis methods (MCDA) are techniques designed to solve this type of problem.

This paper demonstrates an innovative concept based on MCDA methods, including a novel hybrid approach combining COMET with TOPSIS, TOPSIS and VIKOR, used as a tool to support consumer choices in e-commerce systems. The authors performed a comparative analysis of the applied methods using two ranking similarity coefficients: asymmetrical WS and symmetrical r_w. The study was completed with a sensitivity analysis. The results obtained suggest the potentially promising usefulness and suitability of the proposed tool in e-commerce systems.

Keywords: Multi-criteria customer choices · E-commerce · MCDA

I. Woungang et al. (Eds.): ANTIC 2021, CCIS 1534, pp. 546–562, 2022.
https://doi.org/10.1007/978-3-030-96040-7_42

1 Introduction

Nowadays, buying products and equipment requires considering many alternatives available on the market and criteria defining their functionality. It implies that buying decisions for products whose utility is defined by many parameters is a multi-criteria decision problem. Consequently, making an appropriate and satisfactory choice often requires considering opposing criteria and searching for a compromise solution [13]. Compound decision problems are challenging for consumers because an attempt based only on intuition with some criteria is often insufficient. The described situation motivates developing decision support systems based on different methods, both for universal use and specific domains. Multi-criteria decision-making methods (MCDA) are popular and frequently used techniques that allow multiple and conflicting goals to be considered in the decision-making process. The current rapid development of MCDA methods has resulted in the availability of algorithms with different complexity, consideration of criteria and individual preferences of decision-makers, data aggregation, criteria compensation and the possibility of incorporating uncertainty in the data. Thus, while different MCDA methods can improve decision quality, their comparison often yields conflicting results [28]. For this reason, an important stage of research on MCDA methods is a comparative analysis considering several methods [22].

Dynamic digital advancements have now made computers a widely available device and used by most people around the world [10]. Computers simplify people's lives significantly by enabling long-distance communication, storing and providing information, enabling multimedia from anywhere in the world. From a practical point of view, laptops are especially useful devices for work and daily activities because they are mobile, portable and functional [4]. The assortment of laptops available in the market includes many models differing in technical specifications, size, functionality and brand. Objective and complete consideration of all features and parameters that fully satisfy the customer is a complex problem for which an intuitive approach is not enough to solve [1]. Thus, decision support systems seem to be a promising tool to support product purchasing decisions involving electronic devices such as laptops, for example [2].

The aim of this work is to present an innovative approach based on three selected MCDA methods (COMET combined with TOPSIS, TOPSIS and VIKOR), which could be used as a tool to support consumer decisions during multi-criteria problems of purchasing products and devices at e-commerce outlets. In this paper, the authors demonstrate the resolving of a sample multi-criteria problem of choosing the most advantageous laptop model using the concept proposed by the authors. A sensitivity analysis procedure was then performed to determine the robustness of the investigated MCDA models to changes in the criteria weights and to identify the criteria that most strongly affect the final rankings.

The rest of the paper is organised as follows. In Sect. 2 fundamentals and assumptions of the MCDA methods used in the study are provided. Then, in Sect. 3 the problem considered in this article is described. In Sect. 4 final results are presented and discussed. In the last Sect. 5 conclusions and directions for future work are indicated.

2 Preliminaries

2.1 The TOPSIS Method

The algorithm of this method is simple and clear, so this method is popular and widely used in multi-criteria decision-making problems. Furthermore, TOPSIS requires a vector of criteria weights, which can be determined subjectively by the decision-maker or objective techniques. Thus, the TOPSIS algorithm does not require the active involvement of an expert in the computation. A detailed study of the TOPSIS algorithm founded on [3] is given below. This method requires the decision matrix with m alternatives and n criteria represented as $X = (x_{ij})_{m \times n}$.

Step 1. Normalization of the decision matrix. In this article, the authors applied the Max normalization technique. The normalized values r_{ij} are determined by Eq. (1) for profit and (2) for cost criteria.

$$r_{ij} = \frac{x_{ij}}{\max_j (x_{ij})} \tag{1}$$

$$r_{ij} = 1 - \frac{x_{ij}}{\max_j (x_{ij})} \tag{2}$$

Step 2. Computation of the weighted normalized decision matrix v_{ij} as Eq. (3) shows.

$$v_{ij} = w_i r_{ij} \tag{3}$$

Step 3. Determination of Positive Ideal Solution (PIS) and Negative Ideal Solution (NIS) vectors. PIS is represented by maximum values for each criterion (4) and NIS by minimum values (5). There is no necessity to divide criteria into profit and cost because normalization used in step 1 transforms cost criteria into profit criteria.

$$v_j^+ = \{v_1^+, v_2^+, \cdots, v_n^+\} = \{max_j(v_{ij})\} \tag{4}$$

$$v_j^- = \{v_1^-, v_2^-, \cdots, v_n^-\} = \{min_j(v_{ij})\} \tag{5}$$

Step 4. Establishing of distance from PIS and NIS for each alternative as Eqs. (6) and (7) present.

$$D_i^+ = \sqrt{\sum_{j=1}^{n} (v_{ij} - v_j^+)^2} \tag{6}$$

$$D_i^- = \sqrt{\sum_{j=1}^{n} (v_{ij} - v_j^-)^2} \tag{7}$$

Step 5. Calculation of the score for each alternative as Eq. (8) shows. This value is always in the range from 0 to 1. Better alternatives have scores closer to 1.

$$C_i = \frac{D_i^-}{D_i^- + D_i^+} \tag{8}$$

2.2 The VIKOR Method

VIKOR is an acronym in Serbian that means VlseKriterijumska Optimizacija I Kompromisno Resenje, and it was introduced by Opricovic [15]. The VIKOR method aims to choose the closest alternative to the ideal solution with all criteria considered. VIKOR, similarly to TOPSIS, takes into account the proximity to ideal objects, so distance measurement is applied in this algorithm [22]. However, the procedures of the two methods differ at particular stages in their operational approach and consideration of closeness to ideal solutions. The subsequent stages of VIKOR are provided below, according to [22]. As showed in [16], the VIKOR method is defined as follows:

Step 1. Calculation of the best f_i^* and the worst f_i^- values for every criteria functions. Equation (9) is applied for profit criteria and (10) is used for cost criteria.

$$f_j^* = \max_i f_{ij}, \quad f_j^- = \min_i f_{ij} \tag{9}$$

$$f_j^* = \min_i f_{ij}, \quad f_j^- = \max_i f_{ij} \tag{10}$$

Step 2. Computation of the S_i and R_i values according to formulas (11) and (12).

$$S_i = \sum_{j=1}^{n} \left[w_j \frac{(f_j^* - f_{ij})}{(f_j^* - f_j^-)}) \right] \tag{11}$$

$$R_i = \max_j \left[w_j \frac{(f_j^* - f_{ij})}{(f_j^* - f_j^-)} \right] \tag{12}$$

Step 3. Calculation of the Q_i values applying Eq. (13)

$$Q_i = v \frac{(S_i - S^*)}{(S^- - S^*)} + (1 - v) \frac{(R_i - R^*)}{(R^- - R^*)} \tag{13}$$

where
$S^* = \min_i S_i, \quad S^* = \min_i S_i$
$R^* = \min_i R_i, \quad R^* = \max_i R_i$
and v is used as a weight for the strategy named "majority of criteria". Value of $v = 0.5$ was applied in this paper.

Step 4. Ranking alternatives in the procedure of sorting values in S, R, and Q in ascending order. Three ranking lists are provided as a result.

Step 5. S, R and Q ranking lists are considered to suggest the compromise solution or set of compromise solutions, as shown in [15]. In this research, the authors use only Q ranking list.

2.3 The COMET Method

The COMET is an innovative method applied to identify a multi-criteria expert decision-making model for handling decision-making problems [7]. The main advantages of this newly developed method are its complete resistance to the phenomenon known as the rank reversal paradox [12,23,24], accuracy and independence of the complexity of the algorithm from the number of evaluated alternatives. Furthermore, in the innovative hybrid approach used in this work, the time-consuming step of pairwise comparison of a set of characteristic objects by an expert has been replaced by another MCDA method, TOPSIS. This approach was introduced in [12]. This method will be presented in five steps based on [18]:

Step 1. The problem's dimensionality to solve is determined by an expert by choosing number r of criteria, $C_1, C_2, ..., C_r$. Next, the set of fuzzy numbers for every criteria represented by C_i is selected, i.e., $\tilde{C}_{i1}, \tilde{C}_{i2}, ..., \tilde{C}_{ic_i}$. The value of the membership for a given linguistic concept for specific crisp values is determined by each fuzzy number [6,20]. This approach can also be used for non-continuous variables. The result of this step is the result represented by formula (14)

$$\begin{array}{c} C_1 = \left\{\tilde{C}_{11}, \tilde{C}_{12}, ..., \tilde{C}_{1c_1}\right\} \\ C_2 = \left\{\tilde{C}_{21}, \tilde{C}_{22}, ..., \tilde{C}_{2c_2}\right\} \\ \ldots \\ C_r = \left\{\tilde{C}_{r1}, \tilde{C}_{r2}, ..., \tilde{C}_{rc_r}\right\} \end{array} \tag{14}$$

where $C_1, C_2, ..., C_r$ are the ordinates of the fuzzy numbers for every criterion considered.

Step 2. Generation of the characteristic objects (COs), which represent reference points in n-dimensional space. These objects may be real or idealized, which means that they do not exist [17]. (COs)are received using the Cartesian product of fuzzy numbers cores for each criteria [19]. The ordered set of all COs is provided as a result, like formula (15) shows

$$\begin{array}{c} CO_1 = \langle C(\tilde{C}_{11}), C(\tilde{C}_{21}), ..., C(\tilde{C}_{r1})\rangle \\ CO_2 = \langle C(\tilde{C}_{11}), C(\tilde{C}_{21}), ..., C(\tilde{C}_{r2})\rangle \\ \ldots \\ CO_t = \langle C(\tilde{C}_{1c_1}), C(\tilde{C}_{2c_2}), ..., C(\tilde{C}_{rc_r})\rangle \end{array} \tag{15}$$

where t is a number of CO (16):

$$t = \prod_{i=1}^{r} c_i \tag{16}$$

Step 3. The Matrix of Expert Judgement (MEJ) is determined by an expert in the procedure of pairwise comparison of COs. This step depends entirely on the expert's knowledge and opinion in the classical version of the COMET method [11]. The MEJ structure is as follows (17):

$$MEJ = \begin{pmatrix} \alpha_{11} & \alpha_{12} & ... & \alpha_{1t} \\ \alpha_{21} & \alpha_{22} & ... & \alpha_{2t} \\ ... & ... & ... & ... \\ \alpha_{t1} & \alpha_{t2} & ... & \alpha_{tt} \end{pmatrix} \tag{17}$$

where α_{ij} is a result of comparing CO_i and CO_j by the expert. The object that is preferred more gets 1 point, and the object that is preferred less gets 0 points. When the compared objects are equally preferred, they both get 0.5 points [25]. It depends totally on the expert's knowledge and is represented as (18):

$$\alpha_{ij} = \begin{cases} 0.0, & f_{exp}(CO_i) < f_{exp}(CO_j) \\ 0.5, & f_{exp}(CO_i) = f_{exp}(CO_j) \\ 1.0, & f_{exp}(CO_i) > f_{exp}(CO_j) \end{cases} \tag{18}$$

where f_{exp} is an expert mental judgement function. In this study, however, the TOPSIS method was used as an expert function. This approach is presented in [12].

Then, the vertical vector of the Summed Judgements (SJ) is received as formula (19) presents:

$$SJ_i = \sum_{j=1}^{t} \alpha_{ij} \tag{19}$$

The number of query is expressed by $p = \frac{t(t-1)}{2}$ because for each element α_{ij} it can be noticed that $\alpha_{ji} = 1 - \alpha_{ij}$. The vector P is provided as an outcome, where i-th row includes the estimated preference value for CO_i.

Step 4. Each characteristic object is transformed into a fuzzy rule, where the grade of membership to particular criteria is a premise for activating inference in the form of P_i as presented in formula (20). Thus the complete fuzzy rule base is received that estimates the expert mental judgement function $f_{exp}(CO_i)$ [29].

$$IF\ C\left(\tilde{C}_{1i}\right)\ AND\ C\left(\tilde{C}_{2i}\right)\ AND\ \ldots\ THEN\ P_i \tag{20}$$

Step 5. Every alternative A_i is a set of crisp numbers a_{ri} associated to criteria $C_1, C_2, ..., C_r$. It is expressed by formula (21):

$$A_i = \{a_{1i}, a_{2i}, ..., a_{ri}\} \tag{21}$$

2.4 Rankings Similarity Coefficients

In this study, two similarity coefficients, symmetrical r_w (22) and asymmetrical WS (23), were used to check the convergence of the rankings provided by the three MCDA methods applied [21]:

$$r_w = 1 - \frac{6\sum_{i=1}^{N}(R_i - Q_i)^2\left((N - R_i + 1) + (N - Q_i + 1)\right)}{N^4 + N^3 - N^2 - N}, \tag{22}$$

where R_i represents a position in the compared ranking and Q_i means a position in the reference ranking, and N is a number of evaluated alternatives,

$$WS = 1 - \sum_{i=1}^{N} \left(2^{-R_{xi}} \cdot \frac{|R_{xi} - R_{yi}|}{\max\{|1 - R_{xi}|, |N - R_{xi}|\}} \right), \tag{23}$$

where R_{xi} represents a position in the compared ranking, R_{yi} is a position in the reference ranking, and N is a number of alternatives.

2.5 Sensitivity Analysis

A sensitivity analysis procedure is used to identify the most susceptible criteria to weight changes and most significantly affect the final rankings. This technique also allows the identification of tolerable changes in criteria weights to which the rankings are robust. The authors of this paper performed a sensitivity analysis using the two approaches presented in paper [14].

The first stage's goal was an analysis to determine the number of changes in ranking after increasing or decreasing the weights of each criterion. The described procedure allows independent determination of the effect of each criterion under consideration on the rankings provided by MCDA. The technique involves increasing or decreasing the weight of each criterion individually by 5% and by 50%. The results of this procedure are relative sensitivity coefficients that determine the number of changes in rankings caused by a change in criterion weight.

The second step of the sensitivity analysis involves determining the percentage of tolerable weight change for each criterion that does not result in a change in ranking. The sensitivity coefficient of a given criterion C_j is defined as SC_j and is a measure of the sensitivity to changes in the criterion weight. The sensitivity coefficient is given by Eq. (24).

$$SC_j = \frac{1}{D_j}, \; j = 1, 2, \ldots, n \tag{24}$$

where D_j represents the smallest relative change in criterion weight in percentage that causes changes in the ranking.

3 Study Case

This study aimed to answer the multi-criteria problem of choosing the most advantageous laptop model among fifteen available alternatives. Three different MCDA approaches were used to evaluate the alternatives: a novel hybrid approach combining COMET with TOPSIS, TOPSIS and VIKOR. Nine sample criteria were considered in the evaluation procedure: the laptop parameters given in Table 1.

Table 1. Criteria C_1–C_9 with their names, types and units for assessment of alternatives A_1–A_{15}.

C_i	Name	Type	Unit
C_1	Price	Cost	Polish złoty [PLN]
C_2	Hard disk capacity	Profit	Megabyte [MB]
C_3	Random-access memory (RAM)	Profit	Gigabyte [GB]
C_4	Screen size	Profit	Inch [in]
C_5	Battery capacity	Profit	Ampere hour [mAh]
C_6	Refresh rate	Profit	Hertz [Hz]
C_7	Weight	Cost	Kilogram [kg]
C_8	Graphics card memory	Profit	Megabyte [MB]
C_9	Processor cache memory	Profit	Megabyte [MB]

Table 2. Decision matrix containing criteria values for alternatives.

A_i	Alternative name	C_1	C_2	C_3	C_4	C_5	C_6	C_7	C_8	C_9
A_1	HP Pavilion Gaming	4549	1512	32	16.1	4323	144	2.35	4096	8
A_2	Dell Inspiron 3793	3599	256	16	17.3	3500	60	2.79	2048	6
A_3	Lenovo Legion 5-15	4049	512	16	15.6	5350	120	2.46	6144	11
A_4	ASUS TUF Dash F15	4999	512	16	15.6	4940	144	2.06	6144	12
A_5	Lenovo IdeaPad L340-17	3449	512	16	17.3	4000	60	2.78	3072	8
A_6	HP Pavilion 15	3949	512	16	15.6	3440	60	1.73	2048	8
A_7	Dell Inspiron G3	4799	1512	8	15.6	4255	120	2.34	4096	8
A_8	ASUS TUF Dash F15	5199	512	24	15.6	4940	144	2.06	6144	12
A_9	ASUS TUF Gaming FX506IH	3599	512	8	15.6	4240	144	2.04	4096	11
A_{10}	Lenovo Legion Y540-15	3999	512	8	15.6	4670	60	2.30	6144	12
A_{11}	MSI GL65	2999	256	8	15.6	3834	60	2.30	4096	8
A_{12}	MSI GL65	3599	1256	32	15.6	3834	60	2.30	4096	8
A_{13}	Dell Vostro 5301	5299	512	8	13.3	3500	60	1.25	2048	12
A_{14}	MSI GL75	4649	512	16	17.3	3834	144	2.50	4096	8
A_{15}	MSI GL75	5099	1512	32	17.3	3834	144	2.50	4096	8

The advantages of MCDA methods are the individual approach to the solved problem and the possibility of interaction with the user, so there is also the opportunity to choose other criteria expressed in numbers relevant to the customer. Seven of mentioned criteria are of the profit type, and two are of the cost type. The decision matrix that includes the values of each criterion for all alternatives considered is presented in Table 2.

The data was collected from various websites through which laptops can be purchased. These websites provide the technical specifications of the laptops from which the parameters were selected as evaluation criteria. Criteria weights for methods requiring weights such as TOPSIS and VIKOR were determined by the objective equal weights method [26]. The decision matrix was normalized using the Maximum normalization method. The result of the normalization of the matrix conducted for the TOPSIS and VIKOR methods requiring it is displayed in Table 3.

Table 3. Normalized decision matrix.

A_i	C_1	C_2	C_3	C_4	C_5	C_6	C_7	C_8	C_9
A_1	0.1415	1.0000	1.00	0.9306	0.8080	1.0000	0.1577	0.6666	0.6666
A_2	0.3208	0.1693	0.50	1.0000	0.6542	0.4166	0.0000	0.3333	0.5000
A_3	0.2358	0.3386	0.50	0.9017	1.0000	0.8333	0.1182	1.0000	0.9166
A_4	0.0566	0.3386	0.50	0.9017	0.9233	1.0000	0.2616	1.0000	1.0000
A_5	0.3491	0.3386	0.50	1.0000	0.7476	0.4166	0.0035	0.5000	0.6666
A_6	0.2547	0.3386	0.50	0.9017	0.6429	0.4166	0.3799	0.3333	0.6666
A_7	0.0943	1.0000	0.25	0.9017	0.7953	0.8333	0.1612	0.6666	0.6666
A_8	0.0188	0.3386	0.75	0.9017	0.9233	1.0000	0.2616	1.0000	1.0000
A_9	0.3208	0.3386	0.25	0.9017	0.7925	1.0000	0.2688	0.6666	0.9166
A_{10}	0.2453	0.3386	0.25	0.9017	0.8729	0.4166	0.1756	1.0000	1.0000
A_{11}	0.4340	0.1693	0.25	0.9017	0.7166	0.4166	0.1756	0.6666	0.6666
A_{12}	0.3208	0.8306	1.00	0.9017	0.7166	0.4166	0.1756	0.6666	0.6666
A_{13}	0.0000	0.3386	0.25	0.7687	0.6542	0.4166	0.5519	0.3333	1.0000
A_{14}	0.1226	0.3386	0.50	1.0000	0.7166	1.0000	0.1039	0.6666	0.6666
A_{15}	0.0377	1.0000	1.00	1.0000	0.7166	1.0000	0.1039	0.6666	0.6666

For the COMET method, characteristic objects were determined using three characteristic values: the minimum, mean, and maximum values for the criteria of the alternatives studied. In this procedure, the TOPSIS method was used to evaluate a set of characteristic objects. The result of the TOPSIS method, in this case, is the vector SJ, which is used in the next COMET step to construct the vector P. By using the TOPSIS method to evaluate the characteristic objects instead of performing a time-consuming and subjective pairwise comparison by an expert, the procedure is faster, easier and objective. In addition, COMET combined with TOPSIS retains all the advantages of the classical version of the COMET method, such as rank reversal free, accuracy and independence of the algorithm complexity from the number of evaluated alternatives. Furthermore, it is possible because TOPSIS works here on a set of characteristic objects instead of evaluating alternatives directly.

There is no need for labour-intensive filling of the MEJ matrix for this approach because TOPSIS provides the SJ. Nevertheless, it is possible to reconstruct the MEJ matrix for a visualisation based on the P vector. The MEJ matrix is visualised in Fig. 1. Green fields in the MEJ matrix indicate the advantage of the compared object over the other object and have a value of 1. Red fields represent comparisons with a value of 0 in which the compared object is worse than the other object, and blue fields with a value of 0.5 indicate a tie between the compared objects.

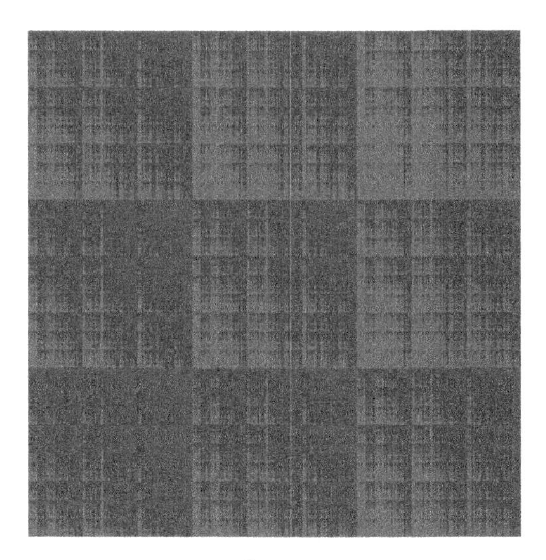

Fig. 1. The MEJ matrix as the stage of the COMET method. (Color figure online)

4 Results and Discussion

Table 4 displays the outcome values of the preference function and the rankings for the evaluated alternatives obtained by the three MCDA methods used in this study. The outcome rankings are visualised in Fig. 2. It can be observed that A_1 (HP Pavilion Gaming) was identified as the leader of all the rankings received. This laptop has a very high value of the profit criteria C_2 (Hard disk capacity), C_3 (Random-access memory (RAM)) and C_6 (Refresh rate). Second place was taken in all rankings by different alternatives. In the COMET ranking, the A_8 (ASUS TUF Dash F15) was ranked second. A_8 has a high value for profit criteria C_3 (Random-access memory (RAM)), C_5 (Battery capacity), C_6 (Refresh rate), C_8 (Graphics card memory) and C_9 (Processor cache memory). In the TOPSIS ranking, A_8 was ranked third, and in the VIKOR ranking, it was ranked sixth. In the TOPSIS ranking, second place was taken by A_{15} (MSI GL75). This model has a very beneficial value of profit criteria C_2 (Hard disk capacity), C_3 (Random-access memory (RAM)), C_4 (Screen size), C_6 (Refresh rate), but cost criteria C_1 (Price) and C_7 (Weight) are quite high, that is, not preferred. In the VIKOR ranking, second place belongs to A_3 (Lenovo Legion 5–15). This laptop has very high values of profit criteria C_5 (Battery capacity), C_8 (Graphics card memory) and C_9 (Processor cache memory). Third place in the COMET and VIKOR rankings was occupied by A_4 (ASUS TUF Dash F15). Third place in the COMET and VIKOR rankings was taken by the A_4 (ASUS TUF Dash F15). This model has a very favourable criterion value of C_6 (Refresh rate), C_8 (Graphics card memory) and C_9 (Processor cache memory).

Table 4. Results including preference values and ranks for alternatives A_1–A_{15} for TOPSIS, VIKOR and COMET method.

A_i	Preference			Rank		
	COMET	TOPSIS	VIKOR	COMET	TOPSIS	VIKOR
A_1	0.7517	0.6556	0.0193	1	1	1
A_2	0.1627	0.2316	1.0000	15	15	15
A_3	0.7122	0.5168	0.1887	4	6	2
A_4	0.7145	0.5412	0.3055	3	5	3
A_5	0.3095	0.3031	0.8463	11	13	11
A_6	0.2096	0.3058	0.9154	13	12	13
A_7	0.4533	0.4939	0.7318	10	7	10
A_8	0.7462	0.5805	0.4239	2	3	6
A_9	0.5865	0.4533	0.6360	6	8	7
A_{10}	0.5037	0.4309	0.6979	8	9	9
A_{11}	0.2537	0.3015	0.8863	12	14	12
A_{12}	0.5344	0.5546	0.6756	7	4	8
A_{13}	0.1982	0.3427	0.9820	14	11	14
A_{14}	0.4579	0.4163	0.4038	9	10	4
A_{15}	0.6812	0.6230	0.4073	5	2	5

The results obtained show that profit criteria and parameters that enhance the usability and capabilities of the laptop have a strong influence on which alternative is selected as a ranking leader. In the case of this study, the alternative that was identified as rankings leader was not cheap compared to the other alternatives. The results demonstrate that a very favourable value for one criterion is not sufficient for an alternative to getting the first ranking. Furthermore, it can be observed that the rankings of the various methods are different. This observation confirms the fact that various MCDA methods can give different results for the same problem.

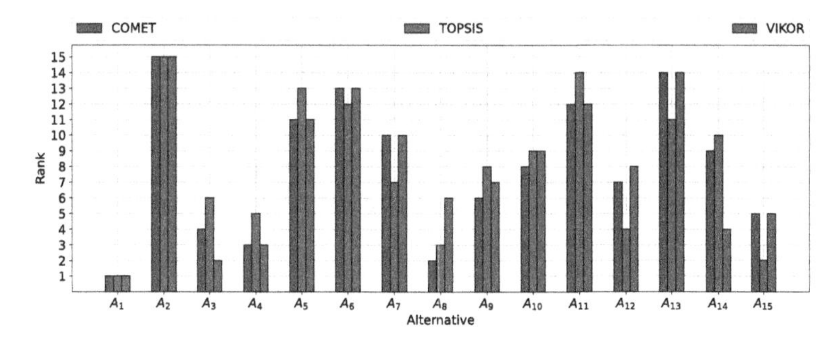

Fig. 2. Column charts illustrating the rankings of alternatives A_1–A_{15} for TOPSIS, VIKOR and COMET methods.

The next stage of the comparative analysis in this work was to compare the convergence of the obtained rankings using two ranking similarity coefficients: asymmetrical WS and symmetrical r_w. The results of this examination are displayed in Fig. 3. It can be observed that the highest convergence occurs for the rankings provided by COMET and TOPSIS, and COMET and VIKOR, and the lowest for VIKOR and TOPSIS.

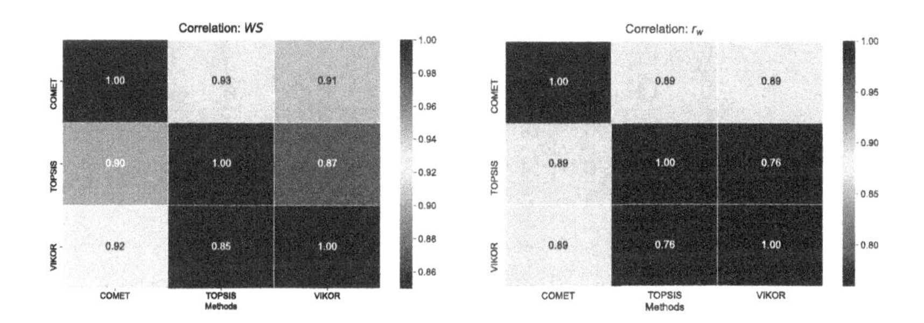

Fig. 3. Visualisation of values of rankings similarity WS and r_w.

4.1 Sensitivity Analysis

Table 5 provides the results of the first stage of sensitivity analysis, which are relative sensitivity coefficients. This step aimed to investigate the effect of individual weights changes for each evaluation criteria on the number of changes in the rankings. The values of individual weights were increased and decreased by a small value (5%) and a large value (50%). As a result, it can be observed that the TOPSIS model has the lowest relative coefficients indicating the fewest changes in ranking and the highest robustness to small (by 5%) and significant changes (by 50%) in weight values in comparison to the other two MCDA models. The VIKOR model showed to be the least resistant to changes in the criteria weights. It is evidenced by the highest values of relative sensitivity coefficients.

For COMET and VIKOR, the results were most strongly affected by changes in the weights of criterion C_3 (Random-access memory (RAM)), while for TOPSIS, the criterion with the most significant effect was C_2 (Hard disk capacity). Both of the criteria mentioned are of the profit type. No criterion was identified for which changes in weight values would leave the rankings unchanged. It means that all nine evaluation criteria used in this study significantly impact the results, so actions such as eliminating any of them or replacing them with another are not recommended.

Table 5. Values of relative sensitivity coefficients in the number of changes in the ranking for modification of criteria weights by 5% and 50%.

Method	COMET				TOPSIS				VIKOR			
Weight modification	Increase		Decrease		Increase		Decrease		Increase		Decrease	
	5%	50%	5%	50%	5%	50%	5%	50%	5%	50%	5%	50%
C_1	2	22	0	10	0	12	0	2	2	44	4	10
C_2	2	16	2	12	0	8	2	20	6	34	0	4
C_3	0	16	4	18	0	10	2	12	10	38	4	18
C_4	0	16	2	10	0	2	0	2	6	6	0	2
C_5	4	12	0	8	0	6	0	2	4	24	0	2
C_6	0	10	4	12	0	12	0	6	8	8	2	12
C_7	0	14	2	10	2	6	4	8	8	44	0	2
C_8	2	10	0	10	4	16	0	6	4	10	0	6
C_9	2	16	0	14	0	8	0	12	4	24	0	4

The second stage of the sensitivity analysis was to investigate the robustness of the rankings provided by the MCDA models applied in this research, measured as the percentage of tolerable change in the weights of each criterion. The results of this study are included in Table 6. The highest percentages of tolerable changes in criteria weights were noticed for the TOPSIS model. It shows the highest robustness of this model to changes in criteria weights. On the other hand, the VIKOR model appeared to be the least resistant to modifications of the criteria weights. Even small changes in the values of the weights caused changes in the ranking.

Table 6. Values of tolerable weights change in % for modification of criteria weights.

Method	COMET		TOPSIS		VIKOR	
Tolerable weights change	Increase [%]	Decrease [%]	Increase [%]	Decrease [%]	Increase [%]	Decrease [%]
C_1	1	5	5	26	0.8	0.1
C_2	1	2	8	4	0.4	13.2
C_3	11	3	19	2	0.4	1.7
C_4	9	1	9	6	0.4	31
C_5	2	11	15	42	0.5	8.6
C_6	10	2	7	14	0.5	1.8
C_7	7	2	4	2	0.1	11.8
C_8	3	18	2	13	0.5	6.4
C_9	2	5	15	18	0.5	6.4

Figure 4 displays, in the form of a column chart, the values of the sensitivity coefficients SC_j for each criterion C_j for the MCDA models studied, when criteria weights are increased. High values of this coefficient mean that even small changes in the weights of these criteria cause changes in the rankings. Rankings provided by the COMET model are most sensitive to changes in C_1 and C_2, TOPSIS in C_8 and C_7, and VIKOR in C_7, C_2, C_3, and C_4.

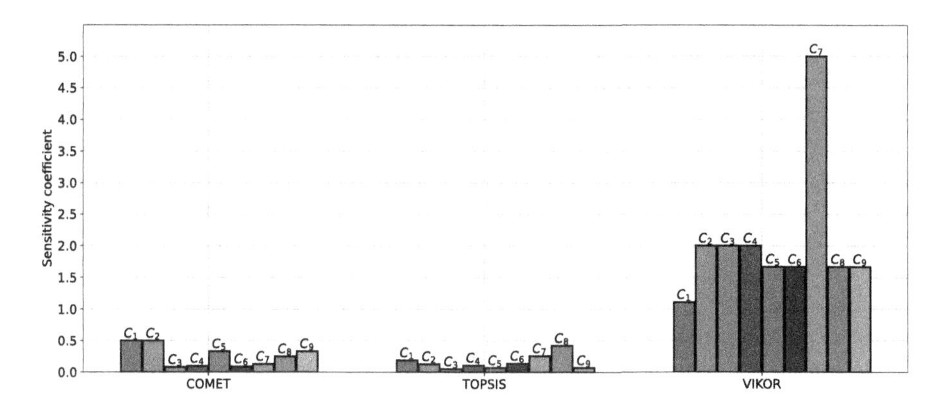

Fig. 4. Values of criteria sensitivity coefficients SCj for increasing of criteria weights.

The values of the similarity coefficients for decreasing the values of the criteria weights are visualised in Fig. 5. The most significant impact on the changes in the COMET rankings is the modification of the weights C_4, C_2, C_6, and C_7. For TOPSIS, these are C_7 and C_3, and for VIKOR, they are C_1, C_3, and C_6.

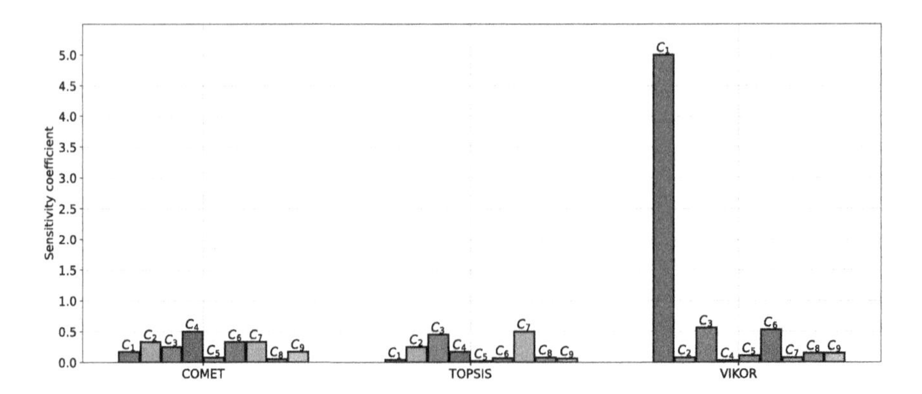

Fig. 5. Values of criteria sensitivity coefficients SCj for decreasing of criteria weights.

5 Conclusions

In this work, the authors present an innovative approach based on MCDA methods as a tool to support consumer decision-making in e-commerce on the illustrative example of the multi-criteria problem of choosing the most advantageous laptop model. The MCDA methods used in the study successfully identified the most favourable alternatives and enabled the identification of the criteria that had the most significant impact on alternatives reaching high ranks. The most important criteria proved to be C_2 (Hard disk capacity), C_3 (Random-access memory (RAM)), C_5 (Battery capacity), C_6 (Refresh rate), C_8 (Graphics card memory) and C_9 (Processor cache memory). These criteria are the parameters that determine the functionality of laptops and the quality of work on them. Thus, it turned out that it is not the low price but the technical parameters that play the most significant role in the rankings in the study performed by the authors.

The observed differences in the rankings obtained by the three MCDA methods used in this study are due to differences in the algorithms of each method and are natural [27]. Because of this, a comparative analysis of the rankings provided by the different MCDA models is helpful in an insightful and critical evaluation of the final results of the decision procedure. The sensitivity analysis performed proved that the sensitivity of the rankings to changes in the criteria weights is dependent on both the MCDA methods used and the criteria.

The study results prove that a tool based on MCDA methods could successfully support consumers in making multi-criteria decisions regarding the purchase of various products and devices at e-commerce sites. In contrast to decisions based on intuition and considering only the main selection criteria, MCDA methods enable objective, quick and fully automated evaluation regardless of the number of alternatives and criteria.

The obtained outcomes encourage further research, including MCDA methods as a basis for tools supporting multi-criteria consumer decisions. Furthermore, due to the notable differences in the rankings provided by the investigated methods, another appropriate direction of research seems to be the comparison of the obtained results with other MCDA methods such as SPOTIS [5,24], COPRAS [8], PROMETHEE II [9].

Acknowledgment. The work was supported by the National Science Centre, Dec. number UMO-2018/29/B/HS4/02725 (A.S., B.K. and W.S.).

References

1. Aytaç Adalı, E., Tuş Işık, A.: The multi-objective decision making methods based on MULTIMOORA and MOOSRA for the laptop selection problem. J. Ind. Eng. Int. **13**(2), 229–237 (2016). https://doi.org/10.1007/s40092-016-0175-5
2. Bączkiewicz, A., Wątróbski, J., Sałabun, W.: Towards MCDA based decision support system addressing sustainable assessment (2021). https://aisel.aisnet.org/isd2014/proceedings2021/sustainable/6/
3. Behzadian, M., Otaghsara, S.K., Yazdani, M., Ignatius, J.: A state-of the-art survey of TOPSIS applications. Expert Syst. Appl. **39**(17), 13051–13069 (2012)
4. Sönmez Çakır, F., Pekkaya, M.: Determination of interaction between criteria and the criteria priorities in laptop selection problem. Int. J. Fuzzy Syst. **22**(4), 1177–1190 (2020). https://doi.org/10.1007/s40815-020-00857-2
5. Dezert, J., Tchamova, A., Han, D., Tacnet, J.M.: The SPOTIS rank reversal free method for multi-criteria decision-making support. In: 2020 IEEE 23rd International Conference on Information Fusion (FUSION), pp. 1–8. IEEE (2020)
6. Faizi, S., Sałabun, W., Rashid, T., Wątróbski, J., Zafar, S.: Group decision-making for hesitant fuzzy sets based on characteristic objects method. Symmetry **9**(8), 136 (2017)
7. Faizi, S., Sałabun, W., Ullah, S., Rashid, T., Więckowski, J.: A new method to support decision-making in an uncertain environment based on normalized interval-valued triangular fuzzy numbers and COMET technique. Symmetry **12**(4), 516 (2020)
8. Goswami, S., Mitra, S.: Selecting the best mobile model by applying AHP-COPRAS and AHP-ARAS decision making methodology. Int. J. Data Netw. Sci. **4**(1), 27–42 (2020)
9. Goswami, S.S., Behera, D.K.: Evaluation of the best smartphone model in the market by integrating fuzzy-AHP and PROMETHEE decision-making approach. DECISION **48**(1), 71–96 (2021). https://doi.org/10.1007/s40622-020-00260-8
10. Kecek, G., Demirağ, F.: A comparative analysis of TOPSIS and MOORA in laptop selection. Res. Humanit. Soc. Sci. **6**(14) (2016). 2225-0484
11. Kizielewicz, B., Kołodziejczyk, J.: Effects of the selection of characteristic values on the accuracy of results in the COMET method. Procedia Comput. Sci. **176**, 3581–3590 (2020)
12. Kizielewicz, B., Shekhovtsov, A., Sałabun, W.: A new approach to eliminate rank reversal in the MCDA problems. In: Paszynski, M., Kranzlmüller, D., Krzhizhanovskaya, V.V., Dongarra, J.J., Sloot, P.M.A. (eds.) ICCS 2021. LNCS, vol. 12742, pp. 338–351. Springer, Cham (2021). https://doi.org/10.1007/978-3-030-77961-0_29
13. Kizielewicz, B., Wątróbski, J., Sałabun, W.: Identification of relevant criteria set in the MCDA process-wind farm location case study. Energies **13**(24), 6548 (2020)

14. Maliene, V., Dixon-Gough, R., Malys, N.: Dispersion of relative importance values contributes to the ranking uncertainty: sensitivity analysis of multiple criteria decision-making methods. Appl. Soft Comput. **67**, 286–298 (2018)
15. Opricovic, S., Tzeng, G.H.: Compromise solution by MCDM methods: a comparative analysis of VIKOR and TOPSIS. Eur. J. Oper. Res. **156**(2), 445–455 (2004)
16. Papathanasiou, J., Ploskas, N.: Multiple Criteria Decision Aid. SOIA, vol. 136. Springer, Cham (2018). https://doi.org/10.1007/978-3-319-91648-4
17. Sałabun, W.: Reduction in the number of comparisons required to create matrix of expert judgment in the COMET method. Manag. Prod. Eng. Rev. **5** (2014)
18. Sałabun, W.: The characteristic objects method: a new distance-based approach to multicriteria decision-making problems. J. Multi-Criteria Decis. Anal. **22**(1–2), 37–50 (2015)
19. Sałabun, W., Karczmarczyk, A.: Using the COMET method in the sustainable city transport problem: an empirical study of the electric powered cars. Procedia Comput. Sci. **126**, 2248–2260 (2018)
20. Sałabun, W., Karczmarczyk, A., Wątróbski, J.: Decision-making using the hesitant fuzzy sets COMET method: an empirical study of the electric city buses selection. In: 2018 IEEE Symposium Series on Computational Intelligence (SSCI), pp. 1485–1492. IEEE (2018)
21. Sałabun, W., Urbaniak, K.: A new coefficient of rankings similarity in decision-making problems. In: Krzhizhanovskaya, V.V., et al. (eds.) ICCS 2020. LNCS, vol. 12138, pp. 632–645. Springer, Cham (2020). https://doi.org/10.1007/978-3-030-50417-5_47
22. Sałabun, W., Wątróbski, J., Shekhovtsov, A.: Are MCDA methods benchmarkable? A comparative study of TOPSIS, VIKOR, COPRAS, and PROMETHEE II methods. Symmetry **12**(9), 1549 (2020)
23. Sałabun, W., Ziemba, P., Wątróbski, J.: The rank reversals paradox in management decisions: the comparison of the AHP and COMET methods. In: Czarnowski, I., Caballero, A.M., Howlett, R.J., Jain, L.C. (eds.) Intelligent Decision Technologies 2016. SIST, vol. 56, pp. 181–191. Springer, Cham (2016). https://doi.org/10.1007/978-3-319-39630-9_15
24. Shekhovtsov, A., Kizielewicz, B., Sałabun, W.: New rank-reversal free approach to handle interval data in MCDA problems. In: Paszynski, M., Kranzlmüller, D., Krzhizhanovskaya, V.V., Dongarra, J.J., Sloot, P.M.A. (eds.) ICCS 2021. LNCS, vol. 12747, pp. 458–472. Springer, Cham (2021). https://doi.org/10.1007/978-3-030-77980-1_35
25. Shekhovtsov, A., Kołodziejczyk, J., Sałabun, W.: Fuzzy model identification using monolithic and structured approaches in decision problems with partially incomplete data. Symmetry **12**(9), 1541 (2020)
26. Shekhovtsov, A., Kozlov, V., Nosov, V., Sałabun, W.: Efficiency of methods for determining the relevance of criteria in sustainable transport problems: a comparative case study. Sustainability **12**(19), 7915 (2020)
27. Shekhovtsov, A., Więckowski, J., Kizielewicz, B., Sałabun, W.: Towards reliable decision-making in the green urban transport domain. Facta Universitatis, Series: Mechanical Engineering (2021)
28. Wątróbski, J., Jankowski, J., Ziemba, P., Karczmarczyk, A., Zioło, M.: Generalised framework for multi-criteria method selection. Omega **86**, 107–124 (2019)
29. Wątróbski, J., Sałabun, W., Karczmarczyk, A., Wolski, W.: Sustainable decision-making using the COMET method: an empirical study of the ammonium nitrate transport management. In: 2017 Federated Conference on Computer Science and Information Systems (FedCSIS), pp. 949–958. IEEE (2017)

Unmasking the Masked Face Using Zero-Shot Learning

Pranjali Singh(✉) and Amritpal Singh

Department of Computer Science and Engineering, Dr B.R. Ambedkar National Institute of Technology, Jalandhar, India
{pranjalis.cs.19,apsingh}@nitj.ac.in

Abstract. Nowadays, people are required to wear masks due to the COVID-19 pandemic. The COVID-19 is an ongoing crisis that has resulted in a large number of fatalities and safety concerns. People also carry masks to cover themselves to effectively prevent the transmission of this virus. In this situation recognizing a face is very challenging. In certain cases, like facial attendance, face access control, facial security, this makes traditional facial recognition technology ineffective, for that urgent requirement to improve this recognition performance and use the technology on the masked face. During the current pandemic, the main objective of researchers is to deal with these problems through quick and accurate approaches. Throughout this report, suggest a clear way centred on removing masked areas and deep learning related techniques to resolve the issues of mask detection. Another way of finding the masked face is to go through TensorFlow, YOLOv5, SSDMNV2, SVM, OpenCV, Keras.

Deep Learning of artificial intelligence (AI) is an exciting future technology with explosive growth. Masked face recognition is a mesmerizing topic that contains several AI technologies including classifications, SSD object detection, MTCNN, FaceNet, data preparation, data cleaning, data augmentation, training skills, etc. Takes two datasets, CASIA-Web face datasets used for training purpose and LFW is used for testing purpose. Here, face alignment has been done by SSD and MTCNN. After face alignment, deleting the misleading image. Then wear a face mask in the image by using Dlib. From Dlib get the facial landmark. So take the mouth part for the face mask. Then detect the face mask. In green box shows that the person is wearing the mask. In red it shows that person is not wearing the mask. After detecting the face, also have to recognize the person in the mask. In training, it gives 99% of accuracy and in testing, it gives 96% of accuracy.

Keywords: Masked face dataset · COVID-19 epidemic · OpenCV · TensorFlow · Zero-shot learning · Python · Deep learning · Convolution neural network · Masked face recognition and detection

1 Introduction

1.1 Background

The COVID-19 outbreak has been major healthcare and social problem in the world since November 2019. As per the WHO, the pandemic is causing a global health emergency,

I. Woungang et al. (Eds.): ANTIC 2021, CCIS 1534, pp. 563–585, 2022.
https://doi.org/10.1007/978-3-030-96040-7_43

making this the most recent human health virus outbreak throughout the last century so wearing a mask is required. Before this pandemic, people used to carry the mask only to protect themselves from air pollution. This pandemic is circulated through the respiratory system which is spreading very fast. Given the fact that many states mandate people to wear masks in public areas, many individuals forget or refuse to wear masks, or they wear them inappropriately [1]. As a result of these realities, the illness will spread faster and place a higher strain on the public health care system. Regarding the effectiveness of several vaccines, wearing masks is among the most efficient and cost-effective strategies to prevent 80% of respiratory illnesses [3, 4]. As a result, many monitoring systems have been established to provide efficient supervision in sporting events, airports, public transit systems, hospitals, and retail locations to detect the mask. This pandemic affects many areas like the institute, organizations [2–4].

The coronavirus illness (COVID-19) is an unprecedented catastrophe that there have been a lot of deaths and security concerns as a result of this. People frequently use masks to safeguard themselves against the spreading of coronavirus. Because significant portions of the face are concealed, facial identification is extremely difficult [3].

The only choice left is to keep away from this disease like keep social distancing, wear a mask, regularly wash hands. Many countries have their regulations to follow guidelines to get prevented from this virus which will minimize the risk of COVID-19.

Face Recognition

Face recognition is one of the most promising fields in computer vision. The facial recognition technology recognizes a person's face and verifies them automatically from photos. Face recognition is very crucial in our daily lives [5]. Face recognition is extensively used to quickly and accurately identify a person in a passport check, smart door, credit card, ATM, criminal or terrorist investigation, voter verification and many more applications. Face recognition is the most used biometric technology because of these factors [1, 3].

During the present coronavirus epidemic, researchers are concentrating their efforts on developing ideas for dealing with the problem in a timely and effective manner [6]. To solve the challenge of masked face recognition, we present an effective approach based on occlusion elimination and deep learning-based features. Face recognition, as a unique Biometric recognition technique in all automated personal identification systems, has gotten a lot of attention. Many sectors around the world are now attempting to apply this authentication process in their organizations to protect their assets [7]. Face recognition systems are being considered by several governments across the globe to safeguard public places such as railway stations, bus stops and airports [8]. However, due to the low recognition rate, real-time recognition is still ineffective. The rate of recognition is determined by the image quality. When an image is noisy or of poor quality, the recognition rate drops. Pre-processing is necessary for a greater recognition rate. Boosting, sharpening, resizing, normalizing, de-noising and cropping are some of the pre-processing techniques utilized in the face recognition process.

Masked Face Recognition

Masked face recognition has become more essential in recent years. The masked face has fewer features than a regular face, which makes it more difficult to recognize [9].

As a result, the rate of recognition accuracy is falling. As a result, the masked face is among the most important considerations in the field of face recognition.

The method of screening requires the identification of someone who does not wear a face mask. Here the dataset is a collection without the mask and with mask images and is using real-time mask detection through a webcam. Mask detection by using the popular Deep Learning technique is useful for figuring out who wears the mask and who does not and that can be used on any common device [4, 5].

Deep transfer learning was used to extract features, and it was paired with three traditional machine learning algorithms. The greatest thing regarding deep learning is there is deep architecture in all the models. Deep architecture has many layers, becoming the biggest opponents of deep architecture that has a few hidden layers [16]. The CNN is used to function extraction from images, and then several hidden layers learn these features. Also SSDMNV2 technology, SVM technology is used to detect mask detection [6]. Here performed a comparison among them to identify the most appropriate algorithm that reached the best accuracy while taking the least amount of time during the training and detection processes [17].

Generate Mask

Generated masked face to train a deep network for facial recognition due to the unavailability of a masked face dataset. Masking the face successfully masks faces, resulting in a massive dataset of masked faces [11]. Cover the Face is software that uses computer vision to mask faces in pictures. It employs a dlib based facial landmarks detector to determine the face tilt and six essential characteristics required for mask application. A matching mask template is chosen from the mask library depending on the facial tilt [16]. The template mask is then modified to fit the face exactly depending on the six main characteristics. The template mask is then modified to fit the face exactly depending on the four main characteristics.

For using face detection/recognition, there are several relevant face datasets accessible, such as CASIA web faces, LFW to mention a few. Masking the face may be used to transform current datasets into masked-face datasets, that can subsequently be said to train a deep network for the underlying technology.

1.2 Traditional Approach

For the masked face recognition, have traditional approaches:

Machine Learning-Based Approach

Facial recognition is a technique that can recognise a person only by looking at them. It uses machine learning techniques to identify, collect, store, and analyse face characteristics so that they can be matched to photos of people in a database.

Face Detection: To begin, the system must find the face in the image or video. Most cameras now include a built-in facial detection feature. Facebook, Snapchat and other social media platforms employ face identification to let users apply effects to images and videos taken using their apps.

Face Alignment: To a computer, faces excluded from the focal point appear completely different. To normalize the face and make it consistent with the faces in the database, an algorithm is necessary. Using a variety of generic face landmarks is one method to do this. The bottom of the chin, the top of the nose, the outsides of the eyes, different places surrounding the eyes and lips, and so on are examples. The next stage is to train a machine learning system to locate these spots on any face and turn it around the middle.

Feature Measurement and Extraction: This phase entails measuring and extracting numerous characteristics from the face so that the algorithm can compare it to other faces in its database. However, it was initially unclear which traits should be collected and retrieved until researchers realised that letting the ML system decide which data to gather on its own was the optimal method. Embedding is a technique that employs deep neural networks to teach itself to create numerous characteristics of a face, enabling it to differentiate it from other faces.

Face Recognition: The machine learning algorithm will compare the measures of each face to known individuals in a database, and use the distinctive characteristics of each face. There will be a match when the face in your database comes the closest to the dimensions of the face under consideration.

Face Verification: Face verification contrasts the distinct characteristics of one face to those of another. To determine if the faces matching or not, the ML algorithm will provide a certain number over the scale number approaching towards least will say low match and number is approaching towards high on the scale will say there is a high match.

Deep Learning-Based Approach

Extract a wide range of characteristics from pictures using CNN. It turns out that the same feature extraction concept may also be applied to face recognition. CNNs has the primary advantage of automatic feature extraction. The requested input data is first forwarded to a feature extraction network, and then the retrieved features are forwarded to a classifier network. There are a lot of convolutional and pooling layer pairings in the feature extraction network. The convolutional layer is made up of a group of digital filters that execute the convolution on the input data. The threshold is determined by the pooling layer, which acts as a dimensionality reduction layer.

- Draw a bounding box around one or more faces throughout the image for face detection.
- For face alignment, equalize the face to match the database's photometric and geometry.
- Extract facial characteristics that can be utilized in the recognition job for facial extraction.
- Align the face with one or more identified faces in a database that has already been generated.

1.3 Zero-Shot Learning (ZSL)

ZSL methods are intended to understand intermediary semantic layers and their characteristics, then use them to forecast a new class of data at inference time. As a result, unknown classes may be predicted, and no training data is necessary for such classes.

2 Literature Review

Many non-masked face recognition algorithms have recently been developed that are widely used and provide better performance. Few contributions have been made in the area of masked face recognition. As a result, a statistical protocol has been chosen for this study, which can be used in both non-masked and masked face recognition techniques. Many areas focused on recognizing the masked face like Principal Component Analysis(PCA), Convolutional Block Attention Module(CBAM) and datasets like SMFRD, CASIA.

2.1 Current Scenario

In today's world, the word "security" is crucial. Face recognition is commonly used in biometric technology to protect any device since it is better than other conventional techniques such as fingerprint, password, PIN and so on, and is the most accurate way to recognize or validate an individual efficiently. Md. Sabbir Ejaz et al. [1] focuses on non-masked and masked facial recognition by using PCA. Face identification has become difficult in recent years due to various masks like the use of sunglasses, scarves, caps, and various forms of make-up products. These types of masks affect face recognition accuracy. PCA is a commonly used statistical method that is more accurate and successful. As a result, the PCA algorithm was chosen for this analysis [1]. Finally, for a deeper understanding, a comparative analysis was conducted. In a PCA-based face recognition method, it has been shown that a face without a mask has a higher recognition score. When an individual wears a mask, however, facial recognition has a low recognition rate. Extracting features from a masked face is found to be less challenging than extracting features from a non-masked face. Due to the extreme lack of distinguishing features when wearing a mask, the identification rate is reduced [1]. Finally, find that the standard statistical algorithm (PCA) is superior to masked face recognition for regular face recognition.

The worldwide COVID-19 outbreak has made people aware that wearing a mask is among the most important ways to defend themselves from infections, posing significant challenges to the current face recognition system [1, 2]. Yande Li et al. [2] suggests two methods in this paper to overcome the problem of recognizing a person in the mask i.e. an attention-based approach and a cropping-based approach. In an attention-based method, the CBAM attention system is used to concentrate on the region around the eyes [2]. Investigate the best cropping for each case in masked face recognition using a cropping-based approach and combines it with the CBAM to resolve the difficulty. With each case, the best cropping is determined, and the CBAM module is used to concentrate on the areas around the eyes [2]. Faces without masks were used for training to identify masked faces and used for training to recognise faces without the mask. Extensive tests on the Extend Yela B, CASIA-Web face and SMFRD datasets demonstrate that the proposed scheme can dramatically increase masked face recognition efficiency when compared to other state-of-the-art methods [2].

J. Deng et al. [3] host the Masked Face Recognition (MFR) challenge, which focuses on evaluating deep face recognition techniques when facial masks are present. The InsightFace track as well as the WebFace260M track are the two primary tracks in the MFR challenge [3]. Here, systematically obtain a large-scale mask face sample test data with 7K identities for the InsightFace track and also gather a children's test data, which contains 14K IDs, and a multi-racial test set, which has 242K identities. Also create an online model testing method utilising these three test sets that may provide a complete review of face detection and recognition models [3].

Fadi Boutros et al. [4] introduces the Masked Face Recognition(MFR) Competitions held as part of the 2021 International Joint Conference on Biometrics (IJCB 2021) are summarised in the paper [4]. The tournament drew a total of ten teams, all of which submitted acceptable entries. Such teams have a wide range of ties, including connections with academia and business in nine various countries. These teams were successful in submitting 18 proposals that were valid. The challenge is intended to encourage innovations aimed at improving the accuracy of MFR [4]. Furthermore, the competition took into account the deployability of the submitted solutions by taking into consideration the efficiency of the face recognition models. The proposed methods are evaluated using an unique dataset that represents a multisession, collaborative real masked capture situation. Ten of the 18 submitted solutions had greater masked face validation accuracy than one of the best performing academic face recognition methods [4].

Weiqiu Wang et al. [5] compares between face recognition and masked face recognition. Face recognition has been greatly improved using deep learning approaches. When wearing a mask, however, the effectiveness of deep learning algorithms degrades [5]. MFR has recently received more attention as a result of the worldwide COVID-19 epidemic [5]. Improving MFR performance is a non-trivial and critical task. MaskOut, a convenient and efficient data augmentation approach, is introduced in this paper. MaskOut masks out original face characteristics by replacing a random area below the nose with such a random mask pattern [5]. Our method is both computationally and memory-efficient, as well as easy to integrate with other approaches. MaskOut significantly improves efficiency in MFR, according to the findings of the experiments. In addition, author create MCPRL-Mask, a real-life masked face dataset, to test the performance of masked face recognition algorithms [5].

2.2 Challenges

Learning Delay: As data is big with many classes, the model takes many epochs to learn parameters.

Embedding: As embeddings are calculated based on partial facial features, it sometimes leads to misclassification as the distance between faces of two different people may come less than the set threshold. To rectify this we use parameters validation rate and false acceptance rate.

3 Problem Statement

As a result of the pandemic, need to create a system that can detect faces in real-world recordings, determine whether or not the identified faces are wearing masks, and also identify people behind the mask. Our ultimate aim after the mask/no mask classification is to identify the person behind the mask in real-world videos.

3.1 Need Analysis

Earlier, the Face, thumb, the eye is used for biometric and security purposes. But nowadays, due to COVID-19, it is difficult to recognize face due to mask. So here, Face recognition is not an easy task for biometric purposes. To overcome this problem, the need to identify the person even in the mask through a webcam will help in any scenario.

3.2 Objectives

- To study and compare the state-of-art techniques available for face recognition.
- To propose and implement a model to recognize the person in a mask, with minimum model training efforts, using Zero-shot learning.
- To test and validate the proposed model in a simulated environment on some real datasets.

4 Methodology

The methodology adopted is described in the following steps:

Knowing the Dataset: First, take the dataset like CASIA datasets as well as LFW datasets.

Model Training and Testing: Use CASIA web face database as training set and create an AI model and then training loop. Set that epoch number and then go for the training loop and optimization to modify the weights again and again. And then output fixed model parameters as the PB file here. Don't use the CASIA dataset to evaluate the model, use LFW datasets as our testing set and to evaluate its model and to get accuracy to know if our model is good or not.

Generate Mask face: By using DLib generate the mask on faces. Here, use D-library(Dlib) to perform this function because the library has another function code facial landmark so that can do face detection. And then wear the artificial face mask on faces and then save images in the output folder.

Face Detection and Recognition: As a final result, have to detect the mask on a person's face and also recognize the person behind the mask which can be evaluated with high accuracy.

5 Implementation Approach

5.1 Datasets

Here, for training purposes choose the CASIA-Web face dataset with 494,414 images and for testing purposes, choose the LFW dataset with 13,233 images (Table 1).

Table 1. The information of CASIA-WebFace and comparison to other scale face datasets.

Dataset	#Subjects	#Images	Availability
LFW [7]	5.749	13,233	Public
WDRef [4]	2,995	99,773	Public (feature only)
CelebFaces	10,177	202,599	Private
SFC	4030	4,400,000	Private
CACD [3]	2,000	163,446	Public (partial annotated)
CASIA-WebFace	10,575	494,414	Public

5.2 Experimental Approach

FaceNet

FaceNet may be used to collect high-quality characteristics from faces, referred to as face embeddings, which can subsequently be used to train a face recognition system.

Authentication of the Face: A one-to-one match between a face and a recognised identity.

Recognition of Face: A one-to-many link exists between a particular face and a database of other people's faces.

Embeddings mean a face is transformed into a sequence of numbers. These numbers are used to describe face characteristics. Here, use 128-dimension embeddings. Here, take embeddings with 12 dimensions as an example as shown in Fig. 1.

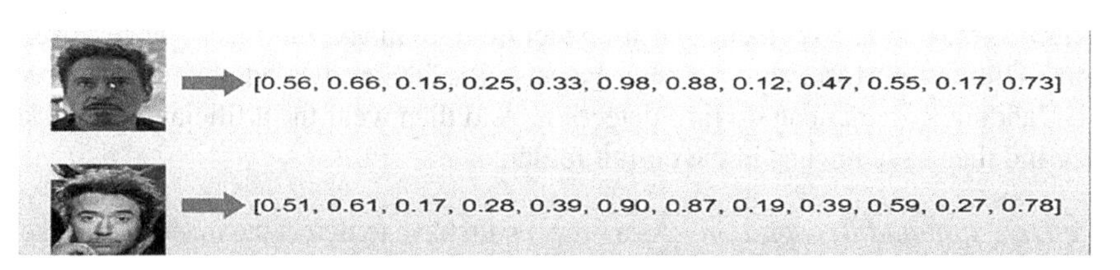

Fig. 1. Embeddings

Steps to Calculate Euclidean distance.

1. Calculate the embedding of each image.
2. Calculate the Euclidean distance (scalar) between embeddings
3. Set the threshold
4. The smallest distance and under the threshold will be the answer.
5. FaceNet needs a face dataset to do face matching.

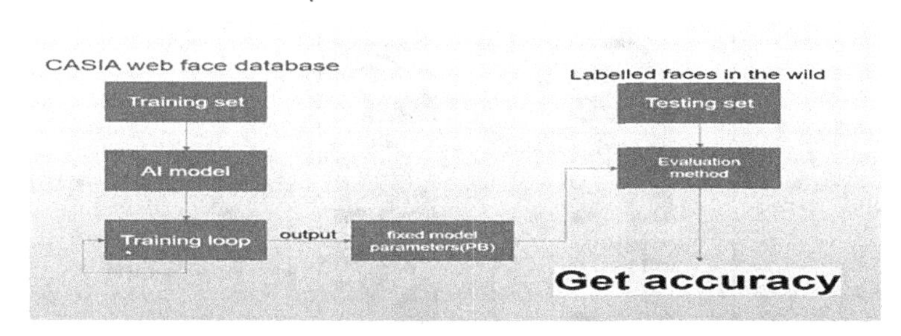

Fig. 2. Face recognition training and evaluation diagram

Face recognition training and evaluation diagram, use CASIA web face database as training set and create an AI model and then training loop. Set that epoch number and then go for the training loop and optimization to modify the weights again and again. And then output fixed model parameters as the PB file here. Don't use the CASIA dataset to evaluate the model, use LFW datasets as our testing set and to evaluate its model and to get accuracy to know if our model is good or not (Fig. 2).

Using SSD Model

There's no need to align your face but when doing real-time facial recognition, it takes time. So, allow the model to learn different angles, including side faces, to make it more robust. Simply perform face detection and cropping to generate a large number of face images from various angles (Fig. 3).

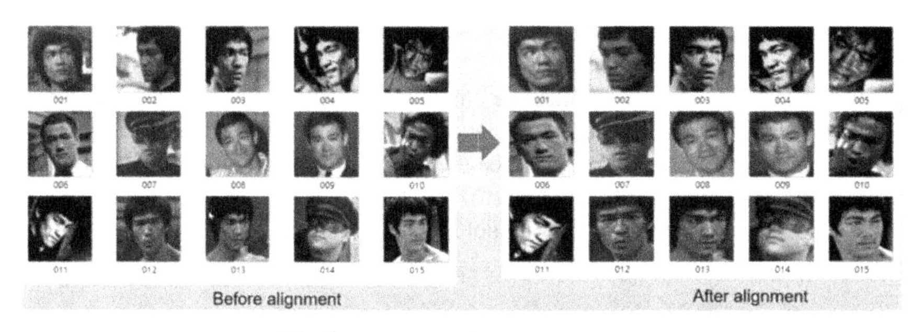

Fig. 3. Face detection and alignment

Face Detection methods are MTCNN, SSD, Dlib, OpenCV, here, MTCNN is normally used, SSD is faster than MTCNN, Dlib and OpenCV are conventional methods. SSD is an object detection method that is famous for its fast interference.

Using MTCNN Model

The first step is to resize the image and there is the scale factor. The scale factor is 0.7. So the image is resized to the smaller one and then do it again and again until the width or height is under 12. So these images are code image pyramids (Fig. 4).

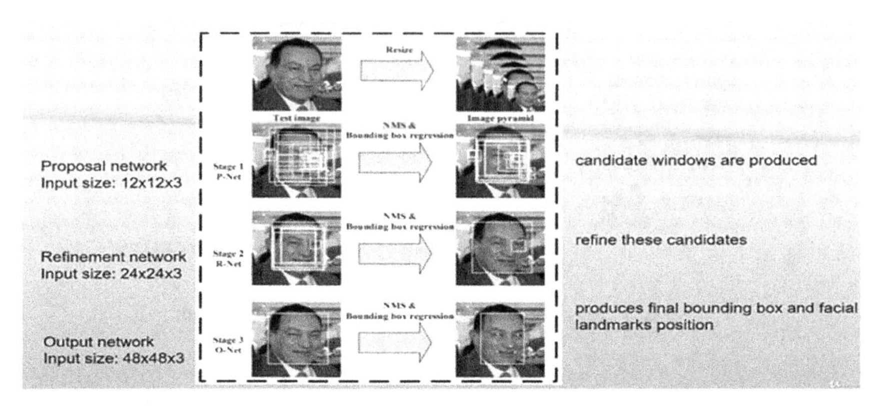

Fig. 4. Working of MTCNN [6].

There are 3 networks:

Proposal Networks

The first one is a coded proposal and that emphasis is $12 \times 12\times 3$. And this network is used to find a candidate window and can see the feature after the network and after the NMS bounding box regression and can find a boundary box.

Refinement Network

The second network is the Refinement network. The emphasis is trying to forward by $24\times 24\times 3$. And this network is used to refine these candidates.

Output Network

The third is called Output Network. The emphasis is a 48x48x3 and after that can find the final boundary boxes and the facial landmark positions, the positions of the facial landmarks are five points. So, one can see the two points on the sides of the mouth and the one point on the tip of the nose and two points on the eyes. But just want to find all the faces, so don't use the five-point model structure (Fig. 5).

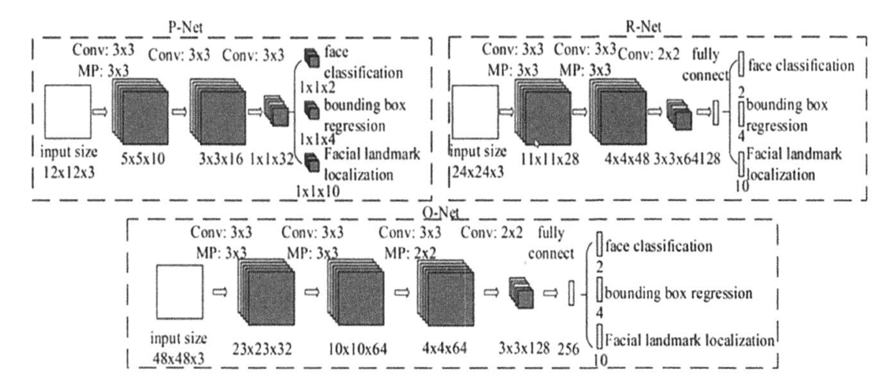

Fig. 5. Model structure [6]

Here, In MTCNN, use this model and don't want to change this model for that use margin difference (Fig. 6).

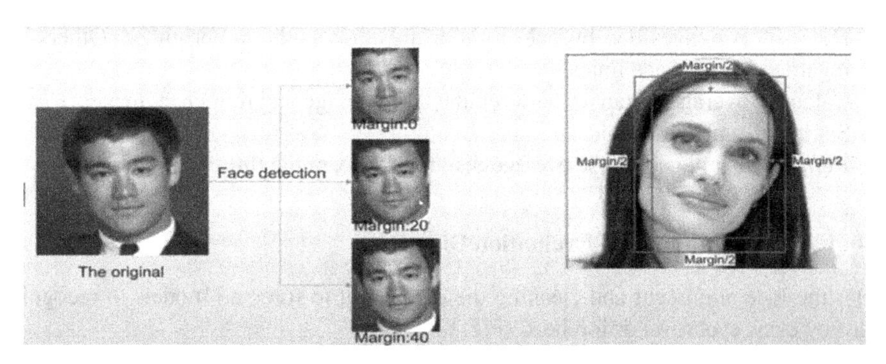

Fig. 6. Different margins

So the margin is used to include more phase features. When compare the margin zero and the margin for 40, the margin 40 has improved features and more in here features and the neck features compared to the margin zero. And in MTCNN and a model, the default value of the margin is 44. Also, set margin 0 when using the default value. That's the result of the phase detection without any margin shown in red. The blue rectangle is the result considering the margin.

5.3 Data Cleaning

There are 10575 classes in CASIA datasets. Now have to clean mislabelled images (Fig. 7).

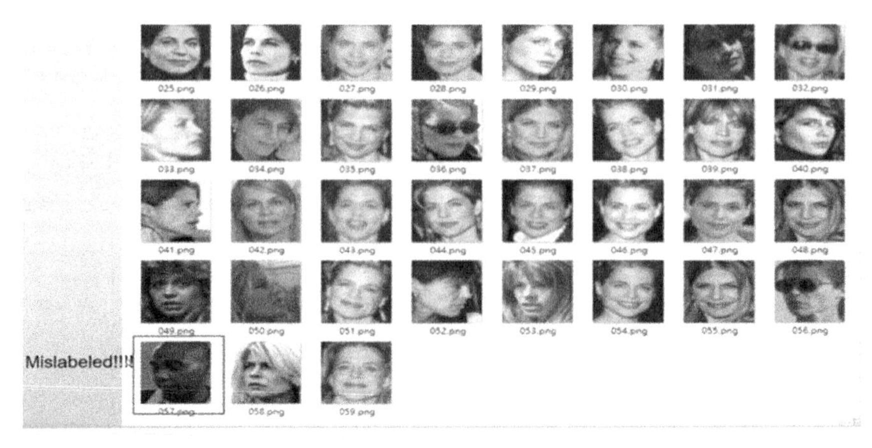

Fig. 7. Mislabelled data

Steps to remove the mislabelled image:

1. Here select images one by one in the same class as the target image. Others are regarded as reference images.
2. Calculate average distances between the target image and reference images.
3. Set the distance threshold.
4. Remove the images whose average distances surpass the threshold.

5.4 Face Recognition and Evaluation Diagram

After the face alignment and cleaning the data, want to force all models to recognize faces by eyes, eyebrows or forehead (Fig. 8).

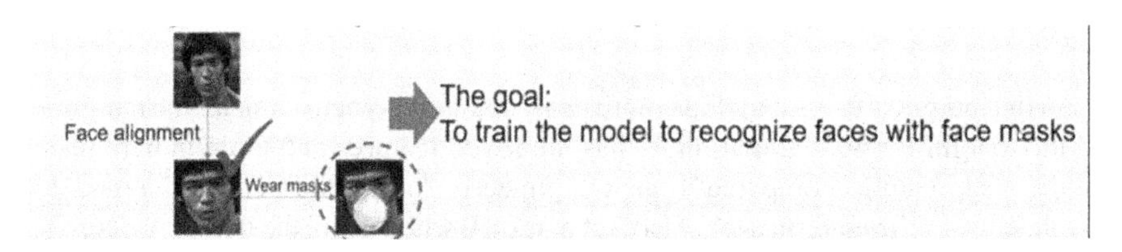

Fig. 8. Face Recognition training and evaluation diagram

The mask image should be in PNG files. The image has 4 channels. The additional 4th channel is used to describe transparency. The background is transparent when a PNG image copy on the PPT (Fig. 9).

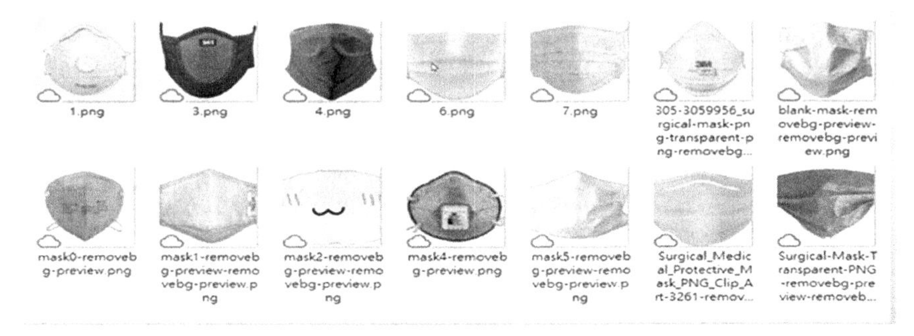

Fig. 9. Variety of face mask

Here, want to change the model to know that people can wear different kinds of face masks and when created the dataset, will use the random selection for each image.

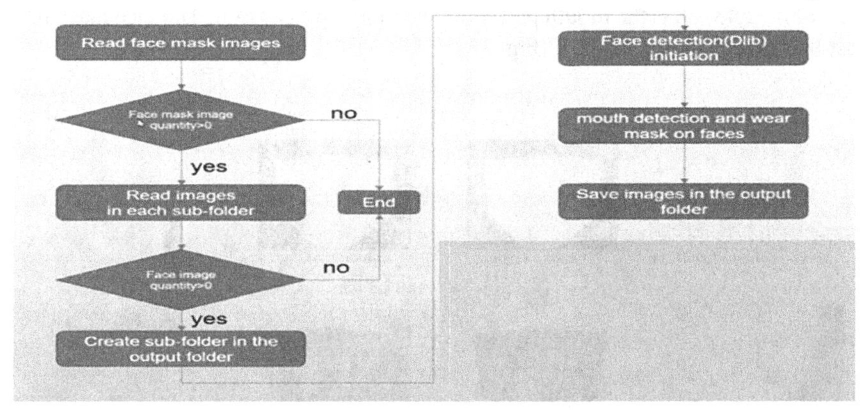

Fig. 10. Program flowchart

So, check the face mask image, the quantity of face masks is zero and not. If not, will come to an end if give yes read images in each subfolder and the have to check the first image is of a zero. If no, the program will come to an end and if yes, created a subfolder in the output folder of the folder is saving new images with face masks and then do the face detection here (Fig.10).

Here, use Dlib to perform this function because the library has another function code facial landmark so that can do face detection. And then wear the artificial face mask on faces and then save images in the output folder. The output sub-folder is obtained from the path information. Dlib and MTCNN produced very similar findings, with MTCNN having a slight advantage, but Dlib is unable to recognise very small faces. Also, if the image size is extremely large and there is a reasonable chance that the lighting will be good, with minimal occlusion and mostly front-facing faces, MTCNN will achieve good performance, as saw when analysing the images (Fig. 11).

Fig. 11. How to find the mouth

Use Dlib face detection to find the face. Use Dlib 68 face landmarks to find the mouth part. Number 48–68 is the mouth part. Find the mouth part as ROI. The face mask image must be resized as same as ROI (Fig. 12).

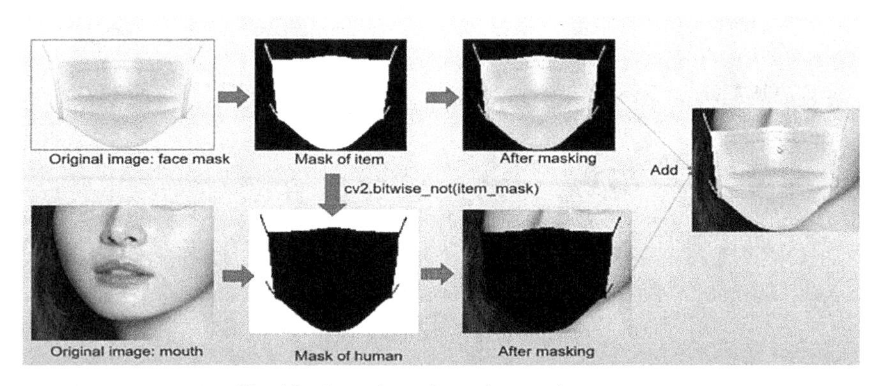

Fig. 12. Overview of wearing mask process

The steps are listed here.

1. The first step is to use a new dataset that is never used in FaceNet training.
2. Select 1000 different-class images randomly. These images are regarded as the face database(ref_data)
3. Wear face masks on selected images. These images are used as test images (tar_data)
4. Calculate embeddings and use them to do face matching.

After done with the Face mask using Dlib, go for the Face recognition using webcam. Here in Fig. 13, shows the flowchart for detection of masks and recognising the person behind the mask.

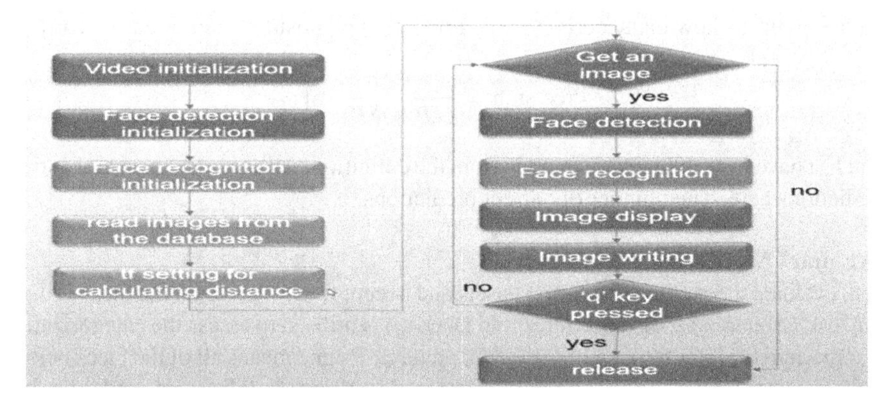

Fig. 13. Face recognition flowchart

5.5 Model Evaluation

Precision and Recall

Here, for evaluation of the model. The total dataset is always divided into the training part and the testing part. Important thing is that datasets don't have identical images. The training part is called a training dataset that's for model training and the testing part is called the testing data set for model evaluation. And use the accuracy as the model performance. The number of right predictions divided by the total number of predictions is the definition of accuracy, and accuracy is used in all types of binary classification. The definition of accuracy is from the terms of negatives and positives, just like this formula, precision, recall and accuracy. Here is my model prediction, which contains positive predictions and negative predictions. And here is the ground truth. It means the answers, which also contains positive and negative answers. When the model prediction is positive and the answer is positive, it means the model prediction is true. So, it is called a true positive (TP). But if the answer is negative, it means the model prediction is wrong. So, it's called false positive (FP). If the model prediction is negative, but the answer is positive. So, the model prediction is wrong. So, it's called a false negative (FN). If the answer is negative, the model prediction is correct is true. So, it's called a true negative (TN).

In all kinds of classification (see Eq. 1).

$$\text{Accuracy} = \frac{(Number\ of\ Correct\ Prediction)}{(Total\ Number\ of\ Prediction)} [6] \tag{1}$$

In a binary classification (see Eq. 2).

$$\text{Accuracy} \frac{(TP + TN)}{(TP + TN + FP + FN)} [6] \tag{2}$$

The number of positive class predictions made out of all positive examples in the dataset is measured by the recall (see Eq. 3).

$$\text{Recall} = \frac{(TP)}{(TP + FN)} [6] \tag{3}$$

Precision means how many correct predictions from all positive predictions (see Eq. 4).

$$\text{Precision} = \frac{(TP)}{(TP + FP)} \text{ [6]} \quad (4)$$

And last have to calculate the accuracy from the definition. An accuracy is several correct predictions here. The number of correct predictions.

VAL and FAR

Here evaluate the methods on four datasets and except for LFW and on the face verification test, a squared L2 distance threshold D(xi, xj) is utilised to assess the categorization of same and different for a pair of two face pictures. Psame means all of the faces belong to the same person. All faces pairs of different identities are Pdifferent, and, two face images of the same people, the embedding L2 distance(the capital D) it's smaller than the threshold, the d is called TA. Two face images of different people. The embedding L2 distance is smaller than the threshold this called F.A.

Define the set of all true accept as.

$$\text{TA(d)} = \{(\text{i, j}) \in P_{same}, \text{ with } \text{D}(x_i, x_j) \le d\} \text{ [6]} \quad (5)$$

These are the face pairs(i, j) that were correctly classified as the same at threshold d, Similarly.

$$\text{FA(d)} = \{(\text{i, j}) \in P_{diff}, \text{ with } \text{D}(x_i, x_j) \le d\} \text{ [6]} \quad (6)$$

Is the set of all pairs that were incorrectly classified as the same (false accept).

The validation rate VAL(d) and the false accept rate FAR(d) for a given face distanced are then defined as.

$$\text{VAL(d)} = \frac{|TA(d)|}{|P_{same}|} \text{ [6]} \quad (7)$$

Validation Rate(VAL) is a true positive rate in binary systems.

$$\text{FAR(d)} = \frac{|FA(d)|}{|P_{diff}|} \text{ [6]} \quad (8)$$

False accept rate(FAR) is a false positive rate in the binary system. Here, make VAL higher and FAR lower.

ROC Curve

In Fig. 14, the colour curves, including the orange one, the green one, the red one and the blue one are formed by shifting the threshold d and then calculate FAR and VAL. With fixed FAR, find out the best curve and get the threshold d.

Use prelogit after the L2 normalization to form the embedding. So here is embedding and then use the pre logit to form the embedding. This is the simple Resnet model. After

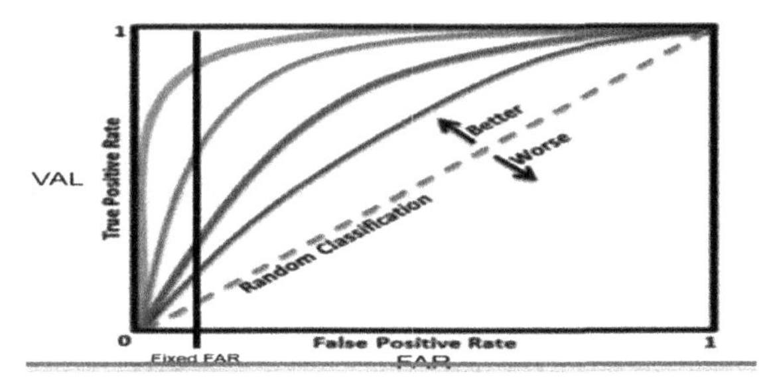

Fig. 14. ROC curve

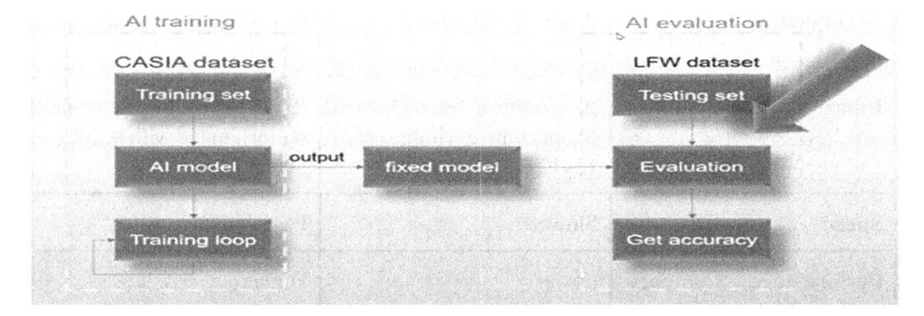

Fig. 15. Training and evaluation of FaceNet

the flatten, the first connection with drop out is preloaded. Here, can train FaceNet using the Simple Resnet model but the training result won't be good. Many famous models such as Efficient Net, SeNet, GoogLeNet, ResNet always use the ImageNet dataset for evaluations. ImageNet datasets have 1000 classes where CASIA-dataset has more than 10,000 classes (Fig. 15).

Consider the CASIA and LFW datasets, where the CASIA dataset is for training and the LFW dataset is utilised for testing. To assess the performance, the AI model weights are fixed after several training cycles (Fig. 16).

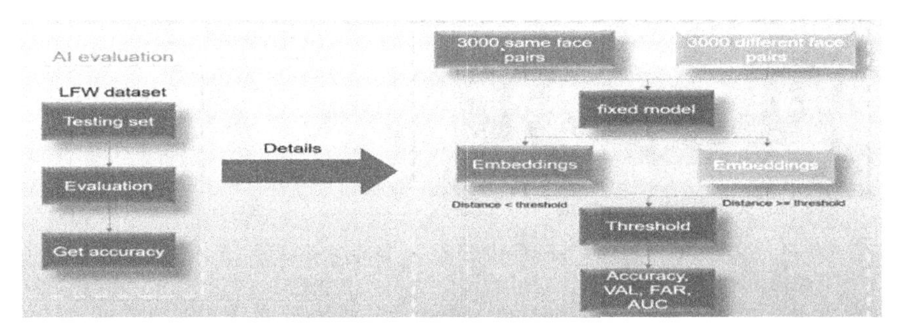

Fig. 16. Evaluation flowchart

6 Result

If accuracy is a top concern, MTCNN should be used, but if speed is a top requirement, SSD should be used. Furthermore, SSD will be unable to locate facial landmarks and will have to rely on OpenCV's Dlib to locate eye positions to align faces. This might have a detrimental impact on output (Fig. 17).

Properties	**MTCNN**	**SSD**
Approach	Bounding box approach	Grid based approach
Architecture	P-Net, R-Net, O-Net	Network input, Backbone network.
Input	It accepts a three-channel (color) image with a size of 12x12.	It accepts a three-channel (color) image with a size of 300x300
Speed	Slower	Faster
Performance Comparison*(mAP)*	Lower	Higher
RAM And GPU memory usage	Higher	Lower

Fig. 17. Comparison between MTCNN and SSD

SSD Model

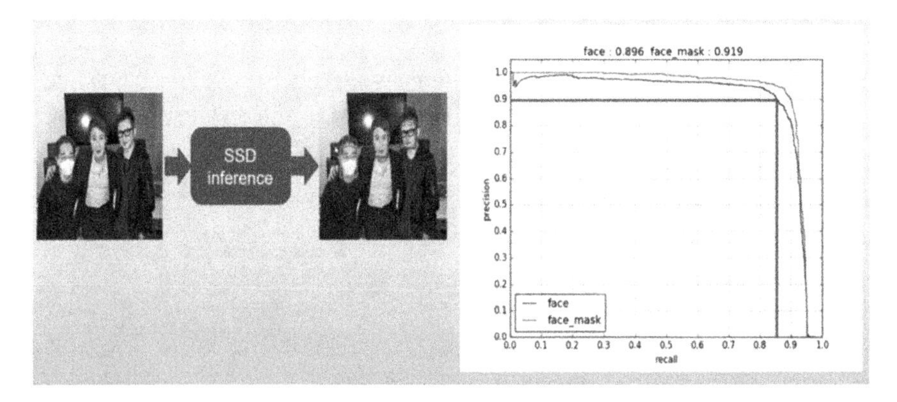

Fig. 18. SSD face detection

It's a face mask detection, it improves the image into the SSD model and after the inference will all pull an image with several bounding boxes. The best performance in a recall is 0.86, precision is 0.9 and if the rebounding bounce colour is green, it means that the person uses the face mask, if red then doesn't. 86% in opposition is around 90%. So the performance is quite good. Want to use to do face detection and phase parts, face mask detection class (Figs. 18 and 19).

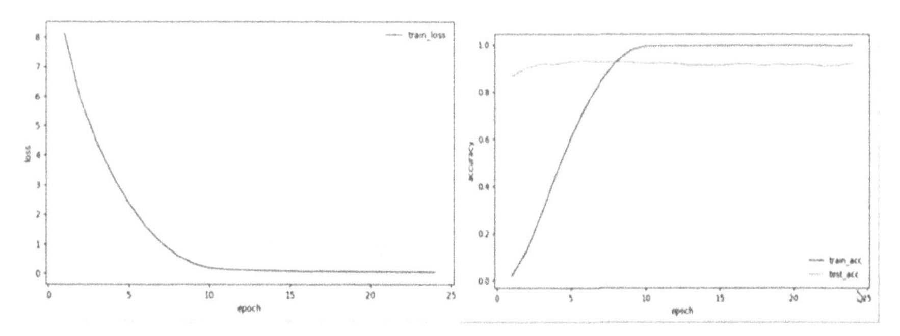

Fig. 19. Graph between Train_acc and Test_acc using SSD model

Result with SSD Face Detection

- The best training set accuracy is 99.77%.
- The best test set accuracy is 93.2%.

MTCNN Model

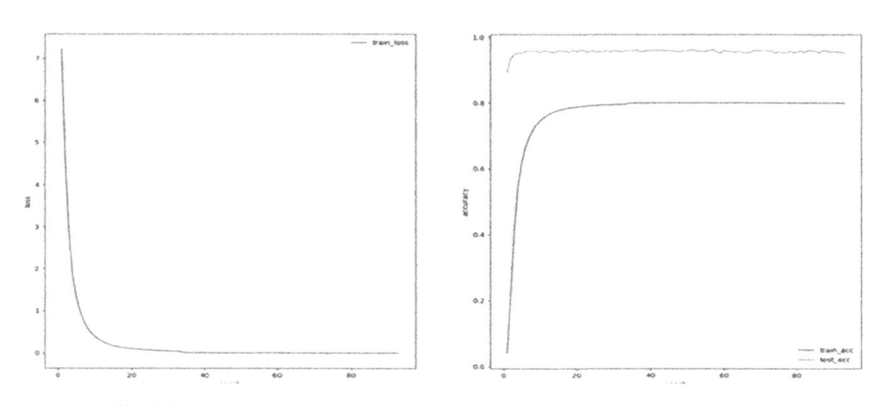

Fig. 20. Graph Between Train_acc and Test_acc using MTCNN model

Result with MTCNN Face Detection

- The best training set accuracy is 96.77% (Fig.20).
- The best test set accuracy is 99.9%.

Masked Face Detection and Recognition

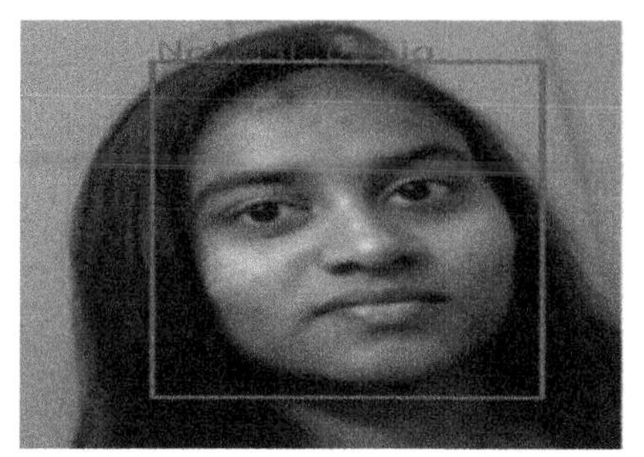

Fig. 21. Result with mask detection and recognition

Here, do LFW dataset evaluation and then can get the accuracy back here. The accuracy is the face recognition without facial masks. Here show faces with the mask as well as identify the person in the mask by using facial features in Fig. 21.

7 Conclusion

The proposed approach in this thesis has helped to provide an optimal solution for the problem of reliably detecting masked faces using current facial recognition systems in this work. As a result, a big dataset of faces is created. The dataset created using this tool may subsequently be used to train a successful facial recognition technology for masked faces using target accuracy. In this study, we used existing facial recognition algorithms to tackle the challenge of consistently recognizing masked faces. Show the tool that can be used to mask people's faces. Therefore, a massive dataset of masked faces is created. The accuracy of the re-trained system was also evaluated on a customized real-world dataset and found to be similar, indicating that it may be used for real-world masked faces.

Applications

- This method may be used in ATMs, to detect duplicate voters, verify passports and visas, verify driving licences, in defence, competitive and other tests, and the public and private sectors.

- Forecasters predict that this technology will expand at a rapid pace and produce significant income in the future years.
- The key areas that will be heavily affected are security and surveillance.
- Private enterprises, public buildings, and schools are among the other places that are now welcome with open arms.
- It is expected that merchants and financial institutions would embrace it in the future years to prevent fraud in debit/credit card transactions and payments, particularly those made online. This technique would plug the gaps in a password system that is widely used yet insecure. Robots using masked facial recognition technology may potentially make an appearance in the future. They can assist humans in completing activities that are unrealistic or impossible for them to perform.

Future Scope

There are numerous issues on which I intend to work:

- Machine Learning is becoming increasingly popular in the realm of mobile deployment. As a result, there are plans to transfer our model to their TensorFlow light versions.
- Our architecture can be considered TensorFlow Run Time (TFRT) suitable, which will improve inference effectiveness on edge devices that make the model's multithreading CPU efficient.

References

1. Ejaz, M.S., Islam, M.R., Sifatullah, M., Sarker, A.: Implementation of principal component analysis on masked and non-masked face recognition. In: 2019 1st International Conference on Advances in Science, Engineering and Robotics Technology (ICASERT), Dhaka, Bangladesh, pp. 1–5 (2019). https://doi.org/10.1109/ICASERT.2019.8934543
2. Li, Y., Guo, K., Lu, Y., Liu, L.: Cropping and attention based approach for masked face recognition. Appl. Intell. **51**(5), 3012–3025 (2021). https://doi.org/10.1007/s10489-020-02100-9
3. Deng, J., Guo, J., An, X., Zhu, Z., Zafeiriou, S.: Masked face recognition challenge: the insightface track report. In: Proceedings of the IEEE/CVF International Conference on Computer Vision, pp. 1437–1444 (2021)
4. Boutros, F., et al.: MFR 2021: masked face recognition competition. In: 2021 IEEE International Joint Conference on Biometrics (IJCB), pp. 1–10 (2021). https://doi.org/10.1109/IJCB52358.2021.9484337
5. Wang, W., Zhao, Z., Zhang, H., Wang, Z., Su, F.: MaskOut: a data augmentation method for masked face recognition. In: Proceedings of the IEEE/CVF International Conference on Computer Vision, pp. 1450–1455 (2021)
6. Khapra, M.M.: Master deep learning in four stage (2019). GUVI | Learn to code in your native language. www.guvi.in/deep-learning
7. Wang, Z., et al.: Masked face recognition dataset and application, arXiv preprint arXiv:2003.09093

8. Loey, M., Manogaran, G., Taha, M.H.N., Khalifa, N.E.M.: A hybrid deep transfer learning model with machine learning methods for face mask detection in the era of the COVID-19 pandemic. Measurement **167**, 108288 (2021)
9. Nagrath, P., Jain, R., Madan, A., Arora, R., Kataria, P., Hemanth, J.: SSDMNV2: a real-time DNN-based face mask detection system using a single shot multibox detector and MobileNetV2. Sustain. cities Soc. **66**, 102692 (2021)
10. Rahman, M.M., Manik, M.M.H., Islam, M.M., Mahmud, S., Kim, J.-H.: An automated system to limit COVID-19 using facial mask detection in smart city network. In: 2020 IEEE International IOT, Electronics and Mechatronics Conference (IEMTRONICS), Vancouver, BC, Canada, pp. 1–5 (2020). https://doi.org/10.1109/IEMTRONICS51293.2020.9216386
11. Chen, Y., et al.: Face mask assistant: detection of face mask service stage based on mobile phone. IEEE Sens. J. https://doi.org/10.1109/JSEN.2021.3061178
12. Dey, S.K., Howlader, A., Deb, C.: MobileNet mask: a multi-phase face mask detection model to prevent person-to-person transmission of SARS-CoV-2. In: Kaiser, M.S., Bandyopadhyay, A., Mahmud, M., Ray, K. (eds.) Proceedings of International Conference on Trends in Computational and Cognitive Engineering. AISC, vol. 1309. Springer, Singapore (2021). https://doi.org/10.1007/978-981-33-4673-4_49
13. Militante, S.V., Dionisio, N.V.: Real-time facemask recognition with alarm system using deep learning. In: 2020 11th IEEE Control and System Graduate Research Colloquium (ICSGRC), Shah Alam, Malaysia, pp. 106–110 (2020). https://doi.org/10.1109/ICSGRC49013.2020.9232610
14. Jignesh Chowdary, G., Punn, N.S., Sonbhadra, S.K., Agarwal, S.: Face mask detection using transfer learning of inceptionV3. In: Bellatreche, L., Goyal, V., Fujita, H., Mondal, A., Reddy, P.K. (eds.) BDA 2020. LNCS, vol. 12581, pp. 81–90. Springer, Cham (2020). https://doi.org/10.1007/978-3-030-66665-1_6
15. Hariri, W.: Efficient masked face recognition method during the COVID-19 pandemic (2020). https://doi.org/10.21203/rs.3.rs-39289/v1
16. Zereen, A.N., Corraya, S., Dailey, M.N., Ekpanyapong, M.: Two-stage facial mask detection model for indoor environments. In: Kaiser, M.S., Bandyopadhyay, A., Mahmud, M., Ray, K. (eds.) Proceedings of International Conference on Trends in Computational and Cognitive Engineering. AISC, vol. 1309, pp. 591–601. Springer, Singapore (2021). https://doi.org/10.1007/978-981-33-4673-4_48
17. Li, C., Ge, S., Zhang, D., Li, J.: Look through masks: towards masked face recognition with de-occlusion distillation. In: Proceedings of the 28th ACM International Conference on Multimedia (MM 2020), pp. 3016–3024. Association for Computing Machinery, New York, (2020). https://doi.org/10.1145/3394171.3413960I
18. Chen, Y., Song, L., Hu, Y., He, R.: Adversarial occlusion-aware face detection. In: 2018 IEEE 9th International Conference on Biometrics Theory, Applications and Systems (BTAS), pp. 1–9 (2018). https://doi.org/10.1109/BTAS.2018.8698572
19. Anwar, A., Raychowdhury, A.: Masked face recognition for secure authentication. arXiv preprint arXiv:2008.11104. 25 August 2020
20. Lin, S., Cai, L., Lin, X., Ji, R.: Masked face detection via a modified LeNet. Neurocomputing **218**, 197–202 (2016)
21. Podbucki, K., Suder, J., Marciniak, T., Dąbrowski, A.: CCTV based system for detection of anti-virus masks. In: 2020 Signal Processing: Algorithms, Architectures, Arrangements, and Applications (SPA), Poznan, Poland, pp. 87–91 (2020)

22. Damer, N., Grebe, J.H., Chen, C., Boutros, F., Kirchbuchner, F., Kuijper, A.: The effect of wearing a mask on face recognition performance: an exploratory study. In: 2020 International Conference of the Biometrics Special Interest Group (BIOSIG), Darmstadt, Germany, pp. 1–6 (2020)
23. Yang, G., et al.: Face mask recognition system with YOLOV5 based on image recognition. In: 2020 IEEE 6th International Conference on Computer and Communications (ICCC), Chengdu, China, pp. 1398–1404 (2020). https://doi.org/10.1109/ICCC51575.2020.9345042

Machine Learning Models for Predicting Indoor Air Temperature of Smart Building

Salam Traboulsi(✉) and Stefan Knauth

HFT Stuttgart - Stuttgart University of Applied Sciences, Stuttgart, Germany
{salam.traboulsi,stefan.knauth}@hft-stuttgart.de

Abstract. The indoor air temperature is one of the key factors to improve the performance of energy efficiency of buildings and quality of life in a very smart IoT environment. Therefore, a periodic and accurate prediction of the minimum and maximum indoor air temperature allows taking necessary precautions to handle the variations' impact and tendencies. During this assessment, we developed minimum and maximum indoor air temperature prediction models using multiple statistical regression (MLR), multilayered perceptron (MLP), and random forest (RF, where Rf is achieved once with tree depth 10 (RFdepth10), and once with tree depth 50 (RFdepth50)). The study was conducted at a building located within the University of Applied Sciences, Stuttgart, in Germany. Sensors were accustomed to aggregate data, which were used because of the input variables for the prediction. The variables are outdoor air temperature, indoor air temperature, humidity, and heating temperature. Performance of the models was evaluated with the coefficient of determination R^2 and therefore the root means square error (RMSE). The simulation results showed that the prediction by the MLP algorithm, based on minimum indoor air temperature models and also maximum indoor air temperature models, provides better accuracy with the very best R^2 and lowest RMSE in the independent test dataset. This survey developed a straightforward and powerful MLP model to predict the minimum and therefore the maximum indoor air temperature, which may integrate into smart building management system technology in the future.

1 Introduction

In recent years, Indoor air quality, user comfort, and energy consumption in buildings have received increasing attention, as they contribute to improving the health conditions and quality of life (QoL) of people in buildings and to reducing energy costs [15]. According to the [2] studies, occupants spend about 60–90% of their time indoors of buildings, thus occupants' behaviors are strongly considered as part of any energy management system of a building. The recent COVID-19 pandemic condition has even further provoked the importance of the topic of energy use and QoL of life in indoor locations.

I. Woungang et al. (Eds.): ANTIC 2021, CCIS 1534, pp. 586–595, 2022.
https://doi.org/10.1007/978-3-030-96040-7_44

To optimize energy use in buildings, the building energy management systems necessitate continuous monitoring and administration of the time series data of parameters affecting the basic environment variables of the building, such as energy system (e.g., the HVAC-Heating, Ventilation, and Air Conditioning-system and lighting systems), and they should be able to monitor indoor temperature, air quality (CO_2), and visual comfort, etc. [23]. At the same time, the recent works and signs of progress in the Internet of Things (IoT) technologies are among the main cores of smart buildings constructions. A smart building is any structure that uses automated processes to control observed indoor parameters, it uses sensors to collect data and makes decisions according to predefined services and functions, This structure helps directly to reduce energy use and the environmental impact of buildings [5].

Meteorologists and many researchers are interested in the maximum and minimum of the surface air temperature estimation, it impacts meteorological events like precipitation, wind flow, humidity, and pressure. They use the maximum air temperature degree to analyze the maximum air temperature during summer, the minimum air temperature during winter, and the maximum and minimum temperature during spring. Whatever, in the research field, there is a lack of such predictions in the indoor environment, on the other hand, the prediction of the maximum indoor air temperature in winter is very important to reducing energy consumption in buildings, and improving the workers' QoL and health conditions. Therefore, indoor air temperature is considered as one of the important parameters that impact directly the energy consumption and life quality in the indoor environment. Thus, to develop an ideal management system, it is essential to consider a relevant model to predict the minimum and maximum indoor air temperature [19]. In previous studies, various static and dynamic models have been explored [8,17]. Existing studies have used different models, such as regression models, time series models, computational intelligence approaches, and hybrid approaches to predict short- and long-term variables based on time series [16]. A thorough evaluation of the performance of different prediction models allows the development of an ideal strategy to improve air quality in the indoor environment. The present research uses Machine Learning (ML) methods to develop minimum and maximum indoor air temperature prediction models. Researchers have devoted different modeling techniques to predict the building's indoor temperature. whatever the adopted techniques, such as the data-driven models, rely on experimental data, show prediction accuracy if training data is available [2,13,22]. From existing work in the literature, a wide variety of ML-based demonstrated methods were improved accuracy as in predicting indoor temperature [1,11]. In this paper, we aim to compare the performance of different models of Machine Learning (ML). It evaluates the performance of the physical model (multiple linear regression (MLR)), computational model (multilayer perceptron (MLR) with backpropagation), and time-series forecasting model (random forest (RF), where Rf is achieved once with tree depth 10 (RFdepth10), and once with tree depth 50 (RFdepth50) with limited input variables. Since our work is part of a smart city project, therefore these predictions models are to evaluate the effect of the high- and low-temperature predictions on the performance of the energy consumption and QoL in smart environments.

2 Data

The current study was conducted at a building located in the University of Applied Sciences, Stuttgart, Germany. The GPS coordinates for the site were 48°54'6.754" N, 9°15'43.84" E. The campus consists of eight buildings from different years of construction, ranging from 2016 back to the 19th century [7]. Building 1, one of the older university buildings, has been selected as a test-bed for the research. It includes staff offices, workshops rooms, library, and cafeteria.

This study carries on previous achieved work published in [20], we aim continually to identify the input parameters for the indoor space temperature predictions model, sensors were used and a time series 5 min interval real-time data of relevant parameters were collected for a period of 11 months starting from 1 January 2016. The aggregated environmental data is represented by the temperature measurements, like outdoor, indoor, and heating temperature. Outdoor temperature is the outside ambient air temperature, the indoor temperature is the room temperature, and the heating temperature is the temperature of the radiator. The data set contains details of Indoor Temperature at different points and different techniques. The observed data-set for the study area are used as parameters to predict the next day's minimum temperature and maximum temperature. In the present study, we apply the concept of lead time. Where Lead time is the latency time from the start of a process until its end. Here, it is named as lead day. Where lead day is the period that passes from the start of data collection until at a pre-defined Time and date, till another pre-defined time and date. To cover the period of one year, and to enhance the accuracy of our predictions, we defined eleven Lead days, which means: lead day 1 is one month out, lead day 2 is 2 months out, lead day 3 is 3 months out, lead day 4 is 4 months out, lead day 5 is 5 months out, lead day 6 is 6 months out, lead day 7 is 7 months out, lead day 8 is 8 months out, lead day 9 is 9 months out, lead day 10 is 10 months out, and lead day 11 is 11 months out. Therefore, eleven models with identical structures were fit for both response variables representing eleven different forecast horizons.

3 Methodology: Machine Learning Algorithms

ML is a part of Artificial Intelligence (AI), has reinterpreted the world in diverse forecasting fields for the past two decades. The ML models are capable of adaptive learning from the data, and they can improve themselves from subsequent training, trends, and pattern identification. These related aspects have brought them to efficiently handle complex studies. The use of these technologies allows large data sets to be analyzed more efficiently and with relative ease than physical or statistical models [1]. ML-based models adopted in indoor environment [1] show high outperform, particularly for determining linear and non-linear variables that adopt time series. In the ML framework, a huge number of learning algorithms are available, including linear regression (LR), decision tree regression (DTR), random forest regression (RFR), support vector regression (SVR), etc. have been developed to deal with regression and classification problems. Some studies have used artificial neural networks (ANNs) and ML models to predict indoor temperature variables specifically related to QoL improvement

3.1 Multiple Linear Regression

Multiple linear regression (MLR) is a statistical algorithm that aims to predict the outcome of the response (dependent) variables by using several explanatory (independent) variables. These models are popular among the fields such as weather prediction, energy consumption, electricity load, and business forecast [24]. According to [24], MLR model is expressed by the following equation:

$$Y = a_0 + a_1X_1 + a_2X_2 + \ldots + a_iX_i + \epsilon \tag{1}$$

where Y is the output variable; X is the predictor or independent variable (from X_0 to X_i); a is the regression coefficient to predict Y (from a_1 to a_i); a_0 is the constant of the model; and the ϵ is the random error or the noise of the model.

3.2 Multilayered Perceptron-Back-Propagation

A multilayer perceptron (MLP) [6,21] along with the back-propagation is a class of feed-forward artificial neural network (ANN). It consists of at least three layers of nodes: an input layer, a hidden layer, and an output layer [4]. Where, each layer includes many neurons. The input layer represents the dimension of the input data and the hidden layer has n neurons, which is the fully connected network to the outputs [3]. The BP is a training technique, which assumes reducing the prediction error of the output layer. The MLP with three layers, mathematically, can be indicated as [6,21]:

$$Y_p = f_0\left[\sum_{i=1}^{n} w_{kj} f_h(\sum_{j=1}^{m} w_{ji}x_i + w_{jb}) + w_{kb}\right] \tag{2}$$

As defined in [6,21]: Y_p is the predicted output; f_0 is the activation function for the output neuron; n is the number of output neurons; w_{kj} is the weight for the connecting neuron of hidden and output layers; f_h is the hidden neuron's activation function; and m is the number of hidden neurons; w_{ji} is the weight for the connecting neuron of input and hidden layers; x_i is the input variable; w_{jb} is the bias for the hidden neuron; and w_{kb} is the bias for the output neuron.

3.3 Random Forest

Random Forest (RF) [10] is an algorithm that learns from multiple randomized decision trees driven. The RF trees work in parallel, there is no interaction when building the trees. RF has an operational process similar to decision trees, nevertheless, RF can be applied for both classification and regression. The procedure of the algorithm consists of three steps [14]. The first is to create bootstrap samples from the data. Particularly, each sample (bag) contains k observations which are uniformly selected (with replacement) out of k original observations using bootstrap. Then for each sample, we grow a decision CART (Classification and Regression Tree) [14]. Instead of using all predictors, at each node of each tree, RF is a collection of DTs where all the trees depend on a collection of random variables. the output is predicted by averaging the output of each ensemble tree. The mathematical expression of RF expressed as the following equation [10]:

$$Y = \frac{1}{M}\sum_{i=1}^{M} H(T_i) \tag{3}$$

where M is the total number of trees, Y is the final prediction; $\mathrm{H(T_i)}$ is a sample in the training set.

A Random Forest Regression model is one of the powerful and accurate ML algorithms. One of Its disadvantages is determining the number of trees included in the model. To test the effect of this disadvantage, we run the method once with depth equal to 10, and once to 50.

4 Machine Learning Workflow

Before we jump right into programming phase for predicting the daily maximum indoor temperature and daily minimum indoor temperature, the following steps represent the adopted process of our ML workflow: As a first step, the data were collected and stored from sensors. At next, the stored data were subjected to the pre-processing methods as missing data analysis, data normalization, and splitting data into training and testing partition. The current study adopts minimum-maximum normalization, which is a pre-processing method for ML modelling, it re-scales the input features in the range of 0 to 1 or −1 to 1 [9,18]. After the normalization applied to the input data, we come to the data partition scales (training:testing), the current study utilized 60:40, 60% of the data for training and the rest of data for testing. Researchers [9,12,18] use 80:20, or 90:10, there is any explained reason. Because, it is random, in this study, the lead days 1, 4, 7, 8 and 9 forms the test set or the validation data-set, and the rest represents the training ensemble.

During the testing phase, the indoor daily maximum temperature and indoor daily minimum temperature were predicted for 40% of untrained data sets using all aforementioned ML algorithms. All results were documented to observe the performance of the models during the training and testing phase. At the final step, the model prediction results were evaluated by using root-mean-square error (RMSE), and coefficient of determination (R^2) methods, which could be expressed by the following equations:

$$RMSE = \sqrt{\frac{\sum_{i=1}^{n}(y_i - p_i)^2}{n}} \quad (4)$$

$$R^2 = 1 - \frac{\sum_{i=1}^{n}(y_i - p_i)^2}{\sum_{i=1}^{n}(y_i - \frac{1}{n}\sum_{i=1}^{n} y_i)^2} \quad (5)$$

5 Results

Two variables were predicted: Daily Maximum Indoor Temperature (InT_{max}) and Daily Minimum Indoor Temperature (InT_{min}). This data came from observation from 2016–01–01 to 2016–11–30, 60% randomized observations cross over the data set were used to train the different ML algorithms chosen, with the rest of the data used to assess predictions accuracy.

Table 1. The assessment of all the models during Maximum Indoor Temperature and Minimum Indoor Temperature predictions.

Models	Max. indoor temperature		Min. indoor temperature	
	RMSE	R^2	RMSE	R^2
MLR	2.284628	**0.9470787**	2.416765	**0.9164277**
MLP	**1.155018**	**0.9412482**	**0.679522**	**0.8797165**
RFdepth10	1.93584	0.7765188	1.493203	0.4206477
RFdepth50	**1.898579**	0.775957	**1.466228**	0.4229851

5.1 Selection of the Best Algorithm

Two variables were predicted: Maximum Indoor Temperature and Minimum Indoor Temperature. Three regression algorithms of ML were adopted to evaluate the predictions' performance: multiple linear regression (MLR), random forest regression(RF), and multilayered perceptron. Eleven models with identical structures were fit for both variables, representing eleven different forecast horizons: lead days (1 till 11).

The accuracy of the ML algorithms for the minimum and the maximum indoor temperature predictions are validated and summarized in Table 1, where RMSE represents the Root mean square error and R^2 is the coefficient of determination, we used Bold fonts to represent the top performed results with the corresponding data set.

5.2 Input Data

During the maximum indoor temperature and minimum indoor temperature predictions, the lead days data-set (1, 4, 7, 8, 9) are used to validate our issue

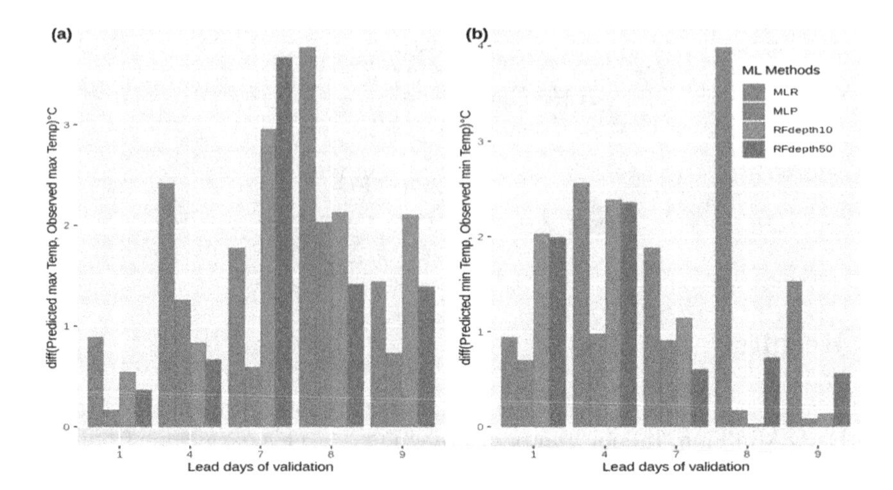

Fig. 1. Training evaluation metric comparison between MLR, MLP, RFdepth10 and RFdepth50 during maximum indoor temperature (a), and minimum indoor temperature (b) predictions

during the performance phase. All ML models outperformed when using designated data. For instance, MLP obtained the best performance (with maximum indoor temperature and minimum indoor temperature testing predictions, provided in Table 1. Since the MLP performed better than other models during indoor temperature predictions testing results, it has been chosen for intercomparison between the minimum and maximum indoor temperature prediction. When compared, we deduce that the RMSE values of MLP and RF ML regression algorithms of Minimum Indoor Temperature set are less than of MLR, the MLR of minimum indoor temperature predictions obtained the least performance among other models during testing the predictions' performance. When compared the results of the maximum indoor temperature data set, we have the same conclusion, where the results were the same that MLP and RF performed better results during temperature prediction.

5.3 Model Performance and Decision

In maximum and minimum indoor temperature predictions, based on the 11 lead days, most of the models performed well during the training time in RMSE and R^2. For instance, the results of all the models RMSE except MLR were less than 2 °C during the training phase, but the MLR model produces over 2 °C. The training accuracy was high in the MLP model with testing data than RF, but the RFdepth50 results are better than RFdepth10.

Though the performance results were less in MLR, therefore MLR performed significantly less accurate predictions than other models. The comparison of evaluation metrics between the ML's models during the validating phase is illustrated in Fig. 1, where the MLP presents better predictions results of minimum indoor temperature 1(b) prediction during lead days 1, 4, and 9; but in the part, RFdepth50 is better during lead day 7, and RFdepth10 is better for lead day 8. Part (a) of the Fig. 1, concerning the maximum indoor temperature predictions' evaluation, presents the same concept only with a little difference, here the MLP is the accurate algorithm for the most lead days (1, 7, 9), the RFdepth50 is better during at the lead days (4, 8), but the RFdepth10 hat any chance in this case.

From the comparison results of predictions, RMSE, which is often used as a verification measure for temperature predictions, gives the least error to the MLP algorithm for the maximum indoor temperature and the minimum indoor temperature. In the second rank, it comes RF with a Tree depth equal to 50. Therefore, this study demonstrated that the MLP model performed better results during the validation phase. Nevertheless, the RF algorithm shows good results also.

6 Conclusion

Despite the availability of present advanced technologies, providing a comfortable environment and improving energy use are still considered challenging phenomenons. Prediction models are necessary for improving environmental control in indoor buildings. The current study successfully predicts Indoor temperature using important and reliable ML regression algorithms. In the end, this study concludes the following points:

- The MLP models performed the most well among all the forecasting models used in this research, most probably.
- This study's results showed the ML algorithms efficiency since the ML algorithms perform better with complex and nonlinear input variables due to the self-adaptive nature. ML algorithms models suppose to be the optimal solver for the performance of the management system of the indoor environment.
- The present study predicted indoor environment predictions without considering the human factor. However, this factor will be considered in the next study.
- Current literature used limited data due to the hindrance to collect indoor climate data for supervised learning. So in the future, big data for many use cases will be used to fortify our solutions for managing and controlling the indoor environment.

Acknowledgement. This work carries on previous achieved work published in [20], all related studies are realized within the context of the i_City project. This project is sponsored by the German Federal Ministry of Education and Research Program "FH-Impuls 2016" under Contract 13FH9I01IA.

References

1. Arulmozhi, E., Basak, J.K., Sihalath, T., Park, J., Kim, H.T., Moon, B.E.: Machine learning-based microclimate model for indoor air temperature and relative humidity prediction in a swine building. Animals **11**(1), 222 (2021). https://doi.org/10.3390/ani11010222
2. ASHRAE: Guideline 10 provides guidance regarding factors affecting indoor environmental conditions acceptable to the comfort and health of human occupants (2016)
3. Basak, J.K., Okyere, F.G., Arulmozhi, E., Park, J., Khan, F., Kim, H.T.: Artificial neural networks and multiple linear regression as potential methods for modelling body surface temperature of pig. J. Appl. Anim. Res. **48**(1), 207–219 (2020)
4. Beghdad, R., Bechar, K., Bouali, M., Haddadi, M.: Neural networks and decision trees for intrusion detections: enhancing detection accuracy. Tech. rep., EasyChair (2020)
5. Chui, K.T., Lytras, M.D., Visvizi, A.: Energy sustainability in smart cities: artificial intelligence, smart monitoring, and optimization of energy consumption. Energies **11**(11), 1–20 (2018). https://ideas.repec.org/a/gam/jeners/v11y2018i11p2869-d177758.html
6. Elanchezhian, A., et al.: Evaluating different models used for predicting the indoor microclimatic parameters of a greenhouse. Appl. Ecol. Environ. Res. **18**, 2141–2161 (2020)
7. Guedey, M., Uckelmann, D.: Exploring smart home and internet of things technologies for smart public buildings. In: Proceedings of the 10th International Conference on the Internet of Things, pp. 1–8 (2020)
8. He, X., Guan, H., Zhang, X., Simmons, C.T.: A wavelet-based multiple linear regression model for forecasting monthly rainfall. Int. J. Climatol. **34**(6), 1898–1912 (2014)
9. Jayalakshmi, T.A.S.: Statistical normalization and back propagation for classification. Int. J. Comput. Theory Eng. (IJCTE) **3**, 89–93 (2011)
10. Vassallo, D., Krishnamurthy, R., Sherman, T., Fernando, H.J.S.: Analysis of random forest modeling strategies for multi-step wind speed forecasting. Energies **13**(20), 5488 (2020). https://doi.org/10.3390/en13205488, https://curate.nd.edu/show/kp78gf09v04
11. Lu, T., Viljanen, M.: Prediction of indoor temperature and relative humidity using neural network models: model comparison. Neural Comput. Appl. **18**, 345–357 (2009). https://doi.org/10.1007/s00521-008-0185-3
12. Mohan, P., Patil, K.: Deep learning based weighted SOM to forecast weather and crop prediction for agriculture application. Int. J. Intell. Eng. Syst. **11**, 167–176 (2018). https://doi.org/10.22266/ijies2018.0831.17
13. Mustafaraj, G., Lowry, G., Chen, J.: Prediction of room temperature and relative humidity by autoregressive linear and nonlinear neural network models for an open office. Energy Build. **43**(6), 1452–1460 (2011)
14. Nguyen, T., Fouchereau, R., Frenod, E., Gerard, C., Sincholle, V.: Comparison of forecast models of production of dairy cows combining animal and diet parameters. Comput. Electron. Agric. **170**, 105258 (2020). https://doi.org/10.1016/j.compag.2020.105258, https://hal.archives-ouvertes.fr/hal-02358044
15. Pérez-Lombard, L., Ortiz, J., Pout, C.: A review on buildings energy consumption information. Energy Build. **40**(3), 394–398 (2008)

16. Qi, C., Chang, N.B.: System dynamics modeling for municipal water demand estimation in an urban region under uncertain economic impacts. J. Environ. Manag. **92**(6), 1628–1641 (2011)
17. Shi, X., Lu, W., Zhao, Y., Qin, P.: Prediction of indoor temperature and relative humidity based on cloud database by using an improved BP neural network in Chongqing. IEEE Access **6**, 30559–30566 (2018)
18. Sola, J., Sevilla, J.: Importance of input data normalization for the application of neural networks to complex industrial problems. IEEE Trans. Nucl. Sci. **44**(3), 1464–1468 (1997). https://doi.org/10.1109/23.589532
19. Tham, K., Ullah, M.: Building energy performance and thermal comfort in Singapore (1993)
20. Traboulsi, S., Knauth, S.: IoT analysis and management system for improving work performance with an IoT open software in smart buildings. J. Ubiquitous Syst. Pervasive Netw. **14**(01), 1–6 (2021)
21. Walker, S., Khan, W., Katić, K., Maassen, W., Zeiler, W.: Accuracy of different machine learning algorithms and added-value of predicting aggregated-level energy performance of commercial buildings. Energy Build. **209**, 109705 (2020)
22. Wargocki, P., Wyon, D.: Effects of HVAC on student performance. ASHRAE J. **48**, 22–28 (2006)
23. Yang, R., Wang, L.: Multi-objective optimization for decision-making of energy and comfort management in building automation and control. Sustain. Cities Soc. **2**(1), 1–7 (2012)
24. Zhao, T., Xue, H.: Regression analysis and indoor air temperature model of greenhouse in northern dry and cold regions. In: Li, D., Liu, Y., Chen, Y. (eds.) CCTA 2010. IAICT, vol. 345, pp. 252–258. Springer, Heidelberg (2011). https://doi.org/10.1007/978-3-642-18336-2_30

Lane Detection for Autonomous Vehicle in Hazy Environment with Optimized Deep Learning Techniques

Bagesh Kumar, Harshit Gupta(✉), Ayush Sinha, and O.P. Vyas

Department of IT, Indian Institute of Information Technology Allahabad, Allahabad, India
{pse2016001,rsi2020501,pro.ayush,opvyas}@iiita.ac.in

Abstract. In this technological era, the devices are getting more intelligent and smarter with the advent of artificial intelligence and related technologies. The autonomous vehicle is one of the emerging and important example using the same machine intelligence for effective and efficient vehicle driving experience. However to make the working of autonomous vehicle in a smooth and responsive, the accurate lane detection is a crucial aspects and other co-variate like weather condition also place an important role in it. The present work explores lane detection in foggy or hazy environment on time series data of a continuous driving scene. The proposed methodology uses Dark Channel Prior (DCP) to make images un-hazy and this is passed as an input to the proposed hybrid architecture by connecting a convolutional neural network (CNN) architecture with a recurrent neural network architecture (RNN). The model uses Long-Short Term Memory (LSTM) for time-series data analysis as it captures and process the time series data well in advance. In this model, the CNN block first extracts the feature maps from each frame which is given as input of the time series data and after this these feature maps are provided to LSTM to obtain the final prediction. The model also uses dark channel prior (Image enhancing technique) to enhance image in case of foggy or hazy environment and then detecting lane with the help of enhanced images. Finally, the accuracy of the model is shown in terms of standard performance metrics like precision, recall and F1 score.

Keywords: Deep learning · Convolution neural networks · LSTM · Driver-less driving · Lane detection · Semantic segmentation · Dark channel prior

1 Introduction

Vehicles were once considered to be the boon of mechanical engineers to the world. However, there have been many advancements in automobiles in recent years. Nowadays, the artificial intelligence embed vehicles are getting more attention due to its self-operable behaviours like self-driving cars where no human

I. Woungang et al. (Eds.): ANTIC 2021, CCIS 1534, pp. 596–608, 2022.
https://doi.org/10.1007/978-3-030-96040-7_45

driver is needed to drive the cars rather the intelligent system of car will automatically drive the car by detecting objects, humans, lanes etc. Keeping the challenges in view for the autonomous vehicles the detection of lanes accurately is very important which needs more exploration. It would help the vehicles to move seamlessly without any collisions or violations of traffic rules. The future of the automobile industry is to make the vehicles autonomous and provide best advance driver assistance system (ADAS) possible [1]. Various environmental factors like fog, dim light etc. are also required to be processed for robust lane detection.

Lane boundaries detection accuracy has been increased using various algorithms over the years. Introduction to deep learning methods has provided a huge improvement in the accuracy and robustness. Now using the deep learning methods with some latest algorithms for image enhancement can improve the scenario even further. As the lanes are continuous line structures, they can be either dashed or solid lines. And these scenes are largely overlapped between two neighbouring frames, so, these neighbouring frames are highly related. Therefore, lanes in the current frame can be generated by using the information from the current frame previous frames as well [1].

Meanwhile, deep learning in recent years have proved that they are better in performance in computer vision problems (i.e. object detection, image classification etc.) and can be relied upon Deep Convulational Neural Networks (DCNN) which do the work of feature abstraction from images. Deep Recurrent Neural Networks (DRNN) are used for information prediction for time-series signals. So, lane detection in continuous frames is a time-series analysis and can be processed using DRNN. As dimensionality of images can be sometimes quite big enough so this makes DRNN network slower. That is why DCNN is used as it can reduce dimensionality keeping intact the important information of the images.

So it can be deduce that there is requirement of hybrid architecture of DCNN and DRNN for lane detection in continuous frames [1]. Furthermore, applying this on foggy images might give low accuracy. Hence, it is required to apply Dark Chanel Prior for foggy or hazy images to remove blurry or hazy effect from the images before forwarding it to above architecture.

The remaining paper is organized as separate sections in which Sect. 2 is Literature Review which provides the previous works related to autonomous vehicles, Sect. 3 is Proposed Model in which the model for lane detection is explained, Sect. 4 Experimental Outcome and Discussions explains the obtained result after experiment. Finally, Sect. 5 is Conclusion and Future Scope in which the entire work is concluded along with the possible works that can be done in future.

2 Literature Review

The advancement of technologies leads to further extensive research for any domain. Similarly in the field of autonomous vehicle industry there are numerous researches has taken place and still going on. In the article [2], the authors have explored the geometrical methods for the detection of lines, curves etc. in the figure. Specifically the Hough Transformation method is taken into consideration for line detection. But it also suffers from some limitations like curved lanes, sharp turns. These limitations reduces the overall efficiency of the entire system.

The introduction of various artificial intelligence techniques have taken the problem rectification way at high level. This can be observed in the paper [3] where a lane detection algorithm is presented which is based on support vector machine (SVM). The SVM pattern recognition is exploited for road surface extraction. The image is transformed into bird-view image by using world coordinate system (WCS) and image coordinate system (ICS) which providesral point of the road from the mid-line of the road. The SVM regression is taken into account for the road shape function. This method is good for dividing a road in two lanes but fails when there are large number of lanes and we need to detect all those lanes at same time. Thus this model is not at all suitable for current needs. In [7], the two group classification problem is also introduced.

The authors in [4] exploited Cascaded CNNs where an end-to-end technique for lane detection related aspects like boundary detection, classification and clustering is done which is based on two CNNs that runs in real-time. The first CNN is used to detect lanes using feature extraction techniques. And the second CNN is used for classifying the type of lanes into 8 categories. This method is not so robust as it takes into account only one image at a time which may not give proper lane detection due to weather conditions and obstruction etc.

In the continuation of findings where most of the works has used single frame for lane detection, the authors in [5] have introduced the similar work by using continuous frames. In this paper, multiple frames are taken from continuous scene captured while driving and proposed a architecture which is hybrid in nature where RNN and CNN is combined. This method although performs good but still produces bad result when weather conditions are not favourable like hazy or foggy environment or when sun flare is very strong resulting in high brightness.

Various approaches is given for the lane detection and in every approach there are some limitations. Similarly, taking the trade-off related to analysis of hazy images a Dark Channel Prior algorithm [6] is proposed which has ability to unhazify an hazy image using single color channel of an image. Using this dark channel and atmospheric light we can get an unhazy image by interpolation and soft-matting techniques.

In [8], the concept utilization of a CNN-RNN network is shown. CNN is used for feature extraction and RNN for doing time series analysis on the features extracted to make the network robust. Using such concept helps in the action recognition which is very important for the autonomous vehicles to self operate. This pipeline of CNN-RNN is useful as it helps to remember useful features and relations in the data like some sequence or patterns.

Finding in [9] gives us a brief about Self-driving and driver relaxing vehicle where the vehicles are focused to be automated to give human driver relaxed driving. The authors in this work has given two insights related to intelligence of automated vehicles. The first is automatically following the other vehicle for being same route, and second is using less resources(breaks, clutches etc.) in case of heavy traffic. From these insights it can be deduced that how much deep research enhancement is to be done for an efficient autonomous vehicle system.

A responsive system of automatic driver assistant with functionality of lane tracking and road signal identification is represented in [10]. Here the important factors are real time video analysis and optimal utilisation of hardware and software. The design of proposed system is like it exploits cheaper cameras. The lane tracking algorithm was implemented on MATLAB in which Hough Transform was applied.

The paper [11] explores another CNN based approach to the lane detection problem wherein a lane segmentation network is used to segment out the visible lane marking pixels before applying a perspective transformation to obtain a bird's eye view which facilitates the ease of voting in clustering and curve fitting that are also applied later. Using this kind of approach provide a base so that the model can perform satisfactory for accuracy and speed.

The paper [12] explored a vision based lane detection approach which can perform responsive operations with higher accuracy for the scenarios like light variation and shadows. The structure of the simulation should be like having cameras mounted on the vehicle and the images obtained from such camera. Then the some algorithms are applied for identification of lane or paths.

A deep-learning-based approach for vehicle detection is represented in [13] that can achieve effective detection performance under extreme conditions. The technique mentioned in the paper can be useful for in-road driver assistance and for autonomous driving.

Perception system design or Insight framework configuration is an essential advance in the improvement of an autonomous vehicle (AV). With the immense determination of accessible off-the-rack plans and apparently unlimited choices of sensor frameworks executed in examination and business vehicles, it tends to be hard to recognize the ideal framework for one's AV application. The [14] presents an exhaustive audit of the best in class AV insight innovation accessible today.

Design of equipment gas pedals for neural networks (NN) applications includes strolling a tightrope in the midst of the imperatives of low-power, high exactness and throughput. NVIDIA's Jetson is a promising stage for implanted AI which looks to accomplish a harmony between the above destinations. In this paper [5], an overview of works that assess and enhance neural organization applications on Jetson stage.

As an essential for independent driving, scene understanding has drawn in broad examination. With the ascent of the convolutional neural organization (CNN)- based profound learning strategy, research on scene understanding has accomplished huge advancement. The paper [15] intends to give an exhaustive review of profound learning-based methodologies for scene understanding in independent driving.

Lane detection is a critical component in increasing driving safety. In this research, a lane detection approach for lane departure warning systems that is both real-time and illumination invariant is presented. The suggested approach [16] works effectively in a variety of lighting circumstances, including bad weather and at night. Lane detection is an essential part of most ebb and ADASs. Countless existing outcomes center around the investigation of vision-based path location techniques because of the broad information foundation and the minimal expense of camera gadgets. In [13], past vision-based path location studies are surveyed as far as three viewpoints, which are path identification calculations, joining, and assessment strategies.

Path discovery in driving scenes is a significant module for independent vehicles and progressed driver help frameworks. Lately, many complex path recognition strategies have been proposed. In any case, most strategies center around identifying the path from one single picture, and regularly lead to inadmissible execution in dealing with some very terrible circumstances like substantial shadow, extreme imprint corruption, genuine vehicle impediment, etc. Truth be told, paths are consistent line structures out and about. Thusly, the path that can't be precisely distinguished in one current casing may conceivably be induced out by joining data of past outlines. To this end, The paper [13] research path recognition by utilizing different casings of a constant driving scene, and propose a half and half profound engineering by joining the convolutional neural organization (CNN) and the intermittent neural organization (RNN).

2.1 Research Gap

Traditional methods including geometric modelling and semantic-segmentation of image works only in the case of single image. This method can under-perform in cases like serious vehicle occlusion, severe mark degradation, heavy shadow and so on [1]. Therefore, to advance in this field we need something more concrete and something we can depend on fully. So we are suggesting a method including a fusion of DRNN and DCNN to detect lanes. This can detect lanes in continuous frame.

3 Proposed Model

In order to integrate images taken from consecutive driving scene with previous images for detecting lanes, we designed a encoder-decoder framework which are fully convolutional network. For this experiment, we used Tu-Simple dataset which is a benchmark dataset for road or lane detection. It contains 3700+ sequences of contiguous road images. Each sequence consisting of 20 images where the 20th image is labeled for ground truth. We changed this dataset to hazy images using image transformation techniques. We made images hazy and some brightened images. Then we used this transformed Tu-simple dataset for

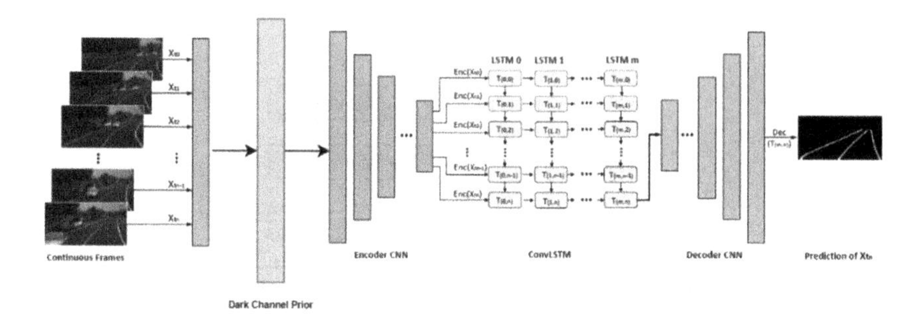

Fig. 1. Architecture of proposed network

this work. Initially. The image is made unhazy by using DCP algorithm. In large number of non-sky patches there are some one color channel whose pixel's intensity is too less and can be said as very near to zero. The dark channel for an arbitrary image **J** is given by

$$J_{dark}(x) = min_{y \in \Omega(x)}(min_{c \in \{r,g,b\}} J^c(y)) \quad (1)$$

The color channel of J is represented by J^x whereas $\Omega(x)$ is denoting local patch which is centered at x. There are two minimum operators which are very important for dark channel $min_{c \in \{r,g,b\}}$ and $min_{y \in (x)}$. These operators are applied on each pixel. Along with the minimum filter these operators are also commutative in nature. Using the concept of a dark channel some facts can be observed like if there is an outdoor image (excluding any sky region image) J which is hazy free then the dark channel intensity of image will be very less and tends towards zero.

$$J_{dark} \rightarrow 0 \quad (2)$$

With many number of continuous frames of un-hazy images as input the encoder CNN convert them into time-series feature maps. Convolution and Pooling are used for image abstraction and feature maps extraction.

For the encoder and decoder CNN we are using the pretrained model of Segnet. It has been successfully trained for semantic segmentation on Imagenet Dataset. The use of pretrained model provides faster and better training for the overall model. It can be seen that lane detection is a semantic segmentation problem and thus we are using encoder-decoder models which will provide an output having the same size of the input. In the encoder phase image operations like convolution and pooling are used to extract feature maps from input images. In the decoder phase deconvolution and unsampling is used to reconstruct the image and highlight the lanes pixelwise. The pretrained segnet model is presented in Fig. 2.

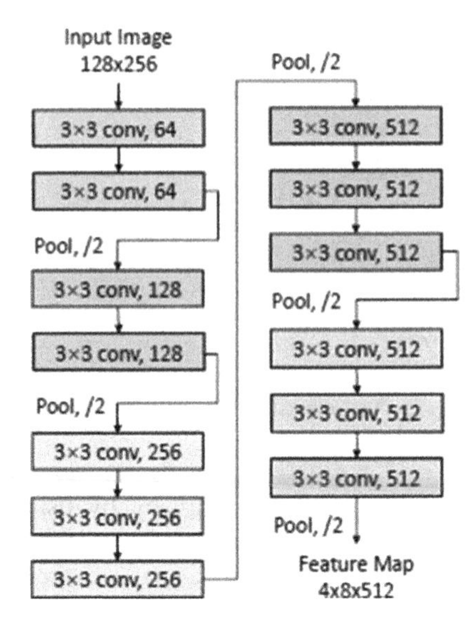

Fig. 2. Segnet encoder model

Now, we pass the feature maps extracted from encoder of CNN for time series analysis. We do this using LSTM (long-short term memory). Modelling the various continuous frames of driving scene as time-series, the RNN block accepts the feature maps of each frame given by CNN encoder as input. LSTM network has the ability to forget the unimportant information and remember the essential ones [1]. Traditional full-connection LSTM is slower as compared to convolutional LSTM. So the convolutional LSTM is used. Convolutional LSTM replaces matrix multiplication in each gate with convolutional operation. Following are the activation functions of a general Convolutional LSTM:-

$$C_t = f_t \circ C_{t-1} + i_t \circ tanh(W_{xc} * X_t + W_{hc} * H_{t-1} + b_c) \tag{3}$$

$$f_t = \sigma(W_{xf} * X_t + W_{hf} * W_{t-1} + W_{cf} \circ C_{t-1} + b_f) \tag{4}$$

$$o_t = \sigma(W_{xo} * X_t + W_{ho} * W_{t-1} + W_{co} \circ C_{t-1} + b_o) \tag{5}$$

$$i_t = \sigma(W_{xi} * X_t + W_{hi} * W_{t-1} + W_{ci} \circ C_{t-1} + b_i) \tag{6}$$

$$H_t = o_t \circ tanh(C_t) \tag{7}$$

The symbols used in all these equations have some specific role and all these can be described in specific manner like X_t is the input feature maps at time t which is extracted by the encoder CNN. C_t, H_t represents memory and output activation at time t and C_{t-1} H_{t-1} represents memory and output activation at time t−1. C_t is cell, i_t is input, f_t is forget and o_t represents output gates. W_{xi}

denotes the weight matrix of the input X_t to the input gate, b_i is the bias of the input gate. The W and b can be inferred from the already provided rules. The sigmoidal operation is represented by $\sigma()$ whereas '$\circ$' and '*' denotes the Hadamard product and the convolution operation respectively.

Next comes the decoder CNN. In the decoder part, deconvolution and upsampling are used to grasp and highlight the information of targets and spatially reconstruct them. For effectively decoding the output of LSTM network the size and number of feature maps should be same as the encoder CNN, while arranged in inverse direction. Accordingly, the up-sampling and convolution in each sub-block of decoder match the corresponding operations in the sub-block of the encoder [1].

4 Experimental Outcome and Discussions

The proposed algorithm is tested against a TuSimple data-set which consists 3700 sequences of 20th image of each sequence has been labeled which are used as ground truth in the this work. The labelling contains the (width, height) of pixels and the belonging lane. So we can easily get the annotated/ground truth image of the labelled image in each sequence. Time interval between consecutive images in each sequence is less than 1 s. This lets us use a time series analysis on the given data-set along with semantic segmentation.

Figure 3 shows a representation of dark channel prior image enhancement technique which is applied on a input image. The Fig. 3(a) is hazy image while the Fig. 3(b) is unhazy image which is obtained after applying the proposed algorithm. It can be clearly seen the clarity of hazy image after applying DCP, so correspondingly the lane detection model will work much better.

The training of proposed model with DCP is done for 10 number of epochs and the result is obtained. The same dataset is also tested without using DCP. The corresponding comparative result for the performance metrices is represented in Table 1.

The average run-time per batch of **100** sequences is **5 s**. So, per sequence average run-time is **50 ms** which is better than most other models.

The Fig. 4 shows the plot between accuracy and epochs where it can be clearly seen that the accuracy increases with number of epoch. The model is able to learn it's parameters with more number of epochs. After a certain number of epochs the model starts over fitting on the data which can be understood at the decrease in accuracy at last. So the training is stopped as soon as it starts over fitting to get best bias-variance tradeoff.

(a)

(b)

Fig. 3. Representation of (a) before (b) after, image processed with proposed algorithm.

Table 1. Comparison of performance matrices "With DCP" and "Without DCP"

Measure	Score with DCP	Score without DCP
Accuracy (in percent)	93.7	91.2
Loss	5.567×10^{-2}	8.3469×10^{-2}
F1 score	0.6331	0.5999
Precision	0.4937	0.4517
Recall	0.9117	0.8931

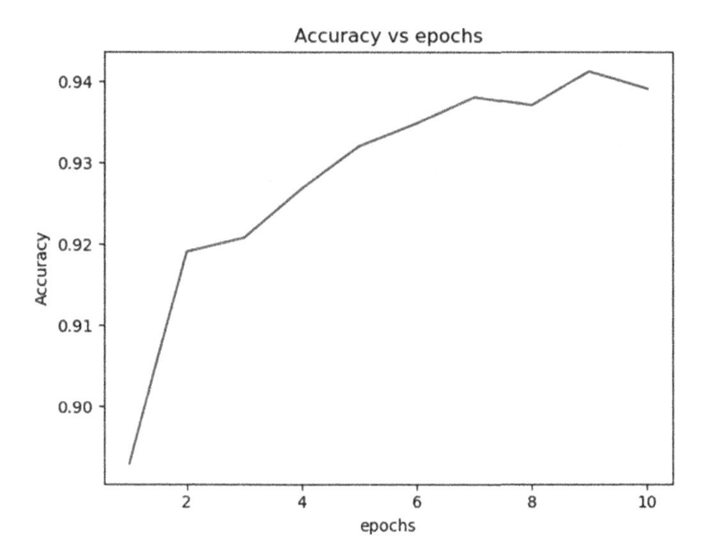

Fig. 4. Accuracy vs epoch plot

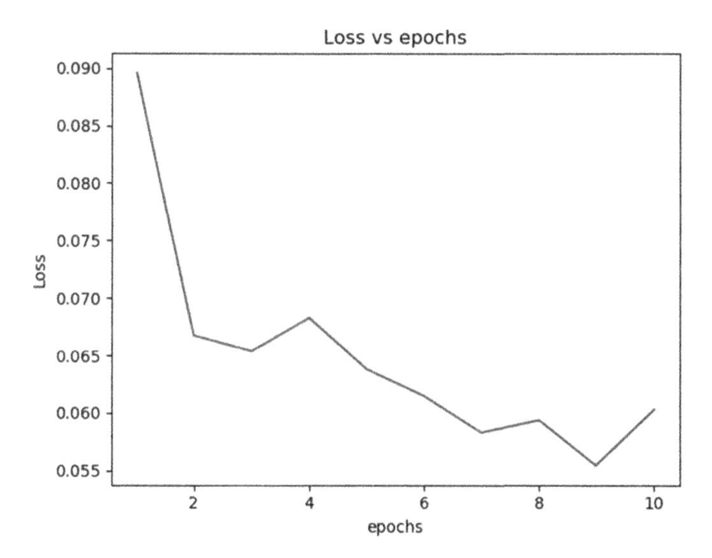

Fig. 5. Loss vs epoch plot

The Fig. 5 shows the plot between Loss and Epochs where it can be clearly observed that the model gets trained with epochs and accuracy increase and the loss decreases until starts over fitting. When it starts over fitting the loss increases and training is stopped to get best fit model.

In the Fig. 7 the experiment on n_frames vs accuracy is performed. n_frames is the number of contiguous frames used for time series analysis. The experiment

Fig. 6. Application on top of the architecture

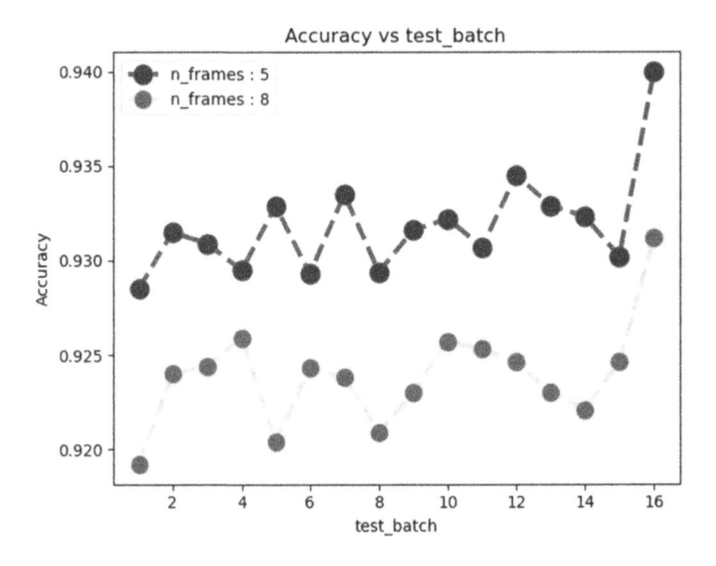

Fig. 7. Accuracy vs epoch plot for different n-frames: 5 and 8

on various n_frames we choose n = 5 as best option as it gives maximum accuracy as shown in Fig. 7. With n = 3 and n = 8 we get less accuracy.

The Fig. 6 shows the use of proposed model in a practical continuous road detection in hazy environment.

5 Conclusion and Future Scope

In this paper, we used a robust hybrid method combining both RNN and CNN for lane detection in driving scenes. The proposed method includes removing haziness from image using dark channel prior method, then using encoder-decoder CNN framework which uses multiple continuous frames as input and using semantic segmentation predict lanes of current frame. In this we first extract feature maps of each frame using encoder CNN. Then, the sequential encoded features of all input frames were processed by a ConvLSTM. Finally, the outputs of the ConvLSTM were fed into the CNN decoder for information reconstruction and lane prediction. To get better results in hazy conditions we also used Dark Channel Prior for image enhancement. The result obtained in this work is able to support the out-performance of the Dark Channel Prior algorithm to detect the lanes efficiently for autonomous vehicles operations.

In the future perspective, the integration of other road segments like vehicles, obstacles and person etc. and create a full fledged ADAS system. Along with this the enhancement in accuracy in different types of environment and roads can be taken into consideration.

References

1. Simonyan, K., Zisserman, A.: Very deep convolutional networks for large-scale image recognition. arXiv preprint arXiv:1409.1556 (2014)
2. Duda, R.O., Hart, P.E.: Use of the Hough transformation to detect lines and curves in pictures. Commun. ACM **15**(1), 11–15 (1972)
3. Zhang, H., Hou, D., Zhou, Z.: A novel lane detection algorithm based on support vector machine. In: Progress in Electromagnetics Research Symposium, pp. 22–26, August 2005
4. Pizzati, F., Allodi, M., Barrera, A., García, F.: Lane detection and classification using cascaded CNNs. arXiv preprint arXiv:1907.01294 (2019)
5. Zou, Q., Jiang, H., Dai, Q., Yue, Y., Chen, L., Wang, Q.: Robust lane detection from continuous driving scenes using deep neural networks. IEEE Trans. Veh. Technol. **69**(1), 41–54 (2019)
6. He, K., Sun, J., Tang, X.: Single image haze removal using dark channel prior. IEEE Trans. Pattern Anal. Mach. Intell. **33**(12), 2341–2353 (2010)
7. Cortes, C., Vapnik, V.: Support-vector networks. Mach. Learn. **20**(3), 273–297 (1995)
8. Zhao, C., Han, J.G., Xu, X.: CNN and RNN based neural networks for action recognition. In: Journal of Physics: Conference Series, vol. 1087, no. 6, p. 062013. IOP Publishing, September 2018
9. Memon, Q., Ahmed, M., Ali, S., Memon, A.R., Shah, W.: Self-driving and driver relaxing vehicle. In: 2016 2nd International Conference on Robotics and Artificial Intelligence (ICRAI), pp. 170–174. IEEE, November 2016
10. Mani, J.R., Gangadhar, N.D., Reddy, V.K.: A real-time video processing based driver assist system. SASTech **9**(1), 9–16 (2010)
11. Chang, D., et al.: Multi-lane detection using instance segmentation and attentive voting. In: 2019 19th International Conference on Control, Automation and Systems (ICCAS), pp. 1538–1542. IEEE, October 2019
12. Assidiq, A.A., Khalifa, O.O., Islam, M.R., Khan, S.: Real time lane detection for autonomous vehicles. In: 2008 International Conference on Computer and Communication Engineering, pp. 82–88. IEEE, May 2008

13. Leung, H.K., Chen, X.Z., Yu, C.W., Liang, H.Y., Wu, J.Y., Chen, Y.L.: A deep-learning-based vehicle detection approach for insufficient and nighttime illumination conditions. Appl. Sci. **9**(22), 4769 (2019)
14. Van Brummelen, J., O'Brien, M., Gruyer, D., Najjaran, H.: Autonomous vehicle perception: the technology of today and tomorrow. Transp. Res. Part C: Emerg. Technol. **89**, 384–406 (2018)
15. Guo, Z., Huang, Y., Hu, X., Wei, H., Zhao, B.: A survey on deep learning based approaches for scene understanding in autonomous driving. Electronics **10**(4), 471 (2021)
16. Son, J., Yoo, H., Kim, S., Sohn, K.: Real-time illumination invariant lane detection for lane departure warning system. Expert Syst. Appl. **42**(4), 1816–1824 (2015)

Review of Extrinsic Plagiarism Detection Techniques and Their Efficiency Comparison

Malya Singh(✉) and Vishal Gupta

University Institute of Engineering and Technology, Panjab University, Chandigarh, India
malyasingh98@gmail.com, vishal@pu.ac.in

Abstract. Plagiarism can be defined as stealing or copying the work of other researchers/authors and presenting it as your own without referencing or acknowledging their work. In this paper various types of extrinsic plagiarism detection techniques have been discussed which are based on Linguistic features, Semantic role labeling, Context matching and Word embeddings, Multi- agents indexing system, Syntactic and Semantic similarity. We have also explored Deep Learning and Machine Learning based plagiarism detection techniques which are based on SVM, Genetic algorithm, CNN & RNN, Logical regression model. These techniques have been implemented on different datasets PAN-PC-(2009–2016), AraPlagDet, Plagiarized Short Answers, SemEval 2015 Twitter and Microsoft Paraphrase corpus. In this paper we have also compared these techniques on the standard measures that are Precision, Recall, F1 Score, PlagDet and Granularity and we have also discussed the pros and cons of these plagiarism detection techniques.

Keywords: Plagiarism · Extrinsic plagiarism detection · Deep learning · Machine learning · Dataset

1 Introduction

Over the years, we have seen a huge growth in the Information Technology world. With the advancement in technology, the internet has expanded and grown so much. Now-a-days we can easily access loads of data online. With the more internet usage, problems like plagiarism gets more complex and harder to detect. In academics, plagiarism has been a major issue, where many students/researchers intentionally/unintentionally copy the work of other authors and present it as their own and it becomes difficult to identify the original work and copied work. There are various types of plagiarism from simple copy and paste (verbatim), to harder forms of plagiarism like Idea plagiarism where the plagiarists copy the idea of the authors work and write it in their own words, which makes it harder to detect. A lot of work and research has been done, and is currently going in this field and with the use of various algorithms and techniques; even the harder forms of plagiarism can be detected.

In Plagiarism detection mainly we have two approaches Intrinsic and Extrinsic [1]. In the Extrinsic plagiarism detection approach we have a reference collection (source

I. Woungang et al. (Eds.): ANTIC 2021, CCIS 1534, pp. 609–624, 2022.
https://doi.org/10.1007/978-3-030-96040-7_46

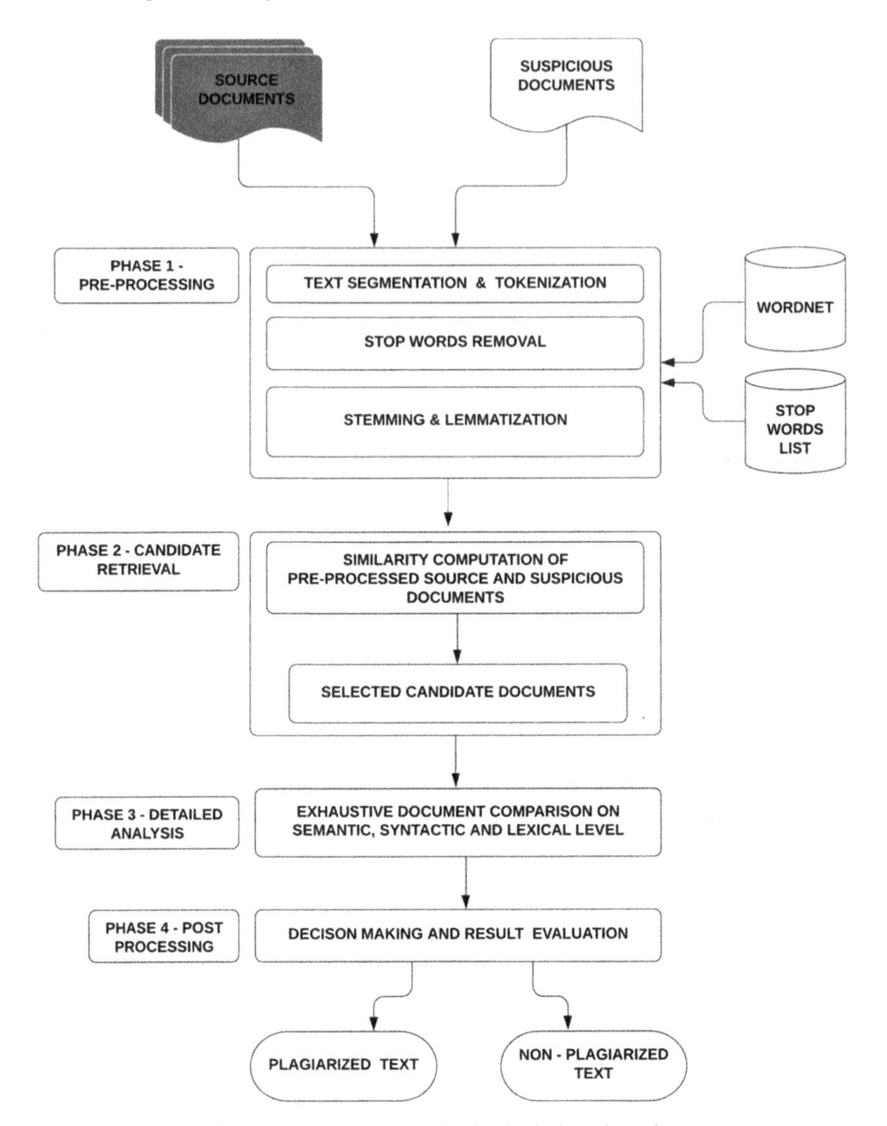

Fig. 1. Basic steps in extrinsic plagiarism detection

documents) to which the suspicious documents are compared, the approach consists of pre-processing of documents which includes tokenization, removing stop words, sentence/word segmentation, stemming/lemmatization. Next is the candidate retrieval stage which includes similarity computation, then is the detailed analysis stage which includes exhaustive comparison on semantic, syntactic and lexical level and lastly post processing stage which includes decision making and filtering out the non plagiarized portion.

In the Intrinsic approach we don't compare the documents to a reference collection, instead we analyze the authors writing style and use stylometric features to detect plagiarism [2]. In this paper we have focused mainly on extrinsic based plagiarism detection techniques [3]. Various Machine learning techniques have also been used in detecting plagiarism based on SVM (Support vector machine) for paraphrase identification [4], Naïve Bayes for detailed analysis task [5], and Genetic algorithm for detection over mail services [6]. Deep learning techniques [7] can also be used to detect various forms of plagiarism. Plagiarism can also be applied in programming languages which is called as Source code plagiarism [8] (Fig. 1).

2 Review of Various Extrinsic Plagiarism Detection Techniques

In this section various extrinsic plagiarism detection techniques have been discussed.

2.1 Detection of Plagiarism Using Semantic-Syntactic Knowledge

In this method [9] different linguistic features are used to compute the similarity, it also uses dice similarity to compute the semantic resemblance. This approach can detect paraphrasing, restructuring, verbatim type plagiarism. The first phase in plagiarism detection is the preprocessing phase which includes text normalization, segmentation, stop-word removal, lemmatization of text. Next phase includes computing the resemblance between pairs of sentences, firstly a joint matrix is formed which includes distinct words from both the source and suspicious sentence pair. This approach also used features like depth estimation and path similarity to calculate the gross resemblance on the given word pairs.

Gross Resemblance score is calculated as:

$$Gross\_Res(Sent\_src, Sent\_sus) = \alpha * sem\_res(Sx, Sy) + (1 - \alpha) * syn\_res(Rx, Ry) \tag{1}$$

The plagiarized sentences are filtered out using threshold value (0.65), if the gross resemblance score is more than the given threshold value, then the sentence is termed as plagiarized. PAN-PC-11 dataset is used to implement this system.

2.2 Novel Approach for Paraphrase Identification

This approach is focused on detecting paraphrase plagiarism, mainly two types of paraphrasing that is word reordering and synonym substitutions. They presented a novel method that uses context-matching and word-embeddings for identifying substitutions and reordering [10]. Paraphrasing type plagiarism is done by copying others work and then obfuscating it by using semantic, syntactic and lexical transformations. Dataset used in this approach is the plagiarized short answers which consist of both source and suspected (paraphrased) sentences (Fig. 2).

The approach consists of three phases, first is the pre-processing phase which includes punctuation removal and converting text to lowercase. Next phase is the identification of word reordering which is done by detecting textual segments from the given pair of sentences which are identical. Finally using contexts, word-embeddings and word-alignment, synonyms substitutions are identified. They used ConceptNet Numberbatch word embeddings and Smith Waterman algorithm to detect these substitutions.

$$\text{Similarity}(\text{K}, \text{L}) = \frac{\vec{\text{K}} . \vec{\text{L}}}{|\vec{\text{K}}| . |\vec{\text{L}}|} \tag{2}$$

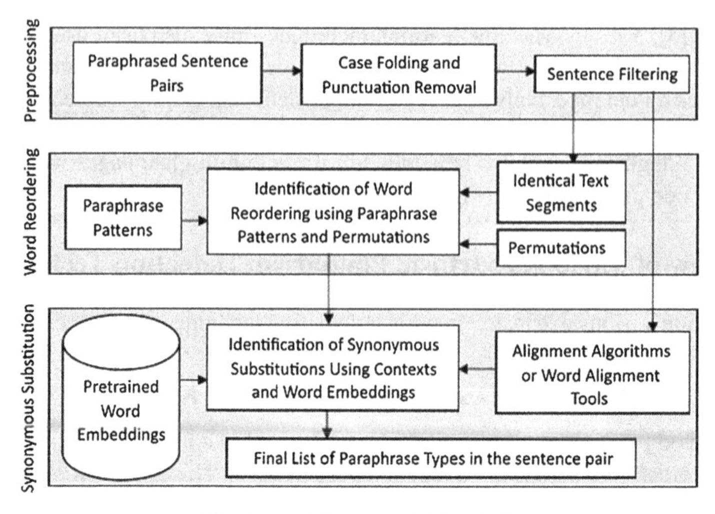

Fig. 2. Architecture of PIPD [10]

The value of Similarity (K, L) is used for sentence filtering; here K and L are vector representations of two given sentences.

2.3 Detecting Plagiarism Using Inverse Path and Depth Estimation (Novel)

This approach [11] considers both the semantic and the syntactic similarity for detecting plagiarism. Novel features like inverse path and depth estimation are used for calculating semantic and syntactic relatedness between source and suspicious sentences by varying weights. This method can detect verbatim, paraphrasing, re-wording and sentence transformation type plagiarism. It also uses the lexical database (WordNet) for finding the relatedness between sentences [12]. There are three stages in this approach, first is the pre-processing of data which includes text segmentation, stop-words removal, stemming. Next the joint vector is formed to compute the syntactic and semantic resemblance between the sentences by using features like depth, local density and two novel features that are inverse path length and estimating depth. The syntactic and semantic relatedness score is calculated on the PAN 2011 dataset and then the final relatedness score is calculated which is given as follows:

$$Overall\_Rel\_score\,(P,L) = \gamma.sem\_rel\,(P,L) + (1-\gamma).syn\_rel\,(P,L) \quad (3)$$

2.4 Novel Approach – Multi-agents Indexing System for Detecting Plagiarism

This approach is in context with the Arabic plagiarism detection system, implemented on the AraPlagDet corpus. SorensenDice method is used for comparing semantic similarity between the given documents. The approach consists of three phases, first is the Natural language processing (NLP) phase which includes removal of numbers, irrelevant characters from the text, followed by stop-words removal, normalization and tokenization of text, lastly stemming is done on the text. Next is the indexing phase in which synonyms are extracted by POS (parts of speech) tags for each given word which is done by using the ShemNet and SorensenDice method [13]. Lastly the evaluation phase is done to evaluate the efficiency of the system using the standard measures. To reduce the processing time for the semantic similarity measure, static and mobile agents are used which can perform tasks and make decisions, these agents are implemented on an open-source platform (JADE) (Fig. 3).

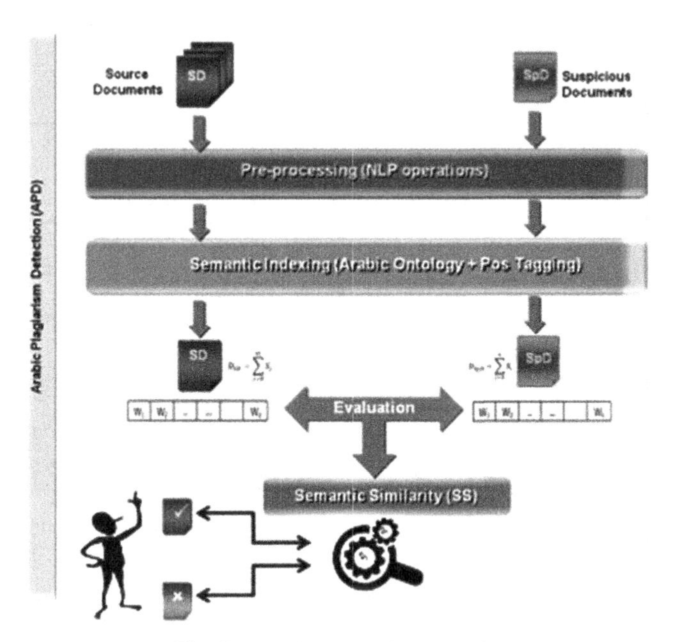

Fig. 3. Architecture of MAIS [13]

2.5 Detecting Plagiarism Based on Semantic-Syntactic Similarity

This method [14] combines the semantic and syntactic similarity to detect plagiarism between source sentence and suspicious sentences. It can detect various types of plagiarism like verbatim, paraphrasing, reordering and restructuring. The approach includes three phases, first is the pre-processing phase which includes sentence segmentation, removal of stop words and stemming. Next is the detailed comparison phase where sentences are compared to check the similarity between documents using the cosine similarity method, if the similarity exceeds the given threshold (0.6), then they are therefore considered as plagiarized. Last phase is the post processing phase where the sentence pairs having the most similarity are considered as plagiarized. The similarity between two given sentences is calculated as:

$$Sim\_measure\,(Sq, Sv) = \alpha.sim\_semantic\,(Sq, Sv) + (1 - \alpha)\,.sim\_wordorder\,(Sq, Sv) \quad (4)$$

2.6 Semantic Similarity Detection Using SRL and Argument Weighting

This approach [15] is based on the SRL (semantic role labeling) which uses argument weighting as a similarity measure to compute the semantic similarity between given pair of sentences. It also uses the lexical database WordNet to analyze the text semantically. The initial phase is the pre processing phase in which text segmentation, stop words removal and stemming is done. Then the sentences are represented as nodes and arguments are extracted and grouped together called the Argument Label Group (ALG), next is the Semantic Term Annotation (STA) phase where concept extraction is done on the PAN-PC-09 dataset (Fig. 4).

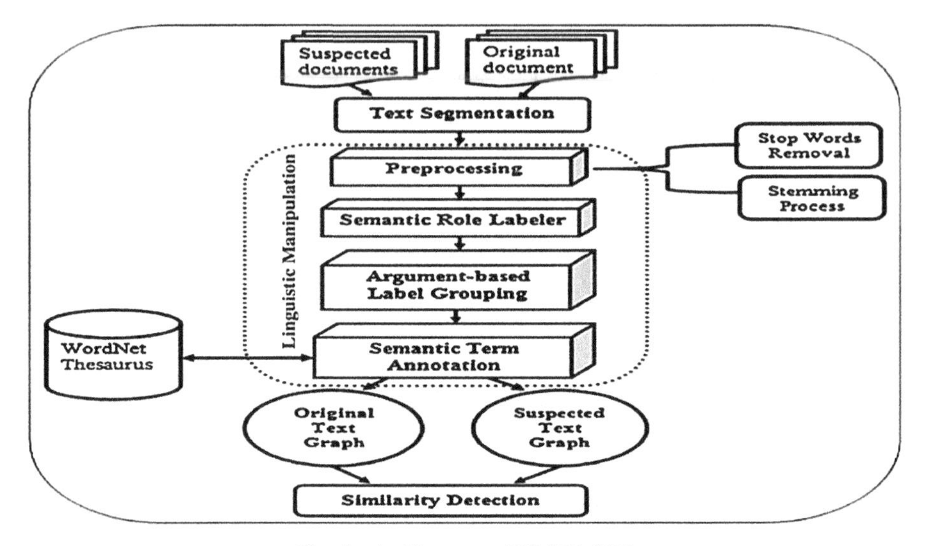

Fig. 4. Architecture of PDSRL [15]

2.7 Text Plagiarism Detection Using Semantic-Syntactic Based NLP Techniques:

In this approach [16] various Natural Language Processing techniques like POS tagging, chunking, SRL and combinations of these techniques like POS + chunking and POS + SRL are analyzed. Initially pre processing of data is done in which lowercasing of text, tokenizing the text, removal of stop words and lastly lemmatization is done. In the next step VSM is used with the tf-idf representations, cosine similarity is also used to compare the similarity between source and suspicious documents. Next shingling is done and tri-grams are formed for suspicious and source documents. The given approach is implemented on all the techniques on various datasets, PAN-PC-2009 to 2014. Parts of Speech tagging performed the best out of all the techniques.

The cosine similarity and overlap coefficient is computed as:

$$Cosine\left(\overrightarrow{K}_{sp}, \overrightarrow{K}_{sc}\right) = \frac{\overrightarrow{K}_{sp}.\overrightarrow{K}_{sc}}{||\overrightarrow{K}_{sp}||\,||\overrightarrow{K}_{sc}||} \tag{5}$$

$$\text{Overlap}(K_{susn}, K_{srcn}) = \frac{|K_{susn} \cap K_{srcn}|}{\text{Min}(|K_{susn}|, |K_{srcn}|} \tag{6}$$

3 Review of Machine Learning and Deep Learning Based Extrinsic Plagiarism Detection Techniques

In this section various machine learning and deep learning based techniques are discussed.

3.1 Paraphrase Detection Using Neural Networks (CNN and RNN)

This approach focuses on identifying text (sentences) which are semantically similar. A hybrid of CNN and RNN model is proposed for detecting paraphrase plagiarism [17]. A new architecture called DeepParaphrase is also developed in this approach. Paraphrase detection can be defined as identifying whether a given sentence is a paraphrase of another one or not [18]. CNN uses n-gram representations for the given text and RNN uses the words in a sequence for the given text. Next the similarity between the given sentences is computed using the matching model and similarity matrix. This approach is tested on both the models which are used in both the coarse and fine grain sentence/word levels, their final model called the AugDeepParaphrase model is implemented on two datasets: SemEval 2015 Twitter & Microsoft Paraphrase (Fig. 5).

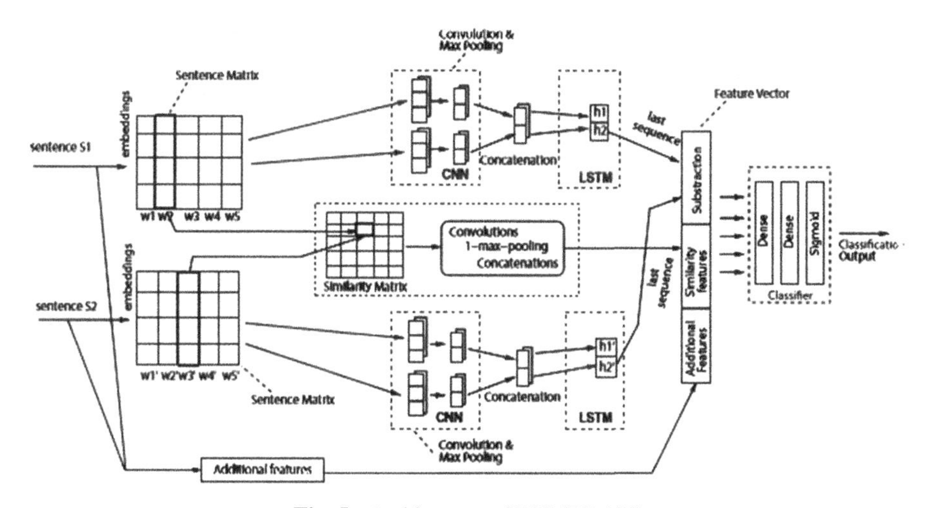

Fig. 5. Architecture of DNMPD [17]

3.2 Idea Plagiarism Detection Using Genetic Algorithm

This approach focused on idea plagiarism where the plagiarist copies the idea of other authors and summarizes it in their own words. In this approach extraction is done using the Genetic algorithm to detect plagiarism within a document at the syntactic and semantic level. The detection is done at both the document and passage/paragraph level [19]. Firstly pre processing is done with includes sentence segmentation, tokenization, POS (parts of speech) tags and lemmatization. Then the documents are represented in VSM with tf-isf weights. At the syntactic level, extraction is done at both sentence and word level. For the semantic level similarity metrics are used with the use of the WordNet database. The approach is implemented on the summary obfuscation PAN-PC-(13 – 14) datasets. Semantic Similarity is calculated (for given threshold values) as follows:

$$Similarity_1 = \sum \frac{MaxSim}{|Ssus| + |Ssrc|} \tag{7}$$

$$Similarity_2 = \frac{|Cnt|}{Max(|Ssus|, |Ssrc|)} \tag{8}$$

3.3 Plagiarism Detection in Persian Texts Using SVM

In this approach [20] they have used Support vector machine (SVM) to detect plagiarism in Persian text. They have also introduced a novel method called 'Index words Replacement to detect the semantic similarity and also used the semantic database names FarsNet.

Firstly pre processing is done which includes normalization, stemming, stop words removal. Next the novel method is implemented to detect the semantic similarity where if there is a match for each of the index words in source and suspicious texts, it will be replaced by its related index word, the index word is chosen out of collection of synonyms. They have also used Dice similarity, Levenshtein similarity, and Longest Common Subsequence as the statistical attributes. The SVM is implemented through the RBF (Radial Base Function) kernel where the extracted sentences from the source and suspicious texts are classified using SVM.

3.4 Hybrid Approach for Plagiarism Detection Using ANN and SVM

In this approach [21] hybrid of SVM and ANN is implemented on the data collected through internet search. Firstly data is validated by experts where the data is collected from the internet with the input document and sentence filtering is also done. Then preprocessing is done which includes tokenization, stemming and stop words removal and then it is represented in VSM. Hybrid machine learning is designed where the output of VSM is the input for the different models of machine learning. For the experiments they have used different combinations of the models that are KNN and ANN, KNN and SVM, SVM and ANN and ANN and SVM. The approach has shown that the hybrid approach has better performance than the pure ANN, SVM and KNN but not in all cases. The combination of ANN and SVM shows the best performance out of all the different combinations.

3.5 High Obfuscation Detection Using Machine Learning - LR Model

In this approach [22] various features are used to detect high obfuscation cases of plagiarism. A classifier based on combination of different features is proposed which uses the Logical Regression (LR) model. It integrates the lexicon, syntactic, semantic and structural features which are extracted from the source and suspected documents to detect the high obfuscation cases. The lexical features are based on the word and character based n-grams, for the syntactic features parts of speech n-gram distance is calculated, for the semantic features the semantic similarity distance is calculated and for the structure features stop-words n-gram distance and word pair order is calculated. All the different types of features are integrated using the LR model. The experiments showed that the lexical features fusion is better than the other type of features fusion, followed by structural and syntactic features fusion. The approach is implemented on the summary obfuscation subset of the PAN@CLEF2013 dataset (Fig. 6).

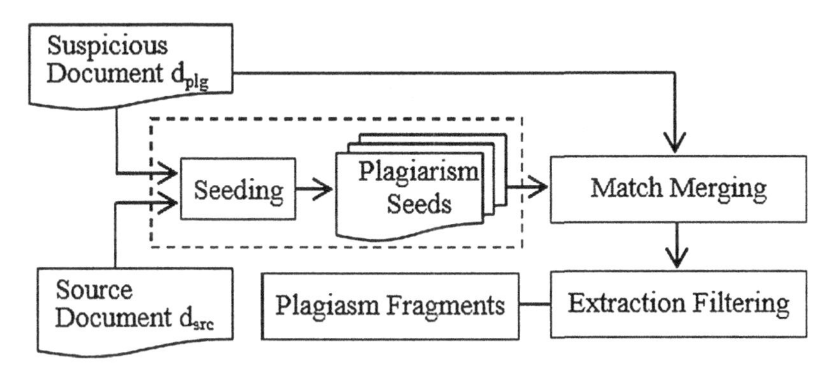

Fig. 6. Architecture for text alignment in HOPML [22]

4 Results and Efficiency Comparison

Various techniques that are discussed in this paper are implemented on different datasets: PAN-PC (2009–2016), AraPlagDet, Corpus of Plagiarized short answers, SemEval 2015 Twitter. Each of the above discussed techniques are analyzed and compared on the standard measures which are Precision, Recall, F1 score, Granularity, PlagDet Score.

$$\text{PRECISION (P)} = \frac{\text{True} - \text{Positives}}{(\text{True} - \text{Positives} + \text{False} - \text{Positives})} \tag{9}$$

$$\text{RECALL (Rec)} = \frac{\text{True} - \text{Positives}}{(\text{True} - \text{Positives} + \text{False} - \text{Negatives})} \tag{10}$$

$$\text{F1} - \text{SCORE} = \frac{2 * (\text{P} * \text{Rec})}{(\text{P} + \text{Rec})} \tag{11}$$

$$\text{GRANULARITY (G)} = \sum_{i=1}^{m} \frac{(No.\,of\ True\,Positives)}{(P + Rec)} \tag{12}$$

$$PLAGDET\ SCORE = \frac{F1 - SCORE}{Log_2(1 + G)} \tag{13}$$

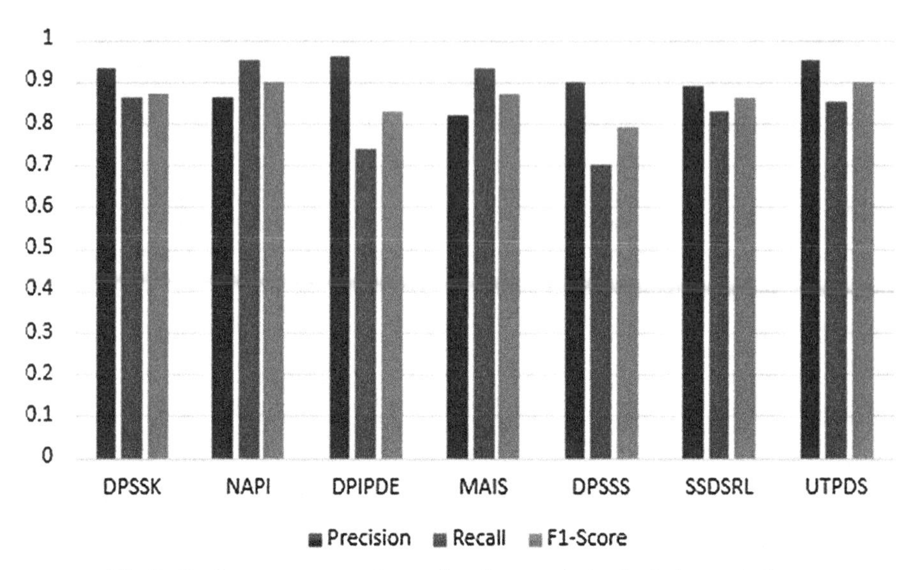

Fig. 7. Performance comparisons of various extrinsic plagiarism techniques

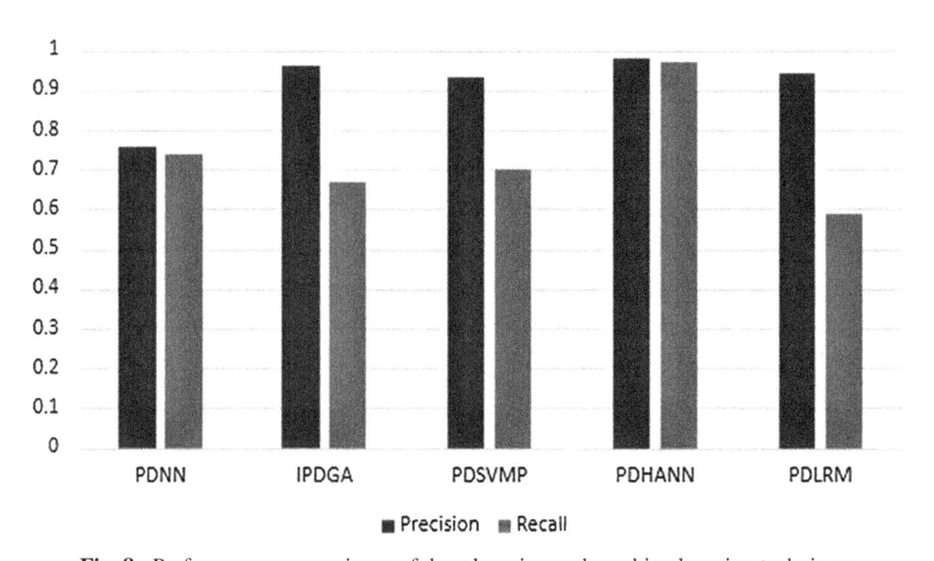

Fig. 8. Performance comparisons of deep learning and machine learning techniques

Table 1. Comparison of various extrinsic plagiarism detection systems

Name	Techniques	Dataset	Results	Pros	Cons
DPSSK [9]	Gross resemblance, Semantic and syntactic resemblance	PAN-PC-11	Precision = 0.93 Recall = 0.86 PlagDet score = 0.87 F1-score = 0.87	It can detect synonyms replacement, paraphrasing and verbatim copying	Not able to detect complex cases like translated or summarized text
NAPI [10]	Context matching and word embeddings	Plagiarized short answers	Precision = 0.86 F1-score = 0.90 Recall = 0.95	It can detect word reordering and synonyms substitution	It cannot detect ghostwriting type plagiarism
DPIPDE [11]	Depth estimation and inverse path estimation	PAN-PC-11	Precision = 0.96 PlagDet score = 0.83 Recall = 0.74 F1-score = 0.83	It can detect rewording in sentences, literal copy paste, sentence transformations	In parts of speech pairs only noun and verb are accepted
MAIS [13]	SorensenDice method and ShemNet	AraPlagDet	Precision = 0.82 Recall = 0.93 F1-score = 0.87	Can detect semantic plagiarism and also reduced the time to analyze the documents	No. of agents used affect the quality of the system and time consumption
DPSSS [14]	Semantic and word order similarity metric	PAN-PC-11	Precision = 0.90 F1-score = 0.79 Recall = 0.70 PlagDet score = 0.79	It can detect semantically similar contexts that are expressed in different wording	It cannot detect active and passive sentences
SSDSRL [15]	Semantic role labeling and weighted arguments	PAN-PC-09	Precision = 0.89 Recall = 0.83 F1-score = 0.86	It can detect verbatim, rewording, semantic plagiarism	Implemented only on a few documents of the dataset

(continued)

Table 1. (*continued*)

Name	Techniques	Dataset	Results	Pros	Cons
UTPDS [16]	Semantic role labeling, POS tagging, semantic and syntactic similarity metric	PAN-PC-10	Precision = 0.95 F1-score = 0.90 Recall = 0.85 PlagDet score = 0.90	It can extract word synsets which are used to identify the sentence pairs (plagiarized)	PlagDet reduced for Summary Obfuscation on PAN-PC-14

Table 2. Comparison of various deep learning and machine learning based extrinsic plagiarism detection systems

Name	Techniques	Dataset	Results	Pros	Cons
PDNN [17]	CNN and RNN, DeepParapharase Model	SemEval 2015 Twitter &Microsoft Paraphrase	Precision = 0.76 Recall = 0.74 F1 score = 0.75	Can be implemented on both user-generated noisy short texts and high-quality clean texts	Focused only on paraphrase detection, not on other complex types
IPDGA [19]	Sentence level concept extraction using Genetic algorithm, VSM with tf-isf	PAN-PC-13 and PAN-PC-14 on SO (Summary Obfuscation) set	Precision = 0.96 Recall = 0.67 Granularity = 1 PlagDet = 0.76	It can detect plagiarism at the document level as well as the paragraph level	Approach implemented only on summary obfuscation set, not on the entire dataset
PDSVMP [20]	SVM, Index Word Replacement	PAN 2016 (Persian)	Precision = 0.93 Recall = 0.70 PlagDet = 0.80 Granularity = 1	Novel approach 'Index Word Replacement is proposed which can detect the semantic similarity	Only Statistical attributes were used to train the SVM
PDHANN [21]	SVM, ANN, k-nearest neighbors	Internet (Search Engines)	Precision = 0.98 Recall = 0.97 Accuracy = 0.97	Data is collected from the internet search so that the documents for detection are up to date	Hybrid approach is not always better than individual machine learning techniques

(*continued*)

Table 2. (*continued*)

Name	Techniques	Dataset	Results	Pros	Cons
PDLRM [22]	Multi feature fusion method, Logical regression model	Summary-Obfuscation sub-corpus of PAN@CLEF2013	Precision = 0.94 Recall = 0.59 Granularity = 1 PlagDet = 0.72	High obfuscation cases can be detected using this approach	Recall is low for the given approach

5 Comparative Analysis of Extrinsic Plagiarism Detection Techniques vs. Machine Learning and Deep Learning Based Plagiarism Detection Techniques

There are various techniques which can be used to detect plagiarism; however each technique has its advantages and disadvantages. In the above given Table 1 and Table 2 we have discussed the pros and cons of each technique. In this paper various techniques which are discussed are novel techniques, also many techniques are a combination (hybrid) of existing techniques, which gives better results and can detect different cases of plagiarism. However results can vary depending on the dataset used. The table given below includes the key features of the plagiarism detection techniques.

Table 3. Key features of different extrinsic plagiarism detection techniques

Extrinsic plagiarism detection techniques	Deep learning and machine learning based plagiarism detection techniques
Can detect Summary, translations and Simple obfuscation cases	Better results in detecting Complex and High Obfuscation cases of Plagiarism
Can be implemented using semantic, syntactic and linguistic features	Higher accuracy and precision values (depending on the dataset used)
Can detect different cases of plagiarism like verbatim, translations and paraphrasing	Better results when implemented on larger datasets and faster computation results

6 Future Work

Future work includes, exploring different merging heuristics to improve the granularity score [23]. It also includes finding a technique that can process documents and its entire content in less time [24]. Also many techniques are implemented on smaller datasets or only on a part of the entire dataset, because these techniques show better results on smaller sets of documents. We need better plagiarism detection systems that can be implemented on larger datasets and also require less time consumption. Large databases with manual plagiarism cases is not easily available, therefore the creation of databases is required to test various plagiarism detection approaches [25]. Also deep learning methods can be used to find similarity between documents as its much more efficient gives better results [26] and it can be used to detect high obfuscation cases which many methods like VSM [27] and n grams [28] aren't able to detect. Future work also includes experimenting in detecting plagiarism in multi-languages [29]. Also the use of ultra fine grained repositories in computer programming can be done to help students dealing with difficulty in learning programming [30].

7 Conclusion

In this paper we have discussed and explored various extrinsic based plagiarism techniques. We have also compared those techniques on the standard measures which can be seen in Fig. 7 and Fig. 8. In terms of Precision DPIPDE performed better but in terms of Recall NAPI had better results. In terms of F1 score NAPI and UTPDS performed better and UTPDS also performed better in terms of PlagDet. Each technique has its advantages and disadvantages which are discussed in Table 1. For deep learning and machine learning based techniques PDHANN performed best in terms of Precision, Recall and Accuracy. PDLRM can detect high obfuscation plagiarism, PDSVMP performed better in terms of PlagDet, also the Pros and Cons of these techniques have been discussed in Table 2. With the advancement in technology, plagiarism has also increased. More and more complex cases of plagiarism are there now-a-days which weren't before and to be able to detect these complex cases we need better plagiarism detection systems that not only can detect simple copy paste plagiarism but can detect complex cases of plagiarism on semantic, syntactic and lexical level.

References

1. Foltýnek, T., Meuschke, N., Gipp, B.: Academic plagiarism detection: systematic literature review. ACM Comput. Surv. **52**(6), 42 (2019)
2. Al-Sallal, M., Iqbal, R., Palade, V., Amin, S., Chang, V.: An integrated approach for intrinsic plagiarism detection. Future Gen. Comput. Syst. **96**, 700-712 (2017)
3. Duarte, F., Caled, D., Xexéo, G.: Minmax circular sector arc for external plagiarisms heuristic retrieval stage. Knowl.-Based Syst. **137**, 1–18 (2017)
4. Brychcín, T., Svoboda, L.: UWB at SemEval-2016 task1: semantic textual similarity using lexical, syntactic,and semantic information (2016)
5. Sánchez-Vega, F., Villatoro-Tello, E., Montes-y-Gómez, M., Villaseñor-Pineda, L., Rosso, P.: Determining and characterizing the reused text for plagiarism detection. Expert Syst. Appl. **40**, 1804–1813 (2013)
6. Bouarara, H.A., Hamou, R.M., Rahman, A., Amine, A.: Machine learning tool and meta-heuristic based on genetic algorithms for plagiarism detection over mail service. In: 2014 IEEE/ACIS 13th International Conference on Computer and Information Science (ICIS), pp. 157–162 (2014)
7. Hambi, E.M., Benabbou, F.: A deep learning based technique for plagiarism detection: a comparative study. Int. J. Artif. Intell. **09**(1), 81–90 (2020)

8. Cheers, H., Lin, Y., Smith, S.P.: Academic source code plagiarism detection by measuring program behavioral similarity. IEEE Access **09**, 50391–50412 (2021)
9. Ahuja, L., Gupta, V., Kumar, R.: A new hybrid technique for detection of plagiarism from text documents. Arab. J. Sci. Eng. **45**(12), 9939–9952 (2020). https://doi.org/10.1007/s13369-020-04565-9
10. Alvi, F., Stevenson, M., Clough, P.: Paraphrase type identification for plagiarism detection using contexts and word embeddings. Int. J. Educ. Technol. High. Educ. **18**(1), 1–25 (2021). https://doi.org/10.1186/s41239-021-00277-8
11. Sahi, M., Gupta, V.: A novel technique for detecting plagiarism in documents exploiting information sources. Cogn. Comput. **9**(6), 852–867 (2017). https://doi.org/10.1007/s12559-017-9502-4
12. Miller, G.A.: WordNet: a lexical database for English. Commun. ACM **38**(11), 39–41 (1995)
13. Zouaoui, S., Rezeg, K.: Multi-agents indexing system (MAIS) for plagiarism detection. J. King Saud Univ. Comput. Inf. Sci. (2020)
14. Abdi, A., Idris, N., Alguliyev, R.M., Alguliyev, R.M.: PDLK: Plagiarism detection using linguistic knowledge. Expert Syst. Appl. (2015)
15. Osman, A.H., Salim, N., Binwahlan, M.S., Alteeb, R., Abuobieda, A.: An improved plagiarism detection scheme based on semnatic role labeling. Appl. Soft Comput. **12**(1568–4946), 1493–1502 (2012)
16. Vani, K., Gupta, D.: Unmasking text plagiarism using syntactic-semantic based natural language processing techniques: comparisons, analysis and challenges. Inf. Process. Manag. **54**, 0306–4573, 408–432 (2018)
17. Agarwal, B., Ramampiaro, H., Langseth, H., Ruocco, M.: A deep network model for paraphrase detection in short text messages. Inf. Process. Manage. **54**(6), 922–937 (2018)
18. Hunt, R. et al.: Machine learning models for paraphrase identification and its applications on plagiarism detection. In: 2019 IEEE International Conference on Big Knowledge (ICBK), pp. 97-104 (2019)
19. Vani, K., Gupta, D.: Detection of idea plagiarism using syntax - semantic concept extractions with genetic algorithm. Expert Syst. Appl. **73**, 11–26 (2017)
20. Esteki, F., Esfahani, F.S.: A plagiarism detection approach based on SVM for persian texts. In: FIRE (2016)
21. Subroto, I.M.I., Selamat, A.: Plagiarism detection through internet using hybrid artificial neural network and support vectors machine. TELKOMNIKA **12**, 209–218 (2014)
22. Kong, L., Lu, Z., Qi, H., Han, Z.: Detecting high obfuscation plagiarism: exploring multi-features fusion via machine learning. Int. J. u- e-Serv. Sci. Technol. **7**(4), 385–396 (2014)
23. Altheneyan, A.S., Menai, M.E.B.: Automatic plagiarism detection in obfuscated text. Pattern Anal. Appl. **23**(4), 1627–1650 (2020). https://doi.org/10.1007/s10044-020-00882-9
24. Umareta, C.F.O., Mariyah, S.: Fuzzy semantic-based string similarity experiments to detect plagiarism in indonesian documents. In: 2019 3rd International Conference on Informatics and Computational Sciences (ICICoS), pp. 1–6 (2019)
25. Vani, K., Gupta, D.: Text plagiarism classification using syntax based linguistic features. Expert Syst. Appl. **88**, 448–464 (2017)
26. JavadiMoghaddam, S., Roosta, F., Noroozi, A.: Weighted semantic plagiarism detection approach based on AHP decision model. Acc. Res. (2021)
27. Ekbal, A., Saha, S., Choudhary, G.: Plagiarism detection in text using vector space model. In: 2012 12th International Conference on Hybrid Intelligent Systems (HIS), pp. 366–371 (2012)
28. Wielgosz, M., Szczepka, P., Russek, P., Jamro, E., Wiatr, K.: Evaluation and implementation of n-gram-based algorithm for fast text comparison. Comput. Inform. **36**, 887–907 (2017)

29. Alfikri, Z.F., Purwarianti, A.: Detailed analysis of extrinsic plagiarism detection system using machine learning approach (Naive Bayes and SVM). TELKOMNIKA Indonesian J. Electr. Eng. **12**(11), 7884–7894 (2014)
30. Ljubovic, V., Pajic, E.: Plagiarism detection in computer programming using feature extraction from ultra-fine-grained repositories. IEEE Access **8**, 96505–96514 (2020)

Autism Detection Using Surface and Volumetric Morphometric Feature of sMRI with Machine Learning Approach

Mayank Mishra(✉) and Umesh C. Pati

Department of Electronics and Communication Engineering, National Institute of Technology Rourkela, Rourkela 769008, India
mmishra1208@gmail.com, ucpati@nitrkl.ac.in

Abstract. Brain imaging has played a very crucial role in the detection of various brain disorders. Among many brain imaging modalities, Magnetic Resonance Imaging (MRI) has proven its importance due to its detailed information regarding the insight of the brain. Autism Spectrum Disorder (ASD) has emerged as a very serious brain disorder due to its late detection among people. It comprises symptoms that are generally ignored, and this creates the urgency for its early detection. This work puts forward the method for the detection of ASD utilizing Machine Learning (ML) with the features extracted from sMRI (Structural Magnetic Resonance Imaging). Surface morphometric and volumetric morphometric features have been utilized for training the machine learning models. The cross-validation approach has been used to avoid overfitting problem occurred during training and testing steps. Machine learning models such as Random Forest (RF), Extra Trees (ET), Linear Support Vector Machine (SVM), Non - Linear SVM, and K- Nearest Neighbors (KNN) have been used for classification between ASD and controls. To evaluate the performance of classification, accuracy, precision, recall, and ROC-AUC score values have been considered.

Keywords: Autism · Brain imaging · Random Forest · Support Vector Machine · sMRI

1 Introduction

Brain imaging eases the process for the insight information of the brain with a noninvasive approach. The information can be further utilized for the detection of various brain disorders [1]. Among various brain imaging modalities, MRI holds its uniqueness for its detailed insight and soft tissue information of the brain [2]. Structural brain imaging such as sMRI (Structural Magnetic Resonance Imaging) provides the anatomical information of the brain which can be further utilized for the detection of brain disorders. Among many disorders, ASD has also affected many lives. It is recognized by its symptoms such as weak social communication, repetition in behavior, etc. [3]. Anatomical information of the brain has played a very important role in past years towards the detection of ASD [4]. The increase in the utilization of machine learning in the field of medical diagnosis

I. Woungang et al. (Eds.): ANTIC 2021, CCIS 1534, pp. 625–633, 2022.
https://doi.org/10.1007/978-3-030-96040-7_47

has simplified the purpose. Many research works have been done towards the detection of various brain disorders utilizing anatomical information of the brain with machine learning [5].

2 Related Works

Lauren E. Libero et al. have presented a comparative study of cortical surface area, volume, thickness, and gyrification index of the brain of ASD and controls. The study concludes with the alteration observed in the anatomy of the social brain region [6]. Gajendra J. Katuwal et al. have presented the machine learning approach for classification between ASD and controls. The work concludes with high accuracy of the individual site compared to a large heterogeneous dataset [7]. Gajendra J. Katuwal et al. have sub-divided the heterogeneous dataset based on autism severity, VIQ (Verbal IQ), and age. Classification performance has been improved for the subdivision process compared to the whole dataset [8].

Osman Altay et al. presented the work for the prediction of ASD using K- Nearest Neighbors (KNN) and Linear Discriminant Analysis (LDA) classifiers. The dataset utilized for training the model includes a variety of questions encountered by ASD and controls during the diagnosis process [9]. Milan N. Parikh et al. have utilized personal characteristic data for the classification between ASD and controls with machine learning models [10]. Kayleigh K. Hyde et al. presented the survey on the supervised ML approach for the detection of ASD. The survey comprises various aspects of data such as behavioral, brain imaging, developmental, genetic data, etc. [11]. Tania Akter et al. presented the machine learning approach for the early detection of ASD in which dataset has been further sub-categorized based on the age factor such as toddlers, children, adolescents, and adults. The result shows that different machine learning models perform differently when dealing with different sub-categories [12].

Shirajul Islam et al. proposed a machine learning approach for the detection of ASD at an early stage. The dataset has been sub-divided as per the medical, health, and social science criteria. The limitation of this work has been found to be model overfitting [13]. F. Catherine Tamilarasi et al. have utilized the gray level co-occurrence matrix for the feature selection from the thermal face images. Among various machine learning models, SVM performs better in their experiment [14]. Our previous work [19] has been focused on the detection of ASD using surface morphometric features with Decision Trees and Random Forest machine learning models. It also presented the comparative analysis between left and right hemispheric surface morphometric features for these models.

In this work, the machine learning approach has been presented for the detection of ASD utilizing surface morphometric features and volumetric morphometric features. Comparative analysis of Random Forest (RF), Extra Trees (ET), Linear Support Vector Machine (SVM), Non - Linear SVM, and K- Nearest Neighbors (KNN) on the final dataset which includes surface morphometric, as well as volumetric morphometric features towards the ASD detection, has been presented in the further section of this work.

3 The Proposed Work

The flow of the proposed work follows the pipelines which include the collection of raw MRI datasets, preprocessing, extraction of surface and volumetric morphometric features, combining of the features, exploratory data analysis, ML model training, and classification.

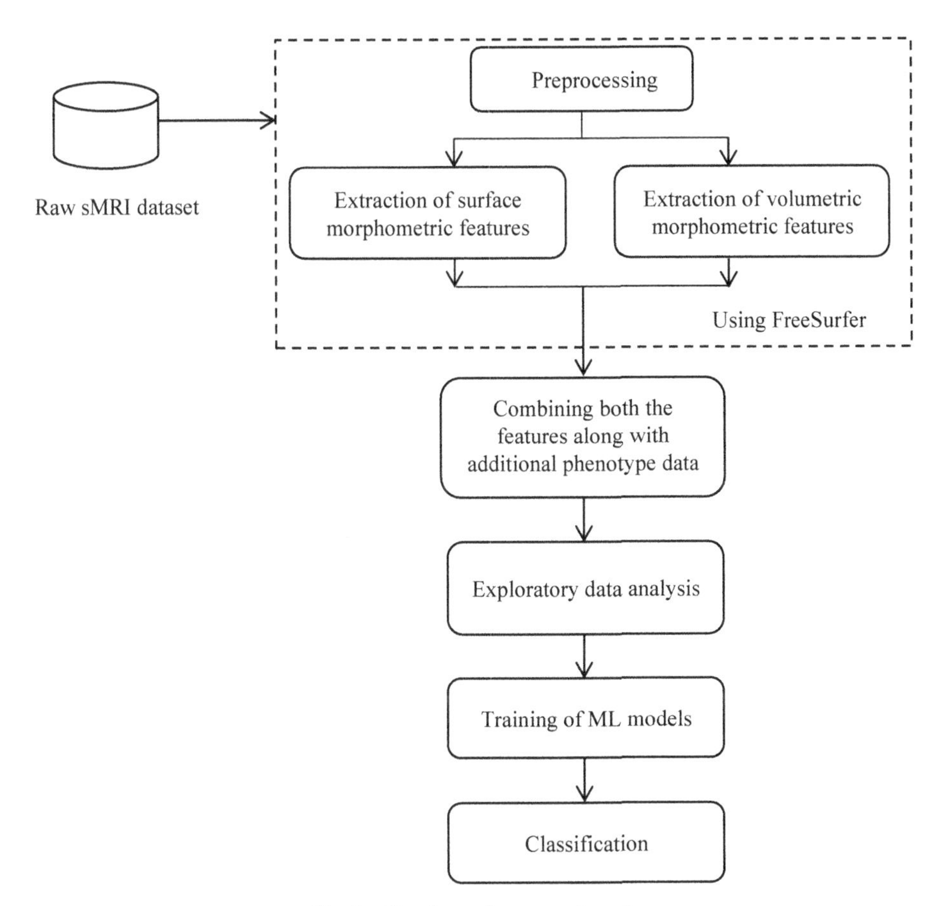

Fig. 1. The flow of proposed work.

3.1 Dataset Collection

The ABIDE-1 [15] dataset collected from COINS [16] has been utilized for the experimentation. Total 100 T1-weighted sMRI data has been taken for this work, which includes 68 ASD and 32 Controls.

3.2 Preprocessing

Once the dataset is collected it goes through the 'recon-all' pipeline of FreeSurfer [17]. It includes stripping off the skull, normalization of intensity, volumetric labeling, surface parcellation, and volumetric segmentation. The thickness of various surface regions measured using FreeSurfer [17] of Desikan-Killiany Atlas [18] has been considered for surface morphometric features as shown in Fig. 2. The volume measured from the segmented regions (such as Left-Lateral-Ventricle, Left-Caudate, Left-Putamen, Left-Hippocampus, etc.) of the sMRI of the brain using FreeSurfer [17] has been considered as volumetric morphometric features. Total 103 features which is the combination of surface as well as volumetric morphometric features have been utilized for the execution of this work. As presented in [7] the performance of the ML model has been improved by adding additional phenotype information to the dataset. The presented approach also utilizes the same approach as mentioned in [7] by combining the additional phenotype data such as age at the time of sMRI scan, VIQ, and performance IQ (PIQ) of the person in the dataset of 103 extracted features.

3.3 Exploratory Data Analysis (EDA)

Once the entire extracted feature along with additional phenotype information is combined, it goes through EDA. It includes finding the missing values in the dataset, removing the incomplete information present in the dataset and lastly finding the Pearson Correlation coefficients [20] between all the features of the dataset. While executing the experiment the threshold of 0.9 (obtained after trial and error approach) has been kept for removing highly correlated features from the dataset.

3.4 Machine Learning Models

For training purposes, Random Forest (total number of decision trees = 150), Extra Trees (total number of decision trees = 150), Linear SVM, Non-Linear SVM (degree = 2), and KNN (number of neighbors = 6) machine models have been used. All the mentioned parameters have been taken into consideration after the 'trial and error approach'.

3.5 Classification

Once the model is trained it is ready for testing purposes. Hence, the classification has been performed for the detection of ASD. To evaluate the classification performance of the models various classification evaluation parameters such as accuracy, precision, recall, and ROC-AUC score have been calculated using the following expressions:

$$\text{Accuracy} = \frac{\text{TP} + \text{TN}}{\text{TP} + \text{TN} + \text{FP} + \text{FN}} \tag{1}$$

$$\text{Precision} = \frac{\text{TP}}{\text{TP} + \text{FP}} \tag{2}$$

$$\text{Recall or True Positive Rate (TPR)} = \frac{\text{TP}}{\text{TP} + \text{FN}} \tag{3}$$

$$\text{False Positive Rate (FPR)} = \frac{\text{FP}}{\text{FP} + \text{TN}} \quad (4)$$

By putting the values of FPR on the x-axis and the values of TPR on the y-axis, ROC is plotted and then the AUC value has been calculated.

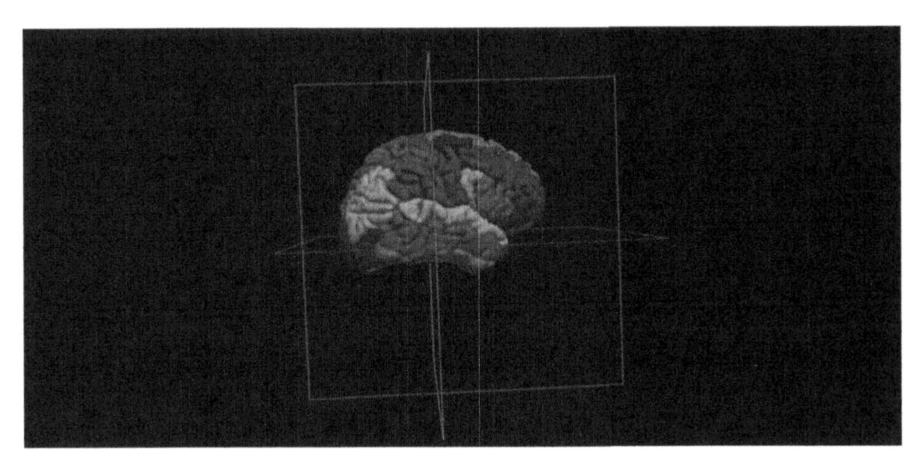

Fig. 2. Surface parcellation using Desikan-Killiany Atlas in FreeSurfer.

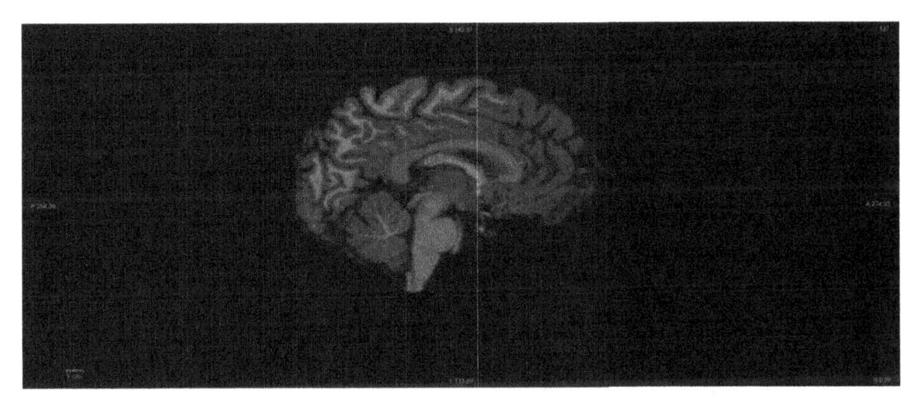

Fig. 3. Sagittal view of volumetric segmentation in FreeSurfer.

4 Results and Discussion

Instead of dividing the complete dataset into training and test data set, this approach utilizes the cross-validation method to overcome the "overfitting" problem. Dataset has been split into 'k' folds. For every iteration, 'k-1' folds are treated as a training dataset and the remaining fold has been taken as a test dataset. The number of iteration depends on the number of 'k-fold' used in cross-validation.

For each iteration, the mentioned classification evaluation parameter has been calculated. After the ‘kth’ iteration, the mean of these parameters has been calculated as shown in Tables 1, 2, 3, 4 and 5 for every ML model.

Table 1. Classification performance evaluation of Random Forest

k	Accuracy	Precision	Recall	ROC-AUC
5	0.7890	0.8397	0.8800	0.8685
10	0.8000	0.8500	0.8761	0.8700
20	0.8100	0.8733	0.8791	0.8708

Table 2. Classification performance evaluation of Extra Trees

k	Accuracy	Precision	Recall	ROC-AUC
5	0.8000	0.8599	0.8956	0.8740
10	0.8099	0.8721	0.8761	0.8761
20	0.8100	0.8775	0.8791	0.8833

Table 3. Classification performance evaluation of Linear-SVM

k	Accuracy	Precision	Recall	ROC-AUC
5	0.7000	0.7399	0.8505	0.7329
10	0.7400	0.7918	0.8666	0.7577
20	0.7300	0.8091	0.8166	0.7416

Table 4. Classification performance evaluation of Non-linear SVM

k	Accuracy	Precision	Recall	ROC-AUC
5	0.7100	0.8189	0.7318	0.7677
10	0.7100	0.8488	0.7095	0.7803
20	0.7900	0.9174	0.7750	0.8041

Table 5. Classification performance evaluation of KNN

k	Accuracy	Precision	Recall	ROC-AUC
5	0.61	0.7217	0.7098	0.5647
10	0.60	0.7063	0.6952	0.5636
20	0.61	0.7308	0.7125	0.6020

As per the results mentioned from Tables 1, 2, 3, 4 and 5, the performance of the ML models has been improved as the value of 'k' for the cross-validation increases. Experimental results show the better performance of models at k = 20.

On comparing the performance of ML models at k = 20 as shown in Fig. 4, the Extra Trees model performs superior to all the models used for the execution of the experiment. After the Extra Trees model, the Random Forest model also performs well on the various classification evaluation parameters. It has been also observed that Non-linear SVM performs superior to other ML models on the scale of 'precision'.

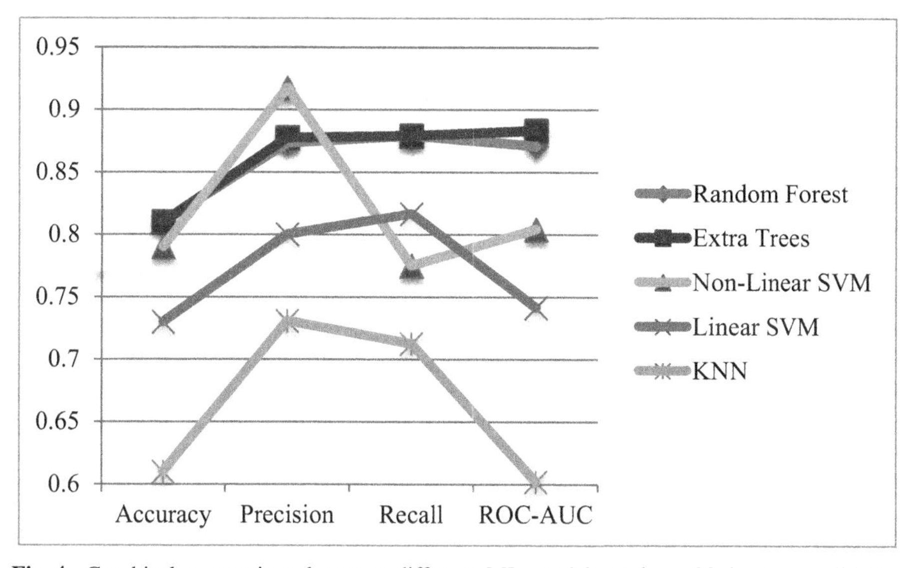

Fig. 4. Graphical comparison between different ML models at k = 20 in cross-validation approach.

5 Conclusion

The presented machine learning approach utilizes the volumetric, surface morphometric features along with additional phenotype information for the detection of ASD. To remove the overfitting problem that arises in the smaller dataset, the cross-validation approach has been utilized in the execution of the experiment. It has been observed that the performance of ML models has improved with the increase in the value of 'k' for the cross-validation. Experimental results show the better performance of Extra Tree and Random Forest. Non-Linear SVM also performs well in terms of precision. The limitation of this work is limited dataset utilization towards the execution of the approach. Hence, the results obtained cannot be taken as generalized findings. Future work will be dedicated to overcoming this limitation.

References

1. Bhatele, K., Bhadauria, S.S.: Brain structural disorders detection and classification approaches: a review. Artif. Intell. Rev. **53**, 3349–3401(2020)
2. Möllenhoff, K., Oros-Peusquens, A.M., Shah, N.J.: Introduction to the basics of magnetic resonance imaging. In: Gründer, G. (ed.) Molecular Imaging in the Clinical Neurosciences. Neuromethods, vol. 71, pp. 75–98. Humana Press, Totowa, NJ (2012)
3. Ecker, C., et al.: MRC AIMS consortium. brain surface anatomy in adults with autism: the relationship between surface area, cortical thickness, and autistic symptoms. JAMA Psychiatry **70**(1), 59–70 (2013)
4. Brašić, J.R., Mohamed, M.: Human brain imaging of autism spectrum disorders. In: Imaging of the Human Brain in Health and Disease. Academic Press (2014)
5. Mateos-Pérez, J.M., Dadar, M., Lacalle-Aurioles, M., Iturria-Medina, Y., Zeighami, Y., Evans, A.C.: Structural neuroimaging as clinical predictor: a review of machine learning applications. NeuroImage: Clin. **20**, 506–522 (2018)
6. Libero, L.E., DeRamus, T.P., Deshpande, H.D., Kana, R.K.: Surface-based morphometry of the cortical architecture of autism spectrum disorders: volume, thickness, area, and gyrification. Neuropsychologia **62**, 1–10 (2014)
7. Katuwal, G.J., Cahill, N.D., Baum, S.A., Michael, A.M.: The predictive power of structural MRI in Autism diagnosis. In: 37th Annual International Conference of the IEEE Engineering in Medicine and Biology Society (EMBC), pp. 4270–4273. Milan, Italy (2015)
8. Katuwal, G.J., Baum, S.A., Cahill, N.D., Michael, A.M.: Divide and conquer: sub-grouping of ASD improves ASD detection based on brain morphometry. PLoS ONE **11**(4), 1–24 (2016)
9. Osman, A., Mustafa, U.: Prediction of the autism spectrum disorder diagnosis with linear discriminant analysis classifier and K-nearest neighbor in children. In: 6th International Symposium on Digital Forensic and Security (ISDFS), pp. 1–4 (2018)
10. Parikh, M.N., Li, H., He, L.: Enhancing diagnosis of autism with optimized machine learning models and personal characteristic data. Front Comput. Neurosci. **13**(9), 1–5 (2019)
11. Hyde, K.K., et al.: Applications of supervised machine learning in autism spectrum disorder research: a review. Rev. J. Autism Dev. Disord. **6**(2), 128–146 (2019). https://doi.org/10.1007/s40489-019-00158-x
12. Akter, T., Satu, S., et al.: Machine learning-based models for early stage detection of autism spectrum disorders. IEEE Access **7**, 166509–166527 (2019)
13. Islam, S., Akter, T.. et al.: Autism spectrum disorder detection in toddlers for early diagnosis using machine learning. In: 2020 IEEE Asia-Pacific Conference on Computer Science and Data Engineering (CSDE), pp. 1–6. Gold Coast, Australia (2020)
14. Tamilarasi, F.C., Shanmugarn, J.: Evaluation of autism classification using machine learning techniques. In: 3rd International Conference on Smart Systems and Inventive Technology (ICSSIT), pp. 757–761 (2020)

15. Di Martino, A., Yan, C-G., Li, Q., Denio, E., Castellanos, F.X., Alaerts, K., et al.: The autism brain imaging data exchange: towards a large-scale evaluation of the intrinsic brain architecture in autism. Mol. Psychiatry **19**, 659–667 (2014)
16. Scott, A., Courtney, W., et al.: COINS: An innovative informatics and neuroimaging tool suite built for large heterogeneous datasets. Front. Neuroinformatics **5**, 33 (2011)
17. Fischl, B.: FreeSurfer. Neuroimage **62**(2), 774–781 (2012)
18. Winkler, A.M., Kochunov, P., et al.: Cortical thickness or grey matter volume? the importance of selecting the phenotype for imaging genetics studies. NeuroImage **53**(3), 1135–1146 (2010)
19. Mishra, M., Pati, U.C.: Autism spectrum disorder detection using surface morphometric feature of sMRI in machine learning. In: 8th International Conference on Smart Computing and Communications (ICSCC), pp. 17–20 (2021)
20. Zhou, H., Deng, Z., et al.: A new sampling method in particle filter based on Pearson correlation coefficient. Neuro. Comput. **216**, 208–215 (2016)

Sign Language Recognition System Using TensorFlow Object Detection API

Sharvani Srivastava, Amisha Gangwar, Richa Mishra, and Sudhakar Singh(✉)

Department of Electronics and Communication, University of Allahabad, Prayagraj, India
{richa_mishra,sudhakar}@allduniv.ac.in

Abstract. Communication is defined as the act of sharing or exchanging information, ideas or feelings. To establish communication between two people, both of them are required to have knowledge and understanding of a common language. But in the case of deaf and dumb people, the means of communication are different. Deaf is the inability to hear and dumb is the inability to speak. They communicate using sign language among themselves and with normal people but normal people do not take seriously the importance of sign language. Not everyone possesses the knowledge and understanding of sign language which makes communication difficult between a normal person and a deaf and dumb person. To overcome this barrier, one can build a model based on machine learning. A model can be trained to recognize different gestures of sign language and translate them into English. This will help a lot of people in communicating and conversing with deaf and dumb people. The existing Indian Sing Language Recognition systems are designed using machine learning algorithms with single and double-handed gestures but they are not real-time. In this paper, we propose a method to create an Indian Sign Language dataset using a webcam and then using transfer learning, train a TensorFlow model to create a real-time Sign Language Recognition system. The system achieves a good level of accuracy even with a limited size dataset.

Keywords: Sign Language Recognition (SLR) · Computer vision · Machine learning · Indian Sign Language

1 Introduction

Communication can be defined as the act of transferring information from one place, person, or group to another. It consists of three components: the speaker, the message that is being communicated, and the listener. It can be considered successful only when whatever message the speaker is trying to convey is received and understood by the listener. It can be divided into different categories as follows [1]: formal and informal communication, oral (face-to-face and distance) and written communication, non-verbal, grapevine, feedback, and visual communication, and the active listening. The formal communication (official communication) is steered through the channels that are pre-determined. The unofficial or grapevine communication is the spontaneous communication between

I. Woungang et al. (Eds.): ANTIC 2021, CCIS 1534, pp. 634–646, 2022.
https://doi.org/10.1007/978-3-030-96040-7_48

individuals in one's profession that does not have any formal protocol or structure. The oral communication (face-to-face and distance) is the communication in which words are exchanged between people who are present in front or at a distance (with the help of technology including voice and video calls, webinars, etc.). The written communication is the communication in which letters, emails, notices, or any other written form is used for communicating. The non-verbal communication is the communication that uses gestures, facial expressions, body language, etc. The feedback communication happens when a person gives feedback on some product or service provided by an individual or a company. The visual communication occurs when a person gets information from a visual source like televisions, social networking, or any other source. Active listening is when a person listens to and understands what the other individual is trying to convey so that the communication becomes more meaningful and effective [1].

Non-verbal communication helps deaf and dumb people to communicate amongst themselves and with others. Deaf is a disability that impairs a person's hearing ability and makes them incapable to hear while dumb is a disability that impairs the speaking ability and makes them incapable to speak. Not being able to speak or listen makes it difficult to establish communication with others. This is where sign languages come into the role, it enables a person to communicate without words. But a problem still exists, not many people possess the knowledge of sign language. Deaf and dumb may be able to communicate amongst themselves using sign languages but it is still difficult for them to communicate with people having normal hearing and vice-versa due to the lack of knowledge of sign languages. This issue can be resolved by the use of a technology-driven solution. By using such a solution, one can easily translate the gestures of sign language into the commonly spoken language, English.

A lot of research has been done in this field and there is still a need for further research. For gesture translation, data gloves, motion capturing systems, or sensors have been used [2]. Vision-based SLR systems have also been developed previously [3]. The existing Indian Sign Language Recognition system was developed using machine learning algorithms with MATLAB [4]. Authors have worked on single-handed and double-handed gestures. They used two algorithms to train their system, K Nearest Neighbours Algorithm and Back Propagation Algorithm. Their system achieved 93–96% accuracy. Though being highly accurate, it is not a real-time SLR system. The objective of this paper is to develop a real-time SLR system using TensorFlow object detection API and train it using a dataset that will be created using a webcam.

The rest of this paper after the introduction is organized as follows. Section 2 presents the related work on the SLR system. Section 3 describes the data acquisition and generation. Section 4 focuses on the methodology of the developed system. Section 5 presents the experimental evaluation of the system, and finally, Sect. 6 concludes the paper with future work.

2 Related Work

Sign languages are defined as an organized collection of hand gestures having specific meanings which are employed from the hearing impaired people to communicate in everyday life [3]. Being visual languages, they use the movements of hands, face, and body as communication mediums. There are over 300 different sign languages available all around the world [5]. Though there are so many different sign languages, the percentage of population knowing any of them is low which makes it difficult for the specially-abled people to communicate freely with everyone. SLR provides a means to communicate in sign language without knowing it. It recognizes a gesture and translates it into a commonly spoken language like English.

SLR is a very vast topic for research where a lot of work has been done but still various things need to be addressed. The machine learning techniques allow the electronic systems to take decisions based on experience i.e. data. The classification algorithms need two datasets – training dataset and testing dataset. The training set provides experiences to the classifier and the model is tested using the testing set [6]. Many authors have developed efficient data acquisition and classification methods [3, 7]. Based on data acquisition method, previous work can be categorized into two approaches: the direct measurement methods and the vision-based approaches [3]. The direct measurement methods are based on motion data gloves, motion capturing systems, or sensors. The motion data extracted can supply accurate tracking of fingers, hands, and other body parts which leads to robust SLR methodologies development. The vision-based SLR approaches rely on the extraction of discriminative spatial and temporal from RGB images. Most of the vision-based methods initially try to track and extract the hand regions before their classification to gestures [3]. Hand detection is achieved by semantic segmentation and skin colour detection as the skin colour is usually distinguishable easily [8, 9]. Though, because the other body parts like face and arms can be mistakenly recognized as hands, so, the recent hand detection methods also use the face detection and subtraction, and background subtraction to recognize only the moving parts in a scene [10, 11]. To attain accurate and robust hands tracking, particularly in cases of obstructions, authors employed filtering techniques, for example, Kalman and particle filters [10, 12].

For data acquisition by either the direct measurement or the vision-based approaches, different devices need to be used. The primary device employed as input process in SLR system is camera [13]. There are other devices available that are used for input such as Microsoft Kinect which provides colour video stream and depth video stream all together. The depth data helps in background segmentation. Apart from the devices, other methods used for acquiring data are accelerometer and sensory gloves. Another system that is used for data acquisition is Leap Motion Controller (LMC) [14, 15] – it is a touchless controller developed by technology company "Leap Motion" now called "Ultraleap" based in San Francisco. Approximately, it can operate around 200 frames per second and can detect and track the hands, fingers, and objects that look alike fingers. Most of the researchers collect their training dataset by recording it from their signer as finding a sign language dataset is a problem [2].

Different processing methods have been used for creating an SLR system [16–18]. Hidden Markov Model (HMM) has been widely used in SLR [12]. The various HMM that have been used are Multi Stream HMM (MSHMM) which is based on the two standard single-stream HMMs, Light-HMM, and Tied-Mixture Density-HMM [2]. The other processing models that have been used are neural network [19–23], ANN [24],

Naïve Bayes Classifier (NBC), and Multilayer Perceptron (MLP) [14], unsupervised neural network Self-Organizing Map (SOM) [25], Self-Organizing Feature Map (SOFM), Simple Recurrent Network (SRN) [26], Support Vector Machine (SVM) [27], 3D convolutional residual network [28]. Researchers have also used self-designed methods like the wavelet-based method [29] and Eigen Value Euclidean Distance [30].

The use of different processing methods or application systems has given different accuracy results. The Light-HMM gave 83.6% accuracy result, the MSHMM gave 86.7% accuracy result, SVM gave 97.5% accuracy result, Eigen Value gave 97% accuracy result, Wavelet Family gave 100% accuracy result [2, 22, 31, 32]. Though different models have given high accuracy results, but the accuracy does not depend only on the processing model used, it depends upon various factors such as size of the dataset, clarity of images of the dataset depending upon data acquisition methods, devices used, etc.

There are two types of SLR systems – isolated SLR and continuous SLR. In isolated SLR, the system is trained to recognize a single gesture. Each image is labelled to represent an alphabet, a digit, or some special gesture. Continuous SLR is different from isolated gesture classification. In continuous SLR, the system is able to recognize and translate whole sentences instead of a single gesture [33, 34].

Even with all the research that has been done in SLR, many inadequacies need to be dealt with by further research. Some of the issues and challenges that need to be worked on are as follows [2, 4, 6, 33].

- Isolated SLR methods need to do strenuous labeling for each word.
- Continuous SLR methods make use of isolated SLR systems as building blocks with temporal segmentation as pre-processing, which is non-trivial and unescapably proliferates errors into subsequent steps, and sentence synthesis as post-processing.
- Devices needed for data acquisition are costly, a cheap method is needed for SLR systems to be commercialized.
- Web camera is an alternative to higher specification camera but the image is blurred so, the quality is compromised.
- Data acquisition by sensors also has some issues e.g., noise, bad human manipulation, bad ground connection, etc.
- Vision-based methodologies introduce inaccuracies due to overlapping of hand and finger.
- Large datasets are not available.
- There are misconceptions about sign languages like sign language is same around the world, while sign language is based upon the spoken language.
- Indian Sign Language is communicated using hand gestures made by a single hand and double hands due to which there are two types of gestures representing the same thing.

In this paper, the dataset that will be used is created using Python and OpenCV with the help of a webcam. The SLR system that is being developed is a real-time detection system.

3 Data Acquisition

A real-time sign language detection system is being developed for Indian Sign Language. For data acquisition, images are captured by webcam using Python and OpenCV. OpenCV provides functions which are primarily aimed at the real-time computer vision. It accelerates the use of machine perception in commercial products and provides a common infrastructure for the computer vision-based applications. The OpenCV library has more than 2500 efficient computer vision and machine learning algorithms which can be used for face detection and recognition, object identification, classification of human actions, tracking camera and object movements, extracting 3D object models, and many more [35].

The created dataset is made up of signs representing alphabets in Indian Sign Language [36] as shown in Fig. 1. For every alphabet, 25 images are captured to make the dataset. The images are captured in every 2 s providing time to record gesture with a bit of difference every time and a break of five seconds are given between two individual signs, i.e., to change the sign of one alphabet to the sign of a different alphabet, five seconds interval is provided. The captured images are stored in their respective folder.

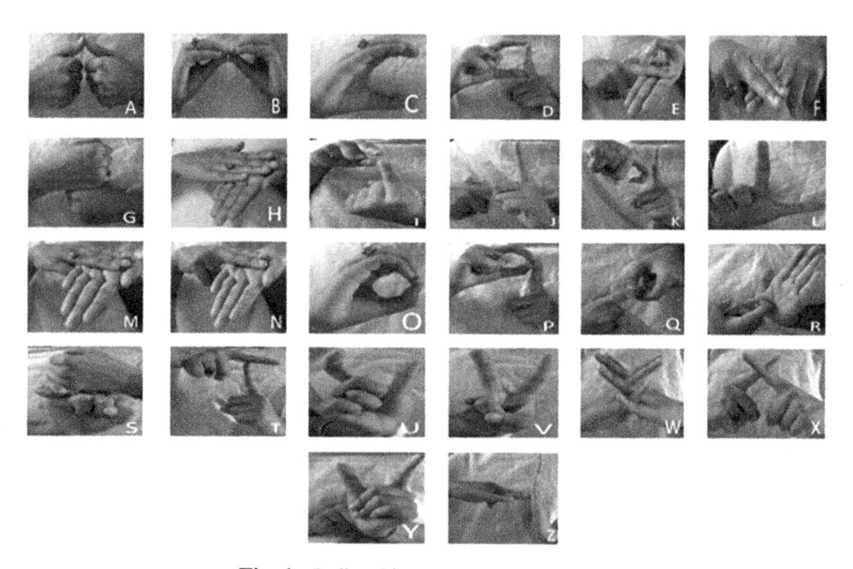

Fig. 1. Indian Sign Language alphabets

For data acquisition, dependencies like cv2, i.e., OpenCV, os, time, and uuid have been imported. The dependency os is used to help work with file paths. It comes under standard utility modules of Python and provides functions for interacting with the operating systems. With the help of the time module in Python, time can be represented in multiple ways in code like objects, numbers, and strings. Apart from representing time, it can be used to measure code efficiency or wait during code execution. Here, it is used to add breaks between the image capturing in order to provide time for hand movements. The uuid library is used in naming the image files. It helps in the generation of random objects of 128 bits as ids providing uniqueness as the ids are generated on the basis of time and computer hardware.

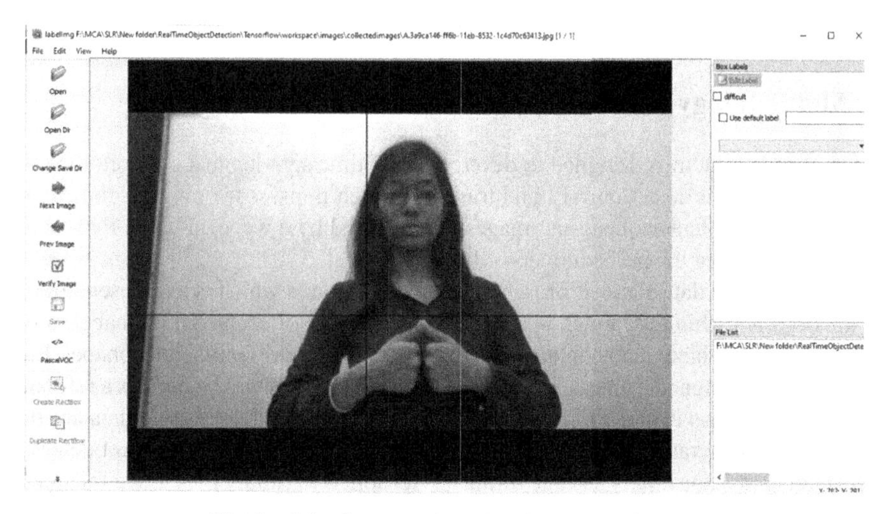

Fig. 2. Selecting a portion of the image to label it

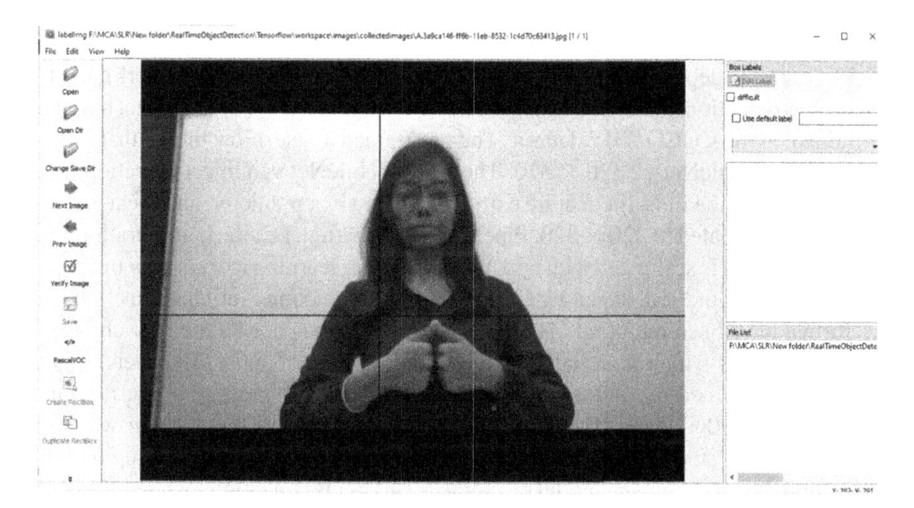

Fig. 3. Labelling the selected portion

Once all the images have been captured, they are then one by one labelled using the LabelImg package. LabelImg is a free open-source tool for graphically labelling images. The hand gesture portion of the image is labelled by what the gesture in the box or the sign represents as shown in Fig. 2 and Fig. 3. On saving the labelled image, its XML file is created. The XML files have all the details of the images including the detail of the labelled portion. After labelling all the images, their XML files are available. This is used for creating the TF (TensorFlow) records. All the images along with their XML files are then divided into training data and validation data in the ratio of 80:20. From 25 images of an alphabet, 20 (80%) of them were taken and stored as a training dataset and the remaining 5 (20%) were taken and stored as validation dataset. This task was performed for all the images of all 26 alphabets.

4 Methodology

The proposed system is designed to develop a real-time sign language detector using a TensorFlow object detection API and train it through transfer learning for the created dataset [37]. For data acquisition, images are captured by a webcam using Python and OpenCV following the procedure described under Sect. 3.

Following the data acquisition, a labeled map is created which is a representation of all the objects within the model, i.e., it contains the label of each sign (alphabet) along with their id. The label map contains 26 labels, each one representing an alphabet. Each label has been assigned a unique id ranging from 1 to 26. This will be used as a reference to look up the class name. TF records of the training data and the testing data are then created using generate_tfrecord which is used to train the TensorFlow object detection API. TF record is the binary storage format of TensorFlow. Binary files usage for storage of the data significantly impacts the performance of the import pipeline consequently, the training time of the model. It takes less space on a disk, copies fast, and can efficiently be read from the disk.

The open-source framework, TensorFlow object detection API makes it easy to develop, train and deploy an object detection model. They have their framework called the TensorFlow detection model zoo which offers various models for detection that have been pre-trained on the COCO 2017 dataset. The pre-trained TensorFlow model that is being used is SSD MobileNet v2 320 × 320. The SSD MobileNet v2 Object detection model is combined with the FPN-lite feature extractor, shared box predictor, and focal loss with training images scaled to 320 × 320. Pipeline configuration, i.e., the configuration of the pre-trained model is set up and then updated for transfer learning to train it by the created dataset. For configuration, dependencies like TensorFlow, config_util, pipeline_pb2, and text_format have been imported. The major update that has been done is to change the number of classes which is initially 90 to 26, the number of signs (alphabets) that the model will be trained on. After setting up and updating the configuration, the model was trained in 10000 steps. The hyper-parameter used during the training was to set up the number of steps in which the model will be trained which was set up to 10000 steps. During the training, the model has some losses as classification loss, regularization loss, and localization loss. The localization loss is mismatched between the predicted bounding box correction and the true values. The formula of the localization loss [38] is given in Eq. (1)–(5).

$$L_{loc}(\mathrm{x}, \mathrm{l}, \mathrm{g}) = \sum_{i \in Pos}^{N} \sum_{m \in \{cx, cy, w, h\}} x_{ij}^{k}\, smooth_{L1}\left(l_i^m - \hat{g}_j^m\right) \tag{1}$$

$$\hat{g}_j^{cx} = (g_j^{cx} - d_i^{cx})/d_i^w \tag{2}$$

$$\hat{g}_j^{cy} = (g_j^{cy} - d_i^{cy})/d_i^h \tag{3}$$

$$\hat{g}_j^w = \log(g_j^w / d_i^w) \tag{4}$$

$$\hat{g}_j^h = \log(g_j^h / d_i^h) \tag{5}$$

where, N is the number of the matched default boxes, l is the predicted bounding box, g is the ground truth bounding box, $\hat{g}$ is the encoded ground truth bounding box and x_{ij}^k is the matching indicator between default box i and ground truth box j of category k.

The classification loss is defined as the softmax loss over multiple classes. The formula of the classification loss [38] is as Eq. (6).

$$L_{conf}(x, c) = -\sum_{i \in Pos}^{N} x_{ij}^p \log\left(\hat{c}_i^p\right) - \sum_{i \in Neg} \log(\hat{c}_i^0) \tag{6}$$

where, $\hat{c}_i^p = \exp\left(c_i^p\right) / \sum_p \exp(c_i^p)$ is the softmax activated class score for default box i with category p, x_{ij}^p is the matching indicator between default box i and the ground truth box j of category p.

The different losses incurred during the experimentation are mentioned in the subsequent section. After training, the model is loaded from the latest checkpoint which makes it ready for real-time detection. After setting up and updating the configuration,

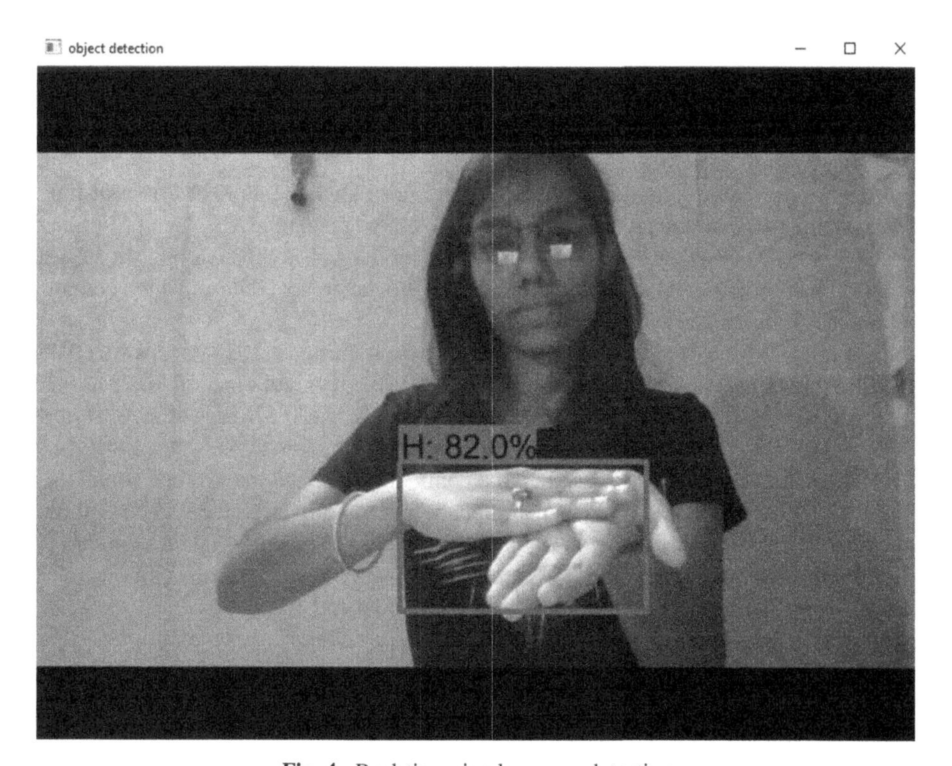

Fig. 4. Real-time sign language detection

the model will be ready for training. The trained model is loaded from the latest checkpoint which is created during the training of the model. This completes the model making it ready for real-time sign language detection.

The real-time detection is done using OpenCV and webcam again. For, real-time detection, cv2, and NumPy dependencies are used. The system detects signs in real-time and translates what each gesture means into English as shown in Fig. 5. The system is tested in real-time by creating and showing it different signs. The confidence rate of each sign (alphabet), i.e., how confident the system is in recognizing a sign (alphabet) is checked, noted, and tabulated for the result.

5 Experimental Evaluation

5.1 Dataset and Experimental Setup

The dataset is created for Indian Sign Language where signs are alphabets of the English language. The dataset is created following the data acquisition method described in Sect. 3.

The experimentation was carried out on a system with an Intel i5 7th generation 2.70 GHz processor, 8 GB memory and webcam (HP TrueVision HD camera with 0.31 MP and 640 × 480 resolution), running Windows 10 operating system. The programming environment includes Python (version 3.7.3), Jupyter Notebook, OpenCV (version 4.2.0), TensorFlow Object Detection API.

5.2 Results and Discussion

The developed system is able to detect Indian Sign Language alphabets in real-time. The system has been created using TensorFlow object detection API. The pre-trained model that has been taken from the TensorFlow model zoo is SSD MobileNet v2 320 × 320. It has been trained using transfer learning on the created dataset which contains 650 images in total, 25 images for each alphabet.

The total loss incurred during the last part of the training, at 10,000 steps was 0.25, localization loss was 0.18, classification loss was 0.13, and regularization loss was 0.10 as shown in Fig. 4. Figure 4 also shows that the lowest lost 0.17 was suffered at steps 9900.

The result of the system is based on the confidence rate and the average confidence rate of the system is 85.45%. For each alphabet, the confidence rate is recorded and tabulated in the result as shown in Table 1. The confidence rate of the system can be increased by increasing the size of the dataset which will boost up the recognition ability of the system. Thus, improving the result of the system and enhancing it.

The state-of-the-art method of the Indian Sign Language Recognition system achieved 93–96% accuracy [4]. Though being highly accurate, it is not a real-time SLR system. This issue is dealt with in this paper. In spite of the dataset being small, our system has achieved an average confidence rate of 85.45%.

```
INFO:tensorflow:Step 9800 per-step time 1.665s
I0828 23:17:04.872565 19444 model_lib_v2.py:700] Step 9800 per-step time 1.665s
INFO:tensorflow:{'Loss/classification_loss': 0.06653614,
 'Loss/localization_loss': 0.014709826,
 'Loss/regularization_loss': 0.10198762,
 'Loss/total_loss': 0.18323359,
 'learning_rate': 0.07380057}
I0828 23:17:04.872565 19444 model_lib_v2.py:701] {'Loss/classification_loss': 0.06653614,
 'Loss/localization_loss': 0.014709826,
 'Loss/regularization_loss': 0.10198762,
 'Loss/total_loss': 0.18323359,
 'learning_rate': 0.07380057}
INFO:tensorflow:Step 9900 per-step time 1.642s
I0828 23:19:49.079613 19444 model_lib_v2.py:700] Step 9900 per-step time 1.642s
INFO:tensorflow:{'Loss/classification_loss': 0.05087668,
 'Loss/localization_loss': 0.014473952,
 'Loss/regularization_loss': 0.10147699,
 'Loss/total_loss': 0.16682762,
 'learning_rate': 0.073662736}
I0828 23:19:49.095237 19444 model_lib_v2.py:701] {'Loss/classification_loss': 0.05087668,
 'Loss/localization_loss': 0.014473952,
 'Loss/regularization_loss': 0.10147699,
 'Loss/total_loss': 0.16682762,
 'learning_rate': 0.073662736}
INFO:tensorflow:Step 10000 per-step time 1.649s
I0828 23:22:34.039046 19444 model_lib_v2.py:700] Step 10000 per-step time 1.649s
INFO:tensorflow:{'Loss/classification_loss': 0.12946561,
 'Loss/localization_loss': 0.01821224,
 'Loss/regularization_loss': 0.1009706,
 'Loss/total_loss': 0.24864845,
 'learning_rate': 0.07352352}
I0828 23:22:34.039046 19444 model_lib_v2.py:701] {'Loss/classification_loss': 0.12946561,
 'Loss/localization_loss': 0.01821224,
 'Loss/regularization_loss': 0.1009706,
 'Loss/total_loss': 0.24864845,
 'learning_rate': 0.07352352}
```

Fig. 5. Loss incurred at different steps

Table 1. Confidence rate of each alphabet

A	B	C	D	E	F	G	H	I
94%	98%	90%	90%	70%	96%	73%	97%	95%
J	**K**	**L**	**M**	**N**	**O**	**P**	**Q**	**R**
57%	87%	93%	91%	55%	78%	95%	95%	83%
S	**T**	**U**	**V**	**W**	**X**	**Y**	**Z**	
86%	81%	87%	86%	87%	88%	90%	80%	

6 Conclusion and Future Works

Sign languages are kinds of visual languages that employ movements of hands, body, and facial expression as a means of communication. Sign languages are important for specially-abled people to have a means of communication. Through it, they can communicate and express and share their feelings with others. The drawback is that not everyone possesses the knowledge of sign languages which limits communication. This limitation can be overcome by the use of automated Sign Language Recognition systems which will be able to easily translate the sign language gestures into commonly spoken language. In this paper, it has been done by TensorFlow object detection API.

The system has been trained on the Indian Sign Language alphabet dataset. The system detects sign language in real-time. For data acquisition, images have been captured by a webcam using Python and OpenCV which makes the cost cheaper. The developed system is showing an average confidence rate of 85.45%. Though the system has achieved a high average confidence rate, the dataset it has been trained on is small in size and limited.

In the future, the dataset can be enlarged so that the system can recognize more gestures. The TensorFlow model that has been used can be interchanged with another model as well. The system can be implemented for different sign languages by changing the dataset.

References

1. Kapur, R.: The types of communication. MIJ. **6** (2020)
2. Suharjito, S., Anderson, R., Wiryana, F., Ariesta, M.C., Kusuma, G.P.: Sign language recognition application systems for deaf-mute people: a review based on input-process-output. Procedia Comput. Sci. **116**, 441–448 (2017). https://doi.org/10.1016/J.PROCS.2017.10.028
3. Konstantinidis, D., Dimitropoulos, K., Daras, P.: Sign language recognition based on hand and body skeletal data. In: 3DTV-Conference, pp. 1–4 (2018). https://doi.org/10.1109/3DTV.2018.8478467
4. Dutta, K.K., Bellary, S.A.S.: Machine Learning techniques for Indian sign language recognition. In: International Conference Current Trends Computing Electric Electronics Communication CTCEEC 2017, pp. 333–336 (2018). https://doi.org/10.1109/CTCEEC.2017.8454988
5. Bragg, D., et al.: Sign language recognition, generation, and translation: an interdisciplinary perspective. In: 21st International ACM SIGACCESS Conference on Computer Accessibility,12, pp.16–31 (2019). https://doi.org/10.1145/3308561
6. Rosero-Montalvo, P.D., et al.: Sign language recognition based on intelligent glove using machine learning techniques. In: 2018 IEEE Third Ecuador Technical Chapters Meeting (ETCM), pp. 1–5 (2018). https://doi.org/10.1109/ETCM.2018.8580268
7. Zheng, L., Liang, B., Jiang, A.: Recent advances of deep learning for sign language recognition. In: DICTA 2017 - 2017 International Conference on Digital Image Computing Techniques and Applications 2017-Decemember, pp. 1–7 (2017). https://doi.org/10.1109/DICTA.2017.8227483
8. Rautaray, S.S.: A real time hand tracking system for interactive applications. Int. J. Comput. Appl. **18**, 975–8887 (2011)
9. Zhang, Z., Huang, F.: Hand tracking algorithm based on super-pixels feature. In: Proceedings - 2013 International Conference on Information Science and Cloud Computing Companion, ISCC-C 2013, pp. 629–634 (2014). https://doi.org/10.1109/ISCC-C.2013.77
10. Lim, K.M., Tan, A.W.C., Tan, S.C.: A feature covariance matrix with serial particle filter for isolated sign language recognition. Expert Syst. Appl. **54**, 208–218 (2016). https://doi.org/10.1016/J.ESWA.2016.01.047
11. Lim, K.M., Tan, A.W.C., Tan, S.C.: Block-based histogram of optical flow for isolated sign language recognition. J. Vis. Commun. Image Represent. **40**, 538–545 (2016). https://doi.org/10.1016/J.JVCIR.2016.07.020
12. Gaus, Y.F.A., Wong, F.: Hidden Markov Model - Based gesture recognition with overlapping hand-head/hand-hand estimated using Kalman Filter. In: Proceedings of - 3rd International Conference on Intelligent Systems Modelling and Simulation, ISMS 2012, pp. 262–267 (2012). https://doi.org/10.1109/ISMS.2012.67

13. Nikam, A.S., Ambekar, A.G.: Sign language recognition using image based hand gesture recognition techniques. In: Proceedings of 2016 Online International Conference Green Engineering Technologies IC-GET 2016, pp. 1–5 (2017). https://doi.org/10.1109/GET.2016.7916786
14. Mohandes, M., Aliyu, S., Deriche, M.: Arabic sign language recognition using the leap motion controller. In: IEEE International Symposium Industrial Electronics, pp. 960–965 (2014). https://doi.org/10.1109/ISIE.2014.6864742
15. Enikeev, D.G., Mustafina, S.A.: Sign language recognition through leap motion controller and input prediction algorithm. J. Phys. Conf. Ser. **1715**, 012008 (2021). https://doi.org/10.1088/1742-6596/1715/1/012008
16. Cheok, M.J., Omar, Z., Jaward, M.H.: A review of hand gesture and sign language recognition techniques. Int. J. Mach. Learn. Cybern. **10**(1), 131–153 (2017). https://doi.org/10.1007/s13042-017-0705-5
17. Wadhawan, A., Kumar, P.: Sign language recognition systems: a decade systematic literature review. Arch. Comput. Meth. Eng. **28**(3), 785–813 (2019). https://doi.org/10.1007/s11831-019-09384-2
18. Camgöz, N.C., Koller, O., Hadfield, S., Bowden, R.: Sign language transformers: Joint end-to-end sign language recognition and translation. In: Proceedings of IEEE Computer Social Conference Computer Vision Pattern Recognition, pp. 10020–10030 (2020). https://doi.org/10.1109/CVPR42600.2020.01004
19. Cui, R., Liu, H., Zhang, C.: A deep neural framework for continuous sign language recognition by iterative training. IEEE Trans. Multimed. **21**, 1880–1891 (2019). https://doi.org/10.1109/TMM.2018.2889563
20. Bantupalli, K., Xie, Y.: American sign language recognition using deep learning and 13 computer vision. In: Proceedings of - 2018 IEEE International Conference on Big Data, Big Data 2018, pp. 4896–4899 (2019). https://doi.org/10.1109/BIGDATA.2018.8622141
21. Hore, S., et al.: Indian sign language recognition using optimized neural networks. In: Balas, V.E., Jain, L.C., Zhao, X. (eds.) Information Technology and Intelligent Transportation Systems. AISC, vol. 455, pp. 553–563. Springer, Cham (2017). https://doi.org/10.1007/978-3-319-38771-0_54
22. Kumar, P., Roy, P.P., Dogra, D.P.: Independent Bayesian classifier combination based sign language recognition using facial expression. Inf. Sci. (Ny) **428**, 30–48 (2018). https://doi.org/10.1016/J.INS.2017.10.046
23. Sharma, A., Sharma, N., Saxena, Y., Singh, A., Sadhya, D.: Benchmarking deep neural network approaches for Indian sign language recognition. Neural Comput. Appl. **33**(12), 6685–6696 (2020). https://doi.org/10.1007/s00521-020-05448-8
24. Kishore, P.V.V., Prasad, M.V.D., Prasad, C.R., Rahul, R.: 4-camera model for sign language recognition using elliptical fourier descriptors and ANN. In: International Conference on Signal Processing Communication Engineering System – Proceedings SPACES 2015, Association with IEEE, pp. 34–38 (2015). https://doi.org/10.1109/SPACES.2015.7058288
25. Tewari, D., Srivastava, S.K.: A Visual recognition of static hand gestures in indian sign language based on kohonen self-organizing map algorithm. Int. J. Eng. Adv. Technol. **2**(2), 165–170 (2012)
26. Gao, W., Fang, G., Zhao, D., Chen, Y.: A Chinese sign language recognition system based on SOFM/SRN/HMM. Pattern Recognit. **37**, 2389–2402 (2004). https://doi.org/10.1016/J.PATCOG.2004.04.008
27. Quocthang, P., Dung, N.D., Thuy, N.T.: A comparison of SimpSVM and RVM for sign language recognition. In: ACM International Conference Proceeding Series, pp. 98–104 (2017). https://doi.org/10.1145/3036290.3036322

28. Pu, J., Zhou, W., Li, H.: Iterative alignment network for continuous sign language recognition. In: IEEE Computer Social Conference Computing Vision Pattern Recognition, pp. 4160–4169 (2019). https://doi.org/10.1109/CVPR.2019.00429
29. Kalsh, E.A., Garewal, N.S.: Sign language recognition system. Int. J. Comput. Eng. Res. **3**(6), 15–21 (2013)
30. Singha, J., Das, K.: Indian sign language recognition using eigen value weighted Euclidean distance based classification technique. Int. J. Adv. Comput. Sci. Appl. **4**(2), 188–195 (2013). https://doi.org/10.14569/IJACSA.2013.040228
31. Liang, Z., Liao, S., Hu, B.: 3D Convolutional neural networks for dynamic sign language recognition. Comput. J. **61**, 1724–1736 (2018). https://doi.org/10.1093/COMJNL/BXY049
32. Pigou, L., Van Herreweghe, M., Dambre, J.: Gesture and sign language recognition with temporal residual networks. In: Proceedings of - 2017 Proceedings of the IEEE International Conference on Computer Vision Workshops. ICCVW 2017. 2018-January, pp. 3086–3093 (2017). https://doi.org/10.1109/ICCVW.2017.365
33. Huang, J., Zhou, W., Zhang, Q., Li, H., Li, W.: Video-based sign language recognition without temporal segmentation. In: Thirty-Second AAAI Conference on Artificial Intelligence (2018)
34. Cui, R., Liu, H., Zhang, C.: Recurrent convolutional neural networks for continuous sign language recognition by staged optimization. In: Proceedings of the IEEE Conference on Computer Vision and Pattern Recognition, CVPR 2017. 2017-January, pp. 1610–1618 (2017). https://doi.org/10.1109/CVPR.2017.175
35. About - OpenCV. 14
36. Poster of the Manual Alphabet in ISL | Indian Sign Language Research and Training Center (ISLRTC), Government of India
37. Transfer learning and fine-tuning | TensorFlow Core
38. Wu, S., Yang, J., Wang, X., Li, X.: IoU-balanced loss functions for single-stage object detection (2020)

APM Bots: An Automated Presentation Maker for Tourists/Corporates Using NLP-Assisted Web Scraping Technique

Shajulin Benedict(✉), Rahul Badami, and M. Bhagyalakshmi

Department of Computer Science and Engineering, Indian Institute of Information Technology Kottayam, Valavoor P.O., Kottayam, Kerala 686635, India
{shajulin,rahulbadami2017,blakshmib.phd2111}@iiitkottayam.ac.in
http://www.iiitkottayam.ac.in, http://www.sbenedictglobal.com

Abstract. Summarizing information and preparing presentation slides remain the biggest measure for corporates, or startup enthusiasts to augment clients. Actively involving clients or convincing tourists with appropriate stories that mimic their interests in a short span of time has been a challenging task for most of the sales communities or governmental agencies. This paper proposed an Automated Presentation Maker (APM) bots, which utilized Natural Language Processing (NLP) and web scraping techniques, to deliver conceivable contents within a limited time frame. APM automated the preparation processes of presentation slides considering the emotions of clients/tourists. Sentiment analyzer and Stochastic Gradient Descent (SGD) classifier with an accuracy of 84% were implemented in APM for organizing presentation slides. Experiments were carried out at the IoT Cloud Research laboratory with specific case studies. Additionally, the performance study of the proposed APM bots, while experimented with different websites, was studied. The proposed approach would be beneficial for business corporates or government agencies to quickly capture the minds of clients/citizens with the automated presentations of APM.

Keywords: Bots · LSTM · Machine learning · NLP · Web scraping

1 Introduction

Information stands in as a tool that steadfastly grow retaining the search engine developers and information seekers with frustration, time-consuming, and inefficient. Corporates, startups, government officials, researchers, and so forth, struggle to get the core message from lengthy websites or search engine results within a short span of time, especially when the web pages were not well written.

In general, preparing presentation slides that focus on the requirements of clients or communal interests wins laurels and businesses albeit of being challenged by the hidden efforts and talents of manual developers. Understanding the market demand, corporates have taken strides to advance technologies that

I. Woungang et al. (Eds.): ANTIC 2021, CCIS 1534, pp. 647–659, 2022.
https://doi.org/10.1007/978-3-030-96040-7_49

optimize the development of presentation slides. In fact, capitalizing on digital marketing by adopting bots, including bots that generate presentation slides, in web/mobile applications could enrich profits if presentations were customized to the tone and need of clients or users.

In the recent past, engaging bots such as voice bots or chatbots in the majority of enterprise applications has been practiced to clarify business queries among pre-defined customers/clients. Obviously, presenting the users with conceivable contents/information has been an arduous task that needs to be addressed by solving the following research questions:

1. How effectively precise information could be scraped from websites?
2. Whether the sentiments of texts could be considered while preparing presentation slides?
3. Whether historical information of sites, especially the cultural heritage information, could be delivered to travelers/tourists?

This paper attempts to address the research questions of these sorts by adopting Natural Language Processing (NLP) based machine learning techniques in the proposed APM bots framework after the websites were scrapped for the analysis. The proposed APM utilizes the web scraping technique to collect data from websites; analyzes the emotional state of information; pre-processes the conceivable information; summarizes using Long Stort Term Memory (LSTM) algorithm; and, prepares the presentation slides for users.

Experiments were carried out using the proposed APM bots at the IoT Cloud research laboratory with two use-cases belonging to the tourism sector and education sector. The use-cases were tailored to prepare presentation slides based on the following titles: i) History of Dinosaurs and ii) Coronavirus Infections. Additionally, the performance of APM bots was analyzed for different web pages and titles. The proposed APM framework could be served in cultural heritage sites for the active participation of tourists or in business endeavors to attract more customers considering the likes/dislikes of clients/users.

The rest of the article is described as follows: Sect. 2 explains the state-of-the-art study of bots and web scraping techniques; Sect. 3 delves into the components and procedures involved in developing automated presentation slides; Sect. 4 highlights the machine learning algorithms and sentiment analysis aspects of the APM bot; Sect. 5 manifests the importance of the proposed framework using case studies; and, finally Sect. 6 provides a few conclusions and outlooks in the direction of the undertaken research work.

2 Related Work

Despite advances in information analysis and allied search-related services, the diligent information retrieval processes from web pages still remain a challenging task in several domains, including cultural heritage tourism and healthcare domains. Researchers have been designing procedures and solutions that offer the synopsis of information from web pages in the past. For instance, authors of

[1] had applied ontologies to extract information about crop diseases from web pages; authors of [4,7] had scraped news content from news-related web pages for providing information or synopsis of the contents. Notably, Gaol et al. [4] had focused on developing an API-based news aggregator system that crawled over news-related webpages as part of the Noox project. Similarly, authors of [17] had attempted to remove noise from web pages while performing the information retrieval processes.

In recent years, the web scraping technique has been modernized to process apt information from web pages using cloud-assisted scalable computing techniques. For instance, Shreya et al. [14] have studied the approach to harvest data from web pages using Amazon cloud elastic EC2 instances. These authors have observed the challenges due to scraping a large amount of data – i.e., limitations due to captcha, storage, and so forth have been studied by them.

In fact, the web scraping techniques require parsers that fetch appropriate information from web pages. These parsers could lead to overheads. Authors of [3] have studied the performance impact of different web parsers during the web crawling processes in recent years. Similarly, Rohmeth et al. [11] have compared the performance efficiency due to different web scraping techniques for web pages.

Applying intelligence while scraping information from web pages in an automated fashion has been an attraction among several researchers and web application developers. For instance, authors of Rizwan et al. [10] have studied the application of bots to automatically feed data to web servers and investigate the crimes that happened on websites. In [5], authors have applied intelligent supervised machine learning algorithms in web page processing methods.

LSTM learning model has been considered as a widely applied learning algorithm for time-series data in several application domains such as emotion classification [8], text analysis [2,6,9], emotion intelligence [18].

In this work, an LSTM-based NLP learning method is adopted in the APM bots framework to automatically create elegant presentation slides considering the emotions of users. The approach fits very well in cultural heritage sites to assist tourists, in corporate sectors to promote businesses by convincing clients, in smart cities to attract more citizens based on their well-wishes.

3 APM Bots

Formulating a synopsis by reading the entire webpage or weblinks is a time-consuming task. Businesses are often driven by presenting apt information to attract the mindsets of clients or customers. In addition, governmental agencies have to convince citizens, including tourists visiting cultural heritage sites, by providing appropriate information orienting to the fervor of citizens/tourists. APM bots are designed to automatically generate presentation slides that customizes content that influence the interests of travels/clients. This section explains APM bots and their processes in detail. The crucial entities that accomplish the goals of generating presentation slides and their functionalities are described below:

1. *Frontend/Backend* – Frontend and backend services of APM enable the users to view presentation slides and input specific requirements for enabling APM services. The frontend of APM could be written in languages such as reactjs, flutter, php, and so forth which improve web layouts and responsiveness of the users; whereas, the backend could be written using nodejs, golang, java, or similar web-based languages to enhance the software engineering functionalities of APM.
2. *Web Scraper* – At edge nodes of APM, web scrapers are designed to capture contents from web pages. Traditionally, while scraping a web page, the scrapers inspect the web pages and identify the information from inbound classes. This traditional approach consumes considerably a large amount of time. In APM, web scrapers are designed such that they could automatically parse the data directly from the web page without inspecting the data using edge nodes. Designing edge for several IoT-enabled applications was carried out earlier [12]. The web scraper of APM outputs data in JSON format for further analysis and for the preparation of slides.
3. *Sentiment Analyzer* – The *Sentiment Analyzer* component of APM is hosted in cloud server instances. This component investigates the emotions of the contents of web pages. It automatically captures the most relevant information that aligns with the sentiments of individuals. It applies the natural language processing technique that analyzes the words and phrases of documents found on the web pages. In the component, the latent sentiment analyzer approach is designed for focusing on a single document of web pages to fetch sentences that scores higher values with respect to the sentiments.
4. *Summarizer* – The major purpose of the *Summarizer* component of APM is to generate a piece of concise information with appropriate flow using neural networks such as LSTM. The contents derived from the LSTM summarizer are applied by the *Presentation Generator* of APM. The *Summarizer* delivers content in the JSON format after they were processed in cloud instances.
5. *Presentation Generator* – This component is located in edge nodes such as mobile devices or laptops of users. The *Presentation Generator* component is often incorporated into the backend services of the framework to generate presentation slides based on the information obtained from the *Summarizer* component of APM. The component could deliver output in ppt or pdf formats for providing elegant/concise views to the users.

Figure 1 illustrates the components involved in the APM bots framework in a pictorial fashion.

The major processes involved in the proposed APM bots framework, which automates in preparing the presentation slides for tourists/corporates, are carried out in three phases, as listed below:

1. *Requisition Phase* – During this phase, requests from clients or users are collected using the *Frontend* component of APM. In addition, the requests need to be pre-processed in this phase for submitting them to the web scraping process of APM. The major objectives of this phase are i) to handle a large number of requests from users. To do so, diligent handling of requests in end

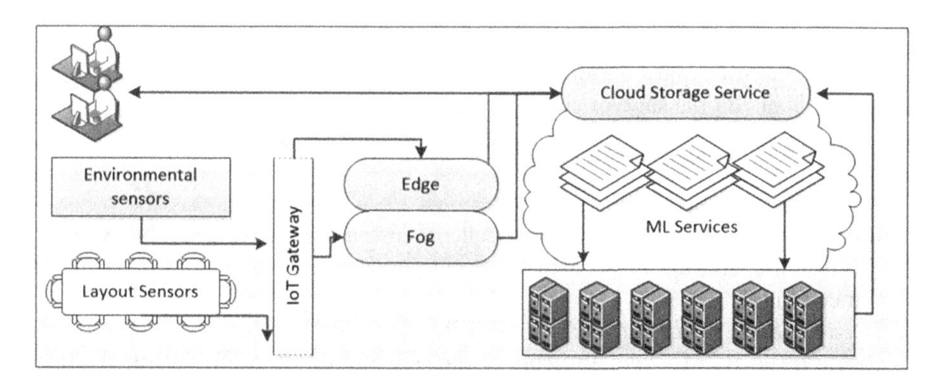

Fig. 1. APM bots framework – A pictorial illustration

devices is required; ii) to pre-process the requests and align them as per the need for extracting information.
2. *Analysis Phase* – In the *Analysis Phase*, web pages are scraped to collect the required information and identify emotions based on the algorithms such as SGD classifiers. These information are collected in *csv* format. Each line of the *csv* file contains one sentence that influenced the requested content of the user. In addition, based on the summarization algorithms, the content necessary for creating presentation slides is collected in a separate file.
3. *Preparation Phase* – Based on the synopsis information observed from the previous *Analysis Phase* of APM, the presentation slides are prepared in this phase. The *Preparation Phase* also controls the number of slides required for presentations. For instance, a user, who wants to quickly understand only a portion of the historical information from tourism locations, could restrict the number of slides to 5 or 10 with sufficient images. In addition, during this phase, users could specify the output format of the presentation slides such as HTML or PPTx.

4 Web Scraping and Presentation Maker Mechanisms

The web scraping methods and procedures accomplished in the APM bots framework for creating presentation slides are described in this section.

4.1 Web Scraping

To scrape information from web pages, APM bots implemented programs based on *Requests* and *BeautifulSoup* packages of python. The purpose of *Requests* package is to collect required information from the specified webpages; and, the purpose of *BeautifulSoup* is to organize data in a structured format using methods given below:

1. *get/text methods* – Once after importing the *Requests* package of python library, the information from web pages are collected using the *get* and *text* methods of the package of the APM framework as shown below:

   ```
   info = requests.get("http://www.domain.com"
   info.text
   ```

   In fact, the output of *info* is not well processed for better readability – i.e., the text information are continually presented without new lines or sufficient punctuation.
2. *BeautifulSoup* – *BeautifulSoup* is utilized in APM bots to vividly parse through the HTML tags and collect the sensible information within the document. For this, HTML tags, which do not contain information, are removed. For instance, HTML tags such as iframe, script, noscript, head, style, and so forth, are removed. Next, heading tags are parsed and collected in JSON format for further emotion analysis.

4.2 Sentiment Analysis

Fetching information from web pages needs to be analyzed for emotions such as bitterness, happiness, sadness, hate, anxiety, and so forth. APM bots framework utilized SGD classifier to detect emotions. The steps undertaken in analyzing the documents are listed as follows:

1. At first, the preprocessed data that were scraped from web pages are fed to the *Sentiment Analyzer* component of APM bots.
2. Next, texts are converted to *wordcount* vectors using the *CountVectorizer* tool. This tool has the feature to process texts to vector words based on tokenization. In this process, the number of words found in the web pages is parsed and counted so that the machine learning algorithm could learn the emotions at ease. Additionally, the tool limits the number of words that have to be processed in the next stage of sentiment analysis.
3. Next, the Term Frequency - Inverse Document Frequency (TF-IDF) method is computed for the vector words. TF-IDF calculates the weights of frequently utilized words using numerical statistical methods – i.e., the value reflects on the importance of words with respect to the collection of documents available on web pages. By this, words such as “a”, “the”, and so forth, will have limited influence scores to proceed further to the next level of analysis – i.e., TF-IDF limits the number of words for further processing.
4. Next, the transformed scores are fed to the Stochastic Gradient Descent (SGD) classifier of APM bots. In general, SGD classifiers are considered as optimization algorithms that have the capability to classify based on the discriminative learning and convex loss functions. SGD classifiers are quite predominantly utilized in text analysis or sentiment analysis works. In APM bots, SGD classifiers are implemented to study the emotions of words of web pages to process information while preparing the presentation slides with utmost accuracy.

4.3 Summarizer Technique

In APM bots, the *Summarizer* technique is implemented so that exact words selected through sentiment analyzer are not utilized to generate presentation slides. Rather, based on the selected words, newer paraphrases are generated using key-value mapping processes. The selected words and sequences are input to the Long Short Term Memory (LSTM) learning model for the accurate development of phrases suitable for the slides.

In general, LSTM is a Recurrent Neural Network (RNN) based artificial intelligence mechanism [15] which suits very well for sequences. LSTM is applicable for almost many time-series-based machine learning problems where there are a large bound of learning rates or input/output biases.

In the APM bots framework, sequences of information from web pages need to be learned for generating sequences based on feed-forward learning mechanisms. Layers of LSTM neural networks are adopted in the APM bots framework to train the sequences of paraphrases of web pages such that the last layer adopts a dense layer with *softmax* activation function [16]. Accordingly, the approach suggests apt paraphrases that could be included in the presentation slides.

Once after the identification of appropriate headings and contextual texts from web pages, the *backend* application, which was developed using nodejs language, prepares the presentation slides in HTML and pptx formats. Figure 2 explains the stages of these phases in detail.

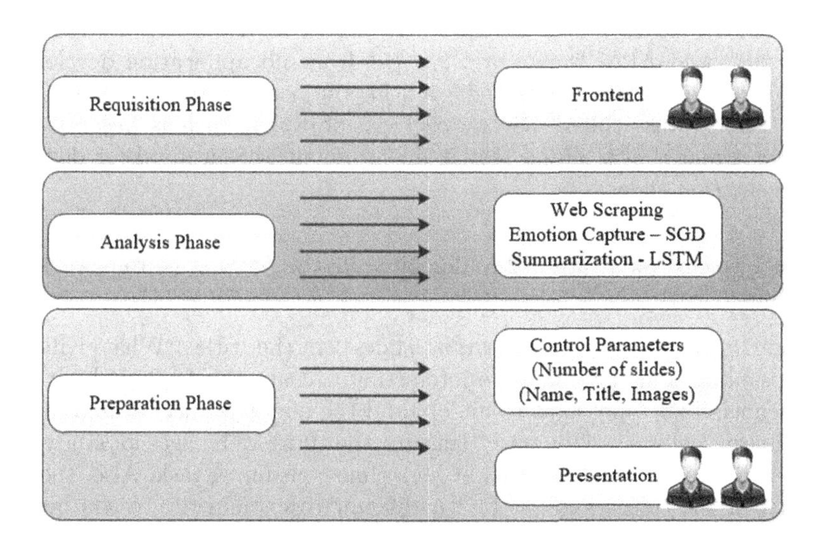

Fig. 2. Phases of APM bots framework

5 Experimental Results

Presenting a gist of information from lengthy web pages is an important task for several domains such as tourism domain, smart city domain, research domain, and so forth.

Tourists would be interested in viewing sites in addition to getting brief information from apt web resources to bolster viewpoints or to associate their visits to cultural heritage. Similarly, smart city officials would be interested to convince their citizens belonging to several communities by inclining their perspectives towards the matching thought processes of communities. Obviously, it is a time-consuming and inaccurate solution to manually surf the internet sites and deliver the objectives.

This section presents the experiments carried out at the IoT Cloud research laboratory to manifest the importance of the proposed APM bots framework. Initially, the experimental setup was discussed; next, the presentation slides created for two cases were discussed; finally, the performance analysis of the implementation was disclosed.

5.1 Experimental Setup

For experimental purposes, the frontend and backend services of the APM bots were developed using Python 3.8 version on ubuntu 20.04 LTS version-based intel i7 processor machine. The system had 64-bit ISA that was processed on a 16 GB RAM machine. This machine was considered as an edge-compute node for the proposed APM framework. For the frontend application development, reactjs version 16.12.0 was utilized; and, for the backend development, NodeJs version 14.17.6 was utilized. Several packages of python such as *Requests, BeautifulSoup, pandas, nltk, pillow, scikit*, and so forth were utilized for developing the presentation slides.

Cases: Two cases were studied to demonstrate the APM bots framework based on the input features:

1. *Case 1:* To prepare presentation slides on the title "What Killed the Dinosaurs?" This title may be interesting to tourists who visit heritages or museums. They may have to understand the non-existence of dinosaurs and the reason behind it. Obviously, learning the entire web pages for studying the non-existence of Dinosaurs might be a time-consuming task. Also, the group leader of the tourists might find it difficult to explain the reason behind it considering the communal information of the visitors;
2. *Case 2:* To prepare presentation slides on the title "Coronavirus". In fact, the nature of Coronavirus depends on different situations and regions. Several research works have progressed to address COVID-19 situations in the recent past [13,19]. According to this work, the presentation slides may be customized for attracting appropriate people through the proposed APM framework. For instance, providing details of COVID-19 cases belonging to overseas

might not be so interesting to the listeners. Rather, the presentations and figures should incline towards the region where it has to be presented for a proper reach.

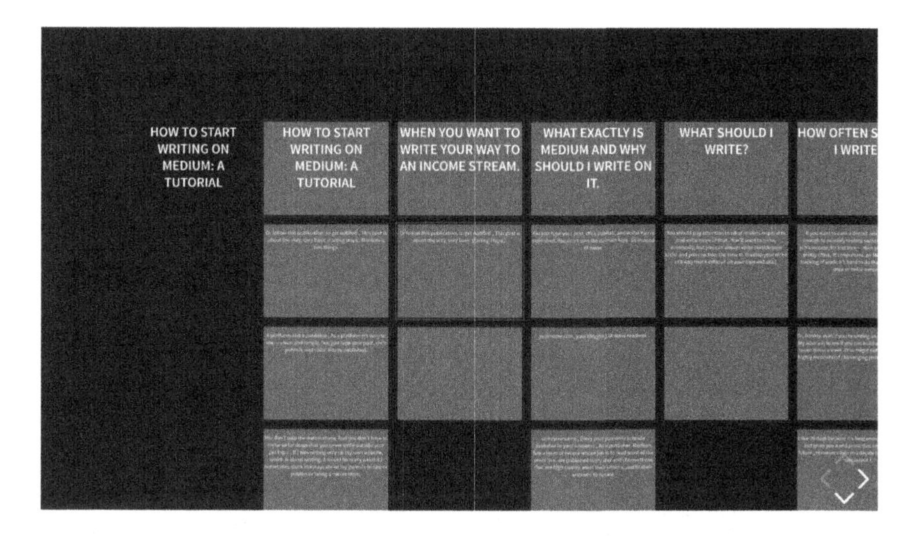

Fig. 3. Presentation slides – case 1

The proposed APM bots framework was designed to customize the presentation slides based on the vicinity and emotions of the listeners/readers.

5.2 Presentation Slides

Albeit any topics could be presented using the APM bots framework, in this work, as discussed above, two cases were studied. The users of the framework were asked to input the following information using the *reactjs* frontend component of the framework so that the presentation slides were generated:

1. *name* – This field generates the name of the presenter in the first slide of the presentation slides;
2. *title* – This field generates the title of the presentation slides in the first slide;
3. *email address* – Email address is included in the first slide of the presentation slides to honor the presenter based on this field;
4. *Type* – The type of the presentation field is required to set the tone of selections from web pages. For instance, the users could select the presentation mode of the slides for either business or education purposes;
5. *Category* – The category of presentations involve personalities, tourism locations, or groups where the slides would orient the selection of words and phrases appropriately; and,

6. *images* – The slides could be excluded from images or included with images based on selecting this field in the frontend of the framework.

A few slides of the presentations that were generated for the case studies are revealed in Figs. 3 and 4.

Fig. 4. Presentation slide – case 2

It could be observed that Case-1 had no images as the images were not selected in the frontend; also, the presentation slides of Case-2 had email addresses and names of the presenter with appropriate images. In addition, the framework opts for selecting the number of slides required for the presentations and the output format of the presentations. In order to study the case studies of the framework, the SGD classifier of the framework was trained using a training dataset at an 80:20 ratio to achieve 84% accuracy in the learning model.

5.3 Performance Study

The automatic creation of presentation slides depends on the web pages utilized in APM bots. In fact, it was interesting to manifest the performance aspects of the proposed APM bots framework owing to utilizing different web pages such as dictionary.com or purdue.com. To do so, different titles were offered to eight different sites to generate the presentation slides. For instances,

```
site 1 -- owl.purdue.edu -- ''University System''
site 2 -- pcworld.com -- ''35 great pc games for linux''
site 3 -- wikipedia.com -- ''Tourism in Agra, India''
site 4 -- dictionary.com -- ''Articles''
```

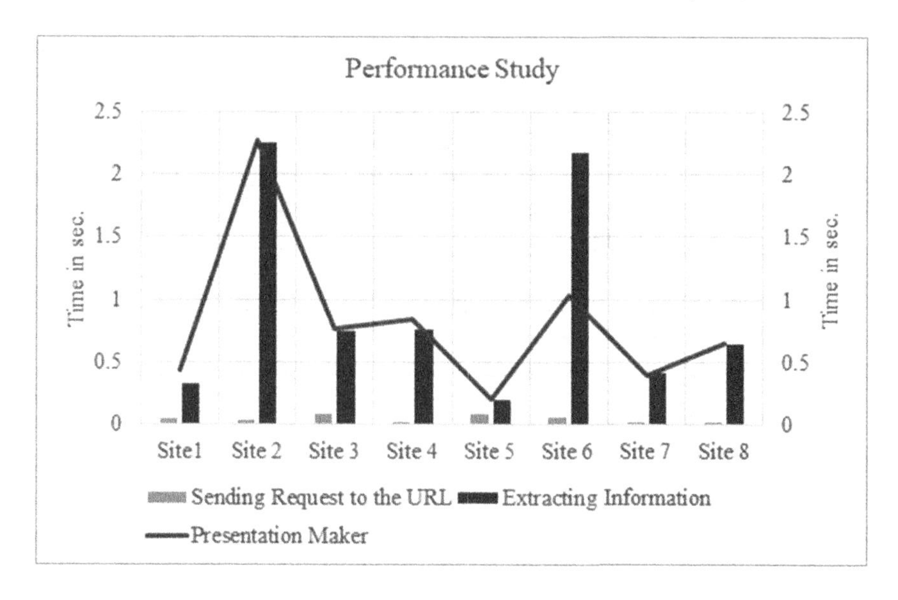

Fig. 5. Performance study for different presentation titles

```
site 5 -- hindustantimes.com -- ''classical music in North India''
site 6 -- medium.com -- ''How To Set Token in Postman?''
site 7 -- theatlantic.com -- ''Aquatic life''
site 8 -- wikihow.com -- ''How to start writing on medium?''
```

The observed times for sending requests to the mentioned URL, extracting information from the URL, and creating presentation slides are illustrated in Fig. 5.

Figure 5 illustrates the following:

1. the time required for extracting information from sites is comparatively higher than submitting URL or generating slides. For instance, see the red bar chart of the figure;
2. a few web pages considerably took more time than the other sites. For instance, creating presentation slides from pcworld.com took 2.263 s when compared to hindustantimes.com (site 5). The reason was due to the fact that sufficient keywords matched with the titles in certain sites for undertaking the sentiment analysis processes based on SGD classifiers.

6 Conclusion and Outlooks

For a decade, the amount of information has surpassed various readability methods/techniques that undermine web page readers. In the modern era of machine-to-machine communications, diligent measures are required to collect the synopsis of web pages within a short span of time, especially when clients were convinced or when travelers were briefed out about the cultural heritage sites. This article proposed an Automated Presentation Maker (APM) bots framework using ML techniques such as LSTM-NLP and sentiment analyzers to tune the contents of a lengthy web page for appeasing the interests of users. The article manifested the framework using use-cases for the tourism and the educational sector. In addition, the performance analysis of experimenting APM for different titles was analyzed. The proposed approach could be implicitly beneficial for tourists, business enablers, and government agencies to prepare conceivable contents in a short span of time considering the emotional mindsets of clients.

Acknowledgment. The authors would like to thank the officials of Indian Institute of Information Technology and the associated funding schemes to undertake the research work at the IoT Cloud research laboratory.

References

1. Jiang, B., Zhu, M., Wang, J.: Ontology-based information extraction of crop diseases on Chinese web pages. J. Comput. **8**(1), 85–90 (2013)
2. Li, D., Qian, J.: Text sentiment analysis based on long short-term memory. In: Proceedings of 2016 First IEEE International Conference on Computer Communication and the Internet (ICCCI), pp. 471–475 (2016). https://doi.org/10.1109/CCI.2016.7778967
3. Uzun, E.: A novel web scraping approach using the additional information obtained from web pages. IEEE Access **8**, 61726–61740 (2020). https://doi.org/10.1109/ACCESS.2020.2984503
4. Gaol, F.L., Trisetyarso, A., Abbas, B.S., Bahana, R., Adinugroho, R., Suparta, W.: Web crawler and back-end for news aggregator system (Noox project). In: Proceedings of 2017 IEEE International Conference on Cybernetics and Computational Intelligence (CyberneticsCom), Phuket, pp. 56–61 (2017). https://doi.org/10.1109/cyberneticscom.2017.8311684
5. Suchack, G., Iwanski, J.: Identifying legitimate Web users and bots with different traffic profiles - an Information Bottleneck approach. Knowl.-Based Syst. **197**, 105875 (2020). https://doi.org/10.1016/j.knosys.2020.105875
6. Elfaik, H., Nfaoui, E.H.: Deep bidirectional LSTM network learning-based sentiment analysis for Arabic text. J. Intell. Syst. **30**(1), 395–412 (2021). https://doi.org/10.1515/jisys-2020-0021
7. Nanba, H., Saito, R., Ishino, A., Takezawa, T.: Automatic extraction of event information from newspaper articles and web pages. In: Urs, S.R., Na, J.-C., Buchanan, G. (eds.) ICADL 2013. LNCS, vol. 8279, pp. 171–175. Springer, Cham (2013). https://doi.org/10.1007/978-3-319-03599-4_21
8. Mahesh, H.G., Sannakki, S.S., Rajpurohit, V.S.: Correction to: Attention-based multimodal contextual fusion for sentiment and emotion classification using bidirectional LSTM. Multimed. Tools Appl. **80**, 13077 (2021). https://doi.org/10.1007/s11042-021-10591-y
9. Mannel, N., Meziane, A.: SAHAR-LSTM: an enhanced model for sentiment analysis of hotels' Arabic reviews based on LSTM. In: Proceedings of 2020 5th International Conference on Cloud Computing and Artificial Intelligence: Technologies and Applications (CloudTech), pp. 1–7 (2021). https://doi.org/10.1109/CloudTech49835.2020.9365921

10. Rahman, R.U., Tomar, D.S.: A new web forensic framework for bot crime investigation. Forensic Sci. Int.: Digit. Investig. **33**, 300943 (2020). https://doi.org/10.1016/j.fsidi.2020.300943
11. Gunawan, R., Rahmatulloh, A., Darmawan, I., Firdaus, F.: Comparison of web scraping techniques: regular expression, HTML DOM and Xpath. In: Atlantis Highlights in Engineering (AHE), vol. 2, pp. 283–287 (2019)
12. Benedict, S.: Energy-aware edge intelligence for dynamic intelligent transportation systems. In: Garg, D., Wong, K., Sarangapani, J., Gupta, S.K. (eds.) IACC 2020. CCIS, vol. 1368, pp. 132–151. Springer, Singapore (2021). https://doi.org/10.1007/978-981-16-0404-1_11
13. Benedict, S.: RandomForest enabled collaborative COVID-19 product manufacturing/fabrications. In: International Semantic Intelligence Conference (ISIC 2021), CEUR-WS (2021)
14. Upadhyay, S., Pant, V., Bhasin, S., Pattanshetti, M.K.: Articulating the construction of a web scraper for massive data extraction. In: Proceedings of 2017 Second International Conference on Electrical, Computer and Communication Technologies (ICECCT), pp. 1–4 (2017). https://doi.org/10.1109/ICECCT.2017.8117827
15. Sundermeyer, M., Schlüter, R., Ney, H.: LSTM neural networks for language modeling (2012). https://doi.org/10.21437/Interspeech
16. Tuske, Z., Tahir, M.A., Schlüter, R., Ney, H.: Integrating Gaussian mixtures into deep neural networks: softmax layer with hidden variables. In: Proceedings of 2015 IEEE International Conference on Acoustics, Speech and Signal Processing (ICASSP), pp. 4285–4289 (2015). https://doi.org/10.1109/ICASSP.2015.7178779
17. Uma, R., Latha, B.: Noise elimination from web pages for efficacious information retrieval. Clust. Comput. **22**(6), 14583–14602 (2018). https://doi.org/10.1007/s10586-018-2366-x
18. Wang, S., Zhu, Y., Gao, W., Cao, M., Li, M.: Emotion-semantic-enhanced bidirectional LSTM with multi-head attention mechanism for microblog sentiment analysis. Information **11**, 280 (2020). https://doi.org/10.3390/info11050280
19. Shi, Y., et al.: An overview of COVID-19. J. Zhejiang Univ. Sci. B **21**, 343–360 (2020). https://doi.org/10.1631/jzus.B2000083

Designing a Residential's Database Management System: A Case Study of Applying Resource, Event and Agent (REA) Approach

Zaini Zainol[1](✉), Siti Noratika Toha[2], Nur Azlinda Azman[2], Rajinah Paling[2], and Mardhiah Abdullah[2]

[1] Department of Accounting, Faculty of Economics and Management Sciences, International Islamic University, 53100 Kuala Lumpur, Malaysia
zzaini@iium.edu.my

[2] A4U Flagship, Department of Accounting, Faculty of Economics and Management Sciences, International Islamic University, 53100 Kuala Lumpur, Malaysia

Abstract. Nowadays, with more and more data is created every second, the dependency on reliable database management systems (DBMS) is so crucial. In the case of a residential college, students' frequent changes every semester have made the process of keeping track of students' data a daunting task. Hence, this paper is conducted to document the proposal of creating a DBMS for the administrative staff in a residential college to record and keep track of the student's necessary information. The database is constructed using the resource, event, and agent (REA) approach, and following the software development life cycle (SDLC) implementation stages. Methods such as interviews, observation, and document review are used to gather information from relevant parties. As a result, a database is designed using Microsoft Access platform. Also, several benefits, limitations and recommendations are acknowledged for future improvement of the database.

Keywords: REA · Database · Malaysia · SDLC

1 Introduction

1.1 Database Management System Concept

Database systems are designed and developed to be built and occupied with data. In addition, any set of electronic records that can be processed to produce useful information can also be defined as a database. Also, the data is usually indexed through rows, columns and tables that enable the processing of workloads and query information to be performed easily. According to Muhammad Raza (2018), DBMS is a software package designed for describing, manipulating, retrieving, and managing data in the database. DBMS manipulates various features such as the data itself, data format, record structure, file structures and field names. The validation and manipulation of data are also made by rules that are defined by the system. Likewise, Mullins and Christiansen (2019) also stated that as DBMS is a system software that functions to create and manage databases,

I. Woungang et al. (Eds.): ANTIC 2021, CCIS 1534, pp. 660–674, 2022.
https://doi.org/10.1007/978-3-030-96040-7_50

it also enables the end users to read, create, update, and delete data in a database. In essence, DBMS acts as an interface between databases and end-users or other application programs, ensuring consistent arrangement and easy accessibility of the data. With appropriate security measures, the database management system can accept the request for data from an application and instructs the operating system to provide specific data.

While among the key components of DBMS are software, data, procedures, database languages, query processor, runtime database manager, database manager, database engine and report generator. Among the characteristics of database management systems is, it offers data security and eliminates redundancy, it enables insulation between programs and data abstraction, it supports multiple views of data, it allows sharing of data, and it supports multi-user environments that allow users to access and manipulate data in parallel. DBMS provides various techniques to store and retrieve data as it uses a variety of functions to do so. In comparison of database management systems with flat-file management systems, the latter does not support multi-user access and it is only limited to smaller DBMS systems. It also does not provide support for complicated transactions. Although it is cheaper compared to DBMS, it is still entangled with redundancy and integrity issues. There is plenty of database programming software available on the market. Among the DBMS software tools are Microsoft Access, MySQL, Oracle, MariaDB and dBASE.

Apart from that, DBMS also provides a wide range of benefits such as data security, data access, data sharing, data integration, uniformity of administration and management. Data security under DBMS allows organizations to impose compliance and security in which the database is available for usage by users according to the policies of the organizations. Meanwhile, data sharing is the advantage of DBMS that enables collaboration between users. Data access on the other hand means that DBMS offers controlled access to the databases. It also collects the logs of activities that are being performed throughout the databases which in return will enable the organizations to perform auditing to check for compliance and security. Another benefit of DBMS as mentioned earlier is data integration which brings about the meaning of having a single interface together with logical and physical relationships that are used to manage databases. DBMS also offers uniform management which is a single consolidated interface used to perform administrative tasks, which makes the job easier for the database administrators and users.

1.2 System Development Life Cycle Concept

SDLC is a conceptual model that involves procedures and policies that aim to develop or alter systems along the period of their life cycles. The process of conceptualizing, developing, implementing, and improving hardware or software or both are part of the SDLC. With the presence of effective computer systems, organizations or companies will be able to produce and deliver high quality products or services to their clients and potential clients as the effectiveness of computers is able to increase general efficiency and ensure logical workflow. High quality systems can be attained through an effective system development life cycle. High quality systems consist of various criteria such as the ability to meet customer expectations and standards, the ability to reach completion within allocated time and cost evaluations, and the ability to operate in current and future

Information Technology (IT) infrastructure effectively and efficiently. It is essential for SDLC to incorporate both the requirements of the end-user and the concerns on security throughout all its phases. System development life cycle is a structured approach that specifically separates the tasks into phases needed for the implementation of either new or modified information systems.

SDLC offers numerous advantages to the organizations that implement and comprehend the system in their organizations. One of them is that it helps an organization to clearly see and track the goals or problems in a particular project that they are involved with. With this ability, the organization can implement a plan or project with full precision and relevancy. Another advantage of SDLC is that the organization will be able to design certain projects or plans that they are interested in with clarity. This is because, under each stage of SDLC, there will be a formal review by the appropriate personnel. Apart from that, any execution of a project or plan will be thoroughly tested before being installed or implemented. This means that the installation of a particular project will need to undergo necessary checks and balances to ensure precision. Finally, the project or plan that is to be implemented can continuously loop around until it is perfect. Hence, SDLC is the best option to ensure optimal control and minimal problems.

2 Background of the Study

Historically, Residential Nusaibah of International Islamic University Malaysia (IIUM) of Gombak campus in Malaysia was officially occupied in February 1997, Semester 2 1997/1998 session. The administration of the Residential is managed by the Residential Office of Nusaibah (MO Nusaibah). The head of the Residential office is Principal, who is assisted by fellows and other staff. The office is under the management of the Residential and Services Department (RSD) of IIUM Gombak. Residential Nusaibah which is one of the female residentials able to accommodate an estimate of about 900 students in six blocks, allocated for undergraduate students while another two blocks allocated for postgraduate students.

The Residential is also occupied by many other facilities for the utilization by the residents of the Residential. Among the facilities are toilets in each level of the building blocks, Residential café, kiosk, prayer room, storage room for the storage of residents' belongings during semester break, a common room for small group activities, multipurpose hall for conducting resident activities, sports playground, and many others. To manage and sustain the administration of Residential, Residential Nusaibah uses both computerized systems and manual systems. For instance, the residents of Residential Nusaibah who wish to report on any defect or malfunction of the Residential's facilities, they can do so by filling in the Report on Maintenance of Facilities Form at the Residential office or by filling in the Google Form of Residential Maintenance Report Form. This shows that the Residential office does not fully use computerized systems and does not fully utilize computerized database management systems.

3 Statement of Problem

As the Residential Office of Nusaibah does not fully use a database management system in their administration and management works, the recording and storing of various information may give rise to several problems as the current system adopted by the Residential office may be unable to capture information effectively and efficiently, especially pertaining to the matters in relation to the residents or issues brought by the residents. The current system also may cause the information to be captured more than once, hence duplicates of data occur which entails more time and costs to detect and fix the issue. Also, the current system may cause the omission of data to happen as some information may be left out from being captured by the system due to non-integrated systems being used in place.

4 Objectives of the Study

A database information system (RNDiS) is suggested to be used by the Residential Office of *Nusaibah* to overcome the issues and problems arise when using the current existing system. Among the objectives of this proposed RNDiS are as described below:

- To ensure adequate record-keeping
  The proposed RNDiS aims to eliminate manual record-keeping of data and install electronic record keeping to ensure sufficient records of information are kept. This also ensures a centralized system in which all necessary data and information can be accessed, tracked, and monitored easily.
- To provide data security
  This proposed RNDiS can provide adequate security measures by giving different access levels to various users of the system.
- To reduce time consumption
  The proposed RNDiS aims to lower the time consumption consumed by the users in capturing or managing data, for instance by enhancing the search facility whereby it can be done through the query wizard function in Microsoft Access.

5 Research Methodology

Data gathering employed are interview, record review and observation or reading.

- Interview
  This method allows the collection of information from individuals or groups of individuals that are generally the users of the existing system and the potential users of the proposed system. This is one of the qualitative information that is able to discover misunderstanding or problematic areas. This method provides information such as the steps or activities involved in the process of allocating rooms to students, steps or activities involved in the process of verifying room availability, steps or activities involved in the process of handling maintenance reports from the residents, steps or activities involved in the process of issuing and recording compound offence to the residents, and others.

- Record review

  This method allows the attainment of facts from documents because the existing system is able to be understood better by examining existing documents, files or records. Among the records can be the standard operating procedure used for Residential Nusaibah, rules and regulation, policy manuals, reports generated from the current existing system and other necessary records.
- Observation/Reading

  This method allows any missing facts or information to be obtained or brings in any other facts that other methods are unable to provide. This method also may enable improvement of existing procedures through the introduction of new procedures or value-added procedures. This method can also provide the opportunity to go and see behind the scenes of how the management of Residential Nusaibah is being conducted as a whole. This method tries to identify the inefficiencies of operations, alternatives procedures, interruptions in normal workflow, obstacles in the usage of documents or files, and informal channels of communication that may exist.

In terms of the scope of study, our focus is on the significant and critical operations that happened within the administrative system of Residential Nusaibah. It includes:

- The student information management
- The Residential staff information management
- The maintenance staff information management
- The room information management
- The maintenance of facilities information management
- The compound offence management

6 Results

6.1 System Planning and Conceptual Design

Following the SDLC stages as discussed earlier, the development of RNDiS is made to overcome the hassle of recording the information of the overall Residential Nusaibah's management, especially to keep track of the number of records and finding a particular data. This is because, this database can help some users to do the following task:

- MO staff are able to allot every student to available rooms.
- MO staff can record the complaints of Residential facilities and assign maintenance staff to do the fixing process.
- MO staff can document the issuance of Residential compounds towards the student
- MO staff and RSD can check the details of the student
- MO staff and RSD are staff able to check the status of payment of the student's Residential fee

A Gantt chart is created to assist and schedule proper system planning in order to identify the exact problems, alternatives, limitations, solutions and so on. With the

Table. 1. REA entities of residential Nusaibah's management.

Resources	Events	Agents
Room	Verify availability Booking room	Mahallah office staff Student
Mahallah facilities	Receive complaint Maintain facilities	Maintenance staff Student
Cash	Execute Mahallah compound Receive payment	Mahallah office staff Student

intention of having proper planning, we use the REA model to specify all information needed to model the database system. Basically, the REA model has three entity types, namely the resources, events, agents and a set of associations that link them together (Hall 2011). Resources are the objects of economic exchange with the trading partners; thus, they give economic values to the organization. Events are the economic phenomena that cause the changes of the resources while agents are internal and external individuals of the organization who are involved in the economic event. The focal point of the REA model is the concept of economic duality, where every economic event is mirrored by another event in the opposite direction which are the give and receive economic events in the organization. The first step taken in planning the RNDiS is identifying the giving and receiving events in the economic cycles. Secondly, the following step in the system planning is to identify the resources, events and internal agents as well as external agents entities for the Residential Nusaibah's management. All entities are demonstrated in Table 1.

Subsequently, we determine the associations and cardinalities between the entities. "Associations are the nature of the relationship between two entities. They are labelled as lines connecting them while cardinalities are the degree of association between the entities that describes the number of possible occurrences in one entity that are associated with a single occurrence in a related entity" (Hall 2011).

The planning process is followed by consolidating all the individual REA diagrams (book room and receive complaint) into a single enterprise-wide diagram as illustrated in Fig. 1. The resources, which are the Residential facilities, room and cash are located in the center of the diagram. The event entities are situated beside the resource entities and lastly, the agent entities are positioned at the edge of the diagram. The cash resource is included in both disbursement of cash and receiving payment events, thus it will be merged as one entity to simplify the diagram. Moreover, the agent entities which are the Residential Office staff, students, and Office of Security Management staff (OSEM) appear more than one time to avoid a confusing diagram as the association lines will cross each other if we choose to merge them as one entity.

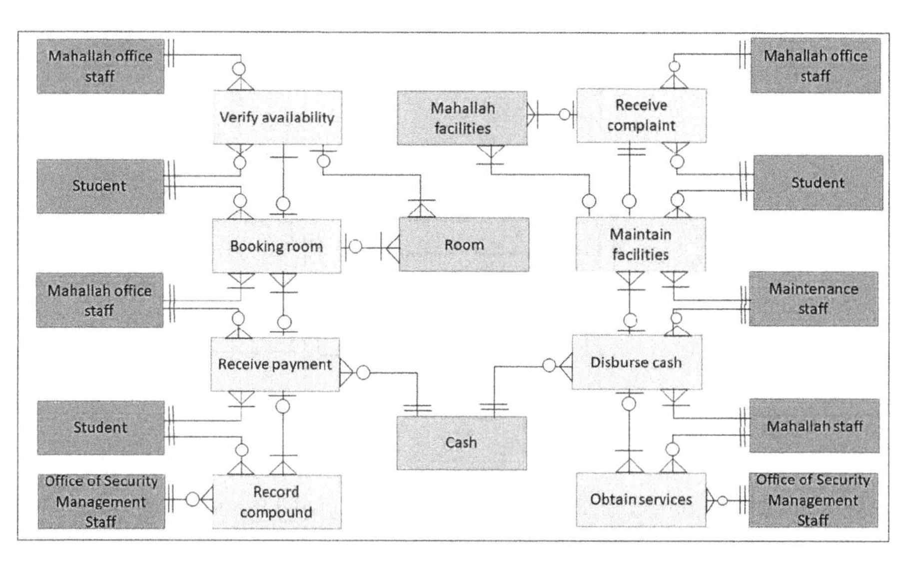

Fig. 1. Enterprise-wide diagram of residential Nusaibah's management

The existence of the enterprise-wide diagram enables us to figure out a clearer picture to build the database system. It helps us to identify and list down all the required items to be recorded with their respective primary key and foreign keys.

After that, we construct the physical database system and produce user-views using Microsoft Access because this software comes at zero cost and it is available on every computer that runs Microsoft operating system. Additionally, it is easy and convenient to customize based on the users' preferences. In Microsoft Access, the first step taken to develop the database is to create the tables with their respective attributes. Secondly, we create the forms with similar names of these tables to record a list of 15 names of students, some names of MO staff, type of maintenance issues and et cetera. After that, we build the relationship from these tables to allow some relational database to link disparate items, store and split data in those tables. This database tool can be done when there is a foreign key in a table that acts as a reference of the primary key of the other table.

Next, we create queries of some items which are compound information, maintenance, and room as the medium to do some data analysis and reviewing purpose. After that, we generate the reports of all the tables to summarize the information in the RNDiS. The final step in the system planning is designing the database according to the specific preference to make the database user-friendly before we truly confirm that the system can be fully implemented.

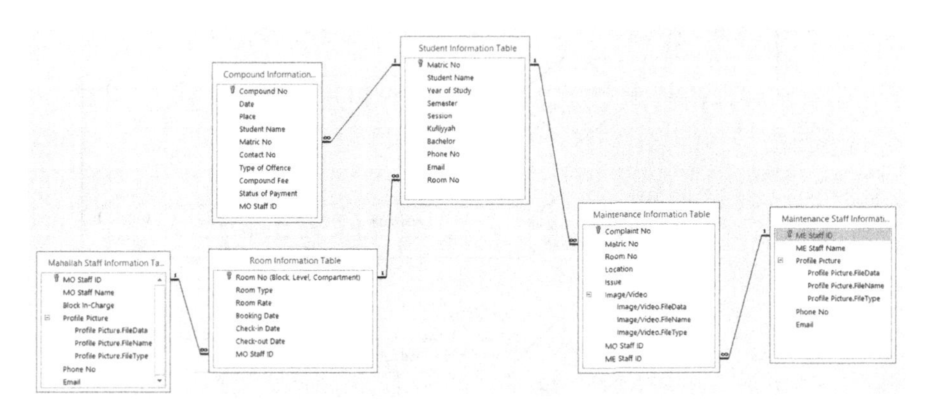

Fig. 2. The relationship of the tables in the microsoft access database

7 System Implementation

7.1 Process Flow

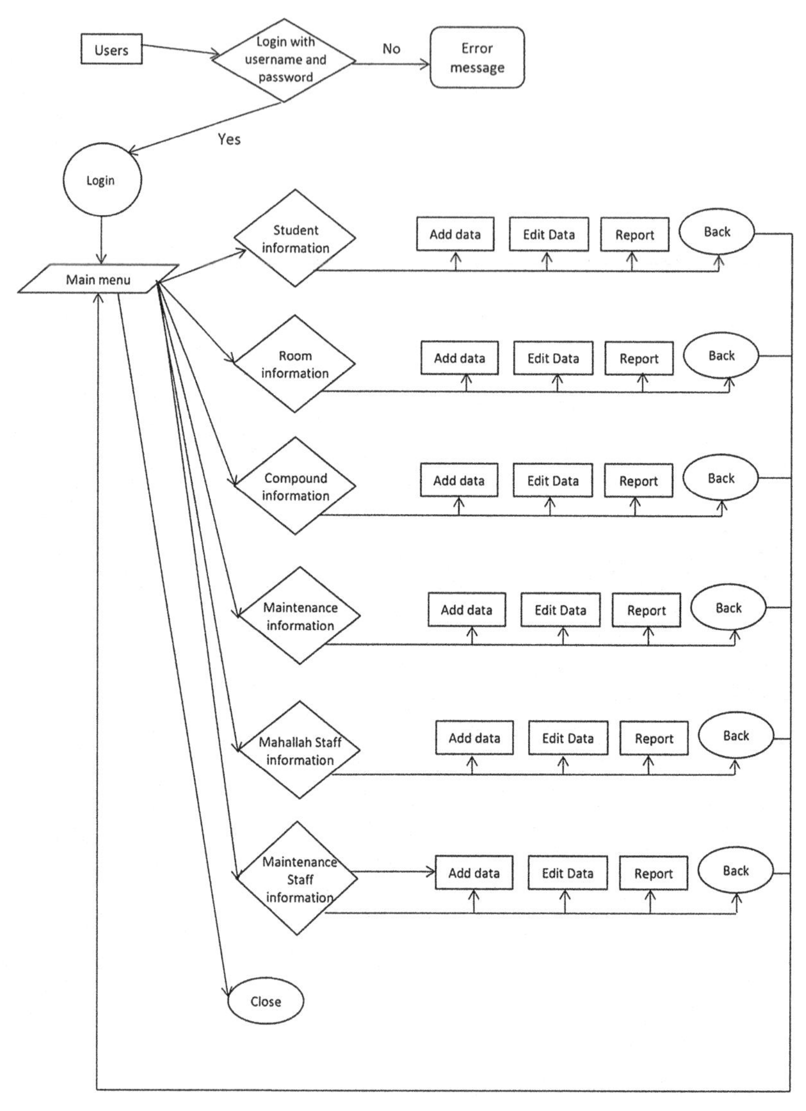

Fig. 3. Process flow of RND is

7.2 System Testing

System testing must be carried out to check or verify whether the system works accurately based on the user's requirements to meet the main objectives. It is also to assure that it is effective and compliant with the required standard or specification before it can be implemented to other users. Every single procedure should be tested throughout the system to find if there is any error or missing in operations. If there is defective or failure found, debugging must be done to solve those problems. There are several types of testing such as functionality testing, recoverability testing, performance testing, documentation testing, regression testing, security testing and so on.

Firstly, security testing should be done in the early phase of the system. It is necessary to make sure that only authorized personnel can access the system, i.e., RNDiS, data or information. In this RNDiS, the authorization is granted to the Residential's staff in Residential Office Nusaibah and also the Principal of Residential. Therefore, they must enter the unique username and password to login into the system to access the data, files and other application features. An error message will be displayed if there is a mistake in data entered in the required fields.

Next, to test the functionality of the RNDiS, test cases can be conducted to the modules or components of the system. For example, users can test to select modules of "Student information" in the main menu to add new information. The system in RNDiS will display a sub-menu whether to add data, edit, display report or back to the main menu. When the user clicks on the add button, the student information form will be displayed. Once the data is keyed in, the system will check on all the data in the required field such as matric number, name, email, faculty and so on. Te matric number is a unique ID which will not be allowed to be duplicate.

Meanwhile, the system also must be able to check the data type in each of the fields. If there is an error occurs, an error message will pop out stating that the data entered is not valid for that specific field. The same test will be conducted on other menus to examine the functionality of the RNDiS.

7.3 System Implementation

After the RNDiS is successfully developed and tested in the earlier phase, the system should be ready to be presented and implemented to the end-users. In this phase, the main objective is to evaluate the success of the system as it is being executed. The end-user will be provided with the system manual and guidance to assist them or training depending on their needs. From this phase, the system developer will gather feedbacks from the end-user regarding the system design, functionality and the effectiveness of the system based on their requirements. The comments or responses will be listed for the future improvements.

Table. 2. Test case for RNDiS login

Action	Input	Required Output	Status
Key in correct Username Key in correct Password Click login button	Username: amirah.asri@iium_nusaibah.my Password: amirasr999	Login successful and will be directed to the switchboard.	Pass
Key in correct Username And leave Password empty Click login button	Username: amirah.asri@iium_nusaibah.my Password:	Display error label "*Incorrect password*"	Pass
Key in incorrect Username Key in correct Password Click login button	Username: amirah@iium_nusaibah.my Password: amirasr999	Display error label "*Incorrect username*"	Pass
Key in correct Username Key in incorrect Password Click login button	Username: amirah.asri@iium_nusaibah.my Password: amirasr000	Display error label "*Incorrect password*"	Pass
Key in incorrect Username Key in incorrect Password Click login button	Username: amirah@iium_nusaibah.my Password: amirasr000	Display error label "*Incorrect username*" and "*Incorrect password*"	Pass

7.4 System Documentation

Database login as in Fig. 4 or encryption of database was made in order to prevent unauthorized use of an access database by simply setting a password. If the developer knows the password for an encrypted database, the developer can also decrypt the database and remove its password. However, it is risky since once the password is forgotten, the database will not be able to be used since it cannot be removed if it does not know the password. The database password or encryption will be removed if users agree with the system concept. The login page for the staff of Nusaibah's Residential Office comprises a dialog box that allows the user to input their Username (using staff email) and password.

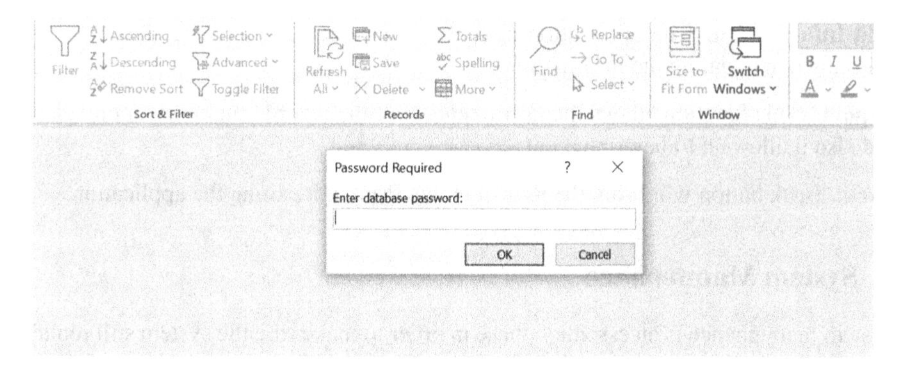

Fig. 4. RNDiS login Interface

The input will be validated when the user keys in a value for either of the two required values and when both are deemed correct or validated it advances to the menu page of the application. The username and password can be changed through the developer version.

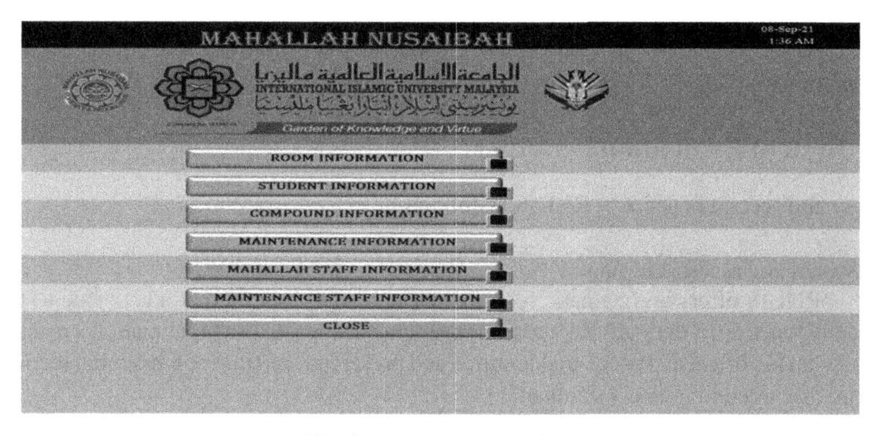

Fig. 5. Home page interface

The Home page interface of the Residential Office of Nusaibah system as shown in Fig. 5 consists of six (6) modules which include room information, student information, compound information, maintenance information, Residential staff information, and maintenance staff information. It also includes the "CLOSE" option exiting the application.

Each module has the same function; add data, edit data, print out reports and also back buttons.

Add Information: It basically allows the user to add new information. All boxes are required to fill in in order to save the data.

Edit Information: The application also allows users to edit existing data, which is needed when updating the information.

Report: This function allows the user to check all the records summary in one glance and also it allows it to be printed out.

Back: Back button will bring the user to Home Page, not exiting the application.

8 System Maintenance

System maintenance is an essential phase in order to make sure the system still remains effective to the user needs and up to date to the changes of environment in the organization. It is an ongoing activity to make sure the system is operated in its maximum capacity to deliver benefits and assistance to the users. The maintenance task could take place in terms of security, system updates, documentation, features and so on. There are several types of maintenance. If the system is bound to adapt to new demands and environments in the organization, the developer will perform adaptive maintenance where it will enhance and upgrade the current system to make sure it will satisfy the user's needs. Apart from that, there is corrective maintenance to repair the errors or defects that happened in the system. It may require some modifications to correct those problems. Next, perfective maintenance also can be carried out to fulfill users' additional demands from time to time.

9 System Evaluation

There are several benefits of RNDiS as follows:

- Systematic record-keeping
  RNDiS offers a systematic records management program that adds value to the daily function of the staff. Records management ensures that an organization's records of vital historical and fiscal are identified and preserved, and that non-essential records are discarded in a timely manner.
- User-friendly (Easy to use)
  A user-friendly system is one that is easy for novices to learn. The user who is not familiar with using the system might not have to worry since the RNDiS we develop for the Residential Office of Nusaibah is easy to handle and can be learned in less than half an hour.
- Reduce the amount of administration work
  The proposed RNDiS will make the administrative work more effective and efficient. It reduces and eliminates redundancies in record-keeping.
- One secure location
  Residential Office staff will not be required to record and save the record manually in the file rack. It will help them in terms of storage where it can reduce the cost for records storage equipment and supplies. Besides, it allows them to increase usable office space through the elimination of unnecessary file storage. In addition, using RNDiS that we developed, it will also provide accountability and timely access to information when needed.

- Manage risk

  As we all know, a system developed in order to help the organization's staff in their daily task. The system we develop can also help the Residential Office in terms of risk reduction. Proper management of records can be helpful in terms of evidence of the organization's actions and decisions. The relationship between records and risks can be seen from two perspectives, risks deriving from bad records management, and records used as a tool to mitigate risks. The organizations can face business risks associated with records management, such as compliance concerns, discrete information, and disaster recovery issues. Whereas the latter is related to the records management system in place to prevent the above-mentioned issues from happening to the Residential Office of Nusaibah.

10 Limitations, Recommendations and Conclusion

We listed below the limitations of our study, recommendations for future RNDiS improvement and finally our conclusion.

10.1 Limitations

As mentioned earlier, the project covers general services provided by the Residential Office of Nusaibah. However, the project has limitations, such as:

- The developed system cannot handle online payment of students' accommodation fees. Students need to pay by themselves and the system will be updated by Residential Office staff in the students' records.
- The developed system cannot handle other Residential issues such as mess activities because the data were not obtained.
- The students' signature is not captured by the developed system. This process might make procedures cumbersome, which is what the study aims to eliminate.However, it is able to capturethe full details of the students.
- Due to time constraints, certain fields were not included; the developed system was therefore reduced to cover only the critical aspects of Residential's management.
- Lack of skills was also one of the most challenging things in developing the system. However, we managed to develop a system that is able to solve the problem faced by the Residential management.

These limitations however were encountered in the course of the study, and appropriate techniques have been applied to ensure the system functions properly thereby eliminating the "stalemate".

10.2 Recommendations

- Biometric measures such as fingerprint and retinal scan should be included in the system to ensure good security of the system thereby avoiding impersonation and unauthorized access to stored data, and also prevent loss of vital information.

- Constant research should be implemented to ensure the effectiveness and efficiency of the system continuously.
- Implementation of more modern online facilities that might help prospective students interact more with the system and the Residential Office in general such as applications to make online transactions.
- Improvement in terms of the accessor of the system. Allow students to access the database by creating another level of security that only allows them to view the room information and their data such as compounds related to Residential and their report of maintenance.

11 Conclusion

Traditionally, the administration and management of the MO Nusaibah are done using the combination of pens and papers, and a computerized system. DBMS using Microsoft Access software is required to assist the management of records and data. The system automates the overall steps throughout the entire process of managing information regarding students, Residential office staff, maintenance staff, rooms, compounds, and complaints. The system will also minimize the required time and costs to perform such tasks. The system can be suited and modified according to the changes of needs of the users from time to time, the system also comes with minimal costs and only consumes minimal resources, thus becoming the best option to be implemented for a prolonged period. Hence, it should be able to cater to the needs of the users for the distant future. It is also believed that the proposed system can reduce the number of tasks that the administration needs to do. The system can also act as a major tool in the efficiency improvement of the Residential Nusaibah management. Therefore, from the above discussion, it can be concluded that the proposed system is a reliable, secure, timely and efficient system that can be implemented in replace of the current existing system that is less reliable to achieve better performance and productivity in the administration matters of Residential Nusaibah as a whole.

References

1. Bmcblogs DBMS: An Intro to Database Management Systems. https://www.bmc.com/blogs/dbms-database-management-systems/. Accessed 21 Nov 2020
2. Guru99: What is DBMS? Application, Types, Example, Advantages. https://www.guru99.com/what-is-dbms.html. Accessed 18 Nov 2020
3. Hall, J.A.: Accounting Information Systems, 10th edn. Cengage Learning, South-Western (2019)
4. International Islamic University Malaysia Official Website. https://www.iium.edu.my/my/mahallah/mahallah-nusaibah. Accessed 20 Nov 2020
5. Microsoft Support: Queries. https://support.microsoft.com/en-us/office/queries-93fb69b7-cfc1-4f3e-ab56-b0a01523bb50?ui=en-US. Accessed 21 Nov 2020
6. SearchSQLServer, Database management system (DMBS). https://searchsqlserver.techtarget.com/definition/database-management-system. Accessed 20 Nov 2020
7. SDLC Tutorial System Implementation and Maintenance. https://www.tutorialspoint.com/sdlc/index.htm. Accessed 22 Nov 2020
8. Techopedia, Database Management System (DBMS), https://www.techopedia.com/definition/24361/databasemanagement-systems-dbms. Accessed 18 Aug 2020

Kannada Dialect Identification from Case-Based Word Utterances Using Gradient Boosting Algorithm

Nagaratna B. Chittaragi[1(✉)] and Shashidhar G. Koolagudi[2]

[1] Department of ISE, Siddaganga Institute of Technology, Tumakuru, Karnataka, India

[2] Department of CSE, National Institute of Technology Karnataka, Surathkal, India

koolagudi@nitk.edu.in

Abstract. Dialects or accents constitute the grammatical variations along with phonological and lexical changes those are commonly observed in the usage of a language with minor and subtle differences. Dialectal variations existing among dialects are mainly due to unique speaking patterns followed among the group of speakers. The dialect processing systems are essential in the development of automatic speech recognition systems (ASRs) for regional and resource-constrained languages in the country like India. Since India is with rich diversity in languages. In this paper, a language-dependent dialect identification system is proposed for Kannada language from words especially with the Kannada language-specific **case** (Vibhakthi Prathyayas) information. Special morphological operations that exist in the Kannada language in terms of various cases commonly called as a grammatical function of a noun or pronoun. These word utterances are used for the classification of five dialects of Kannada. This is a novel idea to use the smaller word utterances that consist of dialect-specific information representing the unique characteristics. In this paper, case-based word utterance dataset is prepared by considering five Kannada dialects from Kannada Dialect Speech Corpus (KDSC). Dynamic and static prosodic features are extracted to capture dialectal variations. Addition to these features, spectral MFCC features are also considered for evaluation of differences among dialects from these word-level units. Initially, multi-class Support vector machine (SVM) technique is used and later effective extreme gradient boosting (XGB) ensemble algorithms are used for the development of an automatic Kannada dialect recognition system. The research findings have demonstrated the words with case information convey dialect specific linguistic cues effectively. The combination of dynamic and static prosodic cues has a significant effect on the characterization of dialects along with spectral features.

Keywords: Kannada dialect identification · Cases (Vibhakthi Prathyayas) · Dynamic and static prosodic features · Gradient boosting algorithm · Support vector machines

I. Woungang et al. (Eds.): ANTIC 2021, CCIS 1534, pp. 675–686, 2022.
https://doi.org/10.1007/978-3-030-96040-7_51

1 Introduction

Nowadays, Automatic Dialect Identification (ADI) systems are gaining more popularity and research attention among active speech researchers due to its implications to ASR systems. Characterization and identification of dialects would be beneficial if linguistic knowledge is available in advance. Dialects of a language can be distinguished based on lexical variations (difference in vocabulary), grammatical variations, and phonological variations. Dialectal differences exists at several levels of linguistic hierarchy (e.g., acoustics, phonetics, prosody) [3,6]. Dialect identification systems are useful in several speech recognition-based applications. There is an extensive increase in usage of speech interactive-based systems nowadays through several electronic smart systems. Interactive ADI-based speech recognition systems can enhance the performance of ASR systems. Spoken dialogue systems, efficient speech-to-text conversion systems, and language translation systems can also be benefited from dialect recognition systems [12]. Nativity identification can be useful in the context of immigration.

Sometimes, certain words, which are used quite often and those words pronunciations differ among various dialects. These variations at the word level play an important role in modeling dialects effectively. It is observed that the written form remains unchanged across dialects of almost all languages except a few [10]. Generally, dialects of a language are said to be mutually intelligible where speakers of the different dialects of the same language can understand and communicate to a larger extent [18]. In this regard, a systematic study of dialect processing systems of any language is in need of a better understanding of linguistic properties of that language [17].

The literature demonstrates several works carried out by considering short utterances such as vowels and consonants with dialect classification. These works can be observed in a few of the languages spoken in India. Kannada is the official language spoken in Karnataka and other parts of India. However, dialect processing systems are found to be comparatively lesser with the Kannada language over other prominent languages spoken in India.

Few studies can be seen where vowels, consonants, and word utterances are used for Kannada dialect classification [6,15]. Kannada language vocabulary is equipped with forty-nine phonemes (letters), these are mainly divided into three groups as; 1. *Swaragalu* (vowels), 2. *Vyanjanagalu* (consonants) and 3. *Yogavahakagalu.* Formation of words happens in the Kannada language as an aggregation of the meaningful combination of *Aksharas*. However, these words are mainly bi and tri-syllabic, many times few may be even larger clusters of vowels and consonants [14]. The **cases** in the Kannada language exhibit morphological peculiarities across the dialects. The grammatical function of a noun or pronoun is called a **case** (Vibhakthi). Vibhakthi's are expressed through appending the suffixes (pratyayas) to the words specifically to nouns or pronouns. These convey the relationship between the words used in the sentences. These cases do not have any independent or unique meaning when used alone. However, they convey a different meaning when combined with the nouns. This distinctive usage of nouns and pronouns has demonstrated variations among five dialectal speakers of the

Kannada language. Based on this observation, in this paper, ADI systems are developed by considering words with case-level utterances.

Most of the time, research activities concerning many regional languages are limited in the context of dialect processing. The prime reason for this is the unavailability of standard datasets for regional languages. In the countries like India, to date, the identification of accurate boundaries between dialects of any language is a major challenge with many regional languages. However, few research teams are working on prominent and languages those are spoken by a large group of people. Because of this, we can find few standard datasets for such languages are available. The dataset used in this work is extracted from KDSC. This speech corpus consists of spontaneous speech recordings from five dialects of Kannada [7]. Five dialects of the Kannada language such as Central Kannada, Coastal (Karavali) Kannada, Hyderabad Kannada, Mumbai Kannada, and Southern Kannada are considered according to KDSC dataset. These dialect classes are identified based on the unique speaking patterns observed among the community of speakers belonging to a specific geographical area across Karnataka state [7].

In this paper, three different ADI systems are discussed by using *case* level information of the Kannada language. Similarly, pitch and energy contours are extracted for modeling dialectal variations in Kannada. Further, Support vector machines and decision tree-based ensemble classification algorithms (boosting concept) are employed for the development of Kannada ADI systems. Experiments are also carried out for evaluation of the contribution of both dynamic and static prosodic and standard spectral features towards recognizing the dialects.

The remaining sections of the paper are organized as follows. Section 2 discusses the implementation details of the case-based dialect recognition system. It includes the procedure followed for the preparation of a word-level dataset with case information. The steps followed to extract dynamic-static prosodic features with details of classification algorithms employed in this work are provided in Sect. 3. Section 4 presents experimental details with SVM and XGB ensemble methods under individual and combinations of features. Analysis of results obtained with proposed static and dynamic prosodic features is also provided in this section. Inference of the dynamic and static variations in pitch and energy features across dialects is provided. Conclusions made out of this study and future works planned are presented in Sect. 5.

2 *Case* Based Dialect Identification System

The Kannada language has complex grammar, and various morphological changes are possible for nouns and verbs when compared to English which has only two variations. For example, "Boy" may change only to "Boys" in the plural. However, in Kannada, the scenario is significantly different as nouns, and verbs take multiple variations based on *cases* in which they are used. For example, the noun word "Mara" in Kannada means a tree in English, and it can be expressed in seven different forms known as *cases* in Kannada. These seven *case* variations are as follows, "Maravu" ("Prathama"), "Maravannu" ("Dviteeya"), "Maradinda" ("Triteeya"), "Marakke" ("Chaturthi"), "Marada-deseyinda" ("Panchami"), "Marada" ("Shashti") and "Maradalli" ("Saptami"). After the careful observation and listening to speaking patterns followed in the Kannada language, it can be clearly noticed that variations exist in usage patterns of this seven textit cases across dialects. In some dialects, it is noticed that the standard form of usage of *case* form is completely different and replaced with other forms only. Detailed information regarding the seven *cases* used in the Kannada language and their English equivalents and functions are presented in Table 1. With this motivation, automatic dialect identification systems are developed in this work by using these *case* level utterances. The workflow diagram of the proposed work is presented in Fig. 1. The details of various stages of the proposed work such as dataset preparation, features extracted, classification algorithms used and experiments carried out are presented in the following subsection.

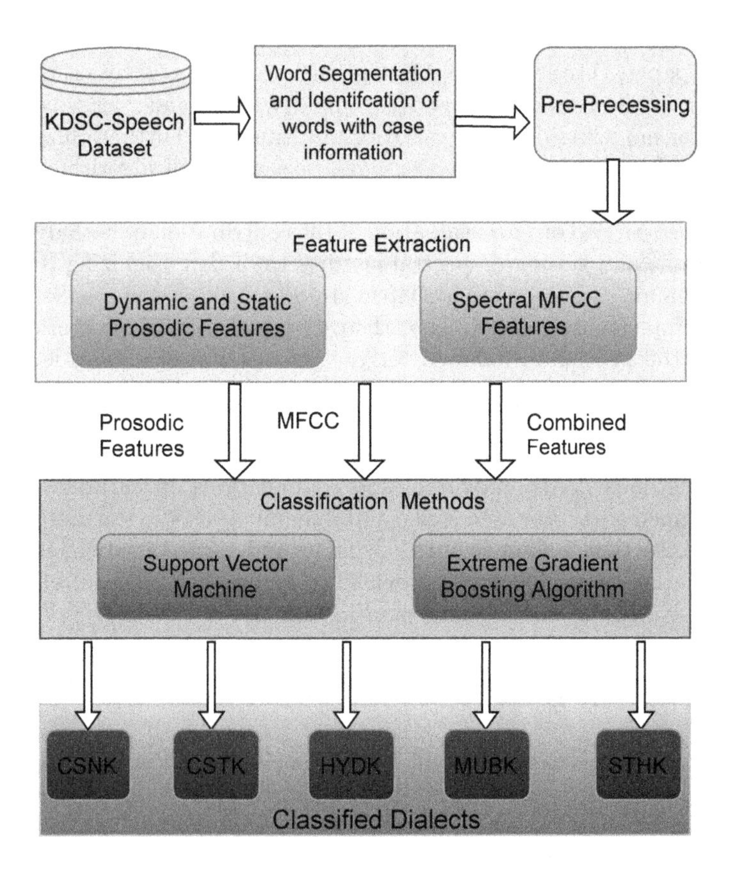

Fig. 1. Workflow diagram of the work proposed

Table 1. Different Kannada *Cases* and their English Equivalents used in this study for dialect characterization

Sl. No.	Kannada	Function	Prepositions	Kannada Suffixes
1	Prathama Nominative	Subject	–	/u/
2	Dvithiya Accusative	Object	To,	/annu/
3	Trithiya Instrumental	Instrument	By/with/through	/inda/
4	Chaturthi Dative	Receiver	To/for	/ge/, /ke/, /goskar/
5	Panchami Ablative	Point of separation	From/than	/deseyinda/
6	Shashti Genitive	Posession/Relation	Of/'s	/a/
7	Sapthami Locative	Location	In/on/at/among	/alli/

Kannada Case Dataset: The grammatical function of a noun or pronoun is called *Case* (Vibhakthi). These are suffixes that get added to the words specifically to nouns or pronouns. These convey the relationship between the words used in the sentences. These cases do not have any independent or unique meaning when used alone. However, they convey and alter the meaning of words when combined with the nouns in different contexts. In this work, an attempt has been made to utilize the dialectal cues present in the usage of this Kannada *cases*. For this purpose, a *case* dataset is prepared from KDSC. This dataset consists of five dialects of Kannada spoken in Karnataka. Dialects considered are Central Kannada (CENK), Costal Kannada(CSTK), Hyderabad Kannada(HYDK), Mumbai Kannada (MUBK), and Southern Kannada (STHK). Details of the dataset with five out of seven *cases* used in the Kannada language are given in Table 2. Complete details of the dataset used in this work can be found in the paper [7].

The word dataset used in this paper consists of a total of 6606 words from all dialects. These words are extracted from the spontaneous Kannada speech. An automatic segmentation algorithm with the help of energy and zero-crossing rate (ZCR) features is used for obtaining them [8]. From this set of words, words are carefully chosen that consist of five different *case* information embedded in it. Every word has listened, and the words with *cases* have been segmented manually using the Praat tool [1]. Out of 6606 words, a total of 1682 specific words with case information are used for the development of ADI systems.

3 Feature Extraction and Classification Algorithms

The literature demonstrates many works indicating how individual word utterances convey the existence of dialect-specific features and their comparative performance analysis [5]. Words embedded with *Case*, the spoken units considered in this work also resemble the word groups specifically with Kannada *case* variations. Hence, the prosodic and spectral features are extracted for analyzing the dialectal differences from *case* information since each word does carry dialect-dependent cues.

3.1 Feature Extraction

This section discusses the extraction procedure of dynamic and static prosodic features and spectral features used in this study.

Table 2. Kannada *Case* dialect dataset, CENK: Central Kannada, CSTK: Coastal (Karavali) Kannada, HYDK: Hyderabad Kannada, MUBK: Mumbai Kannada, STHK: Southern Kannada

Sl. No.	Kannada dialects	Word's with case dataset	Word dataset	Kannada case suffixes considered
1	CENK	404	1426	/u/, /annu/, /inda/, /ge/, /ke/, /ige/, /alli/
2	CSTK	302	1247	/u/, /annu/, /inda/, /ge/, /ke/, /ige/, /alli/
3	HYDK	388	1344	/u/, /annu/, /inda/, /ge/, /ke/, /ige/, /alli/
4	MUBK	194	1240	/u/, /annu/, /inda/, /ge/, /ke/, /ige/, /alli/
5	STHK	394	1349	/u/, /annu/, /inda/, /ge/, /ke/, /ige/, /alli/

Prosodic Features: Prosody can be viewed as one of the important speech features related with longer units of speech such as syllables, words, phrases, and sentences. The prosody correlates with the patterns of duration, intonation (F0 contour), and energy. It is known that prosodic knowledge plays a prominent role in the classification of dialects, emotions, and languages [8]. Similarly, the absence of prosodic cues can easily be perceived from the speech as it will be similar to read speech with just a concatenation of words. Prosodic features are said to add naturalness to a speech by structuring the flow of speech. In this work, both dynamic and static properties of prosodic features are explored for the classification of the Kannada dialects.

Dynamic Prosodic Features: Majority of the times, dialects of languages vary with the use of unique prosodic attributes on the same vocabulary and script. These unique prosodic attributes are reflected through the distinct variabilities existing speaking patterns that includes intonation, stress patterns, prolongation, rhythm and so on. The change in F0 is observed due to the rise and fall in the pitch while speaking, these variations recognizes an unique phonological pattern followed in a sentence or word of a specific dialect. Prosodic features namely, duration, pitch (F0) and energy features are used in this work.

Pitch Contour: Subharmonic-to-harmonic ratio based pitch estimation algorithm with 30 ms frame and 15 ms step size is followed to extract pitch features [16]. These features form the pitch contour for each word.

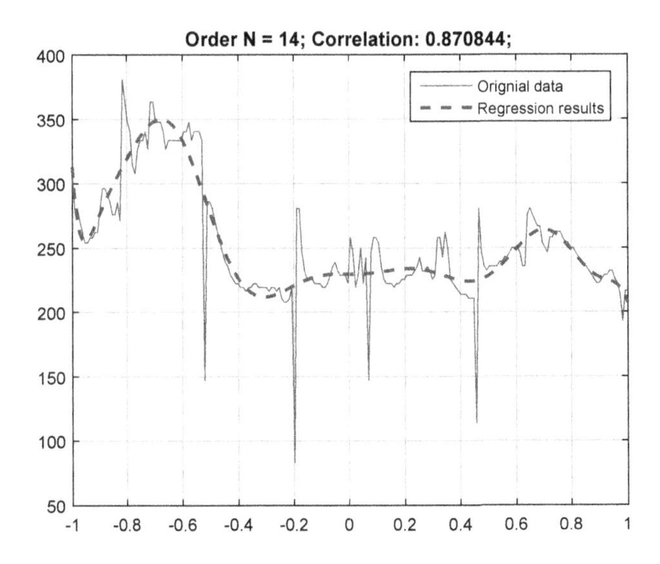

Fig. 2. Legendre polynomial fit for a pitch contour

Further, a Legendre polynomial fit function of order 14 is used to decimate these values. These 14 coefficients of Legendre polynomial have produced a good fit and these form a 14-dimensional feature vector and represent the dynamism of feature contours. The Legendre polynomial fit drawn for the pitch features extracted from the English sentence is shown in Fig. 2.

Energy Contour: The energy contour is also derived by considering the variation in the amplitude values of the signal by using the same method followed in the extraction of pitch contour. Short-term energy features are extracted from the frames of size 30 ms and 15 ms step-sized frame. The number of features obtained from each speech file is different as the lengths of the files are different. Further, a Legendre polynomial function of order 14 is used to fit these values. 14-coefficients resulted in a good fit and are treated in order as a 14-dimensional feature vector which also represents the dynamic stress patterns of different phonemes in a word utterance. These 14 energy features are appended to the 14 pitch values creating the complete feature vector of size 28.

Static Prosodic Features: The 28-dimensional feature vector obtained in the previous section represents the dynamic variations in pitch and energy values in a word concerning time. Apart from these, five pitch and energy values namely, minimum, maximum, mean, standard deviation, and variance are derived from each word unit.

These ten features are considered as the static prosodic evidence of the speech units. The inclusion of these five features forms a 19-dimensional feature vector from pitch contour (14 dynamic+5 static). Similarly, the other 19-dimensional feature vector for energy leads to totally a 38-dimensional feature vector is formed when both dynamic and static parameters are combined.

Spectral Features: Spectral features play a decisive role in capturing the proper dialectal cues existing across languages and dialects. In this work, along with prosodic features, standard Mel frequency spectral coefficients (MFCCs) are also derived from the word utterances to exploit the vocal tract variabilities during the pronunciation of words with case information of different dialects. MFCCs are trusted to be the best available feature representations of the overall characteristics of the vocal tract system. These spectral parameters are widely used in many speech-based applications since they try to imitate the human auditory system [13]. RASTA filter-based processing of speech for finding MFCCs helps in suppressing noisy portions of the speech. These MFCC features are said to capture spectral cues [5]. In this work, RASTA (Relative Spectra) based 13 MFCCs features are extracted from 40 filter banks [11].

3.2 Classification Algorithms

In this paper, a single classifier-based SVM method and multiple classifiers-based gradient boosting algorithms are used. Decision trees are used as the base classifiers in the XG boost algorithm. Automatic dialect identification systems are implemented from *case* level utterances. Dialect-specific cues are captured from various features extracted by using the SVM classification method. The binary classifier SVM is implemented with a one-versus-rest method in order to address the multi-class classification problem. However, one-versus-rest method is preferred over one-versus-one approach as it creates minimum number of binary classification models. Five classes represent the five different dialects of Kannada language considered in this work. Three different kernel functions are tried with SVM, whereas, Radial basis function (RBF) kernel function that is used for separating hyperplane with the maximal margin in a high dimensional feature space has shown better recognition performance [2].

Instead of using only traditional single classifier-based algorithms such as SVM, nowadays, multi-classifier based ensemble techniques are gaining more popularity in learning models. Ensemble algorithms are establish to be powerful prediction and classification techniques. They are performing even better over traditional methods. The boosting technique and learning from smaller-sized many classifiers assists in improving the performance due to model learning happening by using combinations of multiple smaller-sized sub-problems. Always it is better to take opinions from different peoples.

However, usage of ensemble algorithms for solving speech processing tasks is found to be less in literature with respect to dialect identification. Very specifically, dialect recognition system implementation from the speech is rarely seen in the literature [9].

An attempt is made in this paper, for the development of systems by using one of the boosting techniques in a specific XGB learning algorithm. The availability of dialect identification systems using this algorithm is comparatively lesser in literature. Also it is observed that, XGB algorithms provide various advanced features for the purpose of model tuning, computing environments and for the enhancement of algorithms performance. This learning algorithm if found to be robust enough in supporting regularization and fine tuning of parameters. Among all available ensemble-based learning algorithms, three significant steps are followed in the implementation of gradient boosting. These are as follows:

1. Selection of a suitable loss function.
2. Choosing the base classifiers.
3. Procedure for construction of trees.

In the first step, appropriate loss function selection needs to be done based on problem type. Hence, in this work, multi-class *logloss* is used. This function is required to address five different dialect classes. Appropriate selection of a base classifier is the second step. Most commonly decision trees are used for ensembling, a greedy approach is followed for the construction of different smaller trees.

Some of the important parameters such as maximum levels value, number of leaf nodes, and best splits are chosen after sufficient experimentation. Also, these parameter values are fine-tuned with repeated experiments with few values until the best combination was found to yield a recognition performance. In the third step, various trees are constructed by adding one node at a time; a gradient descent procedure is used for the minimization of loss during the addition of trees. The XG boost library is used for implementation to handle five dialect classes [4].

4 Experimental Results and Discussion

Automatic dialect identification systems are developed individually from word-level units using static and dynamic prosodic features along with spectral attributes. Various experiments are carried out to evaluate and analyze the performance of the proposed three various features. Experiments were also carried out with a combination of features.

Analysis of the influence of dynamic and static prosodic features on effective *case*- word-based dialect classification systems is conducted. Since these features have demonstrated their significance on word-level pronunciation in literature [8]. The same features are used for carrying out experiments on *case* based dataset derived from the Kannada dataset. Each such experiment is carried out separately using the SVM and XGB methods. Experiments are done by using dynamic, static, MFCCs, and their combinations. The dialect recognition performances obtained by using these features on *case* dataset are presented in Table 3.

Table 3. 'Case'-based dialect recognition performance (Accuracies in %)

Sl. No.	Features	Kannada word based dialects		Kannada *case* based dialects	
		SVM	XGB	SVM	XGB
1	Prosodic + Static Features	41.25	48.86	56.31	64.96
2	MFCCs	72.94	73.40	75.77	80.83
3	Combination of Features	82.31	**83.06**	84.69	**86.73**

Prosodic and spectral features generally convey different dialect specific features. In this regard, in order to exploit these, both prosodic and spectral features are analyzed individually. It is noticed that both dynamic and static features and their combination considered in these experiments have shown slightly better dialect recognition performance when words with the case are used. The dynamic and static feature set has demonstrated an average recognition rate of 56.31% and 64.96% using SVM and XGB methods respectively when words with case information are used. Similarly, 41.25% and 48.86% when regular utterances are used. Kannada case-level words used in this work have demonstrated significant contributions in the classification of Kannada dialects. These phonological variants are fairly salient markers in the Kannada language, typically for the classification of dialects. It can be observed that the highest accuracy of about 86.73% is achieved with the use of a combination of both types of features. To evaluate the performance of spectral features on dialects, few experiments are carried out by considering 13 MFCCs feature vectors. Comparatively better dialect recognition over the use of only prosodic features is observed with MFCC features on *cases*. From these observations, one can conclude that, from short utterances such as words, spectral features may capture dialect-specific variations more effectively than the prosodic features.

One more important observation was made from the results achieved in the case of word-level utterances and also with words with case information. In the literature, generally, it is said that words and sentences represent the speech from two different dimensions by conveying various dialect-related information from both production and perception points of view. Dynamic and static features along with their statistical parameters are found to be successful in discriminating Kannada dialects even from shorter utterances. Pitch variations are better captured with five statistical parameters with both words and case-based word units. Words are found to be shorter in length to capture intonation features. However, spectral (MFCC) features could identify the dialect-related features from both types of words with almost similar performance. XGB based ensemble classifiers have resulted in somewhat better performance over individual classifier-based SVMs.

It can be observed from the table that, words of every dialect class are highly confused with each other classes. However, the words especially with *case* different information have shown the contribution of these on dialect classification.

The reason may be, word utterances with *case* information makes each word unique and even they are of short duration, dialect recognition performance is comparatively better over just the use of every word for the classification of dialects.

5 Conclusion and Future Work

In this study, characterization, and classification experiments are carried out exclusively for Kannada dialects. For this purpose, a unique grammatical, morphological concept known as *cases* is used for the discrimination of Kannada dialects. Hence, independent studies are carried out concerning the development of ADI systems with words and case-based words. Various experiments are conducted for the classification of five Kannada dialects using word-level utterances.

SVM and XGB algorithms have performed well and resulted in better performances among other algorithms. Case information is known to be the suffixes that get added to nouns in the Kannada language. However, studies conducted in this work have examined the apparent differences across dialects in the use of the words with this *cases*. This work has examined the possible phonetic differences in the way of how the word utterances with *case* are different across dialects and convey unique dialectal cues. In future, performance can be enhanced by exploring excitation source features in addition to prosodic and spectral features. However, the hyper-parameter tuning of XGB algorithm can be performed to enhance system performance. In the future, this work can be extended to apply deep learning models for capturing dialectal cues across longer utterances such as sentences. This work can also be extended to classify dialects of other Dravidian languages such as Telugu and Tamil.

References

1. Boersma, P., Weenink, D., Petrus, G.: Praat, a system for doing phonetics by computer. Glot Int. **5**(9), 341–345 (2002)
2. Chang, C.C., Lin, C.J.: LIBSVM: a library for support vector machines. ACM Trans. Intell. Syst. Technol. (TIST) **2**(3), 27 (2011)
3. Chen, N.F., Shen, W., Campbell, J.P.: A linguistically-informative approach to dialect recognition using dialect-discriminating context-dependent phonetic models. In: IEEE International Conference on Acoustics Speech and Signal Processing (ICASSP), pp. 5014–5017. IEEE (2010)
4. Chen, T., Guestrin, C.: XGBoost: a scalable tree boosting system. In: Proceedings of the 22nd ACM SIGKDD International Conference on Knowledge Discovery and Data Mining, pp. 785–794. ACM (2016)
5. Chittaragi, N.B., Koolagudi, S.G.: Acoustic features based word level dialect classification using SVM and ensemble methods. In: Tenth International Conference on Contemporary Computing (IC3), pp. 1–6 (2017)
6. Chittaragi, N.B., Koolagudi, S.G.: Acoustic-phonetic feature based Kannada dialect identification from vowel sounds. Int. J. Speech Technol. **22**(4), 1099–1113 (2019)

7. Chittaragi, N.B., Koolagudi, S.G.: Automatic dialect identification system for Kannada language using single and ensemble SVM algorithms. Lang. Resource Eval. **54**, 553–585 (2020)
8. Chittaragi, N.B., Koolagudi, S.G.: Sentence-based dialect identification system using extreme gradient boosting algorithm. In: Elçi, A., Sa, P.K., Modi, C.N., Olague, G., Sahoo, M.N., Bakshi, S. (eds.) Smart Computing Paradigms: New Progresses and Challenges. AISC, vol. 766, pp. 131–138. Springer, Singapore (2020). https://doi.org/10.1007/978-981-13-9683-0_14
9. Chittaragi, N.B., Prakash, A., Koolagudi, S.G.: Dialect identification using spectral and prosodic features on single and ensemble classifiers. Arab. J. Sci. Eng. **43**(8), 4289–4302 (2017). https://doi.org/10.1007/s13369-017-2941-0
10. Hansen, J.H.L., Liu, G.: Unsupervised accent classification for deep data fusion of accent and language information. Speech Commun. **78**, 19–33 (2016)
11. Hermansky, H., Morgan, N.: Rasta processing of speech. IEEE Trans. Speech Audio Process. **2**(4), 578–589 (1994)
12. Li, H., Ma, B., Lee, K.A.: Spoken language recognition: from fundamentals to practice. Proc. IEEE **101**(5), 1136–1159 (2013)
13. Liu, G.A., Hansen, J.H.L.: A systematic strategy for robust automatic dialect identification. In: IEEE Nineteenth European Signal Processing Conference, pp. 2138–2141 (2011)
14. Rajapurohit, B.B.: Acoustic characteristics of Kannada, vol. 27. Central Institute of Indian Languages (1982)
15. Rao, K.S., Koolagudi, S.G.: Identification of Hindi dialects and emotions using spectral and prosodic features of speech. IJSCI: Int. J. Syst. Cybern. Inf. **9**(4), 24–33 (2011)
16. Sun, X.: A pitch determination algorithm based on subharmonic-to-harmonic ratio. In: The 6th International Conference of Spoken Language Processing, pp. 676–679 (2000)
17. Tong, R., Ma, B., Zhu, D., Li, H., Chng, E.S.: Integrating acoustic, prosodic and phonotactic features for spoken language identification. In: IEEE International Conference on Acoustics, Speech and Signal Processing, vol. 1 (2006)
18. Wang, D., Ye, S., Hu, X., Li, S., Xu, X.: An end-to-end dialect identification system with transfer learning from a multilingual automatic speech recognition model. In: Proceedings of Interspeech 2021, pp. 3266–3270 (2021)

Efficacy of Transfer Learning Over Semantic Segmentation

Adamya Shyam(✉) and Suresh Selvam

Department of Computer Science, Banaras Hindu University, Varanasi, Uttar Pradesh, India
adamyashyam2016@gmail.com, suresh.selvam@bhu.ac.in

Abstract. Semantic Segmentation is amongst the most difficult and important tasks in the field of computer vision. Used in the field of medical imaging, automated vehicles, geo-sensing, etc., the main idea of semantic segmentation is to link each pixel to a class label. Like other computer vision tasks, it too requires highly powerful and computational resources to produce good results. Also, most of the machine learning algorithms assume that the training, as well as the future data, will be in the same feature space, but, in real-world applications, the assumption does not hold. To resolve this problem, in this paper we have implemented the idea of transfer learning over semantic segmentation to study its efficacy. The paper also proposes some encoder-decoder framework-based models developed using transfer learning that perform at least as good as a fully supervised model. The paper concludes that transfer learning serves as a readily effective solution to enhance supervised learning models when properly carried out.

Keywords: Semantic segmentation · Transfer learning · ResNet · Computer vision · U-Net

1 Introduction

Semantic Segmentation is amongst the most difficult and important tasks in the field of computer vision. The main idea of semantic segmentation is to link each pixel to a class label in a way that pixels sharing some certain features are labeled the same. The task produces a collection of segments that jointly cover the whole digital image or a set of contours obtained from it. A pixel can be classified as similar to other pixels based on the intensity, color, texture, etc. It plays a highly significant and efficient role in different fields like Autonomous Driving, Classification or Categorization of Objects, Geo-Sensing, Precision Agriculture, and mostly in Medical Imaging by helping in the detection of tumors and cancers. The contours obtained in the segmentation process along with interpolation algorithms as marching cubes can help in the creation of 3D reconstruction. Many benchmark models have been developed through the years for this task based on networks like U-Net, AlexNet, etc.

As like every other task of computer vision, semantic segmentation requires highly powerful and computational resources to produce good results. Also, the assumption of same feature space for training and future data by most of the machine learning

I. Woungang et al. (Eds.): ANTIC 2021, CCIS 1534, pp. 687–699, 2022.
https://doi.org/10.1007/978-3-030-96040-7_52

algorithms do not hold, when applied to real-world application. It may happen that for a particular task, there is adequate training data available in one field of interest but not in the one, we might be working on. This problem of insufficient data and lack of high computability resources incepted the idea of transfer learning as a new learning framework. The appropriate implementation of this technique can greatly enhance the performance of learning without many expensive data-labeling efforts. Transfer learning can produce results at least as good as a fully supervised model and this approach can be especially useful when training time, data quantity, and quality are not good.

In the context of semantic image segmentation, many good pre-trained models are likely to enable effective transfer learning. For example – Oxford VGG, Microsoft ResNet are models that have been trained to perform very well on image classification on the ImageNet dataset, which has millions of data points, across 1000 classes. These models are well trained to perform many generic tasks and detect low-level features like edges, corners, ridges, and blobs. An independently developed model may not learn to perform these tasks as efficiently as the pre-trained models due to a much smaller training data set. Transfer learning can be especially useful when training time and data quantity and quality are not good.

Our paper studies the effectiveness of the transfer learning approach on the semantic segmentation task and proposes encoder decoder framework-based models that perform the task at least as well as a fully supervised benchmark model.

2 Background

2.1 Semantic Image Segmentation

Semantic Image Segmentation or pixel-level classification, is the process of clustering the pixels belonging to the same object class together, for an image. The main objective of this task is to segment unknown objects even if the visual concepts are not understood, unlike classification tasks that classify objects into specific labels. As the segmentation task classifies each pixel into its category, hence is known as the pixel-level classification [1, 2].

With the recent developments in the field of semantic segmentation [3], the task can be classified as

Region-Based Semantic Segmentation. In this approach extraction of free-form patches is done from the images which are then described, further followed by the region-based classification [4]. These predictions are further transformed to pixel predictions, generally by determining the highest-scoring patch or region that contains the pixel and labeling it accordingly [5].

FCN-Based Semantic Segmentation. The approach learns pixel-to-pixel mappings and does not extract regional features. Unlike classical CNN, which have fully connected layers, FCNs have only convolutional and pooling layers allowing them to predict outputs for arbitrary-sized inputs with the size of output depending on the input [6, 7].

Weakly Supervised Semantic Segmentation. As semantic segmentation approaches rely on large datasets having segmentation masks and the process of manual annotation of masks is expensive and time-consuming, hence, to tackle this issue, few weakly supervised methods are proposed recently which utilize annotated bounding boxes, or even image-level labels for this task [8].

2.2 Transfer Learning

The concept of transfer learning is to transfer knowledge obtained from a related task or domain to the target task [9]. Transfer learning can be especially useful when training time and data quantity and quality are not good. Relative to Purely Supervised Learning, transfer learning is expected to be at least as good. The disadvantage with a semi-supervised learning approach relative to Transfer Learning is due to the difficulty associated with manually procuring relevant, well-augmented data in very large quantities. Hence Transfer Learning turns out to be the best executable approach [10].

Transfer Learning can be carried out using Developed Model Approach, where a related predictive modeling problem is chosen and a significantly better model is developed or using Pre-Trained Model Approach, which uses one of the many models available that have been trained on large quantities of data and a large number of classes as the starting point for our target task [11]. In the context of semantic image segmentation, many good pre-trained models like Xception [12], VGG-Net [13], UNet [14], ResNet[15] are likely to enable effective transfer learning [16].

2.3 Related Work

Traditionally implemented using the clustering [17] or edge detection techniques [18], the task of Semantic Segmentation has evolved a lot with the inception of neural networks. Long et al., 2015 [19] adapted the traditional convolutional networks into FCNs using the idea of transfer learning and introduced a skip architecture for combining semantic knowledge with appearance data. Their model produced detailed segmentation and achieved improved results of PASCAL VOC but produced low-resolution output feature maps due to the presence of multiple alternated convolutional and pooling layers. Lu et al., 2019 [20] in their work proposed a pretrained-AlexNet based model obtained in order for automatic pathological brain detection in MRIs based on the deep learning structure and transfer learning. The authors of this paper replaced the last 3 layers' parameters with random weights and the rest parameters served as the initial values. Karimi et al., 2020 [21] in their paper assessed the approach of transfer learning for medical image segmentation. It was found by them that the variability among FCNs trained via transfer learning can be as high as that among FCNs trained with random initialization. Also, they found that the reuse of features can be more significant in the deeper layers and is just not restricted to the early encoder layers. Apart from these, most of the approaches focus on a fully supervised model, rather than the transfer learning approach [22].

3 Proposed Approach

This paper gives some insights about the efficacy of transfer learning over semantic segmentation task and proposes some models using the transfer learning approach for the task of Semantic Segmentation. For this purpose, a set of models based on the Encoder Decoder Framework were developed.

3.1 Encoder Decoder Framework

An Encoder-Decoder Framework is made up of an encoder, the hidden state, and the decoder. The encoder is a network like Fully Convolutional or Convolutional Neural Network or Recurrent Neural Network which takes the inputs and further produces a feature map or vector or tensor. The resulting map holds the information i.e., the features representing the input. The second element of the framework is the hidden state which further holds this representation and passes it to the decoder as an input to it. The decoder then finds the best match to the actual input and gives it as the target output. The decoder is also a network, generally the same as the encoder but with opposite orientation. Figure 1 provides a basic visualization of the framework used for the segmentation task.

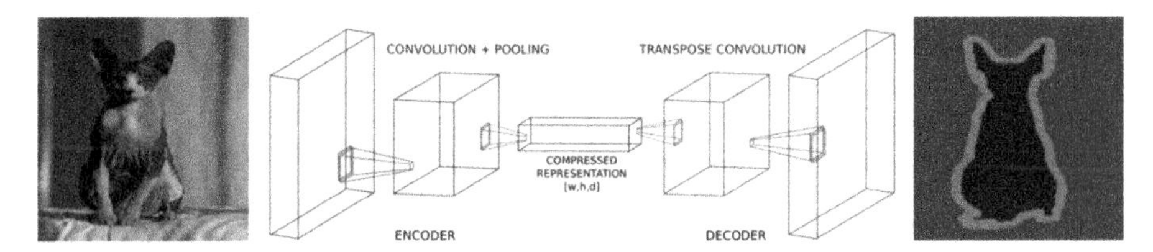

Fig. 1. The encoder-decoder framework

3.2 The Architecture

ResNets were developed to mitigate the effects of the vanishing gradients problem. In ResNet, skip connections are introduced allowing connections from a layer to a few layers ahead. In this way, the loss of signal due to many layers of activation is made up for by stronger signals, by adding the identity term across the residual blocks. There are several ResNet architecture variants available out of which few have been trained on the ImageNet data set and have produced good results for the classification tasks. In this proposed work, we firstly used the ResNet50 Architecture (Fig. 2(a)) as the encoder of the encoder-decoder framework to perform the segmentation task. Further, the best model developed using the ResNet50 Architecture was modified and ResNet152V2, VGG16, and Xception (Fig. 2(b–d)) Architectures were used as Encoder in place of ResNet50 in that model. The fully connected layers of each of the Architectures that are used to make predictions were replaced with dense layers of custom configuration in the proposed models. These Dense layers which are deeply connected layers perform the task of Decoder. The dense layer performs matrix-vector multiplication where the matrix values are trainable parameters that can be updated using backpropagation. This layer is used for modifying the dimensions of the input vector. The output so produced is further reshaped to the required dimension, here into a low-resolution image.

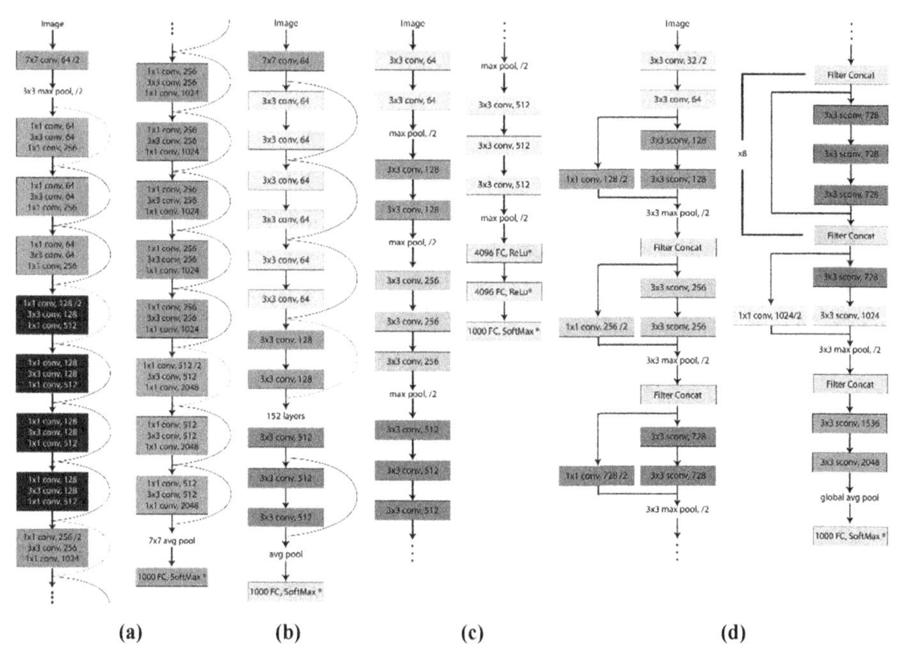

(a) (b) (c) (d)

The FC layers of the architecture was removed from the Encoder used in proposed models.

Fig. 2. (**a**) ResNet50 (**b**) ResNet152V2 (**c**) VGG16 (**d**) Xception

The architecture of models developed vary in four dimensions based on:

The Encoder's Trainability. This was done to measure the sufficiency of the pre-trained models for the particular task. If the pre-trained models have a good generalization ability, then there may be more to lose than benefit out of training the model on the data set.

The Decoder's Depth. The Decoders are made of Dense Layers. The number of dense layers decides the depth of the decoder. This was chosen to determine the right architecture of the decoder.

Weight Initialization. Initialization of weights is done to measure the effectiveness of Transfer Learning versus Purely Supervised Learning for the same architecture. Two different weight initializations viz. Random & ImageNet were used in this proposed work and have been described briefly later in this paper.

The Encoder's Architecture. To study the effect of the Encoder's weight over transfer learning, four different architectures viz. ResNet50, ResNet152V2, VGG16, and Xception were used as an Encoder in this proposed work.

3.3 The Workflow

Each of these proposed models follows the workflow shown in Fig. 3. For an H × W × C input image, semantic segmentation generates a label map of the same dimensions where each pixel is labeled with a category label and H, W, and C represent the height, width, and color channels respectively.

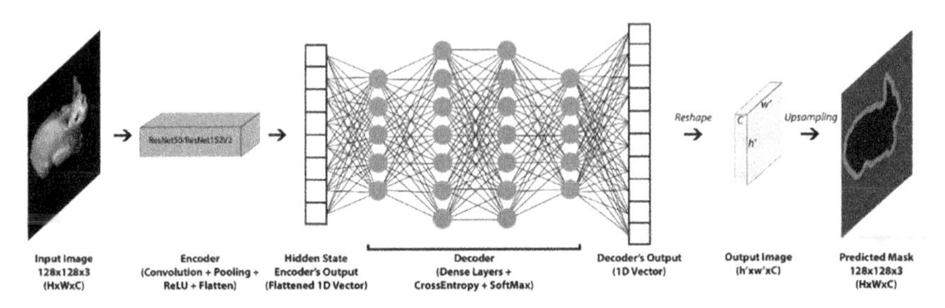

Fig. 3. The workflow of the proposed approach

When the input image is fed to any of the proposed models it goes through the encoder which is a ResNet50 or ResNet152V2. The ResNet, as discussed before, consists of Convolutional and Pooling layers. The ResNet first produces a h × w × c feature map, where h = H/r, w = W/r, and r is the downsampling factor. It is then flattened into a 1D vector by the ResNet. The idea of the work was to use the pretrained weights of ResNet for feature extraction and then feed it to the decoder of our configuration. Hence, the FC layers of the ResNet were removed. The 1D vector so produced is then fed to the decoder which is a collection of dense layers with custom configuration i.e., depth of our choice. The decoder produces a 1D Vector which is then reshaped to h' × w' × C, where h' and w' are respectively the height and width of the low-resolution image obtained. Further, upsampling of the output image to dimensions as of the input image produced the predicted mask.

4 Implementation

4.1 Dataset and Preprocessing

The dataset used in our work is the Oxford-IIIT Pet dataset, available in Tensorflow datasets. It consists of images of pets from 37 different categories with over 200 images per category. In total, the data set consists of over 7000 images varying in scale, pose, and lighting. The data comes with images of the dimension 128 × 128 × 3, along with annotations for each image. Annotations include a bounding box of the head of the pet in each image and ground truth for the segmentation mask. The ground truth data is represented in the form of a tri-map (Fig. 4), with each pixel getting one of the three labels – Interior, Boundary, and Exterior.

The dataset is already split into a ratio of approximately 50:50 between train and test images. Each image is labeled using Class ID ranging from 1 to 37, Species ID viz. 1 and 2, and Breed ID ranging from 1 to 25 and 1 to 12 for Species ID 1 and Species ID 2

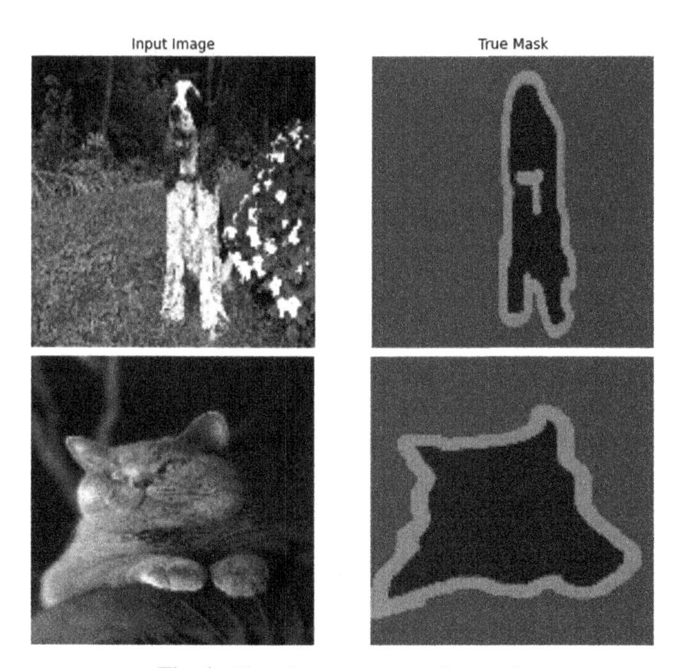

Fig. 4. The tri-map segmentation mask

respectively. In the preprocessing task, data augmentation and normalization were done. Data Augmentation is a technique to manipulate data in such a way that the manipulated data does not lose the core or essence of the original data. It is practiced to increase the dataset by slightly introducing distortion to the images which helps to reduce overfitting during the training step. Normalization is a process that changes the range of pixel intensity values. It is used for noise removal as well as to make the intensity values of the image follow a normal distribution as far as possible, making it look better for the visualizer. In this work, augmentation of data was done by flipping the images and each image was normalized to [0,1]. For convenience, the pixels of the segmentation mask which were labeled as $\{1, 2, 3\}$, were decremented by a unit as $\{0, 1, 2\}$.

4.2 Implementation of Transfer Learning

Transfer learning consists of extracting features learned on one problem and leveraging them on a new, similar problem. The basic idea of transfer learning implemented in our work is as follows:

1. Initiation of a base model with pre-trained weights.
2. Freezing all layers of the base model (trainable = False).
3. Creation of a new model over the output of one or several layers of the base model.
4. Training of the modified model on the desired dataset.

Another step that is practiced in transfer learning is fine-tuning. It is the process of unfreezing the whole (or a portion of) model, and re-training it over another dataset using a lower learning rate. Potential meaningful improvements can be achieved by incrementally adjusting the pretrained features to the new dataset. In this work, the first few models were initiated with non-trainable encoders, and further were unfroze and retrained to get better results.

The models were all trained till epoch 20 with a batch size of 64. The encoder used average pooling for downsampling along with flattening. The pixel-wise loss function used was the Sparse – Categorical Loss.

4.3 Performance Metric

To evaluate the accuracy of the proposed models, the performance metric used in this work is pixel accuracy. Generally, the metric is calculated for each class separately as well as all classes together. For the per-class accuracy, i.e., when the binary mask is evaluated; each predicted pixel class value can be true positive (TP), true negative (TN), false positive (FP), or false negative (FP). The predicted pixel class if is the same as the actual class according to the target mask, it is said to be TP; whereas a pixel when predicted correctly to be not of the given class, then it represents TN. A pixel when is incorrectly classified to a class it doesn't belong to, then it is called FP; whereas an FN signifies a pixel wrongly recognized as belonging to the given class.

$$Pixel\ Accuracy = \frac{TP + TN}{TP + TN + FP + FN} \tag{1}$$

4.4 Results and Discussions

After developing various models and performing the semantic segmentation task with the help of it, the following results were obtained as shown in Table 1. The table represents the series of models developed and the validation accuracy achieved after 20 epochs. As discussed, the models vary on the basis of their initialization, trainability of encoders, and the depth of the decoder.

Table 1. Accuracy of the proposed models

	Encoder	Trainable	Decoder	Initialization	Accuracy	Validation accuracy
1	ResNet50	False	1 (12288 Units)	ImageNet	65.70	64.86
2	ResNet50	False	1 (3072 Units)	ImageNet	67.76	66.67
3	ResNet50	False	2	ImageNet	71.10	70.24
4	ResNet50	False	4	ImageNet	72.22	71.22
5	ResNet50	True	4	ImageNet	87.18	84.37

(continued)

Table 1. *(continued)*

	Encoder	Trainable	Decoder	Initialization	Accuracy	Validation accuracy
6	ResNet50	True	6	ImageNet	89.20	86.76
7	ResNet50	True	6	Random	84.58	77.43
8	ResNet50	True	8	ImageNet	86.83	84.43
9	ResNet50	True	8	Random	83.87	76.54
10	ResNet50	True	10	ImageNet	86.70	84.09
11	ResNet152V2	True	6	ImageNet	86.81	84.20
12	VGG16	True	6	ImageNet	85.86	83.16
13	Xception	True	6	ImageNet	85.30	83.16

Initially, few models were developed with a non-trainable encoder which means by freezing the layers of the base model (ResNet50) with ImageNet initialization. As evident from the table, models 1, 2, 3, and 4 were developed using this configuration and produced modest results with validation accuracy between 65% to 75%. As visible in Fig. 5 these models learned a single structure giving an acceptable result for all the images. These four models though produced some good results but were found to be underfitting due to the non-trainability of the encoder.

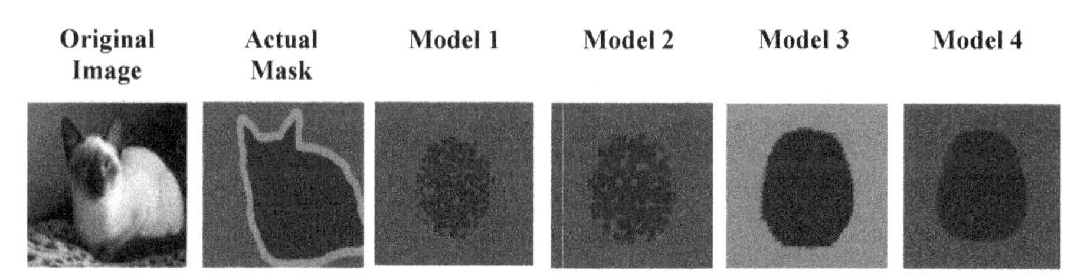

Fig. 5. Segmentation masks produced by Model 1–4

Model 5 was developed the same as Model 4 but with a trainable encoder to study its effect on accuracy. It was found that trainable encoders perform much better than the non-trainable ones for our task as shown in Fig. 6.

Fig. 6. Segmentation masks produced by Model 5

Hence, rest of the models were developed using trainable encoders. On comparing Models 6 and 7 or Models 8 and 9, who differ from each other only on the basis of weight initialization, it was found that models with ImageNet initialization perform better than models with Random initialization.

The difference in validation accuracy of Model 6 and 7 or Model 8 and 9 as shown in Fig. 7, given their values, is both quantitatively and qualitatively significant and serves as strong evidence of the performance benefitting from transfer learning.

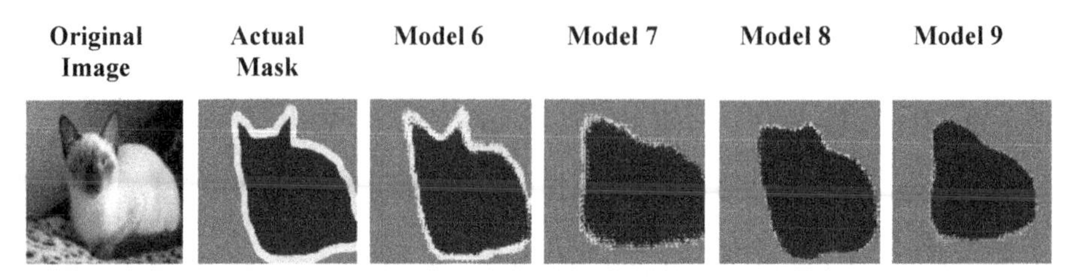

Fig. 7. Segmentation masks produced by Model 6–9

Model 8 and 10 were developed with depth 8 and 10 respectively in order to study the relationship between the decoder depth and validation accuracy. It was found that although they produce good results but the validation accuracy starts to decrease slowly with an increase in depth of the decoder. This serves as evidence of fact that transfer of knowledge plays greater importance than the depth of the decoder. Figure 8 shows the mask produced by Model 10.

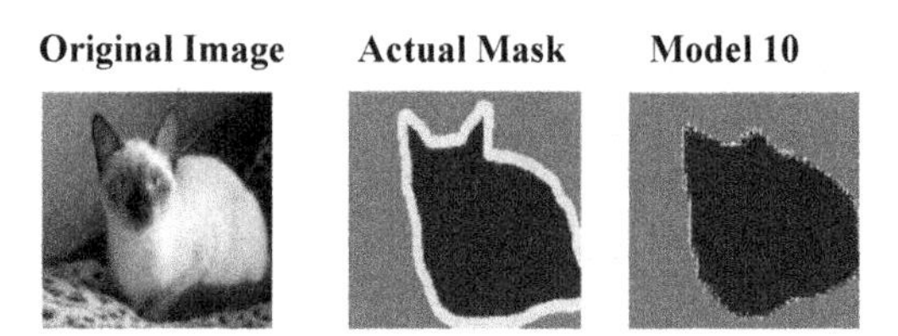

Fig. 8. Segmentation mask produced by Model 10

To check the effect of change in Encoder's Architecture over transfer learning, three more models were developed with ResNet152V2 (Model 11), VGG16 (Model 12) and Xception (Model 13) as the Encoder with the hyperparameters same as Model 6. The results obtained were satisfying, but, showed no significant change in the accuracy of the model reflecting that the change of encoder's architecture has no significant effect on the performance of transfer learning for the task of semantic segmentation. Figure 9 shows the masks produced by Models 11, 12, and 13.

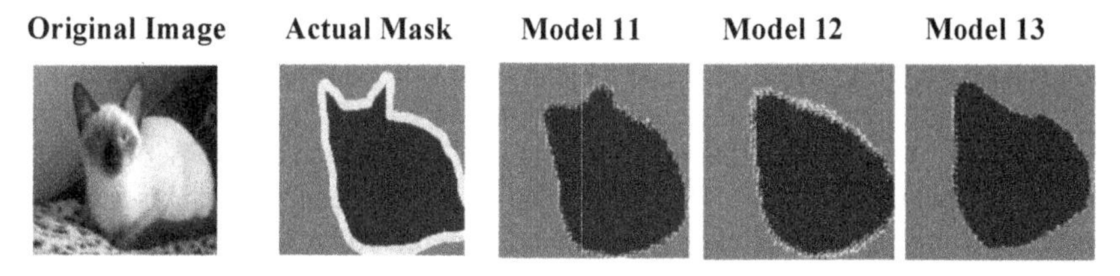

Fig. 9. Segmentation masks produced by Model 11, 12 and 13

After, comparing all the models based on the validation accuracy and masks predicted, Model 6 was found to be the best model developed through transfer learning for the semantic segmentation task. The model had an accuracy of 89.20% and validation accuracy of 86.76% and was an implementation of transfer learning with 'ImageNet' initialization, had a decoder depth of 6, and a trainable encoder with ResNet50 Architecture. The model outperforms the benchmark model based on U-Net architecture (as shown in Fig. 10) which had accuracy of 88.94% and validation accuracy of 85.78% by a satisfactory margin.

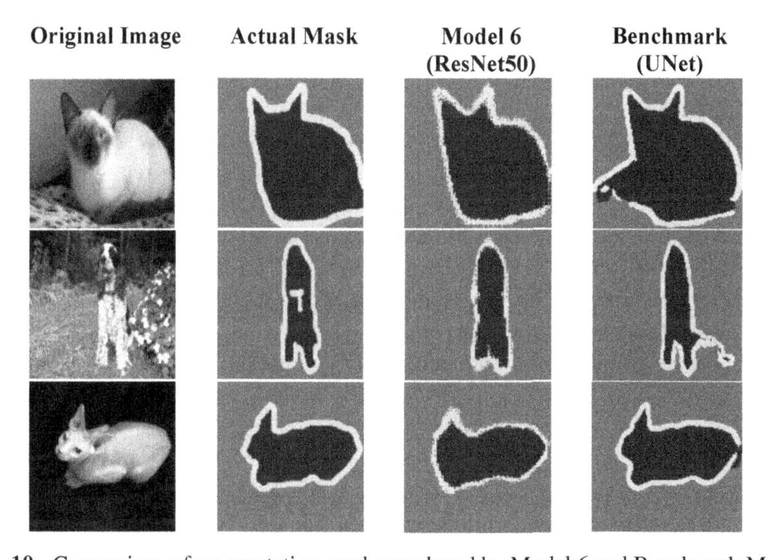

Fig. 10. Comparison of segmentation masks produced by Model 6 and Benchmark Model

5 Conclusion and Future Works

Computer Vision tasks are generally performed on large datasets and require high computational power and resources. Also, like most of the ML algorithms, these tasks also assume that the training as well as the future data will be in the same feature space, but, in real-world applications, the assumption does not hold. It may happen that for a particular task, we have sufficient training data in one domain of interest but not for another domain of interest we might be working on. The idea of transfer learning incepted from this. In our work we tried to perform semantic segmentation with the help of transfer learning and developed a series of models in order to obtain a model which can perform at least as well as a benchmark model. The models varied on the basis of encoder trainability, decoder depth, weight initialization, and encoder architecture. In general, all the models that we developed involving transfer learning proved to be effective, when allowing trainability of the encoder. Hence, we suggest that transfer learning serves as a readily effective solution to enhance supervised learning models when properly carried out.

The best model that we developed outperforms the benchmark that we set out to match. It can also be concluded from the work that transfer of knowledge plays greater importance than the depth of decoder as the accuracy of models increased with decoder depth but then decreased gradually with it. Also, it can be seen by comparing models who differ from each other only on the basis of weight initialization that models with ImageNet initialization perform better than models with Random initialization. The difference in validation accuracy of these models, given their values, is both quantitatively and qualitatively significant and serves as strong evidence of the performance benefitting from transfer learning. Change in the architecture of the encoder has no significant effect on the performance of transfer learning for the semantic segmentation task. Further, transfer learning has been established as a very effective methodology to enhance model performance and more importantly, the quality of the model performance over purely supervised learning for a certain architecture.

To further enhance the performance on this task, we could follow several approaches such as changing the architecture of the decoder, changing the entire architecture of the model like generative adversarial networks, introducing encoder-decoder skip connections, or combining methodologies like transfer learning initialization along with semi-supervised learning, while procuring large volumes of un-labeled and well-augmented data. Also, the other models developed in this work through transfer learning performed well and suggest that the architecture of choice to go along with transfer of knowledge is an area that we could improve on, potentially gaining in the validation accuracy.

References

1. Thoma, M.: A survey of semantic segmentation. arXiv preprint arXiv:1602.06541 (2016)
2. Li, R., et al.: DeepUNet: a deep fully convolutional network for pixel-level sea-land segmentation. IEEE J. Select. Top. Appl. Earth Observ. Remote Sens. **11**(11), 3954–3962 (2018)
3. Liu, X., Deng, Z., Yang, Y.: Recent progress in semantic image segmentation. Artif. Intell. Rev. **52**(2), 1089–1106 (2018). https://doi.org/10.1007/s10462-018-9641-3
4. Lalaoui, L., Mohamadi, T.: A comparative study of image region-based segmentation algorithms. Int. J. Adv. Comput. Sci. Appl. **4**(6), 198–206 (2013)
5. Rajab, M.I., Woolfson, M.S., Morgan, S.P.: Application of region-based segmentation and neural network edge detection to skin lesions. Comput. Med. Imaging Graph. **28**(1–2), 61–68 (2004)
6. Villa, M., Dardenne, G., Nasan, M., Letissier, H., Hamitouche, C., Stindel, E.: FCN-based approach for the automatic segmentation of bone surfaces in ultrasound images. Int. J. Comput. Assist. Radiol. Surg. **13**(11), 1707–1716 (2018). https://doi.org/10.1007/s11548-018-1856-x

7. Guo, Y., Liu, Y., Georgiou, T., Lew, M.S.: A review of semantic segmentation using deep neural networks. Int. J. Multim. Inf. Retriev. **7**(2), 87–93 (2017). https://doi.org/10.1007/s13735-017-0141-z
8. Khoreva, A., Benenson, R., Hosang, J., Hein, M., Schiele, B.: Simple does it: weakly supervised instance and semantic segmentation. In: Proceedings of the IEEE Conference on Computer Vision and Pattern Recognition, pp. 876–885 (2017)
9. Pan, S.J., Yang, Q.: A survey on transfer learning. IEEE Trans. Knowl. Data Eng. **22**(10), 1345–1359 (2009)
10. Van Opbroek, A., Ikram, M.A., Vernooij, M.W., De Bruijne, M.: Transfer learning improves supervised image segmentation across imaging protocols. IEEE Trans. Med. Imaging **34**(5), 1018–1030 (2014)
11. A Gentle Introduction to Transfer Learning for Deep Learning. machinelearningmastery.com. Accessed 21 June 2021
12. Lo, W.W., Yang, X., Wang, Y.: An xception convolutional neural network for malware classification with transfer learning. In: 2019 10th IFIP International Conference on New Technologies, Mobility and Security (NTMS), pp. 1–5. IEEE (2019)
13. Geng, L., Zhang, S., Tong, J., Xiao, Z.: Lung segmentation method with dilated convolution based on VGG-16 network. Comput. Assist. Surg. **24**(sup2), 27–33 (2019)
14. Ronneberger, O., Fischer, P., Brox, T.: U-net: convolutional networks for biomedical image segmentation. In: Navab, N., Hornegger, J., Wells, W.M., Frangi, A.F. (eds.) Medical Image Computing and Computer-Assisted Intervention – MICCAI 2015: 18th International Conference, Munich, Germany, October 5–9, 2015, Proceedings, Part III, pp. 234–241. Springer, Cham (2015). https://doi.org/10.1007/978-3-319-24574-4_28
15. Hong, J., Cheng, H., Zhang, Y.-D., Liu, J.: Detecting cerebral microbleeds with transfer learning. Mach. Vis. Appl. **30**(7–8), 1123–1133 (2019). https://doi.org/10.1007/s00138-019-01029-5
16. Shin, H.C., et al.: Deep convolutional neural networks for computer-aided detection: CNN architectures, dataset characteristics and transfer learning. IEEE Trans. Med. Imaging **35**(5), 1285–1298 (2016)
17. Muthukrishnan, R., Radha, M.: Edge detection techniques for image segmentation. Int. J. Comput. Sci. Inf. Technol. **3**(6), 259 (2011)
18. Sharma, P., Suji, J.: A review on image segmentation with its clustering techniques (2016)
19. Long, J., Shelhamer, E., Darrell, T.: Fully convolutional networks for semantic segmentation. In: Proceedings of the IEEE Conference on Computer Vision and Pattern Recognition, pp. 3431–3440 (2015)
20. Lu, S., Lu, Z., Zhang, Y.D.: Pathological brain detection based on AlexNet and transfer learning. J. Comput. Sci. **30**, 41–47 (2019)
21. Karimi, D., Warfield, S.K., Gholipour, A.: Critical assessment of transfer learning for medical image segmentation with fully convolutional neural networks. arXiv preprint arXiv:2006.00356 (2020)
22. Xing, Y., Zhong, L., Zhong, X.: An encoder-decoder network based FCN architecture for semantic segmentation. Wirel. Commun. Mob. Comput. **2020**, 1–9 (2020)

Performances of Different Approaches for Fake News Classification: An Analytical Study

Md. Abdullah-Al-Kafi, Israt Jahan Tasnova, Md. Wadud Islam, and Sumit Kumar Banshal(✉)

Department of Computer Science and Engineering, Daffodil International University, Dhaka 1207, Bangladesh
{abdullah15-12152,israt15-12932,wadud15-12547, sumit.cse}@diu.edu.bd

Abstract. The penetration of social and online platforms has opened a new substantial domain of Fake news dissemination in the current time. Also, this dynamic form of data opens up new dimensions for researchers to detect Fake news from the ocean of data. Therefore, Fake news detection has attracted both academia and industry indifferently as research or analytical domain in the concurrent time. Due to data availability, the classification tasks have been tested in different sets and types of data. Detecting Fake news evolves as an actual potential domain to explore with more efficient algorithms and parameter-based modified algorithms. In this work, an analytical sketch has been drawn to compare the performances of different classifiers depending on accuracy and time. Seven classifiers of four different types have been implemented and tested namely, Multilayer Perceptron, Sequential Minimal Optimization, Logistic Regression, Decision Tree, J48, Random Forest and Naïve Bayes Classifier. The analytical evaluation process has been designed with three experimental setups, 10-fold cross-validation, 70% split and 80% split. The separate setups show distinctive outcomes across the algorithms. Naïve-Bayes classifier model shows its prominence along with the Random Forest classifier. However, the and Decision Tree-based classifiers perform differently from earlier knowledge. Furthermore, this paper identifies a different aspect of using testing-training splitting in classifier tasks.

Keywords: Fake news detection · Machine learning · Classification

1 Introduction

The term "Fake news" is used to describe false stories spreading on social media. It has been invoked to discredit some news organizations' critical reporting [1]. That means the news that is based on false facts is called Fake news. Fake news became popularized during the 2016 U.S. elections, where the top twenty frequently-discussed false election stories generated 8,711,000 shares, reactions, and comments on Facebook. Ironically, more significant than the total of 7,367,000 for the top twenty most-discussed factually correct election stories posted by 19 major news websites [2]. Fake news can be disseminated in society through different mediums. Sometimes it can be spread through

I. Woungang et al. (Eds.): ANTIC 2021, CCIS 1534, pp. 700–714, 2022.
https://doi.org/10.1007/978-3-030-96040-7_53

people, and sometimes, it can be spread through news mediums. But nowadays, the most prominent medium of spreading Fake News is Social media and online platforms such as Facebook, Twitter, YouTube, and other websites [3].

Fake news is generated to convince its readers to believe in a particularly intended purview. So, Fake News can create mistrust among the people in society. Fake news highlights the erosion of long-standing institutional earthworks against misinformation in the internet age, a global problem. Fake news overlaps with other information disorders like misinformation and disinformation [4]. Misinformation is false or misleading information and disinformation is purposely spread information to deceive people. This misleading approach towards information has transformed Fake news as a political weapon [4]. During the election, voters can be influenced by misleading political statements and claims [5]. Fake news has drawn significant concerns from both industry and academia due to its use in the current era of technology. A massive amount of misleading information is created and displayed on the internet. It hurt the internet activities like online shopping and social marketing [5]. There are countless web pages established to publish fake news and stories. Researchers identified these types of several various pages such as denverguardian.com, wtoe5news.com, ABCnews.com.co, and so on [5]. Due to its speed and potent of spreading misinformation, the Fake news detection topic has gained a great deal of interest from researchers across the globe [3]. A substantial number of academic articles used the term "fake news" between 2003 and 2017 resulted in a typology of types of fake news: news satire, news parody, fabrication, manipulation, advertising, and propaganda [1]. Fake news can be any content that is not truthful and generated to convince its readers to believe in something that is not true. So fake news can create mistrust among the people in society. This news is nearly impossible to verify analytically because of its huge quantity and high dimensions. So, a machine learning approach is better in this situation. In the computer science domain as well, the fake news classification is a well-discussed domain. There are so many papers [6–8] that most of the approaches share the same characteristics. Fake news classification is a Supervised Text classification task. To identify fake news Kareem et al. used seven different supervised learning classifications in their paper and compared results of classification [9] Lui et al. [10] proposed an ensemble framework in their paper to address the fake news classification challenge that was in ACM WSDM Cup 2019. Hakak et al. [11] also proposed an ensemble classification model to detect fake news. In Othman et al. [12] investigate the performance of different classification or clustering methods for a set of large data in their paper. Rubin et al. [13] proposed the SVM classification using five features. LR avoids general-purpose nonlinear optimization algorithms and its works well in text classification [14]. Naive Bayes algorithm is used in text classification because of its simplicity and effectiveness. A Naive Bayesian model is easy to build and it has no complicated iterative parameter estimation which makes it particularly useful for very large datasets [15]. SMO(SVM) is known as the hyperplane, between classes. This hyperplane separates the data into classes. It makes the path easier to get the result of text classification [16]. This work aims to assess the performance of different classification algorithms using the WEKA data mining tool for classifying fake news. WEKA or Waikato Environment for Knowledge Analysis is a data mining and machine learning tool to help users make a wide range of sophisticated learning algorithms available through open-source packages [17].

2 Related Work

The evolution of social and online platforms enables researchers to harvest various data from these dynamic mediums. Which eventually trigger the usage of classification algorithms to detect Fake news from such broadcasting sources. This robust and gigantic data facilitate researchers to train these classifiers with real-time scenarios to investigate the authenticity of the news. Two classifier approaches, the Support Vector Machine (SVM) and Naïve Bayes were noted as better-performing algorithms. The performance was evaluated based on their accuracy of detecting Fake news correctly in this approach [18]. In another work, Kareem et al. [9] noted the K Nearest Neighbors (KNN) as the best performing classifier with 70% accuracy followed by logistic regression with 69% accuracy on their dataset. They used seven different media fake news classification approaches on 344 news articles by scrapping popular news websites. They labeled the data to train their classifiers in two categories: Fake or True. They used two feature extraction techniques: Term Frequency (TF) and Term Frequency-Inverse Document Frequency (TF-IDF).

Beyond traditional direct classification approaches, there were several approaches [3, 4, 8, 9] implemented with an ensemble framework. Liu et al. [3] proposed an ensemble framework to address the fake news classification challenge in ACM WSDM Cup 2019. They regarded this problem as the Natural Language Inference (NLI) task and proposed a novel empirical ensemble framework. This framework performed with more than 85% accuracy. In another approach of ensemble classification model for detection of the fake news, a better level of accuracy was achieved [11]. Their proposed model extracts relevant features from the fake news dataset. Then the extracted features were classified using the ensemble model comprising three popular machine learning models, i.e., Decision Tree, Random Forest, and Extra Tree Classifier. They achieved a training and testing accuracy of 99.8% and 44.15%, respectively, on the ISOT dataset. At the same time, the accuracy went up to 100% for the Liar dataset.

In another work, SVM classification-based approach, researchers proposed five feature-based models to identify satire and humor news articles [13]. In this work, 360 satirical news articles were explored from four domains and achieved 90% accuracy in detecting satire and humor. The final findings were reported based on three (Absurdity, Grammar, and Punctuation) features instead of five. Also, several approaches are implemented in different works where a limited number of instances are used or tested with a limited number of algorithms or used a dataset with fewer events or relied on crowdsourcing for validation, etc. [6, 8, 19]. Furthermore, Twitter threads were also characterized to understand their potential to create fictitious information [20]. And surprisingly enough, this microblogging platform was noted as one of the prominent sources to be used as evidence to produce fabricated news [21]. Though there are several approaches have been implemented to detect fake news. But still, we found this domain to be in scattered implementation with different smaller datasets. Therefore, we tried to fill this well-addressed research problem with a more comprehensive approach. We tested four different types (Function-based, Ensemble, Tree, Bayesian) of seven classifiers on the same dataset [22]. Although these papers have addressed different algorithms on different datasets. But a comparative analysis of the performance of the algorithms depending on Time and Accuracy is missing, In this paper, we tried to bridge the gap by introducing a comparative landscape on different algorithms on different experiment setups.

3 Data and Methodology

3.1 Data

Fake news detection is a text processing technique where the text has been used to detect its validity. In our dataset, we have text data from [23] which had 2 separate file with 4 columns such as "Title, Text, Subject, Date". The news data are collected form social media focusing on three subjects i.e., Politics, world news, and others. The dataset has 21417 instances of true news and 23481 instances of fake news in 2 separate files. A new dataset was created after margining Fake.csv and True.csv [23] at random and reducing the dimensions [22]. The dataset used in this work comprises 7819 instances with four fields, namely 'id', 'title', 'text' and 'label' with some errors and extra characters. The initially collected dataset has been cleaned and processed in multiple steps to remove unwanted features, texts, symbols, unnecessary punctuations, as well as unreadable sentence structures. The text is further processed by removing invalid sequences/characters in Unicode language tools so that WEKA can work fine on the dataset. Then the CSV file was converted to attribute relation file format manually. Because WEKA ARFF loaders work better than CSV converters and are more reliable. After that, the dataset was ready for applying different WEKA Filters for further cleaning and processing. Several WEKA features have been used to perform these preprocessing of data such as, for stemmer "weka.core.stemmers.IteratedLovinsStemmer"; "weka.core.tokenizers.WordTokenizer" for tokenizing purposes. To change the strings to word vector "StringToWordVector" from weka.filters were used. After applying the filters, a dataset with 381 attributes and 254 instances was created [22]. The data cleaning and preprocessing workflow have been depicted in Fig. 1 for better visualization of the approach.

3.2 Experiment

The practical design concerns detecting the Fake news which is enhanced from text classification eventually. Two possible classes exist in the detection process, i.e., 'fake' and 'real'. Therefore, it's a binary classification problem and also non-linear. Algorithms such as Neural Network, Multilayer perceptron, Support Vector machine, Logistic Regression, Tree algorithms such as Decision Tree are ubiquitous in Text and fake news classification [9, 13, 20, 24, 25]. So, in this experiment the aim is to measure the performance in terms of time and correct classification rate of Multilayer Perceptron (MLP), Sequential Minimal Optimization (SMO), Logistic Regression (LR), J48, Random Forest (RF), Decision Tree (DT) and Naïve Bayes (NB) on the same dataset. For

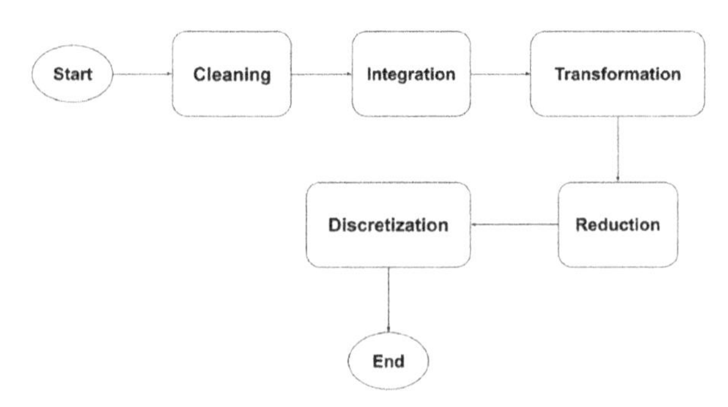

Fig. 1. Data preprocessing workflow diagram

binary classification problems, logistic regression works very efficiently [26] whereas, MLP or Neural network is heavily used in Natural Language Processing (NLP) and text synthesis [27]. Though there might be some issues with the Decision tree classifier as it is based on nodes and our dataset is not the same. But still, we want to keep it along with J48. Random forest is an Ensemble classifier and it can easily overfit to noise in the data. Whereas, it tries to control the variance in the dataset. Naïve Bayes classifier is a well-known classifier using probabilities that eventually performs well in text classification tasks using count vectorization [28]. SVM is a very well-established classifier in various studies and also in fake news detection. It was found to show good results in identifying fake news correctly [18]. While implementing the experiment, all the parameters were set to default for achieving a neutral environment for all the algorithms.

As the study focused on the performance of the algorithms depending on time and correct classification, we used three different approaches namely, 10-fold Cross-Validation, 70% split, and 80% split of the dataset. Though, cross-validation is reported to have the problem of overfitting. Algorithms such as the Decision tree do prune and can face overfitting. We also checked the dataset by splitting it into two parts as training and testing. In 70% split, dataset split 70% as training dataset and 30% as testing dataset. We used a 70% split on the dataset to evaluate the performance more effectively. The most common split ratio is 80:20 that data scientists use. In 80% split, dataset split 80% as training dataset and 20% as testing dataset. We used an 80% split on the dataset to compare 70% split and 80% split in terms of time and accuracy. Ten-fold cross-validation, 70% split, and 80% split were implemented to measure the performance with default parameters for each of these seven classifier algorithms.

3.3 Function Classifiers

Logistic Regression (LR). Logistic regression is an algorithm used to predict the categorical dependent variable using a given set of independent variables. The logistic regression model is susceptible to "bad" data [29]. "Bad" data pointing the outlying responses and extreme points in the design space(X) [29]. The model takes the natural logarithm of the odds as a regression function of the predictors. The fundamental equation of the generalized linear model is,

$$g(E(y)) = \alpha + \beta x1 + yx2$$

It predicts the probability of occurrence of an event by fitting data to a logit function [30].

Multilayer Perceptron (MLP). Artificial neural networks are an alternative to many statistical modeling techniques used across different scientific sectors. A multilayer perceptron is eventually a form of artificial neural network. Most of the applications of MLP are related to classification, prediction, pattern recognition [31]. In general, MLP can be depicted as,

$$y = f(x)$$

Where,

y = [n * j]
x = [n * k]
n is the number of training instances
k is the number of input variables
j is the number of output variables

Backpropagation is used to find the weight optimizes the function y = f(x), where the x and y are training matrices.

Sequential Minimal Optimization (SMO). SMO or Sequential Minimal Optimization algorithm effectively trains support vector machines (SVMs) on classification [32]. Flake et al. [8] express the runtime of a single SMO step as,

$$(p \cdot W \cdot n + (1 - p) \cdot n)$$

Here,

p = the probability that the second Lagrange multiplier is in the working set
W = the size of the working set
n = the input dimensionality

SMO breaks significant quadratic programming problems into a series of most minor possible quadratic problems.

3.4 Tree Classifiers

Decision Tree (DT). A decision tree is widely used as a learning algorithm called Decision Tree Learning [33]. In a decision tree, each node contains a decision rule. Decision tree split based on the condition. A Dataset is assigned a label to characterize its data point [12]. The Model needs to learn features to take and corresponding correct threshold to optimally split the dataset. It is possible by information theory. When the information-theoretic point of view is pursued, the amount of average mutual information is gained at each tree level [34]. Information gain is calculated by comparing the entropy of the dataset. So, the way to quantify:

$$Entropy = \sum -p_i \log(p_i)$$

Here, p_i =probability of class i. Entropy is measured between 0 and 1. The state that gives minimum entropy is a pure node.

J48. J48 is an approach to discover the hidden relationships among data [35]. J48 has been considered the most efficient machine learning algorithm for predicting any crime dataset [35]. J48 is a decision tree algorithm based on ID3 and C4.5 algorithms [36]. It performs better both in performance and execution time. Kaur et al. proposed [37] the modified J48 classifier to increase the accuracy of the data mining procedure. The algorithm is applied by the popular WEKA tool.

3.5 Ensemble Classifier

Random Forest (RF). Random Forest classifier produces multiple decision trees. To decrease the correlation between decision trees, random forest considers controlling the term ρσ2 [33]. ρσ2 is the main part of the variance. Lee et al. introduced average relative importance

$$I_j^2 = \frac{1}{B} \sum_{b=1}^{B} I_j^2(b)$$

Where I_j^2(b) is the relative importance for the b-th decision trees [33]. RF classifiers can successfully handle high data dimensionality and multicollinearity, being both fast and insensitive to overfitting. Random Forest is a type of machine learning called bootstrap aggregation or bagging. Combining results from multiple models is called aggregation (majority votes). By bagging Random Forest algorithms gain better accuracy.

3.6 Bayesian Classifier

Naïve Bayes (NB). The Naive Bayes algorithm is a simple probability classifier. It calculates a set of probabilities by counting the frequency and combinations of values in a given data set [28]. This classifier learns from training data. In this classifier, the conditional probability of each attribute Ai given the class label C [38, 39]. Naive Bayes is applied on some data set and the confusion matrix is generated for class having possible values. For example, in a news dataset the method follows:

$$\text{pr[E/H]} = N! * \prod_{i=1}^{k} \frac{p_i^{n_i}}{n_i!}$$

Here pr[E/H] is the probability of the document/news given its class H. and N is the number of words in the report.n_i is the time of occurrence of the word in the news. pi is the probability of obtaining the word from the news concerning category H.

3.7 Tools and Data Preprocessing

Tools. For this experiment used tools are: Machine Configuration:

1.Cpu: I3 1005 (2 core,4 thread)
2.Storage: SSD 256 GB (R = 465 MB, W = 375 MB)
3.Ram: 8 gb DDR4 2600 MHz
3.No discrete GPU
4.Windows 10 pro 64bit
6.Weka 3.9
7.Python 3.9

4 Result

As discussed in the previous section, to investigate the performance of classification approaches, MLP, SMO, LR, J48, RF, DT, and NB algorithms were selected. WEKA uses ARFF file format which is "Attribute Relation File Format" where all the features are considered attributes and data are viewed as an instance. The experiment is performed in three parts for evaluating each of the algorithms. Those are related to performance at 10-fold cross-validation, 70% split, and 80% split. Hereafter, these three approaches are addressed as ex1, ex2 and ex3, respectively. The experimental results have been elaborated in several tables and figures for better visualization. The performances of these algorithms have been compared using five different measures: Accuracy, Incorrectness, Time taken, Kappa statistic and finally comparing Accuracy. The accuracy has been compared in three possible combinations: cross-validation vis-à-vis 70% split, cross-validation vis-à-vis 80% split and 70% split vis-à-vis 80% split.

Table 1. Accuracy of the algorithms

Algorithm	Cross-validation	70% split	80% split
MLP	78.7402	72.3684	72.5490
SMO	73.6220	73.6842	68.6275
LR	77.5591	80.2632	76.4706
J48	64.9606	67.1053	64.7059
RF	82.2835	78.9474	82.3529
DT	67.3228	67.1053	68.6275
NB	81.8898	85.5263	78.4314

The first experimental result has been represented in terms of the correctness of each of these algorithms in three approaches, i.e., 10-fold Cross-Validation, 70% Split and 80% Split. In Table 1 and Fig. 2, all the classifiers can be seen visualized with their accuracy (in percentage) for all these three approaches. Among all the implementations,

NB classifiers were prominent with more than 85% accuracy for the ex2 approach. This performance is followed by RF having more than 82% accuracy for both ex1 and ex3. However, the performance of RF in ex2 is relatively low but not much behind with more than a 78% accuracy level. Among all the other algorithms, LR only crosses the 80% accuracy level in the ex2.

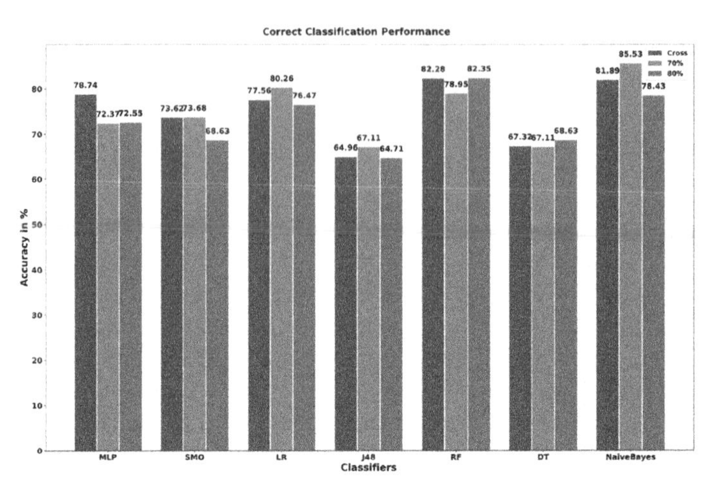

Fig. 2. Accuracy levels of all classifiers (in different experimental setups)

The effective performance of NB is also noted in previous works [9] [24] though the accuracy levels were found lower with 75% and 63%, respectively. In earlier work [9], LR and SVM performed with lower accuracy levels around 60%, but we found both these classifiers performed way ahead in this experimental setup. With more than 80% in the case of LR (in ex2 arrangement), SMO performed way ahead of 70% accuracy in both ex1 and ex2 design. Though the performance of SMO in ex3 is relatively lower with 68% accuracy but still better than previously noted [9]. In the case of DT, the current experimental setup found to perform poorly comparing previous implementation [25]. Where, 70% accuracy level was achieved using DT with N-gram analysis in contrast we found less than 70% accuracy in all three experimental setups. Still the lagging is not much as the highest performance of DT is noted more than 68% in ex3 configuration. For better visualization, we have also plotted the percentages of wrongly classified news by all these algorithms in different experimental setups in Fig. 3. From this representation, it is more apparent that J48 performed with significantly lower accuracy in all three experimental setups. DT follows this in our implementation and obviously NB performed best for ex2 whereas for ex1 and ex3, RF performed best with the lowest wrong classification percentages respectively.

Kappa statistics are used for evaluating the accuracy of classifiers. Kappa is robust and can be used in both nominal and ordinal data. The original intent of Cohen's Kappa was to measure the degree of agreement or disagreement of two or more people observing the same phenomenon [40, 41]. From Fig. 4 we can see that the Random Forest has the

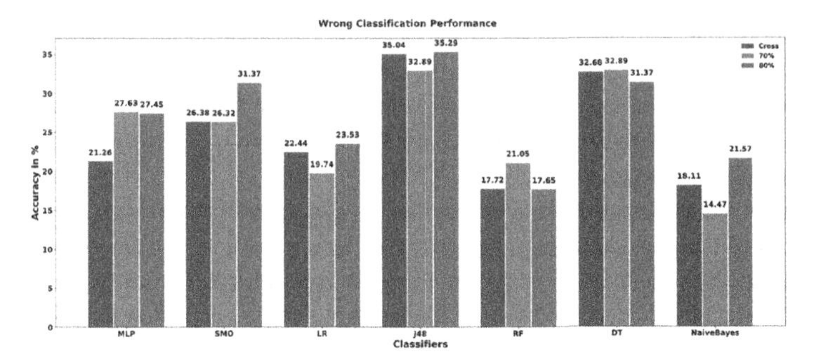

Fig. 3. Statistics of wrongly classified data (in percentage).

highest kappa value. And Naive Bayes has the second-highest value. If we compare Fig. 4 with Landis and Koch [42] interpretation metrices, we can see that most of the MLP is at a "moderate" level as the Kappa value is between 0.41–0.60. SMO belongs to Moderate level for cross validation and 70% split but for ex3, performance reduces and the value is between "0.21–0.40" can be interpreted as a fair level. LR is at moderate level with a value between "0.52–0.602". J48 and DT have the lowest level of "Fair" in all the cases and the performance is inferior concerning other algorithms. RF and NB algorithms have the highest level of significance, which is "Substantial" with values between 0.61–0.75. In the case of RF in ex2 and NB in ex3 the performance reduces which is "moderate". So, kappa Statistics also shows that Naïve Bayes is the top-performing algorithm for this dataset followed by RF and LR.

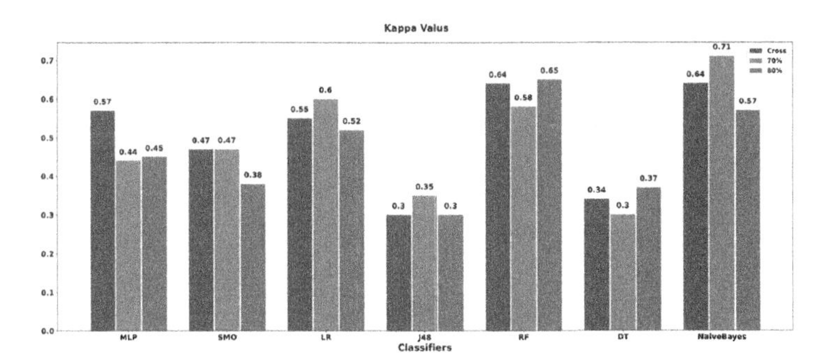

Fig. 4. Kappa statistic for all the classifiers in different experimental setups

Figure 5 represents the time taken by algorithms on logarithmic scale. From the Fig. 5 we can see that for all the three experimental arrangements, the highest time is taken by MLP, and the Lowest time is taken by Naive Bayes. MLP took the highest time at ex3 which is 113.83 s and Naive Bayes took the highest time at ex1 which is 0.04 s. SMO took slightly more time than Naive Bayes in all three experimental setups. In ex1, SMO took 0.1 s which is the highest time among all these three experimental designs. Decision Tree is in third place with a consistent time. It took nearly the same time in all

3 experiments. the highest time taken by Decision Tree is for ex3. Logistic regression is at the fourth position with a highest time of 0.21 s for ex2. J48 is in the fifth position with the highest time of 0.36 s among three experiments. Random forest is in the sixth position with nearly consistent speed. It took 0.45 s in both ex2 and ex3 and with 0.47 s for ex1 placed in the seventh position.

If we consider time with the accuracy, we can see that Naive Bayes took the lowest time and gave the highest accuracy. All the algorithms took more time in Cross-validation among the three experiments except for Logistic Regression. It took significantly less time and the accuracy was 77.5591%. Considering the Time taken by this algorithm Naïve Bayes is still on the first place because it took only 0.04 s, 0.01 s, 0.01 s in three tests. But the Random Forest algorithm took 0.4 s, 0.45 s, 0.45 s in three tests. Logistic Regression took less time than Random Forest. It took 0.17 s, 0.21 s, 0.2 s in three tests. But MLP took the highest time of 104.5 s, 103.69 s, 113.83 s in three tests. From Fig. 5 and Fig. 2 it can be seen that Naive Bayes is clearly on the top position considering time and accuracy Logistic regression is in the 2nd place and Random Forest is in the third position.

In final segment of experimental result representation, the difference of accuracy levels for all these seven classifier approaches have been computed and listed in Table 2. The difference has been computed between two experimental setups which make three combinations. In the first pair of 10-fold cross-validation (ex1) vs. 70%split (ex2), MLP performs better for ex1 comparing ex2 followed by RF which is also performed better in ex1. Other than these two all the negative differences indicate that ex2 is better setup for these algorithms than ex1. Though for SMO and DT it can't be said evidently as the difference is too less to indicate any significance between these setups. Therefore, it can be concluded. These two algorithms are working indifferently irrespective of the experimental design.

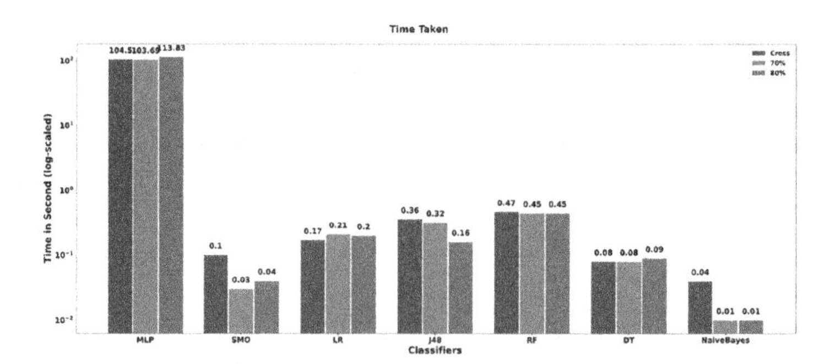

Fig. 5. Time graph

In contrast, ex3 seems to be a less proper setup for almost all the algorithms except DT and RF though the insignificant difference in the case of RF fails to highlight any suitability between ex1 and ex3. Other than these two, for all the algorithms ex1 is found to be more suitable except J48 where the difference is insignificant. For the 80% split, the lower performance has been noted comparing 70% split in four of the algorithms

Table 2. Comparing performances of algorithm in terms of accuracy difference

Algorithm	Cross-validation vis-à-vis 70% split	Cross-validation vis-à-vis 80% split	70% split vis-à-vis 80% split
MLP	6.37	6.19	−0.18
SMO	−0.06	4.99	5.06
LR	−2.7	1.09	3.79
J48	−2.14	0.25	2.4
RF	3.34	−0.07	−3.41
DT	0.22	−1.3	−1.52
NB	−3.64	3.46	7.09

(SMO, LR, J48, NB) with a significant margin. In the MLP also, ex3 is favorable set up with very little margin comparing ex2. These significant margins favoring ex2 (70% splits) show that the commonly used 80% split setup is not much suitable with these algorithms for our dataset. This is a kind of new enlightenment where; it can be said the commonly used 80% split setup should not be considered alone as an excellent setup to evaluate classifiers. As it is a well-known fact that, machine learning algorithms are performed differently on different datasets. Similarly, the split setup gives an altered picture for the different dataset.

5 Conclusion

This study aims to compare the performance of seven algorithms using three different experimental setups for detecting fake news. The study is based on two key aspects of these algorithms, accuracy levels and time consumed to classify. Furthermore, it compares these algorithms in different experimental based on their experimental setups. Also, the significance has been measured Kappa statistics. The top-performing algorithms are Naïve Bayes, followed by the Random Forest and Logistic Regression. But in terms of time Naïve Bayes and Logistic Regression outperformed the Random Forest algorithm. Also, in the experimental setup, 70% split setup is more suitable than an 80% split setup for most of the algorithms. K-fold Cross Validation (here $K = 10$) is an ideal for Perceptron based approach. It is well known that Deep Learning algorithms show great accuracy but comes with huge time and resource overhead. Perceptron based approach in our experiment also shows the same. On the other hand, Algorithms such as Sequential Minimal Optimization, Logistic Regression, Decision Tree, J48, Random Forest, and Naïve Bayes Classifiers are easier to implement and can show better results in some cases and in this paper we compared algorithms such as these on different experiments to compare the performance on fake news dataset.

References

1. Tandoc, E.C., Lim, Z.W., Ling, R.: Defining "Fake News": a typology of scholarly definitions. Dig. J. **6**(2), 137–153 (2017). https://doi.org/10.1080/21670811.2017.1360143
2. Zhou, X., Reza, Z.: A survey of fake news. ACM Comput. Surv. **53**(1), 1–40 (2020). https://doi.org/10.1145/3395046
3. Shu, K., Sliva, A., Wang, S., Tang, J., Liu, H.: Fake news detection on social media: a data mining perspective. ACM SIGKDD Explor. Newslett. **19**(1), 22–36 (2017). https://doi.org/10.1145/3137597.3137600
4. Lazer, D.M.J., et al.: The science of fake news. Science **359**(6380), 1094–1096 (2018). https://doi.org/10.1126/science.aao2998
5. Zhang, X., Ghorbani, A.A.: An overview of online fake news: characterization, detection, and discussion. Inf. Process. Manag. **57**, 102025 (2020). https://doi.org/10.1016/J.IPM.2019.03.004
6. Gupta, A., Kumaraguru, P.: Credibility ranking of tweets during high impact events. ACM Int. Conf. Proceeding Ser. (2012). https://doi.org/10.1145/2185354.2185356
7. Mohd Shariff, S., Zhang, X., Sanderson, M.: User perception of information credibility of news on Twitter. In: de Rijke, M., et al. (eds.) ECIR 2014. LNCS, vol. 8416, pp. 513–518. Springer, Cham (2014). https://doi.org/10.1007/978-3-319-06028-6_50
8. Finding true and credible information on Twitter | IEEE Conference Publication | IEEE Xplore. https://ieeexplore.ieee.org/abstract/document/6915989. Accessed 30 Aug 2021
9. Kareem, I., Awan, S.M.: Pakistani media fake news classification using machine learning classifiers. In: 3rd International Conference Innovative Computing (ICIC 2019). (2019). https://doi.org/10.1109/ICIC48496.2019.8966734
10. Liu, S., Liu, S., Ren, L.: Trust or Suspect? An Empirical Ensemble Framework for Fake News Classification
11. Hakak, S., Alazab, M., Khan, S., Gadekallu, T.R., Maddikunta, P.K.R., Khan, W.Z.: An ensemble machine learning approach through effective feature extraction to classify fake news. Futur. Gener. Comput. Syst. **117**, 47–58 (2021). https://doi.org/10.1016/J.FUTURE.2020.11.022
12. Bin Othman, M.F., Yau, T.M.S.: Comparison of different classification techniques using WEKA for breast cancer. In: Ibrahim, F., Osman, N.A.A., Usman, J., Kadri, N.A. (eds.) 3rd Kuala Lumpur International Conference on Biomedical Engineering 2006. IP, vol. 15, pp. 520–523. Springer, Heidelberg (2007). https://doi.org/10.1007/978-3-540-68017-8_131
13. Rubin, V.L., Conroy, N.J., Chen, Y., Cornwell, S.: Fake News or Truth? Using Satirical Cues to Detect Potentially Misleading News, pp. 7–17 (2016)
14. Komarek, P., Moore, A.: Fast Logistic Regression for Data Mining, Text Classification and Link Detection, pp. 1–8 (2003)
15. Alsari, A.B.A.: Short Text Classification using Machine Learning Techniques (2018)
16. Al Qadi, L., El Rifai, H., Obaid, S., Elnagar, A.: Arabic text classification of news articles using classical supervised classifiers. In: Proceedings of the 2nd International Conference on New Trends in Computing Sciences (ICTCS 2019) (2019). https://doi.org/10.1109/ICTCS.2019.8923073
17. Thornton, C., Hutter, F., Hoos, H.H., Leyton-Brown, K.: Auto-WEKA: combined selection and hyperparameter optimization of classification algorithms. In: Proceedings of the ACM SIGKDD International Conference on Knowledge Discovery and Data Mining, Part F1288, pp. 847–855 (2013). https://doi.org/10.1145/2487575.2487629
18. Ahmed, A.A.A., Aljabouh, A., Donepudi, P.K., Choi, M.S.: Detecting fake news using machine learning: a systematic literature review. Psychol. Educ. J. **58**, 1932–1939 (2021)

19. Castillo, C., Mendoza, M., Poblete, B.: Information credibility on Twitter. In: Proceedings of the 20th International Conference on Companion World Wide Web (WWW 2011), pp. 675–684 (2011). https://doi.org/10.1145/1963405.1963500
20. Buntain, C., Golbeck, J.: Automatically identifying fake news in popular Twitter Threads. In: Proceedings of the 2nd IEEE International Conference on Smart Cloud, SmartCloud 2017, pp. 208–215 (2017). https://doi.org/10.1109/SMARTCLOUD.2017.40
21. Vosoughi, S., Roy, D., Aral, S.: The spread of true and false news online. Science **359**(6380), 1146–1151 (2018). https://doi.org/10.1126/science.aap9559
22. Fakenews Dataset (csv and arff) | Kaggle. https://www.kaggle.com/abkafi/fakenews-datasetcsv-and-arff. Accessed 10 Sept 2021
23. Bisaillon. Clément: Fake and real news dataset | Kaggle. https://www.kaggle.com/clmentbisaillon/fake-and-real-news-dataset. Accessed 24 Aug 2021
24. Granik, M., Mesyura, V.: Fake news detection using naive Bayes classifier. In: Proceedings of the IEEE 1st Ukraine Conference on Electrical and Computer Engineering (UKRCON 2017), pp. 900–903 (2017). https://doi.org/10.1109/UKRCON.2017.8100379
25. Keskar, D., Palwe, S., Gupta, A.: Fake news classification on Twitter using flume, N-gram analysis, and decision tree machine learning technique. In: Bhalla, S., Kwan, P., Bedekar, M., Phalnikar, R., Sirsikar, S. (eds.) Proceeding of International Conference on Computational Science and Applications. AIS, pp. 139–147. Springer, Singapore (2020). https://doi.org/10.1007/978-981-15-0790-8_15
26. Ali, S., Smith, K.A.: On learning algorithm selection for classification. Appl. Soft Comput. **6**(2), 119–138 (2006). https://doi.org/10.1016/j.asoc.2004.12.002
27. Goldberg, Y.: Neural network methods for natural language processing. Synth. Lect. Hum. Lang. Technol. **10**(1), 1–309 (2017). https://doi.org/10.2200/S00762ED1V01Y201703HLT037
28. Saritas, M.M., Yasar, A.: Performance analysis of ANN and Naive Bayes classification algorithm for data classification. Int. J. Intell. Syst. Appl. Eng. **7**, 88–91 (2019). https://doi.org/10.18201//IJISAE.2019252786
29. Pregibon, D.: Logistic regression diagnostics. Ann. Statist. **9**, 705–724 (1981). https://doi.org/10.1214/AOS/1176345513
30. Kleinbaum, D.G., Klein, M.: Logistic Regression. Springer, New York (2010). https://doi.org/10.1007/978-1-4419-1742-3
31. Garner, S.R.: WEKA: the Waikato environment for knowledge analysis. Proc. New Zeal. Comput. Sci. Res. Stud. Conf. 57–64 (1995)
32. Flake, G.W., Flake, G.W., Lawrence, S.: Efficient SVM regression training with SMO. Mach. Learn. **46**, 1–3 (2000)
33. Lee, T.-H., Ullah, A., Wang, R.: Bootstrap aggregating and random forest. In: Fuleky, P. (ed.) Macroeconomic Forecasting in the Era of Big Data. ASTAE, vol. 52, pp. 389–429. Springer, Cham (2020). https://doi.org/10.1007/978-3-030-31150-6_13
34. Safavian, S.R., Landgrebe, D.: A survey of decision tree classifier methodology. IEEE Trans. Syst. Man Cybern. **21**, 660–674 (1991). https://doi.org/10.1109/21.97458
35. Ivan, N., Ahishakiye, E., Omulo, E.O., Taremwa, D.: Crime prediction using decision tree (J48) classification algorithm. Int. J. Comput. Inf. Technol. (2017)
36. Ozbay, F.A., Alatas, B.: Fake news detection within online social media using supervised artificial intelligence algorithms. Phys. A Stat. Mech. Appl. **540**, 123174 (2020). https://doi.org/10.1016/J.PHYSA.2019.123174
37. Kaur, G., Chhabra, A.: Improved J48 classification algorithm for the prediction of diabetes. Int. J. Comput. Appl. **98**, 975–8887 (2014)
38. Bouckaert, R.R.: Properties of Bayesian belief network learning algorithms. Uncertain. Proc. **1994**, 102–109 (1994). https://doi.org/10.1016/B978-1-55860-332-5.50018-3

39. Buntine, W.: Theory refinement on Bayesian networks. Uncertain. Proc. **1991**, 52–60 (1991). https://doi.org/10.1016/B978-1-55860-203-8.50010-3
40. Cohen, J.: A coefficient of agreement for nominal scales. Educ. Psychol. Measur. **20**, 37–46 (2016). https://doi.org/10.1177/001316446002000104
41. Ben-David, A.: Comparison of classification accuracy using Cohen's Weighted Kappa. Expert Syst. Appl. **34**, 825–832 (2008). https://doi.org/10.1016/J.ESWA.2006.10.022
42. Landis, J.R., Koch, G.G.: An application of hierarchical kappa-type statistics in the assessment of majority agreement among multiple observers. Biometrics **33**, 363 (1977). https://doi.org/10.2307/2529786

Image Coding by Count Sample, Motivated by the Mechanisms of Light Perception in the Visual System

V. E. Antsiperov(✉) and V. A. Kershner

Kotelnikov Institute of Radioengineering and Electronics of RAS, Mokhovaya 11-7, Moscow, Russian Federation
antciperov@cplire.ru

Abstract. The paper presents the results of most adequate visual input formation/coding in modern imaging systems. Adequacy here is understood as the maximum relevance between the methods of registration of radiation by material detectors of imaging devices and the methods of coding data in the retina of the human visual system. In this connection, the paper discusses general statistical issues of (photo) counts photoelectric detection and formalizes the concept of an ideal image formation by (ideal) visualization device. The problems arising in practice when working directly with ideal images are discussed and proposed a method of their reduction to the count sample of fixed (controllable) size, which, in fact, constitute the representation (coding) of registered data. Results of computational experiments on count coding of the common digital images given by pixel data are presented. Examples of count samples of different sizes generated for the tested digital image illustrate these results. Based on the given results, the dependence of characteristics of sampling representations on the parameter of sample size is discussed.

Keywords: Image coding · Image representation · SPAD image sensor · Photocounts · Ideal imaging device · Ideal image concept · Sampling representation · Digital DSP camera

1 Introduction

1.1 Human Visual System in Brief

The human visual system registers optical radiation coming from outside with the help of photoreceptor cells that make up the outer layer of the retina, see Fig. 1. The retinal photoreceptors (rods, cones) are responsible for collecting primary data, signals from the world around us. Based on these data the visual (nervous) system performs further interpretation, recognition, assigning semantic meanings to the information about the environment. The presence of useful information in registered radiation and the possibility of its correct interpretation are conditioned by the fact that most of the objects surrounding us, even if they are not themselves light sources, can reflect light from other

I. Woungang et al. (Eds.): ANTIC 2021, CCIS 1534, pp. 715–729, 2022.
https://doi.org/10.1007/978-3-030-96040-7_54

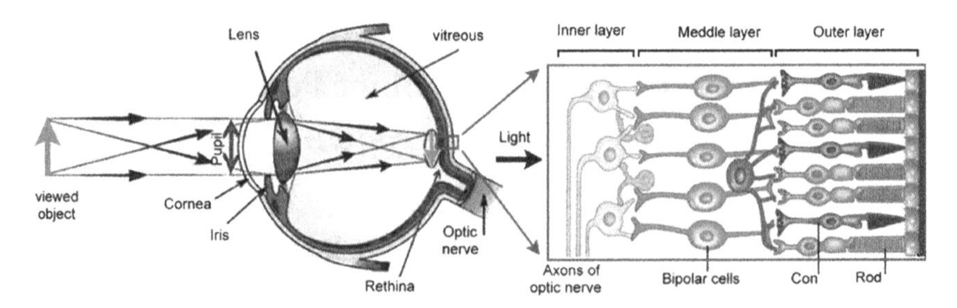

Fig. 1. The structure of the human eye and retina (enlarged).

(artificial/natural) sources according to the shape, structure, chemical composition, etc. of their surface.

Principles of transformation and registration of optical radiation in the visual system were already used in the designs of the photo-cameras. In the process of evolution from the simplest imaging devices based on the camera obscura up to modern digital cameras, the number of borrowed principles of visual system only grew. These include the use of a diaphragm (iris), optical focusing system (lens), and radiation sensing surface in the form of a transparent plate with deposited photo emulsion, or as a matrix of CCD/CMOS photodiodes (photoreceptors), see Fig. 1. The fact that the photosensitive surface of the eye (retina) has the form of the spheric surface, rather than a plane, is not crucial in this case.

1.2 Using the Visual Mechanisms of Light Registration in Imaging Systems

The evolution of the use of visual mechanisms in artificial imaging systems has been most clearly seen in the issue of radiation registration. Indeed, starting from the silver-coated copper plates developed by Daguerre in the first half of the 19th century and their refinement in the 20th century to sheet and roll celluloid films with gelatin-silver emulsion (analog photography), the registering elements are transformed now into matrices of photodiodes - digital-analog integrated circuits (digital photography). This progress is largely due to the emergence of charge-coupled devices (CCD) in the late 1960s and the subsequent invention of light-sensitive matrices on complementary MOS structures (CMOS) in the early 1990s. In practical terms, the switch to new (digital) technologies of radiation registration has made it possible to improve a whole range of camera characteristics, the most important of which is the increase of spatial resolution. This could be achieved by sequentially reducing the size of the photodiodes (photodetectors) in the matrix down to a few microns. Among other achievements, it is necessary to mention an essential increase in the frame rate, an increase of a dynamic range, a decrease in power consumption, etc. It should be noted that the above technological advances are largely due to the development of theoretical physics of the 20th century in the field of interaction of radiation with matter (quantum electrodynamics) [1] and related research in the field of semiconductor science (solid-state theory) [2].

According to modern physical concepts [1, 2], with a gradual decrease in the size of digital matrix photodetectors, the nature of radiation registration necessarily acquires an increasingly pronounced quantum character and, in the limit, proceeds to the registration of single photons. A remarkable circumstance is that this limiting case has already been reached by several technologies. The question is about the technologies of producing the so-called photon-counting sensors [3] – sensors working in the mode of photon counting. As an example, let us mention the electron-multiplication charge-coupled matrices (EMCCD) [4], single-photon avalanche diodes (SPAD) [5], Geiger-mode avalanche photodiodes (GMAPD) [6], see Fig. 2.

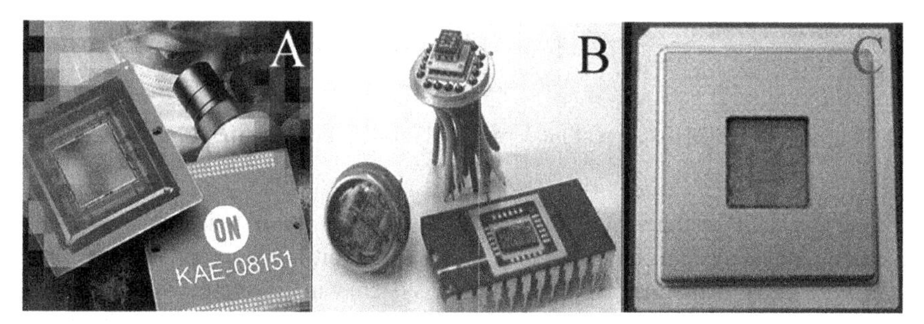

Fig. 2. Modern photon-counting sensor technologies A) EMCCD, B) SPAD, C) GMAPD.

1.3 The Role of Digital Signal-Processing (DSP) in Modern Imaging Systems

Since the result of radiation registration by matrices of photodiodes is the electrical current, they can be easily incorporated into various electronic circuits containing microprocessors, which besides the control functions can contain the applications for digital signal processing (DSP). Given the high performance of modern microprocessors, it is possible to fulfil not only the standard tasks of video signal preprocessing, such as linearization, dark-current compensation, flare compensation, and white balance, but also to solve many more intellectual problems like image classification, object recognition, scene analysis, etc., all online. In other words, conversion of a video signal at the output of a light-sensitive matrix into the digital domain opens completely new opportunities in the image formation and analysis based on DSP methods.

Modern digital image processing theory (iDSP) [7], in its turn, provides digital cameras with an extensive arsenal of image processing methods and tools, ranging from the simplest, linear methods, to complex machine vision algorithms. The variety of iDSP techniques is usually conventionally divided into three main levels – low-, medium- and high-level computerized processes. Low-level processes usually include image preprocessing operations for color correction, smoothing to remove noise, or, conversely, contrast enhancement to improve quality, etc. Mid-level image processing covers tasks such as classification, segmentation, object and scene recognition, compression into a suitable form for further computer processing, etc., see Fig. 3. Finally, the high-level processing includes the semantic analysis of the recognized objects, scenes, etc., in the perspective, it carries out the "understanding" of the reality captured by the image that is associated with the human vision.

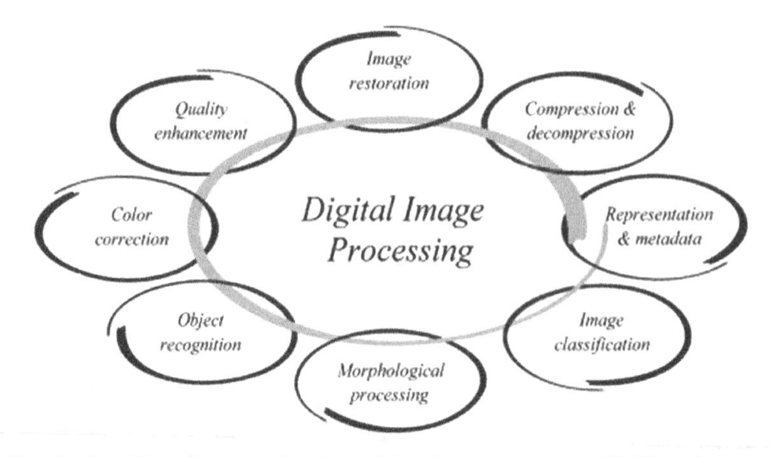

Fig. 3. Simplified diagram of modern digital image processing (iDSP) techniques.

Note that iDSP methods of all three levels also widely use individual mechanisms of the human visual system [8]. Low-level methods exploit image preprocessing mechanisms represented in the periphery of the visual system – in the retina, while middle- and high-level methods use neuro models of visual information processing in the cerebral cortex [9]. Here, first, we should mention machine learning methods [10], such as image classification, morphological processing, object recognition, etc.

Certainly, simulation of visual perception mechanisms is carried out within iDSP with some degree of approximation. Approximate modeling begins already at the level of registered data representation. Indeed, most traditional iDSP methods are focused on bitmap images, which are discrete pixels – the digitized result of the energy of the registered radiation accumulated by the matrix detectors during the exposure time. In other words, pixel values are proportional to the total number of photons falling on the sensitive surfaces of the corresponding detectors during the exposition. On the contrary, photoreceptors in the retina react rather to individual photons of radiation and transmit a signal about their registration to ganglion cells of the inner layer at once, without accumulation [9]. The noted difference in the representation of input data in artificial and natural imaging systems is explained by the fact that until recently there were simply no technological possibilities of implementing methods analogous to image registration in the visual system. However, as was mentioned above, during the last decades considerable progress has been achieved in the development of digital matrices operating in the photon counting mode (see also [11]). In this connection, it seems natural to adapt iDSP procedures to the new opportunities for modeling video registration data closer simulating input data of the visual system formed by the retina. One of the possible approaches to solving this problem - a new method of representing (coding) images by samples of counts due to registered radiation photons is presented below.

2 Image Coding by Count Sample

2.1 Ideal Imaging Device Definition

To substantiate the proposed method of coding (representation) of video data, let us first discuss the used model of recording radiation by the retina of the human eye. This model is motivated by known biophysical facts on the mechanisms of light interaction with photoreceptors of the human retina [9] and fundamental principles of quantum electrodynamics [1] in its semiclassical approximation [12]. Further in the work, this model will be formalized in the form of an ideal image concept [13], which is registered by an ideal imaging device containing a large array (matrix) of point detectors [14].

The prototype model of an ideal imaging device in the human visual system is the retina. As a motivation for the model presented below, let us list a few basic facts concerning it. The human retina contains about 10 million cones and 100 million rods, capable of registering separate photons of visible radiation. Their density reaches values of 100000–160000 receptors per mm^2 (birds of prey that must search for small objects at a large distance have about one million receptors per square millimeter of the retina). By the way, it is worth mentioning that the signals that come to the brain via the optic nerve (neural impulses) are not, strictly speaking, directly recorded by receptors, but are formed on their basis by a complex system of cells in the middle and inner layers of the retina, see Fig. 1. In total, the video information arriving in the brain is transmitted via axons of ganglion cells (optic nerves), the number of which is ~two orders of magnitude less (about a million) than the number of receptors.

A matrix of single-photon avalanche diodes (SPAD) [11] or any of its analogues [3] can be used also as a prototype of an ideal imaging device of artificial origin, which can also be considered as a technical realization of the latter. Now there are achieved serious results in the production of such matrices: $1024 \times 1000 = 1.024.000$ micro-detectors of light radiation (pixels) are placed on an area 11×11 mm^2 with a spacing of 9.4 μm is a fact. Each element has a dynamic binary memory for storing the photo recording data, the memory is updated at a rate of 24.000 frames per second [11]. Accordingly, information flow from SPAD-matrix can reach 25 Gbit/s, which significantly exceeds information flow obtained by the human visual system, which is about 50 Mbit/s [15].

With all the differences of the above two prototypes, they share several common features. Both prototypes contain a finite-sized photosensitive region, which contains a very large number of detectors/photoreceptors. Each photodetector can detect individual photons of incident radiation. All events of registration of photons by detectors are stored by prototypes for some time (frame). The listed properties can be put at the basis of a sufficiently general concept of an ideal imaging device, which is a formal generalization of both presented prototypes and some other imaging systems (photographic plates, photographic films with gelatin-silver emulsion, etc.).

In this regard, let us formulate the following definition. By an ideal imaging device, we will mean some two-dimensional region Ω with coordinates $\vec{x} = (x_1, x_2)$ and overall area S, on which point detectors of radiation with vanishingly small areas ds of sensitive surfaces are located close to each other [14]. Accordingly, the total number of detectors in the region Ω is equal to $N = S/ds$. Under the assumption of $ds \to 0$, the number N is assumed to be arbitrarily large, in the limit $N \to \infty$. Thus, formally, the ideal imaging device Ω is a virtually "continuous" sensitive surface, its points with coordinates $\vec{x}$ define positions of ideal "point" detectors.

2.2 The Statistical Description of Counts Registered by Ideal Imaging Device

When radiation with an intensity $I(\vec{x}), \vec{x} \in \Omega$ (which does not change over the exposure time) is incident on the sensitive surface Ω, some point detectors of the ideal imaging device register individual photons. The event of registration incident photon over the frame time T by some detector will be called a (photo) count and attribute to that count the coordinates $\vec{x}$ of the detector. Within the semi-classical theory of interaction of radiation with matter the counts are random events and at the limit $ds \to 0$ are given by the probabilities $P(\vec{x}) = \alpha TI(\vec{x})ds$ where $\alpha = \eta(h\overline{\nu})^{-1}$, $h\overline{\nu}$ – the mean energy of the photons (h – Planck's constant, $\overline{\nu}$ – the characteristic frequency of radiation), the dimensionless coefficient $\eta < 1$ is the quantum efficiency of the detector material and has the meaning of the count probability per photon (at $w = TI(\vec{x})ds = h\overline{\nu}$). Thus, when the radiation intensity $I(\vec{x})$ is registered, with each point detector $\vec{x} \in \Omega$ can be associated a binary random variable $\sigma \in \{0, 1\}$, taking values $\sigma = 1$ и $\sigma = 0$ depending on whether the detector registers a count. So, the conditional (at a given intensity $I(\vec{x})$) probability distribution of σ has the form of Bernoulli distribution:

$$\begin{aligned} P(\sigma = 1|\vec{x}) &= P(\vec{x}) = \alpha TI(\vec{x})ds, \\ P(\sigma = 0|\vec{x}) &= 1 - \alpha TI(\vec{x})ds. \end{aligned} \tag{1}$$

Note that, according to the distribution (1), the mean number of counts $\overline{\sigma}$ at the point $\vec{x}$ is equal to $\alpha TI(\vec{x})ds$. Accordingly, the integral $\overline{n} = \alpha T \iint_{\Omega} I(\vec{x})ds$ defines the average number of all counts registered at Ω over time T.

Selecting randomly from the array of all N point detectors of the ideal imaging device someone having coordinates $\vec{x}$ (with uniform probability $Q(\vec{x}) = N^{-1}$), it is possible, using (1), to find a joint distribution of random variables $\vec{x}$ and σ:

$$\begin{aligned} P(\sigma = 1, \vec{x}) = P(\sigma = 1|\vec{x})Q(\vec{x}) &= \tfrac{\alpha TI(\vec{x})ds}{N}, \\ P(\sigma = 0, \vec{x}) = P(\sigma = 0|\vec{x})Q(\vec{x}) &= \tfrac{1-\alpha TI(\vec{x})ds}{N}. \end{aligned} \tag{2}$$

Summing the distribution (2) over all N point detectors, we obtain the marginal unconditional distribution of σ, which specifies the probabilities of the appearance/absence of a count for an arbitrary detector:

$$\begin{aligned} P(\sigma = 1) &= \frac{\alpha T}{N} \iint_{\Omega} I(\vec{x})ds = \frac{\overline{n}}{N}, \\ P(\sigma = 0) &= 1 - \frac{\alpha T}{N} \iint_{\Omega} I(\vec{x})ds. \end{aligned} \tag{3}$$

Finally, by dividing the probability of $P(\sigma = 1, \vec{x})$ (2) by the corresponding probability $P(\sigma = 1)$ (3), we obtain the conditional probability of count to have the coordinates $\vec{x}$ in the region Ω:

$$P(\vec{x}|\sigma = 1) = \frac{P(\sigma = 1, \vec{x})}{P(\sigma = 1)} = \frac{I(\vec{x})ds}{\iint_{\Omega} I(\vec{x})ds} = \frac{\overline{\sigma}(\vec{x})}{\overline{n}}, \tag{4}$$

which should not be confused with the count probability $P(\sigma = 1|\vec{x})$ (1) at the point $\vec{x}$.

Further, it is convenient to use the corresponding density instead of the conditional probability of count coordinates (4) – $\rho(\vec{x}|I(\vec{x})) = P(\vec{x}|\sigma = 1)/ds$, which explicitly contains the distribution condition in the form of a given intensity $I(\vec{x})$. Rewriting (4) for the density $\rho(\vec{x}|I(\vec{x}))$, we obtain the following important result:

$$\rho(\vec{x}|I(\vec{x})) = \frac{I(\vec{x})}{\iint_{\Omega} I(\vec{x})ds}, \tag{5}$$

which declares that the probability density of count $\vec{x} \in \Omega$ when the radiation is detected by an ideal imaging device is the same as normalized intensity $I(\vec{x})$ at Ω. Note in this connection the universal character of (5): the conditional probability density does not depend on either the quantum efficiency of the detector material η, or on the spectrum (including the characteristic frequency $\overline{\nu}$) of radiation, or on the time of the frame T. Moreover, it also does not depend on the value of total radiation power $W = \iint_{\Omega} I(\vec{x})ds$ since it is defined only by the shape of intensity – the normalized version of $I(\vec{x})$. Note in this regard that the statistical characteristics of a set of all counts – their average number $\overline{n} = \alpha TW$ and, therefore, the distribution P_n of total counts number n are dependent, however, on the mentioned parameters. So, let us emphasize once more, that for the probability distribution of counts (5) the sufficient statistic is only the normalized intensity of $I(\vec{x})/W$.

2.3 Ideal Image Concept Definition

Basing on the formulated model of an ideal imaging device, the concept of the ideal image can be defined. Namely, the ideal image is the (ordered) set $X = (\vec{x}_1, \ldots, \vec{x}_n), \vec{x}_i \in \Omega$ of all n random counts registered by the ideal device during the frame time T. The ideal image is thus an essentially random object to be distinguished from its realizations. Note that the random nature of the ideal image is determined not only by the random count coordinates $\vec{x}_i$, but also by random number n of them.

Complete statistical description of ideal image in the form of all finite-dimensional densities of probability distributions $\{\rho(\vec{x}_1, \ldots, \vec{x}_n, n|I(\vec{x}))\}, \vec{x}_i \in \Omega, n = 0, 1, \ldots$ can be obtained by assuming conditional independence of all counts $\vec{x}_i$ (under the condition $I(\vec{x})$). The standard derivation of this description based on the coordinate distribution (5) and Poisson approximation (in the limit $ds \to 0, N \to \infty, Nds = S = const$) for P_n – for the probability distribution of their total number n, can be found, for example, in [16]:

$$\begin{aligned}\rho(\vec{x}_1, \ldots, \vec{x}_n, n|I(\vec{x})) &= \rho(\vec{x}_1, \ldots, \vec{x}_n, |n, I(\vec{x})) \times P_n(I(\vec{x})) \\ &= \prod_{i=1}^{n} \rho(\vec{x}_i|I(\vec{x})) \times \tfrac{\overline{n}^n}{n!} \exp(-\overline{n}), \overline{n} = \alpha T \iint_{\Omega} I(\vec{x})ds.\end{aligned} \tag{6}$$

Note that the statistical description (6) corresponds to some two-dimensional point Poisson nonhomogeneous process with intensity $\lambda(\vec{x}) = \alpha TI(\vec{x})$, and the above result reflects the well-known fact that, under the assumptions made, the set of Bernoulli tests $\{(\vec{x}, \sigma)\}$ are well approximated by the Poisson point process [17].

Notice, that in contrast to the one-count probability distribution (5), the densities $\rho(\vec{x}_1, \ldots, \vec{x}_n, n|I(\vec{x}))$ (6) of count set through P_n и $\overline{n} = \alpha TW$ depend on parameters α, T, W, i.e., they partially lose the property of universality. Nevertheless, it is known that if one fixes n, then the conditional densities $\rho(\vec{x}_1, \ldots, \vec{x}_n|n, I(\vec{x}))$ do not depend on these parameters [17]. The latter circumstance often allows simplifying the analysis by separating associated with n "energetic" estimations from "geometric", structural estimations related only to the configuration of coordinate set $X_n = (\vec{x}_1, \ldots, \vec{x}_n)$ [18].

2.4 Reduced Statistical Description of Sampling Representation

The ideal image model and its statistical description (6) are useful in theoretical studies, for example, in the search for optimal image analysis procedures, in particular the recognition or identification of objects in images [18]. In addition, at low intensities $I(\vec{x})$ of registered radiation (with not large $\overline{n} = \alpha TW$), the statistical description (6) is quite successfully used in the fields of fluorescence microscopy, positron emission tomography (PET), single-photon emission computed tomography (SPECT), optical and infrared astronomy, etc. [19].

However, at ordinary radiation intensities, corresponding, for example, to daylight, the practical use of an ideal image model turns out to be problematic. The matter is that fluxes of photons falling on a region with $S \sim 1$ mm^2 per second, for instance, from the sun in usual conditions are enormous - on the Earth's surface on a clear day they are $\sim 10^{15} - 10^{16}$ photons [9]. For devices that allow the photon counting mode, even if they form one count per ~10 photons (with a quantum efficiency of $\eta = 0.1$), the number of counts per second will be $\overline{n} \sim 10^{15}$ (1,000,000 Gbit/s = 1 Pbit/s). Dealing with such data flows would require too many resources in practical tasks. Therefore, it is desirable to develop other approaches to image coding/representation.

Some time ago we proposed the following solution to the above problem of reducing the dimensionality of the representation of ideal images [14]. Let us fix from the beginning some acceptable value of the representation dimension $k \ll \overline{n}$ and, considering the ideal image representation $X = \{\vec{x}_i\}$ as some general population of random counts, let us perform a random sampling from it in k counts $X_k = \{\vec{x}_j\}$. Obviously, in full agreement with the classical statistical theory such a "sampling" representation at dimensions $k \ll \overline{n}$ will still represent the ideal image X. Let's call the sample X_k the image representation by sampling of random counts or, in short, by sampling representation. The statistical description of sampling representation follows easily from (6) by integrating $\rho(\vec{x}_1, \ldots, \vec{x}_n, n|I(\vec{x}))$ over the counts unselected in X_k and summing the result over the number of these unselected counts $l = n - k = 0, 1, \ldots$:

$$\begin{aligned} \rho(X_k|I(\vec{x})) &= \rho(\vec{x}_1, \ldots, \vec{x}_k|n, I(\vec{x})) \times \textstyle\sum_{l=0}^{\infty} P_{k+l}(I(\vec{x})) \\ &= \textstyle\prod_{j=1}^{k} \rho\left(\vec{x}_j|I(\vec{x})\right) \times P_{n \geq k}(I(\vec{x})). \end{aligned} \tag{7}$$

where $P_{n \geq k}(I(\vec{x}))$ denotes the probability that the ideal image contains more than k counts. Given the Poisson character of the probabilities P_n and their asymptotic tendency at $\overline{n} \to \infty$ to the Gaussian distribution (with mean value $\overline{n}$), it is easy to show (see Fig. 4) that in the case of $\overline{n} \gg 1$ probability $P_{n<k}(I(\vec{x}))$ will be less than ε, as soon as $k \leq 2\varepsilon\overline{n}$, respectively, at that $P_{n>k}(I(\vec{x}))$ will differ from unity by less than ε.

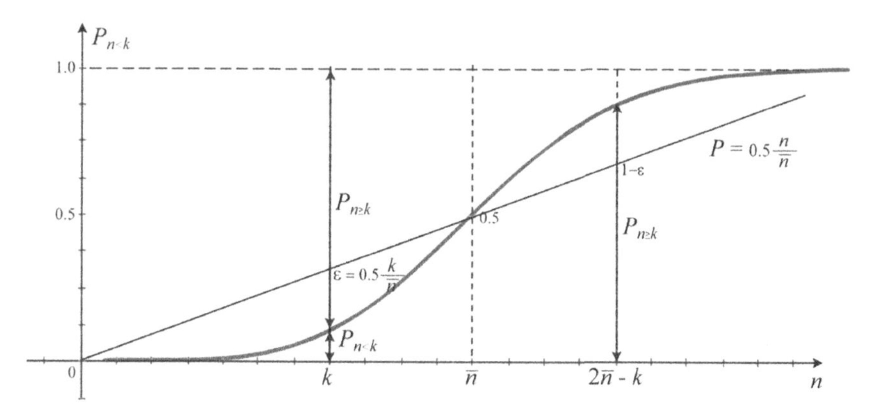

Fig. 4. Graphic justification of the estimates $P_{n<k} < \varepsilon$ and $P_{n\geq k} > 1 - \varepsilon$ at $k = 2\varepsilon\bar{n}$.

So, if we accept that $P_{n>k}(I(\vec{x})) \cong 1$ holds for the representation of the image by a random sample X_k, the size of which satisfies $k < 2\varepsilon\bar{n}$, then with an arbitrarily small accuracy ε we can obtain for statistical description (7) the following remarkable result:

$$\rho(X_k|I(\vec{x})) = \prod_{j=1}^{k} \rho\left(\vec{x}_j|I(\vec{x})\right). \tag{8}$$

Note, if, for example, $\varepsilon = 0.001$ и $\bar{n} \sim 10^{15}$, then we obtain that (8) holds up to $k \sim 2 \times 10^{12}$, however, and this estimate is, most likely, strongly underestimated.

It is easy to see that the statistical description of sampling representations (8) can be obtained at once from (5) under the assumption that the probability $\rho(\vec{x}|I(\vec{x}))$ of any count from X_k does not depend on which and how many counts besides $\vec{x}_j$ are in the sample. As shown above, this assumption of count independence is equivalent to the assumption that the ideal image almost certainly contains (much) more counts than the sampling representation X_k.

The statistical description (8) of sampling representation $X_k = \{\vec{x}_1, \ldots, \vec{x}_k\}$ turns out to be much more convenient than the full statistical description of ideal image (6) not only because of the fixed size of the representation $k \ll \bar{n}$, but also because of several other circumstances. First, it fixes the conditional independence and the same conditional distribution (iid property) of all k counts $\vec{x}_j$. Second, the densities of distributions of individual counts $\rho\left(\vec{x}_j|I(\vec{x})\right)$ are given by the normalized intensity $I(\vec{x})$ in the region Ω. Hence, third, as well as for the conditional densities $\rho(\vec{x}_1, \ldots, \vec{x}_n|n, I(\vec{x}))$ for a description (8) the property of universality is also fulfilled – $\rho(X_k|I(\vec{x}))$ does not depend on the quantum efficiency of detector material η, on a radiation spectrum, or on time of frame T. The enumerated properties of the sampling representations (8) provide a suitable type of input data for many well-developed statistical approaches and machine learning methods, including the naive Bayesian approach [10].

This last note is related to the following circumstance. Since $\rho\left(\vec{x}_j|I(\vec{x})\right)$ (5) does not depend on the absolute values of intensity but is determined only by its normalized version (form) $-I(\vec{x})/\iint_{\Omega} I(\vec{x})ds$, the statistical description of the sampling representations (coded images) (8) also does not depend on $I(\vec{x})$ physical units. In particular, if the intensity of registered radiation is given by pixels $\{n_i\}$ of some raster image obtained by digitization $I(\vec{x})$ with a resolution parameter of digital quantization $Q = \Delta I$ then description (8) will not directly dependent on Q, but only through the parameter of pixel bit depth $\upsilon = \log_2(I_{max}/\Delta I))$ – standard characteristic of digital images. Some examples of count coding of digital images are given in the next section.

3 Experiments on Digital Image Coding by Count Samples

In connection with the last remark, we note that the procedure of count coding (sampling representation) of digital raster images can be essentially reduced to the normalization $\pi_i = n_i/\sum n_i$ of its pixel values $n_i \sim I_i$ and the subsequent sampling k of random counts from the resulting probability distribution $\rho(\vec{x}_j|I(\vec{x})) \approx \pi_j$. Note that in machine learning there is a large arsenal of methods for the organization of sampling procedures, jointly called Monte Carlo methods [20]. It includes well-known methods such as importance sampling, acceptance-rejection, Metropolis-Hastings, Gibbs's sampling algorithms, etc. Having such an arsenal of methods one can optimize sampling procedures from different positions - from positions of computational efficiency, representativeness, specificity of a problem, etc. At the same time, it should be noted, that for some sampling methods even preliminary normalization is not required – it is enough to provide that all pixels are limited by value 2^υ, where υ – is the parameter of image pixel bit depth.

For example, Fig. 5 shows random sampling representations of the standard test image "House", from USC-SIPI Image Database [21] often used in publications on image processing. The "House" image is initially specified in TIFF format, has the dimensions of 512 × 512 pixels, and a (color) pixel depth of $\upsilon = 24$ bits. To reduce the amount of computation, the image was converted to GIF format of the same size $s \times s$, $s = 512$, but in a gray palette with a pixel depth of $\upsilon = 8$ bits (see Fig. 5(A)). A sampling of representations of sizes $k = 100.000$, 500.000, 1.000.000, 2.000.000, and 5.000.000 counts (Fig. 5B–F)) was performed by one of the simplest methods – rejection sampling [20] with a uniform auxiliary distribution $g(\vec{x}) = (s \times s)^{-1} = 512^{-2}$ and the upper likelihood boundary estimation constant $c = 2^\upsilon = 256$. The choice of $g(\vec{x})$ и c is due to the constraints $n_j < 2^\upsilon$ which in the case of the mean pixel value $\overline{m} = \sum n_j/s^2 > 1$ lead to the following majorization of the distribution density of the counts:

$$\rho(\vec{x}_j|I(\vec{x})) \approx \pi_j = \frac{n_j}{\sum n_j} < \overline{m}\frac{n_j}{\sum n_j} = \frac{n_j}{s \times s} < \frac{2^\upsilon}{s \times s} = cg(\vec{x}_j). \quad (9)$$

Note that the algorithmic implementation of this sampling procedure is reduced to the random selection of uniformly distributed in the region Ω (of area $s \times s$, with coordinates – floating-point numbers) random vectors $\vec{x}_j$ and including them into the sample of counts X_k while performing the test equivalent to (9) $u_j < n_j$ where j – index of containing $\vec{x}_j$ image pixel, and u_j – is the realization of a uniformly distributed on $(0, 2^\upsilon)$ random variable (on this point, see [20]). In this implementation, pixel normalization is not required.

Because of the simplicity of formation random samples $X_k = \{\vec{x}_1, \ldots, \vec{x}_k\}$ and universality of their statistical description (8), the related representations are useful

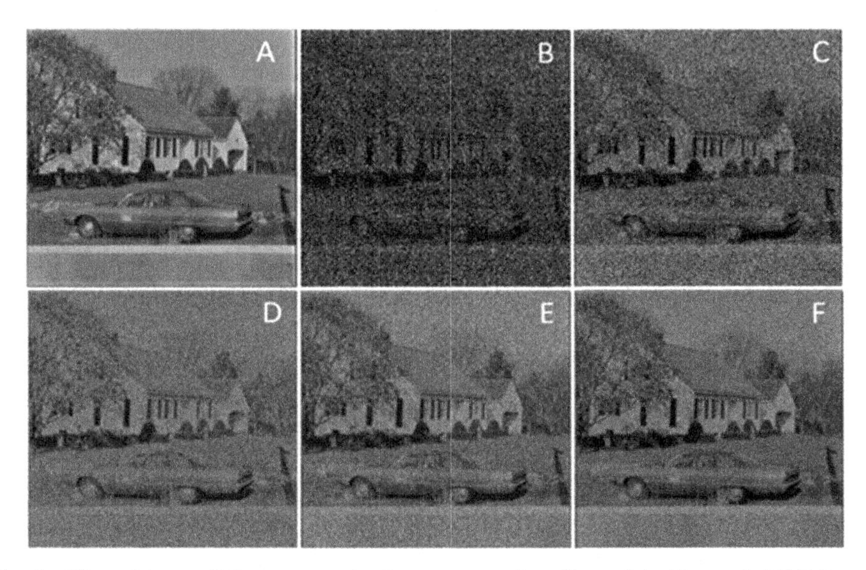

Fig. 5. "House" image [21] representation by count samples: A) – original image in TIFF format, B) – F) sample sizes, respectively k = 100.000, 500.000, 1.000.000, 2.000.000, and 5.000.000 counts.

in many problems of analysis, classification, and identification of images, see [14]. However, in several applications critical to visual perception, sampling representations encounter several problems. Indeed, as can be seen in the fragments of Fig. 5, sampling representations have a less smooth, more granular texture than, for example, conventional raster images.

At the same time, it should not be forgotten that traditional digital images represented in several popular formats, including the TIFF of image "House" in Fig. 5A, taken as an example, are the result of rather complex processing of original images obtained with digital cameras or scanned photographic plates/films, which usually do not possess the required quality. In this regard, like the task of preparing raster representations, the task of improving the quality of visual perception of sampling representations can also be posed. This very important topic, however, is beyond the scope of this paper. Leaving its solution for future research, we shall confine ourselves in conclusion only to the results of the simplest approach to smoothing noisy images based on the Parzen-Rosenblatt window method.

Smoothing by the Parzen-Rosenblatt window method is closely related to the non-parametric reconstruction of probability density distribution based on its kernel estimation [22]. Since in our case the distribution density $\rho(\vec{x}|I(\vec{x}))$ of independent counts in sampling representation $X_k = \{\vec{x}_1, \ldots, \vec{x}_k\}$ is a multiple of intensity $I(\vec{x})$ (5), the kernel estimate

$$\hat{\rho}(\vec{x}|X_k) = \frac{1}{kh}\sum\nolimits_{j=1}^{k} K\left(\frac{\vec{x}-\vec{x}_j}{h}\right). \tag{10}$$

reconstructs both the density of distribution of samples and, to the accuracy of normalization, the intensity $I(\vec{x})$ (intensity shape).

In (10) the kernel $K(\vec{x})$ is assumed to be a non-negative, normalized, symmetric function with a unit second momentum on a two-dimensional plane with coordinates $\vec{x} = (x_1, x_2)$. In other words, the kernel is the simplest smoothing window of unit width. The parameter $h > 0$ is the smoothing parameter, it is often also called the window width [22]. In our experiments, a two-dimensional Gaussian distribution was used as the kernel $K(\vec{x}) = N(\vec{x}|\vec{0}, E)$, the window width parameter h was not used – by default it was assumed $h = 1$. The image sizes $s \times s$, $s = 512$ were chosen as that of the original image, i.e. it was assumed that the parameter h is equal to the pixel size of original image, which here assumed as the unity.

Fig. 6. "House" image [21] reconstruction using random samples: A) –original image in TIFF format, B) – F) – reconstructed images by sample sizes, respectively k = 100.000, 500.000, 1.000.000, 2.000.000, and 5.000.000 counts.

For example, Fig. 6 shows the images obtained from sampling counts (Fig. 5) corresponding to sample representations of size $k = 100.000$, 500.000, 1.000.000, 2.000.000, and 5.000.000 counts (Fig. 5B–F). It is worth noting that the image obtained already from a relatively small sample of counts ($k = 1.000.000$) is visually perceived as an image of acceptable quality despite the presence of noise and blurring. Increasing the number of counts k allows more detailed information and higher quality of image, but also increases the time required for sample formation and image reconstruction.

At the formation of a sampling representation of the image, it is necessary to borne in mind not only the obvious reduction of volume of the initial image, but also other its characteristics. It is worth noting that the time spent on obtaining sampling representations of different sizes (Fig. 7), depends on the number of random counts selected. For example, for the relatively small amount of counts the time of image "House" generation appeared insignificant, for $k = 100.000$ it was on the average $t = 0.56$ s, an increase in the number of counts by 10 times (up to $k = 1.000.000$) causes an increase in the time to

$t = 3.76$ s. It is worth noting that a sampling representation with the number of counts equal to $k = 1.000.000$ has already a sufficiently accurate set of properties allowing sufficiently precise comparison of the objects displayed on it with similar objects on the source image. So, the time required to form the most plausible sampling representation for images with similar initial characteristics (size 512 × 512 pixels, 8-bit pixel depth), will not exceed 5 s. With further sampling with $k = 2.000.000$ and $k = 5.000.000$ the time was about $t = 7.3$ s and $t = 13.9$ s respectively. In Fig. 7, the blue line shows the dependence of the time formation of the sampling representation depending on the number of counts. The dependence of average time is almost linear, however, the more counts we use the greater spread becomes.

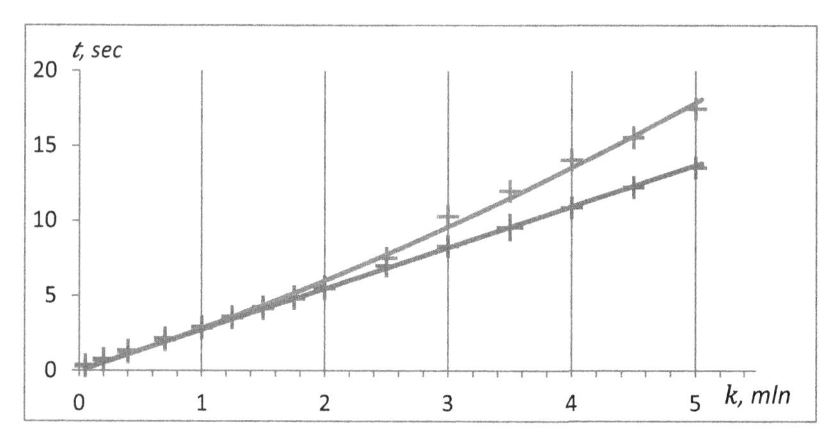

Fig. 7. The dependence of sample formation and image reconstruction times on the number of counts k. The blue line is the time spent on sample formation, the red – spent on its smoothing. (Color figure online)

The time spent on smoothing samples of different sizes was also recorded. For small numbers of counts $k = 100.000$, 500.000, and 1.000.000 the time was $t = 0.53$, 1.51, and 2.88 s respectively. For large samples $k = 2,000,000$ and 5,000,000 it took time $t \sim$ 5.5 and 13.0 s. Also, as well as in the case with the formation of count samples, the best image reconstruction was reached using the samples with the number of counts $k \sim 10^6$. However, it is necessary to note that in the case of small sample sizes it is also possible to receive the recognizable images.

4 Conclusion

The paper presents the coding/representation of images by random count samples of a fixed size. A method for such representation formation is proposed. The proposed sampling representations have a form of a random count sets, with a smaller (controlled) dimension, which can be interpreted as specifical image coding. This coding makes possible not only to compress images, but also to efficiently perform some subsequent stages of their processing. It is possible, in particular, to reconstruct the original images from the obtained representations (with a certain degree of accuracy) by the means, for example, of kernel smoothing methods (Parzen-Rosenblatt method). The degree of smoothness of the reconstructed image, as well as the time of its reconstruction from a representation, linearly depends on the size of the sample used.

The proposed concept of an ideal image also opens the new opportunities in the development and improvement of various imaging devices, for example, modern SPAD cameras, operating in the photon counting mode.

We believe that the approach described in the article, aimed at creating a representation of the image by count samples, will be effectively applied in medical images processing, especially in the case of low radiation intensities like, for example, in the case of photon emission tomography (PET). In the future, we plan to improve the efficiency of approach described and the representation proposed to a wide range of image classes.

Acknowledgements. The authors express their gratitude to the Ministry of Science and Higher Education of Russia for the possibility of using the Unique Science Unit "Cryointegral" (USU #352529) designed for computer simulation and for financial support (Project No. 075-15-2021-667).

References

1. Fox, M.: Quantum Optics: An Introduction. Oxford University Press, New York (2006)
2. Holst, G.C.: CMOS/CCD Sensors and Camera Systems. SPIE Press, Bellingham (2011)
3. Fossum, E.R., Teranishi, N., et al. (eds.): Photon-counting image sensors. In: MDPI (2017)
4. Robbins, M.: Electron-multiplying charge coupled devices-EMCCDs. In: Seitz, P., Theuwissen, A.J. (eds.) Single-Photon Imaging, pp. 103–121. Springer, Berlin (2011). https://doi.org/10.1007/978-3-642-18443-7_6
5. Dutton, N.A.W., Gyongy, I., Parmesan, L., et al.: A SPAD–based QVGA image sensor for single–photon counting and quanta imaging. IEEE Trans. Electron. Devices **63**(1), 189–196 (2016)
6. Aull, B.F., Schuette, D.R., Young, D.J., et al.: A study of crosstalk in a 256x256 photon counting imager based on silicon Geiger–mode avalanche photodiodes. IEEE Sens. J. **15**(4), 2123–2132 (2015)
7. Gonzalez, R.C., Woods, R.E.: Digital Image Processing, 3rd edn. Prentice Hall Inc., New Jersey (2007)
8. Gabriel, C.G., Perrinet, L., et al. (eds.): Biologically Inspired Computer Vision: Fundamentals and Applications. Wiley, Weinheim (2015)
9. Rodieck, R.W.: The First Steps in Seeing. Sinauer, Sunderland (1998)
10. Barber, D.: Bayesian Reasoning and Machine Learning. Cambridge University Press, Cambridge (2012)
11. Morimoto, K., Ardelean, A., et al.: Megapixel time-gated SPAD image sensor for 2D and 3D imaging applications. Optica **7**, 346–354 (2020)
12. Goodman, J.W.: Statistical Optics, 2nd edn. Wiley, New York (2015)
13. Pal, N.R., Pal, S.K.: Image model, poisson distribution and object extraction. Int. J. Pattern Recognit. Artif. Intell. **5**(3), 459–483 (1991)
14. Antsiperov, V.: Maximum similarity method for image mining. In: Del Bimbo, A., et al. (eds.) ICPR 2021. LNCS, vol. 12665, pp. 301–313. Springer, Cham (2021). https://doi.org/10.1007/978-3-030-68821-9_28

15. Koch, K., McLean, J., et al.: How much the eye tells the brain? Curr. Biol. **16**(14), 1428–1434 (2006)
16. Streit, R.L.: Poisson Point Processes. Imaging, Tracking and Sensing. Springer, New York (2010). https://doi.org/10.1007/978-1-4419-6923-1_1
17. Gallager, R.: Stochastic Processes: Theory for Applications. Cambridge University Press, Cambridge (2013)
18. Antsiperov, V.: Machine learning approach to the synthesis of identification procedures for modern photon-counting sensors. In: Proceedings of the 8th International Conference on Pattern Recognition Applications and Methods, ICPRAM, Prague, vol. 1, pp. 814–821 (2019)
19. Bertero, M., Boccacci, P., Desidera, G., Vicidomini, G.: Image deblurring with Poisson data: from cells to galaxies. Inverse Prob. **25**(12), 123006 (2009)
20. Robert, C.P., Casella, G.: Monte Carlo Statistical Methods, 2nd edn. Springer, New York (2004). https://doi.org/10.1007/978-1-4757-4145-2
21. The USC-SIPI Image Database. http://sipi.usc.edu/database/. Accessed 21 Aug 2021
22. Wand, M.P., Jones, M.C.: Kernel Smoothing. Chapman & Hall, Boca Raton (1994)

Analysis of Machine Learning Algorithms for Facial Expression Recognition

Akhilesh Kumar(✉) and Awadhesh Kumar

Department of Computer Science, Banaras Hindu University, Varanasi, UP, India
akhilesh.kumar17@bhu.ac.in

Abstract. People can identify emotion from facial expressions easily, but it is more difficult to do it with a computer. It is now feasible to identify feelings from images because of recent advances in computational intelligence. Emotional responses are those mental states of thoughts that develop without conscious effort and are naturally associated with facial muscles, resulting in different facial expressions such as happy, sad, angry, contempt, fear, surprise etc. Emotions play an important role in nonverbal cues that represent a person's interior thoughts. Intimate robots are expanding in every domain, whether it is completing requirements of elderly people, addressing psychiatric patients, child rehabilitation, or even childcare, as the human–robot interface is grabbing on every day with the increased demand for automation in every industry. We evaluate and test machine learning algorithms on the FER 2013 data set to recognize human emotion from facial expressions, with some of them achieving the highest accuracy and the others failing to detect emotions. Many researchers have used various machine learning methods to identify human emotions during the last few years. In this research article, we analyze eight frequently used machine learning techniques on the FER 2013 dataset to determine which method performs best at categorizing human facial expression. After analyzing the results, it is found that the accuracy of some of the algorithms is quite satisfactory, with 37% for Logistic Regression, 33% for K-neighbors classifier, 100% for Decision Tree Classifier, 78% for Random Forests, 57% for Ada-Boost, 100% for Gaussian NB, 33% for LDA (Linear Discriminant Analysis), and 99% for QDA (Quadratic Discriminant Analysis). Furthermore, the experimental results show that the Decision Tree and Gaussian NB Classifier can correctly identify all of the emotions in the FER 2013 dataset with 100% classification accuracy, while Quadratic Discriminant Analysis can do so with 99% accuracy.

Keywords: Facial expression · Emotion · Machine learning · KNN · QDA · Decision tree · Gaussian NB · Random Forest

1 Introduction

Facial expressions are important indications of individual sentiments since they are linked to emotions. It may be utilized as strong evidence to evaluate if someone is telling the truth or not. Face expressions are the result of a mix of emotions and accompanying

I. Woungang et al. (Eds.): ANTIC 2021, CCIS 1534, pp. 730–750, 2022.
https://doi.org/10.1007/978-3-030-96040-7_55

changes in facial muscles [1]. It gives us insight into a person's mental state and allows us to communicate with them based on their mood. Furthermore, facial expressions aid in determining a person's present emotional and mood state [2]. Facial expression is highly important in nonverbal communication between people. Effective emotion detection algorithms can help systems understand humans better and aid the growth of human-computer interaction (HCI) systems [3]. It displays a person's precise reaction to any situation. As a result, when a robot learns emotion, it will be able to interact with humans and educate or lead them better. To enable a computer to do this task, a training method is required to enable the machine to identify human expression [4]. The universal facial expressions of eight emotions, including neutral joy, sorrow, anger, contempt, disgust, fear, and surprise, have solid evidence. As a result, detecting these emotions on the face is critical, as it has a several applications in computer vision and AI. These disciplines are investigating face expressions in order to mechanically detect human attitudes. Since the robot is capable of reading a human reply, emotion categorization in robotics can be utilized to improve human-robot interactions [5]. Data mining, also known as data or knowledge discovery, is the act of examining data from many angles and distilling it into valuable information. Recognizing emotion from pictures has been a popular research topic in image processing and human-computer interaction applications [6]. Those technologies have spawned a slew of new applications. It may be utilized in a variety of applications, including HMI, robotics, security, and entertainment. The most essential cognitive skill that our brain efficiently accomplishes is facial emotion detection. Facial expression refers to the position of a person's face underneath the skin. Though all emotions are communicated by speech, hand, or body movement, all primary forms of expression are exclusively expressed through the human face [7]. The progress of Artificial Intelligence (AI) has allowed humans and computers to form a symbiotic relationship in which computers can comprehend human emotions and adapt to human emotion reactions [8]. Facial expression detection is one of the most significant directions in the science of computer vision, and it plays an essential part in people's everyday work and lives. The implementation of intelligent human-computer interaction relies heavily on human emotion detection based on facial expressions [9]. Machine learning and computer vision investigates and assesses machine learning applications in computer vision, as well as forecasting future prospects. The machine learning techniques in computer vision are supervised, unsupervised, and semi-supervised [10]. The accomplishment of robotic systems requires smooth device interaction. As a consequence, a robot can extract admissions from a person's face alone, such as detecting their gender or identifying their emotions. Because each challenge's data is so diverse, applying machine learning (ML) techniques to properly understand any of these factors has proven challenging [11]. As a result, frameworks containing millions of characteristics and thousands of observations are constructed [12]. The human brain can also predict how a person will feel depending on their mood. Deep learning aspires to get as similar to the human mind as feasible. Happy, surprise, sad, fear, wrath, disgust, and neutral are the seven fundamental emotions linked with facial expressions [13], each of which is associated with a specific expression. Preprocessing, feature extraction, and classification are three stages of FER [6, 14, 15] systems. To improve image quality, the pre-processing stage uses an image enhancement method by removing noise and artifacts, followed by the extraction of features and classification of emotions using a machine learning classifier. The third stage is classification, in which facial expression classes are used to name features, and then categorization is done using the labelled features. Logistic Regression, KNeighbors, Decision tree, Gaussian NB, Random Forest, AdaBoost, LDA, and QDA are the classifiers used.

Research on the FER 2013 dataset [11] shows that a person's capacity to read other people's emotions is still 95% competent. A number of techniques have been used to study the prediction of human emotion. Real-time accuracy is still a challenge. According to psychological research [16], the actual identification of a person's emotional condition is critical for computational intelligence and interaction. In this research article, we studied and analyzed eight frequently used machine learning techniques on the FER 2013 dataset to determine which method performs best at categorizing human facial expressions. As tested, the accuracy of each algorithm is quite satisfactory: 37% for Logistic Regression, 33% for Kneighbors Classifier, 100% for Decision Tree Classifier, 78% for Random Forests, 57% for AdaBoost, 100% for Gaussian NB, 33% for LDA, and 99% for QDA. Furthermore, the experimental results show that the Decision Tree and Gaussian NB Classifier can correctly identify all of the emotions in the dataset with 100% classification accuracy, while Quadratic Discriminant Analysis can do so with 99% accuracy. Figure 1 depicts the working procedure for recognizing facial expressions using machine learning classifiers. First, we select an image or a real-time image, then apply image segmentation for detecting faces, after that extract the features using statistical methods like mean, standard deviation, and root means square, and finally classify the emotions using the machine learning algorithm like Gaussian NB, Decision Tree, Random Forest, etc. It means the human emotion recognition system is categorized into four stages, i.e. data collection (image selection), segmentation (face detection), feature extraction with the help of statistical techniques, and finally classify the emotion using Machine learning classifier.

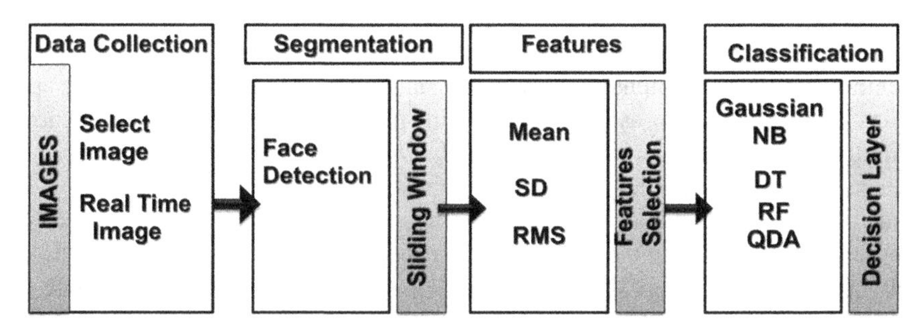

Fig. 1. The architecture of human emotion recognition systems

Section 2 of the research article covered related research which included the state of the art and techniques used to perform emotion recognition. Section 3 narrates the experimental methodology, including dataset information. In Sect. 4, we discussed our experimental environment and Sect. 5 describes the performance analysis and its outcomes. Section 6 contains the conclusions and recommendations for further study.

2 Related Work

This section examines works that are relevant to current research in the fields of human emotion recognition and bio-signal data processing methods. Facial expressions are a common sign for all people to express their present circumstance. Human Behavior Predictor [17, 18], Surveillance System [19], Medical Rehabilitation [20, 21], Crime detection [19, 22], and e-learning [23–25] are some applications that might benefit from facial expression categorization. In [26], the author presented a Multi-Support Vector Neural Network based on the Whale-Grasshopper Optimization algorithm for face emotion detection. The SIFT and the SLDP are used to extract characteristics from a face picture. The Authors present a tree-based genetic program with three functional layers for feature selection, fusion for facial expression recognition and classifications in [27]. The creator of [28] finds a user's social media profile by recognizing their gender, age, and personality attributes. In healthcare, computer vision technologies are used to provide pre-programmed interactions for patient monitoring. A multi-modal visualization analysis approach for enhancing the less complicated processing nature of programmed human-machine interactions (HMI) in health monitoring is proposed in [20]. The suggested technique achieves 95.702% recognition accuracy using CNN, according to the testing data. Based on an automated and more efficient face decomposition into regions of interest, a multiclass SVM classifier is utilized to identify the six fundamental facial expressions and the neutral state [29]. For emotion recognition, a mixture of contour and region harmonics is utilized to create the interconnectedness of sub-local regions in the human face. The study [30] used a multi-class SVM classifier with subject-dependent k-fold cross-validation to classify human feelings into expressions, and then tested the method on three public facial expression datasets for semi-regions of the human face, achieving 94.90%, 93.43%, and 92.57% recognition rates for the CK+, CFEE, and MUG repositories, respectively. In reality, data is frequently acquired from a variety of devices or settings, causing the identification performance to suffer significantly. [31] investigates the topic of cross-dataset facial expression identification and proposes a unique dual-graph regularized transfer sparse coding approach to solve it. The coherence of the DWT is combined with four different algorithms in [32], including the error vector of PCA, the eigen vector of principal component analysis, the eigen vector of LDA, and the CNN. The four results are then combined using the entropy of predictive performance and the Fuzzy system. Compared to earlier studies where separate methods were executed on a specific collection of photos, the combined technique of the study yielded recognition rates of 89.56% for the worst case and 93.34% for the best scenario. Both may be said to be superior.

[33] Uses a bagging algorithm to produce variable training subsets, which are then rebalanced and sized appropriately. The SVM model is then utilized as an individual basis classifier to create a variety of ensemble input members. The goal of [34] is to construct a mapping between facial expressions and the facial muscle contractions that correlate to them, as well as their movement directions. Mathematical symbolic representations are used to demonstrate this mapping. Different normalized facial characteristics and 2D spatial coordinates of a face are used to assess these symbolic representations of facial emotions. Machine learning techniques were used to predict anxiety, sadness, and stress in [35]. The Depression, Anxiety, and Stress Scale questionnaire was used to collect

data from employed and unemployed people from various cultures and groups in order to apply these algorithms. The use of Machine Learning methods in recognizing toddlers' emotions while interacting with edutainment applications is discussed in [36]. The scenarios studied point to the right way to measure children's pleasure using edutainment apps. [2] Investigated the impact of mood on impression generation and individual memory. For individuals in a controlled happy or sad mood, realistic human descriptions with positive and negative features were provided. Facial recognition technology (FRT) has been widely used, but it comes with the risk of abuse owing to technical limits and legal inconsistencies. [37] presents a conceptualization approach for examining public perceptions of FRT and identifying FRT scenarios that might be misused. To train the Support Vector Machine Learning Algorithm, the author [38] used Computer Vision and Haar-Cascade Classifier, as well as feature descriptors, Histogram of Oriented Gradient, and Local Binary Pattern. Gathering and expanding data with Asian/Filipino respondents while being added to the CK+ database at various stages of retraining, refinement, and cross-fold validation resulted in improved outcomes, with the highest average accuracy of 87.14% across the seven emotions. [6] focuses on live webcam pictures and develops an automated face expression detection system for anxious people, giving them music therapy to help them relax [39]. Emotion recognition was done in both real-time and static pictures in [5]. The author utilized the Cohn-Kanade Database and the CK+ Database, both of which include numerous static pictures of 640 $\times$ 400 pixels, as well as a camera for real-time use. Faces must first be detected in static pictures or real-time movies using the HAAR filter from OpenCV. After the face has been recognized, it may be cropped and processed to find more facial landmarks. The datasets are then educated using face landmarks and categorized using a machine learning technique (Support Vector Machine). Using SVM, the author was able to achieve an accuracy of 93.7%. Using the Zernike moments, they present a method to categorize facial expressions into two groups in [40]. Facial feature extraction and facial expression categorization are the two elements of this method. Higher order Zernike moments were used to extract face characteristics, which were then categorized using an ANN-based classifier. On the Cohn-Kanade (CK) dataset, the system achieved 69% accuracy. The performance of three machine learning algorithms to classify human face expressions is compared and analyzed in [4]. The classification procedure uses a total of 23 variables derived from the distance between face characteristics as input, yielding seven categories: angry, disgust, fear, happy, neutral, sad, and surprise. To evaluate the system, some test cases were created, each with a different quantity of training data ranging from 165 to 520. The accuracy of the KNN method is 75.15%, the SVM algorithm is 80%, and the Random Forests strategy is 76.97% when evaluated with the least amount of data. On the FER2013 dataset, the authors obtained the best single-network classification accuracy in [41]. He uses the VGGNet architecture, fine-tunes the hyper parameters, and tries out other optimization approaches. Without the need for additional training data, this achieves an accuracy of 73.28% on FER2013. Real-time student involvement tracking is a critical stage in the educational process. In [25], the author proposes a hybrid architecture that uses student eye gaze movements, head movements, and facial expression to dynamically forecast student attention and engagement levels toward the tutor, and the material is dynamically modified based on the output value. As a result, this notion has a lot of potential in e-learning, classroom training, and human behaviour analysis. He utilized PCA for face expression identification, Haar Cascade for pupil detection, Local Binary Patterns for identifying head motions, and OpenCV for machine learning model development and comparison in the feature extraction stage. Existing advances in the field of Human-Computer Interaction (HCI) aim to develop a more natural interaction between the actors involved. Recently, there has been a lot of interest in automatic and accurate estimates of emotional states, particularly using physiological data. In terms of physiological measurements, emotion evaluation benefits from pure, unmodified experiences as opposed to artificial facial or verbal assessments. The use of biological cues to recognize emotions is critical in the study of psychological states [42]. An electroencephalogram (EEG) and electrocardiogram (ECG)-based multimodal fusion emotion identification technique is based on the Dempster-Shafer evidence theory [43]. He uses the SVM classifier to categorize features in EEG, and he creates the matching Bi-directional LSTM network emotion recognition structure in ECG, which is then merged with the EEG classification results using evidence theory. He chose 25 video clips that represented five emotions (happy, calm, urious, sad, and disgusted), and 20 people took part in our emotional study. Some physiological measure-based categorization methods for measuring emotional state are examined in various circumstances in [42]. In emotion recognition systems, emotion representation is a significant research topic. Some research has used Paul Ekman's [44] categorical representations of emotions, which include six fundamental types: happiness, anger, disgust, sorrow, fear, and surprise. In the study [44] on opinion mining and sentiment analysis, a more generally recognized category of emotions from the standpoint of readers is social emotions. Interconnected representation models for emotions have been used in other investigations. The literature review on facial expression and psychological signals for recognizing the emotion has been discussed in this section. On the benchmark dataset FER 2013, we focused on facial expression recognition using eight machine learning algorithms in this paper.

3 Experimental Methodology

3.1 Dataset

Data Set: FER 2013
In the ICML 2013 Challenges in Representation Learning, the Facial Expression Recognition 2013 (FER-2013) database was launched. Faces were automatically recorded in the database, which was generated using the Google image search API. Any of the six fundamental expressions, as well as the neutral, are assigned to faces. There are 35,887 pictures in the database as a consequence of this process.

Figure 2 shows facial image samples of seven types of emotions, which are labeled by integer values ranging from 0 to 6. The labels 0 to 6 specify happy, angry, disgusted, surprised, fearful, and neutral faces respectively (Table 1).

Table 1. Number of images of each emotions of data set FER 2013

Emotions	Happy	Anger	Disgust	Sad	Surprise	Fear	Neutral
Images#	8989	4953	547	6077	4002	5121	6198

3.2 Emotion Samples of FER Dataset

Fig. 2. Sample images of emotions of data set FER 2013

3.3 ML Classifier

Logistic Regression. It is used to calculate discrete values such as 0/1 or yes/no from a set of independent variables (s). It works by fitting data to a logit function to predict the likelihood of an event occurring. As a result, it's also called logit regression. Its output values are between 0 and 1 since it forecasts probability (as expected). Let's assume you're trying to figure out if a particular face is joyful or not. There are just two possible outcomes: happy face or not happy face. Logistic Regression can help you with this.

$$\text{logit}(p) = \ln(p/(1 - p)) \tag{1}$$

Above, p represents the likelihood of an event occurring, (1 − p represents the chance of an event not occurring, and log (p/1 − p) represents the link function. We can represent a non-linear connection in a linear fashion by using a logarithmic transformation on the result variable. In Logistic Regression, this is the equation that is utilized. The odd ratio here is (p/1 − p).

Kneighbors Classifier. A k-nearest neighbor estimation is an analytical technique that employs the nearest neighbors to estimate the unknown parameters or missing point [45, 46]. The nearest neighbors are generally determined to be the points with the shortest distance to the unknown point from their adjacency. The simplest approach for measuring the distance between neighbors is the Euclidian distance function, which is given in (2).

$$d(x, y) = ||x - y|| = \sqrt{\sum\nolimits_{i=1}^{n}}(x_i - y_i)^2 \quad (2)$$

Where x = (x_1, x_2,... x_n) and y = (y_1, y_2,.. y_n), and n is the vector size. The k neighbor points that have the shortest distance to the unknown point is used to estimate its value using (3).

$$\hat{y}_i = \sum\nolimits_{i=1}^{n} w_i y_i \quad (3)$$

Where w_i is the weight of every single neighbor point y_i to the query point ŷ.

Decision Tree Classifier. The Classification and Regression Tree (CART) is a DT method. CART employs a recursive partitioning approach, in which nodes are produced depending on a set of dividing criteria. The tree is then grown using these nodes (made and split nodes). The optimal split point must be determined before the split criterion can be applied. The quality of the splitting criterion is measured by a function obtained by using the processing variance function. A generated function is applied to every split point [47] to determine the optimal point for splitting. It's a supervised learning method that's commonly used to solve classification issues. It works for both continuous and categorical dependent variables, which is surprising. It employs a variety of methods, including Gini, Information Gain, Chi-square, and entropy, to divide the population into separate heterogeneous groups.

Random Forest Classifier. Random Forest is a classification approach that works by building regression trees out of a collection of decision trees [48]. Breiman's bagging approach is used to merge the trees, which are chosen at random from the given training data. The outcome is predicted by combining the decisions made by the decision trees. Bagging is a strategy for constructing an ensemble of classifiers in which each classifier is constructed from a randomly selected sample of data, resulting in classifiers that are likely to differ from one another.

RF offers certain advantageous features. For starters, RF can handle high-dimensional data. RF may also handle continuous, categorical, and binary data. It can also deal with missing data. Second, RF is capable of dealing with big variable inputs as well as balance faults in imbalanced datasets. Finally, RF is less susceptible to outliers when dealing with training data. Finally, RF produces reliable prediction findings in a variety of domains. The model will be trained when it has been developed. Each data feature's significance may be determined through model training.

The classification performance of RF outperforms a single decision tree [48]. For a variety of datasets, RF has a good prediction accuracy. Because RF employs a random sample and ensemble method, it is able to make more accurate predictions and generalizations.

AdaBoost Classifier. To train a group of weak classifiers into a strong classifier, the AdaBoost method [49] is employed. Each poor classifier isn't expected to do a good job of classifying data, thus building one is simple. The weak classifier employed here is a straightforward function that chooses a feature and a threshold that best classifies the training data. The following is the syntax for this function:

$$h_t(x|k, p, \theta) = \begin{cases} 1, px(k) < p\theta \\ -1, px(k) \geq p\theta \end{cases} \tag{4}$$

Where θ is a threshold, p denotes the inequation's direction, and k denotes that the k'th characteristic of x has been chosen.

Gaussian NB. The Bayes theorem provides the basis for a series of supervised machine learning classification algorithms known as Naïve Bayes. It's a basic categorization approach with a lot of power. They're useful when the inputs' dimensionality is high. The Gaussian Naïve Bayes with Gaussian Normal Distribution is a version of Naïve Bayes that accepts continuous data and follows the Gaussian normal distribution. Naïve Bayes Classifiers are easy to develop and execute, and they may be used in a variety of real-world settings. When working with continuous data, one common assumption is that the continuous values associated with each class follow a normal (or Gaussian) distribution.

Linear Discriminant Analysis. LDA is a technique for finding ideal linear maximizations and, subsequently, grouping and classification in the projected subspace [50]. Ideal projections maximize projected distances across classes while limiting projected distances between topics in the same class to a minimum. Take a p-dimensional case for example; mathematically the ideal projection is the eigenvector b in

$$\sum \mathrm{w}^{-1} \sum \mathrm{Bb} = \lambda \mathrm{b} \tag{5}$$

The inverse of the within-class covariance matrix $\sum \mathrm{w}$ is $\sum \mathrm{w}^{-1}$, and the between-class covariance matrix is$\sum \mathrm{B}$. $\sum \mathrm{w}$ is invertible, under traditional multivariate setting.

Quadratic Discriminant Analysis. QDA is a technique for forecasting the direction of a wave. In this situation, estimating the wave direction is regarded as a classification challenge. Classification is a widely used machine learning approach for categorizing things based on input data. A training dataset is used to train the model, which comprises classes, which are the wave directions that have been chosen. As a result, the output can only be as precise as the data used to train it. QDA is a variation of LDA in which each class of observations has its own covariance matrix calculated. QDA is especially beneficial if you already know that different classes have different covariances. QDA has the drawback of not being able to be utilized as a dimensionality reduction approach. Each class is represented by QDA as follows:

$$f_k(x) = \frac{1}{(2\pi)^{p/2}\left|\sum_k\right|^{1/2}} \exp -\frac{1}{2}(x-\mu_k)^T \sum_k^{-1} (x-\mu_k) \tag{6}$$

For class k, $\sum_k$ is the covariance matrix, μ_k is the mean, and p is the number of dimensions.

3.4 Machine Learning Model Training

After extracting the features from the pre-processed data, the next step is to train the ML models on them. Over the feature vector, we utilized eight different machine learning models. Logistic regression, Kneighbors classifier, Decision Tree, Random Forest, AdaBoost, Gaussian NB, LDA, and QDA are examples of machine learning models. We used a train set to train these eight ML models, and we mentioned all the attributes along with their value in Table 2.

Table 2. Attribute-value for machine learning classifiers

Classifiers	Attributes	Value
Logistic regression	Fit intercept	True (default)
	Maximum no. of iteration	100 (default)
	Intercept scaling	1 (default)
	Multi class	'auto' (default)
	Tolerance criteria	0.0001
	Solver	'lbfgs' (default)
Kneighbors classifier	Leaf size	30 (default)
	No. of neighbors	5 (default)
	Weights	'uniform' (default)
Decision tree	Criterion	'gini' (default)
	Minimum samples for leaf	1 (default)
	Minimum samples for split	2 (default)
Random forest	Maximum features	'auto' (default)
	Minimum samples required for leaf	1 (default)
	Minimum samples required for split	2 (default)
	No of estimators (trees)	100 (default)
AdaBoost	Algorithm	'SAMME.R' (default)
	Rate of learning	1.0 (default)
	Maximum No. of estimators	50 (default)

(continued)

Table 2. (*continued*)

Classifiers	Attributes	Value
Gaussian NB	Smoothing	1e−09 (default)
LDA	Store covariance	False (default)
	Tol (absolute threshold)	0.0001(default)
QDA	Store covariance	False (default)
	Tol (absolute threshold)	0.0001 (default)

3.5 Trained Model Testing and Performance Comparison

Once the ML models are trained over the trained set, the next step is to test these trained models over the test set to evaluate their performance on unseen data. The performance of eight machine learning algorithms for detecting human emotions is evaluated using four commonly used performance testing parameters: precision, recall, accuracy, and F1-score. Afterward, we compared the performance of each trained ML model with all other ML models based on the four prescribed performance parameters. The detailed performance analysis of each ML model and their comparison is given in Sect. 5.2.

4 Experimental Environment

4.1 Google Colab

Google spent years developing Tensor Flow, AI architecture, and Colaboratory, a development platform. Google Colab, or simply Colab, is the new name for Colaboratory. Colaboratory, or "Colab" for short, is a browser-based platform that allows anybody to develop and run any Python code. It is particularly well suited to computer vision, data analysis, and teaching. Colab is a cloud-based notebook platform that is free to use. The utilization of GPU is another appealing feature that Google provides to developers. Colab is a free application that supports the GPU. It's secure, at least as secure as your Google Doc. No one else has access to your personal Colab notebooks.

5 Experiment Result Evaluation and Discussion

The different ML methods were used to classify seven human emotions. As a result, we'll go over the performance metrics we used to compare the performance of eight machine learning algorithms. Following that, we examine the findings of human emotion identification using eight frequently used machine learning classifiers that were trained and evaluated on the FER 2013 dataset.

5.1 Performance Parameters

As previously stated, we evaluated the effectiveness of eight different machine learning algorithms for human emotion identification. On FER 2013 data, we trained and evaluated eight machine learning algorithms in this experiment. Four performance testing criteria are used to evaluate the performance of eight machine learning algorithms for detecting human emotion detection in experiments: precision, recall, accuracy, and F1-score.

Precision. Precision is defined as the ratio of activities that are truly recognized to all activities that are recognized. It may be represented mathematically as:

$$Precision = (TP/(TP + FP)) \times 100$$

Recall. The capacity of the system to accurately recognize all of the actions is described as recall. The real positive rate is another name for it. It's written like this in math:

$$Recall = (TP/(TP + FN)) \times 100$$

Accuracy. The fraction of properly identified actions across all samples determines the accuracy. It may be represented mathematically as:

$$Accuracy = ((TP + TN)/(TP + TN + FP + FN)) \times 100$$

F1-Score. The harmonic mean of accuracy and recall is defined as the F1-score. It may be represented mathematically as:

$$F1 - Score = 2 \times ((Precision \times Recall)/(Precision + Recall))$$

5.2 Performance of ML Classifiers for Recognition of Human Emotion

In the experiment, we trained eight machine learning algorithms (LR, KN, DT, RF, AdaBoost, GNB, LDA, and QDA) over the facial image dataset FER-22013. Table 3 shows the result of these machine learning algorithms for human emotion recognition in the form of a confusion matrix for each algorithm with header columns and rows representing the actual class labels and predicted values, respectively. From the confusion matrices of these algorithms, we found that the DT algorithm and Gaussian NB classifier efficiently recognized emotions with 100% accuracy; QDA recognized emotions with 99% accuracy while algorithms RF, LR, and AdaBoost identified all emotions with accuracy 78%, 37%, and 57% respectively. Figure 3 shows the overall performance of the eight machine learning algorithms for the performance matrices defined in Sect. 5.1. It can be observed that the DT and Gaussian NB algorithms outperformed all other machine learning algorithms with 100% precision, 100% recall, 100% accuracy, and 100% F1-score recognizing human emotions. Tables 4, 5, 6, 7, 8, 9, 10, and 11 display the individual performance analysis with all performance parameters for individual ML classifiers LR, KN, DT, RF, AdaBoost, GNB, LDA, and QDA, respectively, as well as their time-consuming. From Fig. 4 which shows ML classifiers and time consumed by each classifier, we observed that DT and GNB take very little time i.e. 12 s and 3 s respectively, with 100% accuracy in classifications of human emotions.

In order to estimate the likelihood of a given category result based on many explanatory factors, both logistic regression and discriminant analysis are used (predictors). LR should not be used if the number of observations is fewer than the number of features, since it may result in overfitting. It doesn't make any assumptions about class distributions in the feature set. When your classes are strongly correlated or nonlinear, for example, the coefficients of your logistic regression will not correctly determine the gain or loss from each particular feature. If the input is a mix of several classes, KNN

Table 3. Confusion matrices of 8 machine learning algorithm to recognizing human facial expression

LR: Actual Emotions \ Predicted Emotions	0	1	2	3	4	5	6
0	182	0	93	341	125	88	159
1	14	0	16	31	22	9	15
2	102	0	162	304	162	153	186
3	101	0	82	1245	149	103	142
4	123	1	93	364	261	129	243
5	37	0	68	161	72	348	79
6	82	1	82	380	145	97	426

Accuracy: 37%

KN: Actual Emotions \ Predicted Emotions	0	1	2	3	4	5	6
0	269	15	99	238	131	46	190
1	14	41	7	17	11	4	13
2	138	17	284	221	142	78	189
3	184	39	168	822	180	86	343
4	165	26	143	287	292	41	260
5	74	9	92	166	62	258	104
6	140	19	128	316	141	59	410

Accuracy: 33%

DT: Actual Emotions \ Predicted Emotions	0	1	2	3	4	5	6
0	1031	0	0	0	0	0	0
1	0	107	0	0	0	0	0
2	0	0	1030	0	0	0	0
3	0	0	0	1767	0	0	0
4	0	0	0	0	1227	0	0
5	0	0	0	0	0	752	0
6	0	0	0	0	0	0	1264

Accuracy: 100%

RF: Actual Emotions \ Predicted Emotions	0	1	2	3	4	5	6
0	634	0	118	235	38	3	3
1	9	35	23	33	6	1	0
2	90	0	578	277	64	13	8
3	16	0	52	1626	53	18	2
4	5	0	29	164	980	29	20
5	0	0	11	68	119	529	25
6	0	0	1	18	40	13	1192

Accuracy: 78%

AdaBoost: Actual Emotions \ Predicted Emotions	0	1	2	3	4	5	6
0	1031	0	0	0	0	0	0
1	0	0	0	107	0	0	0
2	0	0	0	1030	0	0	0
3	0	0	0	1767	0	0	0
4	0	0	0	1227	0	0	0
5	0	0	0	752	0	0	0
6	0	0	0	0	0	0	1264

Accuracy: 57%

GNB: Actual Emotions \ Predicted Emotions	0	1	2	3	4	5	6
0	1037	0	0	0	0	0	0
1	0	107	0	0	0	0	0
2	0	0	1030	0	0	0	0
3	0	0	0	1767	0	0	0
4	0	0	0	0	1227	0	0
5	0	0	0	0	0	752	0
6	0	0	0	0	0	0	1264

Accuracy: 100%

LDA: Actual Emotions \ Predicted Emotions	0	1	2	3	4	5	6
0	185	15	120	297	197	59	158
1	14	19	16	28	16	6	8
2	102	17	175	258	173	134	171
3	151	17	148	953	207	75	216
4	137	25	143	311	326	66	219
5	56	8	86	123	59	326	94
6	113	10	135	320	226	72	388

Accuracy: 33%

QDA: Actual Emotions \ Predicted Emotions	0	1	2	3	4	5	6
0	1031	0	0	0	0	0	0
1	0	26	81	0	0	0	0
2	0	0	1030	0	0	0	0
3	0	0	0	1767	0	0	0
4	0	0	0	0	1227	0	0
5	0	0	0	0	0	752	0
6	0	0	0	0	0	0	1264

Accuracy: 99%

will fail because it will try to identify k nearest neighbors, but all points will be random. Since KNN is a distance-based algorithm, calculating the distance between a new point and an old point is highly expensive, reducing the algorithm's performance. The LR, KNN, AdaBoost, and LDA methods do not give significantly improved outcomes because of above said limitations. The remaining ML classifiers such as RF, QDA, DT, and GNB perform better with the accuracy of 78%, 99%, 100%, and 100% respectively. Hence from Figs. 3 and 4, we found that the DT and GLB algorithms recognized human emotions with 100% accuracy and also consume less time i.e. 12 s and 3 s respectively.

Table 4. Accuracy parameters for all the seven emotions separately and combined using LR

Logistic regression (LR)						
Emotions	Precision	Recall	F1-score	Support	Accuracy	Time (seconds)
0	0.28	0.18	0.22	988	0.37	46.26
1	0.00	0.00	0.00	107		
2	0.27	0.15	0.19	1069		
3	0.44	0.68	0.54	1822		
4	0.28	0.21	0.24	1214		
5	0.38	0.45	0.41	765		
6	0.34	0.35	0.35	1213		

Table 5. Accuracy parameters for all the seven emotions separately and combined using KN

KNeighbors classifier (KN)						
Emotions	Precision	Recall	F1-score	Support	Accuracy	Time (seconds)
0	0.27	0.27	0.27	988	0.33	2644.46
1	0.25	0.38	0.30	107		
2	0.31	0.27	0.29	1069		
3	0.40	0.45	0.42	1822		
4	0.30	0.24	0.27	1214		
5	0.45	0.34	0.39	765		
6	0.27	0.34	0.30	1213		

Table 6. Accuracy parameters for all the seven emotions separately and combined using DT

Decision tree classifier (DT)						
Emotions	Precision	Recall	F1-score	Support	Accuracy	Time (seconds)
0	1.00	1.00	1.00	1031	1.00	12.51
1	1.00	1.00	1.00	107		
2	1.00	1.00	1.00	1030		
3	1.00	1.00	1.00	1767		
4	1.00	1.00	1.00	1227		
5	1.00	1.00	1.00	752		
6	1.00	1.00	1.00	1264		

Table 7. Accuracy parameters for all the seven emotions separately and combined using RF

Random forest classifier (RF)						
Emotions	Precision	Recall	F1-score	Support	Accuracy	Time (seconds)
0	0.84	0.61	0.71	1031	0.78	106.88
1	1.00	0.33	0.49	107		
2	0.71	0.56	0.63	1030		
3	0.67	0.92	0.78	1767		
4	0.75	0.80	0.78	1227		
5	0.87	0.70	0.78	752		
6	0.95	0.94	0.95	1264		

Table 8. Accuracy parameters for all the seven emotions separately and combined using ADB

AdaBoost classifier (AdaBoost)						
Emotions	Precision	Recall	F1-score	Support	Accuracy	Time (seconds)
0	1.00	1.00	1.00	1031	0.57	206.01
1	0.00	0.00	0.00	107		
2	0.00	0.00	0.00	1030		
3	0.36	1.00	0.53	1767		
4	0.00	0.00	0.00	1227		
5	0.00	0.00	0.00	752		
6	1.00	1.00	1.00	1264		

Table 9. Accuracy parameters for all the seven emotions separately and combined using GNB

Gaussian NB (GNB)						
Emotions	Precision	Recall	F1-score	Support	Accuracy	Time (seconds)
0	1.00	1.00	1.00	1031	1.00	3.33
1	1.00	1.00	1.00	107		
2	1.00	1.00	1.00	1030		
3	1.00	1.00	1.00	1767		
4	1.00	1.00	1.00	1227		
5	1.00	1.00	1.00	752		
6	1.00	1.00	1.00	1264		

Table 10. Accuracy parameters for all the seven emotions separately and combined using LDA

Linear discriminant analysis (LDA)						
Emotions	Precision	Recall	F1-score	Support	Accuracy	Time (seconds)
0	0.24	0.18	0.21	1031	0.33	47.88
1	0.17	0.18	0.17	107		
2	0.21	0.17	0.19	1030		
3	0.42	0.54	0.47	1767		
4	0.27	0.27	0.27	1227		
5	0.44	0.43	0.44	752		
6	0.32	0.31	0.31	1264		

Table 11. Accuracy parameters for all the seven emotions separately and combined using QDA

Quadratic discriminant analysis (QDA)						
Emotions	Precision	Recall	F1-score	Support	Accuracy	Time (seconds)
0	1.00	1.00	1.00	1031	0.99	119.94
1	1.00	0.24	0.39	107		
2	0.93	1.00	0.96	1030		
3	1.00	1.00	1.00	1767		
4	1.00	1.00	1.00	1227		
5	1.00	1.00	1.00	752		
6	1.00	1.00	1.00	1264		

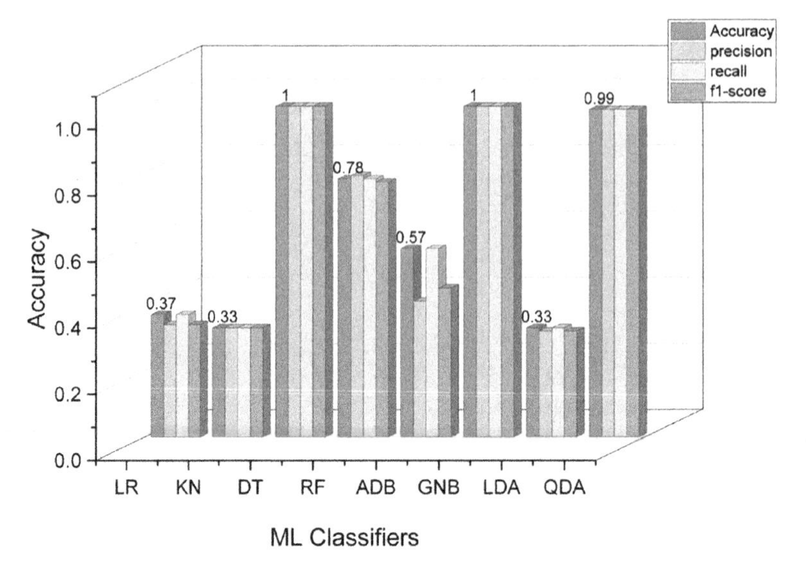

Fig. 3. Performance comparison of 8 machine learning classifiers

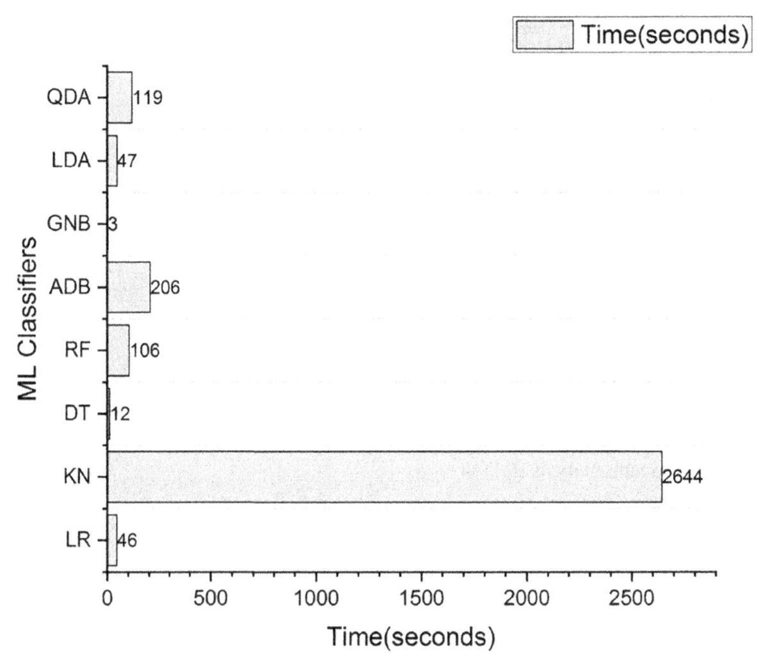

Fig. 4. Time Consuming of 8 machine learning classifiers

6 Conclusion and Future Work

Human emotion recognition in different aspects of our daily life scenarios brings relevant outcomes for novel applications such as identifying people in a crowd, monitor citizens for suspicious behavior, promote public health, e-learning, and many other fields. Moreover, numerous machine learning techniques have been proposed to detect human emotions using facial expressions. This study overviewed eight commonly used machine learning algorithms for recognizing seven emotions. We use a publicly available dataset FER-2013 and trained eight commonly used machine learning algorithms on facial images, evaluating their performance in each case. In this paper, we measure the performance of each of the above machine learning algorithms in various parameters such as precision, recall, F1-score, support, accuracy, and time. The experimental results found that the Decision Tree (DT) and Gaussian NB demonstrated the best performance in recognizing seven human emotions with 100% accuracy and consumed very little time i.e. 12 s and 3 s respectively. Finally, after analyzing the overall results with respect to all the parameters we found that Gaussian NB performs better rather than other discussed algorithms.

Acknowledgement. This research is supported by seed grant under IoE, BHU [grant no. R/Dev/D/IoE/SEED GRANT/2020-21/Scheme No. 6031].

References

1. Ekman, P., Oster, H.: Facial expressions of emotion. Annu. Rev. Psychol. **30**, 527–554 (1979). https://doi.org/10.1146/annurev.ps.30.020179.002523
2. Forgas, J.P., Bower, G.H.: Mood effects on person-perception judgments. J. Pers. Soc. Psychol. **53**, 53–60 (1987). https://doi.org/10.1037/0022-3514.53.1.53
3. Zhao, G., Yang, H., Yu, M.: Expression recognition method based on a lightweight convolutional neural network. IEEE Access **8**, 38528–38537 (2020). https://doi.org/10.1109/ACCESS.2020.2964752
4. Nugrahaeni, R., Mutijarsa, K.: SVM and random forests algorithm for. In: 2016 International Seminar on Application for Technology of Information and Communication, pp. 163–168 (2016)
5. Gupta, S.: Facial emotion recognition in real-time and static images. In: Proceedings of the 2nd International Conference on Inventive Systems and Control (ICISC 2018), pp. 553–560 (2018). https://doi.org/10.1109/ICISC.2018.8398861
6. Machines, S.V.: Machine learning approach summarization. Facial Emot. Recognit. Syst. Mach. Learn. Approach. 272–277 (2017)
7. Kalita, D.: Designing of facial emotion recognition system based on machine learning. In: IEEE 2020 8th International Conference on Reliability, Infocom Technologies and Optimization (Trends and Future Directions) (ICRITO 2020), pp. 969–972 (2020). https://doi.org/10.1109/ICRITO48877.2020.9197771
8. Al-Omair, O.M., Huang, S.: A comparative study of algorithms and methods for facial expression recognition. Proceedings of the 13th Annual IEEE International System Conference (SysCon 2019), pp. 1–6 (2019). https://doi.org/10.1109/SYSCON.2019.8836770
9. Li, B., Lima, D.: Facial expression recognition via ResNet-50. Int. J. Cogn. Comput. Eng. **2**, 57–64 (2021). https://doi.org/10.1016/j.ijcce.2021.02.002

10. Khan, A.I., Al-Habsi, S.: Machine learning in computer vision. Procedia Comput. Sci. **167**, 1444–1451 (2020). https://doi.org/10.1016/j.procs.2020.03.355
11. Goodfellow, I.J., et al.: Challenges in representation learning: a report on three machine learning contests. Neural Netw. **64**, 59–63 (2015). https://doi.org/10.1016/j.neunet.2014.09.005
12. Amodei, D., et al.: Deep speech 2: end-to-end speech recognition in English and Mandarin. In: 33rd International Conference Machine Learning (ICML 2016), vol. 1, pp. 312–321 (2016)
13. Tarnowski, P., Kołodziej, M., Majkowski, A., Rak, R.J.: Emotion recognition using facial expressions. Procedia Comput. Sci. **108**, 1175–1184 (2017). https://doi.org/10.1016/j.procs.2017.05.025
14. Yang, D., Alsadoon, A., Prasad, P.W.C., Singh, A.K., Elchouemi, A.: An emotion recognition model based on facial recognition in virtual learning environment. Procedia Comput. Sci. **125**, 2 (2018). https://doi.org/10.1016/j.procs.2017.12.003
15. Ewees, A.A., Ellaban, H.A., Eleraky, R.M.: Features selection for facial expression recognition. In: 2019 10th International Conference on Computing, Communications and Networking Technologies (ICCCNT 2019), pp. 1–6 (2019). https://doi.org/10.1109/ICCCNT45670.2019.8944459
16. Acharya, D., Billimoria, A., Srivastava, N., Goel, S., Bhardwaj, A.: Emotion recognition using Fourier transform and genetic programming. Appl. Acoust. **164**, 107260 (2020). https://doi.org/10.1016/j.apacoust.2020.107260
17. Park, K.H., Lee, H.E., Kim, Y., Bien, Z.Z.: A steward robot for human-friendly human-machine interaction in a smart house environment. IEEE Trans. Autom. Sci. Eng. **5**, 21–25 (2008). https://doi.org/10.1109/TASE.2007.911674
18. Akhand, M.A.H., Roy, S., Siddique, N., Kamal, M.A.S., Shimamura, T.: Facial emotion recognition using transfer learning in the deep CNN. Electron. **10**, 192–196 (2021). https://doi.org/10.3390/electronics10091036
19. Tiwari, S., Aju, D.: Operating an alert system using facial expression. 2017 Innovative Power Advanced Computing Technology (i-PACT 2017). 1–6 January 2017 (2017). https://doi.org/10.1109/IPACT.2017.8244915
20. Altameem, T., Altameem, A.: Facial expression recognition using human machine interaction and multi-modal visualization analysis for healthcare applications. Image Vis. Comput. **103**, 104044 (2020). https://doi.org/10.1016/j.imavis.2020.104044
21. Amara, K., et al.: Towards emotion recognition in immersive virtual environments: a method for Facial emotion recognition. CEUR Workshop Proc. **2904**, 253–263 (2021)
22. Shehu, H.A., Browne, W., Eisenbarth, H.: An adversarial attacks resistance-based approach to emotion recognition from images using facial landmarks. In: 29th IEEE International Conference on Robot and Human Interactive Communication (RO-MAN 2020), pp. 1307–1314 (2020). https://doi.org/10.1109/RO-MAN47096.2020.9223510
23. Shojaeilangari, S., Yau, W.Y., Nandakumar, K., Li, J., Teoh, E.K.: Robust representation and recognition of facial emotions using extreme sparse learning. IEEE Trans. Image Process. **24**, 2140–2152 (2015). https://doi.org/10.1109/TIP.2015.2416634
24. Nosu, K., Kurokawa, T.: Facial tracking for an emotion-diagnosis robot to support e-Learning. In: Proceedings of the 2006 International Conference on Machine Learning and Cybernetics 2006, pp. 3811–3816 (2006). https://doi.org/10.1109/ICMLC.2006.258689
25. Leelavathy, S., Jaichandran, R., Shantha Shalini, K., Surendar, B., Philip, A.K., Ravindra, D.R.: Students attention and engagement prediction using machine learning techniques. Eur. J. Mol. Clin. Med. **7**, 3011–3017 (2020)
26. Michael Revina, I., Sam Emmanuel, W.R.: Face expression recognition with the optimization based multi-SVNN classifier and the modified LDP features. J. Vis. Commun. Image Represent. **62**, 43–55 (2019). https://doi.org/10.1016/j.jvcir.2019.04.013

27. Ghazouani, H.: A genetic programming-based feature selection and fusion for facial expression recognition. Appl. Soft Comput. **103**, 107173 (2021). https://doi.org/10.1016/j.asoc.2021.107173
28. Ellouze, M., Mechti, S., Belguith, L.H.: Automatic profile recognition of authors on social media based on hybrid approach. Procedia Comput. Sci. **176**, 1111–1120 (2020). https://doi.org/10.1016/j.procs.2020.09.107
29. Lekdioui, K., Messoussi, R., Ruichek, Y., Chaabi, Y., Touahni, R.: Facial decomposition for expression recognition using texture/shape descriptors and SVM classifier. Signal Process. Image Commun. **58**, 300–312 (2017). https://doi.org/10.1016/j.image.2017.08.001
30. Raza Shahid, A., Khan, S., Yan, H.: Contour and region harmonic features for sub-local facial expression recognition. J. Vis. Commun. Image Represent. **73**, 102949 (2020). https://doi.org/10.1016/j.jvcir.2020.102949
31. Chen, D., Song, P.: Dual-graph regularized discriminative transfer sparse coding for facial expression recognition. Digit. Signal Process. A Rev. J. **108**, 102906 (2021). https://doi.org/10.1016/j.dsp.2020.102906
32. Tabassum, F., Imdadul Islam, M., Tasin Khan, R., Amin, M.R.: Human face recognition with combination of DWT and machine learning. J. King Saud Univ. - Comput. Inf. Sci. (2020). https://doi.org/10.1016/j.jksuci.2020.02.002
33. Yu, L., Zhou, R., Tang, L., Chen, R.: A DBN-based resampling SVM ensemble learning paradigm for credit classification with imbalanced data. Appl. Soft Comput. J. **69**, 192–202 (2018). https://doi.org/10.1016/j.asoc.2018.04.049
34. Saha, P., Bhattacharjee, D., De, B.K., Nasipuri, M.: Mathematical representations of blended facial expressions towards facial expression modeling. Procedia Comput. Sci. **84**, 94–98 (2016). https://doi.org/10.1016/j.procs.2016.04.071
35. Priya, A., Garg, S., Tigga, N.P.: Predicting anxiety, depression and stress in modern life using machine learning algorithms. Procedia Comput. Sci. **167**, 1258–1267 (2020). https://doi.org/10.1016/j.procs.2020.03.442
36. Guran, A.M., Cojocar, G.S., DioSan, L.: A step towards preschoolers' satisfaction assessment support by facial expression emotions identification. Procedia Comput. Sci. **176**, 632–641 (2020). https://doi.org/10.1016/j.procs.2020.08.065
37. Lai, X., Patrick Rau, P.L.: Has facial recognition technology been misused? A user perception model of facial recognition scenarios. Comput. Human Behav. **124**, 106894 (2021). https://doi.org/10.1016/j.chb.2021.106894
38. Rosula Reyes, S.J., Depano, K.M., Velasco, A.M.A., Kwong, J.C.T., Oppus, C.M.: Face detection and recognition of the seven emotions via facial expression: integration of machine learning algorithm into the NAO robot. In: 2020 5th International Conference on Control and Robotics Engineering (ICCRE 2020), pp. 25–29 (2020). https://doi.org/10.1109/ICCRE49379.2020.9096267
39. Rahman, M.F.A., Vincent, Giovanni, V.C., Warnars, H.L.H.S., Aryono, G.D.P., Megantoro, B.: Sasmoko: facial recognition development to detect corporate employees stress level. In: 2019 IEEE International Conference on Engineering, Technology and Education (TALE 2019), pp. 5–10 (2019). https://doi.org/10.1109/TALE48000.2019.9225909
40. Mandal, M., Poddar, S., Das, A.: Comparison of human and machine based facial expression classification. Int. Conf. Comput. Commun. Autom. ICCCA **2015**, 1198–1203 (2015). https://doi.org/10.1109/CCAA.2015.7148558
41. Khaireddin, Y., Chen, Z.: Facial Emotion Recognition: State of the Art Performance on FER2013. (2021)
42. Dumitriu, T., Cimpanu, C., Ungureanu, F., Manta, V.I.: Experimental analysis of emotion classification techniques. Proceedings of the 2018 IEEE 14th International Conference on Intelligent Computer Communication and Processing (ICCP 2018), pp. 63–70 (2018). https://doi.org/10.1109/ICCP.2018.8516647

43. Chen, T., Yin, H., Yuan, X., Gu, Y., Ren, F., Sun, X.: Emotion recognition based on fusion of long short-term memory networks and SVMs. Digit. Signal Process. A Rev. J. **117**, 103153 (2021). https://doi.org/10.1016/j.dsp.2021.103153
44. Ngai, W.K., Xie, H., Zou, D., Chou, K.L.: Emotion recognition based on convolutional neural networks and heterogeneous bio-signal data sources. Inf. Fusion. **77**, 107–117 (2022). https://doi.org/10.1016/j.inffus.2021.07.007
45. Zahara, L., Musa, P., Wibowo, E.P., Karim, I., Musa, S.B.: The facial emotion recognition (FER-2013) dataset for prediction system of micro-expressions face using the convolutional neural network (CNN) algorithm based Raspberry Pi. In: 2020 Fifth International Conference on Informatics and Computing (ICIC), pp. 1–9 (2020)
46. Hamed, Y., Ibrahim Alzahrani, A., Shafie, A., Mustaffa, Z., Che Ismail, M., Kok Eng, K.: Two steps hybrid calibration algorithm of support vector regression and K-nearest neighbors. Alexandria Eng. J. **59**, 1181–1190 (2020). https://doi.org/10.1016/j.aej.2020.01.033
47. Subasi, A., Ahmed, A., Alickovic, E.: Effect of flash stimulation for migraine detection using decision tree classifiers. Procedia Comput. Sci. **140**, 223–229 (2018). https://doi.org/10.1016/j.procs.2018.10.332
48. Silitonga, P., Dewi, B.E., Bustamam, A., Al-Ash, H.S.: Evaluation of dengue model performances developed using artificial neural network and random forest classifiers. Procedia Comput. Sci. **179**, 135–143 (2021). https://doi.org/10.1016/j.procs.2020.12.018
49. Lin, G., Zou, X.: Citrus segmentation for automatic harvester combined with AdaBoost classifier and Leung-Malik filter bank. IFAC-PapersOnLine **51**, 379–383 (2018). https://doi.org/10.1016/j.ifacol.2018.08.192
50. Wei, Y., Gu, K., Tan, L.: A positioning method for maize seed laser-cutting slice using linear discriminant analysis based on isometric distance measurement. In: Information Processing in Agriculture, pp. 1–9 (2021). https://doi.org/10.1016/j.inpa.2021.05.002

An Optimize Gene Selection Approach for Cancer Classification Using Hybrid Feature Selection Methods

Sayantan Dass[1](✉), Sujoy Mistry[1], Pradyut Sarkar[1], and Pradip Paik[2]

[1] Department of Computer Science and Engineering, Maulana Abul Kalam Azad University of Technology, Nadia, West Bengal, India

[2] School of Biomedical Engineering, IIT BHU, Varanasi, India

Abstract. In the field of diseases diagnosis and treatment, microarray gene expression data plays a crucial role. But for analysis, expression data are available with a huge number of genes in comparison with few tissue samples. So, the most challenging task is to find out most influential genes from the high-dimensional, noisy and redundant microarray data. To overcome the above-said issues, in this paper, we proposed a two-stage gene subset selection mechanism by a combination of non-parametric Kruskal-Wallis test (KWs test) and Correlation-based Feature Selection (CFS) algorithms. The proposed technique selects most important and significant features (here genes) as well as eliminates insignificant and redundant features (here genes), that have been playing an important role to address this problem. Over three publicly available microarray datasets, proposed technique has been evaluated using two classifiers, namely supported vector machines (SVM) and k-nearest neighbors (k-NN). We also compared experimental outcomes obtained from our proposed model with recently published feature selection and classification models to determine whether or not proposed model is suitable for high-dimensional microarray data analysis. The proposed technique achieves the prediction accuracy rate of 98.61% for leukemia, 90.90% for colon cancer, and 99.60% for ovarian cancer using a support vector machine (SVM). Compared to other existing models, our proposed model shows relatively higher accuracy. Therefore, the proposed model can be used as a reliable framework for gene selection in cancer classification.

Keywords: Gene expression · Feature selection · Microarray · Kruskal-Wallis test · Correlation feature selection · Classification

1 Introduction

Computational biology primarily deals with data on high-dimensional gene expression and is commonly used in many areas, including cancer diagnosis and prognosis. However, cancer detection or classification is very challenging task because of high dimensional data structure contain a huge number of genes in comparison to tissue sample [1]. Generally, in high-dimensional data, certain genes are meaningless and irrelevant

I. Woungang et al. (Eds.): ANTIC 2021, CCIS 1534, pp. 751–764, 2022.
https://doi.org/10.1007/978-3-030-96040-7_56

to some of the diseases and it not only reduce the prediction ability of the classification model but also deteriorates speed of the classifier as well as increases the computational cost [2]. Therefore, the selection of genes, also known as the feature selection process, would improve cancer classification efficiency by picking highly influential or informative genes (here features) from the gene expression microarray data. There are several feature selection techniques namely Information Gain, mRMR, ReliefF, Mutual Information and many more but to select important and significant features by removing redundant and irrelevant features becomes a crucial task [3]. So far, we know this is the first time that this has been used Kruskal-Wallis test (KWs test) and Correlation based Feature Selection (CFS) algorithms to boosts efficiency of cancer classification.

Features are gathered in this study through two levels, which can construct the subset with the lowest redundancy. Through applying the proposed two-stage technique, we first use the Kruskal-Wallis test to pick the relevant features and then create a collection of highly correlated features using CFS filter methods, therefore selected features subsets are more flexible than present gene ratio and prevent the curse of dimensionality or overfitting problems. The paper's key contributions are follows:

- In microarray data analysis, highly correlated features were identified between features and feature-class.
- The proposed method has been applied to microarray datasets to determine what genes are relevant or significant.
- To demonstrate the effectiveness, efficiency, scalability and reliability of the proposed model, conduct detailed experiments on three publicly available gene expression datasets.

The rest of the paper is structured as follows: related work on different approaches to the selection of features is elaborated in Sect. 2. The general procedure of the proposed approach is talked about in Sect. 3. Section 4 gave the experimental results and performance analysis of the work. Conclusion of the article talks about in the Sect. 5.

2 Literature Study

The existing literature has been categorized into different approaches of feature (here gene) selection and classification, and following which each step of the entire research work has been completed. There are three distinct methodologies namely; filter, wrapper and embedded [3] which are followed by the feature selection techniques found in the literature. Filter methods only depend on the intrinsic properties of the data, where a set of features has been chosen according to the high score values of the ranking criterion function. While in the wrapper method, the learning algorithm has wrapped on a search algorithm, where based on highest learning algorithm performance find an optimal subset of feature. On the other hand, a blend of both filter and wrapper approaches has been used in embedded approach. So that, several feature selection methods exist to extract influential features. In this article, our attention is to consider the limitations of previously published literature based solely on the filtering method.

Throughout the past literature many methods have been pursued which were carried out in microarray datasets to resolve the issue of high dimensionality. In reference

[4], researchers tried to introduce a new gene selection technique by combining Kolmogorov–Smirnov test (KS test) and Correlation based Feature Selection (CFS) method where most of the noisy and redundant features (here genes) eliminated and significant and informative genes are selected. In the article [5], Kruskal-Wallis Test along with Bonferroni correction is applied in the microarray datasets. SU filter are used to reduce the number of features from the raw datasets and then applied different attribute selection methods namely CFS, Fast-CFS, GSNR, ReliefF and Minimum Redundancy Maximum Relevance (mRMR) to select the feature subsets are shown in article [6]. In article [7], researchers compare the three different selection methods namely Supervised Weighted Kernel Clustering/SVM (SWKC/SVM), Bhattacharyya distance with SVM (B/SVM) and SVM with Recursive Feature Elimination (SVM-RFE) on different biological datasets where the performance of B/SVM method was better than others. In literature [8], through combining ReliefF and mRMR, a new two-stage gene selection technique has been implemented to find out a significant gene subset by deleting the redundant genes. In reference [9], researchers have identified the informative genes based on mutual information (MI) between the genes and the class level. The performance of the model measured using Leave-one-Out Cross Validation (LOOCV) technique. A new score-based criteria fusion (SCF) feature selection methodology has been introduced in article [10], where two feature ranking approaches have been used to estimate the significance between features and classes. Authors in [11] proposed a two-stage model combining filter methods such as ReliefF, or Chi-square, and genetic algorithm to achieve better performance in microarray datasets for feature selection. In reference [12], researchers analysed 16 high dimensional gene expression datasets using 22 filter methods and different classifier and conclude which filter method perform well on which datasets on the basis of run time and accuracy. Method of Distributed Feature Selection has used to select highly relevant genes from the gene expression dataset, proposed in article [13]. In reference [14] the CFS algorithm analyses the subset of the feature as per the heuristic merit dependent on the correlation. A subset of informative features provides a strong correlation between features and class. Researchers have combining Kendall Correlation (KC) and Filter-based Feature Selection (FS) method to achieve better classification and prediction in literature [15]. In article [16], the authors have been proposed a Correlation-Based Redundancy Multiple Filter Approach (CBRMFA) using three filtering methods to select the genes that have best classification abilities.

3 Proposed Model

The primary objective of this work is to find out the informative and influential features (here genes) by removing noisy and redundant genes from the cancer dataset and to determine the best features subset for correctly identifying cancer. To fulfill this objective by ranking the features using Kruskal-Wallis test and eliminating the redundant features by using CFS algorithm. Then the subset of selected feature is evaluated the classification performance using SVM and k-NN classifier.

3.1 Kruskal-Wallis Test (KWs Test)

Kruskal-Wallis test is a statistical based non-parametric approach introduced by Kruskal and Wallis in 1952 [17]. KWs test compare more than two independent samples which may have the different sample sizes. It compares several populations by determining whether the samples belong to the same distribution, based on independent random samples from each population [17]. The following steps are described below for the Kruskal-Wallis Test.

1. For each entry of expression vector

   i. Sort the expression level in ascending order across all the classes and rank the sorted data. If any tied value present, there then assign the average rank.

   ii. Calculate the K test statistic (see Eq. 1)

$$K = \left[\frac{12}{N(N+1)} \sum_{j=1}^{c} \frac{R_j^2}{n_j} \right] - 3(N+1) \quad (1)$$

   Where, $N =$ Sum of expression level across all the class
   $c =$ Number of different class
   $T_j =$ Sum of ranks of all expression level for class j
   $n_j =$ Number of expression level from class j

   iii. If ties are occurred during ranking the expression level across the classes. For the correction, K which is calculate by Eq. 2, is divided by

$$1 - \frac{\sum T}{N^3 - N} \quad (2)$$

   Where, $T = (t - 1)t(t + 1)$, t is number of tied expression level in class.

   iv. The p-value for the expression level is estimated (see Eq. 3)

$$\Pr\left(\chi_{C-1}^2 \geq K\right) \quad (3)$$

   Where, $\chi_{c\text{-}i}{}^2$ is critical chi-square value.
2. If p-value of the expression is less than a threshold, then it selected.

The lower p-value significance of the higher rank. So, subset of top ranked feature selected by KWs test will enter the next stage. To select a proper threshold value, we have used scientific parameter tuning between the ranges of 0 to 1.

3.2 Correlation Based Feature Selection (CFS)

Correlation based Feature Selection (CFS) is used to determine the 'merit' of the feature subset according to the following hypothesis: "good feature subsets contain features highly correlated with the class, yet uncorrelated with each other" [18]. The evaluation function can efficiently select feature (here gene) subsets if the degree of correlation between the contained features and class is high and the degree of redundancy between each other is low. So, exclude the irrelevant features as they'll have a low degree of correlation with class. Similarly, the redundant features were removed because they will be highly correlated to one or more of the remaining features (see in Fig. 1). The correlation metric is based on the Pearson Correlation coefficient. The CFS's evaluation metric M_S of a feature subset S consisting of k features with c classes are defined as (see Eq. 4)

$$M_s = \frac{k\overline{r_{cf}}}{\sqrt{k + k(k-1)\overline{r_{ff}}}} \tag{4}$$

Where, k = Number of features
$\bar{r}_{cf}$ = is the mean of correlation between features and classes
$\bar{r}_{ff}$ = is the average coefficient of linear correlation between the various features.

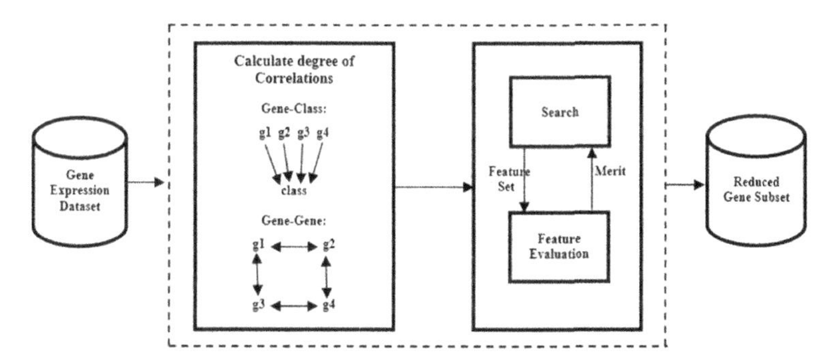

Fig. 1. Correlation-based feature selection (CFS)

3.3 Classification Method

To measuring the performance of proposed KWs-CFG model, different classifier namely Support Vector Machines (SVM) and k-Nearest Neighbour (k-NN) algorithm has been used. The '10-fold cross-validation' test mechanism has been used in our experiment where the datasets are divided 90% of training data (9-fold) and remaining 10% of testing data (1-fold) in each iteration.

Support Vector Machine (SVM) is a function-based classifier introduced in 1992 by Vapnik and co-workers [19]. It creates the decision boundary that used to segregate the data into different classes. Based on the decision boundary unknown data point easily categorized. During this segregation process, many decision boundary will be generated, the best decision boundary is called a hyper-plane. It maximizes the margin between two datasets. We used SVM with RBF Kernel in our experiments.

k-Nearest Neighbor (k-NN) is instance-based learner, introduced in 1978 by Devroye [20]. k-NN is used for classifies the class label of unknown sample on the basis of closest neighbour depending on their Euclidean distance (see Eq. 5).

$$\sqrt{\sum_{i=1}^{k} (a_i - b_i)^2} \quad (5)$$

Where, distance between two objects a and b is calculated.

3.4 Proposed Methodology for Feature Selection

In the first phase of our proposed model (see in Fig. 2), the genes have been ranked according to their p-values using the non-parametric Kruskal-Wallis (KWs) test and selected the subsets of top ranked genes, those p-value below the predetermined appropriate threshold (the range between 0 to 1). So, it eliminates many irrelevant or unimportant genes and the subset of top ranked genes will enter to the next section. In the second phase the CFS selects the features (here genes) which have maximum relevance with target class and minimally similar to each other. So, the integration of Kruskal-Wallis (KWs) test and Correlation-Based Feature Selection (CFS) leads to an efficient feature selection approach.

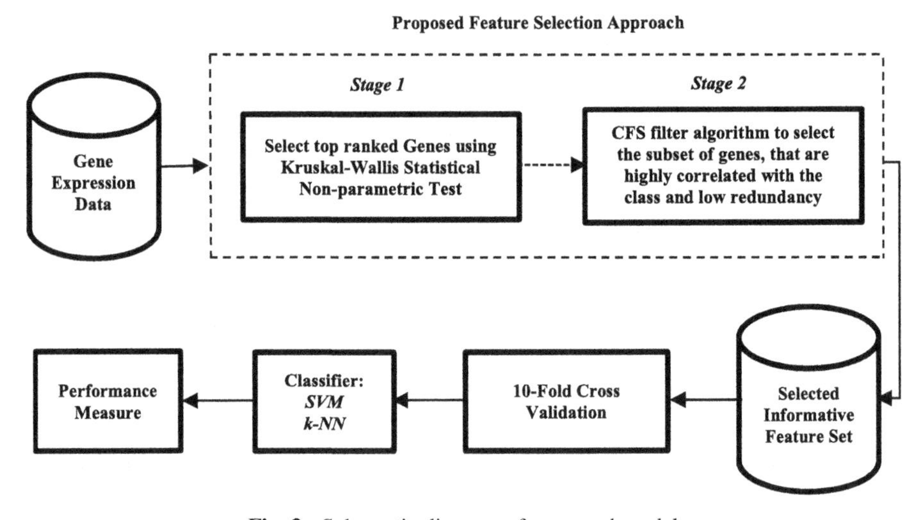

Fig. 2. Schematic diagram of proposed model

4 Result and Discussion

4.1 Experimental Setup

All the experiments have been done on the same machine, having Intel(R) Core(TM) i3-3110 M CPU @ 2.40 GHZ, 6 GB RAM and 64-bit Windows 10 Home operating system.

Kruskal-Wallis test algorithms are implemented using R3.6.3 and CFS and Classifier algorithms are implemented using Weka 3.8.4. The SVM algorithm uses 10-fold cross validation strategy with RBF kernel function.

4.2 Datasets

We tested our proposed feature selection algorithm against 3 publicly available cancer gene expression dataset namely Leukemia, Colon Cancer and Ovarian Cancer. Table 1 lists the descriptions of the datasets.

Table 1. Description of the dataset

Dataset	Gene	Sample	Class	Publication
Leukemia	3571	72	ALL (47) AML (25)	Golub et al. [22]
Colon Cancer	2000	62	Tumor (40) Normal (22)	Alon et al. [23]
Ovarian Cancer	15154	253	Cancer (162) Normal (91)	Petricoin et al. [24]

4.3 Performance Metric

To measure the effectiveness and efficiency of our proposed model are used the following parameters.

Confusion Matrix. The performance of the classification process (i.e. accuracy) is evaluated by the confusion matrix (see in Fig. 3), that compares classifier prediction to the actual value.

		Actual Class: Positive	Actual Class: Negative	Total
Predicted Class	Positive	True Positive (TP)	False Positive (FP)	TP+FP
	Negative	False Negative (FN)	True Negative (TN)	FN+TN
	Total	TP+FN	FP+TN	N

Fig. 3. Confusion matrix

Where,

TP is the number of positive prediction when the observation is positive.
FP is the number of positive prediction when the observation is negative.
FN is the number of negative prediction when the observation is positive.
TN is the number of negative prediction when the observation is negative.

We can conclude some common metrics from confusion Matrix given in Fig. 3.

$$\text{Precision} = \frac{\text{TP}}{(\text{TP} + \text{FP})} \tag{6}$$

$$\text{Sensitivity/Recall} = \frac{\text{TP}}{(\text{TP} + \text{FN})} \tag{7}$$

$$\text{F - Measure} = 2 \times \frac{(\text{Recall } \times \text{Precision})}{(\text{Recall } + \text{Precision})} \tag{8}$$

$$\text{Accuracy} = \frac{(\text{TP} + \text{TN})}{(\text{TP} + \text{TN} + \text{FP} + \text{FN})} \tag{9}$$

4.4 Experimental Result

In this section we have computed the classification accuracy for the leukemia, colon cancer and ovarian cancer microarray gene expression datasets taking all the feature (i.e. without any feature selection mechanism) by using SVM and k-NN classifier. Then our proposed model selects the subsets of informative genes from the above gene expression datasets. Similarly, classification accuracy has been computed for reduced datasets of selected informative genes (i.e. with feature selection mechanism). The proposed model that use integration of Kruskal-Wallis test and CFS method for feature selection, is classified by using 10-fold cross-validation technique on SVM and k-NN classifier algorithm. Then the detailed comparison of performance with or without feature (here gene) selection by using classifiers namely SVM and k-NN, are given in the Tables 2 and 3.

Table 2. Performance comparison with proposed method using SVM classifier

Dataset	Feature selection	Sensitivity	Precision	F-measure	Accuracy (%)
Leukemia	With-out	1	0.8103	0.8952	84.72
	With	1	0.9792	0.9895	**98.61**
Colon Cancer	With-out	0.975	0.6724	0.7959	67.75
	With	0.8333	0.8333	0.8333	**90.90**
Ovarian Cancer	With-out	0.9890	1	0.8333	89.72
	With	0.9890	1	0.9944	**99.61**

Table 3. Performance comparison with proposed method using k-NN classifier

Dataset	Feature selection	Sensitivity	Precision	F-measure	Accuracy (%)
Leukemia	With-out	0.9787	0.92	0.9485	93.05
	With	1	0.9792	0.9895	**98.61**
Colon Cancer	With-out	0.875	0.6481	0.7449	61.29
	With	0.8636	0.8261	0.8444	**88.71**
Ovarian Cancer	With-out	0.8901	0.9310	0.9101	93.68
	With	0.9890	0.9890	0.9890	**99.21**

To evaluate the efficiency and effectiveness, we compared our proposed model with two existing algorithms namely Information Gain and ReliefF. The proposed model that uses integration of Kruskal-Wallis test and CFS method for features (here genes) selection. On the other hand, for both Information Gain and ReliefF algorithm, are calculate score of every feature (here gene) and using forward selection algorithm 50 top most features (here genes) are selected. The subsets of selected features (here genes) for all the models are classified by SVM and k-NN classifier algorithm with 10-fold cross validation mechanism. The details comparison of existing and proposed model is shown in Table 4 and Fig. 4.

Table 4. Performance comparison with 10-fold cross validation using SVM and k-NN

Dataset	Method	SVM	k-NN
Leukemia	Information gain	95.83	93.06
	ReliefF	94.44	93.50
	Proposed	**98.61**	**97.22**
Colon Cancer	Information gain	87.10	82.26
	ReliefF	87.10	80.64
	Proposed	**90.90**	**88.71**
Ovarian Cancer	Information gain	98.81	99.20
	ReliefF	98.41	98.81
	Proposed	**99.61**	**99.21**

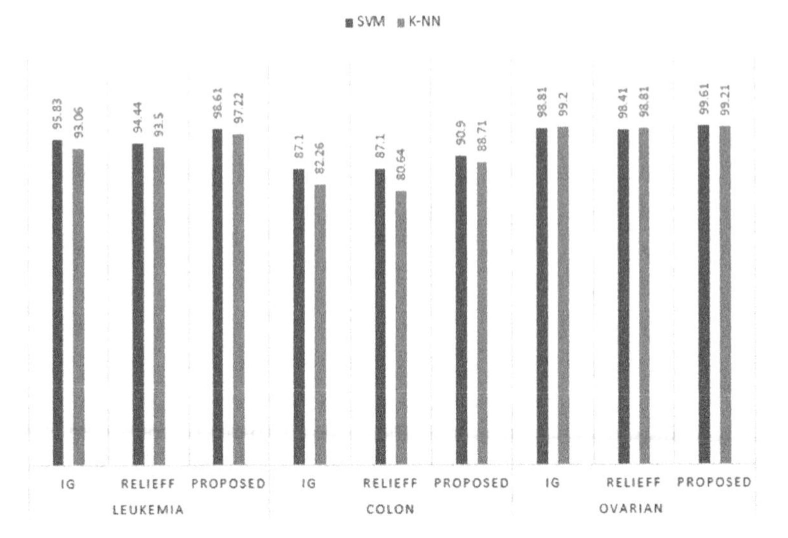

Fig. 4. Performance comparison using SVM and k-NN classifiers

ROC (Receiver operating characteristics) curve is graphical presentation of the relationship between specificity and sensitivity that determine the ability of the binary classifier. The x-axis and y-axis of the curve represents the true positive rate (TPR) and the false positive rate (FPR) respectively. The point on the ROC curve is more closer to the vertical axis (gives more region in ROC space) then the performance of the test is more accurate whereas the point on the ROC curve is more closer to the horizontal axis (gives more region in ROC space) then the performance of the test is less accurate. ROC curve for used three datasets are shown in the Fig. 5 that shows the accuracy of the proposed model are satisfactory.

A relative comparison of classification accuracy of the previous research with proposed method by using SVM classifier for Leukemia, Colon and Ovarian cancer datasets are shown in the Tables 5, 6 and 7. From the comparison table, our model has given the better performance. It has given the accuracy of 98.61%, 90.9% and 99.6% for leukemia, colon cancer and ovarian cancer dataset respectively.

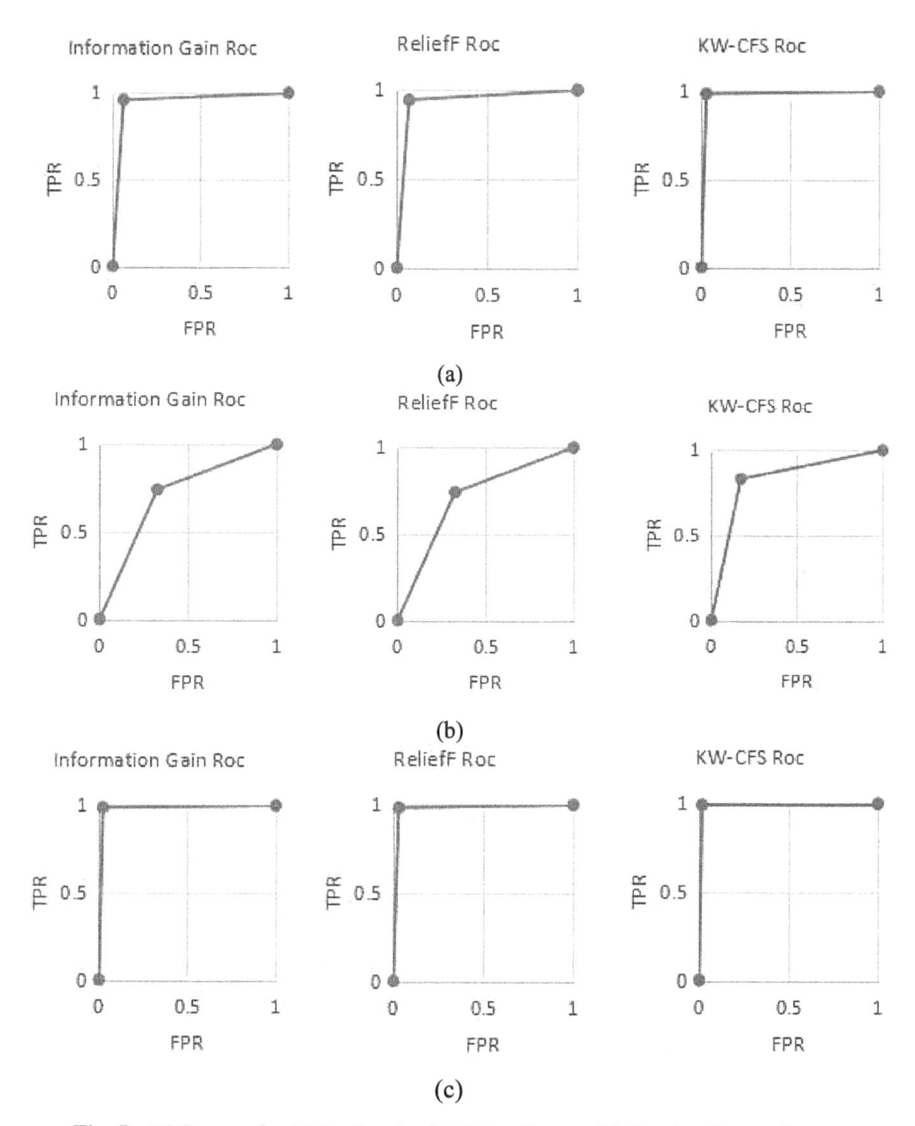

Fig. 5. ROC curve for (a) Leukemia, (b) Colon Cancer, (c) Ovarian Cancer datasets

Table 5. Comparison against performance of different methods for Leukemia dataset

Dataset	Publication	Method	# of selected genes	Accuracy
Leukemia	Wenyan et al. [7]	B/SVM	10	76.9%
	Das et al. [5]	KWs test and Bonferroni Correction	4	91.67%
	Proposed	KWs test and CFS	24	**98.61%**

Table 6. Comparison against performance of different methods for Colon Cancer dataset

Dataset	Publication	Method	# of selected genes	Accuracy
Colon Cancer	Potharaju et al. [21]	DFS	27	83.87%
	Qiang et al. [4]	K-S Test and CFS	10.7	90.1%
	Wenyan et al. [7]	B/SVM	6	90.5%
	Proposed	KWs test and CFS	23	**90.9%**

Table 7. Comparison against performance of different methods for Ovarian Cancer dataset

Dataset	Publication	Method	# of selected genes	Accuracy
Ovarian Cancer	Qiang et al. [4]	K-S Test and CFS	33.2	98.5%
	Proposed	KWs test and CFS	16	**99.6%**

5 Conclusion

In this paper, we have developed a highly informative feature selection strategy in combination with the Kruskal-Wallis (KWs) test and the Correlation-Based Feature Selection (CFS) algorithm, which enables the selection of appropriate and highly related feature (here gene) subsets from high-dimensional real-life datasets. In the first phase of our proposed model, it eliminates irrelevant or unimportant genes and selects the subset of top ranked genes by using non-parametric Kruskal-Wallis (KWs) test. The CFS selects the features (here genes) in the second step which are of maximum relevance with the target class and minimally identical to each other. Thus, combining the Kruskal-Wallis (KWs) method and Correlation-Based Feature Selection (CFS) results in an effective approach to selecting features. We tested the efficiency of the model proposed on two separate classifiers and applied 10-fold cross-validation to test the performance of these classifiers (i.e. SVM and k-NN) in classification. The tests were performed on real data sets, and the two classifiers compared the classification accuracy. The performance of our proposed model shows very promising results compared to other existing feature selection methods.

The proposed method has several limitations, and future directions need to be explored for improvement. Firstly, a small number of well-known filter methods were compared with the proposed framework. Our next steps will be to test the performance of the proposed method against current state-of-the-art methods. Furthermore, the focus of our research is to analyse the performance of the proposed model instead of the investigation of top-ranked genes. The model will be applied to real-world microarray cancer datasets outside of benchmark databases to detect or verify known biomarkers.

References

1. Ferreira, A.J., Figueiredo, M.A.T.: Efficient feature selection filters for high-dimensional data. Pattern Recognit. Lett. **33**, 1794–1804 (2012)

2. Li, Z., Xie, W., Liu, T.: Efficient feature selection and classification for microarray data. PLoS One **13**(8), e0202167 (2018)
3. Chandrashekar, G., Sahin, F.: A survey on feature selection methods. Comput. Electr. Eng. **40**, 16–28 (2014)
4. Su, Q., Wang, Y., Jiang, X., Chen, F., Lu, W.C.: A cancer gene selection algorithm based on the KS test and CFS. Biomed. Res. Int. (2017)
5. Das, U., Hasan, M.A.M., Rahman, J.: Influential gene identification for cancer classification. In: 2019 International Conference on Electrical, Computer and Communication Engineering (ECCE), pp. 1–6. IEEE, Bangladesh (2019)
6. Morovvat, M., Osareh, A.: An ensemble of filters and wrappers for microarray data classification. Mach. Learn. Appl. An Int. J. **3**, 1–17 (2016)
7. Zhong, W., Lu, X., Wu, J.: Feature selection for cancer classification using microarray gene expression data. Biostat. Biometrics Open Access J. **1**(2), 1–7 (2017)
8. Zhang, Y., Ding, C., Li, T.: Gene selection algorithm by combining reliefF and mRMR. BMC Genomics **9**(S2), S27 (2008)
9. Devi Arockia Vanitha, C., Devaraj, D., Venkatesulu, M.: Gene expression data classification using support vector machine and mutual information-based gene selection. Procedia Comput. Sci. **47**, 13–21 (2015)
10. Ke, W., Wu, C., Wu, Y., Xiong, N.N.: A new filter feature selection based on criteria fusion for gene microarray data. IEEE Access **6**, 61065–61076 (2018)
11. Ghosh, M., Adhikary, S., Ghosh, K.K., Sardar, A., Begum, S., Sarkar, R.: Genetic algorithm based cancerous gene identification from microarray data using ensemble of filter methods. Med. Biol. Eng. Compu. **57**(1), 159–176 (2018). https://doi.org/10.1007/s11517-018-1874-4
12. Bommert, A., Sun, X., Bischl, B., Rahnenführer, J., Lang, M.: Benchmark for filter methods for feature selection in high-dimensional classification data. Comput. Statis. Data Anal. **143**, 106839 (2020)
13. Shukla, A.K., Tripathi, D.: Detecting biomarkers from microarray data using distributed correlation based gene selection. Genes Genom. **42**(4), 449–465 (2020). https://doi.org/10.1007/s13258-020-00916-w
14. Lu, X., Peng, X., Liu, P., Deng, Y., Feng, B., Liao, B.: A novel feature selection method based on CFS in cancer recognition. In: 2012 IEEE 6th International Conference on Systems Biology (ISB), pp. 226–231 (2012)
15. Singh, P., Shukla, A., Vardhan, M.: A novel filter approach for efficient selection and small round blue-cell tumor cancer detection using microarray gene expression data. In: 2017 International conference on inventive computing and informatics (ICICI), pp. 827–831. IEEE (2017)
16. Sharifai, A.G., Zainol, Z.: The correlation-based redundancy multiple-filter approach for gene selection. Int. J. Data Min. Bioinform. **23**(1), 62–78 (2020)
17. Kruskal, W.H., Wallis, W.A.: Use of ranks in one-criterion variance analysis. J. Am. Stat. Assoc. **47**(260), 583–621 (1952)
18. Hall, M.A.: Correlation-based feature subset selection for machine learning. Ph. D. dissertation, University of Waikato, Waikato, New Zealand (1999)
19. Boser, B., Guyon, I., Vapnik, V.: A training algorithm for optimal margin classifiers. In: Proceedings of the Fifth Annual Workshop on Computational Learning Theory, Pittsburgh (1992)
20. Devroye, L.: A universal k-nearest neighbor procedure in discrimination. Nearest Neighbor (NN) Norms: NN Pattern Classification Techniques, pp. 101–106 (1978)
21. Potharaju, S.P., Sreedevi, M.: Distributed feature selection (DFS) strategy for microarray gene expression data to improve the classification performance. Clin. Epidemiol. Global Health **7**(2), 171–176 (2019)

22. Golub, T.R., et al.: Molecular classification of cancer: class discovery and class prediction by gene expression monitoring. Science **286**, 531–537 (1999)
23. Alon, U., Barkai, N., Notterman, D., Gish, K., Ybarra, S., Mack, D.: Broad patterns of gene expression revealed by clustering analysis of tumor and normal colon tissues probed by oligonucleotide arrays. Proc. Natl. Acad. Sci. **96**(12), 6745–6750 (1999)
24. Petricoin, E., et al.: Use of proteomic patterns in serum to identify ovarian cancer. Lancet **359**, 572–577 (2002)

Improving Heart Disease Prediction Using Feature Selection Through Genetic Algorithm

Abdul Aleem[1](✉), Gautam Prateek[2], and Naveen Kumar[1]

[1] Siksha 'O' Anusandhan Deemed to be University, Bhubaneswar, Odisha, India
abdulaleem@soa.ac.in

[2] Motilal Nehru National Institute of Technology, Allahabad, India

Abstract. Heart disease is one of the leading causes of fatality. A reliable and robust prediction system is needed for people to take preventive measures and medication beforehand and develop a proactive lifestyle accordingly. Various vital features determine human heart health, and it is important to recognize the critical ones that could be determining the chances of getting heart disease in the future. The various machine learning algorithms based on the critical features could predict heart disease more accurately. This article employs evolutionary algorithms like Genetic Algorithm (GA) and Particle Swarm Optimization (PSO) for the feature selection to improve the accuracy of machine learning algorithms further. GA and PSO combined with Naïve Bayes (NB), Support Vector Machine (SVM), and J48 have been applied for feature selection. After selecting the significant features, the effectiveness of the feature selection algorithm is evaluated by applying machine learning approaches on the complete dataset and reduced dataset. Five different machine learning approaches, viz., NB, SVM, Decision Tree (DT), Logistic Regression (LR), and Random Forest (RF) algorithm, have been used to predict heart disease and thus measure the effectiveness of the feature selection approaches. The results indicate that the GA has been the most effective algorithm for feature selection as it enhances the prediction accuracy most.

Keywords: Feature selection · Genetic Algorithm (GA) · Heart disease prediction · Classification · Particle Swarm Optimization (PSO)

1 Introduction

According to the WHO, cardiovascular diseases (CVDs) are cause of death of 17.9 million people each year, which accounts for 32% of all fatalities worldwide [1]. Heart disease prediction and prevention is one of the major clinical research areas, and even a tiny improvement in this area is significant for medical science. Most of the patients suffering from CVDs are detected when the disease becomes severe. Thus, detecting CVDs at an earlier stage is necessary for saving the

I. Woungang et al. (Eds.): ANTIC 2021, CCIS 1534, pp. 765–776, 2022.
https://doi.org/10.1007/978-3-030-96040-7_57

life of a patient. Manual observation of the patient's record in order to find disease is not only difficult but also takes time. In a study conducted by Richens et al. [2], it is found that machine learning-based approaches can outperform doctors. Predicting whether a person is prone to CVDs depends on many features such as Chest Pain type, Exercise-induced Angina, Thal evolved, etc. Also, using statistical threshold-based approaches for these large number of features may lead to lower accuracy. Therefore, applying machine learning approaches can significantly boost the performance of the prediction.

In this article, we have used the UCI Hungarian dataset [3] to find relevant features vital for detecting CVDs. The selection of features is made using feature selection approaches which select a features' subset by ranking them. After selecting features, the new dataset (reduced dataset) is formed by removing all the other features. Now this reduced dataset can be used for the prediction. The feature selection approach not only minimizes the dataset but also improves the performance. Feature selection is an NP-hard problem; therefore, we have applied metaheuristics for selecting the most appropriate features. This paper uses two metaheuristics approaches, i.e., PSO [4] and GA [5], along with Naïve Bayes [6], SVM [7], and J48 classifier [8] to select relevant features. Here, Naïve Bayes, SVM, and J48 are used as an objective function to compare the performance of two different subsets of features. After selecting relevant features, machine learning based approaches can be applied to evaluate the performance of feature selection approaches.

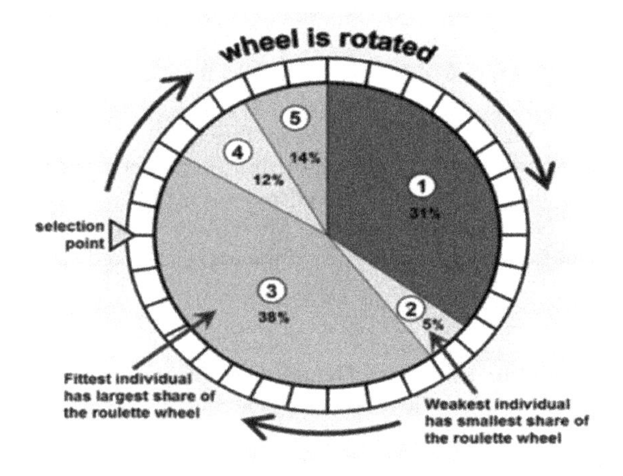

Fig. 1. Roulette wheel

Genetic algorithm for improving feature set is a revolutionary approach and used when the traditional algorithms are not working out well. This type of approach is derived from the biological theory of evolution. GA is generally used for optimization purposes. Every new state of the instance is denoted as a chromosome made up of genes, and using these chromosomes further, a better state can be created. Genetic algorithm have four main functions as follows:

1. **Roulette Wheel Selection**: Its a method for choosing the genomes as the next parent. Figure 1 shows an arbitrary Roulette wheel. All the genomes are given an area on the circle in proportion to their fitness value from the last state. The higher fitness value corresponds higher area for the genome that it will get on the circle/wheel. The wheel is rotated with a fixed pointer to select the genome. In programming, it is implemented with a random number generator and mod function as the pointer to select. Higher the fitness value better probability to be selected as a parent for the next state. Other methods used for selection are Rank selection, Random selection, etc.
2. **Crossover**: It is a process to get the new genomes by crossing over the selected parents on a random basis, cutting feature subset string with a certain probability. Some of the methods are single point crossover, two-point crossover (shown in Fig. 2), uniform crossover, N-point crossover, etc.

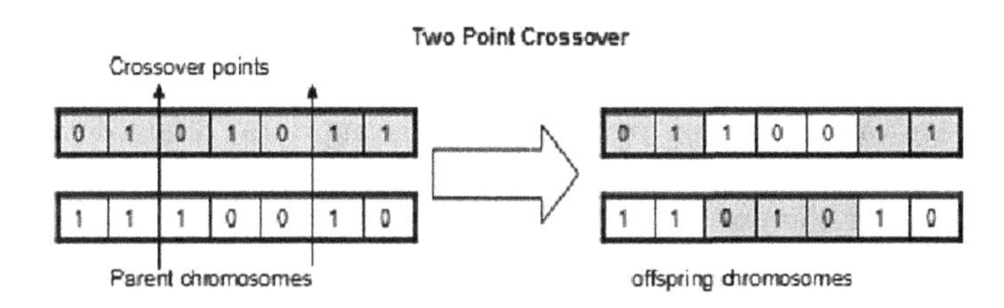

Fig. 2. Two point crossover

3. **Mutation**: It is randomly tweaking the genomes values to get some newer population. It is used to maintain some diversity. Mutation in the genetic algorithm is used with significantly lower probability; otherwise, this algorithm reduces to random searching.
4. **Fitness Model**: Finally, using the various classification methods for the role of judging the fitness of the genomes, which is basically the average of accuracy, precision, and recall for that feature set. Using these values again, parent selection for the next generation is made using the selection method.

This article aims to improve the accuracy of prediction through a suitable feature selection approach. Many feature selection approaches have been applied in the past, but none of them have been done using evolutionary algorithms like GA and PSO. This article uses five machine learning approaches, viz., SVM, DT, NB, LR, and RF, for the CVD prediction. Nonetheless, these approaches have been applied on two datasets, i.e., the original and reduced datasets after feature selection. These approaches are compared via prediction accuracy and the four well-known metrics - precision, recall, ROC area, and accuracy have been computed for the proposed approach. The result shows that a better result is achieved by applying feature selection, and out of all the approaches, the GA-based feature selection approach performs better in terms of prediction accuracy.

This article has been written in five sections. The first section introduced the problem and discussed the probable solution. The rest of the article proceeds as follows. The second section presents the background and related work on heart disease prediction and feature selection. The third section presents the proposed work along with the explanation of the approaches applied. The fourth section presents the experimental work and results along with a brief explanation of the utilized metrics. The fifth section provides the results and analyses it. Finally, the sixth and last section concludes this article and provides direction for future work.

2 Related Work

Many research works have been done for CVD detection using machine learning approaches. Few of the recent and significant contributions have been discussed here to present the background. Kanika and Shah [9] in 2016 proposed an approach to predict CVD considering attributes such as Age, Sex, Weight, Chest pain type, etc., for the prediction. The authors applied preprocessing techniques like noise removal, discarding records with missing data, filling default values, and level-wise classification of attributes to make decisions. SVM and NB had been employed for the prediction, among which SVM is found better.

Mirmozaffari et al. [10] in 2016 used clustering algorithms for feature selection. The authors have applied clustering approaches like K-means, hierarchical clustering, and density-based clustering. The best algorithm has been chosen using multilayer filtering preprocessing and a quantitative evaluation method. The accuracies, error functions, and building times of clusters are compared. Density-Based Clustering and K-Means functioned perform quite well, based on the results.

Jabbar et al. [11] in 2017 applied the genetic algorithm for selecting the optimal feature subset for heart disease prediction. This method works well for pruning redundant and irrelevant features by actually applying every new generation to the test. In this case, KNN is used as a supervised algorithm to check the accuracy for every generation. It is repeated until the performance starts to stabilize. The authors have got 4–5% improved accuracy after applying GA based feature selection approach on various UCI repository datasets.

Gokulnath et al. [12] in 2019 used SVM as the fitness function for the GA, which performs better than the KNN and works on data that has a less linear dependency. The SVM model was 83.70% accurate when classifying CVD with the full features. However, using the framework for feature reduction, an improvement of 5% in the accuracy is seen.

Gárate-Escamila et al. [13] in 2020, the authors have used Principal Component Analysis (PCA), an unsupervised method of feature reduction (filter method), based on a non-parametric statistical technique. PCA had also been utilized by Santhanam and Ephzibah [14] in 2013. The authors used PCA on the

UCI dataset (total of 297 samples, 13 input, and one output attribute), along with the regression technique used for feature reduction and ranking. The features selected using the PCA method were further utilized for classifying and predicting through regression and Feedforward neural network models.

Bashir et al. [15] in 2019 used a hybrid approach of various feature selection methods and ensemble learning with Minimum Redundancy Maximum Relevance Feature (MRMR) selection. Senthil et al. [16] in 2019 also used the hybrid approach of random forest and Linear Model for optimal performance. ANN with backpropagation is used for HRFLM.

All these research works are inspired by some natural phenomena of optimizing the performance in general and have their own sets of advantages as well as limitations. This article goes a step further and involves evolutionary algorithms like GA and PSO for feature reduction in combination with traditional machine learning algorithms. The improvement in prediction accuracy opens the door for the employment of evolutionary algorithms for feature selection before predicting a utility value.

3 Proposed Work

The objective is to improve the accuracy of classification models that predict heart disease when applied to heart datasets. For an accurate prediction model, a dataset is needed that has the best feature set, which has noise and redundancies removed. Wrapper-Method for feature selection is one of the wise choices. It is applied using the genetic algorithm. It could not perform well if proper parameters for the algorithm are not set. A fitness function plays a significant role in choosing the next generation for the algorithm. Choosing the right fitness function could improve GA further. Parameters like crossover and mutation probability values are optimized by trial and test only. The goal is to optimize feature selection using various classification functions as a fitness function in the genetic algorithm for finding a better next state to reach the optimized subset of features. Since the motive is to remove redundant features, Naïve Bayes is one of the stronger candidates. Naïve Bayes also resonates with the same principle because if there is some redundancy left in the feature, the fitness value will be meager as compared to others. Due to this, Naïve Bayes will give more accuracy than other classification methods. The proposed algorithm has been shown in Algorithm 1, and its flow has been explained in Fig. 3. The genomes for the next state will be that features-set only which has a higher fitness value. This hypothesis is further verified with experiments and other classification methods as fitness functions in the analysis section.

4 Experimental Details

WEKA is used as a tool to build a predictive model and further increase the accuracy of models. It is a software tool to analyze and work on different machine learning models. It has all the package to build a classification model based on

Algorithm 1. Proposed Algorithm

INPUT: x_i: Attributes in feature set
N: Total no. of Records

OUTPUT: *Chromosome*: String representing the set of selected features.

1: Calculate the mean (μ) and standard deviation (σ) using equation 1 and equation 2 respectively.

$$\mu = \frac{1}{n}\sum_{i=i}^{n} x_i \tag{1}$$

$$\sigma = \sqrt{\frac{1}{N-1}\sum_{i=1}^{N}(x_i - \mu)^2} \tag{2}$$

2: Normalize the data with z-score method using equation 3.

$$z = (x - \mu)/\sigma \tag{3}$$

3: Generate random population of chromosomes for evaluation.
4: **while** desired accuracy achieved or threshold iterations done **do**
5: Train the model using Naïve Bayes and evaluate the accuracy ($f_1(I)$) for sub-optimal feature sets using equation 4.

$$f_1(I) = \frac{TP + TN}{TP + TN + FP + FN} \tag{4}$$

where TP, TN, FP, FN represent true positive, true negative, false positive and false negative respectively.
6: Select number of genes ($f_2(I)$) using equation 5

$$f_2(I) = 1 - \frac{\text{no. of selected features}}{\text{size of feature set}} \tag{5}$$

7: Fitness function is evaluated for the genomes(feature-sets) using the fitness function of equation 6.

$$f(I) = \alpha f1(I) + (1-\alpha)f2(I) \quad \text{where } 0 < \alpha < 1 \tag{6}$$

8: Perform the genetic operation as crossover and mutation for the next selection.
9: **end while**

the provided data set. In the attribute selection section, various wrapper and filter methods are available to choose an optimal subset. Ranking of attributes can be done with other methods like ReliefF algorithm, Pearson's correlation, etc. Initially, the data set is loaded into the WEKA tool, then various filter methods and selection methods can be applied. The tool allows changing various parameters for the input of algorithms like kernel function for SVM or number of generations for Genetic algorithm. The description of the utilized dataset is provided in Subsect. 4.1 and the flow of experimental activities is discussed in Subsect. 4.2

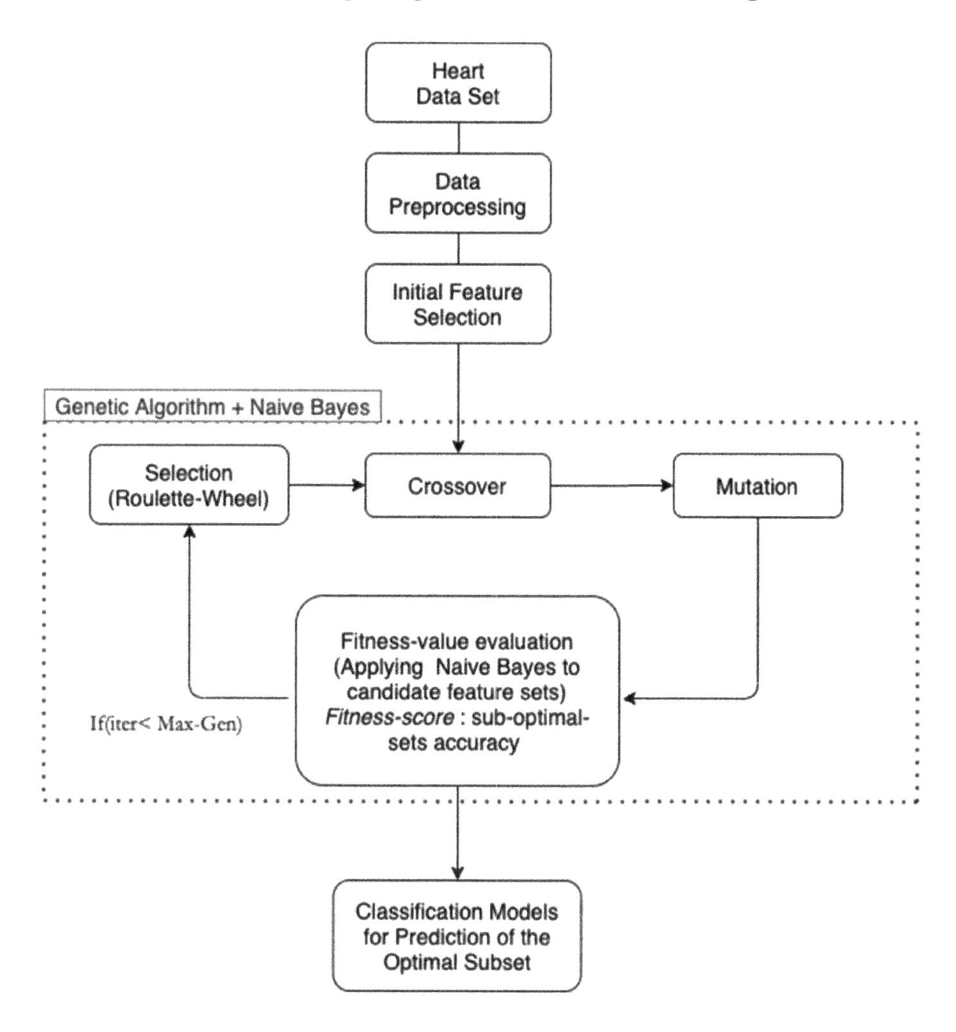

Fig. 3. Flow of proposed algorithm

4.1 Dataset Description

The dataset used is a standard UCI Hungarian dataset which has 14 attributes that describe various factors like age, chest pain, exercise-induced angina, etc. The description of attributes is given in Table 1. It has 294 instances. The motive is to design a framework that gives a better performance in most of the predictive models and produces higher accuracy than others. The data set is split into the training set and testing set. (70% for training) and (30% for testing).

Table 1. Feature description of UCI Hungarian data-set

S.No	*Attribute name*	*Description*
1	Age	Patient's age (years)
2	Sex	Gender(Male/Female)
3	Cp	Type of chest pain
4	Trestbps	Blood pressure while resting
5	Chol	Cholestoral content in Serum (mg/dl)
6	Fbs	Fasting sugar in blood (high/low)
7	Restecg	Electrocardiographic measures while resting
8	Thalach	Maximum count of heart beats
9	Exang	Angina due to exercise (yes/no)
10	OldPeak	Exercise related ST depression (yes/no)
11	Slope	ST segment slope w.r.t. peak exercise
12	Ca	Flourosopy-colored major vessels count (0–3)
13	Thal	Type of defect (normal/fixed/reversible)
14	Target-output	No disease or Heart disease (0/1)

4.2 Flow of Experimental Activities

The experiments have been done on the UCI dataset using the WEKA tool. The value for crossover probability is taken as 0.6, max generation as 20, mutation probability as 0.333, and population size as 20. Classification models like NB, SVM, etc., are deployed one by one as a fitness function to find the feature subset that works best. The flow of experimental activities carried out is as follows:

1. **Data Pre-processing**: Data is normalized to bring every data point to the same scale for fewer errors in the classification.
2. **Feature Selection**: An optimal subset of features is found out using various attribute selector methods, which is less redundant and more relevant for contribution in the model.
3. **Model Building**: Various classification models are build using the curated data subset found in step-2. Models like NB, LR, SVM, DT, and RF are used for classification purposes, as they performed better in comparison to other models.
4. **Performance Comparison**: Comparison of accuracy is made for utilized classification models with and without attribute selection methods.

5 Results and Analysis

Accuracy and performance for the classification models have been compared before and after the feature selection using various methods like ReliefF algorithm, Pearson's coefficient, GA and PSO. For the GA and PSO, accuracy with three "Fitness-functions" have been considered corresponding to NB, SVM, and J48. Table 2 shows the prediction accuracy for various classification models corresponding to different feature selection techniques. The first column shows the feature selection (FS) method applied before executing the classification of the dataset. All the models were tested without feature selection also. The first row depicts the prediction accuracy of models in such a condition. Rest rows show the result for other FS techniques applied, viz., ReliefF algorithm, Pearson's coefficient, GA+NB, GA+SVM, GA+J48, PSO+NB, PSO+SVM, and PSO+J48.

Table 2. Accuracy comparison of models using various FS methods

FS-Methods	NB	LR	SVM	RF	J48
No method	83.67	82.95	82.95	79.54	77.22
Pearson's coef	84.01	83.68	83.54	81.81	78.12
ReliefF algorithm	84.05	83.54	82.86	82.14	77.71
GA + NB	87.36	87.22	87.22	87.22	82.95
GA + SVM	86.09	86.95	86.09	87.00	79.54
GA + J48	84.09	84.09	78.40	84.09	84.09
PSO + NB	86.36	86.36	85.22	84.09	82.95
PSO + SVM	85.22	84.09	85.22	79.54	79.25
PSO + J48	83.88	83.63	78.12	83.27	83.63

Without feature selection, the accuracy recorded is lower. It ranges from 77% to 84%. These results have been recorded using UCI Hungarian dataset, which already has lesser noise. However, when the engaged datasets are just raw data having hundreds of attributes, accuracy will be a lot lesser, and the time for computing the models will be much more. As the number of dimensions increases, the performance would be degraded very quickly due to the overfitting of the model usually caused by unwanted attributes. Naïve Bayes gave the best result in terms of processing time. The highest accuracy value is 83.67% without feature selection, and it improved to 87.36% by just applying the feature selection. The accuracy could be improved further by combining various other optimizations.

Figure 4 shows the bar graph for the accuracy of five models utilized corresponding to the various FS techniques considered. Naïve Bayes is the best performing model in 7 out of 8 cases with all the FS techniques. It can be seen that for all the classification models, the feature selection technique of GA along with Naïve Bayes outperforms others. The best result is achieved using GA+NB as the FS technique and NB as the classification model. The results also infer that the traditional feature selection techniques only marginally improve the classification accuracy. Machine learning makes the feature selection more useful and delivers better results. The improvement in the result is obviously due to the selection of best features that are based on the Fitness-function used in the GA and other machine learning-based techniques.

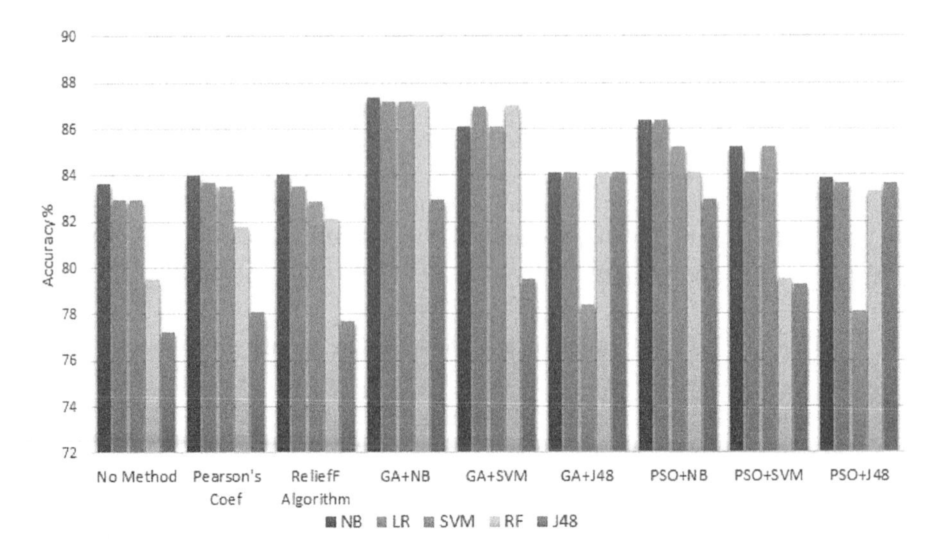

Fig. 4. Graphical comparison of models' accuracy using various FS methods

GA with Naïve Bayes as a fitness function selected the relevant attributes as (3,9,11,13), which is Chest Pain type, Exercise-induced Angina, Slope of Peak exercise ST segment, and Thal. These contribute more than any other attributes, which help build a simple and accurate model with lesser features and decide better for CVD. The results also show an increase in the accuracy from around 1.5% to 6% in different classification models. Various performance measures like precision, recall, F-measure, and ROC curve area for the best-established feature selection technique (GA+NB) corresponding to all the models have been shown in Table 3. The confusion matrices for the testing phase of the NB model using no FS technique, using GA+SVM, and using GA+NB as FS techniques are shown in Fig. 4.

Table 3. Performance measure using GA+NB as FS method

Classifier	Precision	Recall	ROC area	Accuracy %
Naïve Bayes	0.869	0.890	0.898	87.36
Support vector machine	0.874	0.882	0.862	87.22
Logistic regression	0.874	0.882	0.862	87.22
Random forest	0.874	0.882	0.862	87.22
J48 (Decision tree)	0.834	0.830	0.812	82.95

Confusion Matrix without Feature Selection

	Healthy	Unhealthy
Healthy	48	3
Unhealthy	12	25

Confusion Matrix for GA+SVM

	Healthy	Unhealthy
Healthy	49	3
Unhealthy	11	27

Confusion Matrix for GA+NaiveBayes

	Healthy	Unhealthy
Healthy	51	2
Unhealthy	9	28

Fig. 5. Confusion matrices for NB classifier

6 Conclusion and Future Work

This article proposed a feature selection method based on the evolutionary technique of GA. Using the Naïve Bayes as a fitness function, the feature selection method helped enhance the accuracy of CVD prediction. The experiments for prediction have been done using five machine learning techniques - NB, SVM, LR, RF, and J48. Each of the classifications had been done with eight different feature selection mechanisms. The results established NB as the best classifier employed after the feature selection through GA+NB. Hence, Naïve Bayes comes as a decent option for heuristic methods as a next-gen selector and optimizes search even further. It is also established that feature selection in CVD prediction studies has an improving role whenever the right combinations are used. In the future, newer optimizations techniques based on advanced machine learning approaches like deep neural networks could be utilized for feature selection.

References

1. Organization, W.H.: Cardiovascular diseases (2021). www.who.int/health-topics/cardiovascular-diseases/tab/tab/1 Accessed 02 Sept 2021
2. Richens, J.G., Lee, C.M., Johri, S.: Improving the accuracy of medical diagnosis with causal machine learning. Nat. Commun. **11**(1), 1–9 (2020)
3. UCI Machine Learning Repository: Heart disease data set. archive.ics.uci.edu/ml/datasets/heart+disease Accessed 02 Sept 2021
4. Poli, R., Kennedy, J., Blackwell, T.: Particle swarm optimization. Swarm Intell. **1**(1), 33–57 (2007)
5. Mirjalili, S.: Genetic algorithm. In: Evolutionary Algorithms and Neural Networks. Studies in Computational Intelligence, vol. 780. Springer, Cham (2019). https://doi.org/10.1007/978-3-319-93025-1_4
6. Murphy, K.P., et al.: Naive bayes classifiers. Univ. British Columbia **18**(60), 1–8 (2006)

7. Noble, W.S.: What is a support vector machine? Nat. Biotechnol. **24**(12), 1565–1567 (2006)
8. Mathuria, M.: Decision tree analysis on j48 algorithm for data mining. Int. J. Adv. Res. Comput. Sci. Softw. Eng. **3**(6) (2013)
9. Kanikar, P., Shah, D.R.: Prediction of cardiovascular diseases using support vector machine and bayesien classification. Int. J. Comput. Appl. **156**(2) (2016)
10. Mirmozaffari, M., Alinezhad, A., Gilanpour, A.: Heart disease prediction with data mining clustering algorithms. Int. J. Comput. Commun. Instrument. Eng. **4**(1), 16–19 (2017)
11. Deekshatulu, B., Chandra, P., et al.: Classification of heart disease using k-nearest neighbor and genetic algorithm. Proc. Technol. **10**, 85–94 (2013)
12. Gokulnath, C.B., Shantharajah, S.: An optimized feature selection based on genetic approach and support vector machine for heart disease. Cluster Comput. **22**(6), 14777–14787 (2019)
13. Gárate-Escamila, A.K., El Hassani, A.H., Andrès, E.: Classification models for heart disease prediction using feature selection and pca. Inf. Med. Unlocked **19**, 100330 (2020)
14. Santhanam, T., Ephzibah, E.P.: Heart disease classification using PCA and feed forward neural networks. In: Prasath, R., Kathirvalavakumar, T. (eds.) Mining Intelligence and Knowledge Exploration. LNCS, vol. 8284. Springer, Cham (2013). https://doi.org/10.1007/978-3-319-03844-5_10
15. Bashir, S., Khan, Z.S., Khan, F.H., Anjum, A., Bashir, K.: Improving heart disease prediction using feature selection approaches. In: 2019 16th International Bhurban Conference on Applied Sciences and Technology (IBCAST), pp. 619–623. IEEE (2019)
16. Mohan, S., Thirumalai, C., Srivastava, G.: Effective heart disease prediction using hybrid machine learning techniques. IEEE Access **7**, 81542–81554 (2019)

Voronoi Diagrams Based Digital Tattoo for Multimedia Data Protection

Sharmistha Jana[1], Biswapati Jana[2](✉), Prabhash Kumar Singh[2], and Prasenjit Bera[2]

[1] Department of Information Management, Chaoyang University of Technology, Taichung, Taiwan

[2] Department of Computer Science, Vidyasagar University, Midnapore, West Bengal, India

Abstract. The most important parameter in detecting errors in a message stream is the selection of the generator polynomial. To overcome the problem of generating a small degree generator polynomial, we propose a new approach to second generation watermarking using the Voronoi Diagrams (VD) and high degree standard polynomials with particular mathematical properties like Cyclic Redundancy Check (CRC-32, CRC-16 and CRC-8) to generate the watermark. The latter is inserted into each region of the image after decomposition of VD. Voronoi decomposition is used because it has good recovery performance compared to similar geometric decomposition algorithms. The Harris detector is used to extract the points of interest (FPs) considered to be seeds to create a Voronoi decomposition of the image. The proposed method can be applicable in the case where the detection of alteration is critical and only certain regions of interest need to be retransmitted if they are altered, as in the case of medical images. The security aspect of our proposed method is achieved by using the RSA public key encryption system to encrypt FPs. The experimental results show that the proposed schemes obtain good outcomes in terms of standard evaluation metrics and comparison to the existing scheme, ensuring the suggested scheme's uniqueness.

Keywords: Cyclic redundancy check · Digital tattoo · RSA public key encryption · Watermarking · Voronoi diagram

1 Introduction

The development of communication networks and digital media has encouraged the use of computer networks for the transmission of digital information. Several organizations, both public and private, have replaced their files, dispersed and kept manually, with computer systems that give them better access to data. This caused a risk for the data's security. Communicating secret data on the Internet involves several risks including that of sensitive information being easily

I. Woungang et al. (Eds.): ANTIC 2021, CCIS 1534, pp. 777–793, 2022.
https://doi.org/10.1007/978-3-030-96040-7_58

exposed to unauthorized users and hackers. Therefore, along with data communication, data security has become of paramount importance. To satisfy the objective of secure communication, several robust and highly secure techniques have been proposed. Cryptography [1] was a first proposal to secure digital document transfers. Today, modern encryption algorithm [2], with long keys, ensure confidentiality. However, once decrypted, the document is no longer secure and can be fraudulently transmitted or modified. The hiding of information and more particularly the insertion of hidden data can be an answer to this problem. Indeed, the insertion of a watermark in a document makes it possible to authenticate it and to guarantee its integrity. This technique is known by the name "digital watermarking" [3–5]. This technology quickly emerged as a very effective solution for enhancing the security of multimedia documents. Moreover, watermarking [6–8] has evolved as a secure data transmission technique, attracting several industrial applications in the medical, military, multimedia, and other related industries, where secret communications might be useful for internal and external security reasons.

The design of a watermarking system is generally modeled in two phases: in the first phase the watermark is implanted in the document to be protected (called host document or original document). The original document can be a text, image, sound or video file. Then, the watermarked document is transmitted via the network and it can undergo modifications. In the second phase, the watermark is extracted in order to prove the intellectual property of the document. Any digital watermark system must be designed to have certain properties while meeting the functional requirements. In the case of images, the insertion phase must not deteriorate the host image in a perceptible manner, that is the watermarked image must be visually equivalent to the original image. This property is known by the terms imperceptibility, invisibility or even fidelity. Other properties must be taken into consideration: robustness, capacity, safety and complexity. All of these properties depend on the field of application of the digital tattoo. According to Zhao et al. [9], these areas are: copyright protection, hidden annotations, authentication and invisible secret communication. Each of these applications has its own requirements. For example, if watermarking is applied for rights protection, toughness, imperceptibility and security are paramount while capacity is less important.

The principle of this scheme is based on the use of the CRC code to detect the modified pixels [10]. The cyclic redundancy check (CRC) is a powerful and easy to implement data integrity check. It is the main error detection method used in telecommunications. For these reasons, we have chosen to use it in the context of the fragile tattoo with the aim of knowing if the image has undergone any modifications or not. Using CRC [11], binary sequences are treated as polynomials whose coefficient correspond to the binary sequence. We add to the binary sequence with the remainder of a polynomial division by a generator polynomial. On reception, the remainder of the division received and the remainder of the calculated division must be null or there is a transmission error. Our digital tattoo generation scheme consists of three phases. The first phase consists in generating a watermark of size 6 bits which depends on the 18 bits of color component pixels Red, Green and Blue using a secret key K. The second phase is putting two watermark bits into the two LSBs of the colour components pixels Red, Green, and Blue. The last phase consists for detection if the watermarked image has undergone modifications. If the remainder of the division of the received message (18 bits MSB of the three pixels R, G and B concatenated with the 6 bits of the CRC) by the secret key K is equal to zero, then the image is authentic to the original image, otherwise it is not.

In fact, the most important parameter in detecting errors in a message stream is the selection of the generator polynomial. To overcome the problem of generating a small degree generator polynomial, we propose a new approach to second generation watermarking using the Voronoi Diagrams (VD) [12] and high degree standard polynomials with particular mathematical properties like CRC-32, CRC-16 and CRC-8 to generate the watermark. The latter is inserted into each region of the image after decomposition using Voronoi Diagrams VD. Voronoi decomposition is used because it has good recovery performance compared to similar geometric decomposition algorithms. The Harris detector is used to extract the points of interest (FPs) considered being seeds to create a Voronoi decomposition of the image. The proposed method can be applicable in the case where the detection of alteration is critical and only certain regions of interest need to be retransmitted if they are altered, as in the case of medical images. The security aspect of our proposed method is achieved by using the RSA public key encryption system to encrypt FPs.

2 Digital Tattoo

A digital tattoo system [13] is made up of two sub-procedure: an encoding and a decoding scheme. This can be formulated using (O, W, K, Ek, Dk, C_T), where O is original data, W is watermarked and K is secret keys to use. The watermark insertion function (encoder) is described as follows [14]:

$$E_k : O \times W \times K \rightarrow OE_k(I_o, w, k) = I_w \tag{1}$$

This function takes as input an original document I_o, a watermark w and a private or public key k to output a watermarked document I_w (see Fig. 1). The original document can be a text, image, sound or video file. In the rest of this manuscript, we focus on tattooing digital images. The watermark extraction function (decoder) is defined as follows [14]:

$$D_k : O \times K \rightarrow WD_k(I_w^{'}, k) = w^{'} \tag{2}$$

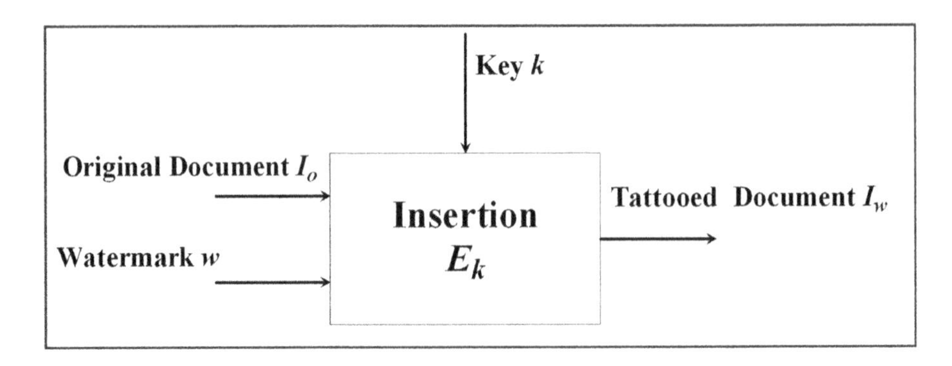

Fig. 1. Insertion function (encoder).

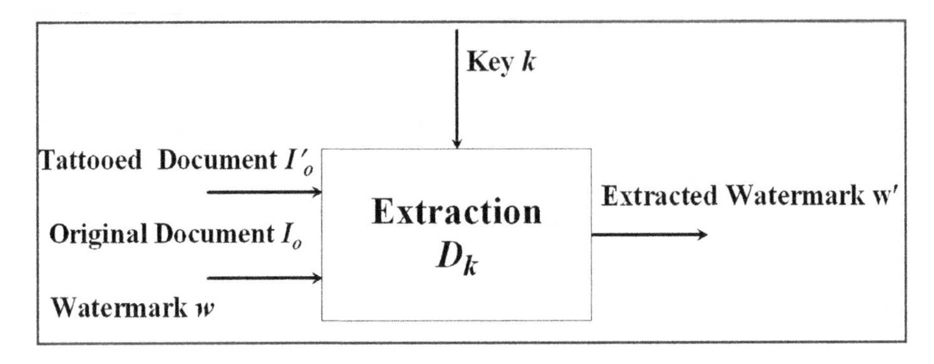

Fig. 2. Extraction function (decoder).

Function D_k takes as input a tattooed document and possibly attacked $I_w^{'}$ and the key to extract the watermark $w^{'}$. Depending on the technique used, the extraction algorithm may also require the original document (Fig. 2). The extracted watermark $w^{'}$ will be different compared with original watermark w due to attacks. It is possible to check these two watermarks; we define the comparison parameter $C\mathcal{T}$ as following:

$$C\mathcal{T} : W^2 \rightarrow \{0, 1\} \tag{3}$$

The $C\mathcal{T}$ allows comparing the extracted watermark with the original watermark, using a comparison threshold τ:

$$C\mathcal{T} = \begin{cases} 1 & \text{if } c \geq x1\tau \\ 0 & \text{if } c < \tau \end{cases} \tag{4}$$

The threshold τ depends on the algorithm used. According to the robustness constraint, we can classify the watermarking algorithms into three categories:

i) Robust tattoo: Robust tattoo techniques have a wide field of theories and results. A watermarking system is said to be robust if it resists possible alterations (compression, filtering, rotation, etc.). This type of tattoo is generally used to protect copyright [16].
ii) Fragile tattoo: In this type of watermark, the watermark is highly sensitive to changes in the watermarked image and should detect any alteration. These techniques are generally used to guarantee an authentication and integrity service for a watermarked file [17].
iii) Semi-fragile tattoo: Semi-fragile tattoo combines fragility against malicious disturbance and robustness with certain classes of mild image degradation, such as compression [18].

3 Cyclic Redundancy Check (CRC)

A cyclic redundancy check (CRC) is an error-detection code that is extensively utilized in telecommunications platforms and storage media to detect unintentional data changes. Various CRC polynomials are used for error detection. CRC is implemented by selecting a polynomial known to both the sender and the receiver, known as the reference generator polynomial, or $G(x)$. The sender applies the encoding technique to the message stream to generate the checksum, which is a set of check bits. This checksum is hidden to the host image that is being sent. The recipient decodes the message upon receipt to ensure that the checksum is correct. The Fig. 3(a) decorates as an example of CRC encoding $(7, 4)$. Assuming $M = \{1101\}$ the information to be moved, and $G(x) = x^3 + 1$, this is equivalent to 1001. M is represented by a polynomial $M(x) = x^3 + x^2 + 1$. These assumptions are depicted in Fig. 3(b) (Table 1).

4 Voronoi Diagram

The Voronoi diagram is a famous structure of computational geometry. The Voronoi diagram is named after Georgy Voronoy, and is also called a Voronoi tessellation, a Voronoi decomposition, a Voronoi partition, or a Dirichlet tessellation.

Let X be a metric space with distance function d. Let K be a set of indices and let $(P_k); k \in K$, be a tuple (ordered collection) of nonempty subsets in the space X. The Voronoi cell, or Voronoi region, R_k, associated with the site P_k is the set of all points in X whose distance to P_k is not greater than their distance to the other sites P_j, where j is any index different from k.

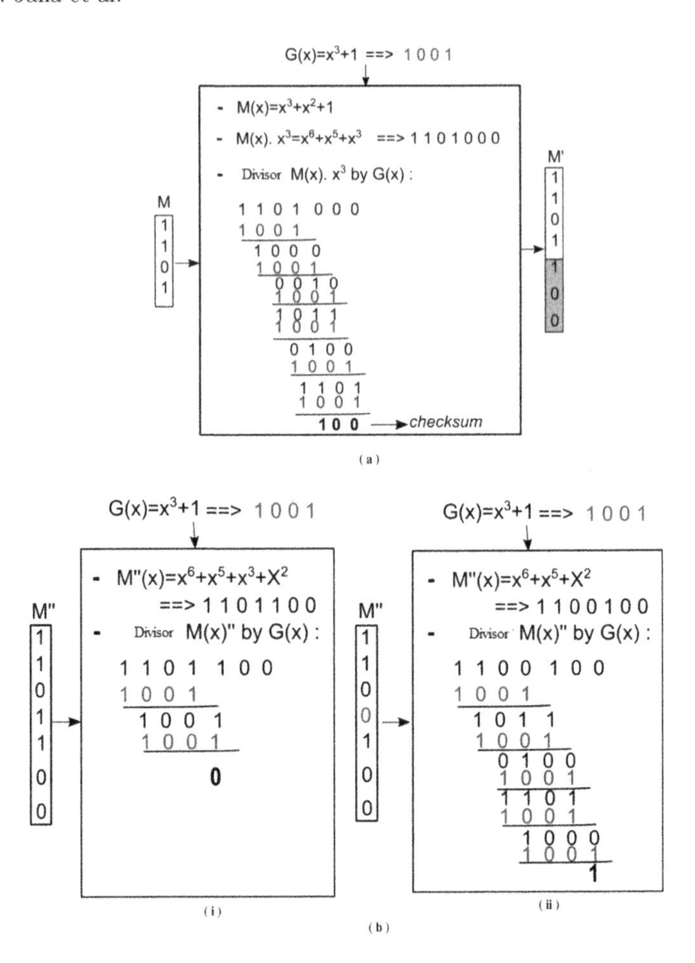

Fig. 3. (a): CRC (7, 4) encoding using $G(x) = x^3 + 1$, (b): CRC (7, 4) decoding using the same generator $G(x) = x^3 + 1$: (i) assumption without errors, (ii) assumption with errors.

In principle, some of the sites can intersect and even coincide but usually they are assumed to be disjoint. Furthermore, the definition allows for an infinite number of sites, yet in many circumstances, only a finite number of sites are examined (Fig. 4).

Table 1. Polynomials generating certain standard CRC codes [19].

Name	Degree	Polynomial
LRCC-8	8	$x^8 + 1$
CRC-12	12	$x^{12} + x^{11} + x^3 + x^2 + x + 1$
CRC-16	16	$x^{16} + x^{15} + x^2 + 1$
CRC-CCITT	16	$x^{16} + x^{12} + x^5 + 1$
LRCC-16	16	$x^{16} + 1$
CRC-32	32	$x^{32} + x^{26} + x^{23} + x^{22} + x^{16} + x^{12} + x^{11} + x^{10} + x^8 + x^7 + x^5 + x^4 + x^2 + x + 1$

Euclidean Distance

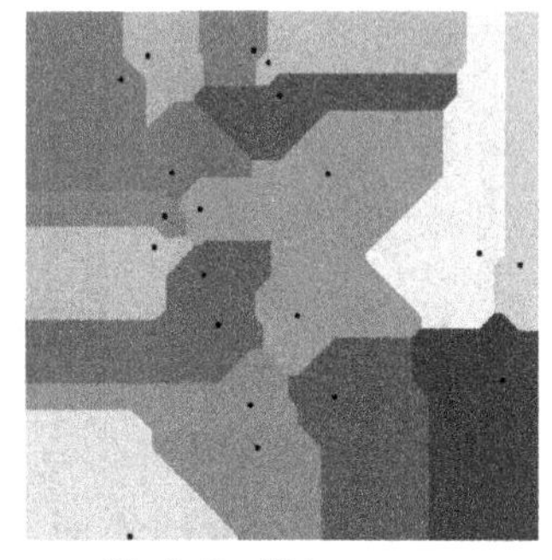

Manhattan Distance

Fig. 4. The corresponding Voronoi diagrams look different for different distance metrics.

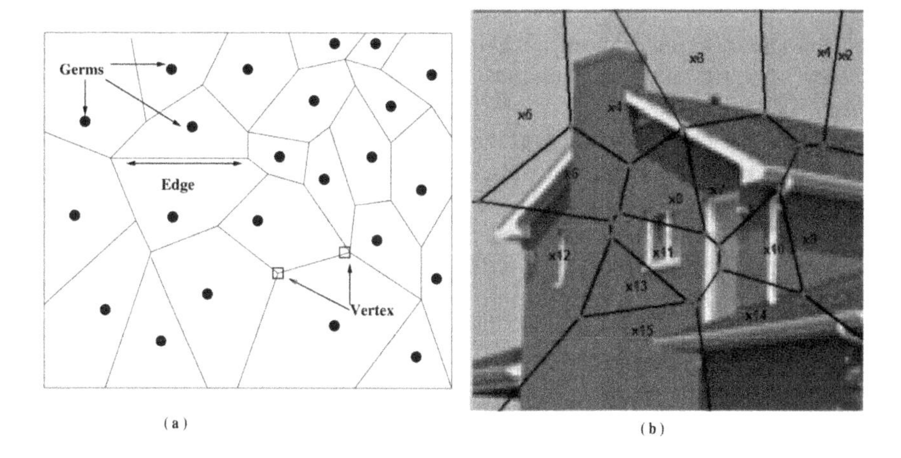

Fig. 5. (a) DV concepts. (b) House image decomposed using DV

The distance between points can be measured using the familiar Euclidean distance or the Manhattan distance. The corresponding Voronoi diagrams look different for different distance metrics. The Voronoi Diagram is defined as a partition of the plane into polygons or regions according to the nearest neighbor principle.

Given a set of points $P = \{p_1, p_2, ..., p_n\}$. The region of Voronoi for a point p_i is defined as the set of all points closest to p_i than of all other points. The points p_i are called Voronoi generators (germs). The borders common to two Voronoi regions are called Voronoi Borders (edge). Vertices where three or more Voronoi edges meet are called Voronoi Vertices (vertices). We say that a Voronoi generator p_i is adjacent to p_j when their Voronoi regions share a common border [20]. Figure 5 illustrates the concept of House image decomposition using DV (Fig. 7).

Fig. 6. Fragile tattoo with CRC-32, impact of the decomposition into blocks on the quality of the tattooed image.

5 The Proposed Model

The approach described above authenticates RGB images using the CRC code. However, the degree of the generator polynomial is very small and does not exceed six.

In fact, the most important parameter in detecting errors in a message stream is the selection of the generator polynomial. To overcome this insufficiency and widen the degree of generator polynomial, we propose a brittle block watermarking technique using a standard polynomial generator $G(x)$ having particular mathematical properties like CRC-32, CRC-16 and CRC-8 to generate the watermark. The latter is inserted into small blocks of the image after decomposition in block (Figs. 8 and 9).

We shown through tests that the deconstruction in blocks causes a reduction in the quality of the watermarked images. For this reason, we have proposed a new decomposition using the Voronoi Diagrams. We first started with an approach based on a decomposition into blocks. The Fig. 6 presents the image Lena tattooed using the CRC-32, different sizes have 128×128, 256×256 and 512×512 were taken into consideration.

It is clear that the decomposition into blocks generates the mosaic effects. To solve this problem, we have proposed a new decomposition using Voronoi Diagrams. Our system generates segments related to intensity values using information from the vertices of the outer border of VD. Thus, the image is decomposed into a set of polygons, where the extracted points of interest (FPs) are the seeds. The generated Watermark is integrated into each pixel of each segment.

The proposed method is described by three algorithms: the watermark generation, insertion and verification algorithms. These algorithms are based on the CRC to generate, insert and extract the watermark.

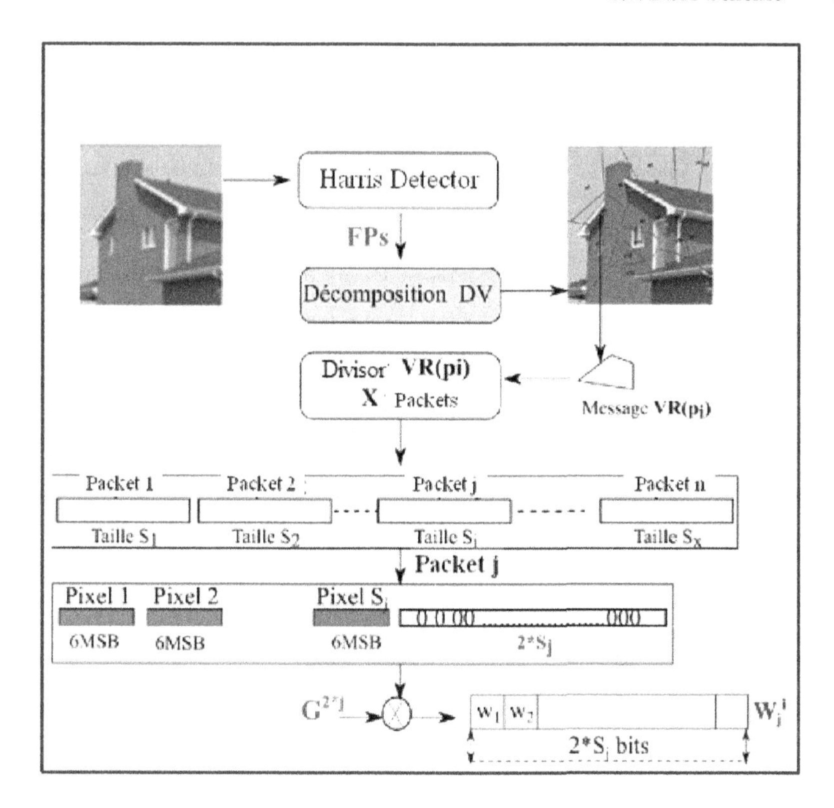

Fig. 7. Watermark generation process using DV decomposition.

Algorithm 1: Watermark generation algorithm using VD

---

input : Original Image (f)
output: Watermark (W), FPCryp : set of digitized FPs

Algorithm `WatermarkGeneration()`:

Step 1: Select a set of N points of interest $FP = \{p_1, p_2, ..., p_{NOT}\}$ using the Harris detector
Step 2: Match this set with RSA
Step 3: Decompose the image f by creating N voronoi region (VR) using FP as seeds. Each VR region (p_i) is considered a message to be transmitted
Step 4: For each VR message (p_i) make:
Step 5: Divide the VR message (p_i) in X packages of size $S_j = \{16, 8, 4, 2or1\}$, where j is the number of packets.
For example, if the first VR message (p_1) is composed of 47 pixels, VR (p_1) $= \{f_1, ..., f_{47}\}$. The X packages are:
$X_1 = \{f_1, ..., f_{16}\}$ size $S_1 = 16 \Rightarrow G(x)$ of degree 32, CRC (32);
$X_2 = \{f_{17}, ..., f_{32}\}$ size $S_2 = 16 \Rightarrow G(x)$ of degree 32, CRC (32);
$X_3 = \{f_{33}, ..., f_{40}\}$ size $S_3 = 8 \Rightarrow G(x)$ of degree, CRC (16);
$X_4 = \{f_{41}, ..., f_{44}\}$ size $S_4 = 4 \Rightarrow G(x)$ of degree, CRC (8);
$X_5 = \{f_{45}, ..., f_{46}\}$ size $S_5 = 2 \Rightarrow G(x)$ of degree, CRC (4);
$X_6 = \{f_{47}\}$ size $S_6 = 1 \Rightarrow G(x)$ of degree, CRC (2);
Step 6: For each package X_j size S_j make:
Step 9: Extract the six MSB bits from each pixel
Step 10: Concatenate these bits together to create the sequence my
Step 11: Add $2 \times S_j$ zeros bits at the end of m to create m^0. This is equivalent to $m^0 = mx^2S_j$
Step 12: The watermark W_{ji} is the remainder of the division of m^0 by a normalized CRC of degree $d = 2S_j$

Algorithm 2: Insertion algorithm based on DV

input : f : original image, W : watermark generated
output: f_w : tattooed images

Algorithm `WatermarkGeneration()`:
Step 1: Decompose f by creating N region of voronoi Region (VR) using FP as seeds.
Step 2: For each VR message (p_i) make:
Step 3: Divide the message into X packets of size $S_j = \{16, 8, 4, 2 \text{ or } 1\}$, where j is the number of packets.
Step 4: For each package P_j size S_j make:
Insert every two bits of the watermark W_{ji} in the two LSB bits of each pixel.
Rebuild the tattooed WP package j using tattoo pixels.
Rearrange the X tattooed packets to reconstruct the tattooed VR message $w(p_i)$.
Step 5: Rearrange the N watermarked messages (segments) VRw to reconstruct the tattooed image f_w.

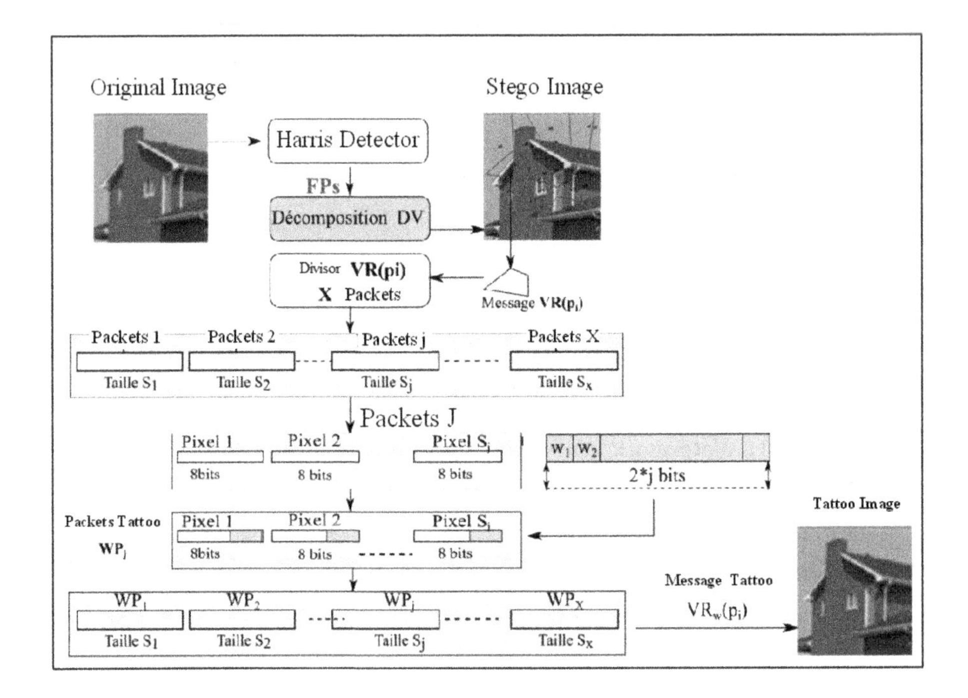

Fig. 8. Watermark insertion process.

Algorithm 3: Verification algorithm based on the decomposition of DV

input : $f_w^\star$: tattooed image and possibly attacked. FP_{Cryp} : set of digitized FPs
output: TDM: the alteration detection card

Algorithm `WatermarkGeneration()`:

Step 1: Create N region (VR_w) using FP as germs.
Step 2: For each VR message $w(p_i)$ make:
Step 3: Divide each tattooed VR message $w(p_i)$ in $X^\star$ pixel packets of size $S_j = \{16, 8, 4, 2or1\}$.
Step 4: For each WP tattooed package $j^\star$ size S_j make:
Extract the two LSB bits from each pixel.
Concatenate these bits together to create the extracted checksum $W_{ji}^\star$.
Extract the six MSB bits from each pixel.
Concatenate these bits together to create the sequence $m_w^\star$.
Add $W_{ji}^\star$ at the end of $m^\star$ to create $m_w^\star$.
Divide $m_w^\star$ by a normalized CRC of degree $d = 2 \times S_j$.
If the remainder of the division is not null then the WP package $j^\star$ is altered.

If one of the packets is corrupted, then the VR message $w(p_i)$ is also. Depending on the degree of tampering and the interest of the region, the receiver sends a negative acknowledgment (NAK) to the sender, requesting that the message be retransmitted.

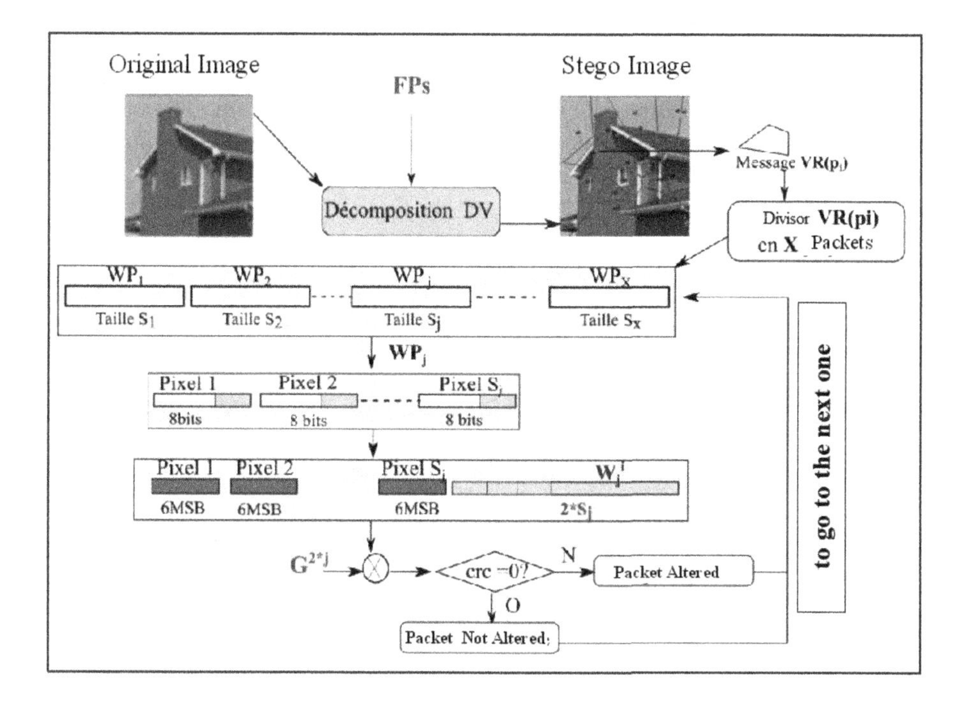

Fig. 9. Alteration detection process.

6 Experimental Results

Some exploratory experiments were conducted in this part to assess the efficacy of our tattooing approach. These tests are based on imperceptibility fragility, capacity and calculation time. We also compared our scheme with a similar fragile tattooing method.

6.1 Analysis of Imperceptibility

Several typical images with different size (64×64, 128×128, 256×256 and 512×512) were tattooed, in order to assess the imperceptibility property of our tattoo method (see Figs. 10). The PSNR and SSIM values achieved by our approach for varied sizes of host images are compared with a similar existing approach in Table 2. The results reported in the Table 2 show satisfactory results of the proposed scheme. In all cases, the PSNR values are greater than 47 dB and the size of the host image is limited to 256×256, because the pixel position (column and row) is converted to an 8-bit binary representation. Also, the SSIM values of the proposed scheme lies within 0.97 to 0.99. If a comparison is made one to one among the proposed scheme and Durgesh et al. 2013 for various sizes of image, the proposed scheme seems to deliver better result both in terms of PSNR and SSIM. The histograms between the original and watermarked images are shown in Fig. 11. From these plots, it is clear that the histograms are similar, which demonstrates the imperceptibility of our diagram. Therefore, the difference between the original images and the integrated images is not noticeable in the human visual system.

Fig. 10. Host images.

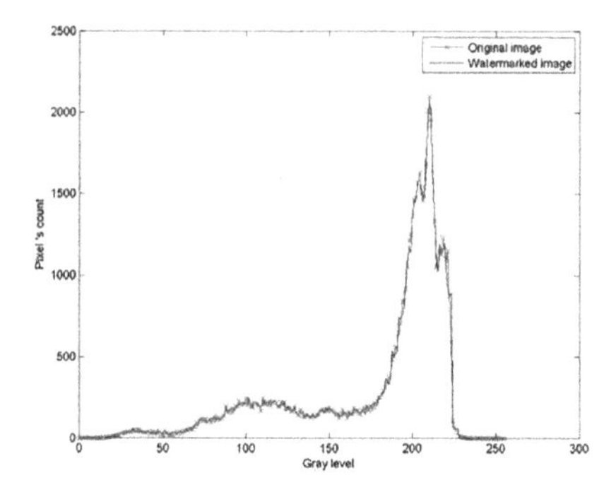

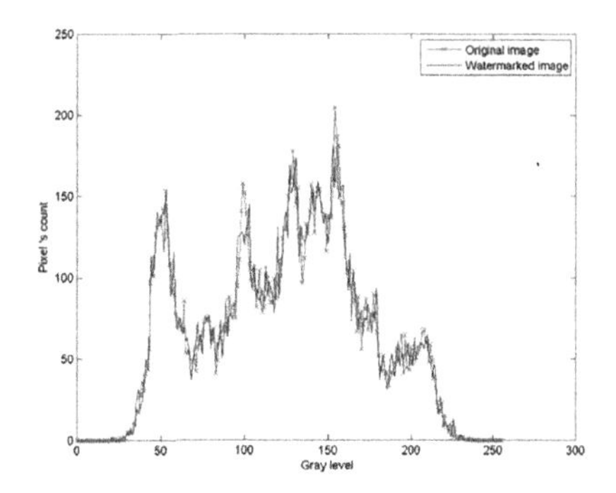

Fig. 11. Presentation of imperceptibility through histograms for Airplane (256 × 256) and Lena (128 × 128) images.

Table 2. Estimation of the quality of watermarked images (PSNR and SSIM).

Host image		Proposed approach				Approach [Durgesh et al. 2013]			
		64 × 64	128 × 128	256 × 256	512 × 512	64 × 64	128 × 128	256 × 256	512 × 512
Airplane	PSNR	47.04	47.02	47.15	47.37	41.51	41.27	40.95	–
	SSIM	0.990	0.986	0.981	0.979	0.970	0.958	0.945	–
Baboon	PSNR	47.61	47.19	47.16	47.17	41.40	41.02	40.99	–
	SSIM	0.995	0.994	0.993	0.993	0.981	0.978	0.975	–
Elaine	PSNR	49.26	47.07	47.12	47.16	41.07	41.07	41.02	–
	SSIM	0.995	0.991	0.985	0.985	0.983	0.968	0.948	–
Lena	PSNR	47.10	47.18	48.18	47.15	41.01	41.04	40.98	–
	SSIM	0.996	0.990	0.984	0.980	0.985	0.968	0.9480	–
Man	PSNR	47.22	47.18	47.23	49.23	41.20	41.18	41.13	–
	SSIM	0.997	0.994	0.990	0.986	0.990	0.976	0.962	–
Peppers	PSNR	47.04	47.08	47.19	47.16	41.48	41.34	40.97	–
	SSIM	0.996	0.990	0.984	0.981	0.987	0.969	0.946	–
Splash	PSNR	48.28	47.22	47.66	49.17	41.13	41.27	41.19	–
	SSIM	0.9888	0.9820	0.9803	0.9752	0.956	0.9413	0.928	–
Tree	PSNR	47.24	47.14	47.23	47.23	41.16	40.93	41.07	–
	SSIM	0.9969	0.991	0.987	0.984	0.9889	0.9725	0.960	–

6.2 Analysis of Accuracy and Fragility

To estimate the accuracy and fragility of the proposed scheme, we use a TDM (Tamper Detection Map) image to indicate the corrupted packets in each region. If there is no attack, the TDM is a black image, otherwise white pixels indicate corrupt pixels. The Fig. 12 illustrates the alteration detection map for various watermarked images in the absence of an attack. From these images we can see that the correction property is satisfied.

Analyzing the ability to detect tampering is similar to assessing the ability to detect errors. At this stage, we study several scenarios of the alterations at the bit level of the watermarked pixels. In addition to the comparison of our method with the Durgesh's method [15], we also compared the proposed scheme with the use of a polynomial generator of degree 3 (CRC-3). Thus, the generated watermark is inserted directly into the three LSB bits as in Durgesh's method.

To highlight the fragility of our method, we have taken into account several types of attacks. The Fig. 13 shows the attacked images and their extracted alteration detection maps.

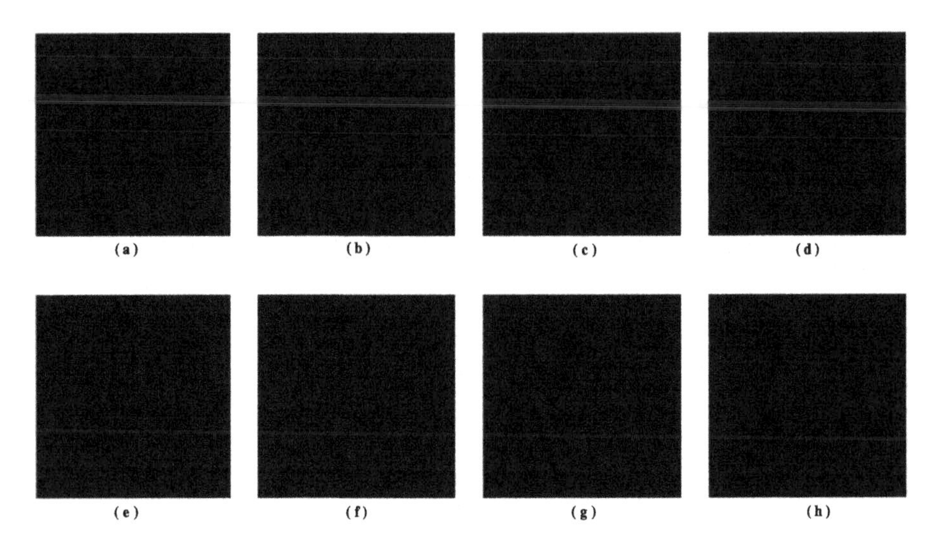

Fig. 12. Alteration detection maps extracted from watermarked images, in case no attack: (a) Airplane, (b) Baboon, (c) Elaine, (d) Lena, (e) Man, (f) Pepper, (g) Splash and (h) Tree.

In our approach, $N_w = 2$ then the insertion capacity is 25% of the size of the original image. This capacity is high. In the diagram [Durgesh et al. 2013], $N_w = 3$ and the capacity is 37.5% of the size of the host image. This capacity is better than the capacity reached by our system, but in [Durgesh et al. 2013] the size of the host image is limited to 256×256.

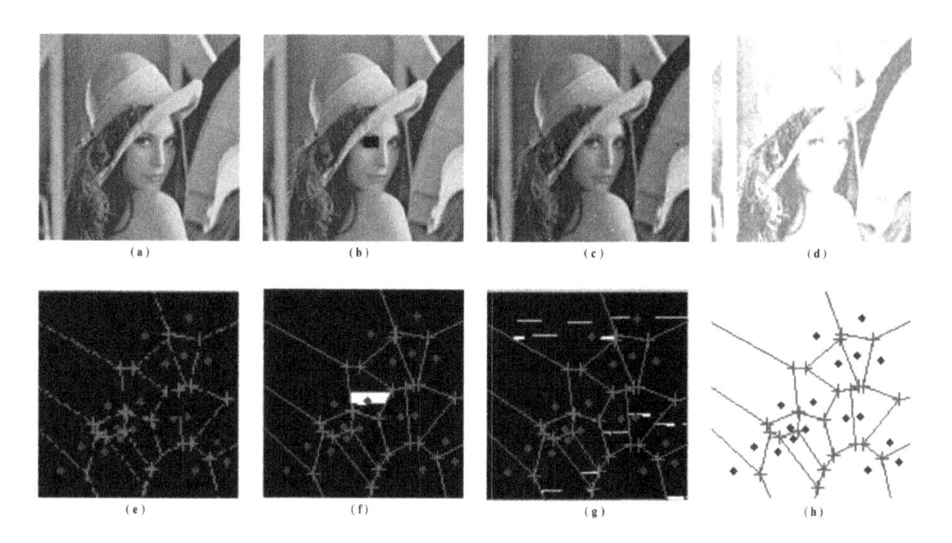

Fig. 13. Fragility against attacks: (a) A corrupted pixel. (b) cropping. (c) Salt and pepper noise (d) Gaussian noise. (e-h) Alteration detection cards: altered packets detected in each region after various attacks.

Attack types	Attacked watermarked image	Extracted watermark	Normalized Correlation (NC)	Attack types	Attacked watermarked image	Extracted watermark	Normalized Correlation (NC)
Salt & Pepper noise(0.03)			0.9153	Gaussian low pass filter(3x3)			0.8420
Gaussian noise(0.01)			0.9209	JPEG Compression(QF:50)			0.9607
Poisson noise			0.9021	Cropping (10%)			0.9506
Speckle noise			0.9146	Cropping (25%)			0.8947
Median filter(2x2)			0.9159	Rotation (clock wise 10^0)			0.8927
Average filter(3x3)			0.8911	Rotation (clock wise 20^0)			0.8606

Fig. 14. Attacked watermarked image, corresponding extracted watermark and NC values.

The watermarked image is compressed, filter and crop with various factors to test the fragility of our method against JPEG compression, Gaussian noise. The Fig. 14 illustrates the result.

7 Time Complexity

The proposed algorithm basically follows three algorithms as mentioned in the above section. Algorithm 1 incorporates Harris detector, RSA algorithm and decomposition with time complexity $\mathcal{O}(N^2)$, $\mathcal{O}(k^2)$ and $\mathcal{O}(N)$ respectively. Here, k is the key size. Similarly, Algorithm 2 takes the maximum time complexity of $\mathcal{O}(N^2)$ for N number of vornoi regions. At last, the Algorithm 3 which verifies based on the decomposition of DV consumes a time of $\mathcal{O}(N)$ for N regions. So, the time overhead for the proposed algorithm can be explicitly defined to be $\mathcal{O}(N^3)$.

8 Conclusion

The objective of this approach is to optimize the two contradictory requirements of tattooing: imperceptibility and robustness. This approach is inspired by network transmission where the use of detector and error correcting codes appeared natural. One of these approaches relies on the use of the error detector code (CRC) in order to guarantee the authentication of the region of interest of the medical image. While, the other uses an error correcting code (RS) to ensure in addition to authentication, integrity. We may proposed a new second generation watermarking approach based on Voronoi VD diagrams and CRC code using high degree standard polynomials with particular mathematical properties like CRC-32, CRC-16 and CRC-8 to generate the watermark. The new tattooing approaches may be applied to medical imaging.

References

1. Buchmann, J.: Introduction to Cryptography, vol. 335. Springer, New York (2004). https://doi.org/10.1007/978-1-4419-9003-7
2. Rivest, R.L.: Cryptography. In: Algorithms and Complexity, pp. 717–755. Elsevier (1990)
3. Jana, M., Jana, B.: A new DCT based robust image watermarking scheme using cellular automata. Inf. Secur. J. Global Perspect. 1–17 (2021)
4. Pal, P., Jana, B., Bhaumik, J.: A secure reversible color image watermarking scheme based on LBP, lagrange interpolation polynomial and weighted matrix. Multimedia Tools Appl. **80**(14), 21651–21678 (2021). https://doi.org/10.1007/s11042-021-10651-3
5. Dey, A., Pal, P., Chowdhuri, P., Jana, B., Jana, S., Singha, A.: Dual image based watermarking scheme using quorum function. In: Mandal, J.K., De, D. (eds.) EAIT 2021. LNNS, vol. 292, pp. 114–123. Springer, Singapore (2022). https://doi.org/10.1007/978-981-16-4435-1_13
6. Jana, B.: High payload reversible data hiding scheme using weighted matrix. Optik - Int. J. Light Electron Opt. **127**(6), 3347–3358 (2016)
7. Jana, B., Giri, D., Mondal, S.K.: Dual image based reversible data hiding scheme using (7, 4) Hamming code. Multimedia Tools Appl. **77**(1), 763–785 (2018)

8. Pal, P., Chowdhuri, P., Jana, B.: Weighted matrix based reversible watermarking scheme using color image. Multimedia Tools Appl. **77**(18), 23073–23098 (2018). https://doi.org/10.1007/s11042-017-5568-y
9. Zhao, J., Koch, E.: Embedding robust labels into images for copyright protection. In: KnowRight, pp. 242–251 (1995)
10. Koopman, P., Chakravarty, T.: Cyclic redundancy code (CRC) polynomial selection for embedded networks. In: International Conference on Dependable Systems and Networks, pp. 145–154. IEEE (2004)
11. Suresh, B., Rath, G.S.: Image encryption and authentication by orthonormal transform with CRC. In: Proceedings of the 2011 International Conference on Communication, Computing & Security, pp. 582–585 (2011)
12. Edelsbrunner, H., Seidel, R.: Voronoi diagrams and arrangements. Discrete Comput. Geom. **1**(1), 25–44 (1986). https://doi.org/10.1007/BF02187681
13. Mitchell, J., Underhill, C.: Learners and digital identity: the digital tattoo project. In: Mastering Digital Librarianship: Strategies, Networking, and Discovery in Academic Libraries, pp. 103–120 (2013)
14. Memon, N., Wong, P.W.: Protecting digital media content. Commun. ACM **41**(7), 35–43 (1998)
15. Singh, D., Shivani, S., Agarwal, S.: Self-embedding pixel wise fragile watermarking scheme for image authentication. In: Agrawal, A., Tripathi, R.C., Do, E.Y.-L., Tiwari, M.D. (eds.) IITM 2013. CCIS, vol. 276, pp. 111–122. Springer, Heidelberg (2013). https://doi.org/10.1007/978-3-642-37463-0_10
16. Chae, J.J., Manjunath, B.S.: Robust embedded data from wavelet coefficients. In: Storage and Retrieval for Image and Video Databases VI, vol. 3312, pp. 308–317. International Society for Optics and Photonics (1997)
17. Lin, E.T., Delp, E.J.: A review of fragile image watermarks. In: Proceedings of the Multimedia and Security Workshop (ACM Multimedia 1999) Multimedia Contents, vol. 1, pp. 25–29 (1999)
18. Lin, C.Y., Chang, S.F.: Semifragile watermarking for authenticating JPEG visual content. In: Security and Watermarking of Multimedia Contents II, vol. 3971, pp. 140–151. International Society for Optics and Photonics (2000)
19. Ramabadran, T.V., Gaitonde, S.S.: A tutorial on CRC computations. IEEE Micro **8**(4), 62–75 (1988)
20. Wang, J., Ju, L., Wang, X.: An edge-weighted centroidal Voronoi tessellation model for image segmentation. IEEE Trans. Image Process. **18**(8), 1844–1858 (2009)

Evaluating Technology Acceptance Model on the User Resistance Perspective: A Meta-analytic Approach

Aygul Donmez-Turan(✉) and Mehmet Tugrul Odabas

Yildiz Technical University, Esenler 34220, Istanbul, Turkey
ayturan@yildiz.edu.tr

Abstract. Technology acceptance model has been researched with many different external variables in the literature. One of the external variables of technology acceptance model is resistance. In present study, we aim to investigate the association between resistance and exogenous variables of technology acceptance model (perceived ease of use and perceived usefulness), on the basis of the findings of prior researches. We achieved 41 papers, which are indexed in SCOPUS database, have correlation scores between resistance and each variable of technology acceptance model. Then we used correlation scores and sample sizes reported in the papers and we conducted meta-analyses with Comprehensive Meta Analysis program. Publication bias was checked first, and no publication bias was found. After that we applied fixed and random effect models to the data and we found that random effect model was appropriate. Results of the analyses showed that average effect size of resistance and perceived ease of use was negative and small level ($\bar{r} = -0{,}225$, $p = 0.00$). Similarly, average effect size of resistance and perceived usefulness was negative and small level ($\bar{r} = -0{,}238$, $p = 0.00$) in this study. So we reached general conclusion about the association between resistance and exogenous variables of technology acceptance model in present study.

Keywords: Technology acceptance model · TAM · Resistance to change · User resistance · Meta-analysis · Perceived ease of use · Perceived usefulness

1 Introduction

Individuals show resistance to the innovations or developing technologies around them, because they do not know how to use them, they consider that the use of the existing technology is not easy and sufficient, or they believe that they cannot use the new technology. This circumstance, which is defined as user resistance, can be explained as the adverse reaction of the users towards the perceived change related to the innovation or technology desired to be realized.

There is growing body of literature about the acceptance and resistance of a technology or system. It began with technology acceptance model [1] and it has been evolving and improving since then. At the further researches, technology acceptance has been explained with different variables [2–4] and some other external variables have been

I. Woungang et al. (Eds.): ANTIC 2021, CCIS 1534, pp. 794–808, 2022.
https://doi.org/10.1007/978-3-030-96040-7_59

added [5–9]. In this technology acceptance literature jungle, there is a need to be done comprehensive researches in order to evaluate the finding of the previous researches and draw some conclusions. Bibliometric and scientiometric methods are the examples for those comprehensive analyses. Similar to those analyses, meta-analysis is entitled as the father of the analysis. It enables to researchers generalize some researchs' findings which was conducted before on many different samples [10]. So we used meta-analysis to reach general conclusion about the relationship between resistance and exogenous variables of technology acceptance model in present study.

2 Theoretical Background

2.1 Technology Acceptance Model (TAM)

There are emerging stream of theories and models relating to acceptance or adoption of individuals towards a new system or technology in the literature. The first and most known one is technology acceptance model (TAM) that was put forth by Davis [1]. TAM claims that if individuals consider that a technological device or system does not require so much effort to use (*perceived ease of use*) and if individuals think that this device or system facilitate or improve their works (*perceived usefulness*), they will adopt or accept aforementioned technological device or system. Adoption or acceptance of a technological device or system will be seen, if individuals has positive attitude towards it or individuals intent to use it. After that they will be eager to use this system or technology. In other words, TAM put forth the causal relationships among the variables of *perceived ease of use (PEOU), perceived usefulness (PU), attitude, intention* and *use* [1]. PEOU and PU have been exogenous variables, attitude, intention and use have been endogenous variables of TAM.

Afterwards, different theories has emerged to explain technology acceptance, such as *diffusion of innovation* [2], *adaptation of information technology innovation* [3], *unified theory of acceptance and use of technology* [4], etc. Each of them has added some other variables to the TAM to lead to better understanding of acceptance of a technology. However, none of them remove any variable from the TAM. Nevertheless, some variables of TAM have been entitled with different concepts. For instance, PEOU has been entitled as *complexity* by Rogers [2] or *effort expectancy* by Venkatesh et al. [4] and PU has been named as *relative advantage* by Rogers [2] or *performance expectancy* by Venkatesh et al. [4].

TAM has been evolved or improved by adding some other external variables since Davis [1]. Generally those external variables have been allocated as influencing the exogenous variables of TAM (PEOU and PU). For instance, personal innovativeness [8, 11], organizational innovativeness [6], system quality, information quality [7], technology readiness [9, 12, 13], user involvement [14, 15], user anxiety [5, 16] and user resistance [17, 18] are some of them.

2.2 Resistance

Kurt Levin [19] put forth the concept of resistance to change. Since then, the concept has been researched to explain individuals' resistance towards many different things in the literature. Furthermore, the concepts of technology resistance [20, 21], innovation resistance [22], user resistance [17, 23–25] have been frequently used to mention that individuals hinder themselves to change. Bhattacherjee and Hikmet [26] have adapted Levin's [19] resistance to change concept to the information system framework. In addition, they have measured user's attitude towards change in order to explain their acceptance of a new technology or system.

Occasionally, innovative products come across customer resistance. Therefore, those innovations result in failure owing to the resistance to change of customers [27]. So, innovation resistance is a resistance towards any innovation or innovative product [22]. Norzaidi et al. [28] has defined the user resistance as users' biases to alter something when they are working with a technology or system.

As seen in the definitions of resistance types, individuals' acceptance or adoption of a new system or technology can be influenced by their level of resistance to change. Furthermore, in the literature, there are many researches indicating the association between resistance and acceptance of any system or technology [17, 20, 29–31]. As mentioned before, resistance has been examined as an external variable of TAM [18, 32]. So we proposed the following hypotheses, on the basis of those researches;

Hypothesis 1: Resistance is associated with PEOU.
Hypothesis 2: Resistance is associated with PU.

3 Methodology

3.1 Literature Search

Figure 1 showed the frequencies of researches conducted between TAM and the other variables in the literature. As seen in the Fig. 1, TAM was studied together with different

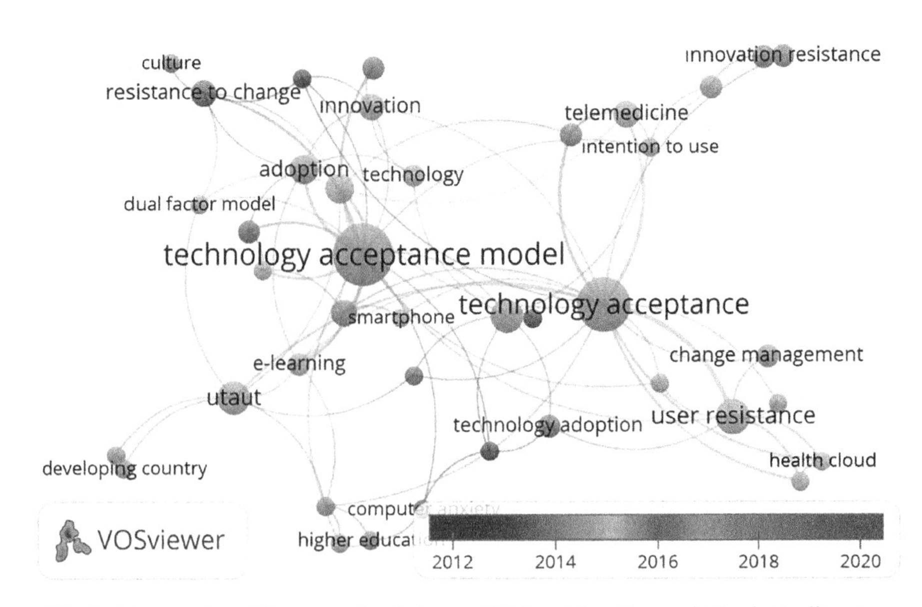

Fig. 1. Frequencies of the researches between TAM and the other variables in the literature

types of resistance (resistance to change, user resistance, innovation resistance etc.). Details of those resistance types were reported in the Appendix. Based on this literature review, it was seen that the association between technology acceptance and resistance could be researched with meta-analysis.

To get data of acceptance and resistance of individuals towards a system or technology, we carried out literature search in October 2021. In this search, we defined the search terms based on the TAM and limited the document type as "peer reviewed journal articles, conference proceedings and book chapters" written in English. As mentioned before resistance is frequently used as an external variable of TAM in the literature [18, 32]. In other words, resistance is evaluated as predicting variable of TAM's exogenous variables, PU and PEOU. In addition PU was named as *relative advantage* and *performance expectancy*; PEOU was entitled as *complexity* and *effort expectancy* in different version of TAM in the literature [2–4].

In SCOPUS database, we searched the term of resistance together with two exogenous variables of TAM (PU/PEOU, with different written forms) in abstracts, keywords and titles of all documents. We used "AND" conjunction between the searching rows. Then we attained 13509 documents. After that we added "technology acceptance" searching term to limit the documents, as follows:

1st searching row: *"perceived ease of use" OR "complexity" OR "effort expectancy" OR "perceived usefulness" OR "relative advantage" OR "performance expectancy"* were written.
2nd searching row: *"resistance"* was typed.
3rd searching row: *"technology acceptance"* was written.

We achieved 113 documents but we only accessed the full-texts of 90 of 113 documents. After evaluation of those articles which have correlation scores between resistance and each variable of TAM (perceived ease of use, perceived usefulness, attitude, intention, use), we attained 41 relevant articles that were reported in the Appendix. 30 of them had the correlation scores between resistance and PEOU and 27 of them had the correlation scores between resistance and PU.

3.2 Meta-analysis Procedures

To prepare data for meta-analysis, exact names of the concepts and measurement instruments of variables (resistance, PU and PEOU) were coded. In addition, countries that samples come from and size of the samples were coded. Furthermore, correlation values between resistance and each exogenous variable of TAM (PEOU/PU) were coded. In this phase, we separated the correlation values between resistance and PEOU from the correlations between resistance and PU. After that we conducted two distinct meta-analyses for both groups of data, based on the Cooper and Hedges's [33] and Lipsey and Wilson's [34] meta-analysis procedures. We used the scores of sample sizes and correlation coefficients in order to reckon effect sizes. *"Ficher's Z-transformed correlations were weighed with sample size, and then weighed coefficients were summed and divided by sum of weights"* [35]. So we obtained true population correlation on the basis of the effect sizes which were calculated by Comprehensive Meta-analysis software program.

Publication Bias. Before conducting meta-analysis, outliers should be checked [36] by evaluating the forest and funnel plots. Figure 1 exhibited Fisher's Z scores of the relation between PEOU and resistance, Fig. 3 represented Fisher's Z scores of the association between PU and resistance.

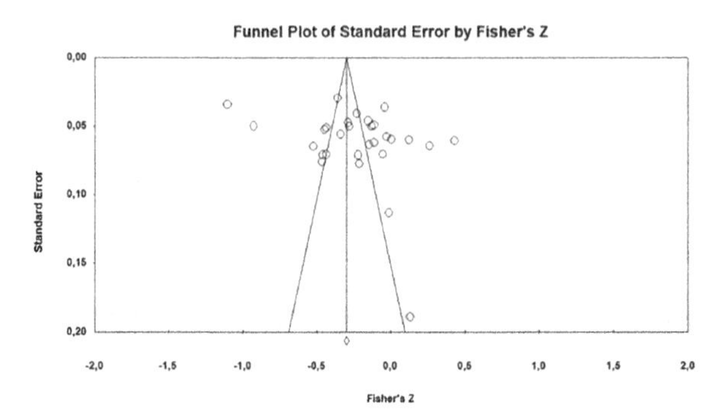

Fig. 2. Funnel Plot of PEOU and Resistance

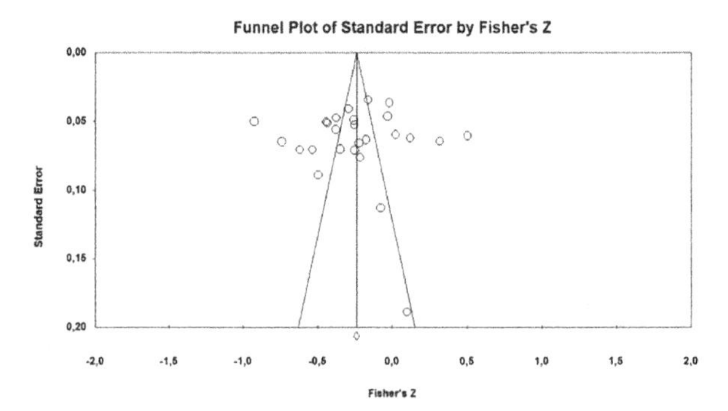

Fig. 3. Funnel Plot of PU and Resistance

An outlier study was seen on the right part of the funnel at the Fig. 2 as well as Fig. 3. Fixed and random effect sizes were calculated in two cases, with the outlier study and without the outlier study. Each effect sizes were reported in Table 1. Both random and fixed effect sizes' scores rose substantially when the outlier study was removed. So this outlier study was removed from the sample of this research.

Table 1. Examining the outlier study removing, $*p < 0.05$

Effect sizes ($\bar{r}$)	PEOU and resistance ($\bar{r}$)	PU and resistance ($\bar{r}$)
Fixed eff. (with an outlier)/No. of eff. size	−0.277*/30	−0.212*/27

(*continued*)

Table 1. *(continued)*

Effect sizes ($\bar{r}$)	PEOU and resistance ($\bar{r}$)	PU and resistance ($\bar{r}$)
Fixed eff. (without outlier)/No. of eff. size	−0.288*/29	−0.237*/26
Random eff. (with an outlier) No. of eff. size	−0.210*/30	−0.201*/27
Random eff. (without outlier) No. of eff. size	−0,225*/29	−0.238*/26

There is no publication bias in the event that each publication should be allocated symmetrically over the zero point line on funnel plot [36]. Evaluating Fig. 2 and Fig. 3, it could be seen publications' symmetrical distributions. So we can say that there is no publication bias of our sample for meta-analysis. Occasionally, it may not be evaluated publication bias, by looking at the points of publications' distribution. In that circumstances, Begg and Mazumdar's [37] rank correlation scores enable researchers to interpret whether any publication bias exists or not [38]. If Kendal's S statistics of Begg and Mazumdar rank correlation analysis is non-significant, it is interpreted that there is no publication bias [34]. Evaluating Kendal's S statistics of present research, it was seen the non-significant results for the relation between PEOU and resistance ($p = 0.69 > 0.05$) and for the relation between PU and resistance ($p = 0.96 > 0.05$), as well. So Kendal Tau's statistics confirmed that there was no publication bias of our sample for meta-analysis.

Q Statistics: Effect Size Variability Across Studies. Meta-analysis program serve researcher both fixed and random effect size scores, based on the correlation scores. Fixed effect model presumes that all effect sizes scores, which are estimated from the other researches' correlation scores, has the same effect to calculate the true effect size. Nevertheless, many different factors (genders, countries, scales etc.) disturb effect size of all studies included meta-analysis. On the other hand, random effect model assumes that true effect sizes are normally distributed [39]. Q statistics should be interpreted to make decision which model is appropriate. *"The Q statistics measure whether data suggest that the effect sizes of different studies estimate the same population effect size"* [34]. If Q statistics produce non-significant results, fixed effect model will be appropriate for the studies included to meta-analysis. On the contrary, whether Q statistics produce significant results, researchers should accept that random effect model is relevant for the data included to meta-analysis.

First, we conducted meta-analysis to attain the effect size of PEOU and resistance, in present study. As seen in Table 2, Q statistics produced significant result ($X^2 = 1193.937/df = 28$). So we interpreted that random effect model is proper and we reported random effect size score to represent the relationship between PEOU and resistance. We found that the score was negative and significant ($\bar{r} = -0.225$, $p = 0.00$). At 95% confidence intervals, the effect size scores changed from −0.343 to −0.101 and the interval had not got zero point. According to Cohen's [40] classification of effect sizes, we interpreted that the effect size of PEOU and resistance was small in present study. Consequently, hypothesis 1 was supported.

Table 2. Overall results of meta-analysis, heterogeneity and publication bias

	PEOU and resistance	*p(sig.)*	PU and resistance	*p(sig.)*
Sample size	13879		13351	
Number of effect sizes	29		26	
Effect Size ($\bar{r}$)				
Fixed effect model	−0.288	0.00	−0.237	0.00
Random effect model*	−0,225	0.00	−0.238	0.00
Heterogeneity				
Q statistics (X/df^2)	1193.937/28	0.00	696.979/25	0.00
Publication Bias				
Tau S Statistics (tau coef.)	0.05429	0.69	−0.00615	0.96
95% CIs				
Lower	−0.343		−0.343	
Upper	−0.101		−0.126	

On the other hand, the effect size of PU and resistance was examined with another meta-analysis. Table 2 indicated the significant Q statistics score ($X^2 = 696.979/df = 25$). Therefore, random effect model was appropriate to investigate effect size of PU and resistance. The effect size score was negative and significant ($\bar{r} = -0.238, p = 0.00$); in addition, it was small level in terms of Cohen's [40] classification. The effect size result differentiated from −0.343 to −0.126 at 95% confidence interval and it had not zero point. So hypothesis 2 was supported.

4 Conclusion

Technology acceptance is defined as the adoption or usage of individuals towards any technology or system. However, resistance to change is described as an adverse reaction of individuals towards a new system, technology or change. In the TAM literature, there could be seen many researches investigating the individuals' acceptance of any technology considering their resistance to change [17, 20, 29–31].

On the basis of the previous researches, that reported the correlation values among resistance and exogenous variables of TAM, we conducted meta-analysis to get effect size of resistance and PEOU, and effect size of resistance and PU. We operated Comprehensive Meta Analysis Program to analyze 41 relevant articles' correlation scores. However, 29 of 41 papers have correlation values between resistance and PEOU; 26 of 41 papers have correlation values between resistance and PU. So we carried out two discrete meta-analyses. Publication bias was checked with Begg and Mazumdar [37] statistics and funnel plot; as a result, there was not any publication bias. Afterwards, Q statistics were employed to find out the most appropriate model for the effect size calculation. So, random effect size model is appropriate for the effect size of resistance and PEOU, for the effect size of resistance and PU, as well.

Results showed that effect size of resistance and PEOU was significantly negative and small level, and effect size of resistance and PU was significantly negative and small level, as well. Based on this result, we interpret that individuals can perceive a new system is not easy to use and the system does not improve their work performance, if they have resistance towards the system. The results come from previous researches in the literature.

Diwivedi et al. [32] claimed that resistance should be evaluated an external variable of Venkatesh et al.'s [4] unified theory of acceptance and use of technology model at their meta-analysis research. So we confirmed the findings of this research. However, Diwivedi et al. [32] did not report any effect size scores of variables of TAM and resistance. So, present study takes this previous research results a stage further.

Present research has some limitations. We got relevant articles, proceedings and book chapters from only SCOPUS database, documents which were indexed other databases could be included for further studies, so that sample size could be increased. The effect size between resistance and endogenous variables of TAM (attitude, intention and use) could be estimated. In addition, it could be explained why random effect model found appropriate by conducting moderator analysis. Moreover, other external variables could be analyzed with meta-analysis based on TAM for further research.

Appendix

See Table 3.

Table 3. Publications which have correlation scores between resistance and variables of TAM

No.	Author/s	Year	Country	Sample size	Resistance type
P1 [41]	Kamal et al.	2020	Pakistan	275	Resistance to technology, user resistance
P2 [42]	Asadi and Jusoh	2019	Espana	401	User resistance
P3 [43]*	Tsai et al.	2020	Taiwan	31	Technology resistance
P4 [43]*	Tsai et al.	2020	Taiwan	81	Technology resistance
P5 [44]	Dai et al.	2019	Saharan Africa	350	User resistance

(continued)

Table 3. (*continued*)

No.	Author/s	Year	Country	Sample size	Resistance type
P6 [45]	Raza et al.	2017	Pakistan	300	Technology resistance, user resistance
P7 [46]	Al-Somali et al.	2008	United Kingdom	400	Technology resistance, user resistance
P8 [26]	Bhattacherjee and Hikmet	2007	USA	129	Technology resistance
P9 [47]*	Oh et al.	2019	USA	241	Innovation resistance
P10 [47] *	Oh et al.	2019	USA	202	Innovation resistance
P11 [48]	Hossain et al.	2018	Portugal		Technology resistance
P12 [49]	Lin et al.	2019	Taiwan	363	Innovation resistance
P13 [50]	Sánchez-Prieto et al.	2019	China	222	Technology resistance, resistance to change
P14 [20]	Yang and Park	2019	Korea	383	Technology resistance, user resistance
P15 [51]	Lallmahomed et al.	2017	Mauritius	247	Technology resistance
P16 [52]	Hoque and Sorwar	2017	Bangladesh	300	Technology resistance, user resistance
P17 [53]	Hsieh	2016	Taiwan	681	User resistance
P18 [54]	Ng et al.	2015	Malaysia	414	Technology resistance, technology adoption
P19 [22]	Im et al.	2014	Korea	845	Innovation resistance, technology resistance

(*continued*)

Table 3. (*continued*)

No.	Author/s	Year	Country	Sample size	Resistance type
P20 [55]	Hsieh et al.	2014	Taiwan	443	Technology resistance, user resistance
P21 [17]	Guo et al.	2012	China	204	Technology resistance
P22 [56]	Shih and Huang	2009	USA	271	Technology resistance, user resistance
P23 [57]	Nov and Ye	2008	USA	170	Technology resistance, user resistance
P24 [58]	Şahin et al.	2021	Turkey	321	Resistance to change
P25 [59]	Talukder et al.	2019	Australia	325	Technology resistance
P26 [60]	Beglaryan et al.	2017	Armenia	233	Technology resistance, user resistance
P27 [61]	Wang et al.	2019	China	175	Technology resistance, user resistance
P28 [62]	Chi et al.	2020	Taiwan	252	User resistance
P29 [63]*	Kim and Bae	2020	Korea	398	Innovation resistance
P30 [63] *	Kim and Bae	2020	Korea	203	Innovation resistance
P31 [64]	Vichitkraivin and Naenna	2020	Thailand	466	Technology resistance
P32 [65]	Alaiad et al.	2019	USA	280	User resistance
P33 [66]	Tsai	2021	Taiwan	593	Resistance to change
P34 [67]	Huang	2015	Taiwan	286	User resistance

(*continued*)

Table 3. (*continued*)

No.	Author/s	Year	Country	Sample size	Resistance type
P35 [68]	Kim and Park	2020	South Korea	1128	Resistant attitude, resistance behaviour
P36 [69]	Halbach and Gong	2011	USA	750	Technology resistance, user resistance
P37 [70]	Keung et al.	2018	Hong Kong	243	Technology resistance, user resistance
P38 [71]	Shu-Fong et al.	2007	Malaysia	200	Resistance to change
P39 [72]	Shahbaz et al.	2020	China	283	Resistance to change
P40 [18]	Donmez-Turan	2020	Turkey	262	User resistance
P41 [73]	Tavera-Mesias et al.	2021	Columbia	397	Resistance to change

*It is reported more than one correlation scores from different samples in this paper.

References

1. Davis, F.D.: Perceived usefulness, perceived ease of use, and user acceptance of information technology. MIS Q. **13**(3), 319–340 (1989)
2. Rogers, E.M.: Diffusion of Innovations, 3rd edn. The Free Press, New York (1983)
3. Moore, G.C., Benbasat, I.: Development of an instrument to measure the perceptions of adopting an information technology innovation. Inf. Syst. Res. **2**(3), 192–222 (1991)
4. Venkatesh, V., Morris, M.G., Davis, G.B., Davis, F.D.: User acceptance of information technology: toward a unified view. MIS Q. **27**(3), 425–478 (2003)
5. Compeau, D.R., Higgins, C.A.: Computer self-efficacy: development of a measure and initial test. MIS Q. **19**(2), 189–211 (1995)
6. Agarwal, R., Prasad, J.: A conceptual and operational definition of personal innovativeness in the domain of information technology. Inf. Syst. Res. **9**(2), 204–215 (1998)
7. DeLone, W.H., McLean, E.R.: Information systems success: the quest for the dependent variable. Inf. Syst. Res. **3**(1), 60–95 (1992)
8. Yi, M.Y., Fiedler, K.D., Park, J.S.: Understanding the role of individual innovativeness in the acceptance of itbased innovations: comparative analyses of models and measures. Decis. Sci. **37**(3), 393–426 (2006)
9. Khatri, V., Samuel, B.M., Dennis, A.R.: System 1 and System 2 cognition in the decision to adopt and use a new technology. Inf. Manag. **55**(6), 709–724 (2018)

10. Humphrey, S.E.: What does a great meta-analysis look like? Organ. Psychol. Rev. **1**(2), 99–103 (2011)
11. Donmez-Turan, A., Zehir, C.: Personal innovativeness and perceived system quality for information system success: the role of diffusability of innovation. Tehnicki vjesnik/Technical Gazette **28**(5), 1717–1726 (2021)
12. Parasuraman, A.: Technology readiness index (TRI) a multiple-item scale to measure readiness to embrace new technologies. J. Serv. Res. **2**(4), 307–320 (2000)
13. Donmez-Turan, A., Oren, B.: Technology readiness as an antecedent of technology acceptance model: a meta-analytic approach. In: Abdalmuttaleb, M.A., Al-Sartawi, M., Razzaque, A., Kamal, M.M. (eds.) Artificial Intelligence Systems and the Internet of Things in the Digital Era: Proceedings of EAMMIS 2021, pp. 513–522. Springer International Publishing, Cham (2021). https://doi.org/10.1007/978-3-030-77246-8_47
14. Hunton, J.E., Beeler, J.D.: Effects of user participation in systems development: a longitudinal field experiment. MIS Q. **21**(4), 359–388 (1997)
15. Turan, A., Tunç, A.Ö., Zehir, C.: A theoretical model proposal: personal innovativeness and user involvement as antecedents of unified theory of acceptance and use of technology. Procedia-Soc. Behav. Sci. **210**, 43–51 (2015)
16. Donmez-Turan, A., Kır, M.: User anxiety as an external variable of technology acceptance model: a meta-analytic study. Procedia Comput. Sci. **158**, 715–724 (2019)
17. Guo, X., Sun, Y., Wang, N., Peng, Z., Yan, Z.: The dark side of elderly acceptance of preventive mobile health services in China. Electron. Mark. **23**(1), 49–61 (2013). [P21]
18. Donmez-Turan, A.: Does unified theory of acceptance and use of technology (UTAUT) reduce resistance and anxiety of individuals towards a new system? Kybernetes **49**(5), 1381–1405 (2019). [P40]
19. Levin, K.: Frontiers in group dynamics: II: channels of group life; social planning and action research. Hum. Relat. **1**(2), 1430–1453 (1947)
20. Yang, H.S., Park, J.W.: A study of the acceptance and resistance of airline mobile application services: with an emphasis on user characteristics. Int. J. Mobile Commun. **17**(1), 24–43 (2019). [P14]
21. Tsai, T.H., Lin, W.Y., Chang, Y.S., Chang, P.C., Lee, M.Y.: Technology anxiety and resistance to change behavioral study of a wearable cardiac warming system using an extended TAM for older adults. Plos One **15**(1), e0227270 (2020)
22. Im, H., Jung, J., Kim, Y., Shin, D.H.: Factors affecting resistance and intention to use the smart TV. J. Media Bus. Stud. **11**(3), 23–42 (2014). [P19]
23. Doll, W.J., Torkzadeh, G.: The measurement of end-user computing satisfaction. MIS Q. **12**(2), 259–274 (1988)
24. Rafaeli, S.: The electronic bulletin board: a computer-driven mass medium. Soc. Sci. Micro Rev. **2**(3), 123–136 (1984)
25. Igbaria, M., Chakrabarti, A.: Computer anxiety and attitudes towards microcomputer use. Behav. Inf. Technol. **9**(3), 229–241 (1990)
26. Bhattacherjee, A., Hikmet, N.: Physicians' resistance toward healthcare information technology: a theoretical model and empirical test. Eur. J. Inf. Syst. **16**(6), 725–737 (2007). [P8]
27. Zaltman, G., Wallendorf, M.: Consumer Behavior: Basic Findings and Management Implications. Wiley, New York (1983)
28. Norzaidi, M.D., Salwani, M.I., Chong, S.C., Rafidah, K.: A study of intranet usage and resistance in Malaysia's port industry. J. Comput. Inf. Syst. **49**(1), 37–47 (2008)
29. Rose, J.: The problem of technological barriers. Kybernetes **38**(1/2), 25–41 (2009)
30. Lian, J.W., Yen, D.C.: Online shopping drivers and barriers for older adults: age and gender differences. Comput. Hum. Behav. **37**, 133–143 (2014)

31. Lwoga, E.T., Komba, M.: Antecedents of continued usage intentions of web-based learning management system in Tanzania. Educ. Train. **57**(7), 738–756 (2015)
32. Dwivedi, Y.K., Rana, N.P., Chen, H., Williams, M.D.: A Meta-analysis of the unified theory of acceptance and use of technology (UTAUT). In: Governance and Sustainability in Information Systems: Managing the Transfer and Diffusion of IT, pp. 155–170. Springer, Heidelberg (2011)
33. Cooper, H., Hedges, L.V.: The Handbook of Research Synthesis. Russell Sage Foundation, New York (1993)
34. Lipsey, M.W., Wilson, D.B.: Practical Meta-Analysis: Applied Social Research Methods Series, vol. 49. SAGE publications, Londra (2001)
35. Witherspoon, C.L., Bergner, J., Cockrell, C., Stone, D.N.: Antecedents of organizational knowledge sharing: a meta-analysis and critique. J. Knowl. Manag. **17**(2), 250–277 (2013)
36. Geyskens, I., Krishnan, R., Steenkamp, J.B.E., Cunha, P.V.: A review and evaluation of meta-analysis practices in management research. J. Manag. **35**(2), 393–419 (2009)
37. Begg, C.B., Mazumdar, M.: Operating characteristics of a rank correlation test for publication bias. Biometrics **50**(4), 1088–1101 (1994)
38. Borenstein, M., Hedges, L.V., Higgins, J.P., Rothstein, H.R.: Introduction to Meta-analysis. Wiley, New Jersey (2011)
39. Borenstein, M., Hedges, L.V., Higgins, J.P., Rothstein, H.R.: Fixed-effect versus random-effects models. Introduct. Meta-anal. **77**, 85 (2009)
40. Cohen, J.: Statistical Power Analysis for the Behavioral Sciences, 2nd edn. Erlbaum, Hillsdale (1988)
41. Kamal, S.A., Shafiq, M., Kakria, P.: Investigating acceptance of telemedicine services through an extended technology acceptance model (TAM). Technol. Soc. **60**, 101212 (2020). [P1]
42. Asadi, S., Abdullah, R., Jusoh, Y.Y.: An integrated SEM-neural network for predicting and understanding the determining factor for institutional repositories adoption. In: Bi, Y., Bhatia, R., Kapoor, S. (eds.) Intelligent Systems and Applications: Proceedings of the 2019 Intelligent Systems Conference (IntelliSys) Volume 2, pp. 513–532. Springer, Cham (2020). https://doi.org/10.1007/978-3-030-29513-4_38
43. Tsai, T.H., Lin, W.Y., Chang, Y.S., Chang, P.C., Lee, M.Y.: Technology anxiety and resistance to change behavioral study of a wearable cardiac warming system using an extended TAM for older adults. Plos One **15**(1), 0227270 (2020). [P3]–[P4]
44. Dai, B., Larnyo, E., Tetteh, E.A., Aboagye, A.K., Musah, A.A.I.: Factors affecting caregivers' acceptance of the use of wearable devices by patients with dementia: an extension of the unified theory of acceptance and use of technology model. Am. J. Alzheimer's Dis. Other Dement. **35**, 1533317519883493 (2020). [P5]
45. Raza, S.A., Umer, A., Shah, N.: New determinants of ease of use and perceived usefulness for mobile banking adoption. Int. J. Electron. Custom. Relationsh. Manag. **11**(1), 44–65 (2017). [P6]
46. Al-Somali, S.A., Gholami, R., Clegg, B.: An investigation into the acceptance of online banking in Saudi Arabia. Technovation **29**(2), 130–141 (2009). [P7]
47. Oh, Y.J., Park, H.S., Min, Y.: Understanding location-based service application connectedness: model development and cross-validation. Comput. Hum. Behav. **94**, 82–91 (2019). [P9]–[P10]
48. Hossain, A., Quaresma, R., Rahman, H.: Investigating factors influencing the physicians' adoption of electronic health record (EHR) in healthcare system of Bangladesh: an empirical study. Int. J. Inf. Manag. **44**, 76–87 (2019). [P11]
49. Lin, C.W., Lee, S.S., Tang, K.Y., Kang, Y.X., Lin, C.C., Lin, Y.S.: Exploring the users behavior intention on mobile payment by using TAM and IRT. In: Proceedings of the 2019 3rd International Conference on E-Society, E-Education and E-Technology, pp. 11–15 (2019). [P12]

50. Sánchez-Prieto, J.C., Huang, F., Olmos-Migueláñez, S., García-Peñalvo, F.J., Teo, T.: Exploring the unknown: the effect of resistance to change and attachment on mobile adoption among secondary pre-service teachers. Br. J. Educ. Technol. **50**(5), 2433–2449 (2019). [P13]
51. Lallmahomed, M.Z., Lallmahomed, N., Lallmahomed, G.M.: Factors influencing the adoption of e-Government services in Mauritius. Telemat. Inf. **34**(4), 57–72 (2017). [P15]
52. Hoque, R., Sorwar, G.: Understanding factors influencing the adoption of mHealth by the elderly: an extension of the UTAUT model. Int. J. Med. Inf. **101**, 75–84 (2017). [P16]
53. Hsieh, P.J.: An empirical investigation of patients' acceptance and resistance toward the health cloud: the dual factor perspective. Comput. Hum. Behav. **63**, 959–969 (2016). [P17]
54. Ng, S.N., Matanjun, D., D'Souza, U., Alfred, R.: Understanding pharmacists' intention to use medical apps. Electron. J. Health Inf. **9**(1), 7 (2015). [P18]
55. Hsieh, P.J., Lai, H.M., Ye, Y.S.: Patients' acceptance and resistance toward the health cloud: an integration of technology acceptance and status quo bias perspectives, PACIS 2014 Proceedings. p. 230 (2014). [P20]
56. Shih, Y.Y., Huang, S.S.: The actual usage of ERP systems: an extended technology acceptance perspective. J. Res. Pract. Inf. Technol. **41**(3), 263–276 (2009). [P22]
57. Nov, O., Ye, C.: Users' personality and perceived ease of use of digital libraries: the case for resistance to change. J. Am. Soc. Inf. Sci. Technol. **59**(5), 845–851 (2008). [P23]
58. Şahin, F., Doğan, E., İlic, U., Şahin, Y.L.: Factors influencing instructors' intentions to use information technologies in higher education amid the pandemic. Educ. Inf. Technol. **26**(4), 4795–4820 (2021). https://doi.org/10.1007/s10639-021-10497-0
59. Talukder, M.S., Sorwar, G., Bao, Y., Ahmed, J.U., Palash, M.A.S.: Predicting antecedents of wearable healthcare technology acceptance by elderly: a combined SEM-Neural Network approach. Technol. Forecast. Soc. Change **150**, 119793 (2020). [P25]
60. Beglaryan, M., Petrosyan, V., Bunker, E.: Development of a tripolar model of technology acceptance: hospital-based physicians' perspective on EHR. Int. J. Med. Inf. **102**, 50–61 (2017). [P26]
61. Wang, G., Wang, P., Cao, D., Luo, X.: Predicting behavioural resistance to BIM implementation in construction projects: an empirical study integrating technology acceptance model and equity theory. J. Civil Eng. Manag. **26**(7), 651–665 (2020). [P27]
62. Chi, W.C., Lin, P.J., Chang, I.C., Chen, S.L.: The inhibiting effects of resistance to change of disability determination system: a status quo bias perspective. BMC Med. Inf. Decis. Mak. **20**(1), 1–8 (2020). [P28]
63. Kim, D., Bae, J.K.: The effects of protection motivation and perceived innovation characteristics on innovation resistance and innovation acceptance in internet primary bank services. Glob. Bus. Financ. Rev. **25**(1), 1–12 (2020). [P29]–[P30]
64. Vichitkraivin, P., Naenna, T.: Factors of healthcare robot adoption by medical staff in Thai government hospitals. Heal. Technol. **11**(1), 139–151 (2020). https://doi.org/10.1007/s12553-020-00489-4
65. Alaiad, A., Alsharo, M., Alnsour, Y.: The determinants of m-health adoption in developing countries: an empirical investigation. Appl. Clin. Inf. **10**(05), 820–840 (2019). [P32]
66. Tsai, L.L.: Why college students prefer typing over speech input: the dual perspective. IEEE Access **9**, 119845–119856 (2021). [P33]
67. Huang, T.K.: The role of user resistance in the adoption of screenshot annotation for computer software learning. In: 2015 48th Hawaii International Conference on System Sciences, pp. 101–110. IEEE (2015). [P34]
68. Kim, J., Park, E.: Understanding social resistance to determine the future of Internet of Things (IoT) services. Behav. Inf. Technol. 1–11 (2020). [P35]

69. Halbach, M., Gong, T.: What predicts commercial bank leaders' intention to use mobile commerce?: the roles of leadership behaviors, resistance to change, and technology acceptance model. In: E-commerce for Organizational Development and Competitive Advantage, pp. 151–170. IGI Global (2013). [P36]
70. Keung, K.L., Lee, C., Ng, K.K.H., Leung, S.S., Choy, K.L.: An empirical study on patients' acceptance and resistance towards electronic health record sharing system: a case study of Hong Kong. Int. J. Knowl. Syst. Sci. **9**(2), 1–27 (2018). [P37]
71. Shu-Fong, L., Yin, F.M., Ming, S.K., Ndubisi, N.O.: Attitude towards internet banking: a study of influential factors in Malaysia. Int. J. Serv. Technol. Manag. **8**(1), 41–53 (2007). [P38]
72. Shahbaz, M., Gao, C., Zhai, L., Shahzad, F., Arshad, M.R.: Moderating effects of gender and resistance to change on the adoption of big data analytics in healthcare. Complexity 2020, 1–13 (2020). [P39]
73. Tavera-Mesias, J.F., van Klyton, A., Zuñiga Collazos, A.: Social stratification, self-image congruence, and mobile banking in Colombian cities. J. Int. Consum. Mark. (2021). https://doi.org/10.1080/08961530.2021.1955426. [P41]

The Online k-Taxi Problem

Kapil[1(✉)], Dharmendra Prasad Mahato[1,2(✉)], and Van Huy Pham[2]

[1] National Institute of Technology Hamirpur, Hamirpur 177 005, Himachal Pradesh, India
{cs16mi549,dpm}@nith.ac.in

[2] Faculty of Information Technology, Ton Duc Thang University, Ho Chi Minh City, Vietnam
phamvanhuy@tdtu.edu.vn

Abstract. The online k-taxi problem, a generalization of the k-server problem, involves k taxis providing a metric space series of requests. A request is made up of two coordinates, a and b, representing a customer wishing to be transported by taxi from coordinate a to coordinate b. The goal is to satisfy all demands and at the same time minimize the overall trip distance for all cabs. The problem is classified into two types: easy k-taxi problems and hard k-taxi problems. Easy problem of k-taxi is defined as the cost is the whole distance covered by the cabs during satisfy request of present customer present in cab, hard problem of k-taxi is defined as the it just takes the distance of vacant runs when cab is free from customer and go for coordinate from where next customer made request and vacant run distance add on until cab is full.

In the k-server and the k-taxi problem the distance that the entities cover is the main and sole cost source. In this paper we propose a new approach that takes both the driving distance as well as the find the good cost for taxi driver while not carrying a passenger in hard k-taxi problem. For this new cost-function which is defined by as probability density function we will present an algorithm and discuss its performance.

Keywords: K-taxi problem · K-server problem · Stochastic model

1 Introduction

1.1 Overview

Let us assume a city with community of cars, pedestrians, sidewalks, and intersections. Customers who want to go from one part of the city to another call a taxi company, but the company only has a limited number of taxis. This is the aim of the original k - Taxi problem, which is to reduce the total distance travelled by taxis. It falls under the category of online problems. An online algorithm gets its data in bits, and some of the information it needs isn't accessible right away. The algorithm must continuously change its decisions in response to new data packages.

The k-taxi problem is a general form of the fundamental k-Server problem, which was first proposed by Karloff and accepted by Fiat et al. [1]. In this

I. Woungang et al. (Eds.): ANTIC 2021, CCIS 1534, pp. 809–817, 2022.
https://doi.org/10.1007/978-3-030-96040-7_60

situation, k taxis are dispersed throughout a metric space and must fulfil a sequence of demands. A request is made up of pair of two metric space points (a, b) that indicate a passenger who has decided to travel from coordinate a to coordinate b. In response to the request, a cab travels to a and then to b. The goal is to satisfy all demands in queue they were received and at the same time minimize the overall trip distance for all cabs. We look at the online problem of this issue, in which requests are displayed one at a time, with a new request showing only after the preceding one has been served.

Since this distance between a and b must be travelled without concern for the algorithm's evaluations, and the exclusion of this from the costs is logical, and focusing on overhead journeys depending on evaluations of the algorithm. There are different ways to describe hard k - taxi problem and we describe it as: it just takes the distance of vacant runs when cab is free from customer and go for coordinate from where next customer made request and vacant run distance add on until cab is full, that is, the overhead distance travelled in addition to the distances between the starting coordinate and destination coordinate. In the hard version, when we talked about cost, it is described in a way as we describe hard k - taxi problem which means the distance travelled without a client, that is, the overhead distance traversed in addition to the distances between the beginning and destination coordinates. In the simple version, however, the total distance driven by the cabs is considered in cost. As a result, the cost of any cab schedule differs between the two versions by the exact number of a-b-distances, and for both, the best offline solutions are the same. Because of the numerous cost functions, estimating the optimal solution value in the hard version is more challenging. The Uber problem [2], which explored the simple version of the problem with stochastic input, was recently reintroduced. We also refer to Ma, W., Xu, Y., Wang, K. [3] which give the brief overview of Online k - Truck Problem.

We compare the output of various online algorithms using competitive analysis [4], focusing on how well the algorithms respond to the input rather than their runtime. One of these online problems is the well-known k-server problem. The whole k server activity which executes an on-line series of requests must be minimized in case of the k server problem. We are given T independent distributions $P_1, P_2, \ldots, P_T$ in advance in the stochastic setting, and a request is drawn from P_i at each time stage i. An online algorithm must travel k servers to satisfy requests in the k-server problem. The servers are dispersed around the graph's vertices. Requests appear on the graph's vertices as input progresses. To fulfil requests, an algorithm must shift servers along the edges of the network to their optimal position. To minimize the total length between all servers is the main goal of k - Server problem. The fundamental objective of k - server problem is to minimize the net distance between all servers.

Consider the matter of online 2 - Taxi problem, in which a customer request appears to an algorithm and requests a ride from s to t. The algorithm now selects one of the taxis, and the taxi arrives to pick up the customer. The taxi is now full and cannot accommodate another passenger. The customer's request is met as soon as the taxi arrives at the destination, and the taxi is free for a new customer. In the cost model of the original k-taxi problem, a customer may have to wait a long time for his request to be scheduled, and requests may be cancelled.

***k* - Server Problem.** The k-server problem came into play as a natural generalization of many online problems and a foundation for additional issues, such as metrical task systems which is defined by Manasse, McGeoch, and Sleator [5]. It was thought that the adversary model did not know about the future demands in the online algorithm. They employ competitive analyses to measure how well Sleator and Tarjan [5] perform an online algorithm. In this model when we compare an online algorithm (OA) with an offline optimum algorithm (OOA) so, online algorithm is aware from input in past.

Let the net cost of OA & OOA is given as $|OA\ (\rho)|\&|OOA\ (\rho)|$ for fulfilling a sequence of requests ρ. If for every, $|OA(\rho)| \leq c|OOA(\rho)| + c_0$, where c_0 is independent of, an algorithm is c - competitive.

A lower limit k for the competitiveness of any deterministic algorithm in any metric space have equal and greater than $k+1$ point is established by Manasse et al. [6]. This limit is tight for general metrics, according to the well-known k-server prediction. In past all k which is in present in exponential is known as the upper limits then Koutsoupias and Papadimitriou [7] made a challenge and gave working function algorithm which is $(2k-1)$ competitive and complete their challenge with respect to this algorithm known for its best result. The tight competitive ratio has become a "holy grail" field in the previous two decades. As a result of this dilemma the uniform (also known as the paging) metric was examined in special areas, including the line, the circle and tree metric (see [8, 9] and references therein).

In a well - organized and competent way moving k number of servers in the metric space in order to serve request is one of challenge for k - server problem. The k number of servers are at first configured with a predetermined set of k points. Then a request occurs somewhere in a metric space at the end of each stage, and some server must travel to that place to fulfil it. The goal is to reduce the entire journey of the servers. From research, we assuming that requests made by client in present will be correct and until this request is completed by server a new request will not accepted by client.

Uber Problem. The rapid rise of online network transportation services is said to be the cause of the Uber problem. Two coordinates in the metric: a source coordinate and a destination coordinate made each request means the location from where client request a cab. Serving a request entails moving a server to its source coordinate and after picking client in cab moving to its destination coordinate. The total movement while client is present in cab and when cab make a vacant run should be minimized is the main goal. We research the stochastic Uber problem because demands are normally strongly associated with time. In research they show that uber problem is depend upon k - server problem which means if we can get for k - server a α -approximation algorithm then we can get for uber problem a $(\alpha + 2)$ - approximation algorithm.

Stochastic Model. We analyze the stochastic k-server problem [10], in which the input is drawn from given probability distributions rather than being chosen adversarial. This issue has a wide range of applications, including network transportation and data centers equipment replacement. In today's giant data centers, there are huge numbers of servers and controllers with short lifespans. Robots are programmed to do maintenance activities including repairs and manual server operations, and automation is the most effective approach to scale out data center maintenance. This issue can also be applied to physical networks. Consider the following scenario: A organization offering service of shopping (like amazon, flipkart and another online shopping application) is modelled as a k-server problem. An online sequence of request of client for placing order for various merchant consider in this. We have a fleet of k-shopping cars (i.e. servers) that can meet customer requirements in the shops. It is only logical to conclude that the requests come from a distribution that can be found by observing the past at a certain time of the week/day.

2 Problem Formulation

We consider an arbitrary space environment in which the server receives k taxi requests and must perform a job scheduling task of assigning taxis to customers while reducing constraints such as distance travelled and overhead cost. In addition, we apply intelligence and data-based approaches to eliminate error when considering a real-world situation, allowing us to achieve the goal of statistical decision making.

The client and the cab driver are satisfied in the case of hard k - Taxi problem and thus the request of the client is not denied and a good price for an empty run is given to the taxi driver by means theoretical computer science approach.

Definition 1. *An instance of the k-taxi problem is a tuple (G, Z, K). Here G is an undirected, connected, and weighted graph with a set of vertices V, a set of edges E, and a weighted edge function w:*

$$\forall e \in E, \exists\ w(e) \in \mathbb{R}^{+}$$

K is a list of vertices that contains the starting positions of the k taxis, and Z is a customer request list (see Definition 3. It is worth noting that the property connected is only included for convenience. A separated graph instance can be divided into sub-graphs and solved independently. The weight of each edge reflects the amount of time required for a taxi to move from one end to the other. We implement an additional variable to calculate the waiting cost that any customer can incur.

Definition 2. *A time t where $t \in \mathbb{R}^{+\nvdash}$ reflects the amount of time that has elapsed since the start of an algorithm's execution.*

Definition 3. *A customer request is a tuple* $c = (v_{arr}, v_{dest}, t_{arr})$ *and the three components are defined as follows:*

Where,

- v_{arr}: The arriving vertex is a vertex in G that represents where the customer will appear and where the taxi must travel to and pick him up.
- v_{dest}: The destination vertex is a vertex in G that represents the location to which the customer wishes to travel.
- t_{dest}: The arrival time is the time when the customer request is available to the algorithm.

Definition 4. *The set of all customers is called the customer request list, and its cardinality is* n. *The algorithm has access to the entire customer request list in the offline problem, but only the customer requests with arrival times that are less than or equal to the current time point are accessible.*

A configuration is the current state of an algorithm's execution on a specific instance of the k-taxi problem at a given time. The taxi locations, details about the customer is actually in which taxi, and a list of satisfied customer requests make up a setup. If the time point value is 0, the configuration is called initial configuration. If an online algorithm is currently configured with time point t, it will not be able to enter a configuration with a time point value less than t.

Definition 5. *A taxi is said to be free if it has no passengers in the current configuration. When a taxi has a passenger, it is said to be occupied.*

The following commands can be executed by an algorithm on a k-taxi problem instance in any configuration:

- **Pickup (Taxi):** If the taxi from the argument is on the same vertex as a customer, this customer would occupy the taxi.
- **Drop (Taxi):** If the taxi is full and the customer's destination is the taxi's current position, the taxi has transported the customer and is now empty.
- **Move (Taxi, Vertex):** This determines the taxi's destination vertex. The taxi now travels to this vertex using the most efficient path. This command can be used when the taxi is in motion. The taxi would simply drive to the new destination vertex in that situation.

Definition 6. *The costs of running an online algorithm on a specific instance of the online k-taxi problem is defined as the amount of all taxis' driving time and all customers' waiting times. The difference between the time the customer is picked up by the taxi and the* t_{arr} *of that customer request is the waiting time for that customer request.*

$$cost = \sum_{i=1}^{k} t_{arr} + \sum_{j=1}^{k} t_{waiting}$$

We compare several online algorithms with our concept of competitive ratio. The competitive ratio compares costs for the online algorithm to the best offline. An offline algorithm has all the information regarding incoming requests of the customers and can use it to find the best solution. The competitive relationship between the cost of the online algorithm performance in a specific instance and the cost of the optimal solution of the instance is the highest ratio. This ratio must be of at least 1. It's vital to highlight.

Definition 7. *If the competitive ratio of all online algorithms is even or higher, an online algorithm for the k-taxi issue is regarded as an ideal online algorithm.*

3 Objectives

- Propose a new solution for the hard k - taxi problem, which is a type of the online k - taxi problem.
- To examine various approaches for online k taxi problem and k- server problem.
- Propose a pseudo algorithm to validate the result.

4 Proposed Approach and Algorithm

The data set we incorporate here is the density function of population of people or simply customers. At any location who are willing to book a cab or are probable to book a cab in the time to come. This density function is represented by $[F(P)] \in [0, 1]$ which means that it takes values between 0 and 1.

Assuming max booking rate for cabs in a span of 1 day at any point in its 2 km radius is 1000. We assumed this value is rarely exceeded. We take values for all distinct points in space at 2km separation and assign value of total bookings every day = 'n'. The definition of F[P] is now

$$[F(P)] = n/1000$$

Value of $F[P]$ is now known which is less than 1 always according to our assumptions. Value of $F[P]$ will be updated on the map grid periodically indicating the density function. As any request for cab is recorded by the server, decision for job ordering is made by referring to total distance value D and $F[P]$. Suppose the server gets request for n cabs at an instant the values of D i.e. total distance parameter and value of $F[P]$ is also known. Below are the basic set of rules that will be followed.

In case of single request

The server will simply assign the nearest available driver to the cab.

In case of multiple requests

1. If the number of bookings is less than the number of cabs available The values of D will compare for every taxi and the one request with minimum value of D is chosen for assigning the cab to the nearest customer.
2. In case of bookings greater than the number of cabs available, i.e. non availability of a nearby driver.
   - In case of conflict we will also refer to $F[P]$, a driver from far off location will be assigned to the passenger going to point 't' if value of $F[P]$ there is greater or equal to 0.75 and his estimated time to reach the customer will be calculated.
   - Also, we will check the local rides that are about end and from their location how much time is needed to reach customer.
   - Which so ever takes less time will be assigned to the customer.
3. If all value of D is higher in a series of requests at an instant
   - Then the taxi with minimum value of D can be sent to a place where $F[P]$ is lowest among others since he is getting to complete a ride nearby in less time.
   - Similarly, the request with max value of D will be assigned to a nearby place indicating high density value, in this way even though he has to travel more distance, but he will have high chances of getting a customer for return.

The presented method can be useful for making the taxi server decision autonomous and for the welfare of everyone involved.

4.1 Proposed Algorithm

Algorithm 1. Proposed Algorithm

1: Start the open list
2: Start the closed list
3: **while** The open list is not open **do**
4: Find the minimum f node in the open list, call it "B"
5: From open list pop B off
6: Create 8 successors of B and set B to their parents
7: **for** Each successor **do**
8: **if** successor is the aim **then**
9: Stop search
10: $successor.g = B.g +$ distance between $successor$ and B
11: $successor.h =$ distance from goal to successor (This can be done using by Manhattan Heuristics)
12: $successor.f = successor.g + successor.h$
13: **end if**
14: **if** a node in the open list has lower f than its successor with the same position as the successor **then**
15: Skip it

```
16:         end if
17:         if There is a node in the closed list with the same position as the
    successor, which has a f lower than the successor then
18:             Skip it, add a node to the open list
19:         end if
20:     end for
21:     From closed list push B
22: end while
23: Get the all request list (Starting point) from server
24: while The list not ends do
25:     if no.request ≤ no.taxi then Find  the nearest cab Find the time for it
    to reach Pop request off the list
26:     end if
27:     if no.request > no.taxi then
28:         for Elements in list till no.taxi - 1 do
29:             Find the nearest cab
30:             Find the time for it to reach
31:             Pop request off the list
32:         end for
33:     end if
34:     for Elements in list after no.taxi - 1 do
35:         Find the nearest cab in nearby areas
36:         Find the time to reach customer (T_x)
37:         Find all the rides in local area which will end before T_z
38:         Find the time (T_y) which local cab will take to reach customer
39:
40:         Get probability of ride booking from end location [P(i)]
41:         if T_x < T_y + Bias then
42:             Sort rides on basis of increasing probability
43:             Allot cabs to farther the distance greater probability
44:             Pop request off the list
45:         end if
46:         if T_x > T_y + Bias then
47:             Find all the rides in local area which will end before T_z
48:             Find the time (T_y) which local cab will take to reach customer
49:             Pop request off the list
50:         end if
51:     end for
52: end while
53:
```

5 Conclusions and Future Scope

We introduced a probability density function for the current hard k-taxi problem in this thesis. We devised an algorithm to solve this new problem. We also demonstrated how the algorithm would behave in a real-world situation. Another purpose could be to see how the probability density function affects the many k-taxi adaptations that have been proposed in the community. By adding statistical data and removing probability uncertainty, our solution will make server systems in taxi booking systems autonomous, ensuring that nothing is left to chance. We believe that our approach for hard k-taxi problem on the line can serve as the foundation for a more general solution to the problem. Since this problem has not been tested in the field of high-performance computing, we are especially interested in working with high-performance computing to generate predictions from high performance computation systems. In the mean - time, we have developed a new computational model for the problem, which is derived from the literature. Our initial hypothesis, that an adaptation would have to be an extreme feature of a taxiing system (because of the complexity of the problem or other factors), was based on the fact that the expected distribution of our data was relatively uniform across all of the inputs. However, the distribution of such data can vary with the complexity of the taxiing system. We found that our data can be easily and efficiently partitioned by all input variables.

In this paper, we apply the $A*$ algorithm, which employs a one-directional approach, to determine the shortest path. In future, we will use the shortest path algorithm to find by utilizing a bidirectional approach in $A*$ algorithm. We employ the Probability Density Function, and its value is updated on a regular basis, so that in the future, storing and updating data on a regular basis is as simple as possible.

References

1. Fiat, A., Rabani, Y., Ravid, Y.: Competetive k-server algorithms (extended abstract). In: 31st Annual Symposium on Foundations of Computer Science (1990)
2. Dehghani, S., Ehsani, S., Hajiaghayi, M., Liaghat, V., Seddighin, S.: Stochastic K-server: how should Uber Woek? In: 44th International Colloquium on Automata, Lamguages & Programming (ICALP 2017). Leibniz International Proceedings in Informatics, vol. 80 (2017)
3. Ma, W., Xu, Y., Wang, K.: Online K-truck problem its competitive algorithms. J. Glob. Optim. (2001)
4. Manasse, M., McGeoch, L., Selator, D.: Competitive algorithms for online problems. In: Proceedings 20th ACM Symposium on Theory of Computing (1988)
5. Manasse, M., McGeoch, L., Sleator, D.: Competitive algorithms for server problems. J. Algorithms **11**(2), 208–230 (1990)
6. Sleator, D.D., Tarjan, R.E.: Amortized efficiency of list update and paging rules. Commun. ACM **28**(2), 202–208 (1985)
7. Koutsoupias, E., Papadimitriou, C.H.: On the K-server conjecture. J. ACM (JACM) **42**(5), 971–983 (1995)

A Lossless Compression Algorithm Based on High Frequency Intensity Removal for Grayscale Images

Sangeeta Sharma, Nishant Singh Hada, Gaurav Choudhary(✉), and Syed Mohd. Kashif

Computer Science and Engineering Department, National Institute of Technology Hamirpur, Hamirpur 177005, Himachal Pradesh, India

Abstract. In the current world where data is one of the most important entities in our life, we cannot afford to save everything in its full size. Compression is a technique to represent the data in fewer bits as compared to the original data. Compression has been in practice since the early days of the information age and researchers continue to build better algorithms to increase its efficiency. But, the problem with widely used compression techniques is that they are designed to work on all types of data and generalize the algorithm. Having a specialized compression algorithm for a particular type of data will present more compression efficiency by using the intrinsic properties of that data. Therefore, we propose a fully lossless compression algorithm made for grayscale images that applies various pre-processing steps and exploits pixel properties of the images to produce a high compression ratio. The algorithm runs in multiple rounds to remove the pixel intensity that occurs the most number of times in the image and stores its position for the decompression. After the rounds are completed, we combine all files into one compressed file using Lempel–Ziv–Markov chaining algorithm. The decompression process is the reverse of the compression process. We test the algorithm over 100s of images and present the results generated by our algorithm. We compare the results for standard test images with existing algorithms to demonstrate the high efficiency of our algorithm. In addition, we suggest the future scope of our work as well.

Keywords: Multi round compression · Selective pixel position storage · MRI images

1 Introduction

Since the dawn of the information age and with the advancement of technology, information sharing has become more frequent and efficient than ever imagined before. It has led to the evolution of society to a level that societal growth, in any case, is not stunted by lack of means of transmission of knowledge. Infrastructure has been developed to transmit knowledge in any form [1, 2] required be it text, images, graphic objects, audio,

I. Woungang et al. (Eds.): ANTIC 2021, CCIS 1534, pp. 818–831, 2022.
https://doi.org/10.1007/978-3-030-96040-7_61

or video. With the digitalization of printed works, all information would be soon available digitally and in a transmittable form. People have developed systems [3] to transmit all types of information in any format required and continue to make advances in the fields of data storage capacity and data transmission speed. To be able to efficiently store and transmit a large amount of information over a network we need to decrease the size of the data by using alternative representations which use less space. To achieve this purpose, we need well-defined algorithms to convert the data and retrieve it back in its original form as and when required.

Data compression is the process wherein fewer bits are used than the original representation to encode information [4]. In information transmission, we perform compression at the information source before storing it or transmitting it, and therefore compression in this context is also known as source coding [5]. Decompression is the process of retrieving original information from compressed information. The compression process can result in loss of information wherein it is called lossy compression [6]. In the case of lossless compression [7], redundancy in the given data is identified and removed while making sure all the information could be retrieved back when decompression is done. However, in lossy compression [8], we reduce the size of data by removing information that is of less importance and therefore cannot recover all the information by decompression. We use lossless compression where the difference between original data and decompressed data cannot be tolerated such as in discrete data [9]. Lossy compression is used where information loss is acceptable such as in images, audio and video.

Compression is important, as it is crucial for efficient utilization of system resources. It improves storage efficiency and decreases data transmission cost and time. Once data is compressed, we can store more data on the same storage and therefore eradicate the need of buying more storage space. Compressed information can be transmitted at a faster rate as there is significant size reduction, which in turn helps to cut down network costs by reducing bandwidth requirements. Improving storage efficiency is a crucial step to allow individuals and organizations to collect and process large amounts of data for data analytics. Every year with more users connecting to the internet there has been an increase in the use of computer algorithms to interpret large amounts of data to extract valuable information. If we take the example of medical institutions, billions of patient test reports are generated every year which are then stored in databases [10]. These reports include both text and medical images which may be required to be processed further for diagnosis and predictions. The results of this processing would also be stored and hence the total size of storage space required is huge. Therefore, data compression has become extremely important in defining the amount of information one can store and process.

Grayscale images have a significant amount of redundant pixel intensities and hence compression techniques that involve the removal of those redundancies can prove to be highly efficient. The problem with widely used compression techniques such as Lempel–Ziv–Welch (LZW) [11] and Direct Cosine Transform and Discrete Wavelength Transform [12] is that they are designed to work on all types of data and do not take account of properties specific to grayscale images. Hence, they can be less efficient when compared to the algorithms designed for compressing specific types of data such as grayscale images. Therefore, we propose a lossless grayscale image compression algorithm that applies various pre-processing steps and exploits properties of grayscale images and the pixel relationships to produce a high compression ratio. The proposed algorithm takes the input image as binary in RGB order to avoid any extension related restrictions. The user can easily convert their image into binary and feed it into our system. We process the image using the properties of grayscale images and pass the processed image through multiple compression rounds. Each round aims to remove the pixel intensity that occurs the most number of times in the image. The compression information for each round is stored within binary configuration files that later help in the decompression of the image. Finally, we use Lempel–Ziv–Markov chaining algorithm (LZMA) [13] to combine all files as one to get the compressed image. The decompression process starts with the same chaining algorithm to get all configuration files. Using the configuration files the decompression rounds run and we can achieve the original image back. We calculated experimentally that five rounds give us a good compression ratio but this can differ based on the image being used.

The rest of the paper is organized as follows: Sect. 2 shares the background of compression algorithms. Section 3 discusses the proposed method in detail. Section 4 introduces the packages for our proposed work. Section 5 contains the efficacy results of our algorithm based on various parameters. Section 6 addresses the conclusions we drew from the proposed work.

2 Background Work

In this paper, we propose an efficient compression algorithm specifically for grayscale images. There have been various research works in the field of compression in the past that present the background of our work. J. Ziv et al. [14] proposed an algorithm for compressing data sequentially. Non-probabilistic model of constrained sources was used to investigate its performance. The universal code proposed achieved compression ratio, which approached lower bounds on compression ratios of other codes uniformly which were designed for matching specified sources.

Again, in 1978, J. Ziv et al. [15] proposed a compression method using variable-rate coding for compressing individual sequences in which a class of lossless encoders investigated the compressibility of individual sequences. Compressibility is defined for every sequence as lower bound attainable asymptotically by any finite-state encoder for the sequence on compression ratio.

N. Manglani et al. [12] proposed a compression technique in which they used discrete cosine transformation, discrete wavelet transformation and code book vector. This technique was proposed for scanned documents and images.

J. Vaisey et al. [16] proposed image compression wherein image is segmented into blocks of variable size to achieve compression. Segmentation is performed by isolating important sections of an image after identifying random large texture blocks. A quadtree data structure has been used for segmentation. The quality of the results obtained by the proposed technique is dependent upon the image selected for testing.

Setia A. et al. [17] proposed an algorithm by enhancing the existing LZW algorithm. The proposed algorithm eliminates spaces from the data to achieve a high compression factor. LZW achieves an excellent compromise between speed of execution and compression performance. The algorithm compresses the data in a manner, which makes sure that all the information is preserved and no information is lost in the compression process. Dictionary is created while scanning the data, which helps to look out for repeated data.

M. Sun et al. [18] proposed a near-lossless image compression algorithm. The effects that the distribution of pixels has on the compression ratio have been leveraged by the algorithm. Row by row classification of pixels is done and the results are recorded in a mask image. Two sequences are then made by decomposing the image data and the mask image is hidden in them. The LZW algorithm is then used to encode the sequences. The algorithm claims to produce a higher compression ratio as compared to Run-length encoding (RLE), LZW and Huffman encoding.

T. D. Gedeon et al. [19] proposed progressive image compression in which a network of processing units arranged in multiple layers is assumed. All the units are connected to the immediate unit above them simply via a weighted connection. The hidden layer in the algorithm as compared to the input layer consists of a lesser number of units that compresses the image. The compressed image obtained is recovered by using the output layer.

M. J. Weinberger et al. [20] proposed a paper discussing the principles fundamental to the design of Low Complexity Lossless Compression for Images (LOCO-I), and its standardization into JPEC-LS. The new ISO/ITU standard has at its core LOCO-I algorithm for lossless and near-lossless compression of continuous-tone images. LOCO-I has better or nearly the same compression ratios when compared with some modern schemes. This comparison is based on arithmetic coding. In comparison to various schemes with the best available compression ratios, it has a significantly lower complexity level.

Al-Dmour et al. [21] proposed a scheme wherein at first for each symbol, its frequency of occurrence in the original image is calculated. The symbols are then sorted according to their corresponding frequencies in descending order. These symbols are then replaced by weighted 8-bit codes of fixed length. Finally, LZW compression algorithm has been applied to the binary file generated.

Shuqin Zhu et al. [22] proposed a scheme to encrypt and compress images. The proposed scheme used sparse transform for the purpose of image compression. The algorithm stores the compressed data after changing the data range to 0–255 using the sigmoid function. To counter data expansion, this transformed data is then stored as an 8-bit binary.

Muhammad Ali Qureshi et al. [23] proposed a compression algorithm for images that is claimed to outperform numerous compressive sensing based algorithms. The algorithm uses a normalized random Gaussian matrix for compression. This matrix is used further for compressed sampling and serves as an important factor for this algorithm to deliver better results as compared to other algorithms. For the purpose of decoding the algorithm uses l-minimization. The results are obtained by testing the algorithm on natural images.

Miao Zhang et al. [24] proposed a Non-uniform Discrete Cosine Transform (NDCT) and nearest-neighbouring coupled-map lattices (NCML) based algorithm for compressing and encrypting images. The proposed algorithm uses a cross chaotic map and compresses the image using NDCT and Huffman coding. Blocks of compression data are made and then they simultaneously undergo diffusion and permutation. The permutation is performed between the blocks and diffusion is performed inside the blocks.

3 Proposed Methodology

The proposed algorithm to compress the image takes the input image as a binary file making the algorithm extension-independent. The users can simply convert their image data in binary where the Red, Green and Blue components are arranged in the order where all the red intensities come first then green and then blue. The algorithm takes in another parameter called *degree*, which specifies the number of rounds of compression to be applied to the dataset. The *degree* varies with the dataset as it is analogous to the number of maximum frequency pixel intensities in an image. Maximum frequency pixel intensity means the pixel intensity (Red, Green and Blue combined) that occurs the most number of times in an image. The algorithm follows various steps before returning the compressed image dataset or file as a ZIP or the decompressed image as a binary. The steps followed are explained in the following sections of the paper.

3.1 Compression Process

In this process, the image is compressed in several rounds specified by *degree* and various intermediate binary files are generated which store the information related to respective rounds. All these intermediate files and the final compressed image are stored in a folder and zipped using LZMA. LZMA was applied through libraries provided within Python and JavaScript. The compression process comprises different steps that are explained in the following sections.

Stripping of Colour Component

The first step we perform is to exploit the properties of a grayscale image. We know that the red, green and blue components in a grayscale image are equal. Therefore, we can just keep one and remove the other two. We remove the green and blue components of the image as shown in Algorithm 1 and keep the red component of the image for further steps. We use *bytearray* in Python to convert the raw binary in the *image-file* to an array of 8-bit integers.

Algo. 1

1. initialize *(image-file)* array and set values to contain binary of the RGB image
2. initialize *(length)* variable as length of *(image-file)* array divided by 3
3. modify *(image-file)* array to keep only first *(length)* values in it
4. initialize bytearray *(image)* from *(image-file)*

Round-Wise Compression

After the stripping of green and blue components from the image, the image is passed through multiple compression rounds. In these rounds, we find the pixel intensity that occurs the most number of times in the *image*. After that, create a 1D array of the same size as the *image* called *bit-array* and set all values to 0. Now, at whatever index in the 1D image array the maximum frequency pixel intensity appears, we set 1 at the same index in *bit-array* as shown in Algorithm 2. To save the *bit-array* we again use the concept of *bytearray* and pad 0s in front of the bit array to make its size a multiple of 8. Finally, we create a binary config file which is separate for each round named as *config{round_number}.bin* and share three details regarding the particular round stored as *bytearray* (8 bits integer array). Firstly, the maximum frequency pixel intensity is stored in it as 8-bit integer. Secondly, the number of zeros padded to bit-array are stored in it as 8-bit integers. Finally, the *bit-array* (size a multiple of 8) is converted to corresponding 8-bit integers and are stored in it. The final step in the round is to remove the maximum frequency pixel intensity from the *image* as shown in Algorithm 2.

Algo. 2

1. create function named *(handler)* with parameter *(image)*
   a. initialize *(intensity)* to the pixel intensity that occurs maximum times in *(image)*
   b. initialize *(img-len)* to the length of *(image)*
   c. initialize *(bit-array)* array of length *(img-len)* and set all values to 0
   d. loop from 0 to *(img-len)*
      i. set value at index i in *(bit-array)* to 1, if index i in *(image)* has intensity equal to *(intensity)*
   e. initialize *(new-image)* array that contains modified *(image)* array containing no occurrence of *(intensity)* value in it
   f. return *(intensity, new-image, bit-array)*

This completes the steps in one round. The degree number of rounds are run to compress and image as shown in Algorithm 3. The number of rounds can differ based on the input image. More rounds signify more number of pixels with same intensity. Finally, after all the rounds are completed we are left with *image* which is the most compressed version of itself and various *config* files which are already stored on the disk. We save the image back to the disk and zip all the binary files using LZMA to complete the compression process.

Algo. 3

1. create function named *(compress)* with parameter *(image, degree)*
   a. loop from 1 to *(degree)*
      i. initialize *(intensity, image, bit-array)* and get values from *(handler)* function
      ii. initialize *(bytes)* as bytearray
      iii. initialize *(zero_count)* and set to *(8 - len(bit-array) % 8)*
      iv. append *(intensity)* to *(bytes)* array
      v. append *(zero_count)* to *(bytes)* array
      vi. modify *(bit-array)* by appending *(zero_count)* number of 0s in front of it
      vii. convert *(bit-array)* to 8 bit integer array and set to *(int-array)* array
      viii. loop in *(int-array)*
         1. append values to *(bytes)* array
      ix. initialize *(file)* as a file with name *config{i}.bin*
      x. write *(bytes)* on *(file)*
   b. save the final *(image)* to a binary file *(image-file)*
   c. zip all binary files using LZMA and save as *(final-zip)*
   d. return *(final-zip)*

3.2 Decompression Process

The decompression process is similar to compression but is totally reversed. We have a zip file as an input and we need to unzip to get all the *config* files and the *image* file. The rounds happen in reverse order to how they happened while compression. The round one's compression will be recovered in the end and the last round's compression will be recovered first. This means that *config* file of round one will be used in the end and the last round's *config* will be used first. Like compression, decompression is also made up of various steps which are explained in the following sections.

Unzipping Files

We use the LZMA algorithm to unzip the files and get a list of files. The *number of files*-1 determines the count of the rounds performed in the compression process. This value will become our new *degree*. We extract and separate the *image-file* from *config* files. The *image-file* is in binary so we use the *bytearray* array again to convert it into readable 8-bit integers that demonstrate the intensities. The process is shown in Algorithm 4.

Algo. 4

1. create function named *(unzip)* with parameter *(zip-file)*
  a. initialize *(files)* as unzipped *(zip-file)* using LZMA algorithm
  b. initialize *(degree)* to count of *(files)* - 1
  c. loop in *(files)*
    i. if "config" substring is not in *(file)*
      1. initialize *(image-file)* to *(file)*
      2. remove *(file)* from *(files)*
      3. break from loop
  d. initialize *(image)* as bytearray of *(image-file)*
  e. return *(image, files, degree)*

Round-Wise Decompression

In each round of decompression, we read the *config* file for that round specified by *config{round_number}.bin* as *bytearray*. The first 8-bit integer gives us the maximum frequency pixel intensity that we need to add to our *image*. The second 8-bit integer tells us the number of zeros that we padded in the *bit-array* to convert it to an array with a size of multiple of 8. Finally, the remaining integers are the 8-bit integers that were converted from the *bit-array*.

To begin the process, we convert the 8-bit integers to binary leaving the first two and create a *bit-array*. Using the second integer, we remove those many zeros from the *bit-array* from the beginning. Finally, using the *bit-array* we insert the maximum frequency pixel intensity in the *image* at the places where 1 appears in the *bit-array* as shown in Algorithm 5.

The recovered *image* from one round is passed on to the other round for recovery. After all the rounds are completed we get the original recovered *image* but with only red components. So, we append the *image* with itself twice to add green and blue components behind the red component. This completes the decompression process.

Algo. 5

1. create function named *(decompress)* with parameter *(zip-file)*
    a. initialize *(image, configs, degree)* and set value from *(unzip)* function with *(zip-file)* parameter
    b. loop from (*degree*) to 1 by decrementing by 1
        i. initialize (*config)* and read config{i}.bin file from (*configs)* as bytearray
        ii. initialize (*intensity)* and set to 0th index value of *(config)*
        iii. initialize (*zero-count)* and set to 1st index value of *(config)*
        iv. initialize (*int-array)* as subarray from 2nd index to last index of *(config)*
        v. initialize (*bit-array*) as empty list
        vi. loop in (*int-array)* as val
            1. append binary bits of val in (*bit-array)*
        vii. modify (*bit-array)* as subarray of *(bit-array)* starting from index *(zero_count)*
        viii. initialize (*new-image)* as empty list
        ix. loop in (*bit-array*) as j
            1. if j = 1, append *(intensity)* to *(new-image)*
            2. else, append *(image[index])* to *(new-image)* and increment *(index)* by 1
        x. modify (*image)* to *(new-image)*
    c. modify *(image)* to (*image + image + image)*
    d. save (*image)* as binary file (*image-file*)
    e. return (*image-file)*

4 Library and Packages

We created a Python Package to allow easy access of our algorithm to where it is needed the most and make use of its full potential. To provide easy integration in the existing software without breaking any old functionality, we created these packages. Python Package provides compression and decompression functions, which return a zip file after the compression and the original image binary file after the decompression. The Python Package can be installed using the **pip install greycomp** command in the terminal.

5 Experimental Results and Discussion

In this section, we analyse our algorithm implemented in Python on a system with a 1.8 GHz Dual-Core Intel Core i5 processor, 8 GB RAM and Python 3.8. To test the efficacy of our proposed algorithm, we used the Brain MRI Dataset for Brain Tumour Detection [31]. This section is divided into two subsections. The first subsection runs the compression algorithm over the dataset and calculates the compression ratio. The second subsection demonstrates the time taken by each file to compress.

5.1 Compression Ratio

In this section, we converted the images from the dataset into binary files. We used the binary files as input to our algorithm. The compression ratio is calculated using the following formula:

$$Compression\ ratio = size\ before\ compression/size\ after\ compression$$

We ran the code with different degrees of compression and the result is displayed in Table 1. A good compression algorithm has more compression ratio. In our algorithm, we achieved an average compression ratio of 5.6, a maximum compression ratio of 11.84 at compression degree 2 and a minimum compression ratio of 3.76 at compression degree 5.

Table 1. Compression ratio of 10 random images for four degrees of compression

Image	Ratio (Degree 2)	Ratio (Degree 3)	Ratio (Degree 4)	Ratio (Degree 5)
Image 1	11.8465824	11.7488603	11.5690675	11.4261861
Image 2	9.64614812	9.55073382	9.38076795	9.29478189
Image 3	9.39605546	9.19964257	9.01539284	8.84927043
Image 4	9.37873358	9.19677415	9.03555435	8.87946795
Image 5	9.156343	8.95216041	8.77269244	8.62070214
Image 6	9.07172142	9.00963495	8.90620729	8.8208142
Image 7	9.07172142	9.00963495	8.90620729	8.8208142
Image 8	9.02968267	8.86334452	8.7115414	8.56324845
Image 9	7.95337795	7.81862017	7.70571708	7.59811987
Image 10	7.89844614	7.74983117	7.61520414	7.47731156

5.2 Compression Time

Time taken to run an algorithm is very important for its consideration to be a good algorithm. In this section, we calculate the time (in seconds) taken by 154 images to compress. The results for 10 random images are shown in Table 2 with varying degrees of compression. The algorithm took an average of 0.74 s for compression degree 2, 1.60 s for compression degree 3, 1.88 s for compression degree 4 and 2.16 s for compression degree 5.

Table 2. Compression time of 10 random images for four degrees of compression

Image	Time (Degree 2)	Time (Degree 3)	Time (Degree 4)	Time (Degree 5)
Image 1	0.14875817	0.32626319	0.36713028	0.58458901
Image 2	0.14932489	0.2747469	0.59300017	0.30414605
Image 3	0.15318179	0.33963394	0.35277796	0.34705901
Image 4	0.15563798	0.33706999	0.43521881	1.06318283
Image 5	0.15623498	0.29073882	0.46481299	0.55236101
Image 6	0.16242003	0.41520405	0.41097403	0.83742499
Image 7	0.16467571	0.38026023	0.46279693	0.77180004
Image 8	0.16493511	0.35359597	0.44375491	0.41154194
Image 9	0.17093301	0.50922799	0.46202493	0.30229402
Image 10	0.17238879	0.33173919	0.93450284	0.4975481

5.3 Comparison with Other Algorithms

In literature we have various well-known compression algorithms such as Location Based Approach [26], Huffman Coding [28, 30] and Discrete Cosine Transform (DCT) [27, 29]. We have compared the compression ratio of our algorithm with these techniques. We compressed the generic grayscale images used in the literature of these techniques

(a) Lena (b) Cameraman

(c) Mandrill (d) Grey Leaf

Fig. 1. (a–d) shows the grayscale images

to compare the performance without any bias. The images used in the experiments are of 512 × 512 dimension and are shown in Fig. 1.

The proposed algorithm achieves a better compression ratio compared to other algorithms for all four standard grayscale images (Lena, Cameraman, Mandrill and Grey Leaf) as shown in Table 3. The proposed method gave best results when degree was chosen to be 2. The optimal value of degree in the proposed method can be calculated experimentally by observing how better the compression ratio gets when degree is increased/decreased by 1.

Table 3. Compression ratio of various methods

Image	Location based approach [25, 26]	Huffman coding [25, 28, 30]	DCT [25, 27, 29]	Proposed method
Lena	1.751	1.851	2.139	4.271
Cameraman	1.789	1.887	2.107	3.541
Mandrill	1.829	1.930	2.208	2.911
Grey Leaf	2.093	2.440	2.968	3.011

6 Conclusion and Future Work

In this work, we proposed a lossless compression algorithm for grayscale images that used the properties of these images and the pixel relationships to produce a high compression ratio. It ran in multiple rounds where the pixel intensity that occurred the most number of times in the image was removed. After the completion of compression rounds, Lempel–Ziv–Markov chaining algorithm helped in combining all files as one. We ran the algorithm over the MRI dataset and achieved an average compression ratio to be 5.6, maximum compression ratio to be 11.84 and minimum compression ratio to be 3.76. This ratio can increase and decrease based on the compression degree and the image type. Achieving such high ratios shows the high efficacy of the algorithm. The algorithm being highly effective runs fast as well. Our algorithm took an average of 0.74 s for degree 2, 1.60 s for degree 3, 1.88 s for degree 4 and 2.16 s for degree 5 compression. After comparison with other known compression techniques, we saw a better compression ratio of our proposed algorithm on standard grayscale images. From these results, we can conclude that the algorithm performs good compression at high speed. The decompression process is the reverse of the compression process. The future scope of our algorithm is to devise the pre-processing steps for the coloured images to use their pixel relationship to achieve a high compression ratio.

Declarations.

Funding This research did not receive any specific grant from funding agencies in the public, commercial, or not-for-profit sectors.

Availability of Data and Material. The datasets analysed during the current study are available from [Brain MRI Images for Brain Tumor Detection], [https://www.kaggle.com/navoneel/brain-mri-images-for-brain-tumor-detection].

Code Availability The code that supports the findings of this study is available in [Grey-Image-Compression], [https://github.com/ThisIsNSH/Grey-Image-Compression].

Conflicts of Interest/Competing Interests. The authors have no relevant financial or non-financial interests to disclose.

References

1. Saleh, Z., Abu Baker, A., Mashhour, A.: Evaluating the effectiveness of using the internet for knowledge acquisition and students' knowledge retention. In: Cherifi, H., Zain, J.M., El-Qawasmeh, E. (eds.) DICTAP 2011. CCIS, vol. 167, pp. 448–455. Springer, Heidelberg (2011). https://doi.org/10.1007/978-3-642-22027-2_36
2. Burgos, D.: Online technology in knowledge transfer. In: Burgos, D. (ed.) Radical Solutions and Open Science. LNET, pp. 91–103. Springer, Singapore (2020). https://doi.org/10.1007/978-981-15-4276-3_6
3. Alavi, M., Leidner, D.: Review: knowledge management and knowledge management systems: conceptual foundations and research issues. MIS Q. (2001). https://doi.org/10.2307/3250961
4. Mahdi, O.A., Mohammed, M.A., Mohamed, A.J.: Implementing a novel approach an convert audio compression to text coding via hybrid technique. Int. J. Comput. Sci. Issu. **9**(6(3)), 53–59 (2012)
5. Salomon, D.: A Concise Introduction to Data Compression. Springer, Berlin (2008)
6. Pujar, J.H., Kadlaskar, L.M.: A new lossless method of image compression and decompression using Huffman coding techniques. J. Theor. Appl. Inf. Technol. **15**(1), 18–23 (2010)
7. Patel, R.A., Zhang, Y., Mak, J., Davidson, A., Owens, J.D.: Parallel lossless data compression on the GPU. Innov. Parall. Comput. (2012). https://doi.org/10.1109/InPar.2012.6339599
8. Kontoyiannis, I.: Pointwise redundancy in lossy data compression and universal lossy data compression. IEEE Trans. Inf. Theory (2000). https://doi.org/10.1109/18.817514
9. Sayood, K.: Introduction to Data Compression (Fifth Edition). Morgan Kaufmann, Burlington (2017)
10. Institute of Medicine. Committee on Regional Health Data Networks. In: Donaldson, M.S., Lohr, K.N. (eds.) Health Data in the Information Age: Use, Disclosure, and Privacy. The National Academies Press, Washington, DC (1994). https://doi.org/10.17226/2312
11. Singh, S., Pandey, P.: Enhanced LZW technique for medical image compression. In: 2016 3rd International Conference on Computing for Sustainable Global Development (INDIACom), pp. 1080–1084 (2016)
12. Manglani, N., Singh, S.: Gray scale image compression. Int. J. Adv. Res. Comput. Commun. Eng. **5**(3) (2016). https://doi.org/10.17148/IJARCCE.2016.5384
13. Wyner, A.D., Ziv, J.: Fixed data base version of the Lempel-Ziv data compression algorithm. IEEE Trans. Inf. Theory (1991). https://doi.org/10.1109/18.79955
14. Ziv, J., Lempel, A.: A universal algorithm for sequential data compression. IEEE Trans. Inf. Theory (1977). https://doi.org/10.1109/TIT.1977.1055714
15. Ziv, J., Lempel, A.: Compression of individual sequences via variable-rate coding. IEEE Trans. Inf. Theory (1978). https://doi.org/10.1109/TIT.1978.1055934

16. Vaisey, J., Gersho, A.: Image compression with variable block size segmentation. IEEE Trans. Signal Process. (1992). https://doi.org/10.1109/78.150005
17. Setia, A., Ahlawat, P.: Enhanced LZW algorithm with less compression ratio. In: Kumar, M.A.R.S., Kumar, T. (eds.) Proceedings of International Conference on Advances in Computing. Advances in Intelligent Systems and Computing, vol. 174. Springer, New Delhi (2013). https://doi.org/10.1007/978-81-322-0740-5_41
18. Sun, M. -Y., Xie, Y.-H., Tang, X.-A., Sun, M.-Y.: Image compression based on classification row by row and LZW encoding. In: 2008 Congress on Image and Signal Processing (2008). https://doi.org/10.1109/CISP.2008.302
19. Gedeon, T.D., Harris, D.: Progressive image compression. Int. Joint Conf. Neural Netw. (1992). https://doi.org/10.1109/IJCNN.1992.227311
20. Weinberger, M.J., Seroussi, G., Sapiro, G.: The LOCO-I lossless image compression algorithm: principles and standardization into JPEG-LS. IEEE Trans. Image Process. (2000). https://doi.org/10.1109/83.855427
21. Al-Dmour, A., Abuhelaleh, M., Musa, A., Al-Shalabi, H.: An efficient bit-level lossless grayscale image compression based on adaptive source mapping. J. Inf. Process. Syst. (2016). https://doi.org/10.3745/JIPS.03.0051
22. Zhu, S., Zhu, C.: A new image compression-encryption scheme based on compressive sensing and cyclic shift. Multim. Tools Appl. **78**(15), 20855–20875 (2019). https://doi.org/10.1007/s11042-019-7405-y
23. Qureshi, M.A., Deriche, M.: A new wavelet based efficient image compression algorithm using compressive sensing. Multim. Tools Appl. **75**(12), 6737–6754 (2016). https://doi.org/10.1007/s11042-015-2590-9
24. Zhang, M., Tong, X.: A new algorithm of image compression and encryption based on spatiotemporal cross chaotic system. Multim. Tools Appl. **74**(24), 11255–11279 (2015). https://doi.org/10.1007/s11042-014-2227-4
25. Anantha Babu, S., Eswaran, P., Senthil Kumar, C.: Lossless compression algorithm using improved RLC for grayscale image. Arab. J. Sci. Eng. **41**(8), 3061–3070 (2016). https://doi.org/10.1007/s13369-016-2082-x
26. Hasan, M., Nur, K.: A lossless image compression technique using location based approach. Int. J. Sci. Technol. Res. **1**(2) (2012)
27. Alfiah, F., Setiadi, A., Saepudin, Supriadi, A., Maulana, I.: Discrete cosine transform DCT methods on compression RGB and grayscale image. Int. J. Comput. Techniq. **4**(6) (2017)
28. Patil, R.B., Kulat, K.D.: Image and text compression using dynamic Huffman and RLE coding. In: Deep, K., Nagar, A., Pant, M., Bansal, J.C. (eds.) Proceedings of the International Conference on Soft Computing for Problem Solving (SocProS 2011) December 20–22, 2011. AISC, vol. 131, pp. 701–708. Springer, New Delhi (2012). https://doi.org/10.1007/978-81-322-0491-6_64
29. Ponomarenko, N., Lukin, V., Egiazarian, K., Astola, J.: DCT based high quality image compression. In: Kalviainen, H., Parkkinen, J., Kaarna, A. (eds.) SCIA 2005. LNCS, vol. 3540, pp. 1177–1185. Springer, Heidelberg (2005). https://doi.org/10.1007/11499145_119
30. Ranjan, R.: Canonical Huffman coding based image compression using wavelet. Wireless Pers. Commun. **117**(3), 2193–2206 (2021). https://doi.org/10.1007/s11277-020-07967-y
31. Chakrabarty, N.: Brain MRI Images for Brain Tumor Detection. Kaggle (2019). https://www.kaggle.com/navoneel/brain-mri-images-for-brain-tumor-detection. Accessed 10 May 2021

Author Index

Printed in the United States
by Baker & Taylor Publisher Services